The CRC Press Laser and Optical Science and Technology Series

Editor-in-Chief: Marvin J. Weber

A.V. Dotsenko
L. B. Glebov
V. A. Tsekhomsky
Physics and Chemistry of Photochromic Glasses

Andrei M. Efimov
Optical Constants of Inorganic Glasses

Alexander A. Kaminskii
Crystalline Lasers: Physical Processes and Operating Schemes

Valentina F. Kokorina
Glasses for Infrared Optics

Sergei V. Nemilov
Thermodynamic and Kinetic Aspects of the Vitreous State

Piotr A. Rodnyi
Physical Processes in Inorganic Scintillators

Michael C. Roggemann and Byron Welsh
Imaging Through Turbulence

Hiroyuki Yokoyama and Kikuo Ujihara
Spontaneous Emission and Laser Oscillation in Microcavities

Marvin J. Weber, Editor
Handbooks of Laser Science and Technology
Volume I: Lasers and Masers
Volume II: Gas Lasers
Volume III: Optical Materials, Part 1
Volume IV: Optical Materials, Part 2
Volume V: Optical Materials, Part 3
Supplement I: Lasers
Supplement II: Optical Materials

HANDBOOK OF LASER WAVELENGTHS

Marvin J. Weber, Ph.D.
Lawrence Berkeley National Laboratory
University of California
Berkeley, California

CRC Press
Boca Raton Boston London New York Washington, D.C.

Library of Congress Cataloging-in-Publication Data

Weber, Marvin John, 1932–
Handbook of laser wavelengths / Marvin John Weber.
p. cm. -- (CRC Press laser and optical science and technology series)
Includes bibliographical references.
ISBN 0-8493-3508-6 (alk. paper)
1. Lasers--Handbooks, manuals, etc. 2. Laser materials--Handbooks, manuals, etc. 3. Light--Wave-length--Handbooks, manuals, etc. I. Title. II. Series.
TA1683.W43 1999
621.36′6—dc21 98-20095
CIP

International Standard Book Number 0-8493-3508-6
Library of Congress Card Number 98-20095
Printed in the United States of America 1 2 3 4 5 6 7 8 9 0
Printed on acid-free paper

Foreword

It is really amazing how many laser transitions and how many laser wavelengths have been discovered. They cover nearly every class of material, from free electrons through gases, liquids, and solids. It is perhaps even more amazing that the comprehensive listing in this book could be compiled through the collaboration of leading experts in each of the fields.

Forty years ago, when Charles Townes and I were first trying to discover how lasers might be made, it seemed very difficult. We had always been taught that the world was pretty close to being in equilibrium, even though masers had shown that you could sometimes get away from it. As Ali Javan pointed out then, when discussing possible gas lasers, there are many processes tending to restore equilibrium. Moreover, since nobody had ever made a laser, we thought it might be very difficult. There might be some hidden problem that we had overlooked. But that turned out to be wrong and some kinds of lasers are quite easy to make once you know how.

When thinking of possible laser materials, I for one had plenty of blind spots and poorly based prejudices. For instance, I knew that the optical gain, for a given excess of excited atoms, would be inversely proportional to the spectral linewidth. Thus I felt that narrow lines were essential, overlooking the fact that some broad bands in things like organic dyes have large oscillator strengths and so make up for their large width. Also, for a time I couldn't see why anyone would want to use a laser to pump another, thereby compounding their inefficiencies.

Fortunately, lasers attracted the interest and stimulated the imagination of large numbers of very clever people. Some of them had specialized knowledge of things like crystal growing, very hot plasmas, or semiconductor luminescence. From their work have come the very many types of lasers listed in this book. Some of the discoveries resulted from careful study and planning, while others were serendipitous.

Many lasers have been discovered but never put to any practical use. In some cases, gases are too corrosive or too easily adsorbed on the walls. In others, crystalline materials are too difficult to grow in useful sizes, or are too hygroscopic. Sometimes, there just isn't any obvious need for that kind of laser. Perhaps someone browsing in this book will find something for a new use, or will think of ways to overcome the apparent difficulties.

Perhaps also in the future, or even now, someone will recognize other blind spots and will see new approaches to yield still more types of useful lasers.

Arthur L. Schawlow
Stanford University

Preface

Although we are well into the fourth decade since the advent of the laser, the number and type of lasers and their wavelength coverage continue to expand. One seeking a photon source is now confronted with an enormous number of possible lasers and laser wavelengths. In addition, various techniques of frequency conversion—harmonic generation, optical parametric oscillation, sum- and difference-frequency mixing, and Raman shifting—can be used to enlarge the spectral coverage.

This volume seeks to provide a comprehensive compilation of the wavelengths of lasers in all media in a readily accessible form for scientists and engineers searching for laser sources for specific applications. The compilation also indicates the state of knowledge and development in the field, provides a rapid means of obtaining reference data, is a pathway to the literature, contains data useful for comparison with predictions and/or to develop models of processes, and may reveal fundamental inconsistencies or conflicts in the data. It serves both an archival function and as an indicator of newly emerging trends.

The *Handbook of Laser Wavelengths* is derived from data evaluated and compiled by the contributors of Volumes I and II and Supplement 1 of the *CRC Handbook Series of Laser Science and Technology*. In most cases it was possible to update these tabulations to include more recent additions and new categories of lasers. For semiconductor lasers where in some instances the lasing wavelength may not be a fundamental property but the result of material engineering and the operating configuration, an effort was made to be representative rather than exhaustive in the coverage of the literature. The number of gas laser transitions is huge; they constitute nearly 80% of the over 15,000 laser wavelengths in this volume. Laser transitions in gases are well covered through the late 1980s in the above volumes. An electronic database of gas lasers prepared from the tables in Volume II and Supplement 1 by John Broad and Stephen Krog (Joint Institute of Laboratory Astrophysics) was used for this volume, but does not cover all recent developments.

In Section 1, a brief description of various types of lasers is given. Lasers are divided by medium—solid, liquid, and gas—each one of which is further subdivided, as appropriate, into distinctive types. Thus there are sections on crystalline paramagnetic ion lasers, glass lasers, color center lasers, semiconductor lasers, polymer lasers, liquid and solid-state dye lasers, rare earth liquid lasers, and neutral atom, ion, and molecular gas lasers. A separate section on "other" lasers covers lasers having special operating conditions or nature. These include extreme ultraviolet and soft x-ray lasers, free electron lasers, nuclear-pumped lasers, lasers in nature, and lasing without inversion. Brief descriptions of each type of laser are given followed by tables listing reported lasing wavelength, lasing element or medium, host, other experimental conditions, and primary literature citations. All lasers are listed in order of increasing wavelength.

The realm of tunable lasers has expanded and includes liquid and solid-state dye lasers, lanthanide and transition-metal crystalline lasers, color center lasers, and semiconductor and polymer lasers. Tuning ranges, when reported, are given for these broadband lasers. For most types of lasers, lasing—light amplification by stimulated emission of radiation—includes, for completeness, not only operation in a resonant cavity but also single-pass gain or amplified spontaneous emission (ASE).

The wavelengths of lasing transitions are of primary concern. No detailed descriptions of laser structure, operation, or performance are provided. These properties are covered in Volumes I and II and Supplement 1 of the *CRC Handbook Series of Laser Science and Technology.* Although laser performance data are not tabulated, a special section on commercially available lasers is included to provide a perspective on the current state-of-the-art and performance boundaries (although these are expected to change due to advances in technology). Further background information about lasers in general and about specific types of lasers in particular can be obtained from the books and articles listed under Further Reading in each section.

To cope with the continuing and bewildering proliferation of acronyms, abbreviations, and initialisms that range from the clever and informative to the amusing or annoying, two appendices are included—one for types and structures of lasers and amplifiers and one for solid-state laser materials. A third appendix provides a list of fundamental physical constants of interest to laser scientists and engineers.

Because lasers now cover such a large wavelength range and because researchers in different fields are frequently accustomed to using different units, there is also a "Rosetta stone for spectroscopists" on the inside back cover.

I wish to acknowledge the valuable help and expertise of the Advisory Board for this volume who reviewed the material, made suggestions about the contents, and in several cases contributed material (the Board, however, is not responsible for the accuracy nor thoroughness of the tabulations). We are all indebted to the contributors to Volumes I and II and Supplement 1 of the *CRC Handbook Series of Laser Science and Technology* who compiled the data from which most of this volume was derived. Others who have provided helpful comments, suggestions, and data include Eric Bründermann, Federico Capasso, Henry Freund, Claire Gmachl, Victor Granatstein, Eugene Haller, Stephen Harris, John Harreld, Thomas Hasenberg, Alan Heeger, Heonsu Jeon, George Miley, Michael Mumma, Dale Partin, Maria Petra, Jin-Joo Song, and Riccardo Zucca. Finally I appreciate the help of the CRC Press staff during the preparation of this volume—Tim Pletscher, Acquiring Editor for Engineering, Felicia Shapiro, Suzanne Lassandro, Gerry Axelrod—and especially Mimi Williams for her careful and excellent editing of the manuscript.

Marvin John Weber
Danville, California

The Author

Marvin John Weber received his education at the University of California, Berkeley, and was awarded the A.B., M.A., and Ph.D. degrees in physics. After graduation, Dr. Weber continued as a postdoctoral Research Associate and then joined the Research Division of the Raytheon Company where he was a Principal Scientist working in the areas of spectroscopy and quantum electronics. As Manager of Solid State Lasers, his group developed many new laser materials including rare-earth-doped yttrium orthoaluminate. While at Raytheon, he also discovered luminescence in bismuth germanate, a scintillator crystal widely used for the detection of high energy particles and radiation.

During 1966 to 1967, Dr. Weber was a Visiting Research Associate with Professor Arthur Schawlow's group in the Department of Physics, Stanford University.

In 1973, Dr. Weber joined the Laser Program at the Lawrence Livermore National Laboratory. As Head of Basic Materials Research and Assistant Program Leader, he was responsible for the physics and characterization of optical materials for high-power laser systems used in inertial confinement fusion research. From 1983 to 1985, he accepted a transfer assignment with the Office of Basic Energy Sciences of the U.S. Department of Energy in Washington, DC where he was involved with planning for advanced synchrotron radiation facilities and for atomistic computer simulations of materials. Dr. Weber returned to the Chemistry and Materials Science Department at LLNL in 1986 and served as Associate Division Leader for condensed matter research and as spokesperson for the University of California/National Laboratories research facilities at the Stanford Synchrotron Radiation Laboratory. He retired from LLNL in 1993 but continues as a Participating Guest in the Physics and Space Technology Department. He presently does consulting and is a physicist in the Center for Functional Imaging at the Lawrence Berkeley National Laboratory.

Dr. Weber is Editor-in-Chief of the multi-volume *CRC Handbook Series of Laser Science and Technology*. He has also served as Regional Editor for the *Journal of Non-Crystalline Solids*, as Associate Editor for the *Journal of Luminescence* and the *Journal of Optical Materials*, and as a member of the International Editorial Advisory Boards of the Russian journals *Fizika i Khimiya Stekla* (Glass Physics and Chemistry) and *Kvantovaya Elektronika* (Quantum Electronics).

Among several honors he has received are an Industrial Research IR-100 Award for research and development of fluorophosphate laser glass, the George W. Morey Award of the American Ceramics Society for his basic studies of fluorescence, stimulated emission and the atomic structure of glass, and the International Conference on Luminescence Prize for his research on the dynamic processes affecting luminescence efficiency and the application of this knowledge to laser and scintillator materials.

Dr. Weber is a Fellow of the American Physical Society, the Optical Society of America, and the American Ceramics Society and has been a member of the Materials Research Society and the American Association for Crystal Growth.

Advisory Board

Contents of previous volumes on lasers from the

CRC HANDBOOK OF LASER SCIENCE AND TECHNOLOGY

VOLUME I: LASERS AND MASERS

VOLUME II: GAS LASERS

SUPPLEMENT 1: LASERS

HANDBOOK OF LASER WAVELENGTHS

TABLE OF CONTENTS

Section 1: Introduction

Section 1

INTRODUCTION

The laser has become an invaluable tool for mankind. The ubiquitous presence of lasers in our lives is evident from their use in such diverse applications as science and engineering, communications, medicine, manufacturing and materials processing, art and entertainment, data processing, environmental sensing, defense, energy, astronomy, and metrology. It is difficult to imagine state-of-the-art physics, chemistry, biology, and medicine research without the use of radiation from various laser systems.

Laser action occurs in all states of matter—solids, liquids, gases, and plasmas. In this volume lasers are categorized based on the active medium. The spectral output ranges of solid, liquid, and gas lasers are shown in Figure 1.1 and extend from the soft x-ray and extreme ultraviolet regions to millimeter wavelengths, thus overlapping masers. In addition to lasers operating at one or more discrete wavelengths, some are tunable over broad wavelength bands. Using various frequency conversion techniques—harmonic generation, parametric oscillation, sum- and difference-frequency mixing, and Raman shifting—the wavelength of a given laser can be extended to longer and shorter wavelengths. Frequently a laser is used as an excitation source for a second medium that generates new laser wavelengths. The medium in essence acts as a wavelength shifter.

Within each category of lasing medium there may be differences in the nature of the active lasing ion or center, the composition of the medium, and the excitation and operating techniques. For some lasers, the periodic table has been extensively explored and exploited; for others—solid-state lasers in particular—the compositional regime of hosts continues to expand. In the case of semiconductor lasers the ability to grow special structures one atomic layer at a time by liquid phase epitaxy, molecular beam epitaxy, and metal-organic chemical vapor deposition has led to numerous new structures and operating configurations, such as quantum wells and superlattices, and to a proliferation of new lasing wavelengths.

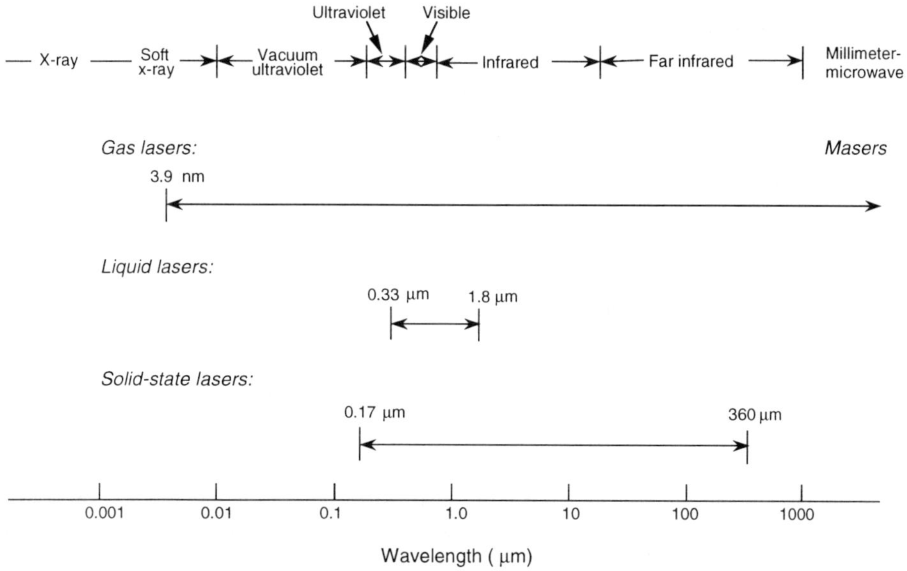

Figure 1.1 Reported ranges of output wavelengths for various laser media.

As will be evident from the brief descriptions below of the different types of lasers covered in this volume, the vitality of the field of lasers is stunning. Furthermore, recent announcements such as those of a single-atom laser,[1] lasing without inversion,[2] and the use of Bose-Einstein condensates for an atom laser[3] continue to extend our understanding of atomic coherence and interference effects in laser physics and quantum optics.

Solid State Lasers

This group includes lasers based on paramagnetic ions, organic dye molecules, and color centers in crystalline or amorphous hosts. Semiconductor lasers are also included in this section because they are a solid state device, although the nature of the active center—recombination of electrons and holes—is different from the dopants or defect centers used in other lasers in this category. The recently emerging field of conjugated polymer lasers is also covered in this section. Solid-state excimer lasers, for which the number of reported cases of lasing is insufficient to warrant a tabulation, are noted at the end of this section.

Reported ranges of output wavelengths for various types of solid-state lasers are shown in Figure 1.2. The differences in the ranges of spectral coverage arise in part from the dependence on host properties, in particular the range of transparency and the rate of nonradiative decay due to multiphonon processes.

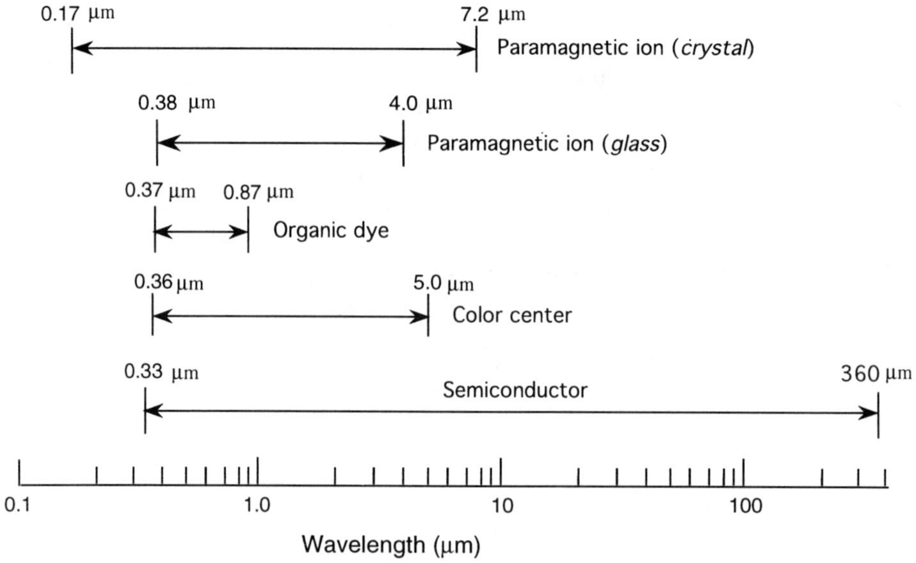

Figure 1.2 Reported ranges of output wavelengths for various types of solid state lasers.

Crystalline Paramagnetic Ion Lasers

The elements that have been reported to exhibit laser action as paramagnetic ions (incompletely filled electron shells) in crystalline hosts are indicated in the periodic table of the elements in Figure 1.3. These are mainly transition metal and lanthanide group ions. Also included are several elements (in italics) for which only gain has been reported (see Table 2.1.3). Typical concentrations of the lasing ion are ≤1%, however for some hosts and ions concentrations up to 100%, so-called stoichiometric lasers, are possible.

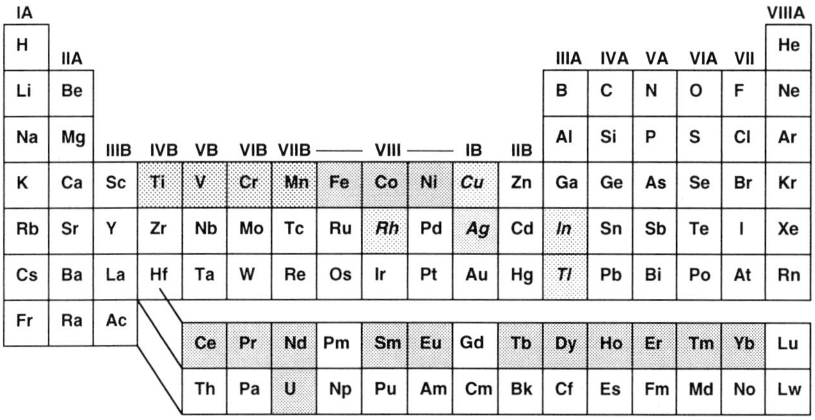

Figure 1.3 Periodic table of the elements showing the elements (shaded) that have been reported to exhibit laser action as paramagnetic ions in crystalline hosts. Gain has been reported for elements shown in italics.

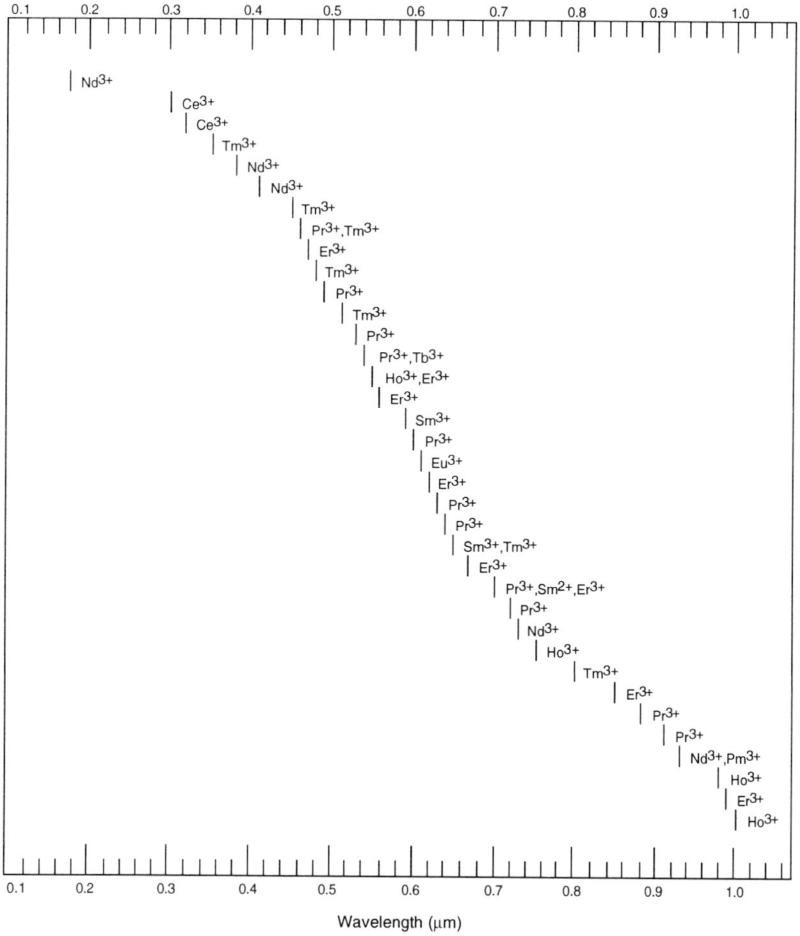

Figure 1.4a Approximate wavelengths of crystalline lanthanide-ion lasers; exact wavelengths are dependent on the host and temperature and the specific Stark levels involved (see Table 2.1.4).

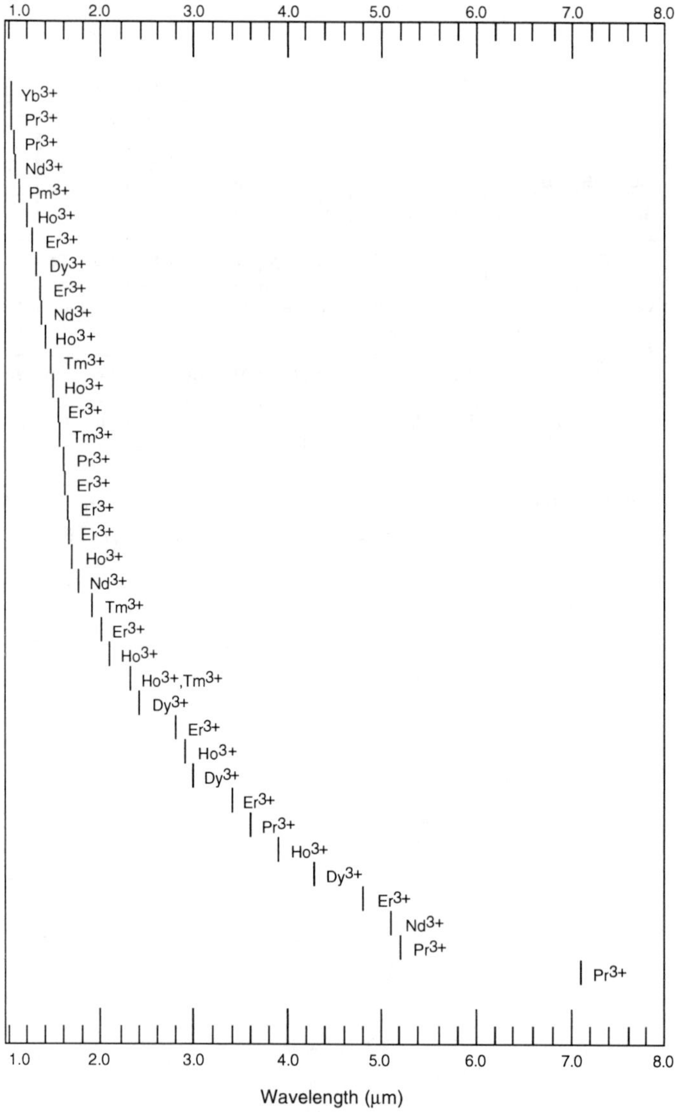

Figure 1.4b Approximate wavelengths of crystalline lanthanide-ion lasers; exact wavelengths are dependent on the host and temperature and the specific Stark levels involved (see Table 2.1.4).

The general operating wavelengths of crystalline lanthanide-ion lasers are given in Figure 1.4 and range from 0.17 μm for the 5d→4f transition of Nd^{3+} to 7.2 μm for the 4f→4f transition transition between J states of Pr^{3+}. Whereas f→f transitions of the lanthanide ions have narrow linewidths and discrete wavelengths, d→f transitions of these ions and transitions of many iron group ions have broad emission and gain bandwidths and hence provide a degree of tunability. The tuning ranges of several paramagnetic laser ions in different hosts are shown in Figure 1.5; the ranges for explicit host crystals are included in Table 2.1.4. As evident from Figure 1.5, tunable lasers are based almost exclusively on iron transition group elements. Whereas the narrow emission lines of Cr^{3+} in Al_2O_3 (ruby) were used for the first demonstration of laser action, the broadband emissions of divalent, trivalent, and tetravalent chromium now provide tunable laser radiation throughout much of the region from 0.7 to 2.5 μm.

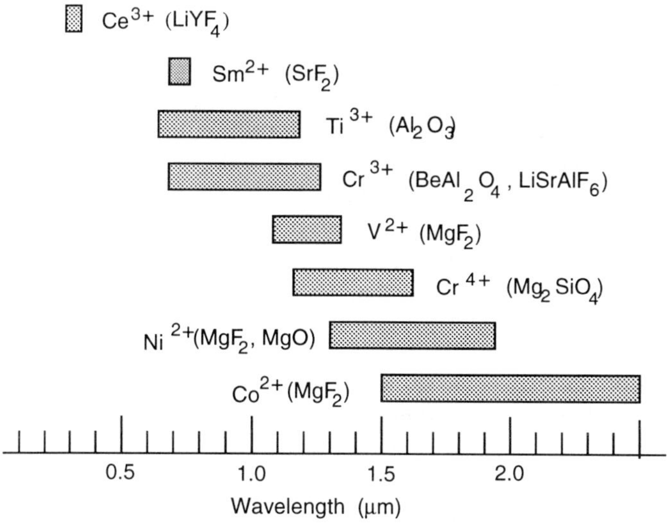

Figure 1.5 Reported wavelength ranges of representative tunable crystalline lasers operating at room temperature (see Table 2.1.4 for details).

Over 300 ordered and disordered crystals have been used as hosts for laser ions.[4] These include oxide, halide, and, recently, chalcogenide compounds. That so many different crystals of sufficient size and quality necessary to demonstrate laser action have been prepared is testimony to the crystal growers' art and capabilities. Codopant ions are sometimes added to the hosts to improve optical pumping efficiency. These sensitizer ions are included in Table 2.1.4.

The field of solid state lasers is large and still amazingly vital. These lasers have been operated pulsed, Q-switched, mode-locked, or cw. Picosecond pulses can be obtained from broadband lasers using various mode-locking techniques; femtosecond pulses can be obtained using saturable absorbers. The population inversion necessary for laser action in solid-state lasers has been achieved by optical pumping with flashlamps, cw arc lamps, the sun, or other lasers (electron beam pumping has also been reported). Recent advances in diode laser pumping now provide all solid-state devices that are rugged, compact, and have long lifetimes. As a result, diode-pumped solid-state lasers combined with nonlinear crystals are replacing gas and liquid dye lasers in a number of applications.

Upconversion, a concept promoted initially in the late sixties for phosphor displays and demonstrated for solid state lasing in 1971,[5] has witnessed a rebirth of interest with the resurgence of diode pumping and has made possible many new lasing transitions and excitation schemes.[4]

With one or more pulsed lasers as the pumping source, one can establish a population inversion between almost any pair of energy levels of interest and, provided excited state absorption is not dominant, lasing should be achievable, although the result may be neither efficient nor practical. In the case of the thirteen trivalent lanthanide ions, there are 1639 free-ion J states and 192,177 possible transitions between them, yet to date less than 70 have been used, thus one may anticipate the demonstration of many additional lasing transitions and hosts.

Glass Lasers

The past two decades have also witnessed increased activity in glass lasers, both in the form of bulk materials and of fiber and planar waveguides. The former include large neodymium-doped glasses for amplifiers used in lasers for inertial confinement fusion research. Fibers, with their long interaction region, and heavy metal fluoride glasses, with their low vibrational frequencies and hence reduced probabilities for decay by nonradiative processes, have made possible many new lasing transitions and operation at longer wavelengths. These include erbium- and praseodymium-doped fibers for telecommunications and erbium- and thulium-doped lasers for medical applications. Upconversion techniques have also been actively exploited for glass lasers.

The wavelengths of glass lasers are shown in Figure 1.6. The wavelength range is less than that of crystals at both the long and short wavelength extrema. The lasing wavelength could be extended to shorter wavelengths using glassy hosts with larger energy gaps such as beryllium fluoride and silica. Extension further into the infrared is limited by the vibrational frequencies associated with the glass network formers and nonradiative decay processes.

Unlike crystals, which have a unique composition and structure, changes in glass network formers (e. g., silicate, phosphate, borate) and network modifier ions (e. g., alkali, alkaline earths) affect the stimulated emission cross sections, rates of radiative and nonradiative transitions, crystalline field splittings, and inhomogeneous broadening.[6] Although trivially small compositional changes might technically constitute a new host material, those listed in Table 2.2.3 are generally characterized by either different compositions or different operating properties. Commercial glasses are identified by their company's designation. The glass type is generally known but the detailed compositions are usually proprietary.

Because of site-to-site variations in the local fields in glass, there is a distribution of energy levels and transition frequencies which appear as inhomogeneous broadening and provide tunability. In the small signal regime, laser action can be obtained by tuning across the inhomogeneous linewidth, whereas in the large signal or saturated gain regime spectral hole burning may occur. Examples of reported tuning ranges of lanthanide-ion glass lasers are shown in Figure 1.7.

Solid State Dye Lasers

Lasing media based on fluorescing organic dyes may be in the form of solids, liquids, or gases. Although the liquid state is the most familiar and commonly used form, numerous dye-doped solid materials have been reported to lase or exhibit gain in a spectral range extending from the near ultraviolet (376 nm) to the near infrared (865 nm). As shown in Table 2.3.1, a wide diversity of host materials have been utilized. These include various plastics and polymers, organic single crystals, and organic and inorganic glasses. Solid state dye lasers also include—in a somewhat more exotic vein—edible lasers[7] and lasing in animal tissue.[8]

Although the first reports of solid state dye lasers date back to the 1960s, photo-degradation of the dye has been a serious limitation to the utilization of these lasers. Recently there has been a revival of interest in these lasers, principally because materials exhibiting useful lifetimes and tunable laser action have been identified. A solid state dye laser is now offered commercially (see Table 6.1.1).

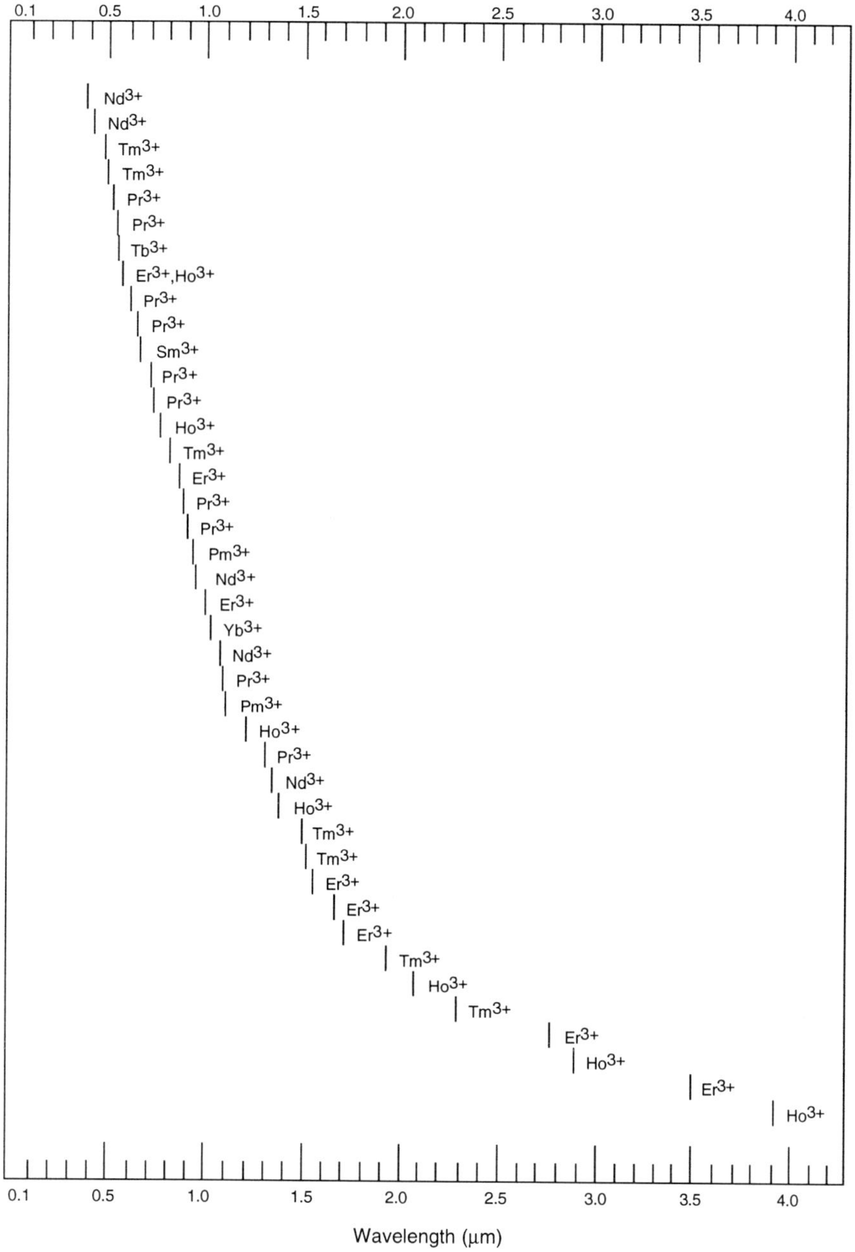

Figure 1.6 Approximate wavelengths of lanthanide-ion glass lasers; exact wavelengths are dependent on the glass composition and temperature and the specific Stark levels involved (see Table 2.2.3).

Color Center Lasers

Color center lasers have been reported that operate in the wavelength range from approximately 0.4 to 5 μm. The optically active centers in these lasers are various types of point defects (i. e., color centers) in alkali halide and oxide crystals. The color centers are

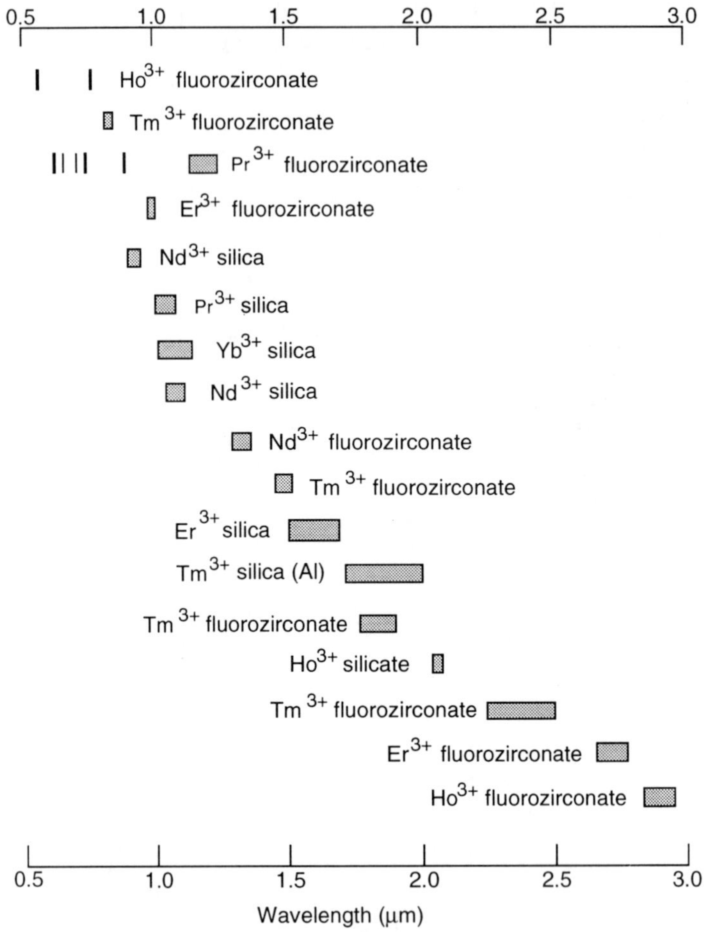

Figure 1.7 Reported tuning ranges of lanthanide-ion glass lasers (see Table 2.2.3).

generally produced by ionizing radiation or are thermally induced. Additional ions may be present to stabilize the defect center and are included in the description of the active center in Table 2.4.1. Other lasers in this category are based on vibrational transitions of molecular defects, such as CN^-.

Color center lasers are usually excited by optical pumping with broadband or laser radiation. Lasing involves allowed transitions between electronic energy levels, hence the gain can be high. Due to their large homogeneous emission bandwidths, color center lasers have varying degrees of tunable. The tuning ranges of some of the longer-lifetime color center lasers are shown in Figure 1.8.

The output of color center lasers may be cw or pulsed. As in the case of paramagnetic ion lasers, picosecond pulses can be obtained using various mode-locking techniques and femtoseconds pulses using saturable absorbers. The operative lifetimes of the color centers in these lasers depend on the temperature and can vary from hours to months. Many color center lasers require operation at low temperatures.

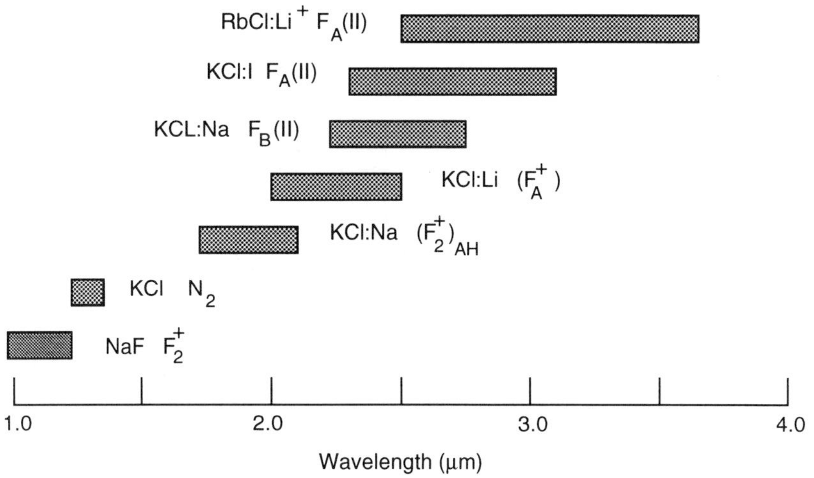

Figure 1.8 Reported tuning ranges of representative color center lasers (see Table 2.4.1).

Semiconductor Lasers

Laser action in semiconductor diode lasers, in contrast to other solid state lasers, is associated with radiative recombination of electrons and holes at the junction of a *n*-type material (excess electrons) and a *p*-type material (excess holes). Excess charge is injected into the active region via an external electric field applied across a simple *p-n* junction (homojunction) or in a heterostructure consisting of several layers of semiconductor materials that have different band gap energies but are lattice matched. Heterostructure enables highly efficient radiative recombination of electrons and holes by confining them into the smaller band gap material sandwiched between higher band gap materials. This has been the most important step in achieving cw operation of diode lasers at room temperature. Excitation of semiconductor lasers has also been achieved by optical pumping and electron beam pumping.

The ability to grow special structures one atomic layer at a time by liquid phase epitaxy (LPE), molecular bean epitaxy (MBE) and metal-organic chemical vapor deposition (MOCVD) has led to an explosive growth of activity and numerous new laser structures and configurations. When dimensions become <100 nm, quantum effects enter that modify the band gap. Quantum wells result from confinement in one dimension, quantum wires from confinement in two dimensions, and quantum dots or boxes from confinement in three dimensions. The wavelength of quantum well lasers can be changed by varying the quantum well thickness or the composition of the active material. If materials of different lattice constants are used, thereby effectively straining the materials, one can further engineer the band gap. Strain-layer technology has led to combinations of various direct band gap materials which have extended the laser wavelength possibilities.

The lasing material may be elemental, but more generally is a compound semiconductor. Figure 1.9 shows the elements that have been used as constituents to achieve laser action in elemental and compound semiconductor materials.

The wavelength ranges of various types of semiconductor lasers are shown in Figure 1.10. III-V compound lasers, including antimonide-based III-V compounds, emit in the visible and the near- and mid-infrared regions. II-VI compound lasers generally emit at shorter wave-

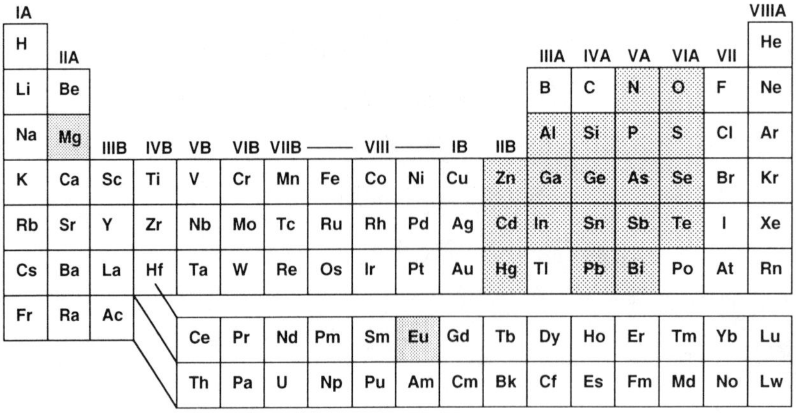

Figure 1.9 Periodic table of the elements showing the elements (shaded) that have been components of semiconductor laser materials.

lengths; mercury-based II-VI compounds extend the coverage over the range 1.9–5.4 μm. Longer wavelength diode lasers are based on IV-VI compounds (lead salts) and can be tuned by changing the temperature or current. The wavelength of quantum cascade lasers, unlike that of diode lasers, is determined by the active layer thickness rather than the band gap of the material. Multiple quantum well lasers have been tailored to operate in the range ~3–13 μm, thereby extending the range of III-V compound lasers. Germanium intervalence band lasers have thus far been operated in the range 75–360 μm.

In edge-emitting lasers the light output is in the plane of the gain medium; in surface-emitting lasers the light output is normal to the axis of the gain medium. The lasing wavelength is determined by the equivalent laser cavity thickness which can be varied by changing the thickness of either the wavelength spacer or the distributed Bragg reflector layers. Vertical-cavity surface-emitting lasers (VCSELs) can be prepared in two-dimensional arrays of independently modulated lasers.

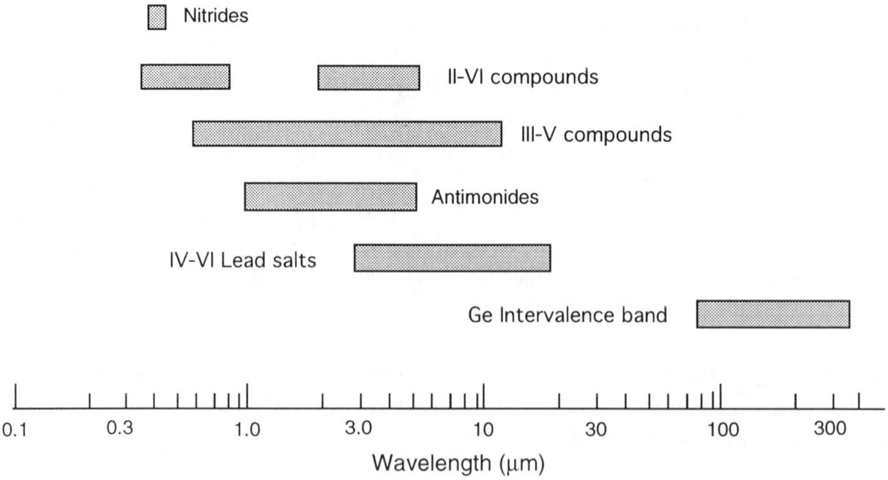

Figure 1.10 Reported ranges of output wavelengths of various types of semiconductor lasers (see Table 2.5.1).

The many different geometries of semiconductor lasers have spawned a proliferation of abbreviations and acronyms for the resultant laser structures, for example, BVSIS (buried V-groove substrate inner strip), GRINSCH (graded-index separate confinement heterojunction) and SELDA (surface emitting laser diode array). Numerous additional terms used to describe semiconductor laser structures are defined in Appendix I.

Because it is possible to vary the constituent elements and tailor the laser emission, the wavelength of semiconductor lasers is a less fundamental property than for other lasers involving transitions between specific atomic levels. Thus the tabulation in Table 2.5.1 includes early pioneering papers and representative examples of different structures and operating conditions rather than an exhaustive listing of all reported semiconductor lasers.

Polymer Laser

Recently a new class of solid-state laser materials based on conjugated polymers has been the subject of increasing activity.[9] Unlike the dye-doped organic solid-state lasers covered in Section 2.3, the active media for these lasers are neat, undiluted, highly purified polymers. These materials have broad optical bands with large cross sections, high radiative quantum efficiencies, and large Stokes shifts of the absorption and emission bands, thereby providing the potential for high-gain, tunable laser action. Transient gain narrowing has been observed in optically pumped neat films of conjugated polymers with the aid of simple planar waveguiding structures or microcavities. The reported observations in Table 2.6.1 are indicative of lasing or amplified spontaneous emission. Thus far all experiments have involved pulsed optical excitation with the material are room temperature. These lasers have operated over a modest wavelength range from 390 to 640 nm.

Electroluminescence is a well-known property of conducting conjugated polymers. The question remaining—the holy grail—is whether it is possible to demonstrate an electrically driven polymer injection laser.

In addition to the pure polymer lasers above and the dye-doped polymer lasers in Section 2.3, amplified spontaneous emission has recently been reported from a Nd^{3+}-doped polymer optical fiber [poly(methyl methacrylate)].[10]

Excimer Lasers

Using matrix isolation techniques, it is possible to grow large, doped, rare-gas crystals. Xenon fluoride molecules can thus be formed by photodissociation of F_2 in Xe-F_2-Ar crystals. Optically pumped solid-state excimer laser action has been reported for XeF in Ar crystals at 286, 411, and 540 nm and for XeF in Ne crystals at 269 nm.[11-13]

Liquid Lasers

Organic Dye Lasers

The most common and familiar liquid lasers are those based on strongly absorbing organic dye molecules in an organic solvent involving allowed transitions of conjugate π electrons. By the selection of the active dye and solvent, laser action spanning a wavelength range from the near-ultraviolet through the near-infrared has been achieved. Laser action has been reported for over 500 different dyes in Table 3.1.1 and extends over a range from 0.336 to 1.8 microns. General categories of dyes and their spectral ranges are shown in Figure 1.11.

Because of the coupling of the electron with molecular vibrations, fluorescing dye emission occurs over a broad wavelength band. The very broad emission and gain spectra of organic dyes lead to tunable laser output—typically over several tens of nanometers. Because of this property, dye lasers are used extensively in wavelength-selective spectroscopy. Tuning curves for various commercial dyes and pumping sources are shown in Section 6.3.

Dye lasers are excited either by linear or coaxial cylindrical flashlamps or other lasers and can be operated in either pulsed, mode-locked, or cw modes. The broad bandwidth of dye lasers is used to advantage in mode-locking schemes to generate ultra-short pulses with durations extending down to a few femtoseconds.

In addition to standard dye laser configurations, organic dye laser action in strongly scattering media consisting of titania nanoparticles in solution has been reported.[16] Lasing from dye-doped micrometer-size liquid droplets[17,18] and from evaporating layered microdroplets in the form of a glass core covered by liquid of dye in solution[19] has also been observed.

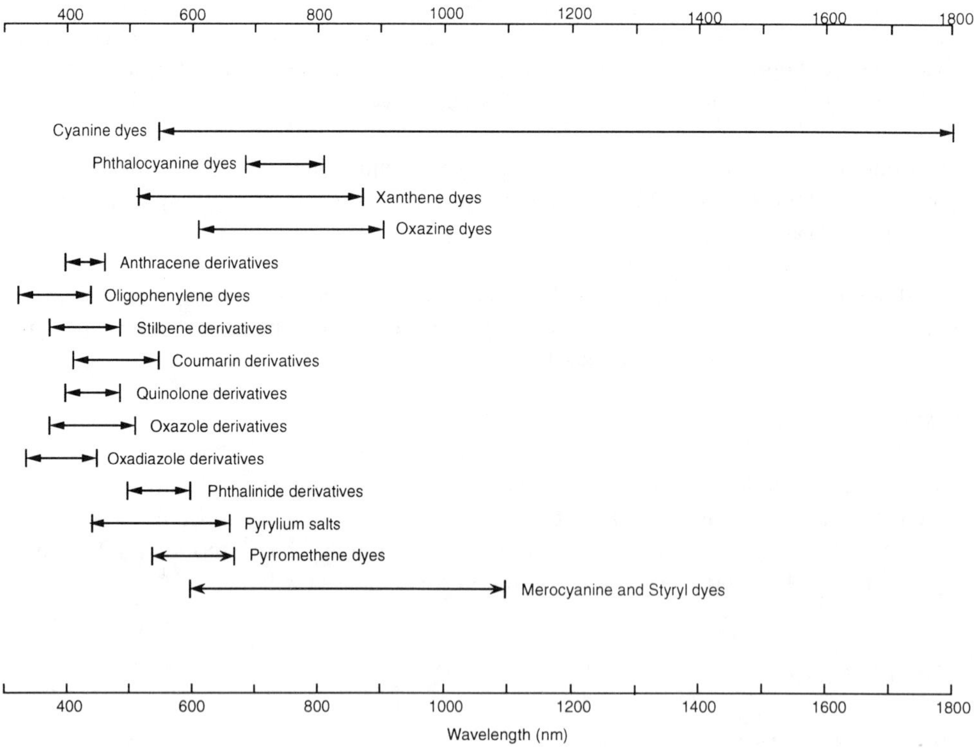

Figure 1.11 Reported ranges of output wavelengths of various types of organic dyes used in liquid lasers (adapted in part from reference 15).

Rare Earth Lasers

Liquid lasers based on lanthanide ions rather than organic molecules have been of two types. In rare-earth chelate lasers, the rare earth is complexed with a chelating agent. Optical pump energy is absorbed by the chelating ligand and transferred to the rare earth ion. Because high frequency vibrations of the organic molecules can give rise to nonradiative quenching of the rare earth fluorescence, only a few ions of the lanthanide series (Nd^{3+}, Eu^{3+}, Tb^{3+}) have been successfully used for this type of laser. These lasers are tabulated in Table 3.2.1

To reduce fluorescence quenching due to vibrations of the ion environment, rare earths have also been incorporated into inorganic aprotic solvents (no hydrogen anions). Thus far these have been oxyhalides or halides of the heavier elements such as phosphorous, sulfur, selenium, zirconium, tin, etc. Neodymium has been the active laser ion although other ions could undoubtedly be used. Neodymium aprotic liquid lasers and amplifiers have been operated in various pulsed modes; operating wavelengths are given in Tables 3.2.2 and 3.2.3. Because of the toxic and corrosive nature of most of these materials, these lasers have found little practical application.

Gas Lasers

Of the spectral ranges for lasers shown in Figure 1.1, by far the largest range is that of gas lasers. The number of elements reported to exhibit laser action in the gas phase has reached 53 and are indicated on the periodic table shown in Figure 1.12. The number of gas laser wavelengths is especially large, totaling more than twelve thousand.

Gas lasers may be categorized as atomic, ionic, or molecular and can be further divided or characterized by the nature of the transitions involved in the stimulated emission process; that is, the transitions may be between electronic, vibrational, or rotational energy levels.

In Figure 1.13 the extremes of the wavelength ranges of different types of gas lasers are shown. The rare gas halide, carbon dioxide, and short-wavelength (200–350 nm) ion lasers are still the dominant sources in their respective wavelength ranges, however the use of gas lasers is being challenged by the compactness, reliability, efficiency, tunability, and spectral

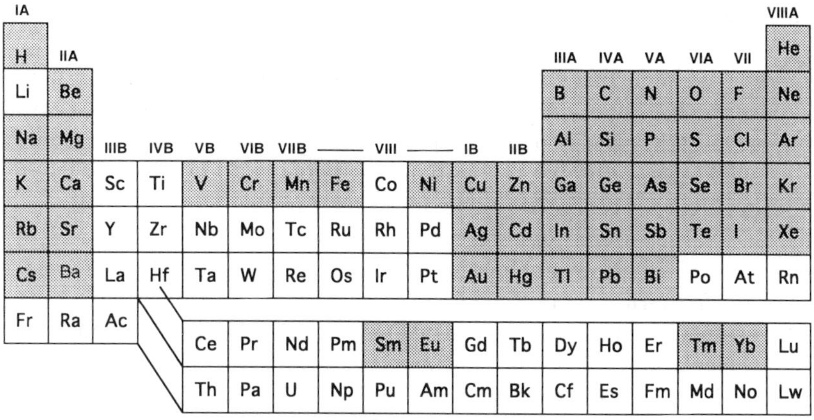

Figure 1.12 Periodic table of the elements showing those atoms or ions (shaded) that have been reported to exhibit laser action in the gas phase.

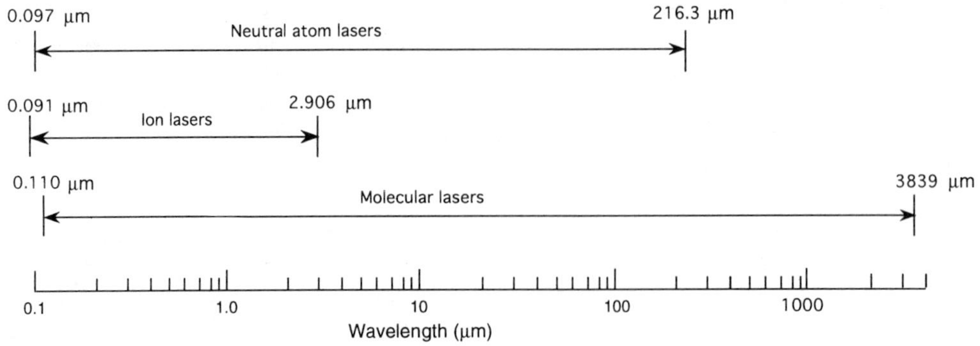

Figure 1.13 Reported ranges of output wavelengths of various types of gas lasers (excluding soft x-ray and far infrared lasers).

quality of all-solid-state lasers, notably in the spectral range between 0.5 and 3 μm. For very short wavelengths, however, only gaseous or plasma sources are possible because of the approximate 100-nm transparency limit of solids.

Gas lasers have been pumped by a variety of processes, several of which are not available for other laser media: (1) direct electron impact, (2) excitation transfer from an ion or an excited atomic or molecular species, (3) highly exothermic chemical reactions, and (4) optical excitation. All have yielded examples of efficient systems and are partially responsible for the wide array of gas lasers that are available. Vacuum ultraviolet and ultraviolet lasers originate from states having short lifetimes and hence are generally pulsed lasers. Optical pumping is used extensively for molecular lasers operating in the mid- to far-infrared regions where CO_2 or other powerful molecular lasers are used for excitation. Far-infrared (FIR) lasers operate cw or quasi-cw and they are used in numerous applications.

Gas pressures range from those in conventional lasers, measured in Torr, to the near vacuum of free electron lasers and of planetary atmospheres and interstellar space (natural lasers), to the high-density, high-temperature plasmas of x-ray lasers. These latter lasers are discussed in Section 5: Other Lasers.

Gas lasers covered in Section 4 are divided into two sections: one on neutral atom, ion, and molecular lasers and one on optically pumped far infrared and millimeter wave lasers. In doing this one must decide on a definition of "far infrared". A perusal of numerous texts and handbooks will reveal a variety of definitions beginning at 10 to 25 μm and extending to 300–1000 μm. Here we use 20 μm as the lower limit for the far infrared. By so doing we avoid the task of separating out the extremely numerous CO_2, N_2O, and other laser transitions in the 10–20 μm region that may be optically pumped. (Others have skirted this issue by using the term "submillimeter" and beginning the tabulations at 10 μm, but omitted the extremely numerous CO_2 and N_2O transitions in the 10 μm region.)[20] We have extended the tabulations of lasing transitions to wavelengths of a few millimeters, thus overlapping millimeter wave masers. It should be noted that the section on neutral atom, ion, and molecular lasers also includes a number of molecular gas lasers that are optically pumped.

Neutral Atom, Ionized, and Molecular Gas Lasers

The wavelength range of neutral atom, ionized, and molecular gas lasers extends from the vacuum ultraviolet through the submillimeter region. (Extreme ultraviolet and soft x-ray lasers are covered separately in Section 5.1.) Individual gas lasers may lase at several different wavelengths and have narrow spectral linewidths. The output of argon lasers, for example, may consist of more than 30 lines of varying intensities from 275 to 686 nm. Neutral atom lasers emit in the visible and near infrared. Ion lasers emit in the ultraviolet and visible, the most important of which are based on the noble gas ions (argon, krypton, xenon). These are operated in various states of ionization and in either pulsed or cw modes.

Metal vapor lasers, which may be either neutral atoms or ions, emit in the near-ultraviolet and visible and operate either pulsed or cw. Of these, copper, gold, and lead are the most important examples with the first two being available commercially.

Excimer lasers are based upon the formation in the gas phase of transient molecules such as XeCl, ArF, KrF, most of which emit in the ultraviolet or vacuum ultraviolet. Produced by collisions between rare gas ions or neutrals in excited states and halogen-containing molecular precursors, these molecules have strongly bound upper laser levels but dissociative or weakly bound ground states. These diatomic molecules are generally produced in fast electrical discharges but can also be pumped optically or by intense electron or proton beams. Since the excited state lifetimes are short, ~10 ns, these lasers can emit powerful ultraviolet pulses of nanosecond duration.

Molecular lasers encompass a wide variety of molecules, operating conditions, and output wavelengths ranging from electronic transitions of the nitrogen (N_2) laser in the near ultraviolet, to the widely used vibrational-rotational transitions of the carbon dioxide (CO_2) laser in the mid-infrared, to the rotational transitions of various halide molecules lasing in the far infrared-submillimeter region. Electrically excited lasers such as H_2O, HCN, and DCN have transitions that extend well into the far infrared region.

In chemical lasers an inverted population is achieved by a chemical reaction (for example, the exothermic reaction of H_2 and F_2 to yield vibrationally excited HF). In the oxygen-iodine laser, excited oxygen transfers electronic energy to metastable levels of iodine. These lasers operate in the near to middle infrared and have been operated pulsed and cw. Several examples of chemical lasers are listed in Table 6.4.1.

The wavelengths of all of the above types of gas lasers are included in Table 4.1.1.

Optically Pumped Far Infrared and Millimeter Wave Lasers

These lasers are optically pumped by narrowband pump sources to excite molecules into a specific rotational state of an excited vibrational state and are usually operated cw or quasi-cw. Organic molecules used for far infrared and millimeter wave lasers now number more than one hundred, the most prominent ones being CH_3OH, $C_2H_2F_2$, and CH_3F. Combined with the use of several different isotopes and the possibility of transitions between many different vibrational and rotational levels, reported lasing transitions are now numbered in the thousands. The wavelengths of these transitions are given in Table 4.2.1. Pump transitions are not included in the table but can be found in the references cited in the introduction to the table.

Laser output powers depend not only on molecular properties but also on factors that vary with the design of the experiment. For those interested in the most intense far infrared and millimeter wave laser lines, a table of calibrated power measurements of over 150 lines between 40 μm and 2 mm having output powers of 1 mW or more is given in reference 20.

Optically pumped far infrared and millimeter wave lasers have found numerous applications in atomic and molecular spectroscopy, heterodyne sources for FIR spectroscopy, atmospheric spectroscopy, plasma diagnostics, and in frequency metrology connecting the microwave and visible regimes.

Other Lasers

Section 5 covers other types of lasers distinguished by their nature or specific characteristics, such as less conventional lasing configurations, extreme spectral regions, or specific operating conditions. In general they constitute many of the most interesting realms of lasing. Most are exploratory laboratory lasers rather than commercial lasers.

Extreme Ultraviolet and Soft X-Ray Lasers

In the decade since the first demonstration of lasing in the extreme ultraviolet (EUV) region, extensive efforts have been devoted to exploring lasing of various ions and in extending the wavelength range of laser action. Lasers have now been observed to operate from the extreme ultraviolet to well out into the soft x-ray region. (87.4 to 3.56 nm). These wavelengths are of particular interest for photolithography and biological imaging instrumentation. Progress in extreme ultraviolet lasers has been made possible in part by advances in multilayer techniques for producing the requisite optics for these wavelengths.

As shown in Figure 1.14, lasing has been reported for more than one-half of the elements in the periodic table. Lasing is achieved in highly ionized plasmas, generally produced by pulses from large, high-power lasers incident on solid targets, by electron collisional excitation from an ionic ground state into the upper laser level, or by sequential ionization followed by recombination into the upper laser level. Recently extreme ultraviolet lasing has

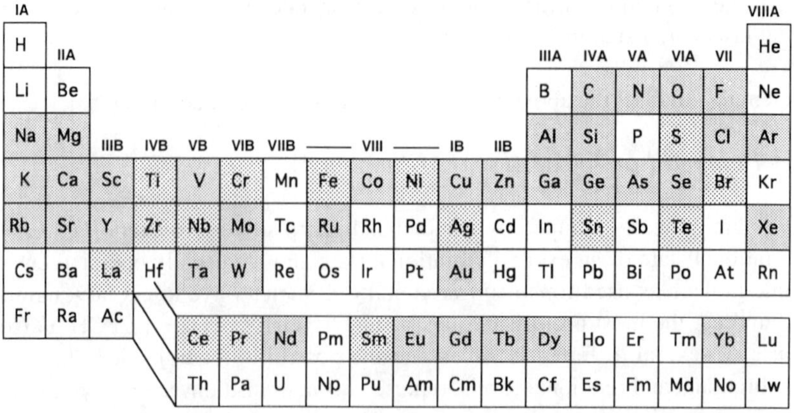

Figure 1.14 Periodic table of the elements showing the elements (shaded) that have been reported to exhibit laser action in the extreme ultraviolet and soft x-ray regions.

been reported from so-called table-top lasers in which fast electric excitation of the plasma is confined in a narrow capillary channel. The discharge plasma is formed by a low-induction, high-voltage current pulse.[21]

The electron configurations of the highly ionized states are similar to those of neutral atoms with the same number of electrons. Thus Al^{10+}, for example, which has only three electrons is described as "Li-like", Se^{24+} (with ten electrons) is described as "Ne-like", and Ta^{45+} (with 28 electrons) is described as "Ni-like". The wavelengths and transitions of extreme ultraviolet and soft x-ray lasers are given in Table 5.1.1.

Free Electron Lasers

Free electron lasers (FEL) provide laser radiation over a large wavelength range spanning six orders of magnitude—from the ultraviolet to millimeter waves (248 nm to 8 mm). They are based on a beam of high energy electrons traversing a spatially varying magnetic field (wigglers, undulators) which cause the electrons to oscillate and emit radiation. Because there is a continuum of states rather than discrete states as for atoms or molecules, a FEL is not a quantum device; almost all features of a FEL can be described classically.

Free electron laser configurations include (1) oscillators with reflectors at the ends of the magnet array, thus providing multi-pass, low-gain operation, (2) amplifiers in which electrons are injected into an undulator in synchronism with a signal derived from a conventional laser source, and (3) self-amplified spontaneous emission amplified by a single pass through a wiggler.

Various types of electron accelerators are used for FELs: storage rings, rf and induction linacs, electrostatic and pulse-line accelerators (operating in single-shot mode), microtrons, modulators, and ignition coils. RF linacs have been the dominant accelerator used. The wavelength ranges of FELs for different types of accelerators are shown in Figure 1.15. Storage rings provide the highest energy electrons and should be exploitable to achieve wavelengths of less than 100 nm.

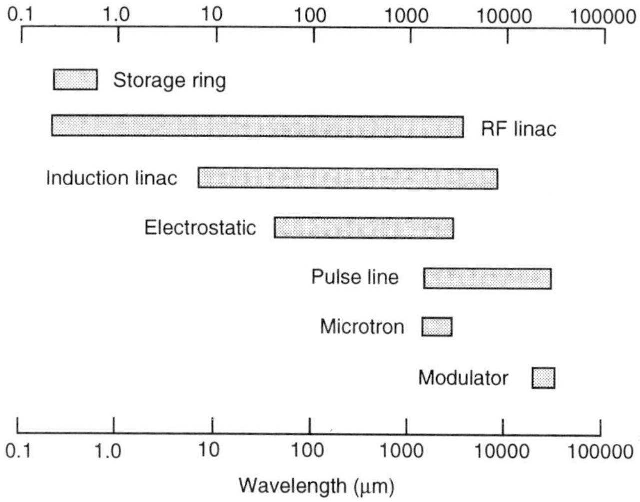

Figure 1.15 Reported ranges of output wavelengths of free electron lasers for various types of electron accelerators (see Table 5.2.1).

Flexibility of design and of operation configuration are characteristic of FELs. They can be operated both cw and pulsed with peak powers of ~1 GW. The spectral output is generally that of the Fourier transform of the optical pulse length. Because the resonance condition depends on the electron velocity component along the undulator axis and the wavelength range is determined by the energy range of the electrons, it is possible to build a continuously tunable source. Such FELs are potentially useful for medical applications and condensed matter research. Current research and development is aimed at obtaining high average power with good overall efficiency, broad bandwidth, and in compact systems. There is also promising work in the area of designing an x-ray FEL.

Nuclear Pumped Lasers

Nuclear pumped lasers (NPL) are gas lasers excited directly or indirectly by high energy particles or gamma rays resulting from nuclear reactions (fission, fusion, radioisotope). This may occur either in a reactor or a nuclear explosion, thus NPLs can be grouped into two broad categories: reactor pumped lasers and nuclear device pumped lasers. Both types provide direct conversion of nuclear energy to directed optical energy. Nuclear pumped lasers have been demonstrated to operate pulsed or steady-state over a wavelength range extending from the vacuum ultraviolet (173 nm) to the infrared (3 μm) using a variety of gases and molecules. These results are summarized in Table 5.3.1.

Natural Lasers

Naturally occurring maser action is frequently found in clouds of molecular gases in our galaxy where water or other molecules amplify radiation from stars. Whereas natural masers operating at microwave and millimeter wavelengths have been known for several decades and now number in the hundreds,[22,23] shorter wavelength natural lasers are much rarer. Those discovered thus far operate at infrared wavelengths of CO_2 in the mesosphere and thermosphere of Mars and Venus and, more recently, at the submillimeter wavelengths of hydrogen in interstellar and circumstellar sources. The tabulated data for these lasers in Table 5.4.1 are necessarily limited, however comprehensive references are provided.

Several years ago C. W. Townes made the thought provoking observation " . . . that if radio astronomy had been sponsored more strongly in the United States we probably would have discovered masers and lasers sooner. Masers could have been detected long ago in the sky—probably as early as the 1930s, and certainly immediately after or during World War II they could easily have been detected, but nobody was looking."[24]

Inversionless Lasers

It is possible to extract energy from a medium even if there are more atoms in the lower level than the upper level. Lasing without inversion (LWI) has been the subject of considerable interest in the past decade and although results to date are not extensive they are included in this volume for completeness. Two schemes, Λ and V, have been used to obtain gain. In the first scheme, the optical fields have a common upper level and LWI is achieved via a coherence between the two lower levels. In the V scheme, the fields have a common lower level and LWI is achieved via a coherence between two upper levels. Gain and cw laser oscillation have now been observed in metal vapors using both Λ and V schemes at wavelengths in the visible-near visible region. These experiments are summarized in Table 5.5.1.

Commercial Lasers

Of the over 15,000 lasing transitions reported in this volume and the many types of lasers that have been demonstrated, only a comparatively few lasers are available commercially. These include numerous gas lasers (noble gases, molecular gases, metal vapors, excimers), liquid lasers (using a multitude of organic dyes to cover the entire visible and near infrared spectral regions), and solid state lasers (paramagnetic ions in crystals and glasses, color centers, organic dyes, and semiconductor materials of various structures and configurations).

Commercial lasers are listed in order of increasing wavelength in Table 6.1 and include (1) gas lasers (atomic: helium-neon, helium-cadmium; ionic: Ar^+, Kr^+), excimer lasers (KrF, ArF, XeF, XeCl), molecular gas lasers (CO_2 and CO), metal vapor lasers (principally Cu and gold), and molecular lasers extending into the submillimeter and millimeter wavelength range, (2) solid state lasers of both lanthanide and iron group ions in crystals and glasses and color center lasers, (3) dye lasers using many organic dyes in liquid solvents, and (4) semiconductor lasers, both single-element and multi-element arrays.

General output properties of representative solid state, semiconductor, dye, and gas lasers are given in Tables 6.2.1, 6.3.1 6.4.1 and 6.5.1, respectively. These data are taken from recent (1995–1997) buyers guides and manufacturer's literature and are representative rather than exhaustive. Also, performance figures may be expected to change due to advances in technology. Ranges of output wavelengths of selected tunable commercial lasers are shown in Figure 1.16.

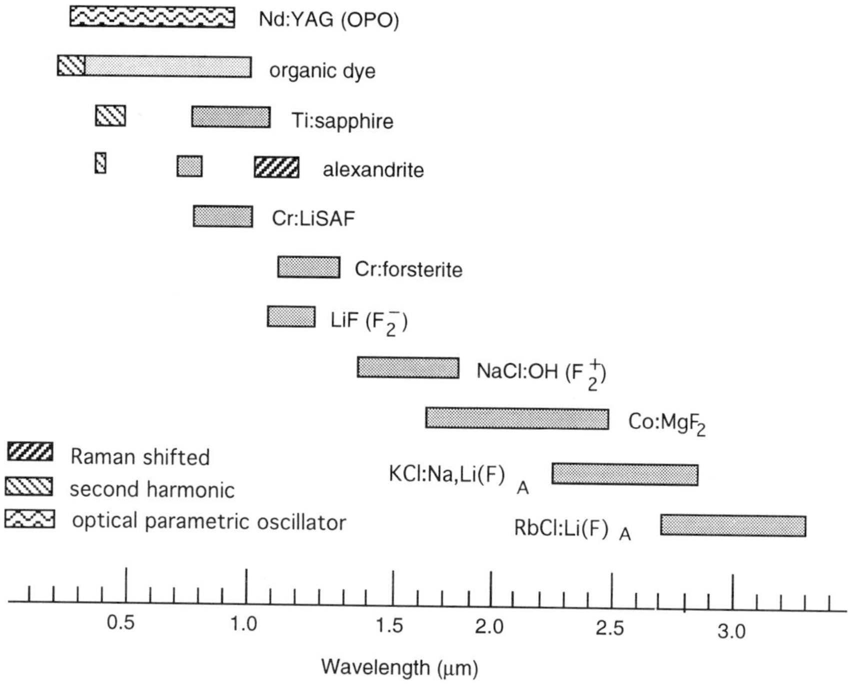

Figure 1.16 Ranges of output wavelengths of selected tunable commercial lasers (1995–1997 data).

Commercial lasers are now available with an astonishingly large range of properties—wavelengths from the ultraviolet through the infrared, pulse lengths from femtoseconds to microseconds to continuous wave, operating with predicted lifetimes in some cases of greater than 10^6 hours, highly monochromatic or tunable, and with peak powers of many joules and cw powers of tens of kilowatts. These lasers are used both for special research and for numerous commercial applications.

Because of their use in local and long distance communications, barcode readers, compact disk (CD) players for entertainment and computer applications, and as pump sources for solid-state lasers, the market quantities of semiconductor diode lasers (~10^8 units per year) generally are many orders of magnitude larger and unit prices are many orders of magnitude smaller than those for most gas or solid state lasers.

The market share of various categories of lasers can change with changing technology. Somewhat in the same way that the vacuum tube in electronics gave way to solid state transistors and integrated circuits, gas lasers—which have been predominant in many applications—are being replaced in a number of applications by all-solid-state lasers pumped by diode lasers.

References

1. An, K., Childs, J. J., Dasari, R. R. and Feld, M. S., Microlaser: a laser with one atom in an optical resonator, *Phys. Rev. Lett.* 73, 3375 (1994).
2. Padmabamdi, G. G., Welch, G. R., Shubin, I. N., Fry, E. S., Nikonov, D. E., Lukin, M. D. and Scully, M. O., Laser oscillation without population inversion in a sodium atomic bean, *Phys. Rev. Lett.* 76, 2053 (1996).
3. Andrews, M. R., Townsend, C. G., Miesner, H.-J., Durfee, D. S., Kurn, D. M. and Ketterle, W., Observation of interference between two Bose condensates, *Science* 275, 637 (1997).
4. Kaminskii, A.A., *Crystalline Lasers: Physical Processes and Operating Schemes*, CRC Press, Boca Raton, FL (1996).
5. Johnson, L. F. and Guggenheim, H. J., Infrared-pumped visible laser, *Appl. Phys. Lett.*, 19, 44 (1971).
6. Weber, M. J., Science and technology of laser glass, *J. Non-Cryst. Solids* 123, 208 (1991).
7. Hänsch, T. W., Pernier, M. and Schawlow, A. L., Laser action of dyes in gelatin, *IEEE J. Quantum Electron.* QE-7, 45 (1971).
8. Siddique, M., Yang, L., Wang, Q. Z. and Alfano, R. R., Mirrorless laser action from optically pumped dye-treated animal tissues, *Optics Commun.* 117, 475 (1995).
9. Dodabalapur, A., Chandross, E. A., Berggren, M. and Slusher, R. L., Organic solid-state lasers: past and future, *Science* 277, 1787 (1997).
10. Zhang, Q. J., Wang, P., Sun, X. F., Zhai, Y., Dai, P., Yang, B., Hai, M. and Xie, J. P., Amplified spontaneous emission of an Nd^{3+}-doped poly(methyl methacrylate) optical fiber at ambient temperature, *Appl. Phys Lett.* 72, 407 (1998).
11. Schwentner, N. and Apkarian, V. A., A solid state rare gas halide laser: XeF in crystalline argon, *Chem. Phys. Lett.* 154, 413 (1989).
12. Zerza, G., Sliwinski, G. and Schwentner, N., Threshold and saturation properties of a solid-state XeF (C-A) excimer laser, *Appl. Phys. B* 55, 331(1992).
13. Zerza, G., Sliwinski, G. and Schwentner, N., Laser investigations at 269 nm for XeF (D-X) in Ne crystals, *Appl. Phys. A* 56, 156 (1993).
14. Tessler, N., Denton, G. J. and Friend, R. H., Lasing from conjugated-polymer microcavities, *Nature* 382, 695 (1996).

15. Maeda, M., *Laser Dyes*, Academic, New York (1984).
16. See, for example, Sha, W. L., Liu, C.-H. and Alfano, R. R., Specral and temporal measurements of laser action of Rhodamine 640 dye in strongly scattering media, *Opt. Lett.* 19, 1922 (1994) and Siddique, R., Alfano, R. R., Berger, G. A., Kempe, M. and Genack, A. Z., Time-resolved studies of stimulated emission from colloidal dye solutions, *Opt. Lett.* 21, 450 (1996) and references cited therein.
17. Biswas, A., Latifi, H., Armstrong, R. L. and Pinnick, R. G., Time-resolved spectroscopy of laser emission from dye-doped droplets, *Opt. Lett.* 14, 214 (1989).
18. Taniguchi, H. and Tomisawa, H., Suppression and enhancement of dye lasing and stimulated Raman scattering from various dye-doped liquid spheres, *Opt. Lett.* 19, 1403 (1994) and references cited therein.
19. Essien, M., Armstrong, R. L. and Gillespie, J. B., Lasing emission from an evaporating layered microdroplet, *Opt. Lett.* 18, 762 (1993).
20. Douglas, N. G., *Millimetre and Submillimetre Wavelength Lasers*, Springer Verlag, Berlin (1987).
21. Rocca, J. J., Shlyaptsev, V., Tomasel, F. G., Cortazar, O. D., Hartshorn, D. and Chilla, J. L. A., Demonstration of a discharge pumped table top soft x-ray laser, *Phys. Rev. Lett.* 73, 2192 (1994); Rocca, J. J., Tomasel, F. G., Marconi, M. C., Shlyaptsev, V. N., Chilla, J. L. A., Szapiro, S. T. and Guidice, G., Discharge-pumped soft x-ray laser in neon-like argon, *Phys. Plasmas* 2, 2547 (1995).
22. Moran, J. M., Maser action in nature, in *Handbook of Laser Science and Technology, Suppl. 1: Lasers,* CRC Press, Boca Raton, FL (1991), p. 579.
23. Elitzur, M., *Astronomical Masers*, Kluwer, New York (1992).
24. Townes, C.H., The early years of research on astronomical masers, in *Astrophysical Masers*, Clegg, A. W. and Nedoluha, G. E., Eds., Springer-Verlag, Berlin (1993), p. 3.

Further Reading

Bertolotti, M., *Masers and Lasers: An Historical Approach*, Hilger, Bristol (1983).

Davis, C. C., *Lasers and Electro-Optics: Fundamentals and Engineering*, Cambridge University Press, New York (1996).

Hecht, J., *Understanding Lasers*, (second edition), IEEE Press, New York (1994).

Svelto, O., *Principles of Lasers*, Plenum, New York (1998).

Meyers, R. A., Ed., *Encyclopedia of Lasers and Optical Technology*, Academic Press, San Diego (1991).

Siegman, A. E., *Lasers*, University Science, Mill Valley, CA (1986).

Silfvast, W. T, Ed., *Selected Papers on Fundamentals of Lasers*, SPIE Milestone Series, Vol. MS 70, SPIE Optical Engineering Press, Bellingham, WA (1993).

Weber, M. J. (Ed.), *Handbook of Laser Science and Technology, Vol. 1 : Lasers and Masers* (1982)*; Vol. II: Gas Lasers* (1982)*; and Supplement 1: Lasers* (1991), CRC Press, Boca Raton, FL.

Yariv, A., *Quantum Electronics*, John Wiley & Sons, New York (1989).

Section 2: Solid State Lasers

Section 2.1
CRYSTALLINE PARAMAGNETIC ION LASERS

Introduction to the Tables

Crystalline paramagnetic ion lasers have used lanthanide and other transition group ions as the active ion. The operative transitions and general wavelengths for these laser ions are given in Tables 2.1.1 and 2.1.2. Although the exact wavelength varies slightly with the crystal host and temperature, these tables can generally be used to identify the transitions involved in laser action.

A few ion-crystal systems in which only gain has been reported are listed in Table 2.1.3.

Crystalline lasers based on transition metal and lanthanide ions are arranged in order of increasing wavelength in Table 2.1.4. Lasers that have been tuned over a range of wavelengths are listed in order of the lowest lasing wavelength reported; the tuning range given is that for the configuration and conditions used and may not represent the extremes possible. A range of wavelengths in brackets denotes a number of discrete lines. The lasing ion is given in the second column and the host crystal in the third column. If codopants or sensitizing ions are added to the host, they are listed following the colon. The operating temperature and reference(s) are given in the final two columns.

A range of wavelengths is also listed if the host consists of mixed elements or a corresponding range of temperatures is given. In such cases the references should be consulted to determine the wavelengths for the specific compositions and temperatures used.

Further Reading

Caird, J. and Payne, S. A., Crystalline Paramagnetic Ion Lasers, in *Handbook of Laser Science and Technology*, *Suppl. 1*: *Lasers*, CRC Press, Boca Raton, FL (1991), p. 3.

Cheo, P. K., Ed., *Handbook of Solid-State Lasers*, Marcel Dekker Inc., New York (1989).

Kaminskii, A. A., *Crystalline Lasers: Physical Processes and Operating Schemes*, CRC Press, Inc., Boca Raton, FL (1996).

Kaminskii, A. A., *Laser Crystals, Their Physics and Properties*, Springer-Verlag, Heidelberg (1990).

Koechner, W., *Solid-State Laser Engineering* (fourth edition), Springer Verlag, Berlin (1996).

Moulton, P., Paramagnetic Ion Lasers, in *Handbook of Laser Science and Technology, Vol. I: Lasers and Masers,* CRC Press, Boca Raton, FL (1995), p. 21

Powell, R. C., *Physics of Solid State Laser Materials*, Springer-Verlag, Berlin (1997).

Powell, R. C., Ed., *Selected Papers on Solid State Lasers*, SPIE Milestone Series, Vol. MS31, SPIE Optical Engineering Press, Bellingham, WA (1991).

See, also, Tunable Solid-State Lasers, *Selected Topics in Quantum Electronics* 1 (1995), Diode-Pumped Solid-State Lasers, *Selected Topics in Quantum Electronics* 3(1) (February 1997), and the following proceedings of the Advanced Solid State Laser Conferences, all published by the Optical Society of America, Washington, DC:

OSA Trends in Optics and Photonics on Advanced Solid State Lasers, Vol. 1, Payne, S. A. and Pollack, C. R., Eds., (1996)

Chai, B. H. T. and Payne, S. A., Eds., Proceedings Vol. 24 (1995).

Fan, T. Y. and Chai, B., Eds., Proceedings Vol. 20 (1994).

Pinto, A. A. and Fan, T. Y., Eds., Proceedings Vol. 15 (1993).

Chase, L. L. and Pinto, A. A., Eds., Proceedings Vol. 13 (1992).

Dubé, G. and Chase, L. L, Eds., Proceedings Vol. 10 (1991).

Jenssen, H. P. and Dubé, G., Eds., Proceedings Vol. 6 (1990).

Table 2.1.1
Transition Metal Ion Laser Transitions in Solids

Wavelength* (μm)	Ion	Transition
0.66–1.18	Ti^{3+}	$^2E \rightarrow {}^2T_2$
0.68–0.70	Cr^{3+}	$^2E \rightarrow {}^4A_2$
0.74–0.89	Cr^{3+}	$^4T_2 \rightarrow {}^4A_2$
1.05–1.33	V^{2+}	$^4T_2 \rightarrow {}^4A_2$
1.17–1.18	Mn^{5+}	$^1E \rightarrow {}^3A_2$
1.17–1.63	Cr^{4+}	$^3T_2 \rightarrow {}^3A_2$
1.31–1.94	Ni^{2+}	$^3T_2 \rightarrow {}^3A_2$
1.62–2.50	Co^{2+}	$^4T_2 \rightarrow {}^4T_1$
2.2–2.9	Cr^{2+}	$^5E \rightarrow {}^5T_2$
3.5	Fe^{2+}	$^5T_2 \rightarrow {}^5E$

* Wavelengths of transitions are only approximate; exact wavelengths and ranges are depend on the host and temperature.

Table 2.1.2
Lanthanide and Actinide Laser Ion Transitions in Solids

Wavelength* (μm)	Ion	Transition
0.17	Nd^{3+}	$5d \rightarrow {}^4I_{11/2}$
0.30	Ce^{3+}	$5d \rightarrow {}^2F_{5/2}$
0.32	Ce^{3+}	$5d \rightarrow {}^2F_{7/2}$
0.35	Tm^{3+}	${}^1G_4 \rightarrow {}^3H_5$
0.38	Nd^{3+}	${}^4D_{3/2} \rightarrow {}^4I_{11/2}$
0.41	Nd^{3+}	${}^2P_{3/2} \rightarrow {}^4I_{11/2}$
0.45	Tm^{3+}	${}^1D_2 \rightarrow {}^3F_4$
0.46	Tm^{3+}	${}^1I_6 \rightarrow {}^3H_4$
0.46	Pr^{3+}	${}^3P_1 \rightarrow {}^3H_4$
0.47	Er^{3+}	${}^2P_{3/2} \rightarrow {}^4I_{11/2}$
0.48	Tm^{3+}	${}^1G_4 \rightarrow {}^3H_6$
0.49	Pr^{3+}	${}^3P_0 \rightarrow {}^3H_4$
0.51	Tm^{3+}	${}^1D_2 \rightarrow {}^3H_5$
0.53	Pr^{3+}	${}^3P_1 \rightarrow {}^3H_5$
0.54	Pr^{3+}	${}^3P_0 \rightarrow {}^3H_5$
0.54	Tb^{3+}	${}^5D_4 \rightarrow {}^7F_5$
0.55	Er^{3+}	${}^4S_{3/2} \rightarrow {}^4I_{15/2}$
0.55	Ho^{3+}	${}^5S_2 \rightarrow {}^5I_8$
0.56	Er^{3+}	${}^2H_{9/2} \rightarrow {}^4I_{13/2}$
0.59	Sm^{3+}	${}^4G_{5/2} \rightarrow {}^6H_{7/2}$
0.60	Pr^{3+}	${}^3P_0 \rightarrow {}^3H_6$
0.61	Eu^{3+}	${}^5D_0 \rightarrow {}^7F_2$
0.62	Er^{3+}	${}^4G_{11/2} \rightarrow {}^4I_{11/2}$
0.62	Er^{3+}	${}^2P_{3/2} \rightarrow {}^4F_{9/2}$
0.63	Pr^{3+}	${}^3P_2 \rightarrow {}^3F_4$
0.64	Pr^{3+}	${}^3P_0 \rightarrow {}^3F_2$
0.65	Sm^{3+}	${}^4G_{5/2} \rightarrow {}^6H_{7/2}$
0.65	Tm^{3+}	${}^1G_4 \rightarrow {}^3F_4$
0.67	Er^{3+}	${}^4F_{9/2} \rightarrow {}^4I_{15/2}$
0.70	Pr^{3+}	${}^3P_0 \rightarrow {}^3F_3$

Table 2.1.2—*continued*
Lanthanide and Actinide Laser Transitions in Solids

Wavelength* (μm)	Ion	Transition
0.70	Er^{3+}	$^2H_{9/2} \rightarrow {}^4I_{11/2}$
0.70	Sm^{2+}	5d, $^5D_0 \rightarrow {}^7F_1$
0.72	Pr^{3+}	$^3P_0 \rightarrow {}^3F_4$
0.73	Nd^{3+}	$^2P_{3/2} \rightarrow {}^4F_{5/2}$
0.75	Ho^{3+}	$^5S_2 \rightarrow {}^5I_7$
0.80	Tm^{3+}	$^3H_4 \rightarrow {}^3H_6$
0.80	Tm^{3+}	$^1G_4 \rightarrow {}^3H_5$
0.85	Er^{3+}	$^4S_{3/2} \rightarrow {}^4I_{13/2}$
0.88	Pr^{3+}	$^3P_1 \rightarrow {}^1G_4$
0.91	Pr^{3+}	$^3P_0 \rightarrow {}^1G_4$
0.93	Nd^{3+}	$^4F_{3/2} \rightarrow {}^4I_{9/2}$
0.93	Pm^{3+}	$^5F_1 \rightarrow {}^5I_5$
0.98	Ho^{3+}	$^5F_5 \rightarrow {}^5I_7$
0.99	Er^{3+}	$^4I_{11/2} \rightarrow {}^4I_{15/2}$
0.99	Pr^{3+}	$^1D_2 \rightarrow {}^3F_3$
1.01	Ho^{3+}	$^5S_2 \rightarrow {}^5I_6$
1.03	Yb^{3+}	$^2F_{5/2} \rightarrow {}^2F_{7/2}$
1.04	Pr^{3+}	$^1G_4 \rightarrow {}^3H_4$
1.05	Pr^{3+}	$^1D_2 \rightarrow {}^3F_4$
1.06	Nd^{3+}	$^4F_{3/2} \rightarrow {}^4I_{11/2}$
1.1	Pm^{3+}	$^5F_1 \rightarrow {}^5I_6$
1.2	Ho^{3+}	$^5I_6 \rightarrow {}^5I_8$
1.26	Er^{3+}	$^4S_{3/2} \rightarrow {}^4I_{11/2}$
1.26	Er^{3+}	$^4F_{9/2} \rightarrow {}^4I_{13/2}$
1.3	Dy^{3+}	$^6H_{9/2} \rightarrow {}^6H_{15/2}$
1.34	Pr^{3+}	$^1G_4 \rightarrow {}^3H_5$
1.35	Nd^{3+}	$^4F_{3/2} \rightarrow {}^4I_{13/2}$
1.40	Ho^{3+}	$^5S_2,{}^5F_4 \rightarrow {}^5I_5$
1.48	Tm^{3+}	$^3H_4 \rightarrow {}^3F_4$
1.49	Ho^{3+}	$^5F_5 \rightarrow {}^5I_6$

Table 2.1.2—*continued*
Lanthanide and Actinide Laser Transitions in Solids

Wavelength* (μm)	Ion	Transition
1.56	Er^{3+}	$^4I_{13/2} \rightarrow {}^4I_{15/2}$
1.58	Tm^{3+}	$^1G_4 \rightarrow {}^3F_3$
1.61	Pr^{3+}	$^3F_3 \rightarrow {}^3H_4$
1.66	Er^{3+}	$^2H_{11/2} \rightarrow {}^4I_{9/2}$
1.67	Er^{3+}	$^4S_{3/2} \rightarrow {}^4I_{9/2}$
1.68	Er^{3+}	$^4I_{9/2} \rightarrow {}^4I_{13/2}$
1.7	Ho^{3+}	$^5I_5 \rightarrow {}^5I_7$
1.8	Nd^{3+}	$^4F_{3/2} \rightarrow {}^4I_{15/2}$
1.9	Tm^{3+}	$^3F_4 \rightarrow {}^3H_6$
2.0	Er^{3+}	$^4F_{9/2} \rightarrow {}^4I_{11/2}$
2.0	Tm^{3+}	$^1G_4 \rightarrow {}^3F_2$
2.1	Ho^{3+}	$^5I_7 \rightarrow {}^5I_8$
2.3	Ho^{3+}	$^5F_5 \rightarrow {}^5I_5$
2.3	Tm^{3+}	$^3H_4 \rightarrow {}^3H_5$
2.35–2.83	U^{3+}	$^4I_{11/2} \rightarrow {}^4I_{9/2}$
2.4	Dy^{2+}	$^5I_7 \rightarrow {}^5I_8$
2.8	Er^{3+}	$^4I_{11/2} \rightarrow {}^4I_{13/2}$
2.9	Ho^{3+}	$^5I_6 \rightarrow {}^5I_7$
3.0	Dy^{3+}	$^6H_{11/2} \rightarrow {}^6H_{15/2}$
3.4	Ho^{3+}	$^5S_2 \rightarrow {}^5F_5$
3.5	Er^{3+}	$^4F_{9/2} \rightarrow {}^4I_{9/2}$
3.6	Pr^{3+}	$^1G_4 \rightarrow {}^3F_4$
3.9	Ho^{3+}	$^5I_5 \rightarrow {}^5I_6$
4.3	Dy^{3+}	$^6H_{11/2} \rightarrow {}^6H_{13/2}$
4.8	Er^{3+}	$^4I_{9/2} \rightarrow {}^4I_{11/2}$
5.1	Nd^{3+}	$^4I_{13/2} \rightarrow {}^4I_{11/2}$
5.2	Pr^{3+}	$^3F_3 \rightarrow {}^3H_6$
7.2	Pr^{3+}	$^3F_3 \rightarrow {}^3F_2$

*Wavelengths of transitions are only approximate; exact wavelengths are dependent on the host and temperature and the specific Stark levels involved.

Table 2.1.3
Paramagnetic Ions and Crystals Exhibiting Gain

Wavelength (μm)	Ion	Crystal	Temperature (K)	Reference
0.186[(a)]	Nd^{3+}	$LiYF_4$	300	1
0.219[(b)]	F(2p)–Ba(5p)	BaF_2	300	2,3
~0.337	Ag^+	RbBr, KI	5	4,5
0.388–0.524	Ti^{4+}	Li_2GeO_3	300	6
0.392	biexciton	CuCl:NaCl	77–108	7,8
0.407	Tl^+	CsI	—	9
0.420	Tl^+	KI	—	10
0.442	In^+	KCl	—	11
0.500–0.550	$UO_2{}^{2+}$	$Ca(UO_2)(PO_4){\cdot}H_2O$	—	12
0.5145	Cu^+	Na–β"–alumina[(c)]	300	13
0.6328	Cu^+	Ag–β"–alumina	300	13
0.700–0.720	Rh^{2+}	$RbCaF_3$	300	14
0.910	Cr^{3+}	$LiNbO_3$	300	15
0.927	Cr^{3+}	$LiNbO_3$	300	16
1.061[(d)]	Nd^{3+}	$Gd_3(Sc,Ga)_5O_{12}$:Cr	300	17
1.064	V^{2+}	$KMgF_3$	—	18
~1.080[(e)]	Nd^{3+}	ZnS film	77	19
1.15	Mn^{5+}	Ca_2PO_4Cl	300	20
1.15–1.5	Cr^{4+}	$CaGd_4(SiO_4)_3O$	300	21
~1.2	Mn^{5+}	$Sr_5(PO_4)_3Cl$	300	20
1.319	Dy^{3+}	$LaCl_3$	300	22
1.338	Dy^{3+}	$LaCl_3$	300	22

(a) Because of the presence of excited state absorption, a negative net-induced gain coefficient was measured.
(b) Core-valence crossover transition: $F^-(2p) \rightarrow Ba^{3+}(5p)$.
(c) Typical composition: $Na_{1.67}Mg_{0.67}Al_{10.33}O_{19}$.
(d) X-ray induced optical gain.
(e) Direct current electroluminescence (DCEL) and cathodoluminescence.

References—Table 2.1.3

1. Cashmore, J. S., Hooker, S. M. and Webb, C. E., Vacuum ultraviolet gain measurements in optically pumped $LiYF_4:Nd^{3+}$, *Appl. Phys. B* 64, 293 (1997).

2. Itoh, M. and Itoh, H., Stimulated ultraviolet emission from BaF_2 under core-level excitation with undulator radiation, *Phys. Rev. B* 46, 15 509 (1992).
3. Liang, J., Yin, D., Zhang, T. and Xue, H., Amplified spontaneous emission of a BaF_2 crystal, *J. Lumin.* 46, 55 (1990).
4. Schmitt, K., Stimulated C^+-emission of Ag^--centers in KI, RbBr, and CsBr, *Appl. Phys. A* 38, 61 (1985).
5. Boutinaud, P., Monnier, A. and Bill, H., Ag^+ center in alkaline-earth fluorides: new UV solid state lasers?, *Rad. Eff. Def. Solids* 136, 69 (1995).
6. Loiacono, G. M., Shone, M. F., Mizell, G., Powell, R. C., Quarles, G. J. and Elonadi, B., Tunable single pass gain in titanium-activated lithium germanium oxide, *Appl. Phys. Lett.* 48, 622 (1986).
7. Masumoto, Y. and Kawamjra, T., Biexciton lasing in CuCl quantum dots, *Appl. Phys. Lett.* 62, 225 (1993).
8. Masumoto, Y., Luminescence and lasing of CuCl nanocrystals, *J. Lumin.* 60/61, 256 (1994).
9. Pazzi, G. P, Baldecchi, M. G., Fabeni, P., Linari, R., Ranfagni, A., Agresti, A., Cetica, M. and Simpkin, D. J., Amplified spontaneous emission in doped alkali-halides, *SPIE* 369, 338 (1982).
10. Nagli, L. E. and Plyovin, I. K., Induced recombination emission of activated alkali halide crystals, *Opt. Spectrosc.* (USSR) 44, 79 (1978).
11. Shkadarevich, A. P., Recent advances in tunable solid state lasers, in *Tunable Solid State Lasers,* Shand, M. L. and Jenssen, H. P., Eds., Optical Society of America, Washington, DC (1989), p. 66.
12. Haley, L. V. and Koningstein, J. A., Time resolved stimulated fluorescence of the uranyl ion in the mineral metaautunite, *J. Phys. Chem. Solids* 44, 431 (1983).
13. Barrie, J. D., Dunn, B., Stafsudd, O. M. and Nelson, P., Luminescence of Cu^+-β-alumina, *J. Lumin.* 37, 303 (1987).
14. Powell, R. C., Quarles, G. L., Martin, J. J., Hunt, C. A. and Sibley, W. A., Stimulated emission and tunable gain from Rh^{2+} ion lin $RbCaF_3$ crystals, *Opt. Lett.* 10, 212 (1985).
15. Zhou, F., De La Rue, R. M., Ironside, C. N., Han, T. P. J., Hendersen, B. and Ferguson, A. I., Optical gain in proton-exchanged $Cr:LiNbO_3$ waveguides, *Electron. Lett.* 28, 204 (1992).
16. Almeida, J. M., Leite, A. P., De La Rue, R. M., et al., Spectroscopy and optical amplification in Cr doped $LiNbO_3$, *OSA Trends in Optics and Photonics on Advanced Solid State Lasers*, Vol. 1, Payne, S. A. and Pollack, C. R., Eds., Optical Society of America, Washington, DC (1996), p. 478.
17. Brannon, P. J., X-ray induced optical gain in Cr,Nd:GSGG, *OSA Proc. Adv. Solid State Lasers*, Chai, B. H. T. and Payne, S. A. (eds) 24, 232 (1995).
18. Moulton, P. F., Recent advances in solid-state laser materials, in *Materials Research Society Symposium Proceedings* 24, 393 (1984).
19. Zhong, G. Z. and Bryant, F. J., Laser phenomena in DCEL of ZnS:Cu:Nd:Cl thin films, *Solid State Commun.* 39, 907 (1981).
20. Capobianco, J. A., Cormier, G., Moncourgé, R., Manaa, H. and Bettinelli, M., Gain measurements of Mn^{5+} ($3d^2$) doped $Sr_5(PO_4)_3Cl$ and Ca_2PO_4Cl, *Appl. Phys. Lett.* 60, 163 (1992).
21. Moncorgé, R., Manaa, H, Deghoul, F., Borel, C. and Wyon, Ch., Spectroscopic study and laser operation of Cr^{4+}-doped $(Sr,Ca)Gd_4(SiO_4)_3O$ single crystal, *Opt. Commun.* 116, 393 (1995).
22. Schaffers, K. I., Page, R. H., Beach, R. J., Payne, S. A. and Krupke, W. F., Gain measurements of Dy^{3+}-doped $LaCl_3$: a potential 1.3 μm optical amplifier for telecommunications, *OSA Trends in Optics and Photonics on Advanced Solid State Lasers*, Vol. 1, Payne, S. A. and Pollack, C. R., Eds., Optical Society of America, Washington, DC (1996), p. 469.

Table 2.1.4
Paramagnetic Ion Lasers Arranged in Order of Wavelength

Wavelength (μm)	Ion	Host: sensitizer ion(s)	Temp. (K)	References
0.172	Nd^{3+}	LaF_3	300	1–4
0.285–0.297	Ce^{3+}	$LiSrAlF_6$	300	5,6
0.286	Ce^{3+}	LaF_3	300	7
0.288	Ce^{3+}	$LiCaAlF_6$	300	6,8
0.306–0.315	Ce^{3+}	$LiYF_4$	300	7,9
0.3146[(a)]	Gd^{3+}	$Y_3Al_5O_{12}$	300	10
0.323–0.328	Ce^{3+}	$LiYF_4$	300	7,9,11
0.323–0.335	Ce^{3+}	$LiLuF_4$	300	12
0.335	Ag^{+}	KI, RbBr, CsBr	5	13
0.345	Ce^{3+}	BaY_2F_8	300	14
0.38006	Nd^{3+}	LaF_3	20–77	15
0.38052	Nd^{3+}	LaF_3	20–77	15
0.413	Nd^{3+}	$LiYF_4$	12	77
0.4502	Tm^{3+}	$LiYF_4$	≤70	18
0.45020	Tm^{3+}	$LiYF_4$	75	19
0.4526	Tm^{3+}	$LiYF_4$	300	20
~0.455[(b)]	Nd^{3+}	$YAl_3(BO_3)_4$	300	1018
0.456	Tm^{3+}	BaY_2F_8:Yb	300	21
0.4697	Er^{3+}	$LiYF_4$	35	22
0.479	Pr^{3+}	$LiYF_4$	300	23
0.48[(b,c)]	Cr^{3+}	$La_3Ga_5SiO_{14}$	300	1017
0.482	Tm^{3+}	BaY_2F_8:Yb	300	21
0.4835	Tm^{3+}	$LiYF_4$	≤160	18
0.4879	Pr^{3+}	$Y_3Al_5O_{12}$	4-32	657
0.488	Pr^{3+}	$SrLaGa_3O_7$	≤230	24
0.4892	Pr^{3+}	$LaCl_3$	5.5–14	25
0.4892	Pr^{3+}	$(La,Pr)Cl_3$	14	26–28
0.5[(b,c)]	Cr^{3+}	$Ca_2Ga_2SiO_7$	300	1017
0.5[(b,c)]	Cr^{3+}	$Sr_3Ga_2GeO_{14}$	300	1017
0.512	Tm^{3+}	BaY_2F_8:Yb	300	21
0.52[(b,c)]	Cr^{3+}	$La_3Ga_5GeO_{14}$	300	1017
0.522	Pr^{3+}	$GdLiF_4$	300	29
0.5236[(b)]	Nd^{3+}	$LiYF_4$	300	316
0.5241[(b)]	Nd^{3+}	$LaBGeO_5$	300	1013–1015
0.5298	Pr^{3+}	$(La,Pr)Cl_3$	12	26-28

Table 2.1.4—*continued*
Paramagnetic Ion Lasers Arranged in Order of Wavelength

Wavelength (μm)	Ion	Host: sensitizer ion(s)	Temp. (K)	References
~0.53[(b)]	Nd^{3+}	$YAl_3(BO_3)_4$	300	1019,1020
0.53[(b,c)]	Cr^{3+}	$La_3Ga_{5.5}Nb_{0.5}O_{14}$	300	1017
0.530[(b)]	Nd^{3+}	$YAl_3(BO_3)_4$	300	47
0.530[(b)]	Nd^{3+}	$Ca_4GdO(BO_3)_3$	300	489
0.532	Pr^{3+}	$LaBr_3$	<300	30
0.5323[(b)]	Nd^{3+}	$La_3Ga_{5.5}Nb_{0.5}O_{14}$	300	1016
0.5335[(b)]	Nd^{3+}	$La_3Ga_5SiO_{14}$	300	1016,1022,1023
0.5337[(b)]	Nd^{3+}	$La_3Ga_5GeO_{14}$	300	1016
0.5344[(b)]	Nd^{3+}	$Ca_2Ga_2SiO_7$	300	1016
0.5347[(b)]	Nd^{3+}	$Sr_3Ga_2GeO_{14}$	300	1017
0.536	XeF	Ar	20	16,17
0.5378	Pr^{3+}	$LiYF_4$	110	31
0.5380	Pr^{3+}	$LiLuF_4$	110	32,33
0.54[(b,c)]	Cr^{3+}	$La_3Ga_{5.5}Ta_{0.5}O_{14}$	300	1017
0.5425[(b)]	Nd^{3+}	$LiNbO_3$:MgO	408–417	1006
0.544075	Er^{3+}	$LiYF_4$	20	34
0.5441[(b)]	Nd^{3+}	$LiNbO_3$:MgO	300	1009–1011
0.5445	Tb^{3+}	$LiYF_4$	300	35
0.545	Pr^{3+}	$GdLiF_4$	300	29
0.5457[(b)]	Nd^{3+}	$LiNbO_3$:MgO	300	1009–1011
~0.546[(b)]	Nd^{3+}	$LiNbO_3:Sc_2O_3$	300	1012
0.5464[(b)]	Nd^{3+}	$LiNbO_3$:MgO	300	1009–1011
0.5465[(b)]	Nd^{3+}	$LiNbO_3$:MgO	~425	36,1008
0.5467[(b)]	Nd^{3+}	$LiNbO_3$:MgO	~360	1005
~0.547	Nd^{3+}	$LiNbO_3$:MgO	300	855
0.547	Nd^{3+}	$LiNbO_3$:MgO	300	81
0.5496	Er^{3+}	$(Y,Er)AlO_3$	30–77	37
0.550965	Er^{3+}	$LiYF_4$	49	34
0.551	Er^{3+}	$LiYF_4$	300	38
0.551	Er^{3+}	$LiYF_4$	300	40,41
0.551	Er^{3+}	$LiYF_4$	40	42
0.551	Er^{3+}	$LiYF_4$	60	863
0.551	Er^{3+}	$LiYF_4$:Yb	300	1030
0.5512	Ho^{3+}	CaF_2	77	44
0.551251	Nd^{3+}	$LiYF_4$	49	38

Table 2.1.4—*continued*
Paramagnetic Ion Lasers Arranged in Order of Wavelength

Wavelength (μm)	Ion	Host: sensitizer ion(s)	Temp. (K)	References
0.5515	Ho^{3+}	$B(Y,Yb)_2F_8$	77	45
0.5540	Er^{3+}	$Ba(Y,Er)_2F_8$	77	46
0.5606	Er^{3+}	$LiYF_4$	<40	47
0.560660	Er^{3+}	$LiYF_4$	20	34
0.561	Nd^{3+}	$Y_3Al_5O_{12}$	300	39
0.561	Nd^{3+}	$Lu_3Al_5O_{12}$	300	39
0.5617	Er^{3+}	$Ba(Y,Er)_2F_8$	77	46
0.5932	Sm^{3+}	TbF_3	116	48
0.5984	Pr^{3+}	PrF_3	15	49
0.5985	Pr^{3+}	LaF_3	110	50,51
0.5985	Pr^{3+}	LaF_3	77	52
0.6001	Pr^{3+}	LaF_3	110	50,51
0.6042	Pr^{3+}	$LiLuF_4$	110	32,33
0.6044	Pr^{3+}	$LiYF_4$	300	53
0.6045	Pr^{3+}	$GdLiF_4$	300	29
0.6048	Pr^{3+}	$LiPrP_4O_{14}$	300	54
0.607	Pr^{3+}	$GdLiF_4$	300	29
0.6071	Pr^{3+}	BaY_2F_8	110, 300	55,56
0.6071	Pr^{3+}	$LiLuF_4$	110	32,33
0.6071	Pr^{3+}	$LiYF_4$	110–200	31
0.6073	Pr^{3+}	$LiYF_4$	300	53
0.6092	Pr^{3+}	$LiYF_4$	300	53
0.6105	Pr^{3+}	$Ca(NbO_3)_2$	110	57
0.6113	Eu^{3+}	Y_2O_3	220	58
0.6116	Ti^{3+}	$YAlO_3$	300	59
0.616	Pr^{3+}	$Y_3Al_5O_{12}$	4–140	657
0.6130	Pr^{3+}	$LiYF_4$	300	53
0.6139	Pr^{3+}	$YAlO_3$	110	60–62
0.6139	Pr^{3+}	$YAlO_3$	300	63
0.6155	Pr^{3+}	$LuAlO_3$	110	61,62
0.6158	Pr^{3+}	$LiYF_4$	300	53
0.6164	Pr^{3+}	$(Pr,La)Cl_3$	65, 300	26–28
0.6164	Pr^{3+}	$LaCl_3$	8, 65	64
0.617	Pr^{3+}	$PrCl_3$	300	30
0.6180	Pr^{3+}	$LiYF_4$	300	53

Table 2.1.4—*continued*
Paramagnetic Ion Lasers Arranged in Order of Wavelength

Wavelength (μm)	Ion	Host: sensitizer ion(s)	Temp. (K)	References
0.6193	Eu^{3+}	YVO_4	90	66
0.620	Pr^{3+}	$PrCl_3$	300	65
0.6201	Pr^{3+}	$LiYF_4$	300	53
0.621	Pr^{3+}	$LaBr_3$	<300	30
0.6213	Pr^{3+}	$YAlO_3$	110	61,62
0.6216	Pr^{3+}	$YAlO_3$	300	63
0.622	Pr^{3+}	$PrCl_3$	300	30
0.632	Pr^{3+}	$LaBr_3$	< 70	30
0.637	Pr^{3+}	$(La,Pr)P_5O_{14}$	300	68
0.6374	Pr^{3+}	PrP_5O_{14}	300	69–71
0.6388	Pr^{3+}	$LiYF_4$	300	53
0.6388	Pr^{3+}	BaY_2F_8	300	72
0.6395	Pr^{3+}	$LiYF_4$	110, 300	31,75
0.6396	Pr^{3+}	$LiPrP_4O_{12}$	300	74
0.6399	Pr^{3+}	$LiLuF_4$	110	32,33
0.64	Pr^{3+}	$LiYF_4$	300	73
0.6401	Pr^{3+}	$LiLuF_4$	300	32,33
0.6444	Pr^{3+}	$LiYF_4$	300	53
0.645	Pr^{3+}	$SrLaGa_3O$	300	24
0.6451	Pr^{3+}	$(Pr,La)Cl_3$	65, 300	26–28
0.6451	Pr^{3+}	$PrBr_3$	300	26
0.6452	Pr^{3+}	$LaCl_3$	300	25
0.647	Pr^{3+}	$LaBr_3$	<300	30
0.647	Pr^{3+}	$PrCl_3$	300	26
0.649	Tm^{3+}	BaY_2F_8:Yb	300	78
0.6490	Tm^{3+}	BaY_2F_8	300	79
0.6497	Pr^{3+}	$CaWO_4$	110	57
0.650	Tm^{3+}	$LiYF_4$:Yb	300	80
0.6571[(b)]	Nd^{3+}	$LaBGeO_5$	300	1013–1015
~0.66[(b)]	Nd^{3+}	$YAl_3(BO_3)_4$	300	1021
0.66–1.06	Ti^{3+}	Al_2O_3	300	82–112
0.6624	Pr^{3+}	$YAlO_3$	300	63
0.67	Er^{3+}	$BaYb_2F_8$	300	78
0.6700	Er^{3+}	$Ba(Y,Yb)_2F_8$	77	45
0.6700	Er^{3+}	$BaYb_2F_8$	110	45,115

Table 2.1.4—*continued*
Paramagnetic Ion Lasers Arranged in Order of Wavelength

Wavelength (μm)	Ion	Host: sensitizer ion(s)	Temp. (K)	References
0.6703	Pr^{3+}	$LiYF_4$	300	53
0.6709	Er^{3+}	$Ba(Y,Er)_2F_8$	77	46
0.6709	Er^{3+}	$Ba(Y,Yb)_2 F_8$	77	45
0.671	Er^{3+}	$LiYF_4$	300	40
0.6799	Cr^{3+}	$BeAl_2O_4$	77	119
0.6803	Cr^{3+}	$BeAl_2O_4$	77, 300	118,121,122
0.685	Cr^{3+}	$Be_3Al_2Si_6O_{18}$	300	123
0.6874	Cr^{3+}	$Y_3Al_5O_{12}$	~77	124
0.6929(R_2)	Cr^{3+}	Al_2O_3	300	125
0.6934	Cr^{3+}	Al_2O_3	77	126–128
0.6935–8	Pr^{3+}	BaY_2F_8	110, 190	55,56
0.6943(R_1)	Cr^{3+}	Al_2O_3	300	131–135,138,297
0.6954	Pr^{3+}	$LiYF_4$	110–180	31
0.6958	Pr^{3+}	$LiLuF_4$	110	32,33
0.6969	Sm^{2+}	SrF_2	4.2	139
0.6977	Pr^{3+}	$LiYF_4$	300	53
0.6977	Pr^{3+}	$LiLuF_4$	110	32
0.6994	Pr^{3+}	$LiYF_4$	300	53
0.70–0.82	Cr^{3+}	$BeAl_2O_4$	300–583	140–152
0.7009(N_2)	$2Cr^{3+}$	Al_2O_3	77	153
0.701–0.818	Cr^{3+}	$BeAl_2O_4$	300	121,154
0.7015	Er^{3+}	$LiYF_4$	300	39
0.7037	Er^{3+}	$Ba(Y,Er)_2F_8$	77	46
0.7041(N_1)	$2Cr^{3+}$	Al_2O_3	77	153
0.7055	Pr^{3+}	$LiYF_4$	300	53
0.7082	Pr^{3+}	$LiYF_4$	300	53
0.7083	Sm^{2+}	CaF_2	65–90	157
0.7085	Sm^{2+}	CaF_2	20	156–160
0.717	Pr^{3+}	$(La,Pr)P_5O_{14}$	300	68
0.7190	Pr^{3+}	$LiYF_4$	180–250	31
0.7191	Pr^{3+}	BaY_2F_8	300	73
0.7192	Pr^{3+}	$LiLuF_4$	110	32,33
0.7194	Pr^{3+}	LaF_3	110	50,51
0.7195	Pr^{3+}	$YAlO_3$	300	63
0.7195	Pr^{3+}	$YAlO_3$	110	60,62

Table 2.1.4—*continued*
Paramagnetic Ion Lasers Arranged in Order of Wavelength

Wavelength (μm)	Ion	Host: sensitizer ion(s)	Temp. (K)	References
0.7197	Pr^{3+}	$YAlO_3$	300	61,62
0.7198	Pr^{3+}	LaF_3	300	50,51,161
0.72–0.84	Cr^{3+}	$LiCaAlF_6$	300	162,163
0.720–0.842	Cr^{3+}	$Be_3Al_2Si_6O_{18}$	300	164–166
0.7204	Pr^{3+}	$LiPrP_4O_{14}$	300	54
0.7206	Pr^{3+}	$LiYF_4$	300	168
0.7207	Sm^{2+}	CaF_2	85–90	157
0.7209	Pr^{3+}	$LiYF_4$	300	53
0.7215	Pr^{3+}	$LiLuF_4$	300	32,33
0.722	Pr^{3+}	$YAlO_3$	300	169
0.7222	Pr^{3+}	$LiYF_4$	300	53
0.7287	Sm^{2+}	CaF_2	110–130	157
0.72952	Nd^{3+}	$LiYF_4$	12	77
0.73–0.95	Ti^{3+}	$BeAl_2O_4$	300	170–172
0.7310	Sm^{2+}	CaF_2	155	157
0.74	Cr^{3+}	$Y_3Ga_5O_{12}$	300	173
0.742–0.842	Cr^{3+}	$Gd_3Sc_2Ga_3O_{12}$	300	174–178
0.7437	Pr^{3+}	$YAlO_3$	300	63
0.744–0.788	Cr^{3+}	$BeAl_2O_4$	300	179
0.745	Sm^{2+}	CaF_2	210	157
0.7469	Pr^{3+}	$YAlO_3$	300	60,61,63
0.748–0.832	Cr^{3+}	$Na_3Ga_3Li_3F_{12}$	300	180
0.7496	Pr^{3+}	$LuAlO_3$	300	169
0.7498	Ho^{3+}	$LiYF_4$	90, 300	167,181
0.75–0.81	Cr^{3+}	$Gd_3Sc_2Al_3O_{12}$	300	174,175,182–185
0.750–0.950	Cr^{3+}	$LiSr_{0.8}Ca_{0.2}AlF_6$	300	186
0.7501	Ho^{3+}	$LiLuF_4$	110	32
0.7505	Ho^{3+}	$LiYF_4$	300	187
0.7505	Ho^{3+}	$LiLuF_4$	110	32
0.7516	Ho^{3+}	$LiYF_4$	116	181
0.753–0.946	Ti^{3+}	$BeAl_2O_4$	300	189
0.7537	Pr^{3+}	$YAlO_3$	300	63
0.7555	Ho^{3+}	$LiYF_4$	116	181
0.7577	Ho^{3+}	$YAlO_3$	110–300	190
0.76	Cr^{3+}	$Y_3Sc_2Ga_3O_{12}$	300	173

Table 2.1.4—*continued*
Paramagnetic Ion Lasers Arranged in Order of Wavelength

Wavelength (μm)	Ion	Host: sensitizer ion(s)	Temp. (K)	References
0.7610	Ho^{3+}	$YAlO_3$	110–300	190
0.766–0.865	Cr^{3+}	$KZnF_3$	300	191–195
0.767	Cr^{3+}	$Y_3Sc_2Al_3O_{12}$	300	196
0.7670	Cr^{3+}	Al_2O_3	300	197
0.769	Cr^{3+}	$Gd_3Ga_5O_{12}$	300	174
0.773–0.814	Cr^{3+}	$(Gd,Ca)_3(Ga,Mn,Zr)_5O_{12}$	300	198
0.775–0.816	Cr^{3+}	$KZnF_3$	80	191,192
0.78–0.92	Cr^{3+}	$LiSrAlF_6$	300	196,199,200
0.780–1.010	Cr^{3+}	$LiSrAlF_6$	300	201
0.787–0.892	Cr^{3+}	$ScBO_3$	300	162,202,203
0.79–0.87	Cr^{3+}	$BeAl_6O_{10}$	300	204
0.790–0.825	Cr^{3+}	$KZnF_3$	200	191,192
0.792	Tm^{3+}	$LiYF_4$:Yb	300	80
0.792	Cr^{3+}	$ScBeAlO_4$	300	206
0.799	Tm^{3+}	BaY_2F_8:Yb	300	207
0.815–1.22	Cr^{3+}	$La_3Ga_5SiO_{14}$	300	208,209,1017
0.80	Cr^{3+}	$Al_2(WO_4)_3$	300	210
0.810	Tm^{3+}	$LiYF_4$:Yb	300	80
0.820	Cr^{3+}	$LiSrGaF_6$	300	212,1025
0.83	Cr^{3+}	$(La,Lu)_3(La,Ga)_2Ga_3O_{12}$	300	174
0.8425	Er^{3+}	$BaEr_2F_8$	110	115
0.8430	Er^{3+}	CaF_2-YF_3	77	215,216
0.8446	Er^{3+}	CaF_2	300	235
0.8456	Er^{3+}	CaF_2-YF_3	77	215,216
0.8456	Er^{3+}	CaF_2-HoF_3-ErF_3	77	217
0.8456	Er^{3+}	CaF_2-HoF_3-ErF_3-TmF_3	77	217
0.8467	Er^{3+}	$KGd(WO_4)_2$	110	218
0.8468	Er^{3+}	$KGd(WO_4)_2$	300	218–220
0.8471	Er^{3+}	$Ca_3Ga_2Ge_3O_{12}$	110	221
0.8474	Er^{3+}	$KY(WO_4)_2$	110	218
0.8479	Er^{3+}	$KLu(WO_4)_2$	110	218
0.84965	Er^{3+}	$YAlO_3$	77	129,130
0.84975	Er^{3+}	$YAlO_3$	300	129,130
0.85	Er^{3+}	$LiYF_4$	300	220
0.85	Er^{3+}	$KLu(WO_4)_2$	300	207

Table 2.1.4—*continued*
Paramagnetic Ion Lasers Arranged in Order of Wavelength

Wavelength (μm)	Ion	Host: sensitizer ion(s)	Temp. (K)	References
0.8500	Er^{3+}	$LiYF_4$	300	222
0.8501	Er^{3+}	$LiY_{0.5}Er_{0.5}F_4$	113, 161	181
0.8503	Er^{3+}	$LiYF_4$:Pr	110, 300	224,226,329
0.8506	Er^{3+}	$LiLuF_4$	110	32
0.8507	Er^{3+}	$LiLuF_4$	300	32,33
0.85165	Er^{3+}	$YAlO_3$	77	129,130
0.852–1.005	Cr^{3+}	$SrAlF_5$	300	227,228
0.853	Er^{3+}	$Bi_4Ge_3O_{12}$	77	229
0.8535	Er^{3+}	$LiYF_4$	110	226
0.8535	Er^{3+}	$LiY_{0.5}Er_{0.5}F_4$	113, 163	181
0.8537	Er^{3+}	$LiYF_4$	116, 300	181,231
0.8538	Er^{3+}	$BaEr_2F_8$	104–123	232
0.8540	Er^{3+}	$LiErF_4$	110	233
0.8542	Er^{3+}	$LiLuF_4$	300	32,33
0.8543	Er^{3+}	$BaEr_2F_8$	110	115
0.8543	Er^{3+}	$LiLuF_4$	110	32
0.8548	Er^{3+}	CaF_2	77	235
0.8594	Er^{3+}	$YAlO_3$	300	129,130
0.8610	Er^{3+}	$KGd(WO_4)_2$	110	218
0.8615	Er^{3+}	$Ca_3Ga_2Ge_3O_{12}$	110	221
0.8621	Er^{3+}	$KY(WO_4)_2$	300	218,236
0.8621	Er^{3+}	$KY(WO_4)_2$	110	218
0.8624	Er^{3+}	$KEr(WO_4)_2$	110	218
0.8624	Er^{3+}	$Y_3Al_5O_{12}$	77	237
0.8625	Er^{3+}	$(Lu,Er)_3Al_5O_{12}$	300	218
0.8627	Er^{3+}	$Y_3Al_5O_{12}$	77, 300	129,238
0.8628	Er^{3+}	$Er_3Al_5O_{12}$	110	218
0.8631	Er^{3+}	$KLu(WO_4)_2$	110	218
0.8631	Er^{3+}	$Lu_3Al_5O_{12}$	110	218
0.86325	Er^{3+}	$KLu(WO_4)_2$	110	218
0.86325	Er^{3+}	$Lu_3Al_5O_{12}$	77	239
0.8633	Er^{3+}	$KLu(WO_4)_2$	300	218
0.87–1.21	Cr^{3+}	$Ca_3Ga_2Ge_4O_{14}$	300	241,1017
0.88–1.22	Cr^{3+}	$La_3Ga_5GeO_{14}$	300	241,242,1017
0.890	Cr^{3+}	$LiSrCrF_6$	300	243

Table 2.1.4—*continued*
Paramagnetic Ion Lasers Arranged in Order of Wavelength

Wavelength (μm)	Ion	Host: sensitizer ion(s)	Temp. (K)	References
0.8910	Nd^{3+}	$Y_3Al_5O_{12}$	300	244
0.895	Cr^{3+}	$Sr_3Ga_2Ge_4O_{14}$	300	1017
0.8999	Nd^{3+}	$Y_3Al_3O_{12}$	300	244
0.90–1.15	Cr^{3+}	$Sr_3Ga_2Ge_4O_{14}$	300	241,242
0.9010	Pr^{3+}	BaY_2F_8	110	245
0.9066	Pr^{3+}	$LiYF_4$	110	241,246
0.9068	Pr^{3+}	LiLuF4	110–250	246
0.9069	Pr^{3+}	$LiYF_4$	300	246
0.9–1.1	Cr^{3+}	$Ca_3Ga_2Ge_4O_{14}$	300	240,241
0.9–1.25	Cr^{3+}	$La_3Ga_{5.5}Nb_{0.5}O_{14}$	300	240,1017
0.911	Nd^{3+}	$SrGdGa_3O_7$	31	247
0.912	Nd^{3+}	Y_2SiO_5	300	248
0.9106	Nd^{3+}	$Ba_3LaNb_3O_{12}$	300	249
0.9137	Nd^{3+}	$KY(WO_4)_2$	77	250–252
0.9145	Nd^{3+}	$CaWO_4$	77	253
0.9145	Pr^{3+}	$LiYF_4$	110	246
0.9148	Pr^{3+}	BaY_2F_8	110	245
0.9150	Pr^{3+}	BaY_2F_8	300	245
0.9185	Pr^{3+}	LaF_3	110	245
0.9190	Pr^{3+}	LaF_3	300	245
0.925–1.24	Cr^{3+}	$La_3Ga_{5.5}Ta_{0.5}O_{14}$	300	240,241
0.9266[(b)]	Tm^{3+}	$LiNbO_3$	77	65
0.930	Nd^{3+}	$YAlO_3$	300	256
0.9308	Pr^{3+}	$YAlO_3$	110	62
0.9312	Pr^{3+}	$YAlO_3$	300	62
0.936	Nd^{3+}	$Gd_3Sc_2Ga_3O_{12}$	300	258,259
0.9385	Nd^{3+}	$Y_3Al_5O_{12}$	300	244
0.9395	Pr^{3+}	$YAlO_3$	300	63
0.941	Nd^{3+}	$CaY_2Mg_2Ge_3O_{12}$	300	265
0.946	Nd^{3+}	$Y_3Al_5O_{12}$	260–300	262–263
0.9460	Nd^{3+}	$Y_3Al_5O_{12}$	300	266
0.9473	Nd^{3+}	$Lu_3Al_5O_{12}$	77	267
0.9660	Pr^{3+}	$YAlO_3$	300	63
0.979	Ho^{3+}	$LiHoF_4$	90	268,269
0.9794	Ho^{3+}	$LiYF_4$	90, 300	67

Table 2.1.4—*continued*
Paramagnetic Ion Lasers Arranged in Order of Wavelength

Wavelength (μm)	Ion	Host: sensitizer ion(s)	Temp. (K)	References
0.98–1.09	Cr^{3+}	$ZnWO_4$	77	271
0.985	Yb^{3+}	$Ca_4Sr(PO_4)_3F$	300	17
0.9960	Pr^{3+}	$YAlO_3$	300	62
1.008	Yb^{3+}	$LiNbO_3$	300	999
1.0143	Ho^{3+}	$LiYF_4$	90, 300	67,181
1.0183	Ho^{3+}	$LiLuF_4$	110	32
1.0230	Yb^{3+}	$Lu_3Ga_5O_{12}$	77	276
1.0232	Yb^{3+}	$Gd_3Ga_5O_{12}$	77	276
1.0233	Yb^{3+}	$Y_3Ga_5O_{12}$:Nd	77	276
1.025	Yb^{3+}	$KY(WO_4)_2$	300	277
1.025	Yb^{3+}	$KGd(WO_4)_2$	300	277
1.0293	Yb^{3+}	$Y_3Al_5O_{12}$	77	276
1.0293	Yb^{3+}	$(Y,Yb)_3Al_5O_{12}$	77	276
1.0294	Yb^{3+}	$Lu_3Al_5O_{12}$	77	276
1.0296	Yb^{3+}	$Y_3Al_5O_{12}$	77	278
1.0297	Yb^{3+}	$Lu_3Al_5O_{12}$	77	276
1.0297	Yb^{3+}	$Y_3Al_5O_{12}$:Nd	200	276
1.0298	Yb^{3+}	$Y_3Al_5O_{12}$:Nd	210	276
1.0299	Yb^{3+}	$Gd_3Sc_2Al_3O_{12}$	77	279
1.0299	Yb^{3+}	$Lu_3Sc_2Al_3O_{12}$	77	279
1.03	Yb^{3+}	$Y_3Al_5O_{12}$	300	280,281
1.03	Yb^{3+}	$Y_3Al_5O_{12}$:Nd	300	282
1.03	Yb^{3+}	$Lu_3Al_5O_{12}$	300	283
1.030	Yb^{3+}	$LiNbO_3$	300	999
1.031	Yb^{3+}	$Y_3Al_5O_{12}$	300	311
1.0311	Ho^{3+}	$YAlO_3$	110–300	190,241
1.0336	Yb^{3+}	CaF_2	120	285
1.034	Yb^{3+}	$BaCaBO_3F$	300	275
1.0369	Nd^{3+}	CaF_2-SrF_2	300	286
1.0370	Nd^{3+}	CaF_2	300	137
1.0370–1.0395	Nd^{3+}	SrF_2	300–530	1024
1.04	Pr^{3+}	$Ca(NbO_3)_2$	77	753
1.04	Pr^{3+}	$SrMoO_4$	—	290
1.0400	Nd^{3+}	LaF_3	77	291–293
1.0404	Nd^{3+}	CeF_3	77	294,295

Table 2.1.4—*continued*
Paramagnetic Ion Lasers Arranged in Order of Wavelength

Wavelength (μm)	Ion	Host: sensitizer ion(s)	Temp. (K)	References
1.04065–1.0410	Nd^{3+}	LaF_3	300–430	137,985
1.0410	Nd^{3+}	CeF_3	300	294,295
1.0412	Nd^{3+}	KYF_4	300	559,573
1.042–1.075	Nd^{3+}	$Na_{0.4}Y_{0.6}F_{2.2}$	300	298
1.043	Yb^{3+}	$Ca_5(PO_4)_3F$	300	284,311
1.0437	Nd^{3+}	SrF_2	77	155
1.044	Yb^{3+}	$Sr_5(VO_4)_3F$	300	17
1.0445	Nd^{3+}	SrF_2	300	155
1.0446	Nd^{3+}	SrF_2	500–550	155
1.0448	Nd^{3+}	CaF_2	50	300–302
1.0451	Nd^{3+}	LaF_3	77	303–305
1.0456	Nd^{3+}	CaF_2	50	301,302
1.0457	Nd^{3+}	CaF_2	77	301,302
1.046	Yb^{3+}	$Ca_4Sr(PO_4)_3F$	300	17
1.046	Yb^{3+}	$Ca_3Sr_2(PO_4)_3F$	300	17
1.046–1.064	Nd^{3+}	$LiNdP_4O_{12}$	300	306,307
1.0461	Nd^{3+}	CaF_2-YF_3	300	308,309
1.0461–1.0468	Nd^{3+}	CaF_2	300–530	137
1.0466	Nd^{3+}	CaF_2	50	301,302
1.0467	Nd^{3+}	CaF_2	77	301,302
1.0468	Pr^{3+}	$CaWO_4$	77	310
1.047	Yb^{3+}	$Sr_5(PO_4)_3F$	300	311,312
1.047	Nd^{3+}	$LiGdF_4$	300	313
1.047	Nd^{3+}	$LiNdP_4O_{12}$	300	317,319
1.047	Nd^{3+}	$LiYF_4$	300	314–316
[1.047–1.078]	Nd^{3+}	NdP_5O_{14}	300	307
1.0471	Nd^{3+}	$LiYF_4$	300	33,314,321,322, 327
1.0472	Nd^{3+}	$LiLuF_4$	300	32,33,983
1.0475	Nd^{3+}	$LaBGeO_5$	300	330
1.0477	Nd^{3+}	$Li(Nd,La)P_4O_{12}$	300	331
1.0477	Nd^{3+}	$Li(Nd,Gd)P_4O_{12}$	30	331
1.048	Nd^{3+}	$Li(Bi,Nd)P_4O_{12}$	300	332
1.048	Nd^{3+}	$Li(Nd,La)P_4O_{12}$	300	333–355
1.048	Nd^{3+}	$Li(Nd,Gd)P_4O_{12}$	300	273
1.048	Nd^{3+}	$K_5Nd,Ce)Li_2F_{10}$	300	357

Table 2.1.4—*continued*
Paramagnetic Ion Lasers Arranged in Order of Wavelength

Wavelength (μm)	Ion	Host: sensitizer ion(s)	Temp. (K)	References
1.0480	Nd^{3+}	CaF_2	50	301,302
1.0481	Nd^{3+}	$LiKYF_5$	300	358,359
1.0482	Nd^{3+}	$LaBGeO_5$	300	330
1.0482	Nd^{3+}	$NaLa(MoO_4)_2$	300	361
1.0486	Nd^{3+}	LaF_3-SrF_2	300	362
1.049–1.077	Nd^{3+}	$NaNdP_4O_{12}$	300	307
1.0491	Nd^{3+}	$SrAl_{12}O_{19}$	300	364,365
1.0493	Nd^{3+}	BaY_2F_8	77	366
1.0493	Nd^{3+}	$Sr_2Y_5F_{19}$	300	367
1.0495	Pr^{3+}	CaWO4	300	73
1.0495	Nd^{3+}	BaY_2F_8	300	366
1.0495	Nd^{3+}	GdF_3-CaF_2	300	369
1.0497	Nd^{3+}	$SrAl_2O_4$	300	370
1.0498	Nd^{3+}	$Ca_2Y_5F_{19}$	300	137
1.0498	Nd^{3+}	$SrAl_{12}O_{19}$	300	313
1.05	Nd^{3+}	$KNdP_4O_{12}$	300	317
1.05	Nd^{3+}	NdP_5O_{14}	300	376
1.05	Nd^{3+}	LaP_5O_{14}	300	376
1.05	Nd^{3+}	$LiLuF_4$	300	32,33
1.0500	Nd^{3+}	CaF_2-ScF_3	300	374
1.0505	Nd^{3+}	$(Nd,La)P_5O_{14}$	300	378
1.0505	Nd^{3+}	$5NaF$-$9YF_3$	300	379
1.0506	Nd^{3+}	$5NaF$-$9YF_3$	300	379
1.0507	Nd^{3+}	CaF_2	50	301,302
1.0507	Nd^{3+}	CdF_2-ScF_3	300	380
1.051	Nd^{3+}	$NaNdP_4O_{12}$	300	335,381,382
1.051	Nd^{3+}	YP_5O_{14}	300	384
1.051	Nd^{3+}	$(La,Nd)P_5O_{14}$	300	347,385
1.051	Nd^{3+}	CeP_5O_{14}	300	384
1.051	Nd^{3+}	GdP_5O_{14}	300	384
1.051	Nd^{3+}	NdP_5O_{14}	300	386
1.0510	Nd^{3+}	CdF_2-ScF_3	≤200	380
1.0511	Nd^{3+}	$(Nd,La)P_5O_{14}$	300	378
1.0512	Nd^{3+}	NdP_5O_{14}	300	378
1.0512	Nd^{3+}	$(Nd,La)P_5O_{14}$	300	390–398

Table 2.1.4—*continued*
Paramagnetic Ion Lasers Arranged in Order of Wavelength

Wavelength (μm)	Ion	Host: sensitizer ion(s)	Temp. (K)	References
1.0512	Nd^{3+}	$(Y,Nd)P_5O_{14}$	300	399–403
1.0513	Nd^{3+}	NdP_5O_{14}	300	387
1.0515	Nd^{3+}	YP_5O_{14}	300	384
1.0515	Nd^{3+}	NdP_5O_{14}	300	355,388,389
1.052	Nd^{3+}	$(Nd,La)P_5O_{14}$	300	390–398,617,621
1.052	Nd^{3+}	$KNdP_4O_{12}$	300	289,335,382, 404–406
1.052	Nd^{3+}	$K_5NdLi_2F_{10}$	300	357
1.052	Nd^{3+}	$Y_3Al_5O_{12}$	300	407,412
1.0521	Nd^{3+}	BaF_2-YF_3	300	413
1.0521	Nd^{3+}	$Y_3Al_5O_{12}$	300	414
1.0521	Nd^{3+}	NdP_5O_{14}	300	387
1.0521	Nd^{3+}	YF_3	300	415
1.0523	Nd^{3+}	LaF_3	77	291,416–418
1.0525	Nd^{3+}	YP_5O_{14}	300	419
1.0526	Nd^{3+}	BaF_2-GdF_3	300	413
1.0528	Nd^{3+}	SrF_2-GdF_3	300	421
1.0528	Nd^{3+}	$LiYF_4$	77	32,33
1.0529	Nd^{3+}	NdP_5O_{14}	300	378
1.0529	Nd^{3+}	BaY_2F_8	77	422
1.0529	Nd^{3+}	$LiLuF_4$	110	32,33
1.053	Nd^{3+}	$LiYF_4$	300	32,33,314–316
1.053	Nd^{3+}	$(La,Nd)P_5O_{14}$	300	429,430
1.053–1.062	Nd^{3+}	$Ca_3(Nb,Ga)_2Ga_3O_{12}$	300	428
1.0530	Nd^{3+}	$LiYF_4$	300	321,322
1.0530	Nd^{3+}	BaY_2F_8	300	366
1.0530	Nd^{3+}	CaF_2-LuF_3	300	425
1.0530–1.059	Nd^{3+}	$LaMgAl_{11}O_{19}$	300	432
1.0531	Nd^{3+}	$LiLuF_4$	300	32,33
1.0532	Nd^{3+}	$LiKYF_5$	300	358,359
1.0534–1.0563	Nd^{3+}	BaF_2-LaF_3	300–920	435
1.0535	Nd^{3+}	$Lu_3Al_5O_{12}$	300	436
1.0535–1.0547	Nd^{3+}	CaF_2-SrF_2-BaF_2-YF_3-LaF_3	300–550	137,438
1.0536	Nd^{3+}	CaF_2-YF_3	110	439
1.0537	Nd^{3+}	BaF_2-CeF_3	300	440
1.0538	Nd^{3+}	BaF_2-LaF_3	77	435

Table 2.1.4—*continued*
Paramagnetic Ion Lasers Arranged in Order of Wavelength

Wavelength (μm)	Ion	Host: sensitizer ion(s)	Temp. (K)	References
1.0539–1.0549	Nd^{3+}	α-$NaCaYF_6$	300–550	137,439,504,983
1.054	Nd^{3+}	$Gd_3Ga_5O_{12}$	300	443
1.054	Nd^{3+}	$LaAl_{11}MgO_{19}$	300	313,326
1.054–1.086	Nd^{3+}	$LaAl_{11}MgO_{19}$	300	431,444,1025
1.0540	Nd^{3+}	CaF_2-YF_3	300	137,308,309
1.0540	Nd^{3+}	BaF_2	300	445
1.0540	Nd^{3+}	CaF_2-YF_3	300	439
1.0543	Nd^{3+}	BaF_2-CeF_3	300	413
1.0543	Nd^{3+}	SrF_2-ScF_3	300	446
1.05436	Nd^{3+}	$Ba_2MgGe_2O_7$	300	447
1.05437	Nd^{3+}	$Ba_2ZnGe_2O_7$	300	448
1.0547	Nd^{3+}	$LaMgAl_{11}O_{19}$	300	318,450,451
1.05499	Nd^{3+}	CsY_2F_7	300	452
1.055	Nd^{3+}	$Na_3Nd(PO_4)_2$	300	373
1.055	Nd^{3+}	$K_3(La,Nd)(PO_4)_2$	300	373,454
1.055	Nd^{3+}	$Na_3(La,Nd)(PO_4)_2$	300	373
1.0550	Nd^{3+}	$LaMgAl_{11}O_{19}$	77	439,455,546
1.0551	Nd^{3+}	$Pb_5(PO_4)_3F$	300	456
1.0552	Nd^{3+}	$LaMgAl_{11}O_{19}$	300	439,455,546
1.0554	Nd^{3+}	$LiNdP_4O_{12}$	300	349
1.0554	Nd^{3+}	KY_3F_{10}	300	461
1.0555	Nd^{3+}	$CsGd_2F_7$	300	462
1.0555	Nd^{3+}	$Ba_5(PO_4)_3F$	300	1027
1.0556	Nd^{3+}	SrF_2-LuF_3	300	463
1.0560	Nd^{3+}	SrF_2-LuF_3	300	464,465
1.0566	Nd^{3+}	$SrAl_4O_7$	77	473
1.0566	Nd^{3+}	$La_2Si_2O_7$	300	466
1.0566	Nd^{3+}	$NdGaGe_2O_7$	110	467
1.0567	Nd^{3+}	SrF_2-YF_3	300	468
1.0567	Nd^{3+}	$GdGaGe_2O_7$	77	467
1.0568	Nd^{3+}	$SrAl_4O_7$	77	473
1.0569	Nd^{3+}	$NdGaGe_2O_7$	300	467,470
1.0570	Nd^{3+}	$GdGaGe_2O_7$	300	467,472
1.0572	Nd^{3+}	$LaSr_2Ga_{11}O_{20}$	300	234,439,546
1.0573	Nd^{3+}	$CaMoO_4$	295	291,474,479

Table 2.1.4—*continued*
Paramagnetic Ion Lasers Arranged in Order of Wavelength

Wavelength (μm)	Ion	Host: sensitizer ion(s)	Temp. (K)	References
1.0574	Nd^{3+}	$SrWO_4$	77	291
1.0575	Nd^{3+}	$CsLa(WO_4)_2$	300	241,475
1.0575	Nd^{3+}	$Y_3Sc_2Ga_3O_{12}$	77	427
1.05755	Nd^{3+}	$Gd_3Sc_2Ga_3O_{12}$	77	477
1.0576	Nd^{3+}	$SrAl_4O_7$	300	473
1.0576	Nd^{3+}	$SrMoO_4$	295	291,479
1.0576	Nd^{3+}	$La_2Si_2O_7$	300	466
1.0580	Nd^{3+}	BaF_2-LaF_3	77	435
1.058	Nd^{3+}	$Gd_3Sc_2Ga_3O_{12}$	300	358
1.0580	Nd^{3+}	$Gd_3Sc_2Ga_3O_{12}$	77	477
1.0580	Nd^{3+}	$KLa(MoO_4)_2$	110	439
1.0581	Nd^{3+}	$CsLa(WO_4)_2$	110	241,475
1.0582	Nd^{3+}	$Ca_3Ga_4O_9$	300	482
1.0582–1.0597	Nd^{3+}	$CaWO_4$	300–700	137,483–487
1.0583	Nd^{3+}	LaF_3	77	416–418
1.0583	Nd^{3+}	$Y_3Sc_2Ga_3O_{12}$	300	427
1.0583	Nd^{3+}	$Y_3Ga_5O_{12}$	77	490,491
1.0583	Nd^{3+}	$Ca_3(Nb,Ga)_2Ga_3O_{12}$	110	439,458
1.0584	Nd^{3+}	$Gd_3Ga_5O_{12}$	77	490,492
1.0584	Nd^{3+}	$Y_3Sc_2Ga_3O_{12}$:Cr	300	342,343
1.0584	Nd^{3+}	$Y_3Sc_2Ga_3O_{12}$	300	493
1.0584	Nd^{3+}	$CaY_2Mg_2Ge_3O_{12}$	300	494
1.0585	Nd^{3+}	$YAlO_3$	300	495
1.0585	Nd^{3+}	$LiLa(MoO_4)_2$	300	496
1.0585	Nd^{3+}	$Sr_5(PO_4)_3F$	300	497
1.0585	Nd^{3+}	CaF_2-SrF_2-BaF_2-YF_3-LaF_3	300–700	137
1.0585	Nd^{3+}	$KLa(MoO_4)_2$	300	498
1.0586	Nd^{3+}	$Sr_5(PO_4)_3F$	300	499
1.0586	Nd^{3+}	$PbMoO_4$	300	291,500
1.0587	Nd^{3+}	$KLa(MoO_4)_2$	300	439,498,501
1.0587	Nd^{3+}	$CaWO_4$	300	17
1.0587	Nd^{3+}	$Y_3Sc_2Al_3O_{12}$	77	476
1.0587	Nd^{3+}	$Lu_3Ga_5O_{12}$	77	505
1.0588	Nd^{3+}	$Ca_3(Nb,Ga)_2Ga_3O_{12}$	300	439,458,506
1.0589	Nd^{3+}	SrF_2-CeF_3-GdF_3	300	421

Table 2.1.4—*continued*
Paramagnetic Ion Lasers Arranged in Order of Wavelength

Wavelength (μm)	Ion	Host: sensitizer ion(s)	Temp. (K)	References
1.0589	Nd^{3+}	$Y_3Ga_5O_{12}$	300	490,492
1.05895	Nd^{3+}	$CaAl_4O_7$	77	510
1.05896	Nd^{3+}	$CaMg_2Y_2Ge_3O_{12}$	300	265,511
1.059	Nd^{3+}	$(La,Sr)(Al,Ta)O_3$	300	512
1.059	Nd^{3+}	$BaLaGa_3O_7$	300	513
1.059	Nd^{3+}	$SrMoO_4$	77	291,497
1.059	Nd^{3+}	$NaY(WO_4)_2$	300	516
1.059	Nd^{3+}	$Na_{1+x}Mg_xAl_{11-x}O_{17}$	300	517
1.059	Nd^{3+}	$Na_2Nd_2Pb_6(PO_4)_6Cl_2$	300	520,521
1.0590	Nd^{3+}	SrF_2-CeF_3	300	518
1.0590	Nd^{3+}	$Ca_3Ga_2Ge_3O_{12}$	77	519
1.0590	Nd^{3+}	SrF_2-CeF_3	300	439
1.0591	Nd^{3+}	$Gd_3Ga_5O_{12}$	300	522
1.0591	Nd^{3+}	$Lu_3Sc_2Al_3O_{12}$	300	279
1.0591	Nd^{3+}	$Ba_3LaNb_3O_{12}$	77	249
1.0591	Nd^{3+}	$LaGaGe_2O_7$	300	525,526
1.05915	Nd^{3+}	$Gd_3Sc_2Al_3O_{12}$	77	279,527
1.0592	Nd^{3+}	$NaGaGe_2O_7$	110	470
1.0592	Nd^{3+}	$Nd_3Ga_5O_{12}$	110	470
1.0593	Nd^{3+}	$Sr_4Ca(PO_4)_3$	300	499
1.0594	Nd^{3+}	$Lu_3Ga_5O_{12}$	300	529
1.0594	Nd^{3+}	SrF_2-CeF_3	110	439
1.0595	Nd^{3+}	5NaF-$9YF_3$	300	530
1.0595	Nd^{3+}	$Y_3Sc_2Al_3O_{12}$	300	427,504
1.0595	Nd^{3+}	$NaLa(MoO_4)_2$	300	439,533,534,536
1.0595	Nd^{3+}	$BaLaGa_3O_7$	300	249
1.0595	Nd^{3+}	$LiGd(MoO_4)_2$	110	439,458
1.0595–1.0613	Nd^{3+}	LaF_3	380–820	137
1.0596	Nd^{3+}	$CaAl_4O_7$	300	510
1.0596	Nd^{3+}	$Ca_3Ga_2Ge_3O_{12}$	300	519
1.0596	Nd^{3+}	$NaLuGeO_4$	77	539,540
1.0596	Nd^{3+}	SrF_2	300	541
1.0597	Nd^{3+}	$Ca_3Ga_2Ge_3O_{12}$	300	675
1.0597–1.0583	Nd^{3+}	SrF_2-LaF_3	300–800	137
1.0597–1.0629	Nd^{3+}	α-$NaCaYF_6$	1000–300	137

Table 2.1.4—*continued*
Paramagnetic Ion Lasers Arranged in Order of Wavelength

Wavelength (μm)	Ion	Host: sensitizer ion(s)	Temp. (K)	References
1.05975	Nd^{3+}	$Y_3Ga_5O_{12}$	77	490,491
1.0599	Nd^{3+}	$Gd_3Ga_5O_{12}$	77	490,492
1.0599	Nd^{3+}	$Lu_3Sc_2Al_3O_{12}$	300	279
1.0599	Nd^{3+}	$LiGd(MoO_4)_2$	300	252,257
1.05995	Nd^{3+}	$Gd_3Sc_2Al_3O_{12}$	300	279,527
~1.06	Nd^{3+}	$(Gd,Ca)_3(Ga,Mg,Zr)_5O_{12}$:Cr	300	502,503
~1.06	Nd^{3+}	$CaGd_4(SiO_4)_3O$	300	551
~1.06	Nd^{3+}	$GdVO_4$	300	550
1.06	Nd^{3+}	$Gd_3Sc_2Al_3O_{12}$	300	279,527
1.06	Nd^{3+}	$NaLa(MoO_4)_2$	300	536
1.06	Nd^{3+}	$NaGd(WO_4)_2$	300	476
1.06	Nd^{3+}	$Na(Nd,Gd)(WO_4)_2$	77	476
1.06	Nd^{3+}	$NdAl_3(BO_3)_4$	300	317
1.06	Nd^{3+}	$YAl_3(BO_3)_4$	300	553
1.06	Nd^{3+}	$Gd_3Ga_5O_{12}$	300	490,492,554–556
1.06	Nd^{3+}	$Gd_3Ga_5O_{12}$	~120	490,492
1.060	Nd^{3+}	$Gd_3Ga_5O_{12}$:Cr	300	557
1.060	Nd^{3+}	BaF_2	77	463
1.060	Yb^{3+}	$LiNbO_3$	300	999
1.060	Nd^{3+}	$Ca_4GdO(BO_3)_3$	300	1033
1.0600	Nd^{3+}	$Gd_3Ga_5O_{12}$	300	490,492
1.0601	Nd^{3+}	$CaWO_4$	77	437
1.0601	Nd^{3+}	$GdGaGe_2O_7$	300	467,472
1.06025	Nd^{3+}	$Lu_3Ga_5O_{12}$	77	505
1.0603	Nd^{3+}	$Y_3Ga_5O_{12}$	300	505,560
1.0603	Nd^{3+}	$Gd_2(WO_4)_3$	300	562
1.0603–1.0632	Nd^{3+}	CaF_2-YF_3	95–300	137
1.0604	Nd^{3+}	HfO_2-Y_2O_3	300	563,564
1.0604	Nd^{3+}	$NaLuGeO_4$	300	539,540,985
1.06045	Nd^{3+}	$Gd_3Sc_2Ga_3O_{12}$	77	477
1.0605	Nd^{3+}	$Lu_3Al_5O_{12}$	77	267
1.0605	Nd^{3+}	$Ca_3(Nb,Ga)_2Ga_3O_{12}$	110	439,458
1.0605	Nd^{3+}	SrF_2-ScF_3	300	446
1.0606	Nd^{3+}	$Gd_2(MoO_4)_3$	300	568,569
1.0606	Nd^{3+}	$Gd_3Ga_5O_{12}$	300	505

Table 2.1.4—*continued*
Paramagnetic Ion Lasers Arranged in Order of Wavelength

Wavelength (μm)	Ion	Host: sensitizer ion(s)	Temp. (K)	References
1.0607	Nd^{3+}	$SrWO_4$	77	291
1.0607	Nd^{3+}	$Sr_3Ca_2(PO_4)_3$	300	499
1.0607	Nd^{3+}	CaF_2-ScF_3	300	374
1.0608	Nd^{3+}	$(Y,Lu)_3Al_5O_{12}$	77	597
1.0608	Nd^{3+}	ZrO_2-Y_2O_3	300	563
1.0608	Nd^{3+}	$Nd_3Ga_5O_{12}$	300	563
1.0608	Nd^{3+}	$NaGaGe_2O_7$	300	470
1.0609	Nd^{3+}	$Lu_3Ga_5O_{12}$	300	505
1.0609	Nd^{3+}	$Ca_2Ga_2SiO_7$	77	439,536,1002
1.0609	Nd^{3+}	$NaYGeO_4$	300	576,577
~1.061	Nd^{3+}	$CaLa(SiO_4)_3O$	300	1007
1.061	Nd^{3+}	$Ca_2Al_2SiO_7$	300	314
1.061	Nd^{3+}	$BaGd_2(MoO_4)_4$	300	584
1.061	Nd^{3+}	$CaMoO_4$	300	291,474,479
1.061	Nd^{3+}	$Gd_3Sc_2Ga_3O_{12}$:Cr	300	531,536,542,581, 583,645,646,650, 653,659,730,732
1.061	Nd^{3+}	$CaLa_4(SiO_4)_3O$	300	497,586,590,608, 644
1.0610	Nd^{3+}	$Ca_2Ga_2SiO_7$	300	439,544–546,1002
1.0610	Nd^{3+}	$7La_2O_3$-$9SiO_2$	300	466
1.0610–1.0627	Nd^{3+}	$Y_3Al_5O_{12}$	77–600	137
1.0611	Nd^{3+}	$SrMoO_4$	77	291,479
1.0612	Nd^{3+}	$Y_3Al_5O_{12}$	77	591,592
1.0612	Nd^{3+}	$Ca(NbO_3)_2$	77	588
1.0612	Nd^{3+}	$Gd_3Sc_2Ga_3O_{12}$	300	477
1.0612	Nd^{3+}	$CaLa_4(SiO_4)_3O$	300	590
1.0612	Nd^{3+}	$Ca_3(Nb,Ga)_2Ga_3O_{12}$	300	439,506,546
1.0613	Nd^{3+}	$Ca_4La(PO_4)_3O$	300	497
1.0613	Nd^{3+}	$Ba_2NaNb_5O_{15}$	300	595
1.0613	Nd^{3+}	$Gd_3Sc_2Ga_3O_{12}$	300	596
1.0614	Nd^{3+}	$Y_3Ga_5O_{12}$	77	435
1.0615	Nd^{3+}	$Ca(NbO_3)_2$	77, 300	588
1.0615	Nd^{3+}	$Ca(NbO_3)_2$	300	598,600
1.0615	Nd^{3+}	$Y_3Al_5O_{12}$	300	407
1.0615	Nd^{3+}	$Gd_3Ga_5O_{12}$	~120	490,492

Table 2.1.4—*continued*
Paramagnetic Ion Lasers Arranged in Order of Wavelength

Wavelength (μm)	Ion	Host: sensitizer ion(s)	Temp. (K)	References
1.0615	Nd^{3+}	$Lu_3Al_5O_{12}$	300	505
1.0615	Nd^{3+}	$Y_3Sc_2Ga_3O_{12}$	300	427
1.0615	Nd^{3+}	$Ba_{0.25}Mg_{2.75}Y_2Ge_3O_{12}$	300	603
1.0615	Nd^{3+}	HfO_2-Y_2O_3	110	439,546
1.0615	Nd^{3+}	$NaGdGeO_4$	300	539,604,605,985
1.0615	Nd^{3+}	$Y_3Al_5O_{12}$	300	414
1.0615–1.0625	Nd^{3+}	$Ca(NbO_3)_2$	300–650	994
1.0618	Nd^{3+}	$Sr_2Ca_3(PO_4)_3$	300	499
1.0618	Nd^{3+}	$SrAl_{12}O_{19}$	300	313
1.0618	Nd^{3+}	CaF_2-ScF_3	300	374
1.0618	Nd^{3+}	$LaNbO_4$	300	611
1.062	Nd^{3+}	$Ca_{0.25}Ba_{0.75}(NbO_3)_2$	295	291
1.062	Nd^{3+}	$LaSc_3(BO_3)_4$	300	612
1.0620	Nd^{3+}	$Gd_3Sc_2Al_3O_{12}$	300	527
1.0620	Nd^{3+}	$Lu_3Sc_2Al_3O_{12}$	300	460
1.0621	Nd^{3+}	$SrAl_{12}O_{19}$	300	615
1.0621	Nd^{3+}	$Gd_3Ga_5O_{12}$	300	490,492
1.0622	Nd^{3+}	$Y_3Sc_2Al_3O_{12}$	300	476,504,609
1.0623–1.10585	Nd^{3+}	CaF_2-SrF_2-BaF_2-YF_3-LaF_3	300	438
1.0623	Nd^{3+}	$Lu_3Ga_5O_{12}$	300	505
1.0623	Nd^{3+}	CaF_2-LuF_3	300	425
1.0623–1.0628	Nd^{3+}	CaF_2	560–300	137
1.0624	Nd^{3+}	$LaNbO_4$	300	619
1.0625	Nd^{3+}	$Y_3Ga_5O_{12}$	300	490,491
1.0625	Nd^{3+}	YVO_4	300	620
1.0626	Nd^{3+}	$Ca(NbO_3)_2$	77	588
1.06265	Nd^{3+}	$SrWO_4$	77	57
1.0627	Nd^{3+}	$SrAl_4O_7$	77	473
1.0627	Nd^{3+}	$SrMoO_4$	77	291,497
1.0627	Nd^{3+}	$SrWO_4$	77	291
1.0628	Nd^{3+}	$SrWO_4$	300	57
1.0629	Nd^{3+}	$Ca_5(PO_4)_3F$	300	487,499
1.0629	Nd^{3+}	α-$NaCaYF_6$	300	379,504
1.0629	Nd^{3+}	$Bi_4Si_3O_{12}$	77, 300	625
1.0629–1.0656	Nd^{3+}	CdF_2-YF_3	600–300	626,627

Table 2.1.4—*continued*
Paramagnetic Ion Lasers Arranged in Order of Wavelength

Wavelength (μm)	Ion	Host: sensitizer ion(s)	Temp. (K)	References
1.063	Nd^{3+}	$SrWO_4$	295	291,373
1.063	Nd^{3+}	$Na_5(Nd,La)(WO_4)_4$	300	373,628
1.063	Nd^{3+}	$NdAl_3(BO_3)_4$	300	641–643
1.063	Nd^{3+}	$(La,Nd)P_5O_{14}$	300	621,634–640
1.063	Nd^{3+}	$NdAl_3(BO_3)_4$	300	630,631
1.0630	Nd^{3+}	$Ca_5(PO_4)_3F$	300	632,633
1.0632	Nd^{3+}	CaF_2-YF_3-NdF_3	300	238
1.0632	Nd^{3+}	CaF_2-YF_3	300	308,309,439
1.0632–1.0642	Nd^{3+}	LaF_3	400–700	137
1.0633	Nd^{3+}	$Ca_3Ga_2Ge_3O_{12}$	77	519
1.0633–1.0653	Nd^{3+}	α-$NaCaCeF_6$	920–300	137
1.06335–1.0638	Nd^{3+}	LaF_3	300–650	137,985
1.0634	Nd^{3+}	$CaWO_4$	77	137
1.0634	Nd^{3+}	YVO_4	300	778
1.0635	Nd^{3+}	LaF_3-SrF_2	300	362
1.0635	Nd^{3+}	$NdAl_3(BO_3)_4$	300	630,652
1.0635	Nd^{3+}	$NaLa(WO_4)_2$	300	654
1.0635	Nd^{3+}	$(Nd,Gd)Al_3(BO_3)_4$	300	405,640,656,658
1.0635	Nd^{3+}	$Bi_4(Si,Ge)_3O_{12}$	300	356
1.0635	Nd^{3+}	CaF_2-LuF_3	300	425
1.0636	Nd^{3+}	$(Y,Lu)_3Al_5O_{12}$	77	597
1.0636	Nd^{3+}	SrF_2-CeF_3	110	439
1.0637–1.0670	Nd^{3+}	$Y_3Al_5O_{12}$	170–900	137
1.06375–1.0672	Nd^{3+}	$Lu_3Al_5O_{12}$	120–900	267
1.0638	Nd^{3+}	$Bi_4Ge_3O_{12}$	77	291
1.0638	Nd^{3+}	CeF_3	300	662,663
1.0638	Nd^{3+}	$CaAl_4O_7$	300	510
1.0638	Nd^{3+}	$NaBi(WO_4)_2$	300	665
1.0638	Nd^{3+}	NdP_5O_{14}	300	394
1.0638	Nd^{3+}	$Ca_3Ga_2Ge_3O_{12}$	300	510
1.0638	Nd^{3+}	$La_3Ga_{5.5}Nb_{0.5}O_{14}$	77	667,668
1.0638–1.0644	Nd^{3+}	$(Y,Ce)_3Al_5O_{12}$	300	669
1.0639	Nd^{3+}	CeF_3	77	662,663
1.0639	Nd^{3+}	$Ca_3Ga_2Ge_3O_{12}$	300	675
1.064	Nd^{3+}	$Y_3Al_5O_{12}$	300	676–694

Table 2.1.4—*continued*
Paramagnetic Ion Lasers Arranged in Order of Wavelength

Wavelength (μm)	Ion	Host: sensitizer ion(s)	Temp. (K)	References
1.064	Nd^{3+}	$Y_3Al_5O_{12}$:Fe	300	345
1.064	Nd^{3+}	$Y_3Al_5O_{12}$:Ti	300	670
1.064	Nd^{3+}	$Y_3Al_5O_{12}$:Cr,Ce	300	671
1.064	Nd^{3+}	$Y_3Al_5O_{12}$:Ho	300	673
1.064	Nd^{3+}	$Y_3Al_5O_{12}$:Er	300	672
1.064	Nd^{3+}	YVO_4	300	684
1.064	Nd^{3+}	LaF_3	300	326
1.064–1.065	Nd^{3+}	$SrGdGa_3O_7$	31	247
1.0640	Nd^{3+}	$SrMoO_4$	77	291,497
1.0640	Nd^{3+}	$La_3Ga_5SiO_{14}$	300	439,476,546,667, 695–698,1000
1.0640–1.0657	Nd^{3+}	CaF_2-CeF_3	700–300	137
1.06405–1.0654	Nd^{3+}	$YAlO_3$	77–500	626,702
1.0641	Nd^{3+}	$Y_3Al_5O_{12}$:Cr	300	592
1.0641	Nd^{3+}	YVO_4	300	620
1.0641	Nd^{3+}	$La_3Ga_{5.5}Ta_{0.5}O_{14}$	300	705,1000
1.06415	Nd^{3+}	$Y_3Al_5O_{12}$	300	25,407–411, 492,610,984
1.0642	Nd^{3+}	$(Y,Lu)_3Al_5O_{12}$	295	597
1.0642	Nd^{3+}	$Ca_3Ga_2Ge_3O_{12}$	300	519
1.0642	Nd^{3+}	$NaBi(WO_4)_2$	300	665
1.06425	Nd^{3+}	$Lu_3Al_5O_{12}$	300	267,610
1.06425	Nd^{3+}	$Bi_4Ge_3O_{12}$	77	291,661
1.0643	Nd^{3+}	$SrMoO_4$	295	291,497
1.0644	Nd^{3+}	$YAlO_3$	300	712
1.0644	Nd^{3+}	Y_2SiO_5	77	713
1.0644	Nd^{3+}	$Bi_4Ge_3O_{12}$	300	291,497
1.0644	Nd^{3+}	$YAlO_3$	300	712
1.0645	Nd^{3+}	CaF_2-LaF_3	300	427
1.0645	Nd^{3+}	$KLa(MoO_4)_2$	110	439
1.0645	Nd^{3+}	$La_3Ga_{5.5}Nb_{0.5}O_{14}$	300	668,1000
1.0645	Nd^{3+}	$La_3Ga_5SiO_{14}$	300	695,697,698
1.0645	Nd^{3+}	$YAlO_3$	300	716–718,729
1.0645	Nd^{3+}	$YAlO_3$:Cr	295	715
1.0646	Nd^{3+}	$Y_3Al_5O_{12}$	300	375
1.0646	Nd^{3+}	$KLa(MoO_4)_2$	300	361

Table 2.1.4—*continued*
Paramagnetic Ion Lasers Arranged in Order of Wavelength

Wavelength (μm)	Ion	Host: sensitizer ion(s)	Temp. (K)	References
1.0647	Nd^{3+}	$CeCl_3$	300	720,721
1.0648	Nd^{3+}	CaF_2	50	301,302
1.0648	Nd^{3+}	YVO_4	300	620
1.0648	Nd^{3+}	$Ca_3(Nb,Ga)_2Ga_3O_{12}$	110	439,540
1.0649	Nd^{3+}	$CaWO_4$	77	137
1.0649	Nd^{3+}	$CaY_2Mg_2Ge_3O_{12}$	300	494
1.065	Nd^{3+}	$GdVO_4$	300	725
1.065	Nd^{3+}	$(Nd,Gd)Al_3(BO_3)_4$	300	405,640,648,656, 658,729–734
1.065	Nd^{3+}	$Sr_5(VO_4)_3Cl$	300	728
1.065	Nd^{3+}	$Sr_5(VO_4)_3F$	300	727
1.0650	Nd^{3+}	$CaWO_4$	77	735
1.0650	Nd^{3+}	$La_3Ga_5GeO_{14}$	300	704,736
1.0650	Nd^{3+}	$NaLa(MoO_4)_2$	110	439
1.0650	Nd^{3+}	$RbNd(WO_4)_2$	300	737
1.0652	Nd^{3+}	$CaWO_4$	300	483–486
1.0652	Nd^{3+}	$SrMoO_4$	77	291,497
1.0652	Nd^{3+}	CdF_2-LuF_3	300	742
1.0652–1.0659	Nd^{3+}	$YAlO_3$	310–500	626,702
1.0653–1.0633	Nd^{3+}	α-$NaCaCeF_6$	300–920	137
1.0653	Nd^{3+}	$NaLa(MoO_4)_2$	300	532–534
1.0653–1.0665	Nd^{3+}	$NaLa(MoO_4)_2$	300–750	137
1.0654	Nd^{3+}	CaF_2-GdF_3	300	427
1.0654	Nd^{3+}	$NdGaGe_2O_7$	300	470
1.0656	Nd^{3+}	CdF_2-YF_3	300	626,627
1.0657–1.0640	Nd^{3+}	CaF_2-CeF_3	300–700	137
1.0657	Nd^{3+}	CaF_2	300	541
1.0658	Nd^{3+}	$LiLa(MoO_4)_2$	300	496
1.0658	Nd^{3+}	$CsNd(MoO_4)_2$	300	737
1.06585	Nd^{3+}	$CaAl_4O_7$	77	510
1.0659	Nd^{3+}	$GdGaGe_2O_7$	300	467,472
1.066	Nd^{3+}	$Nd(Ga,Cr)_3(BO_3)_4$	300	631
1.066	Nd^{3+}	$K_5Nd(MoO_4)_4$	300	758
1.066	Nd_3^+	$K_5Bi(MoO_4)_4$	300	758
1.0661	Nd^{3+}	CaF_2	300	300
1.0662	Pr^{3+}	$Ca(NbO_3)_2$	~110	73

Table 2.1.4—*continued*
Paramagnetic Ion Lasers Arranged in Order of Wavelength

Wavelength (μm)	Ion	Host: sensitizer ion(s)	Temp. (K)	References
1.0663	Nd^{3+}	$NaGd(MoO_4)_2$	110	439
1.0663	Nd^{3+}	$NaY(MoO_4)_2$	110	439
1.0664	Pr^{3+}	$Ca(NbO_3)_2$	300	73
1.0664–1.0672	Nd^{3+}	YVO_4	300–690	626,628
~1.0665	Nd^{3+}	CdF_2-LaF_3	300	627
1.0666	Na^{3+}	CdF_2	300	763
1.0667	Nd^{3+}	CdF_2-CeF_3	300	763
1.0667	Nd^{3+}	$NaGd(MoO_4)_2$	300	439,699
1.0668	Nd^{3+}	CdF_2-LaF_3	300	381
1.0669	Nd^{3+}	$KY(MoO_4)_2$	300	765,766
1.067	Nd^{3+}	$Ca_3(VO_4)_2$	300	767
1.067	Nd^{3+}	LaF_3	77	291–293
1.0670	Nd^{3+}	CdF_2-LaF_3	≤200	380
1.0670	Nd^{3+}	$La_3Ga_5SiO_{14}$	300	439,476,667, 695–698,1000
1.0671	Nd^{3+}	$LuAlO_3$	77	74
1.0671	Nd^{3+}	CaF_2-YF_3	110	439
1.0672	Nd^{3+}	$CaY_4(SiO_4)_3O$	300	497
1.0672	Nd^{3+}	CdF_2-GdF_3	300	439,742
1.0672	Nd^{3+}	$KGd(WO_4)_2$	300	76,117,136,139, 536,610
1.0672	Nd^{3+}	$La_3Ga_5SiO_{14}$	300	695,697,698
1.0673	Nd^{3+}	$CaMoO_4$	77, 300	291,474,479
1.0673	Nd^{3+}	$La_3Ga_5SiO_{14}$	300	695,697,698
1.0674	Nd^{3+}	$NaY(MoO_4)_2$	300	439,699
1.0675	Nd^{3+}	$LuAlO_3$	300	74
1.0675	Nd^{3+}	$Na_2Nd_2Pb_6(PO_4)_6Cl_2$	300	520,521
1.0675	Nd^{3+}	$La_3Ga_5GeO_{14}$	77	1000
1.0675	Nd^{3+}	$La_3Ga_5SiO_{14}$	77	695,697
1.0675	Nd^{3+}	$Nd_3Ga_5SiO_{14}$	300	695,697,698
1.0675	Nd^{3+}	$Nd_3Ga_5GeO_{14}$	300	667,736
1.0676	Nd^{3+}	CdF_2-GdF_3	110	439
1.068	Nd^{3+}	$Na_2Nd_2Pb_6(PO_4)_6Cl_2$	300	520,521
1.0680	Nd^{3+}	$Nd_3Ga_5GeO_{14}$	300	667,736
1.0682	Nd^{3+}	$Y_3Al_5O_{12}$	300	116
1.0687–1.0690	Nd^{3+}	$KY(WO_4)_2$	77–600	250–252,984

Table 2.1.4—*continued*
Paramagnetic Ion Lasers Arranged in Order of Wavelength

Wavelength (μm)	Ion	Host: sensitizer ion(s)	Temp. (K)	References
1.0688	Nd^{3+}	$Ca_3Ga_2SiO_7$	300	1016
1.0688	Nd^{3+}	$Ca_2Ga_2Ge_4O_{14}$	300	188,667,1000
1.0688	Nd^{3+}	$KY(WO_4)_2$	300	250–252
1.0688	Nd^{3+}	$Sr_3Ga_2Ge_4O_{14}$	77	188,667
1.0689	Nd^{3+}	$GdAlO_3$	77	439,546
1.0689	Nd^{3+}	$NdGaGe_2O_7$	300	467,470
1.069	Nd^{3+}	YVO_4	~90	987
1.0690	Nd^{3+}	$GdAlO_3$	300	439,546
1.0690	Nd^{3+}	$Ca_3Ga_2Ge_4O_{14}$	300	118,667
1.0690	Nd^{3+}	$LaSr_2Ga_{11}O_{20}$	110	234,439,546
1.0694	Nd^{3+}	CdF_2-GdF_3	110	439
1.0694	Nd^{3+}	$Sr_3Ga_2Ge_4O_{14}$	300	188,667,1000
1.0698	Nd^{3+}	$La_2Be_2O_5$	300	254,255,260, 270,274,314
1.070	Nd^{3+}	$La_2Be_2O_5$	300	299
1.0700	Nd^{3+}	$LaBGeO_5$		330
1.0701	Nd^{3+}	$Gd_2(MoO_4)_3$	300	568,569
1.0701–1.0706	Nd^{3+}	$KLu(WO_4)_2$	77–600	257,610
1.0706	Nd^{3+}	$KY(WO_4)_2$	300	205
1.0706	Nd^{3+}	$LaSr_2Ga_{11}O_{20}$	300	234,439,546
1.07–1.16	V^{2+}	MgF_2	80	261,303–305
1.0710	Nd^{3+}	Y_2SiO_5	77	320
1.0711	Nd^{3+}	Y_2SiO_5	300	466
1.0714	Nd^{3+}	$KLu(WO_4)_2$	300	257
1.0714–1.0716	Nd^{3+}	$KLu(WO_4)_2$	550–77	272
1.0715	Nd^{3+}	Y_2SiO_5	300	320
1.0716–1.0721	Nd^{3+}	$KLu(WO_4)_2$	550–77	257
1.0720	Nd^{3+}	$CaSc_2O_4$	300	323
1.0721	Nd^{3+}	$KLu(WO_4)_2$	300	211
1.0725	Nd^{3+}	$LaSr_2Ga_{11}O_{20}$	110	234,439,546
1.07255–1.0730	Nd^{3+}	$YAlO_3$	77–490	626,702
1.0726	Nd^{3+}	$YAlO_3$	300	495
1.0726	Nd^{3+}	$(Y,Lu)_3Al_5O_{12}$	77	597
1.0729	Nd^{3+}	$YAlO_3$	300	712,717,718
1.073	Nd^{3+}	Y_2O_3	77	325,593
1.0730	Nd^{3+}	$CaSc_2O_4$	77	323

Table 2.1.4—*continued*
Paramagnetic Ion Lasers Arranged in Order of Wavelength

Wavelength (μm)	Ion	Host: sensitizer ion(s)	Temp. (K)	References
1.0737	Nd^{3+}	$Y_3Al_5O_{12}$	300	407,414
~1.074	Nd^{3+}	Y_2O_3-ThO_2-Nd_2O_3	300	781
1.074	Nd^{3+}	$SrAl_{12}O_{19}$	300	313
1.0740	Nd^{3+}	Y_2SiO_5	77	713
1.0741	Nd^{3+}	Gd_2O_3	300	372
1.0741	Nd^{3+}	Y_2SiO_5	300	466
1.0742	Nd^{3+}	Y_2SiO_5	300	320,360
~1.0746	Nd^{3+}	Y_2O_3	300	325,593
1.0746	Nd^{3+}	$LaSr_2Ga_{11}O_{20}$	110	234,439,546
1.075	Nd^{3+}	La_2O_2S	300	371
1.0755	Nd^{3+}	$CaSc_2O_4$	77	323
1.0757	Nd^{3+}	$Sr_3Ga_2Ge_4O_{14}$	300	188,667
1.0759	Nd^{3+}	$GdAlO_3$	77	213,214
1.0759	Nd^{3+}	$LuAlO_3$	300	75
1.0760	Nd^{3+}	$GdAlO_3$	300	213,214
1.07655	Nd^{3+}	$CaAl_4O_7$	77	510
1.0770	Nd^{3+}	$YScO_3$	77	363
1.0772	Nd^{3+}	$CaAl_4O_7$	77	510
1.0774	Nd^{3+}	$YScO_3$	130	439
1.0775–1.0845	Nd^{3+}	$CaYAlO_4$	300	287
1.0776	Nd^{3+}	Gd_2O_3	77	372
1.0778	Nd^{3+}	$LaSr_2Ga_{11}O_{20}$	110	234,439,546
1.078	Nd^{3+}	Y_2O_3	77	325,593
1.0780	Nd^{3+}	$Y_3Al_5O_{12}$	300	375
1.0780	Nd^{3+}	$Ca_2Ga_2SiO_7$	77	439,544-546,1002
1.0780–1.086	Nd^{3+}	$LaMgAl_{11}O_{19}$	300	431
1.0781	Nd^{3+}	Y_2SiO_5	77	320
1.0782	Nd^{3+}	Y_2SiO_5	300	320,466
1.0782–1.0787	Nd^{3+}	$LiNbO_3$	590–450	626
1.0782–1.0815	Nd^{3+}	$YAlO_3$	300	718
1.0785	Nd^{3+}	$LuScO_3$	300	363
1.0785	Nd^{3+}	$La_2Be_2O_5$	77	270
1.0786	Nd^{3+}	$CaAl_4O_7$	300	510
1.0788	Nd^{3+}	$Ca_2Ga_2SiO_7$	300	544–546,1002
1.0789	Nd^{3+}	Gd_2O_3	77, 300	372

Table 2.1.4—*continued*
Paramagnetic Ion Lasers Arranged in Order of Wavelength

Wavelength (μm)	Ion	Host: sensitizer ion(s)	Temp. (K)	References
1.0789	Nd^{3+}	$Pb_5Ge_3O_{11}$	77	417,418
1.079	Nd^{3+}	La_2O_3	77	296
1.079	Nd^{3+}	$La_2Be_2O_5$	300	254,255,299,416
1.0790	Nd^{3+}	$La_2Be_2O_5$	300	270
1.0790	Nd^{3+}	Lu_2SiO_5	300	466
1.07925	Nd^{3+}	Lu_2SiO_5	300	420
1.0795	Nd^{3+}	$YAlO_3$	300	356,377,423,424, 437,441,442,481, 716–718,995
1.0795–1.0802	Nd^{3+}	$YAlO_3$	77–600	620,702
1.0796	Nd^{3+}	$YAlO_3$	300	712
1.0796–1.0803	Nd^{3+}	$YAlO_3$	600–700	718
1.0799	Nd^{3+}	$Pb_5Ge_3O_{11}$	77	417,418
1.08	Nd^{3+}	Y_2O_3	300	427
1.0804	Nd^{3+}	$LaAlO_3$	300	213
1.0806	Nd^{3+}	$CaYAlO4$	300	422
1.0812	Nd^{3+}	$LaMgAl_{11}O_{19}$	77	439,482,546
1.0812–4	Nd^{3+}	Sc_2SiO_5	300	453,466
1.08145	Nd^{3+}	Sc_2SiO_5	300	420,466
1.0817	Nd^{3+}	$LaMgAl_{11}O_{19}$	300	439,482,546
1.082	Nd^{3+}	$LaMgAl_{11}O_{19}$	300	313
1.082–1.084	Nd^{3+}	$LaMgAl_{11}O_{19}$	300	444
1.0828	Nd^{3+}	$SrAl_4O_7$	300	473
1.0829–1.0859	Nd^{3+}	$LiNbO_3$	300	988,996
1.083	Nd^{3+}	$YAlO_3$	300	666
1.0831	Nd^{3+}	$LuAlO_3$	120	74
1.0832	$Nd3^{+}$	$LuAlO_3$	300	74
1.0832–1.0855	Nd^{3+}	$YAlO_3$	300	666,718
1.0837	Nd^{3+}	$YScO_3$	130	439
1.0840	Nd^{3+}	$LiNbO_3$	77	457
1.0840	Nd^{3+}	$GdScO_3$	200	459
1.0843	Nd^{3+}	$YScO_3$	300	363,439
1.0845	Nd^{3+}	$YAlO_3$	300	718
1.0846	Nd^{3+}	$LiNbO_3$	300	383,449,626
1.0847	Nd^{3+}	$YAlO_3$	530	626,702
1.085	Nd^{3+}	$LiNbO_3$:Mg	300	36,471,782

Table 2.1.4—*continued*
Paramagnetic Ion Lasers Arranged in Order of Wavelength

Wavelength (μm)	Ion	Host: sensitizer ion(s)	Temp. (K)	References
1.08515	Nd^{3+}	$GdScO_3$	300	363
1.0867	Nd^{3+}	$CaSc_2O_4$	77	323
1.0868	Nd^{3+}	$CaSc_2O_4$	300	323
~1.0885	Nd^{3+}	CaF_2-CeO_2	300	523
1.0885–1.0889	Nd^{3+}	CaF_2	300–420	137,344
1.0909	Nd^{3+}	$YAlO_3$	300	717,718
1.0913	Nd^{3+}	$YAlO_3$	530	607,702
1.0921	Nd^{3+}	$YAlO_3$	300	718
1.0922–1.0933	Nd^{3+}	$LiNbO_3$	620–300	994
1.093	Nd^{3+}	$LiNbO_3$	300	36,471,480,503
1.0933	Nd^{3+}	$LiNbO_3$	300	348,383,449
~1.094	Nd^{3+}	$LiNbO_3$:MgO	300	1030
1.0989	Nd^{3+}	$YAlO_3$	300	717,718
1.0991	Nd^{3+}	$YAlO_3$	500	607,702
1.1054	Nd^{3+}	$Y_3Al_5O_{12}$	300	414
1.110	Yb^{3+}	$Ca_4Sr(PO_4)_3F$	300	17
1.1119	Nd^{3+}	$Y_3Al_5O_{12}$	300	414
1.1158	Nd^{3+}	$Y_3Al_5O_{12}$	300	414
1.116	Tm^{2+}	CaF_2	4.2	524,525
1.1213	V^{2+}	MgF_2	77	488
1.1225	Nd^{3+}	$Y_3Al_5O_{12}$	300	414
1.167–1.345	Cr^{4+}	Mg_2SiO_4	300	507,706–709
1.18–1.29	Cr^{4+}	Y_2SiO_5	77	508
1.1810	Mn^{5+}	$Ba_3(VO_4)_2$	300	509
1.190	Ho^{3+}	$BaYb_2F_8$	300	469
1.2–1.32	Cr^{4+}	Mg_2SiO_4	77	514
1.2085	Ho^{3+}	$Gd_3Ga_5O_{12}$	~110	722
1.2155	Ho^{3+}	$Y_3Al_5O_{12}$	110	993
1.2160	Ho^{3+}	$Lu_3Al_5O_{12}$	110	993
1.2195	Er^{3+}	$LiYF_4$	116, 300	222,231
1.2196	Er^{3+}	$LiLuF_4$	110	32
1.2198	Ho^{3+}	$YAlO_3$	110	241
1.228	Er^{3+}	$LiErF_4$	110	233
1.2290	Er^{3+}	$LiYF_4$	110	502
1.2292	Er^{3+}	$LiErF_4$	90–102	223

Table 2.1.4—*continued*
Paramagnetic Ion Lasers Arranged in Order of Wavelength

Wavelength (μm)	Ion	Host: sensitizer ion(s)	Temp. (K)	References
1.2292	Er^{3+}	$LiLuF_4$	110	32
1.2294	Er^{3+}	$LiYF_4$	120, 300	224
1.2295	Er^{3+}	$LiLuF_4$	300	32,33
1.2308	Er^{3+}	$LiYF_4$	~110,300	226
1.2312	Er^{3+}	$BaEr_2F_8$	100–112	232
1.2320	Er^{3+}	$BaEr_2F_8$	110	215
1.234	Er^{3+}	$LiYF_4$:Yb	300	734
1.2342	Er^{3+}	$(Y,Er)AlO_3$	110	190
1.236–1.300	Cr^{4+}	Mg_2SiO_4	300	1032
1.2390	Er^{3+}	$(Y,Er)AlO_3$	110	190
1.2392	Er^{3+}	$(Y,Er)AlO_3$	300	190
1.24–1.33	V^{2+}	$CsCaF_3$	80	582
1.244	Cr^{4+}	Mg_2SiO_4	300	552,558,559
1.245	Er^{3+}	$Y_3Al_5O_{12}$	77	237
1.26	Er^{3+}	CaF_2	77	614
1.26	Er^{3+}	$BaYb_2F_8$	300	587
1.2805	Nd^{3+}	$Ca_3Ga_2Ge_3O_{12}$	110	519,571
1.3	Cr^{4+}	Y_2SiO_5	300	567
1.3	Nd^{3+}	$KNdP_4O_{12}$	300	317
1.3	Nd^{3+}	$NdAl_3(BO_3)_4$	330	317
1.302	Nd^{3+}	KYF_4	300	573
[1.304–1.372]	Nd^{3+}	$(La,Nd)P_5O_{14}$	300	306,566
1.3065	Nd^{3+}	$SrAl_{12}O_{19}$	300	478,324
1.307	Nd^{3+}	KYF_4	300	573
1.3070	Nd^{3+}	$5NaF\text{-}9YF_3$	300	537
1.3077	Nd^{3+}	$Gd_3Ga_5O_{12}$	77	490,492
1.309–1.628	Cr^{4+}	$Y_3Sc_xAl_{5-x}O_{12}$ (x=0–1.7)	300	572
1.311–1.334	Nd^{3+}	$NaNdP_4O_{12}$	300	307
1.3125	Nd^{3+}	LaF_3	77	537
1.3130	Nd^{3+}	CeF_3	77	538
1.313	Nd^{3+}	$LiYF_4$	300	32,33,314
1.3133	Nd^{3+}	$LiLuF_4$	300	32,33
1.3144	Ni^{2+}	MgO	77	535
1.3150	Nd^{3+}	$Ca_3Ga_2Ge_3O_{12}$	300	519
1.316	Ni^{2+}	MgO	82	662

Table 2.1.4—*continued*
Paramagnetic Ion Lasers Arranged in Order of Wavelength

Wavelength (μm)	Ion	Host: sensitizer ion(s)	Temp. (K)	References
1.316–1.340	Nd^{3+}	$LiNdP_4O_{12}$	300	307
1.3160	Nd^{3+}	SrF_2-LaF_3	77	537
1.3165	Nd^{3+}	CdF_2-YF_3	77	538
1.3165	Nd^{3+}	α-$NaCaCeF_6$	77	538
1.317	Nd^{3+}	$Li(La,Nd)P_4O_{12}$	300	105,108,109,112, 113,350,353,354, 382,576
1.317	Nd^{3+}	$LiNdP_4O_{12}$	300	306,560,561
1.3170	Nd^{3+}	CeF_3	300	538
1.3170	Nd^{3+}	LaF_3-SrF_2	77	549
1.3170	Nd^{3+}	BaY_2F_8	77	366
1.3172	Nd^{3+}	$LiLuF_4$	110	519
1.3175	Nd^{3+}	BaF_2	300	541
1.318	Ni^{2+}	MgO	80	261
1.318	Nd^{3+}	$Y_3Al_5O_{12}$	300	346,570,779
1.318	Nd^{3+}	BaY_2F_8	300	366
1.3185	Nd^{3+}	CaF_2-GdF_3	300	434
1.3185	Nd^{3+}	BaF_2-LaF_3	300	602
1.3185	Nd^{3+}	KY_3F_{10}	300	461
1.3187	Nd^{3+}	$Y_3Al_5O_{12}$	300	594,599
1.3188	Nd^{3+}	$Y_3Al_5O_{12}$	300	375
1.319	Nd^{3+}	$LiNdP_4O_{12}$	300	317,319,780
1.319	Nd^{3+}	$Y_3Ga_5O_{12}$	300	613
1.319–1.325	Nd^{3+}	$(Y,Nd)P_5O_{14}$	300	400
1.3190	Nd^{3+}	$Ca_2Y_5F_{19}$	300	606
1.3190	Nd^{3+}	CaF_2-LaF_3	300	434
1.3190	Nd^{3+}	CaF_2-CeF_3	300	537
1.3190	Nd^{3+}	$Sr_2Y_5F_{19}$	300	367
1.3190	Nd^{3+}	α-$NaCaCeF_6$	300	538
1.32	Nd^{3+}	$Gd_3Sc_2Ga_3O_{12}$:Cr	300	565
1.32	Nd^{3+}	NdP_5O_{14}	300	386
1.32	Nd^{3+}	$(La,Nd)P_5O_{14}$	300	386
1.32	Nd^{3+}	$K(Nd,Gd)P_4O_{12}$	300	404
1.32–1.43	Cr^{4+}	$Ca_3Ga_2Ge_3O_{12}$	300	433
1.32–1.53	Cr^{4+}	$Y_3Al_5O_{12}$	300	585,589
1.320	Nd^{3+}	$NaNdP_4O_{12}$	300	382

Table 2.1.4—*continued*
Paramagnetic Ion Lasers Arranged in Order of Wavelength

Wavelength (μm)	Ion	Host: sensitizer ion(s)	Temp. (K)	References
1.3200	Nd^{3+}	$Ca_2Y_5F_{19}$	77	606
1.3200	Nd^{3+}	SrF_2-LuF_3	300	427
1.3200	Nd^{3+}	BaF_2-YF_3	300	413
1.3200	Nd^{3+}	$Y_3Al_5O_{12}$	300	375
1.3200	Nd^{3+}	YP_5O_{14}	300	601
1.3208	Nd^{3+}	$LiLuF_4$	300	32,33
1.3209	Nd^{3+}	$Lu_3Al_5O_{12}$	300	538
1.3209	Nd^{3+}	$Ba_5(PO_4)_3F$	300	1027
1.3212	Nd^{3+}	$LiYF_4$	300	32,33
1.3225	Nd^{3+}	SrF_2-YF_3	77	537
1.3225	Nd^{3+}	CaF_2	300	541
1.323	Nd^{3+}	NdP_5O_{14}	300	656
1.323	Nd^{3+}	$(La,Nd)P_5O_{14}$	300	574,575
1.3235	Nd^{3+}	LaF_3	77	537
1.3235	Nd^{3+}	SrF_2-LaF_3	77	537
1.3240	Nd^{3+}	CeF_3	77	538
1.324	Nd^{3+}	$(La,Nd)P_5O_{14}$	300	566
1.3245	Nd^{3+}	CdF_2-YF_3	300	538
1.3250	Nd^{3+}	SrF_2-LaF_3	300	537
1.3250	Nd^{3+}	SrF_2-GdF_3	77	421
1.3250	Nd^{3+}	SrF_2	300	541
1.3255	Nd^{3+}	CaF_2-YF_3	77	538
1.3255	Nd^{3+}	SrF_2-CeF_3	300	518
1.3256	Nd^{3+}	$LiYF_4$	110	32,33
1.3257	Nd^{3+}	$LiLuF_4$	110	32
1.3260	Nd^{3+}	SrF_2-GdF_3	300	421
1.3260	Nd^{3+}	α-$NaCaYF_6$	77	538
1.3270	Nd^{3+}	CaF_2-YF_3	300	538
1.3270	Nd^{3+}	BaF_2	300	541
1.3270	Nd^{3+}	$Ca_3(Nb,Ga)_2Ga_3O_{12}$	300	506
1.3275	Nd^{3+}	LaF_3-SrF_2	77	549
1.328	Ni^{2+}	MgO	131	662
1.328	Nd^{3+}	$Sr_5(PO_4)_3$	300	515
1.3280	Nd^{3+}	BaF_2-LaF_3	300	602
1.3285	Nd^{3+}	α-$NaCaYF_6$	300	538

Table 2.1.4—*continued*
Paramagnetic Ion Lasers Arranged in Order of Wavelength

Wavelength (μm)	Ion	Host: sensitizer ion(s)	Temp. (K)	References
1.3285	Nd^{3+}	SrF_2-ScF_3	300	446
1.3290	Nd^{3+}	BaF_2-LaF_3	77	602
1.3298	Nd^{3+}	$CsLa(WO_4)_2$	300	241,475
1.3298	Nd^{3+}	$GdGaGe_2O_7$	300	467,472
1.3	Nd^{3+}	$NdAl_3(BO_3)_4$	300	317
1.330	Nd^{3+}	CaF_2-LuF_3	300	425
1.3300	Nd^{3+}	$Gd_3Ga_5O_{12}$	300	618
1.3300	Nd^{3+}	SrF_2-YF_3	77	537
1.3300	Nd^{3+}	$SrMoO_4$	77	616
1.3300	Nd^{3+}	CdF_2-ScF_3	300	360
1.3300	Nd^{3+}	$NaLuGeO_4$	77	539,540
1.3303	Nd^{3+}	$NdGaGe_2O_7$	300	467,470
1.3305	Nd^{3+}	LaF_3	77	537
1.3305	Nd^{3+}	HfO_2-Y_2O_3	300	563,564
1.3305	Nd^{3+}	$Y_3Ga_5O_{12}$	300	490,491
1.3310	Nd^{3+}	LaF_3	300	537
1.3310	Nd^{3+}	CeF_3	77	538
1.3310	Nd^{3+}	$CaWO_4$	77	617
1.3310	Nd^{3+}	$Y_3Sc_2Ga_3O_{12}$	300	427
1.3310	Nd^{3+}	$NaLuGeO_4$	300	539,540
1.3315	Nd^{3+}	LaF_3-SrF_2	300	549
1.3315	Nd^{3+}	$Lu_3Ga_5O_{12}$	300	505
1.3315	Nd^{3+}	$Gd_3Ga_5O_{12}$	300	593
1.3315	Nd^{3+}	$Ca_3Ga_2Ge_3O_{12}$	300	519
1.3317	Nd^{3+}	$Ca_3Ga_2Ge_3O_{12}$	300	675
1.3319	Nd^{3+}	$Lu_3Al_5O_{12}$	77	538
1.3320	Nd^{3+}	CeF_3	300	538
1.3320	Nd^{3+}	SrF_2-YF_3	77	537
1.3320	Nd^{3+}	$SrAl_4O_7$	77	478
1.3320	Nd^{3+}	$PbMoO_4$	77	655
1.3320	Nd^{3+}	ZrO_2-Y_2O_3	300	564
1.3320	Nd^{3+}	$Ca_3Ga_4O_9$	300	482
1.3325	Nd^{3+}	LaF_3-SrF_2	77	549
1.3325	Nd^{3+}	$SrMoO_4$	300	655
1.3325	Nd^{3+}	$NaYGeO_4$	300	539,577

Table 2.1.4—*continued*
Paramagnetic Ion Lasers Arranged in Order of Wavelength

Wavelength (μm)	Ion	Host: sensitizer ion(s)	Temp. (K)	References
1.3326	Nd^{3+}	$Lu_3Al_5O_{12}$	300	538
1.3333	Nd^{3+}	$Lu_3Al_5O_{12}$	77	538
1.3334	Nd^{3+}	$NaGdGeO_4$	300	539,604,605
1.3335	Nd^{3+}	$Ca_3Ga_2Ge_3O_{12}$	110	519,571
1.3338	Nd^{3+}	$Y_3Al_5O_{12}$	300	375
1.3340	Nd^{3+}	$CaWO_4$	300	253,549,980
1.3340	Nd^{3+}	$PbMoO_4$	300	655
1.3342	Nd^{3+}	$KLa(MoO_4)_2$	300	501
1.3342	Nd^{3+}	$Lu_3Al_5O_{12}$	300	538
1.3342	Nd^{3+}	$NaBi(WO_4)_2$	300	665
1.3345	Pr^{3+}	BaY_2F_8	110	241
1.3345	Nd^{3+}	$CaWO_4$	77	253,549,660
1.3345	Nd^{3+}	$SrAl_4O_7$	300	478
1.3345	Nd^{3+}	$Ca_5(PO_4)_3F$	77	606
1.3347	Nd^{3+}	$Ca_5(PO_4)_3F$	300	606
1.3347	Pr^{3+}	$BaYb_2F_8$	110	241,623
1.3347	Nd^{3+}	$SrWO_4$	300	57
1.3350	Nd^{3+}	$KLa(MoO_4)_2$	77, 300	498
1.3350	Nd^{3+}	$Y_3Al_5O_{12}$	300	375
1.3354	Nd^{3+}	$CaLa_4(SiO_4)_3O$	300	644
1.3355	Nd^{3+}	SrF_2-LaF_3	77	537
1.3355	Nd^{3+}	$NaLa(WO_4)_2$	300	655
1.3360	Nd^{3+}	$Y_3Sc_2Al_3O_{12}$	300	465
1.3360	Nd^{3+}	$Gd_3Sc_2Al_3O_{12}$	300	476
1.3360	Nd^{3+}	$Lu_3Sc_2Al_3O_{12}$	300	460
1.3360	Nd^{3+}	$Y_3Sc_2Al_3O_{12}$	300	504
1.3360	Nd^{3+}	CdF_2-CeF_3	300	763
1.3365	Nd^{3+}	$Ca_2Ga_2SiO_7$	300	439,544–546
1.3365	Nd^{3+}	CdF_2-GaF_3	300	742
1.3365	Nd^{3+}	CdF_2-LaF_3	300	380
1.3370	Nd^{3+}	CaF_2-YF_3	300	538
1.3370	Nd^{3+}	$Ca(NbO_3)_2$	77	607
1.3370	Nd^{3+}	$CaWO_4$	300	253,549,980
1.3370	Nd^{3+}	$Gd_3Ga_5O_{12}$	300	279,527
1.3370	Nd^{3+}	$LiLa(MoO_4)_2$	300	655

Table 2.1.4—*continued*
Paramagnetic Ion Lasers Arranged in Order of Wavelength

Wavelength (μm)	Ion	Host: sensitizer ion(s)	Temp. (K)	References
1.3372	Nd^{3+}	$CaWO_4$	77	253,549,980
1.3375	Nd^{3+}	α-$NaCaYF_6$	300	538
1.3375	Nd^{3+}	$PbMoO_4$	77	616
1.3375	Nd^{3+}	$LiLa(MoO_4)_2$	77	655
1.3375	Nd^{3+}	$Ca_2Ga_2SiO_7$	110	544–546
1.3376	Nd^{3+}	$Lu_3Al_5O_{12}$	77	602
1.338	Nd^{3+}	$Y_3Ga_5O_{12}$	300	314,613
1.3380	Nd^{3+}	CaF_2-YF_3	77	538
1.3380	Nd^{3+}	$Ca(NbO_3)_2$	300	607
1.3380	Nd^{3+}	$NaLa(MoO_4)_2$	77, 300	655
1.3381	Nd^{3+}	$Y_3Al_5O_{12}$	300	594
1.3382	Nd^{3+}	$Y_3Al_5O_{12}$	300	375
1.3385	Nd^{3+}	$NaGd(MoO_4)_2$	300	655
1.3387	Nd^{3+}	$Lu_3Al_5O_{12}$	300	267
1.339	Nd^{3+}	YF_3	300	415
1.3390	Nd^{3+}	α-$NaCaYF_6$	77	538
1.3390	Nd^{3+}	$CaWO_4$	300	253,549,980
1.3391	Nd^{3+}	$YAlO_3$	77	549,660,717
1.3393	Nd^{3+}	$YAlO_3$	77	599,717
~1.34	Nd^{3+}	$GdVO_4$	300	550
1.34	Nd^{3+}	YVO_4	300	778
1.3400	Nd^{3+}	$CaAl_4O_7$	77	478
1.3400	Nd^{3+}	$LiGd(MoO_4)_2$	77, 300	655
1.3400	Nd^{3+}	$YAlO_3$	300	549,717,980
1.3407	Nd^{3+}	$Bi_4Si_3O_{12}$	300	625
1.341	Nd^{3+}	$NdAl_3(BO_3)_4$	300	656
1.3410	Nd^{3+}	$Y_3Al_5O_{12}$	300	375
1.3410	Nd^{3+}	$Lu_3Al_5O_{12}$	300	267
1.3410	Nd^{3+}	$YAlO_3$	300	253,599,717
1.3410	Nd^{3+}	$Y_3Al_5O_{12}$	300	375
1.3413	Nd^{3+}	$YAlO_3$	300	712
1.3414	Nd^{3+}	$YAlO_3$	300	995
1.3415	Nd^{3+}	$Ca(NbO_3)_2$	77	607
1.3415	Nd^{3+}	YVO_4	77	537
1.3416	Nd^{3+}	$YAlO_3$	300	549,660,717

Table 2.1.4—*continued*
Paramagnetic Ion Lasers Arranged in Order of Wavelength

Wavelength (μm)	Ion	Host: sensitizer ion(s)	Temp. (K)	References
1.3418	Nd^{3+}	$Bi_4Ge_3O_{12}$	300	291,661
1.3420	Nd^{3+}	$CaAl_4O_7$	300	478
1.3425	Nd^{3+}	$Ca(NbO_3)_2$	300	607
1.3425	Nd^{3+}	YVO_4	300	537
1.3425	Nd^{3+}	$PbMoO_4$	300	616
1.3425	Nd^{3+}	CdF_2-YF_3	300	538
1.3430	Nd^{3+}	$NaLa(MoO_4)_2$	77	655
1.3437	Nd^{3+}	$LuAlO_3$	300	74
1.3440	Nd^{3+}	$SrMoO_4$	77	616,655
1.3440	Nd^{3+}	$LiLa(MoO_4)_2$	77	655
1.3440	Nd^{3+}	$NaLa(MoO_4)_2$	300	655
1.345	Nd^{3+}	$NdAl_3(BO_3)_4$	300	648
1.3450	Nd^{3+}	$PbMoO_4$	77	655
1.3455	Nd^{3+}	$LiGd(MoO_4)_2$	77	655
1.3459	Nd^{3+}	$CaWO_4$	77	253,549,660
1.3465	Pr^{3+}	$LiYbF_4$	110	241
1.3468	Pr^{3+}	$LiYbF_4$	110	241,246
1.3475	Nd^{3+}	$CaWO_4$	300	253,549,980
1.348–1.482	Cr^{4+}	Ca_2GeO_4	300	629
1.3482	Nd^{3+}	$KLu(WO_4)_2$	300	211,272
1.3485	Nd^{3+}	$KY(MoO_4)_2$	300	549,765
1.3493	Nd^{3+}	$Ca_3Ga_2Ge_4O_{14}$	300	188,667
1.3499	Nd^{3+}	$Lu_3Al_5O_{12}$	77	602
1.35–1.5	Cr^{4+}	$Y_3Al_5O_{12}$	300	981,982
1.350–1.560	Cr^{4+}	$Y_3Al_5O_{12}$	300	647
1.3500	Nd^{3+}	CdF_2-LuF_3	300	742
1.3505	Nd^{3+}	CaF_2-ScF_3	300	374
1.351	Nd^{3+}	$La_2Be_2O_5$	300	314,649
1.3510	Nd^{3+}	$KGd(WO_4)_2$	300	76
1.3510	Nd^{3+}	$La_2Be_2O_5$	300	651
1.3510	Nd^{3+}	$Sr_3Ga_2Ge_4O_{14}$	300	188,667
1.3512	Nd^{3+}	$YAlO_3$	300	549
1.3514	Nd^{3+}	$YAlO_3$	300	549,717,980
1.3515	Nd^{3+}	$KY(WO_4)_2$	77	250–252
1.3520	Nd^{3+}	CdF_2-GdF_3	300	742

Table 2.1.4—*continued*
Paramagnetic Ion Lasers Arranged in Order of Wavelength

Wavelength (μm)	Ion	Host: sensitizer ion(s)	Temp. (K)	References
1.3525	Nd^{3+}	$Ca_2Y_5F_{19}$	300	606
1.3525	Nd^{3+}	$KY(WO_4)_2$	300	250–252
1.3525	Nd^{3+}	$Lu_3Al_5O_{12}$	300	267
1.3530	Nd^{3+}	$SrAl_4O_7$	77	478
1.3532	Nd^{3+}	$Lu_3Al_5O_{12}$	300	496
1.3533	Nd^{3+}	$Y_3Al_5O_{12}$	300	660
1.3533	Nd^{3+}	$KLa(MoO_4)_2$	300	501
1.3533	Nd^{3+}	$KLu(WO_4)_2$	300	257
1.354	Nd^{3+}	$La_2Be_2O_5$	300	151
1.3545	Nd^{3+}	$KY(WO_4)_2$	77, 300	250–252
1.3545	Nd^{3+}	$Ca_3Ga_2Ge_3O_{12}$	110	519
1.3550	Nd^{3+}	$LiNbO_3$	300	603
1.3550	Nd^{3+}	$KLu(WO_4)_2$	300	257
1.3564	Nd^{3+}	$Y_3Al_5O_{12}$	77	375
1.3565	Nd^{3+}	$CaSc_2O_4$	300	323,504
1.3572	Nd^{3+}	$Y_3Al_5O_{12}$	300	660
~1.358	Nd^{3+}	Y_2O_3	300	325,593
1.3585	Nd^{3+}	CaF_2-YF_3	300	538
1.3585	Nd^{3+}	Y_2SiO_5	300	466
1.3585	Nd^{3+}	Lu_2SiO_5	300	466
1.3595	Nd^{3+}	LaF_3	300	537
1.3600	Nd^{3+}	CaF_2-YF_3	77	538
1.3600	Nd^{3+}	α-$NaCaYF_6$	300	538
1.3628	Nd^{3+}	$LaSr_2Ga_{11}O_{20}$	300	234,439,564
1.3630	Nd^{3+}	$KLa(MoO_4)_2$	300	501
1.3630–2	Nd^{3+}	Sc_2SiO_5	300	453,466
1.3644	Nd^{3+}	$YAlO_3$	77	549,660,717
1.365	Nd^{3+}	$La_2Be_2O_5$	300	663
1.3657	Nd^{3+}	$KLa(MoO_4)_2$	300	501
1.3665	Nd^{3+}	$SrAl_4O_7$	300	478
1.3670	Nd^{3+}	LaF_3	77	537
1.3675	Nd^{3+}	LaF_3	300	537
1.3675	Nd^{3+}	CeF_3	77	538
1.3675	Nd^{3+}	$CaAl_4O_7$	77	478
1.3680	Nd^{3+}	$SrAl_4O_7$	300	602

Table 2.1.4—*continued*
Paramagnetic Ion Lasers Arranged in Order of Wavelength

Wavelength (μm)	Ion	Host: sensitizer ion(s)	Temp. (K)	References
1.369	Ni^{2+}	MgO	153	662
1.3690	Nd^{3+}	CeF_3	300	602
1.37	Cr^{4+}	$CaGd_4(SiO_4)_3O$	300	701
1.37–1.51	Cr^{4+}	$Y_3Al_5O_{12}$	300	664
1.3707	Nd^{3+}	$La_3Ga_{5.5}Nb_{0.5}O_{14}$	300	667,668
1.3710	Nd^{3+}	$CaAl_4O_7$	300	478
1.3730	Nd^{3+}	$La_3Ga_5GeO_{14}$	300	667,668
1.3730	Nd^{3+}	$La_3Ga_5SiO_{14}$	300	695,697,698
1.3730	Nd^{3+}	$La_3Ga_{5.5}Ta_{0.5}O_{14}$	300	667,705
1.3745	Nd^{3+}	$LiNbO_3$	300	383,538
1.3755	Nd^{3+}	$NaLa(MoO_4)_2$	77	655
1.3760	Nd^{3+}	$LaMgAl_{11}O_{19}$	300	455
1.3780	Nd^{3+}	$PbMoO_4$	77	655
1.3790	Nd^{3+}	$SrMoO_4$	77	655
1.3806	Ho^{3+}	$YAlO_3$	300	190,241
1.3840	Nd^{3+}	$NaLa(MoO_4)_2$	77	655
1.3849	Nd^{3+}	$YAlO_3$	77	549
1.3865	Ho^{3+}	$BaYb_2F_8$	110	703
1.3868	Nd^{3+}	$LaBGeO_5$	300	330
1.3870	Nd^{3+}	$LiNbO_3$	300	383,538
1.3870	Nd^{3+}	$CaSc_2O_4$	77	323,504
1.3880	Nd^{3+}	$CaWO_4$	77	253,549,980
1.3885	Nd^{3+}	$CaWO_4$	300	549
1.390–1.475	Cr^{4+}	Ca_2GeO_4	300	1032
1.3900	Ho^{3+}	$YAlO_3$	110	190,241
1.3908	Ho^{3+}	$KY(WO_4)_2$	~110	993
1.3918	Ho^{3+}	$LiLuF_4$	110	32
1.392	Ho^{3+}	$LiYF_4$	300	719
1.3920	Ho^{3+}	$LiLuF_4$	300	32
1.3950	Ho^{3+}	$YAlO_3$	110	190
1.3960	Ho^{3+}	$LiYF_4$	116, 300	167,181
1.3982	Ho^{3+}	$KGd(WO_4)_2$	110	993
1.4	Ho^{3+}	$KLa(MoO_4)_2$	110	501
1.4003	Ho^{3+}	$YAlO_3$	110	190
1.4026	Nd^{3+}	$YAlO_3$	~110	549

Table 2.1.4—*continued*
Paramagnetic Ion Lasers Arranged in Order of Wavelength

Wavelength (μm)	Ion	Host: sensitizer ion(s)	Temp. (K)	References
1.4028	Ho^{3+}	$YAlO_3$	110	993
1.4040	Ho^{3+}	$Gd_3Ga_5O_{12}$	~110	722
1.4058	Ho^{3+}	$YAlO_3$	110–300	190
1.4072	Ho^{3+}	$Y_3Al_5O_{12}$	~110	993
1.4085	Ho^{3+}	$Lu_3Al_5O_{12}$	~110	993
1.409	Ni^{2+}	MgO	235	662
1.414	Ni^{2+}	MgO	235	375
1.4150	Nd^{3+}	$Y_3Al_5O_{12}$	300	723
1.44	Nd^{3+}	$SrGd_4(SiO_4)_3O$	300	701
1.4440	Ni^{2+}	MgO	235	375
1.4444	Nd^{3+}	$Y_3Al_5O_{12}$	300	724
1.449–1.455	Tm^{3+}	$LiYF_4$:Tb	300	700
1.46	Ni^{2+}	$CaY_2Mg_2Ge_3O_{12}$	80	261,704
1.464	Tm^{3+}	$LiYF_4$:Yb	300	80
1.48	Tm^{3+}	$Gd_3Sc_2Ga_3O_{12}$	300	710
1.482	Tm^{3+}	$BaYb_2F_8$	300	711,714
1.486	Ho^{3+}	$LiYF_4$	90	268,269
1.4862	Ho^{3+}	$LiYF_4$	90, 116	749,750
1.4912	Ho^{3+}	$LiYF_4$	190	181
1.5–2.3	Co^{2+}	MgF_2	80	114,543,704, 762,768–773
1.50	Tm^{3+}	$LiYF_4$	300	743
1.500	Tm^{3+}	$LiYF_4$:Yb	300	80
1.5298	Er^{3+}	CaF_2	77	674
1.530	Er^{3+}	$Ca_2Al_2SiO_7$:Yb	300	744
1.5308	Er^{3+}	CaF_2	77	674
1.532	Er^{3+}	$LiNbO_3$	300	747
1.547	Er^{3+}	CaF_2-YF_3	77	215,216
1.550	Er^{3+}	$Ca_2Al_2SiO_7$:Yb	300	733,744
1.5500	Er^{3+}	$CaAl_4O_7$	77	510
1.554	Er^{3+}	$SrY_4((SiO_4)_3O$:Yb	300	426
1.5542	Er^{3+}	$Er(Y,Gd)AlO_3$	77	783
1.555	Er^{3+}	$Ca_2Al_2SiO_7$:Yb,Ce	300	744
1.5554	Er^{3+}	$YAlO_3$	77	783
1.5578	Er^{3+}	$Bi_4Ge_3O_{12}$	77	229
1.563	Er^{3+}	$LiNbO_3$	300	752

Table 2.1.4—*continued*
Paramagnetic Ion Lasers Arranged in Order of Wavelength

Wavelength (μm)	Ion	Host: sensitizer ion(s)	Temp. (K)	References
1.5646	Er^{3+}	$GdAlO_3$	77	784
1.568	Tm^{3+}	$LiYF_4$:Yb	300	80
1.576	Er^{3+}	$LiNbO_3$	300	752
1.5808	Tm^{3+}	BaY_2F_8	300	79
1.5815	Er^{3+}	$CaAl_4O_7$	77	510
1.59	Ni^{2+}	$KMgF_3$	80	774
1.6(a)	Ni^{2+}	MnF_2	77	535
1.61	Er^{3+}	$Ca(NbO_3)_2$	77	753
1.61–1.74	Ni^{2+}	MgF_2	80	261,543,704,754, 771,774,775
1.6113	Er^{3+}	LaF_3	77	810
1.612	Er^{3+}	$CaWO_4$	77	757
1.617	Er^{3+}	CaF_2	77	760
1.620	Er^{3+}	$LiYF_4$	300	755
1.620	Er^{3+}	ZrO_2-Er_2O_3	77	764
1.62–1.90	Co^{2+}	$KMgF_3$	80	192
1.622–3	Er^{3+}	$YAlO_3$	300	777,787
1.623	Ni^{2+}	MgF_2	77	535,739
1.63–2.11	Co^{2+}	MgF_2	80	754
1.632	Er^{3+}	$Y_3Al_5O_{12}$	300	738
1.636	Ni^{2+}	MgF_2	77–82	740
1.64	Er^{3+}	$Y_3Ga_5O_{12}$	300	976
1.64	Er^{3+}	$Y_3Al_5O_{12}$	300	756
1.643	Er^{3+}	$Y_3Sc_2Ga_3O_{12}$	300	1026
1.6437	Er^{3+}	$YScO_3$:Gd	77	783
1.644	Er^{3+}	$Y_3Al_5O_{12}$	300	954
1.644	Pr^{3+}	LaF_3	130	761
1.6449	Er^{3+}	$Y_3Al_5O_{12}$	295	741
1.6452	Er^{3+}	$Y_3Al_5O_{12}$	77	651
1.6455	Er^{3+}	$BaEr_2F_8$	110	115
1.6459	Er^{3+}	$Y_3Al_5O_{13}$	295	793
1.646	Er^{3+}	$(Y,Er)_3Al_5O_{12}$	300	766
1.6470	Er^{3+}	$LiYF_4$	110	226
1.65–2.15	Co^{2+}	$KZnF_3$	27	192,785,786,1003
1.6525	Er^{3+}	$Lu_3Al_5O_{12}$	77	794
1.6596	Er^{3+}	$(Y,Er)_3Al_5O_{12}$	77	237

Table 2.1.4—*continued*
Paramagnetic Ion Lasers Arranged in Order of Wavelength

Wavelength (μm)	Ion	Host: sensitizer ion(s)	Temp. (K)	References
1.66	Er^{3+}	$YAlO_3$	300	788,789
1.6600	Er^{3+}	$YAlO_3$:Gd	300	783
1.6602	Er^{3+}	$Y_3Al_5O_{12}$	77	651
1.6615	Er^{3+}	$Yb_3Al_5O_{12}$	77	726
1.6615	Er^{3+}	CaF_2-ErF_3	110	791
1.662–3	Er^{3+}	$YAlO_3$	300	788,789
1.6628	Er^{3+}	$(Y,Er)AlO_3$	110	788,789,792
1.6630	Er^{3+}	$Lu_3Al_5O_{12}$	77	794
1.6631	Er^{3+}	$(Y,Er)AlO_3$	110	788–790
1.6632	Er^{3+}	$YAlO_3$	300	119,129,716,797
1.6632	Er^{3+}	$ErAlO_3$	110	925
1.6640	Er^{3+}	$LiYF_4$	138–300	223,231
1.6645	Er^{3+}	$Bi_4Ge_3O_{12}$	77	229
1.667	Er^{3+}	$(Y,Er)AlO_3$	300	777
1.6675	Er^{3+}	$LuAlO_3$	~90	795
1.6682	Er^{3+}	$YScO_3$:Gd	77	783
1.6714	Er^{3+}	$GdAlO_3$	77	784
1.673	Ho^{3+}	$LiYF_4$	300	719
1.674–1.676	Ni^{2+}	MgF_2	82–100	535,739
1.6776	Er^{3+}	$YAlO_3$	300	797
1.696	Er^{3+}	CaF_2	77	592
1.7036	Er^{3+}	$LiYF_4$	116, 250	223,231
1.7042	Er^{3+}	$LiErF_4$	110	233
1.706	Er^{3+}	$(Y,Er)AlO_3$	300	777
1.7061	Er^{3+}	$YAlO_3$	300	797
1.715	Er^{3+}	CaF_2	77	592
1.7155	Er^{3+}	$KGd(WO_4)_2$	300	219,796
1.7178	Er^{3+}	$KY(WO_4)_2$	300	236
1.726	Er^{3+}	CaF_2	77	592
1.726	Er^{3+}	$(Y,Er)AlO_3$	300	777
1.7280	Er^{3+}	$KLa(MoO_4)_2$	110	501
1.7296	Er^{3+}	$YAlO_3$	300	797
1.73	Er^{3+}	$KLa(MoO_4)_2$	300	501
1.730	Er^{3+}	$LiYF_4$	300	808
1.731–1.756	Ni^{2+}	MgF_2	100–192	535,739

Table 2.1.4—*continued*
Paramagnetic Ion Lasers Arranged in Order of Wavelength

Wavelength (μm)	Ion	Host: sensitizer ion(s)	Temp. (K)	References
1.7312	Er^{3+}	$LiYF_4$	116, 300	224
1.732	Er^{3+}	$LiErF_4$	~90	268,269
1.7320	Er^{3+}	$LiYF_4$	110, 300	226
1.7322	Er^{3+}	$LiErF_4$	90	223
1.7325	Er^{3+}	$KGd(WO_4)_2$	300	219,796
1.7330	Er^{3+}	$KGd(WO_4)_2$	300	809
1.7343	Er^{3+}	$LiLuF_4$	110	32
1.7345	Er^{3+}	$LiLuF_4$	300	32,33
1.7350	Er^{3+}	$BaEr_2F_8$	102–112	232
1.7355	Er^{3+}	$BaEr_2F_8$	110	115
1.7355	Er^{3+}	$KY(WO_4)_2$	300	236
1.7360	Er^{3+}	$BaYb_2F_8$	110	115
1.7370	Er^{3+}	$KLu(WO_4)_2$	300	300
1.7372	Er^{3+}	$KEr(WO_4)_2$	300	809
1.7372	Er^{3+}	$KY_{0.5}Er_{0.5}(WO_4)_2$	300	809
1.7383	Er^{3+}	$KLu(WO_4)_2$	300	809
1.7390	Er^{3+}	$KLu(WO_4)_2$	300	809
1.7410	Er^{3+}	$Ca(NbO_3)_2$	110	57
1.75	Ni^{2+}	MgF_2	200	806
1.750	Co^{2+}	MgF_2	77	535,801
1.7757	Er^{3+}	$Y_3Al_5O_{12}$	300	790,802
1.776	Er^{3+}	$(Y,Er)_3Al_5O_{12}$	77–110	237
1.7762	Er^{3+}	$Lu_3Al_5O_{12}$	300	790,794
1.7762	Er^{3+}	$Er_3Al_5O_{12}$	110	790
1.7767	Er^{3+}	$(Lu,Er)_3Al_5O_{12}$	110	790
1.785–1.797	Ni^{2+}	MgF_2	190–240	535,739
1.800–2.450	Co^{2+}	MgF_2	300	799,1003
1.8035	Co^{2+}	MgF_2	77	535,801
1.821	Co^{2+}	$KMgF_3$	77	535,1003
1.833	Nd^{3+}	$Y_3Al_5O_{12}$	293	807
1.85–2.14	Tm^{3+}	$Y_3Sc_2Ga_3O_{12}$	300	800
1.8529	Tm^{3+}	$GdAlO_3$	77	804
1.8532	Tm^{3+}	$LiNbO_3$	77	370
1.856	Tm^{3+}	$YAlO_3$:Cr	90	803
1.8580	Tm^{3+}	α-$NaCaErF_6$	150	805

Table 2.1.4—*continued*
Paramagnetic Ion Lasers Arranged in Order of Wavelength

Wavelength (μm)	Ion	Host: sensitizer ion(s)	Temp. (K)	References
1.860	Tm^{3+}	CaF_2-ErF_3	77	798
1.861	Tm^{3+}	$(Y,Er)AlO_3$	77	130
1.862	Tm^{3+}	$Y_3Sc_2Ga_3O_{12}$	300	979
1.865	Ni^{2+}	MnF_2	20	535
1.87–2.16	Tm^{3+}	$Y_3Al_5O_{12}$	300	800
1.872	Tm^{3+}	$ErAlO_3$	77	819
1.88	Tm^{3+}	$LiYF_4$	300	811
1.880	Tm^{3+}	$(Y,Er)_3Al_5O_{12}$	77	80
1.883	Tm^{3+}	$YAlO_3$	90	803
1.8834	Tm^{3+}	$Y_3Al_5O_{12}$	77	278
1.884	Tm^{3+}	$(Y,Er)_3Al_5O_{12}$	77	794
1.8845	Tm^{3+}	$(Er,Lu)AlO_3$	77	840
1.8850	Tm^{3+}	$(Er,Yb)_3Al_5O_{12}$	77	726
1.8855	Tm^{3+}	$Lu_3Al_5O_{12}$	77	794
1.8885	Tm^{3+}	α-$NaCaErF_6$	77, 150	805
1.8890	Tm^{3+}	$LiYF_4$	110	812
1.894	Tm^{3+}	CaF_2:Er	~100	790,836,936
~1.896	Tm^{3+}	ZrO_2-Er_2O_3	77	764
~1.9	Tm^{3+}	CaF_2:Er	77	755
1.9–2.0	Tm^{3+}	$Er_3Al_5O_{12}$	77	823
1.9060	Tm^{3+}	$CaMoO_4$:Er	77	821
1.9090	Tm^{3+}	$LiYF_4$	110	812
1.91	Tm^{3+}	$Ca(NbO_3)_2$	77	753
1.911	Tm^{3+}	$CaWO_4$	77	291
1.9115	Tm^{3+}	$CaMoO_4$	77	821
1.915	Ni^{2+}	MnF_2	77	535
1.916	Tm^{3+}	$CaWO_4$	77	291,822
1.9190	Tm^{3+}	$BaYb_2F_8$	300	813
1.922	Ni^{2+}	MnF_2	77	535
1.929	Ni^{2+}	MnF_2	85	535
1.9335	Tm^{3+}	$YAlO_3$:Cr	90	803
1.934	Tm^{3+}	Er_2O_3	77	816
1.939	Ni^{2+}	MnF_2	85	535
1.94	Tm^{3+}	YVO_4	300	827
1.94	Tm^{3+}	$CaY_4(SiO_4)_3O$	300	826

Table 2.1.4—*continued*
Paramagnetic Ion Lasers Arranged in Order of Wavelength

Wavelength (μm)	Ion	Host: sensitizer ion(s)	Temp. (K)	References
1.945–2.014	Tm^{3+}	$Y_3Al_5O_{12}$:Cr	300	828
1.96	Er^{3+}	$BaYb_2F_8$	300	953,1004
1.965	Er^{3+}	$BaYb_2F_8$	300	817,1004
1.9654-5	Er^{3+}	$BaYb_2F_8$	300	548,549
1.972	Tm^{3+}	SrF_2	77	463
1.99	Co^{2+}	MgF_2	77	80,535
1.9925	Ho^{3+}	GdAlO3	90	825
1.9925	Er^{3+}	$BaYb_2F_8$:Tm,Yb	110	813
1.9965	Er^{3+}	$BaYb_2F_8$	110	115
1.9975	Er^{3+}	$BaEr_2F_8$	110	115
~2.0	Tm^{3+}	Y_2O_3	300	867
~2.0	Ho^{3+}	$SrY_4(SiO_4)_3O$	77	497
~2.0	Tm^{3+}	YVO_4	77	832
2	Ho^{3+}	$Y_3Al_5O_{12}$:Tm	300	814,815,818
2.0	Ho^{3+}	$Li(Y,Er)F_4$	77	841
2.0005	Er^{3+}	$LiErF_4$	110	233
2.0010	Ho^{3+}	$(Er,Lu)AlO_3$	77	840
2.0025	Er^{3+}	$LiYbF_4$	300	813
2.0132	Tm^{3+}	$Y_3Al_5O_{12}$:Cr	77	278
2.014	Tm^{3+}	$(Y,Er)_3Al_5O_{12}$	85	278
2.014	Tm^{3+}	$Y_3Al_5O_{12}$:Cr	300	977
2.015	Tm^{3+}	$Y_3Al_5O_{12}$:Cr	300	978
2.018	Tm^{3+}	$Y_3Sc_2Ga_3O_{12}$	300	829
2.019	Tm^{3+}	$Y_3Al_5O_{12}$:Cr	300	278
2.0195	Tm^{3+}	$(Er,Yb)_3Al_5O_{12}$	77	820
2.02	Tm^{3+}	$Ca(NbGa)_2Ga_3O_{12}$	300	830
2.0240	Tm^{3+}	$Lu_3Al_5O_{12}$	77	794
2.030	Ho^{3+}	CaF_2-ErF_2	77	835
2.0312	Ho^{3+}	α-$NaCaErF_6$	77	805
2.0318	Ho^{3+}	CaF_2-YF_3	77	834
2.0345	Ho^{3+}	α-$NaCaErF_6$	150	805
2.0377	Ho^{3+}	α-$NaCaErF_6$	77	805
2.0412	Ho^{3+}	YVO_4	77	833
2.0416	Ho^{3+}	$ErVO_4$:Tm	77	833
2.046	Ho^{3+}	$CaWO_4$	77	291,820

Table 2.1.4—*continued*
Paramagnetic Ion Lasers Arranged in Order of Wavelength

Wavelength (μm)	Ion	Host: sensitizer ion(s)	Temp. (K)	References
2.047	Ho^{3+}	$Ca(NbO_3)_2$	77	753
2.048–2.071	Ho^{3+}	$LiYF_4$:Er,Tm	77	842
2.049	Ho^{3+}	$GdVO_4$:Tm	300	578
2.0490–2.0559	Ho^{3+}	$LiErF_4$	300	843
2.0496	Ho^{3+}	SrF_2-(Y,Er)F_3	120, 300	844
2.05	Ho^{3+}	CaF_2-ErF_3-TmF_3-YbF_3	100	836,837
2.05	Co^{2+}	MgF_2	77	80,535
2.050	Ho^{3+}	$NaLa(MoO_4)_2$:Er	90	839
2.0505	Ho^{3+}	$LiYF_4$	300	847,1001
2.053	Ho^{3+}	SrF_2-(Y,Er)F_3	120, 300	844,845
2.0534	Ho^{3+}	$LiYF_4$	300	847
2.055	Ho^{3+}	$LiLuF_4$:Tm	300	831
2.0555	Ho^{3+}	BaY_2F_8:Er,Tm	20	838
2.0556	Ho^{3+}	$CaMoO_4$	77	821
2.0560	Ho^{3+}	$BaTm_2F_8$	110–230	813
2.0563	Ho^{3+}	$BaYb_2F_8$	295	846
2.0565	Ho^{3+}	$KY(WO_4)_2$	110	848
2.059	Ho^{3+}	$CaWO_4$	77	291,820
2.06	Ho^{3+}	Li(Y,Er)F_4	300	849
2.06	Ho^{3+}	$LiLuF_4$	300	849
2.060	Ho^{3+}	CaF_2-ErF_3-TmF_3YbF_3	298	836
2.060	Ho^{3+}	$CaY_4(SiO_4)_3O$	77	497
2.060	Ho^{3+}	$Y_3Al_5O_{12}$	300	858
2.0610–2.0650	Ho^{3+}	$LiErF_4$	300	843
2.0644	Ho^{3+}	BaY_2F_8:Er,Tm	85	152
2.065	Ho^{3+}	BaY_2F_8	77	152
2.065	Ho^{3+}	$LiYF_4$:Er,Tm	220	787
2.065	Ho^{3+}	Li(Y,Er)F_4	220–300	787
2.065	Ho^{3+}	$LiYbF_4$	300	850
2.065	Ho^{3+}	$Y_3Al_5O_{12}$	300	858
2.0654	Ho^{3+}	Li(Y,Er)F_4	300	853
2.0656	Ho^{3+}	Li(Y,Er)F_4	300	847
2.066	Ho^{3+}	$LiYF_4$:Er	77	854
2.0665	Ho^{3+}	$BaYb_2F_8$	110	703,848
2.067	Ho^{3+}	$LiYF_4$:Tm	300	340

Table 2.1.4—*continued*
Paramagnetic Ion Lasers Arranged in Order of Wavelength

Wavelength (μm)	Ion	Host: sensitizer ion(s)	Temp. (K)	References
2.068	Ho^{3+}	$LiLuF_4$:Tm	300	831
2.0672	Ho^{3+}	$LiYF_4$	~90	852
2.07	Ho^{3+}	$LaNbO_4$	~90	839
2.0707	Ho^{3+}	$CaMoO_4$	77	821
2.0715	Ho^{3+}	$BaYb_2F_8$	110	703,848
2.0720	Ho^{3+}	$K(Y,Er)(WO_4)_2$:Tm	110–220	855
2.0725	Ho^{3+}	$LaNbO_4$	110	851
2.074	Ho^{3+}	BaY_2F_8	20	838
2.074	Ho^{3+}	$CaMoO_4$	77	821
2.0740	Ho^{3+}	$KGd(WO_4)_2$	110	851
2.0746–2.076	Ho^{3+}	BaY_2F_8:Er,Tm	85	838
2.0765	Ho^{3+}	$KY(WO_4)_2$	110	848,851
2.0786	Ho^{3+}	$LiNbO_3$	77	370
~2.079	Ho^{3+}	$Ca_5(PO_4)_3F$	77	497
2.0790	Ho^{3+}	$KLu(WO_4)_2$	110	887
2.080–2.089	Ho^{3+}	$Y_3Sc_2Al_3O_{12}$	300	859
2.085	Ho^{3+}	Er_2SiO_5	77	856
2.086	Ho^{3+}	$Y_3Fe_5O_{12}$	77	857
2.086	Ho^{3+}	$Y_3Ga_5O_{12}$	77	857
2.086	Ho^{3+}	$Y_3Ga_5O_{12}$:Fe	77, >140	860
2.086	Ho^{3+}	$Ho_3Ga_5O_{12}$	77	713
2.086	Ho^{3+}	$BaEr_2F_8$:Tm	300	870–872
2.086	Ho^{3+}	$Y_3Sc_2Al_3O_{12}$	300	869
2.086	Ho^{3+}	$Y_3Sc_2Ga_3O_{12}$	300	776,869
2.0866	Ho^{3+}	BaY_2F_8:Er,Tm	77	838
2.087	Ho^{3+}	$Bi_4Ge_3O_{12}$	77	229
2.088	Ho^{3+}	$Gd_3Sc_yGa_{5-y}O_{12}$	300	341,818,873
2.088	Ho^{3+}	$Y_3Sc_2Ga_3O_{12}$	300	874–876
2.0885	Ho^{3+}	$Gd_3Ga_5O_{12}$	110	722,851
2.089	Ho^{3+}	$Y_3Fe_5O_{12}$	77	857
2.089–2.102	Ho^{3+}	$ErAlO_3$	77	878
2.0895	Ho^{3+}	$BaYb_2F_8$	110	703,848
2.09	Ho^{3+}	$Gd_3Sc_2Al_3O_{12}$	300	861
2.090	Ho^{3+}	$(Er,Ho)F_3$	77	863
2.091	Ho^{3+}	$Y_3Al_5O_{12}$:Tm	300	776

Table 2.1.4—*continued*
Paramagnetic Ion Lasers Arranged in Order of Wavelength

Wavelength (μm)	Ion	Host: sensitizer ion(s)	Temp. (K)	References
2.0914	Ho^{3+}	$Y_3Al_5O_{12}$	77	868
2.0917	Ho^{3+}	$(Y,Er)_3Al_5O_{12}$	77	278
2.092	Ho^{3+}	CaF_2	77	463
2.092	Ho^{3+}	Y_2SiO_5	~110	856
2.0960	Ho^{3+}	$Yb_3Al_5O_{12}$	77	866
2.097	Ho^{3+}	$(Ho,Y)_3Al_5O_{12}$	77	893
2.0974	Ho^{3+}	$Y_3Al_5O_{12}$:Tm	300	746,879,880
2.0975	Ho^{3+}	$Y_3Al_5O_{12}$	77	278,868
2.0977	Ho^{3+}	$(Y_3Al_5O_{12}$	~77	579
2.0978	Ho^{3+}	$Y,Er)_3Al_5O_{12}$	77	865
2.0979	Ho^{3+}	$(Y,Er)_3Al_5O_{12}$	77	278
2.0982	Ho^{3+}	$(Y,Er)_3Al_5O_{12}$:Tm	77	817
2.0982	Ho^{3+}	$Y_3Al_5O_{12}$:Er,Tm	110	817,881
2.0983	Ho^{3+}	$Y_3Al_5O_{12}$:Er,Tm,Yb	110	881
2.0985	Ho^{3+}	$Er_3Sc_2Al_3O_{12}$	77	279
2.0985	Ho^{3+}	$Er_3Al_5O_{12}$	110	848
2.0985–2.0997	Ho^{3+}	$ErAlO_3$	110	848,882
2.0990	Ho^{3+}	$(Y,Er)_3Al_5O_{12}$:Tm	77	817
2.0995	Ho^{3+}	$Tm_3Al_5O_{12}$	110	848
2.0998	Ho^{3+}	$Yb_3Al_5O_{12}$:Er,Tm	110	881
2.1	Ho^{3+}	$Y_3Al_5O_{12}$:Tm,Yb	110	883,888
2.1	Ho^{3+}	$(Y,Er)_3Sc_2Ga_3O_{12}$:Tm	77	841,848
2.1	Ho^{3+}	CaF_2-ErF_3-TmF_3-YbF_3	77	836
2.1	Ho^{3+}	$Li(Y,Er)F_4$	77–124	884
2.1	Ho^{3+}	$LiYF_4$:Tm	77	340
2.1	Ho^{3+}	$(Y,Er,)_3Al_5O_{12}$	77	885,886
2.1000	Ho^{3+}	$Yb_3Al_5O_{12}$:Tm,Yb	110	848
2.1004	Ho^{3+}	$Lu_3Al_5O_{12}$:Tm	300	862
2.1005	Ho^{3+}	$Lu_3Al_5O_{12}$:Er	110	881
2.1005	Ho^{3+}	$Lu_3Al_5O_{12}$:Er,Tm,Yb	110	887
2.1008	Ho^{3+}	$Lu_3Al_5O_{12}$:Tm	110	881
2.101	Ho^{3+}	$Yb_3Al_5O_{12}$	300	858
2.1010	Ho^{3+}	$(Er,Tm,Yb)_3Al_5O_{12}$	77	726
2.1020	Ho^{3+}	$Lu_3Al_5O_{12}$:Er,Tm	77	794
2.1020	Ho^{3+}	$Lu_3Al_5O_{12}$	110	889

Table 2.1.4—*continued*
Paramagnetic Ion Lasers Arranged in Order of Wavelength

Wavelength (μm)	Ion	Host: sensitizer ion(s)	Temp. (K)	References
2.105	Ho^{3+}	Y_2SiO_5	~110, 220	856
2.107	Ho^{3+}	$Y_3Fe_5O_{12}$	77	857
2.1110	Ho^{3+}	CaF_2-HoF_3	110	791
2.1135	Ho^{3+}	$Ho_3Ga_5O_{12}$	77	713
2.114	Ho^{3+}	$Y_3Ga_5O_{12}$:Fe	77, >140	860
2.114	Ho^{3+}	$Y_3Ga_5O_{12}$	77	857
2.115	Ho^{3+}	ZrO_2-Er_2O_3	77	764
2.1170	Ho^{3+}	$Ho_3Sc_2Al_3O_{12}$	77	713
2.1185	Ho^{3+}	$YAlO_3$	110	887
2.1189	Ho^{3+}	$YAlO_3$	110	190,241
2.119	Ho^{3+}	(Y,Er)AlO_3:Tm	300	130
2.1193	Ho^{3+}	$YAlO_3$	110	190
2.12	Ho^{3+}	(Y,Er)AlO_3	233	787
2.12	Ho^{3+}	$Y_3Al_5O_{12}$:Tm	215–330	814,890,891
2.1205	Ho^{3+}	$ErAlO_3$	77	840
2.1205	Ho^{3+}	(Er,Lu)AlO_3)	77	840
2.121	Ho^{3+}	Er_2O_3	145	893
2.122	Ho^{3+}	$Ho_3Al_5O_{12}$	90	894
2.1223	Ho^{3+}	$Y_3Al_5O_{12}$:Tm	77	881
2.1223	Ho^{3+}	$Y_3Al_5O_{12}$	300	278
2.1224	Ho^{3+}	$Ho_3Al_5O_{12}$	~90	892
2.1227	Ho^{3+}	$(Y,Er)_3Al_5O_{12}$:Tm	77	817
2.1227	Ho^{3+}	$Ho_3Al_5O_{12}$	77	713
2.123	Ho^{3+}	(Y,Er)AlO_3:Tm	300	130
2.123	Ho^{3+}	$(Y,Ho)_3Al_5O_{12}$	77	278,864
2.1241	Ho^{3+}	$Lu_3Al_5O_{12}$:Tm	300	824
2.1250	Ho^{3+}	$Lu_3Al_5O_{12}$	110	887
2.1285	Ho^{3+}	$(Y,Er)_3Al_5O_{12}$:Tm	77	895
2.1285	Ho^{3+}	$Ho_3Sc_2Al_3O_{12}$	77	713
2.129	Ho^{3+}	$Ho_3Al_5O_{12}$	90	894
2.1294	Ho^{3+}	$Ho_3Al_5O_{12}$	~90	892
2.1295	Ho^{3+}	$Y_3Al_5O_{12}$	300	881
2.1297	Ho^{3+}	$Ho_3Al_5O_{12}$	77	713
~2.13	Ho^{3+}	$(Y,Er)_3Al_3O_{12}$:Tm	300	896,897
2.1300	Ho^{3+}	$Lu_3Al_5O_{12}$	110	887

Table 2.1.4—*continued*
Paramagnetic Ion Lasers Arranged in Order of Wavelength

Wavelength (μm)	Ion	Host: sensitizer ion(s)	Temp. (K)	References
2.1300	Ho^{3+}	$YAlO_3$	110	190,887
2.1303	Ho^{3+}	$Lu_3Al_5O_{12}$	300	881
2.134–2.799	Cr^{2+}	ZnSe	300	914
2.1348	Ho^{3+}	$LuAlO_3$	~90	795
2.165	Co^{2+}	ZnF_2	77	535,801
2.171	Ho^{3+}	BaY_2F_8:Tm	295	838
2.20–2.46	Tm^{3+}	$LiYF_4$	300	912,913
2.234	U^{3+}	CaF_2	77	898
2.274	Tm^{3+}	$YAlO_3$	300	904
2.2845	Tm^{3+}	$BaYb_2F_8$	300	711
2.286–2.530	Cr^{2+}	ZnS	300	914
2.30	Tm^{3+}	$LiYF_4$	300	811
2.303	Tm^{3+}	$LiYF_4$	110	812
2.318	Tm^{3+}	$YAlO_3$:Cr	300	904
2.324	Tm^{3+}	$Y_3Al_5O_{12}$	300	904
2.335	Tm^{3+}	$Gd_3Sc_2Ga_3O_{12}$	300	710,873
~2.34	Tm^{3+}	$YAlO_3$:Cr	90, 300	803
2.3425	Tm^{3+}	$Lu_3Al_5O_{12}$	110	889
2.348	Tm^{3+}	$YAlO_3$	300	905
2.349	Tm^{3+}	$YAlO_3$	300	905
~2.35	Cr^{2+}	ZnS	300	915
~2.35	Cr^{2+}	ZnSe	300	915
2.352	Ho^{3+}	$LiHoF_4$	90	51,124
2.3520	Ho^{3+}	$LiYF_4$	116	181
2.3524	Ho^{3+}	$LiYF_4$	116	750
2.353	Tm^{3+}	$YAlO_3$:Cr	300	904
2.354	Tm^{3+}	$YAlO_3$:Cr	300	904
2.355	Tm^{3+}	$YAlO_3$:Cr	300	904
2.358	Dy^{2+}	CaF_2	4.2, 77	877,906–911
2.362	Ho^{3+}	BaY_2F_8	77	838
2.363	Ho^{3+}	BaY_2F_8	77	838
2.3659	Dy^{2+}	SrF_2	20	903
2.375	Ho^{3+}	BaY_2F_8	77	838
2.407	U^{3+}	SrF_2	20–90	902
2.439	U^{3+}	CaF_2	77	899

Table 2.1.4—*continued*
Paramagnetic Ion Lasers Arranged in Order of Wavelength

Wavelength (μm)	Ion	Host: sensitizer ion(s)	Temp. (K)	References
2.511	U^{3+}	CaF_2	77	899,900
2.515	Cr^{2+}	$Cd_{0.85}Mn_{0.15}Te$	300	1031
2.556	U^{3+}	BaF_2	20	901
2.571	U^{3+}	CaF_2	77	900
2.6	U^{3+}	CaF_2	4.2	932
2.613	U^{3+}	CaF_2	77–90	933
2.62–2.94	Er^{3+}	$Y_3Al_5O_{12}$:Cr,Tm	300	934
2.66	Er^{3+}	$LiYF_4$	300	935
2.6887	Er^{3+}	$K(Y,Er)(WO_4)_2$	300–150	855
2.69	Er^{3+}	CaF_2-ErF_3-TmF_3	298	936
2.6930	Er^{3+}	$(Y,Er)_3Al_5O_{12}$	77	237
2.6970	Er^{3+}	$Er_3Al_5O_{12}$	110	51,882
2.6975	Er^{3+}	$Y_3Al_5O_{12}$:Tm,Yb	300	938
[2.6975–2.6979]	Er^{3+}	$(Y,Er)_3Al_5O_{12}$	110	918
2.6987	Er^{3+}	$(Lu,Er)_3Al_5O_{12}$	110	918
2.699	Er^{3+}	$(Lu,Er)_3Al_5O_{12}$	300	51,919,920
2.6990	Er^{3+}	$(Er,Lu)_3Al_5O_{12}$	300	239
2.6990	Er^{3+}	$Lu_3Al_5O_{12}$:Ho,Tm	300	239
2.7	Er^{3+}	$SrLaGa_3O_7$:Pr	300	733
2.7	Er^{3+}	$Y_3Al_5O_{12}$:Cr	300	776
2.7	Er^{3+}	$Y_3Sc_2Al_3O_{12}$	300	776
2.7034	Er^{3+}	$Gd_3Ga_5O_{12}$	110	877
2.707	Er^{3+}	$Y_3Sc_2Ga_3O_{12}$	300	776
2.71	Er^{3+}	$Ca_3(NbLiGa)_5O_{12}$	300	929
[2.71–2.86]	Er^{3+}	$(Y,Er)AlO_3$	300	916
2.7118	Er^{3+}	$(Y,Er)AlO_3$	300	922
2.7126	Er^{3+}	$LuAlO_3$	110	792
2.7140–43	Er^{3+}	$(Er,Lu)_3Al_5O_{12}$	100	918
2.7170	Er^{3+}	$LiYF_4$	110	788
2.7175	Er^{3+}	$Ca(NbO_3)_2$	110	57
2.7188	Er^{3+}	$Gd_3Ga_5O_{12}$	110	877
2.7220	Er^{3+}	$KLa(MoO_4)_2$	110	501
2.7222	Er^{3+}	$KGd(WO_4)_2$	300	796
2.7285	Er^{3+}	SrF_2-ErF_3	300	921
2.7290	Er^{3+}	CaF_2-ErF_3	300	791

Table 2.1.4—*continued*
Paramagnetic Ion Lasers Arranged in Order of Wavelength

Wavelength (μm)	Ion	Host: sensitizer ion(s)	Temp. (K)	References
2.7295	Er^{3+}	CaF_2-ErF_3	300	791
[2.73–2.92]	Er^{3+}	$(Y,Er)AlO_3$	300	917
2.73	Er^{3+}	$YAlO_3$	290–330	968
2.7305–2.7307	Er^{3+}	$YAlO_3$	300	788,789,922
2.7307[(a)]	Er^{3+}	CaF_2-ErF_3	300	997
2.7307	Er^{3+}	CaF_2	300	998
2.7309	Er^{3+}	$YAlO_3$	300	129
2.7310	Er^{3+}	$(Y,Er)AlO_3$	110	788,789,792,930
2.7398	Er^{3+}	$(Y,Er)AlO_3$	110	788,789,792,930
2.7417	Er^{3+}	$BaEr_2F_8$	110	115,927
2.7450	Er^{3+}	SrF_2-ErF_3	300	921
2.7460	Er^{3+}	CaF_2-ErF_3	300	791
2.747	Er^{3+}	$LiYF_4$	300	776
2.7490	Er^{3+}	CaF_2-ErF_3	300	791
2.75	Er^{3+}	CaF_2-ErF_3	300	923–926
2.7575	Er^{3+}	$KLa(MoO_4)_2$	300	501
2.7595	Er^{3+}	$BaEr_2F_8$	110	115,927
2.7608	Er^{3+}	$(Y,Er)AlO_3$	110	788,789,792,930
2.766	Er^{3+}	$Y_3Al_5O_{12}$	300	928
2.7645	Er^{3+}	$(Y,Er)AlO_3$	300	922
2.7698	Er^{3+}	$(Y,Er)AlO_3$	110	788,789
2.77	Er^{3+}	$LiYF_4$	300	943
2.79	Er^{3+}	$(Y,Er)AlO_3$	290–330	968
2.791	Er^{3+}	$Y_3Sc_2Ga_3O_{12}$	300	942,956,962
2.7930	Er^{3+}	SrF_2-ErF_3	300	921
2.795	Er^{3+}	$Y_3Al_5O_{12}$	300	928
2.7953	Er^{3+}	$(Lu,Er)_3Al_5O_{12}$	300	964
2.7953–2.7958	Er^{3+}	$(Y,Er)_3Al_5O_{12}$	300	918
2.7955	Er^{3+}	CaF_2-ErF_3	300	971
2.7955	Er^{3+}	$(Y,Er)AlO_3$	300	788,930
2.7969	Er^{3+}	$(Y,Er)AlO_3$	110	788,789
2.7973	Er^{3+}	$Lu_3Al_5O_{12}$	300	918
2.7980	Er^{3+}	$BaEr_2F_8$	110	115,927
2.7985	Er^{3+}	CaF_2-ErF_3	300	944,971
2.799	Er^{3+}	$(Lu,Er)_3Al_5O_{12}$	300	51

Table 2.1.4—*continued*
Paramagnetic Ion Lasers Arranged in Order of Wavelength

Wavelength (µm)	Ion	Host: sensitizer ion(s)	Temp. (K)	References
2.7990	Er^{3+}	$KGd(WO_4)_2$	300	796
2.7998	Er^{3+}	$Lu_3Al_5O_{12}$	110	918
2.8	Er^{3+}	$LiYF_4$	300	961
2.80	Er^{3+}	CaF_2-ErF_3	300	923–926
2.80	Er^{3+}	$(Y,Er)_3Al_5O_{12}$	300	969–972
2.80	Er^{3+}	SrF_2-ErF_3	300	926
2.80	Er^{3+}	$Er_{2.7}Gd_{0.3}Al_5O_{12}$	110	367
2.8070	Er^{3+}	$KY(WO_4)_2$	300	236,328
2.8070	Er^{3+}	$KEr(WO_4)_2$	300	339
2.8085	Er^{3+}	$LiYF_4$	110	788
2.8092	Er^{3+}	$KLu(WO_4)_2$	300	938
2.81	Er^{3+}	$LiYF_4$	300	926,935
2.810	Er^{3+}	$LiYF_4$	300	755,926
2.8128	Er^{3+}	$Gd_3Ga_5O_{12}$	300	938
2.8218	Er^{3+}	$Gd_3Ga_5O_{12}$	300	942
2.8230	Er^{3+}	$(Y,Er)AlO_3$	300	922
2.827	U^{3+}	$LiYF_4$	300	945
2.8297–2.8302	Er^{3+}	$Lu_3Al_5O_{12}$	300	918
2.8298	Er^{3+}	$(Er,Lu)_3Al_5O_{12}$:Yb	300	51,941
2.8298	Er^{3+}	$(Er,Lu)_3Al_5O_{12}$	300	918
2.83	Ho^{3+}	$LiYbF_4$	300	850
2.830	Er^{3+}	$(Lu,Er)_3Al_5O_{12}$	300	947,986
2.8302	Er^{3+}	$Y_3Al_5O_{12}$	300	940
2.8302	Er^{3+}	$(Y,Er)_3Al_5O_{12}$	300	964
2.8302	Er^{3+}	$(Lu,Er)_3Al_5O_{12}$	300	964
2.8302	Er^{3+}	$Y_3Al_5O_{12}$:Tm	300	938
2.84	Er^{3+}	$LiYF_4$	300	935
2.8400	Er^{3+}	$(Y,Er)AlO_3$	300	922
2.8415	Ho^{3+}	$KLa(MoO_4)_{12}$	110	501
2.8484	Ho^{3+}	$GdVO_4$:Tm	300	578
2.85	Er^{3+}	$LiYF_4$	300	935
2.850	Ho^{3+}	$LiYF_4$	300	181
2.8500	Er^{3+}	$LiErF_4$	110	233
2.8510	Ho^{3+}	$LaNbO_4$:Er	90	958
2.8549	Er^{3+}	$Gd_3Ga_5O_{12}$	110	877

Table 2.1.4—*continued*
Paramagnetic Ion Lasers Arranged in Order of Wavelength

Wavelength (μm)	Ion	Host: sensitizer ion(s)	Temp. (K)	References
2.8552–2.8590	Er^{3+}	$Lu_3Al_5O_{12}$	110	918
2.8575	Ho^{3+}	$BaYb_2F_8$:Yb	300	203,848
2.8578	Ho^{3+}	$YAlO_3$	300	952
2.8595	Er^{3+}	$Er_3Al_5O_{12}$	110	882
2.86	Er^{3+}	$Er_3Al_5O_{12}$	110	465
2.86	Er^{3+}	$Er_{2.7}Gd_{0.3}Al_5O_{12}$	110	368
2.8637	Ho^{3+}	$YScO_3$:Gd	77	783
2.8665	Er^{3+}	$(Y,Er)AlO_3$	300	922
2.870	Er^{3+}	$LiYF_4$	110, 300	939
2.8700	Er^{3+}	$Lu_3Al_5O_{12}$	110	918
2.8748–2.8752	Er^{3+}	$Lu_3Al_5O_{12}$	110	918
2.8750	Er^{3+}	$Er_3Al_5O_{12}$	110	882
2.8756	Er^{3+}	$(Y,Er)AlO_3$	300	922
2.8758	Er^{3+}	$(Y,Er)AlO_3$	300	922
2.8760	Er^{3+}	$Lu_3Al_5O_{12}$	110	918
2.8868	Er^{3+}	$Er_3Al_5O_{12}$	110	918
2.8967–2.8979	Er^{3+}	$Lu_3Al_5O_{12}$	110	918
2.8970	Er^{3+}	$(Y,Er)_3Al_5O_{12}$	77	237
2.9	Ho^{3+}	$Gd_3Ga_5O_{12}$	300	918
2.9	Ho^{3+}	$BaYb_2F_8$:Yb	300	51,161,703,964
2.9054	Ho^{3+}	$BaYb_2F_8$:Yb	300	966
2.9073	Ho^{3+}	$BaYb_2F_8$	~300	966
2.9073	Ho^{3+}	$BaYb_2F_8$	293	989
2.9155	Ho^{3+}	$YAlO_3$	300	783
2.9180	Ho^{3+}	$YAlO_3$	300	129,792,930
2.9185	Ho^{3+}	$YAlO_3$	110–300	190
2.9200	Ho^{3+}	$YAlO_3$	110	241,792,930
2.9230	Ho^{3+}	$ErAlO_3$	110	792
2.9342	Ho^{3+}	$KGd(WO_4)_2$	300	796
2.936	Er^{3+}	$Y_3Al_5O_{12}$		731
2.9362–2.9366	Er^{3+}	$(Y,Er)_3Al_5O_{12}$	300	918,964
2.9364	Er^{3+}	$(Y,Er)_3Al_5O_{12}$	300	225,237,926, 973–975
2.9365	Er^{3+}	$Y_3Al_5O_{12}$:Tm,Yb	300	938,942
2.9365	Er^{3+}	$(Y,Er)_3Al_5O_{12}$	300	964
2.9365	Er^{3+}	$(Lu,Er)_3Al_5O_{12}$	300	964

Table 2.1.4—*continued*
Paramagnetic Ion Lasers Arranged in Order of Wavelength

Wavelength (μm)	Ion	Host: sensitizer ion(s)	Temp. (K)	References
2.9366	Er^{3+}	$Er_3Al_5O_{12}$	300	918
2.9367	Er^{3+}	$Er_3Al_5O_{12}$	300	882
2.937	Er^{3+}	$(Lu,Er)_3Al_5O_{12}$	300	926
2.937	Er^{3+}	$(Y,Er)_3Al_5O_{12}$	300	926
2.939	Er^{3+}	$Y_3Al_5O_{12}$	300	928
2.9395	Er^{3+}	$Lu_3Al_5O_{12}$:Yb	300	239,941
2.9395	Er^{3+}	$Lu_3Al_5O_{12}$	300	239
2.9395	Er^{3+}	$Er_3Al_5O_{12}$	110	882
2.9395	Ho^{3+}	$KY(WO_4)_2$:Er,Tm	300	959
2.9395	Er^{3+}	$Er_3Al_5O_{12}$	110	882
2.9395–2.9397	Er^{3+}	$Lu_3Al_5O_{12}$	110	918
2.94	Er^{3+}	$Er_{3\text{-}x}Gd_{0.x}Al_5O_{12}$	300	367
2.94	Er^{3+}	$Y_3Al_5O_{12}$:Cr	300	957,992
2.94	Er^{3+}	$(Lu,Er)_3Al_5O_{12}$	300	918
2.94	Er^{3+}	$Er_3Al_5O_{12}$	110	882
2.94	Er^{3+}	$Y_3Al_5O_{12}$:Nd	300	672
2.940	Ho^{3+}	$Y_3Al_5O_{12}$:Ho	300	673
2.9401	Er^{3+}	$Lu_3Al_5O_{12}$	110	918
2.9403	Ho^{3+}	$Y_3Al_5O_{12}$	300	129
2.9403	Er^{3+}	$(Lu,Er)_3Al_5O_{12}$	300	918
2.9403	Er^{3+}	$Y_3Al_5O_{12}$	300	129
2.9405	Er^{3+}	$Lu_3Al_5O_{12}$:Yb	300	941
2.9408	Er^{3+}	$Lu_3Al_5O_{12}$	300	239
2.943	Er^{3+}	$(Y,Er)_3Al_5O_{12}$	110	359
2.9445	Ho^{3+}	$KLu(WO_4)_2$	300	965
2.9460	Ho^{3+}	$Lu_3Al_5O_{12}$:Cr,Yb	300	129,941
2.952	Ho^{3+}	$LiYF_4$	300	750
2.952	Ho^{3+}	$LiYF_4$	116–300	189
2.955	Ho^{3+}	$LiYF_4$	300	967
2.9619	Ho^{3+}	$Gd_3Ga_5O_{12}$	110	887
2.97	Dy^{3+}	LaF_3	300	759
2.9700	Ho^{3+}	$KLa(MoO_4)_2$	300	501
3.011	Ho^{3+}	$Y_3Al_5O_{12}$	300	673
3.0132	Ho^{3+}	$YAlO_3$	300	129,930,948
3.0157	Ho^{3+}	$YAlO_3$	110	792,910

Table 2.1.4—*continued*
Paramagnetic Ion Lasers Arranged in Order of Wavelength

Wavelength (µm)	Ion	Host: sensitizer ion(s)	Temp. (K)	References
3.0165	Ho^{3+}	$YAlO_3$	110–300	190
3.0177	Ho^{3+}	$YAlO_3$	110	792
3.02	Dy^{3+}	BaY_2F_8	300	965
3.022	Dy^{3+}	$Ba(Y,Er)_2F_8$	77	960
3.369	Ho^{3+}	$LiYF_4$	300	967
3.377	Ho^{3+}	BaY_2F_8	20	838
3.40	Dy^{3+}	BaY_2F_8	300	1029
3.41	Er^{3+}	$LiYF_4$	300	951
3.53	Fe^{2+}	n-InP	2	748
3.6050	Pr^{3+}	$BaYb_2F_8$:Yb	110	241,623
3.893	Ho^{3+}	$LiYF_4$	300	967
3.914	Ho^{3+}	$LiYF_4$	300	915
4.34	Dy^{3+}	$LiYF_4$	300	990
4.75	Er^{3+}	$YAlO_3$	110	958–960,965
5.15	Nd^{3+}	BaF_2-LaF_3	300	161,959,991
5.242	Pr^{3+}	LaF_3	130	760
7.141	Pr^{3+}	$LaCl_3$	~300	955
7.152	Pr^{3+}	$LaCl_3$	~30	955
7.24	Pr^{3+}	$LaCl_3$	148	950
7.244	Pr^{3+}	$LaCl_3$	130	955

(a) Stimulated emission and/or wavelength requires more accurate definition.
(b) Self-frequency-doubled emission.
(c) Center wavelength of lasing tuning range.

References

1. Waynant, R. W., Vacuum ultraviolet laser emission from Nd^{+3}:LaF_3, *Appl. Phys. B* 28, 205 (1982).
2. Waynant, R. W. and Klein, P. H., Vacuum ultraviolet laser emission from Nd^{3+}:LaF_3, *Appl. Phys. Lett.* 46, 14 (1985).
3. Dubinskii, M. A., Cefalas, A. C. and Nicolaides, C. A., Solid state LaF_3:Nd^{3+} vuv laser pumped by a pulsed discharge F_2-molecular laser at 157 nm, *Optics Commun.* 88, 122 (1992).
4. Dubinskii, M. A., Cefalas, A. C., Sarantopouou, E., Spyrou, S. M., Nicolaides, C. A., Abdulsabirov, R. Yu., Korableva, S. L. and Semashjko, V. V., Efficient LaF_3:Nd^{3+}-based vacuum-ultraviolet laser at 172 nm, *J. Opt. Soc. Am. B* 9, 1148 (1992).
5. Pinto, J. F., Rosenblat, G. H., Esterowitz, L., Castillo, V. and Quarles, G. J., Tunable solid-state laser action in Ce^{3+}:$LiSrAlF_6$, *Electron. Lett.* 30, 240 (1994).
6. Marshall, C. D., Speth, J. A., Payne, S. A., Krupke, W. F., Quarles, G. J., Castillo, V. and Chai, B. H. T., Ultraviolet laser emission properties of Ce^{3+}-doped $LiSrAlF_6$ and $LiCaAlF_6$, *J. Opt. Soc. Am. B* 11, 2054 (1994).
7. Ehrlich, D. J., Moulton, P. F., and Osgood, R. M., Jr., Optically pumped Ce:LaF, laser at 286 nm, *Opt. Lett.* 5, 339 (1980).
8. Dubinskii, M. A., Semashko, V. V., Naumov, A. K., Abdulsabirov, R. Yu., and Korableva, S. L., Ce^{3+}-doped colquiriite: a new concept of all-solid-state tunable ultraviolet laser, *J. Mod. Optics* 40, 1 (1993).
9. Ehrlich, D. J., Moulton, P. F., and Osgood, R. M., Jr., Ultraviolet solid-state Ce:YLF laser at 325 nm, *Opt. Lett.* 4, 184 (1978) and unpublished data.
10. Azamatov, Z. T., Arsen'yev, P. A., and Chukichev, M. V., Spectra of gadolinium in YAG single crystals, *Opt. Spectrosc.* 28, 156 (1970).
11. Okada, F., Togawa, S. and Ohta, K., Solid-state ultraviolet tunable laser: a Ce^{3+} doped $LiYF_4$ crystal, *J. Appl. Phys.* 75, 49 (1994).
12. Sarukura, N., Liu, Z., Segawa, Y., Edamatsu, K., Suzuke, Y., Itoh, T. et al., Ce^{3+}:$LiLuF_4$ as a broadband ultraviolet amplification medium, *Optics Lett.* 20, 294 (1995).
13. Schmitt, K., Stimulated C'-emission of Ag^+-centers in KI, RbBr, and CsBr, *Appl. Phys. A* 38 (1985).
14. Kaminskii, A. A., Kochubei, S. A., Naumochkin, K. N., Pestryakov, E. V., Trunov, V. I. and Uvarova, T. V., Amplification of ultrviolet radiation due to the 5d-4f configurational transition of the Ce^{3+} ion in BaY_2F_8, *Sov. J. Quantum Electron.* 19, 340 (1989).
15. Macfarlane, R. M., Tong, F., Silversmith, A. J., and Lenth, W., Violet CW neodymium upconversion laser, *Appl. Phys. Lett.* 52, 1300 (1988).
16. Kaminskii, A. A., Demchuk, M. I., Zhavaronkov, N. V. and Mikhailov, V. P., *Phys. Status Solidi* (a) 113, K257 (1989).
17. Payne, S. A., Smith, L. K., DeLoach, L. D., Kway, W. L., Tassano, J. B. and Krupke, W. F., Ytterbium-doped apatite-structure crystals: A new class of laser materials, *J. Appl. Phys.* 75, 497 (1994).
18. Hebert, T., Wannemacher, R., Macfarlane, R.M. and Lenth, W., Blue continuously pumped upconversion lasing in Tm:$YLiF_4$, *Appl. Phys. Lett.* 60, 2592 (1992).
19. Nguyen, D.C., Faulkner, G.E. and Dulick, M., Blue-green (450-nm) upconversion Tm^{3+}:YLF laser, *Appl. Optics*, 28, 3553 (1989).
20. Baer, J. W., Knights, M. G., Chicklis, E. P., and Jenssen, H. P., XeF-pumped laser operation of Tm:YLF at 452 nm, in *Proceedings Topical Meeting on Excimer Lasers*, IEEE-OSA, Charleston, SC (1979).

21. Thrash, R. J. and Johnson, L. F., Upconversion laser emission from Yb^{3+}-sensitized Tm^{3+} in BaY_2F_8, *J. Opt. Soc. Am. B* 11, 881 (1994)
22. Hebert, T., Wannemacher, R., Lenth, W. and Macfarlane, R.M., Blue and green cw upconversion lasing in Er:$YLiF_4$, *Appl. Phys. Lett.* 57, 1727 (1990).
23. Esterowitz, L., Allen, R., Kruer, M., Bartoli, F., Goldberg, L. S., Jensen, H. P., Linz, A. and Nicolai, V. O., Blue light emission by Pr:$LiYF_4$ laser operated at room temperature, *J. Appl. Phys.* 48, 650 (1977).
24. Malinowsdki, M., Pracka, I., Surma, B., Lukasiewicz, T., Wolinski, W. and Wolski, R., Spectroscopic and laser properties of $SrLaGa_3O_7$:Pr^{3+} crystals, *Opt. Mater.* 6, 305 (1996).
25. Smith, R. G., New room temperature CW laser transition in YAlG:Nd, *IEEE J. Quantum Electron.* QE-4, 505 (1968).
26. Varsanyi, F., Surface lasers, *Appl. Phys. Lett.*, 19, 169 (1971).
27. German, K. R., Kiel, A., and Guggenheim, H., Stimulated emission from $PrCl_3$, *Appl. Phys.* Lett., 22, 87 (1973).
28. German, K. R. and Kiel, A., Radiative and nonradiative transitions in $LaCl_3$: Pr and $PrCl_3$, *Phys. Rev. B* 8, 1846 (1973).
29. Danger, T., Sandrock, T., Heumann, E., Huber, G. and Chai, B., Pulsed laser action of Pr:$GdLiF_4$ at room temperature, *Appl. Phys. B* 57, 239 (1993).
30. German, K. R., Kiel, A., and Guggenheim, H. J., Radiative and nonradiative transitions of Pr^{3+} in trichloride and tribromide hosts, *Phys. Rev. B*, 11, 2436 (1975).
31. Kaminskii, A. A., Visible lasing on five intermultiplet transitions of the ion Pr^{3+} in $LiYF_4$, *Sov. Phys. Dokl.* 28, 668 (1983).
32. Kaminskii A. A., Stimulated emission spectroscopy of Ln^{3+} ions in tetragonal $LiLuF_4$ fluoride, *Phys. Status Solidi A* 97, K53 (1986).
33. Kaminskii, A. A., Markosyan, A. A., Pelevin, A. V., Polyakova, Yu. A., Sarkisov, S. E. and Uvarova, T. V., Luminescence properties and stimulated emission from Pr^{3+}, Nd^{3+} and Er^{3+} ions in tetragonal lithium-lutecium fluoride, *Inorg. Mater. (USSR)* 22, 773 (1986).
34. McFarlane, R.A., High-power visible upconversion laser, *Optics Lett.* 16, 1397 (1991).
35. Jenssen, H. P., Castleberry, D., Gabbe, D., and Linz, A., Stimulated emission at 5445 Å in Tb^{3+}: YLF, in *Digest of Technical Papers CLEA* (1973), IEEE/OSA, Washington, DC (1971), p. 47.
36. Fan, T. Y., Cordova-Plaza, A., Digonnet, M. J. F., Byer, R. L., and Shaw, H. J., Nd:MgO:$LiNbO_3$ spectroscopy and laser devices, *J. Opt. Soc. Am. B* 3, 140 (1986).
37. Silversmith, A. I.. Lenth, W. and Macfarlane, R. M., Green infrared-pumped erbium upconversion laser, *Appl. Phys. Lett.* 51, 1977 (1987).
38. Brede, R., Danger, T., Heumann, E. and Huber, G., Room temperature green laser emission of Er:$LiYF_4$, *Appl. Phys. Lett.* 63, 729 (1993).
39. Xie, P. and Rand, S. C., Continuous-wave, fourfold upconversion laser, *Appl. Phys. Lett.* 63, 3125 (1993).
40. McFarlane, R. A., Dual wavelength visible upconversion laser, *Appl. Phys. Lett.* 54, 2301 (1989).
41. Heine, F., Heumann, E., Danger, T., Schweizer, T., Koetke, J., Huber, G. and Chai, B. H. T., Room temperature continuous wave upconversion Er:YLF laser at 551 nm, *OSA Proc. Adv. Solid State Lasers*, Fan, T. Y. and Chai, B. H. T. (eds) 20, 344 (1995).
42. Tong, F., Risk, W. P., Macfarlane, R. M. and Lenth, W., 551 nm diode-laser-pumped upconversion laser, *Electron. Lett.* 25, 1389 (1989).
43. Stephens, R. R. and McFarlane, R. A., Diode-pumped upconversion laser with 100-mW output power, *Opt. Lett.* 18, 34 (1993).

44. Voron'ko, Yu. K., Kaminskii, A. A., Osiko, V. V., and Prokhorov, A. M., Stimulated emission from Ho^{3+} in CaF_2 at 5512 Å, *JETP Lett.* 1, 3 (1965).
45. Johnson, L. F. and Guggenheim, H. J., Infrared-pumped visible laser, *Appl. Phys. Lett.* 19, 44 (1971).
46. Johnson, L. F. and Guggenheim, H. J., New laser lines in the visible from Er^{3+} ions in BaY_2F_8, *Appl. Phys. Lett.* 20, 474 (1972).
47. Lu, B., Wang, J., Pan, I., and Jiang, M., Excited emission and self-frequency-doubling effect of $Nd_xY_{1-x}Al_3(BO_3)_4$ crystal, *Chin. Phys. Lett. (China)* 3, 413 (1986).
48. Kazakov, B. N., Orlov, M. S., Petrov, M. V., Stolov, A. L., and Takachuk, A. M., Induced emission of Sm^{3+}- ions in the visible region of the spectrum, *Opt. Spectrosc. (USSR)* 47, 676 (1979).
49. Hegarty, J. and Yen, W. M., Laser action in PrF_3, *J. Appl. Phys.* 51, 3545 (1980).
50. Kaminskii, A. A., Stimulated radiation at the transitions $^3P_0 \rightarrow {}^3F_4$ and $^3P_0 \rightarrow {}^3H_6$ of Pr^{3+} ions in LaF_3 crystals, *Izv. Akad. Nauk. SSSR* 17, 185 (1981), (in Russian).
51. Kaminskii, A A., Some current trends in physics and spectroscopy of laser crystals, in *Proc. Int. Conf. Lasers '80*, Collins, C. B., Ed., STS Press, Mclean, VA (1981), p. 328.
52. Solomon, R. and Mueller, L., Stimulated emission at 5985 Å from Pr^{3+} in LaF_3, *Appl. Phys. Lett.* 3, 135 (1963).
53. Sutherland, J. M., French, P. M. W., Taylor, J. R. and Chai, B. H. T., Visible continuous-wave laser transitions in Pr^{3+}:YLF and femtosecond pulse generation. *Opt. Lett.* 21, 797 (1996).
54. Szafranski, C., Strek, W., and Jezowqka-Trzebiatowska, B., Laser oscillation of a $LiPrP_4O_{12}$ single crystal, *Opt. Commun.* 47, 268 (1983).
55. Kaminskii, A. A., Sobolev, B. P., Uvarova, T. V., and Chertanov, M. I., Visible stimulated emission of Pr^{3+} ions in BaY_2F_8, *Inorg. Mater. (USSR)* 20, 622 (1984).
56. Kaminskii, A. A. and Sarkisov, S. E., Stimulated-emission spectroscopy of Pr^{3+} ions in monoclinic BaY_2F_8 fluoride, *Phys. Status Solidi A* 97, K163 (1986).
57. Kaminskii, A. A., Petrosyan, A. G., and Ovanesyan, K. L., Stimulated emission of Pr^{3+}, Nd^{3+}, and Er^{3+} ions in crystals with complex anions, *Phys. Status. Solidi A* 83, K159 (1984).
58. Chang, N. C., Fluorescence and stimulated emission from trivalent europium in yttrium oxide, *J. Appl. Phys.* 34, 3500 (1963).
59. Kvapil, J., Koselja, M., Kvapil, J., Perner, B., Skoda, V., Kubeika, J., Hamal, K., and Kubecek, V., Growth and stimulated emission of YAP:Ti, *Czech. J. Phys. B.* 38, 237 (1988).
60. Kaminskii, A. A., Petrosyan, A. G., Ovanesyan, K. L., and Chertanov, M. I., Stimulated emission of Pr^{3+} ions in $YAlO_3$ crystals, *Phys. Status Solidi A*: 77, K173 (1983).
61. Kaminskii, A. A., Petrosyan, A. G., and Ovanesyan, K. L., Stimulated emission spectroscopy of Pr^{3+} ions in $YAlO_3$ and $LuAlO_3$, *Sov. Phys. Dokl.* 32, 591 (1987).
62. Kaminskii, A. A., Kurbanov, K., Ovanesyan, K. L., and Petrosyan, A. G., Stimulated emission spectroscopy of Pr^{3+} ions in orthorhombic $YAlO_3$ single crystals, *Phys. Status Solidi A* 105, K155 (1988).
63. Bleckman, A., Heine, F., Meyn, J. P., Danger, T., Heumann, E. and Huber, G., CW-lasing of Pr:$YAlO_3$ at room temperature, in *Advanced Solid-State Lasers*, Pinto, A. A. and Fan, T. Y., Eds., Proceedings Vol. 15, Optical Society of America, Washington, DC (1993), p. 199.

64. Luo, Z., Jiang, A., Huang, Y. et al., *Chin. Phys. Lett.* 6, 440 (1989).
65. Johnson, L. F. and Ballman, A. A., Coherent emission from rare earth ions in electro-optic crystals, *J. Appl. Phys.* 40, 297 (1969).
66. O'Connor, J. R., Optical and laser properties of Nd^{3+}- and Eu^{3+}-doped YVO_4, *Trans. Metallurg. Soc. AIME* 239, 362 (1967).
68. Szymanski, M., Simultaneous operation at two different wavelengths of an $(Pr,La)P_5O_{14}$ laser, *Appl. Phys.* 24, 13 (1981).
69. Borkowski, B., Crzesiak, E., Kaczmarek, F., Kaluski, Z., Karolczak, J., and Szymanski, M., Chemical synthesis and crystal growth of laser quality praseodymium pentaphosphate, *J. Crystal Growth* 44, 320 (1978).
70. Szymanski, M., Karolczak, I., and Kaczmarek, F., Laser properties of praseodymium pentaphosphate single crystals, *Appl. Phys.* 19, 345 (1979).
71. Dornauf, H. and Heber, J., Fluorescence of Pr^{3+}-ions in $La_{1-x}Pr_xP_5O_{14}$, *J. Lumin.* 20, 271 (1979).
72. Kaminskii, A. A., New room-temperature stimulated-emission channels of Pr^{3+} ions in anisotropic laser crystals, *Phys. Status Solidi A* 125, K109 (1991).
73. Knowies, D. S., Zhang, Z., Gabbe, D., and Jenssen, H.B., Laser action of Pr^{3+} in $LiYF_4$ and spectroscopy of Eu^{3+}- sensitized Pr in BaY_2F_8, *IEEE J. Quantum Electron.* 24, 1118 (1988).
74. Kaminskii, A. A., Ivanov, A. O., Sarkisov, S. E. et al., Comprehensive investigations of the spectral and lasing characteristics of the $LuAlO_3$ crystal doped with Nd^{3+}, *Sov. Phys.-JETP* 44, 516 (1976).
75. Kaminskii, A. A., Eichler, H. J., Liu, B. and Meindl, P., $LiYF_4:Pr^{3+}$ laser at 639.5 nm with 30 J flashlamp pumping and 87 mJ output energy, *Phys. Status Solidi A* 138, K45 (1993).
76. Kaminskii, A. A., Pavlyuk, A. A., Klevtsov, P. V. et al., Stimulated radiation of monoclinic crystals of $KY(WO_4)_2$, and $KGd(WO_4)_2$ with Ln^{3+} ions, *Inorg. Mater. (USSR),* 13. 482 (1977).
77. Macfarlane, R. M., Silversmith, A. J., Tong, F., and Lenth, W., CW upconversion laser action in neodymium and erbium doped solids, in *Proceedings of the Topical Meeting on Laser Materials and Laser Spectroscopy*, World Scientific, Singapore, (1988), p. 24.
78. Antipenko, B. M., Voronin, S. P., and Privaiova, T. A., Addition of optical frequencies by cooperative processes, *Opt. Spectrosc. (USSR)* 63, 768 (1987).
79. Antipenko, B. M., Voronin, S. P. and Privalova, T. A., New laser channels of the Tm^{3+} ion, *Opt. Spectrosc.* 68, 164 (1990).
80. Heine, F., Ostroumov, V., Heumann, E., Jensen, T., Huber, G. and Chai, B. H. T., CW Yb,Tm:$LiYF_4$ upconversion laser at 650 nm, 800 nm, and 1500 nm, *OSA Proc. Adv. Solid State Lasers*, Chai, B. H. T. and Payne, S. A. (Eds.) 24, 77 (1995).
81. Li, R., Xie, C., Wang, J., Liang, X., Peng, K. and Xu, G., CW Nd:MgO:$LiNbO_3$ self-frequency-doubling laser at room temperature, *IEEE J. Quantum Electron.* 29, 2419 (1993).
82. Albers, P., Stark, E., and Huber, G., Continuous-wave laser operation and quantum efficiency of titanium-doped sapphire, *J. Opt. Soc. Am. B*, 3, 134, 1986
83. Moulton, P.F., Spectroscopic and laser characteristics of Ti:Al_2O_3, *J. Opt. Soc. Am. B* 3, 125 (1986).
84. Kruglik, G. S., Skripko, G. A., Shkadarevich, A. P., Kondratyuk, N. V., Urbanovich, V. S., and Nazarenko, R N., Output of $Al_2O_3:Ti^{3+}$ crystals in the continuous and quasicontinuous regimes, *J. Appl. Spectrosc. (USSR).* (Eng. Transl.) 45, 1031 (1986).

85. Birnbaum, M. and Pertica, A. J., Laser material characteristics of Ti:Al_2O_3, *J. Opt. Soc. Am. B.* 4, 1434 (19870.
86. Sanchez, A., Strauss, A. J., Aggarwal, R. L., and Fahey, R. E., Crystal growth, spectroscopy, and laser characteristics of Ti:Al_2O_3, *IEEE J. Quantum Electron.* 24, 995 (1988).
87. Sanchez, A., Fahey, R. E., Strauss, A. J., and Aggurwal, R. L., Room-temperature continuous-wave operation of a Ti:Al_2O_3 laser, *Opt. Lett.* 11, 363 (1986).
88. Sanchez, A., Fahey, R. E., Strauss, A. J., and Aggarwal, R. L., Room-termperature cw operation of the Ti:Al_2O_3 laser, in *Tunable Solid-State Lasers* 11, Vol. S2 Budgor, A. B., Esterowitz, L. and DeShazer, L. G., Eds., Springer-Verlag, New York (1986), p. 202.
89. Albers, P., Jenssen, H. P., Hube, G., and Kokta, M., Continuous wave tunable laser operation of Ti^{3+}- doped sapphire at 300 K, in *Tunable Solid-State Lasers II*, Vol. 52, Budgor, A. B., Esterowitz, L. and DeShazer, L. G., Eds., Springer-Verlag, New York (1986), p. 208.
90. Albrecht, G. F., Eggleston, J. M., and Ewing, J. J., Measurements of Ti^{3+}:Al_2O_3 as a lasing material, in *Tunable Solid State Lasers*, Proc. Int. Conf., Vol. 47, Hammerling, P, Budgor, A. B. and Pinto, A., Eds., Springer-Verlag, New York (1985), p. 68.
91. Sevast'ganov, B. K., Bagdasarov, Kh. S., Fedorov, E. A., Semenov, V. B., Tsigler, I. N., Chirkina, K. P., Starostina, L. S., Chirkin, A. P., Minaev, A. A., Orekhova, V. P., Seregin, V. F., Koierov, A. N., and Vratskii, A. N., Tunable laser based on Al_2O_3:Ti^{3+} crystal, *Sov. Phys. Crystallogr.* 29, 566 (1984).
92. Kruglik, C. S., Skripko, G. A., Shkadurevich, A. P., Kondratyuk, N. V., and Zhdanov, E. A., Output characteristics of a coherently pumped laser utilizing an Al_2O_3:Ti^{3+} crystal, *Sov. J. Quantum Electron.* 16, 792 (1986).
93. Sevast'yanov, B. K., Budasarov, Kh. S., Fedorov, E. A., Semenov, V. B., Tsigler, I. N., Chirkina, K. P., Starostin, L. S., Chirkin, A. P., Minaev, A. A., Orekhova, V. P., Seregin, V. F., and Kobro, A. N., Spectral and lasing characteristics of corundum crystals activated by Ti^{3+} ions (Al_2O_3:Ti^{3+}), *Sov. Phys. Dokl.* 30, 508 (1985).
94. Muller, C. H., III, Lowenthal, D. D., Kangus, K. W., Hamil, R. A., and Tisone, G. C., 2.0-J Ti sapphire laser oscillator, *Opt. Lett.* 13, 380 (1988).
95. Schmid, F. and Khattak, C. P., Growth of Co MgF_2 and Ti:Al_2O_3 crystals for solid state laser applications, in *Tunable Solid State Lasers,* Proc. Int. Conf., Vol. 47, Hammerling, P., Budgor, A. B. and Pinto, A., Eds., Springer-Verlag, New York (1985), p. 122.
96. Moulton, P. F., Tunable paramagnetic-ion lasers, in *Laser Handbook*, Vol. 5, Bass, M. and Stitch, M. L., Eds., North-Holland, Amsterdam, The Netherlands (1985), p. 203.
97. Moulton, P. F., Recent advances in transition metal-doped lasers, in *Tunable Solid State Lasers*, Proc. Int. Conf., Vol 47, Hammerling, P., Budgor, A. B. and Pinto, A., Eds., Springer-Verlag, New York (1985), p. 4.
98. Moulton, P. F., Spectroscopic and laser characteristics of Ti:Al_2O_3. *J. Opt. Soc. Am.* B 3, 125 (1986).
99. Barnes, N. P., Williams, J. A., Barnes, J. C., and Lockard, G. E., A self-injection locked, Q-switched, line-narrowed Ti:Al_2O_3 laser, *IEEE J Quantum Electron.* 24, 1021 (1988).
100. Moncorge, R., Boulon, G., Vivien, D., Lejus, A. M., Collongues, R., Djevahirdjian, V., Djevahirdjian, K., and Cagnard, R., Optical properties and tunable laser action of Verneuil-grown single crystals of Al_2O_3:Ti^{3+}, *J. IEEE Quantum Electron.* 24, 1049 (1988).

101. Bagdasarov, Kh. S., Krasilov, Yu. I., Kuznetsov, N. T., Kuratev, I. I., Potemkin, A. V., Shestakov, A. V., Zverev, G. M., Siyuchenko, O. G. and Zhitnnyuk, V. A., Laser properties of α-Al_2O_3:Ti^{3+} crystals, *Sov. Phys. Dokl.* 30, 473 (1985).
102. Rapoport, W. R. and Khattak, C. P., Titanium sapphire laser characteristics, *Appl. Opt.* 27, 2677 (1988).
103. DeShazer, L. G., Albrecht, C. F., and Seamans, J. F., Tunable titanium sapphire lasers, in Proc. Int. Soc. Opt. Eng., *High Power and Solid State Lasers*, Vol. 622, Simmons, W. W., Ed., SPIE, Bellingham, WA (1986), p. 133.
104. Eggleston, J. M., DeShazer, L. G., and Kangas, K. W., Characteristics and kinetics of laser-pumped Ti:Sapphire oscillators, *IEEE J. Quantum Electron.* 24, 1009 (1988).
105. Rapoport, W. R. and Khattak, C. P., Efficient tunable Ti:sapphire laser, in *Tunable Solid-State Lasers II*, Vol. 52, Budgor. A. B., Esterowitz. L., and DeShazer, L. G., Eds., Springer-Verlag, New York (1986), p. 212.
106. DeShazer, L. G., Eggleston, J. M., and Kangas, K. W., Oscillator and amplifier performance of Ti:sapphire, in *Tunable Solid-State Lasers II,* Vol. 52, Budgor, A. B., Esterowitz, L., and DeShazer L. G., Eds., Springer-Verlag, New York (1986), p. 228.
107. Schepier, K. L., Laser performance and temperature-dependent spectroscopy of titanium-doped crystals, in *Tunable Solid-State Lasers II*, Vol. 52, Budgor, A. B., Esterowitz, L. and DeShazer, L. G., Eds., Springer-Verlag, New York (1986), p. 235.
108. DeShazer, L. G. and Kangas, K. W., Extended infrared operation of a titanium sapphire laser presented at *Conf. on Lasers and Electrooptics*, Baltimore, MD, Apr. 26–May 1 (1987), p. 296.
109. Bagdasarov, Kh. S., Danilov, V. P., Murina, T. M., Novikov, E. G., Prokhorov, A. M., Semenov, V. B., and Fedorov, E. A., Tunable flashlamp-pumped Al_2O_3:Ti^{3+} laser, *Sov. Tech. Phys. Lett.* 13, 152 (1987).
110. Schulz, P. A., Single-frequency Ti:Al_2O_3 ring laser, *IEEE J. Quantum Electron.* 24, 1039 (1988).
111. Lacovara, P., Esterowitz, L., and Allen, R., Flash-lamp-pumped Ti:Al_2O_3 laser using fluorescent conversion, *Opt. Lett.* 10, 273 (1985).
112. Esterowitz, L. and Allen, R., Stimulated emission from flashpumped Ti:Al_2O_3, in *Tunable Solid State Lasers, Proc. Int. Conf.*, Vol. 47, Hammerling, P., Budgor, A. B., and Pinto, A., Eds., Springer-Verlag, New York (1985), p. 73.
113. Alshuler, G. B., Karasev, V. B., Kondratyuk, N. V., Krugiik, G. S., Okishev, A. V., Skripko, G. A., Urbanovich, V. S., and Shkadarevich, A. P., Generation of ultrashort pulses in a synchronously pumped Ti^{3+} laser, *Sov. Tech. Phys. Lett.* 13, 324 (1987).
114. Fox, A. M., Maciel, A. C., and Ryan, J. F., Efficient CW performance of a Co:MgF_2 laser operating at 1.5—2.0 μm, *Opt. Commun.* 59, 142 (1986).
115. Kaminskii, A. A., Sobolev, B. P., Sarkisov, S. E., Denlsenko, G. A., Ryabchenkov, V. V., Fedorov, V. A., and Uvarova, T. V., Physicochemical aspects of the preparation spectroscopy, and stimulated emission of single crystals of $BaLn_2F_8$-Ln^{3+}, *Inorg. Mater. (USSR)* 18, 402 (1982).
116. Marin, V. I., Nikitin, V. I., Soskin, M. S., and Khizhnyak, A. I., Superluminescence emitted by YAG:Nd^{3+} crystals and stimulated emission due to weak transitions, *Sov. J. Quantum Electron.* 5, 732 (1975).
117. Nebdaev, N. Ya., Petrenko, R. A., Piskarskas, A. S., Sirutajtis, V. A., Smil'gyavichyus, V. I., and Yuozapavlchus, A. S., Peculiarities of stimulated emission of picosecond pulses from the potassium gadolinium tungstate laser, *Ukr. Fiz. Zh.* 33, 1165 (1988) (in Russian).

118. Bukin, G. V., Volkov, S. Yu., Matrosov, V. N., Sevast'yanov, B. K., and Timoshechkin, M. I., Stimulated emission from alexandrite ($BeAl_2O_4:Cr^{3+}$), *Sov. J. Quantum Electron.* 8, 671 (1978).
119. Weber, M. J., Bass, M., and Demars, G. A., Laser action and spectroscopic properties of Er^{3+} in $YAlO_3$, *J. Appl. Phys.* 42, 301 (1971).
120. Sevastyanov, B. K., Remigailo, Yu. I., Orekhova, V. P., Matrosov, V. P., Tsvetkov, E. G. and Bukin, G. V., Spectroscopics and lasing properties of alexandrite ($BeAl_2O_4:Cr^{3+}$), *Sov. Phys. Dokl.* 26, 62 (1981).
121. Walling, J. C., Jenssen, H. P., Morris, R. C., O'Dell, E. W., and Peterson, O. G., Tunable-laser performance in $BeAl_2O_4:Cr^{3+}$, *Opt. Lett.* 4, 182 (1979).
122. Walling, J. C. and Peterson, O. G., High gain laser performance in alexandrite, *IEEE J. Quantum Electron.* QE-16, 119 (1980).
123. Buchert, J., Katz, A., and Alfano, R. R., Laser action in emerald, *IEEE J. Quantum Electron.* QE-19, 1477 (1983).
124. Sevast'yanov, B. K., Bagdasarov, Kh. S., Pasternak, L. B., Volkov, S. Yu., and Drekhova, V. P., Stimulated emission from Cr^{3+} ions in YAG crystals, *JETP Lett.* 17, 47 (1973).
125. McClung, F. J., Schwarz, S. E., and Meyers, F. J., R_2 line optical maser action in ruby, *J. Appl. Phys.* 33, 3139 (1962).
126. Collins, R. J., Nelson, D. F., Schawlow, A. L., Bond, W., Garrett, C. G. B., and Kaiser, W., Coherence, narrowing, directionality, and relaxation oscillations in the light emission from ruby, *Phys. Rev. Lett.* 5, 303 (1960).
127. Nelson, D. F. and Boyle, W. S., A continuously operating ruby optical maser, *Appl. Optics*, 1, 181 (1962).
128. Birnbaum, M., Tucker, A. W., and Fincher, C. L., CW ruby laser pumped by an argon ion laser, *IEEE J. Quantum Electron.* QE- 13, 808 (1977).
129. Kaminskii, A. A., Butaeva, T. I., Ivanov, A. O. et al., New data on stimulated emission of crystals containing Er^{3+} and Ho^{3+} ions, *Sov. Tech. Phys. Lett.* 2, 308 (1976).
130. Weber, M., Bass, M., Varitimos, T., and Bua, D., Laser action from Ho^{3+}, Er^{3+} and Tm^{3+} in $YAlO_3$, *IEEE J. Quantum Electron.* QE-9, 1079,1973.
131. Maiman, T. H., Stimulated optical radiation in ruby, *Nature* 187, 493 (1960).
132. Maiman, T. H., Optical maser action in ruby, *Br. Commun. Electron.* 7, 674 (1960).
133. Roess, D., Analysis of room temperature CW ruby lasers, *IEEE J. Quantum Electron.* QE-2, 208 (1966).
134. Evtuhov, V. and Kneeland, J. K., Power output and efficiency of continuous ruby laser, *J. Appl. Phys.* 38, 4051 (1967).
135. Burrus, C. A. and Stone, J., Room-temperature eontinuous operation of a ruby fiber laser, *J. Appl. Phys.* 49, 3118 (1978).
136. Galkin, S. L., Zakgeim, A. L., Markhonov, V. M., Nikolaev, V. M., Pavlyuk, A. A., Petrovich, I. P., Petrun'kin, V. Yu., Shkadarevich, A. P., and Yarzhemkovskil, V. D., Crystalline $KGd(WO_4)_2$ laser with a semiconductor pump system, *J. Appl. Spectrosc. (USSR).* (Engl. Transl.) 37, 886 (1982).
137. Kaminskii, A., High-temperature spectroscopic investigation of stimulated emission from lasers based on crystals and glass activated with Nd^{3+} ions, *Sov. Phys.-JETP* 27, 388 (1968).
138. Kirkin, A. N., Leontovich, A. M., and Mozharovskii, A. M., Generation of high power ultrashort pulses in a low temperature ruby laser with a small active volume, *Sov. J. Quantum Electron.* 8, 1489 (1978).
139. Ivanyuk, A. M., Shakhverdov, P. A., Belyaev, V. D., Ter-Pogosyan, M. A., and Ermolaev, V. L., A picosecond neodymium laser on potassium-gadolinium tungstate

with passive mode locking operating in the repetitive regime, *Opt. Spectrosc. (USSR)* 58, 589 (1985).
140. Walling, J. C., Peterson, O. G., and Morris, R. C., Tunable CW alexandrite laser, *IEEE J. Quantum Electron.* QE-16, 120 (1980).
141. Samelson, H., Walling, J. C., Wernikowski, T., and Harter, D. J., CW arc-lamp-pumped alexandite lasers, *IEEE J. Quantum Electron.* 24, 1141 (1988).
142. Walling, J. C., Heller, D. F., Samelson, H., Harter, D. J., Pete, J. A., and Morris, R. C., Tunable alexandrite lasers: development and performance, *IEEE J. Quantum Electron.* QE-21, 1568 (1985).
143. Lai, S. T. and Shand, M. L., High efficiency cw laser-pumped tunable alexandrite laser, *J. Appl. Phys.* 54, 5642 (1983).
144. Rapoport, W. R. and Samebon, H., Alexandrite slab laser, in *Proc. Int. Conf. Lasers 85*, Wang, C. P., Ed., STS Press, McLean, VA (1986), p. 744.
145. Zhang, S. and Zhang, K., Experiment on laser performance of alexandrite crystals, *Chin. Phys.*, 4, 667 (1984).
146. Guch, S., Jr. and Jones, C. E., Alexandrite-laser performance at high temperature, *Opt. Lett.* 7, 608 (1982).
147. Walling, J. C., Peterson, O. G., Jenssen, H. P., Morris, R. C., and O'Dell, E. W., Tunable alexandrite lasers, *IEEE J. Quantum Electron.* QE-16, 1302 (1980).
148. Shand, M. L., Progress in alexandrite lasers, in *Proc. Int. Conf. Lasers 85*, Wang, C. P., Ed., STS Press, McLean, VA (1986), p. 732.
149. Zhang, G. and Ma, X., Improvement of lasing performance of alexandrite crystals, *Chin. Phys. Lasers*, 13, 816 (1986).
150. Jones, J. E., Dobbins, J. D., Butier, B. D. and Hinsley, R. J., Performance of a 250-Hz, 100-W alexandrite laser system, in *Proc. Int. Conf. Lasers 85*, Wang, C. P., Ed., STS Press, McLean, VA (1986), p. 738.
151. Lisitsyn, V. N., Matrosov, V. N., Pestryakov, E. V., and Trunov, V. I., Generation of picosecond pulses in solid-state lasers using new active media, *J. Sov. Laser Res.* 7, 364 (1986).
152. Jones, J. E., Dobbins, J. D., Butier, B. D., and Hinsley, R. J., Performance of a 250-Hz, 100-W alexandrite laser system, in *Proc. Int. Conf. Lasers 85*, Wang, C. P., Ed., STS Press, McLean, VA (1986), p. 738.
153. Schawlow, A. L. and Devlin, G. E., Simultaneous optical maser action in two ruby satellite lines, *Phys. Rev. Lett.* 6, 96 (1961).
154. Walling, J. C. and Sam, C. L., unpublished data (1980).
155. Kaminskii, A. A. and Li, L., Spectroscopic studies of stimulated emission in an $SrF_2:Nd^{3+}$ crystal laser, *J. Appl. Spectrosc. (USSR)* 12, 29 (1970).
156. Sorokin, P. P., Stevenson, M. J., Lankard, J. R., and Pettit, G. D., Spectroscopy and optical maser action in $SrF_2:Sm^{2+}$, *Phys. Rev.* 127, 503 (1962).
157. Vagin, Yu. S., Marchenko, V. M., and Prokhorov, A. M., Spectrum of a laser based on electron-vibrational transitions in a $CaF_2:Sm^{2+}$ crystal, *Sov. Phys.-JETP* 28, 904 (1969).
158. Sorokin, P. P. and Stevenson, M. J., Solid-state optical maser using divalent samarium in calcium fluoride, *IBM J. Res. Dev.* 5, 56 (1961).
159. Kaiser, W., Garrett, C. G. B., and Wood, D. L., Fluorescence and optical maser effects in CaF_2,:Sm^{2+}, *Phys. Rev.* 123, 766 (1961).
160. Anan'yev, Yu. A., Grezin, A. K., Mak, A. A., Sedov, B. M., and Yudina, Ye. N., A fluorite:samarium laser, *Sov. J. Opt. Technol.* 35, 313 (1968).
161. Kaminskii, A. A., Achievements in the fields of physics and spectroscopy of insulating laser crystals, in *Lasers and Applications, Pt. I, Proc.*, Ursu, I. and Brokhorov, A. M., Eds., CIP Press, Bucharest, Romania (1983), p. 97.
162. Payne, S. A., Chase, L. L., Newkirk, H. W., Smith, L. K., and Krupke, W. F., $LiCaAlF_6:Cr^{3+}$: a promising new solid-state laser material, *IEEE J. Quantum Electron.* 24, 2243 (1988).

163. Chase, L. L., Payne, S. A., Smith, L. K., Kway, W. L., Newkirk, H. W., Chai, B. H. T., and Long, M., Laser performance and spectroscopy of Cr^{3+} in $LiCaAlF_6$ and $LiSrAlF_6:Cr^{3+}$, in *Tunable Solid StateLasers*, Shand M. L. and Jenssen, H. P., Eds., Optical Society of America, Washington, DC (1989), p. 71.
164. Lai, S. T., Highly efficient emerald laser, *J. Opt. Soc. Am. B.*, 4, 1286 (1987).
165. Shand, M. L. and Lai, S. T., CW laser pumped emerald laser, *IEEE J. Quantum Electron.* QE-20, 105 (1984).
166. Shand, M. L. and Walllng, J. C., A tunable emerald laser, *IEEE J. Quantum Electron.* QE-18, 1829 (1982).
167. Podkolzina, I. G., Tkachuk, A. M., Fedorov, V. A., and Feofilov, P. P., Multifrequency generation of stimulated emission of Ho^{3+} ion in $LiYF_4$ crystals, *Opt. Spectrosc.* 40, 111 (1976).
168. Kaminskii, A. A. and Pelevin, A. V., Low-threshold lasing of $LiYF_4:Pr^{3+}$ crystals in the 0.72 μm range as a result of flashlamp pumping at 300 K, *Sov. J. Quantum Electron.* 21, 819 (1991).
169. Kaminskii, A. A. and Petrosyan, A. G., New laser crystal for the excitation of stimulated radiation in the dark-red part of the spectrum at 300 K, *Sov. J. Quantum Electron.* 21, 486 (1991).
170. Pestryakov, E. V., Trunov, V. I., and Alimpiev, A. I., Generation of tunable radiation in a $BeAl_2O_4:Ti^{3+}$ laser subjected to pulsed coherent pumping at a high repetition frequency, *Sov. J. Quantum Electron.* 17, 585 (1987).
171. Alimpiev, A. I., Bukin, G. V., Matrosov, V. N., Pestryakov, E. V., Soinbev, V. P., Trunov, V. I., Tsvetkov, E. G., and Chebobev, V. P., Tunable $BeAl_2O_4:Ti^{3+}$ laser, *Sov. J. Quantum Electron.* 16, 579 (1986).
172. Segawa, Y., Sugimoto, A., Kim, P. H., Namba, S., Yamagishi, K., Anzai, Y., and Yamaguchi, Y., Optical properties and lasing of Ti^{3+} doped $BeAl_2O_4$, *Jpn. J. Appl. Phys.* 26, L291 (1987).
173. Huber, G. and Petermann, K., Laser action in Cr-doped garnet and tungstates, in *Tunable Solid-State Lasers*, Vol. 47, Hammerling, P., Budgor, A. B., and Pinto, A., Eds., Springer-Verlag, New York (1985), p. 11.
174. Struve, B. and Huber, G., Laser performance of Cr^{3+}:Gd(Sc,Ga) garnet, *J. Appl. Phys.* 57, 45 (1985).
175. Huber, G., Drube, J., and Struve, B., Recent developments in tunable Cr-doped garnet lasers, in *Proc. Int. Conf. Lasers 83*, Powell, R. C., Ed., STS Press, McLean, VA (1983), p. 143.
176. Struve, B., Huber, C., Laptev, V. V., Shcherbakov, I. A., and Zharikov, E. V., Tunable room-temperature cw laser action in Cr^{3+} :GdScGa-garnet, *Appl. Phys. B* 30, 117 (1983).
177. Zharikov, E. V., Il'ichev, N. N., Kalitin, S. P., Laptev, V. V., Malyutin, A. A., Osiko, V. V., Ostroumov, V. G., Pashinin, P. P., Prokhorov, A. M., Smirnov, V. A., Umyskov, A. F., and Shcherbakov, I. A., Tunable laser utilizing an electronic-vibrational transition in chromium in a gadolinium scandium gallium garnet crystal, *Sov. J. Quantum Electron.* 13, 1274 (1983).
178. Payne, M. J. P. and Evans, H. W., Laser action in flashlamp-pumped chromium:GSG-garnet, in *Tunable Solid-State Lasers II,* Vol. 52, Budgor, A. B., Esterowitz, L., and DeShazer, L., Eds., Springer-Verlag New York (1986), p. 126.
179. Walling, J. C., Peterson, D. G., and Morris, R. C., Tunable CW alexandrite laser, *IEEE J. Quantum Electron.* QE-16, 120 (1980).
180. Caird, J. A., Payne, S. A., Staver, P. R., Ramponi, A. J., Chase, L. L., and Krupke, W. F., Quantum electronic properties of the $Na_3Ga_2Li_3F_{12}:Cr^{3+}$ laser, *IEEE J. Quantum Electron.* 24, 1077 (1988).

181. Petrov, M. V., Tkachuk, A. M., and Feofilov, P. P., Multifrequency and cascade production of induced emission from Ho^{3+} and Er^{3+} in $LiYF_4$ crystals and delayed induced afterglow from Ho^{3+}, *Bull. Acad. Sci. USSR, Phys. Ser.* 45, 167 (1981).
182. Drube, J., Struve, B., and Huber, G., Tunable room-temperature CW laser action in Cr^{3+}:GdScAl-garnet, *Opt. Commun.*, 50, 45 (1984).
183. Drube, J., Huber, G., and Mateika, D., Flashlamp-pumped Cr^{3+}:GSAG and Cr^{3+}:GSGG: slope efficiency, resonator design color centers, and tunability in *Tunable Solid-State Lasers II*, Vol. 52, Budgor, A. B., Esterowitz, L., and DeShazer, L. G., Eds., Springer-Verlag, New York (1986), p. 118.
184. Meier, J. V., Barnes, N. P., Remelius, D. K., and Kokta, M. R., Flashlamp-pumped Cr^{3+} :GSAG laser, *IEEE J. Quantum Electron.* QE-22, 2058 (1986).
185. Struve, B., Fuhrberg, P., Luhs, W., and Litfin, G., Thermal lensing and laser operation of flashlamp-pumped Cr:GSAG, *Opt. Commun.* 65, 291 (1988).
186. Chai, B. H. T., Lefaucheur, J., Stalder, M. and Bass, M., Cr:$LiSr_{0.8}Ca_{0.2}AlF_6$ tunable laser, *Optics Lett.* 17, 1584 (1992).
187. Chicklis, E. P., Naiman, C. S., Esterowitz, L., and Allen, R., Deep red laser emission in Ho:YLF, *IEEE J. Quantum Electron.* QE-13, 893 (1977).
188. Kaminskii, A. A., Belokoneva, E. L., Mill, B. V., Pisarevskii, Yu. V., Sarkisov, S. E., Silvestrova, I. M., Butashin, A. V. and Khodzhabagyan, G. G., Pure and Nd^{3+}-doped $Ca_3Ga_2Ge_4O_{14}$ and $Sr_3Ga_2Ge_4O_{14}$ single crystals, their structure, optical, spectral luminescence, electromagnetic properties, and stimulated emission. *Phys. Status Solidi A* 86, 345 (1984).
189. Sugimoto, A., Segawa, Y., Anzai, Y., Yamagishi, K., Kim, P.H. and Namba, S., Flash-lamp-pumped tunable Ti:$BeAl_2O_4$ laser, *Japan. J. Appl. Phys.* 29, 1136 (1990).
190. Kaminskii, A. A., Luminescence and multiwave stimulated emission of Ho^{3+} and Er^{3+} ions in orthorhombic $YAlO_3$ crystals, *Sov. Phys. Dokl.*, 31, 823 (1986).
191. Brauch, U. and Dürr, U., $KZnF_3$:Cr^{3+}—a tunable solid state NIR-laser, *Opt. Commun.* 49, 61 (1984).
192. Dürr, U., Brauch, U., Knierim, W., and Weigand, W., Vibronic solid state lasers: transition metal ions in perovskites. in *Proc. Int. Conf. Lasers* 83, Powell, R. C., Ed., STS Press, McLean, VA (1983), p. 42.
193. Brauch, U. and Dürr, U., Room-temperature operation of the vibronic $KZnF_3$:Cr^{3+} laser, *Opt. Lett.* 9, 441 (1984).
194. Dubinskii, M. A., Kolerov, A. N., Mityagin, M. V., Silkin, N. I., and Shkadarevich, A. P., Quasi-continuous operation of a $KZnF_3$:Cr^{3+} laser, *Sov. J. Quantum Electron.* 16, 1684 (1986).
195. Abdulsabirov, R. Yu., Dubinskii, M. A., Korableva, S. L., Mityagin, M. V., Silkin, N. I., Skripko, C. A., Shkadarevich, A. P., and Yagudin, Sh. I., Tunable laser based on $KZnF_3$:Cr^{3+} crystal with nonselective pumping, *Sov. Phys. Crystallogr.* 31, 353 (1986).
196. Payne, S. A., Chase, L. L., Smith, L. K., Kway, W. L., and Newkirk, H. W., Laser performance of $LiSrAlF_6$:Cr^{3+}, *J. Appl. Phys.* 66, 1051 (1989).
197. Woodbury, E. J. and Ng, W. K., Ruby laser operation in the near IR, *Proc. IRE* 50, 2367 (1962).
198. Bazylev, A. G., Voitovich, A. P., Demidovich, A. A., Kalinov, V. S., Timoshechkin, M. I. and Shkadarevich, A. P., Laser performance of Cr^{3+}:$(Gd,Ca)_3(Ga,Mg,Zr)_2Ga_3O_{12}$, *Opt. Commun.* 94, 82 (1992).
199. Beaud, P., Chen, Y.-F., Chai, B. H. T. and Richardson, M. C., Gain properties of LiSrAlF6:Cr^{3+}, *Optics Lett.* 17, 1064 (1992).

200. Dymott, M. J. P., Botheroyd, I. M., Hall, G. J., Lincoln, J. R. and Ferguson, A. J., All-solid-state actively mode-locked Cr:LiSAF laser, *Optics Lett.* 19, 634 (1994).
201. Stalder, M., Chai, B.H.T. and Bass, M., Flashlamp pumped $Cr:LiSrAlF_6$ laser, *Appl. Phys. Lett.* 58, 216 (1991).
202. Lai, S. T., Chai, B. H. T., Long, M., and Morris, R. C., $ScBO_3$:Cr—a room temperature near-infrared tunable laser, *IEEE J. Quantum Electron.*, QE-22, 1931 (1986).
203. Lai, S. T., Chai, B. H. T., Long, M., Shinn, M. D., Caird, J. A., Marion, J. E., and Staver, P. R., A $ScBO_3$:Cr laser, in *Tunable Solid-State Lasers II*, Vol. 52, Budgor, A. B., Esterowitz, L., and DeShazer, L.G., Eds., Springer-Verlag, New York (1986), p. 145.
204. Alimpiev, A. I., Pestryakov, E. V., Petrov, V. V., Solntsev, V. P., Trunov, V. I., and Matrosov, V. N., Tunable lasing due to the T_2-$^4A_2$ electronic-vibrational transition in Cr^{3+} ions in $BeAl_2O_4$, *Sov. J. Quantum Electron.* 18, 323 (1988).
205. Kaminskii, A. A., Sarkisov, S. E., Pavlyuk, A. A., and Lyubchenko, V. V., Anisotropy of luminescence properties of the laser crystals $KGd(WO_4)_2$ and $KY(WO_4)_2$ with Nd^{3+} ions, *Inorg. Mater. (USSR)* 16, 501 (1980).
206. Chai, B. H. T., Shinn, M. D., Long, M. N., Lai, S. T., Miller, H. H., and Smith, L. K., Laser and spectroscopic properties of Cr -doped $ScAlBeO_4$, *Bull. Am. Phys. Soc.* 33, 1631 (1988).
207. Kaminskii, A. A., Pavlyuk, A. A., Agamalyan, N. P., , Bobovich, L. I., Lukin, A. V. and Lyubchenko, V. V., Stimulated radiation of $KLu(WO_4)_2$-Er^{3+} crystals at room temperature, *Inorg. Mater. (USSR)* 15, 1182 (1979).
208. Kaminskii, A. A., Shkadarevich, A. P., Mill, B. V., Koptev, V. G., and Demidovich, A. A., Wide-band tunable stimulated emission from a $La_3Ga_5SiO_{14}$-Cr^{3+} crystal, *Inorg. Mater. (USSR)* 23, 618 (1987).
209. Lai, S. T., Chai, B. H. T., Long, M. and Shinn, M. D., Room temperature near-infrared tunable $Cr:La_3Ga_5SiO_{14}$ laser, *IEEE J. Quantum Electron.* 24, 1922 (1988).
210. Petermann, K. and Mitzscherlich, P., Spectroscopic and laser properties of Cr^{3+}-doped $Al_2(WO_4)_3$ and $Sc_2(WO_4)_3$, *IEEE J. Quantum Electron.* QE-23, 1122 (1987).
211. Chicklis, E. P. and Naiman, C. S., A review of near-infrared optically pumped solid-state lasers, in *Proceedings of the First European Electro-Optics Markets and Technology Conference*, IPC Science and Technology Press. (1973), p. 77.
212. Smith, L. K., Payne, S. A., Krupke, W. F., Kway, W. L., Chase, L. L. and Chai, B. H. T., Investigation of the laser properties of $Cr^{3+}:LiSrGaF_6$, *IEEE J. Quantum Electron.* 28, 2612 (1992).
213. Bagdasarov, Kh. S., Bogomolova, G. A., Gritsenlco, M. M., Kaminskii, A. A., and Kervorkov, A. M., Spectroscopic study of the $LaAlO_3:Nd^{3+}$ laser crystal, *Sov. Phys.-Crystallogr.* 17, 357 (1972).
214. Arsen'yev, P. A. and Bienert, K. E., Synthesis and optical properties of neodymium-doped gadolinium aluminate ($GdAlO_3$) single crystals, *J. Appl. Spectrosc. (USSR)* 17, 1623 (1972).
215. Kaminskii, A. A., Cascading lasers based on activated crystals, *Inorg. Mater. (USSR)* 7, 802 (1971).
216. Kaminskii, A. A., Spectroscopic studies of stimulated emission from Er^{3+} ions in CaF_2-YF_3 crystals, *Opt. Spectrosc.* 31, 507 (1971).
217. Kaminskii, A. A., Cascade laser schemes based on activated crystals, *Inorg. Mater. (USSR)* 7, 802 (1971).

218. Kaminskii, A. A., Stimulated-emission spectroseopy of EB+ ions in cubic (Y,Ln)$_3$Al$_5$O$_{12}$ and monoclinic K(Y,Ln)(WO$_4$)$_2$ single crystals, *Phys. Status Solidi A* 96 K175 (1986).
219. Kaminskii,, A. A., Pavlyuk, A. A., Butaeva, T. I., Fedorov, V. A., Balashov, I. F., Berenberg, V. A., and Lyubchenko, V. V., Stimulated emission by subsidiary transitions of Ho^{3+} and Er^{3+} ions in KGd(WO$_4$)$_2$ crystals, *Inorg. Mater. (USSR)* 13, 1251 (1977).
220. Pollock, S. A., Chang, D. B., and Birnbaum, M., Threefold upconversion laser at 0.85, 1.23 and 1.73 μm in Er:YLF pumped with a 1.53 μm Er glass laser, *Appl. Phys. Lett.* 54, 869 (1989).
221. Kaminskii, A. A., Butashin, A. V., Markabaev, A. K., Mill', B. V., Knab, G. G., and Ursovskaya, A. A., Garnet Ca$_3$Ga$_2$Ge$_3$O$_{12}$:optical properties, microhardness and stimulated emission of Er^{3+} ions, *Sov. Phys. Crystallogr.* 32, 413 (1987).
222. Chicklis, E. P., Naiman, C. S., and Linz, A., Stimulated emission at 0.85 μm in Er^{3+}:YLF, in Digest of Technical Papers Vll International Quantum Electronics Conference, Montreal (1972), p. 17.
223. Kubodera, K. and Otsuka, K., Spike-mode oscillations in laser-diode pumped LiNdP$_4$O$_{12}$ lasers, *IEEE J. Quantum Electron.* QE-17, 1139 (1981).
224. Tkachuk, A. M., Petrov, M. V., Linmov, L. D. and Korablova, S. L., Pulsed-periodic 0.8503-μm YLF:Er^{3+}, Pr^{3+} laser, *Opt. Spectrosc. (USSR)*, 54, 667 (1983).
225. Zhekov, V. I., Lobachev, V. A., Murina, T. M., and Prokhorov, A. M., Efficient cross-relaxation laser emitting at $\lambda = 2.94$ μm, *Sov. J. Quantum Electron.* 13, 1235 (1983).
226. Petrov, M. V. and Tkachuk, A. M., Optical spectra and multifrequency stimulated emission of LiYF$_4$-Er^{3+} crystals, *Opt. Spectrosc.* 45, 81 (1978).
227. Jenssen, H. P. and Lai, S. T., Tunable-laser characteristics and spectroscopic properties of SrAlF$_5$:Cr, *J. Opt. Soc. Am. B* 3, 115 (1986).
228. Caird, J. A., Staver, P. R., Shinn, M. D., Guggenheim, H. J., and Bahnak, D., Laser-pumped laser measurements of gain and loss in SrAlF$_5$:Cr crystals, in *Tunable Solid-State Lasers II*, Vol. 52, Budgor, A. B., Esterowitz, L. and DeShazer, L., Eds., Springer-Verlag, New York (1986), p. 159.
229. Kaminskii, A. A., Sarkisov, S. E., Butaeva, T. I., Denisenko, G. A., Hermoneit, B., Bohm, J., Grosskreutz, W., and Schultze, D., Growth spectroscopy, and stimulated emission of cubic Bi$_4$Ge$_3$O$_{12}$ crystals doped with Dy^{3+}, Ho^{3+}, Er^{3+}, Tm^{3+}, or Yb^{3+} ions, *Phys. Status Solidi A*: 56, 725 (1979).
230. Andryunas, K., Vishchakas, Yu., Kabelka,V., Mochalov, I.V., Pavlyuk, A. A., and Syrus,V., Picosecond lasing in KY(WO$_4$)$_2$Nd^{3+} crystals at pulse repetition frequencies up to 10 Hz, *Sov. J. Quantum Electron.* 15, 1144 (1985).
231. Korableva, S, L., Livanova, L. D., Petrov, M. V. and Tkachuk, A. M., Stimulated emission of Er^{3+} ions in LiYF$_4$ crystals, *Sov. Phys. Tech. Phys.* 26, 1521 (1981).
232. Tkachuk, A. M., Petrov, M. V., Podkolzina, I. G., and Semenova, T. S., Generation of stimulated emission in a concentrated barium-erbium double-fluoride erystal, *Opt. Spectrosc. (USSR)* 53, 235 (1982).
233. Kaminskii, A. A., Sarkisov, S. E., Seiranyan, K. B., and Fedorov, V. A., Generation of stimulated emission for the waves of five ehannels of Er^{3+} ions in a self-activated LiErF$_4$ crystal, *Inorg. Mater. (USSR)* 18, 527 (1981).
234. Kaminskii, A. A., Mill', B. V., Belokoneva, E. L., Butashin, A. V., Sarkisov, S. E., Kurbanov, K., and Khodzhabagyan, G. G., Crystal structure intensity characteristics of luminescence and stimulated emission of the disordered gallate LaSr$_2$Ga$_{11}$O$_{20}$-Nd^{3+}, *Inorg. Mater. (USSR)* 22, 1635 (1986).

235. Voronko, Yu. K. and Sychugov, V. I., The stimulated emission of Er^{3+} ions in CaF_2 at λ = 8456 Å and λ = 8548 Å, *Phys. Status Solidi* 25, K 119 (1968).
236. Kaminskii, A. A., Pavlyuk, A. A., Balashov, I. F. et al., Stimulated emission by $KY(WO_4)_2$-Er^{3+} crystals at 0.85, 1.73 and 2.8 μm at 300 K, *Inorg. Mater. (USSR)* 14, 1765 (1978).
237. Zhekov, V. I., Zubov, B. V., Lobchev, V. A., Murina, T. M., Prokhorov, A. M., and Shevd', A. F., Mechanism of a population inversion between the $^4I_{11/2}$ and $^4I_{13/2}$ levels of the Er^{3+} ion in $Y_3Al_5O_{12}$ crystals, *Sov. J. Quantum Electron.* 10, 428 (1980).
238. Kaminskii, A. A., Butaeva, T. I., Kevorkov, A. M. et al. New data on stimulated emission by crystals with high concentrations of Ln^{3+} ions, *Inorg. Mater. (USSR)* 12, 1238 (1976).
239. Kaminskii, A. A., Butaeva, T. I., Fedorov, V. A., Bagdasarov, Kh. S., and Petrosyan, A. G., Absorption, luminescence and stimulated emission investigations in $Lu_3Al_5O_{12}$-Er^{3+} crystals, *Phys. Status Solidi* 39a, 541 (1977).
240. Kaminskii, A. A., Shkadarevich, A. P., Mill, B. V., Koptev, V. G., Butashin, A. V., and Demidovich, A. A., Wide-band tunable stimulated emission of Cr^{3+} ions in the trigonal crystal $La_3Ga_{5.5}Nb_{0.5}O_{14}$, *Inorg. Mater. (USSR)* 23, 1700 (1987).
241. Kaminskii, A. A., Stimulated-emission spectroscopy of activated laser crystals with ordered and disordered structure: seven selected experimental problems, private communication (1988).
242. Kaminskii, A. A., Shkadarevich, A. P., Mill, B. V., Koptev, V. G., Butashin, A. V., and Demidovich, A., Tunable stimulated emission of Cr^{3+} ions and generation frequency self-multipication effect in acentric crystals of Ca-gallogermanate structure, *Inorg. Mater. (USSR)* 24, 579 (1988).
243. Smith, L. K., Payne, S. A., Kway, W. L. and Krupke, W. F., Laser emission from the transition-metal compound $LiSrCrF_6$, *Optics Lett.* 18, 200 (1993).
244. Birnbaum, M., Tucker, A. W., and Pomphrey, P. I., New Nd:YAG laser transition $^4F_{3/2} \rightarrow {^4I_{9/2}}$. *IEEE J. Quantum Electron.* QE-8, 502 (1972).
245. Kaminskii, A. A., Kurbanov, K., Peievin, A.V., Bobakova, Yu. A. and Uvarova, T. V., Intermultiplet transition $^3P_0 \rightarrow {^1G_4}$—new channel for stimulated Pr^{3+}-ion emission in anisotropic fluoride crystals, *Inorg. Mater. (USSR)* 24, 439 (1988).
246. Kaminskii, A. A., Kurbanov, K., Peievin, A. V., Bobakova, Yu. A., and Uvarova, T. V., New channels for stimulated emission of Pr^{3+} ions in tetragonal fluorides $LiRF_4$ with the structure of scheelite. *Inorg. Mater. (USSR)* 23, 1702 (1987).
247. Hanson, F., Dick, D., Versun, H. R. and Kokta, M., Optical properties and lasing of Nd:$SrGdGa_3O_7$, *J. Opt. Soc. Am. B* 8, 1668 (1991).
248. Beach, R., Albrecht, G., Solarz, R., Krupke, W., Mitchell, S., Comaskey, B., Brandle, C., and Berkstresser, G., Q-switched laser at 912 nm using ground state depleted neodymium in yttrium Orthosilicate, *Opt. Lett.* 15, 1020 (1990).
249. Antonov, V. A., Arsenev, P. A., Evdokimov, A. A., Koptsik, E. K., Starikov, A. M. and Tadzhi-Aglaev, Kh. G., Spectral-luminescence properties of $Ba_3LaNb_3O_{12}$:Nd^{3+} single crystals. *Opt. Spectrosc. (USSR)* 60, 57 (1986).
250. Kaminskii, A. A., Klevtsov, P. V., Li, L., and Pavlyuk, A. A., Spectroscopic and stimulated emission studies of the new $KY(WO_4)_2$:Nd^{3+} laser crystal, *Inorg. Mater. (USSR)* 8, 1896 (1972).
251. Kaminskii, A. A., Klevtsov, P. V., Li, L., and Pavlyuk, A. A., Laser $^4F_{3/2} \rightarrow {^4I_{11/2}}$ and $^4F_{3/2} \rightarrow {^4I_{13/2}}$ transitions in $KY(WO_4)_2$, *IEEE J. Quantum Electron.* QE-8, 457 (1972).

252. Kaminskii, A. A., Klevtsov, P. V., Bagdasarov, Kh. S., Mayyer, A. A., Pavlyuk, A. A., Petrosyan, A. G. and Provotorov, M. V., New cw crystal lasers, *JETP Lett.* 16, 387 (1972).
253. Johnson, L. F. and Thomas, R. A., Maser oscillations at 0.9 and 1.35 microns in $CaWO_4:Nd^{3+}$, *Phys. Rev.* 131, 2038 (1963).
254. Morris, R. C., Cline, C. F., and Begley, R. F., Lanthanum beryllate: a new rare-earth ion host, *Appl. Phys. Lett.* 27, 444 (1975).
255. Jenssen. H. P., Begley, R. F., Webb, R., and Morris, R. C., Spectroscopic properties and laser performance of Nd^{3+} in lanthanum beryllate, *J. Appl. Phys.* 47, 1496 (1976).
256. Birnbaum, M. and Tucker, A. W., Nd-YALO Oscillation at 0.95 μm at 300 K, *IEEE J. Quantum Electron.* QE-9, 46 (1973).
257. Kaminskii, A. A., Agamalyan, N. R., Pavlyuk, A. A., Bobovich, L. I., and Lyubchenko, V. V., Preparation and luminescence-generation properties of KLu(WO4)2-Nd^{3+}, *Inorg. Mater. (USSR)* 19, 885 (1983).
258. Prokhorov, A. M., A new generation of solid-state lasers, *Sov. Phys. Usp.* 29, 3 (1986).
259. Zharikov, E. V., Zhltnyuk, V. A., Kuratev, I. I., Lapaev, V. V., Smirnov, V. A., Shestakov, A. V., and Shcherbakov, I A., Laser based on a GSGG:Cr^{3+}, Nd^{3+} crystal, which operates on the $^4F_{3/2} \rightarrow {}^4I_{9/2}$ transition at room temperature, in *Bull. Acad. Sci. USSR Phys. Ser.* 48, 98 (1984).
260. Scheps, R., Myers, J., Schimitschek, E. J., and Heller, D. F., End-pumped Nd:BEL laser performance, *Opt. Engineer.* 27, 830 (1988).
261. Moulton, P. F., Tunable paramagnetic-ion lasers, in *Laser Handbook*, Vol. 5, Bass, M. and Stitch, M. L., Eds., North-Holland, Amsterdam, The Netherlands (1985), p. 203
262. Fan, T. Y. and Byer, R. L., Continuous-wave operation of a room-temperature diode-laser-pumped 946-nm Nd:YAG laser, *Opt. Lett.* 12, 809 (1987).
263. Risk, W. P. and Lenth, W., Room-temperature, continuous-wave, 946-nm Nd:YAG laser pumped by laser-diode arrays and intracavity frequency doubling to 473 nm, *Opt. Lett.* 12, 993 (1987).
264. Fan, T. Y. and Byer, R. L., Modeling and CW operation of a quasi-three-level 946-nm Nd:YAG laser, *IEEE J. Quantum Electron.* QE-23, 605 (1987).
265. Birnbaum, M., Tucker, A. W., and Fincher, C. L., CW room-temperature laser operation of Nd:CAMGAR at 0.941 and 1.059 μ, *J. Appl. Phys.* 49, 2984 (1978).
266. Wallace, R. W. and Harris, S. E., Oscillation and doubling of the 0.946 μm line in Nd^{3+}:YAG, *Appl. Phys. Lett.* 15, 111 (1969).
267. Kaminskii, A. A., Bogomolova, G. A., Bagdasarov, Kh. S., and Petrosyan, A. C., Luminescence. absorption and stimulated emission of $Lu_3Al_5O_{12}$-Nd^{3+}- crystals, *Opt. Spectrosc.* 39, 643 (1975).
268. Morozov, A. M., Podkolzina, I. A., Tkachuk, A. M., Fedorov, V. A., and Feofilov, P. P., Luminescence and induced emission lithium-erbium and lithium-holmium binary fluorides, *Opt. Spectrosc.* 39, 338 (1975).
269. Christensen, H. P., Spectroscopic analysis of $LiHoF_4$ and $LiErF_4$, *Phys. Rev. B* 19, 6564 (1979).
270. Kaminskii, A. A., Ngoc, T., Sarklsov, S. E., Matrosov, V. N., and Timoshechkin, M. I., Growth, spectral and laser properties of $La_2Be_2O_5:Nd^{3+}$ crystals in the $^4F_{3/2} \rightarrow {}^4I_{11/2}$ and $^4F_{3/2} \rightarrow {}^4I_{13/2}$ transitions, *Phys. Status Solidi A*: 59, 121 (1980).
271. Kolbe, W., Petermann, K., and Huber, G., Broadband emission and laser action of Cr^{3+} doped zinc tungstate at 1 μm wavelength, *IEEE J. Quantum Electron.* QE-21, 1596 (1985).

272. Kaminskii, A. A., Pavlyuk, A. A., Agamalyan, N. R., Sarkisov, S. E., Bobovich, L. I., Lukin, A. V., and Lyubchenko, V. V., Stimulated radiation of Nd^{3+} and Ho^{3+} ions in monoclinic $KLu(WO_4)_2$ crystals at room temperature, *Inorg. Mater. (USSR)* 15, 1649 (1979).
273. Otsuka, K., Nakano, J., and Yamada, T., Laser emission cross section of the system $LiNd_{0.5}M_{0.5}P_4O_{12}$ (M = Gd,La), *J. Appl. Phys.* 46, 5297 (1975).
274. Golubev, P. G., Kandaurov, A. S., Lazarev, V. V., and Safronov, E. K., Lasing properties of neodymium-activated lanthanum beryllate, *Sov. J. Quantum Electron.* 15, 1213 (1985).
275. Schaffers, K. I., DeLoach, L. D. and Payne, S. A., Crystal growth, frequency doubling, and infrared laser performance of Yb^{3+}:$BaCaBO_3F$, *IEEE J. Quantum Electron.* 32, 741 (1996).
276. Bogomolova, G. A., Vylegzhanin, D. N., and Karninskii, A. A., Spectral and lasing investigations of Barnets with Yb^{3+} ions, *Sov. Phys. -JETP* 42, 440 (1976).
277. Kuleshov, N. V., Lagatsky, A. A. and Mikhailov, V. P., High efficiency cw lasing of Yb-doped tungstates, *Proceedings Advanced Solid State Lasers* (1997), to be published.
278. Johnson, L. F., Geusic, J. E., and Van Uitert, L. G., Coherent oscillations from Tm^{3+}, Ho^{3+}, Yb^{3+} and Er^{3+}- ions in yttrium aluminum garnet, *Appl. Phys. Lett.*, 7, 127 (1965).
279. Bagdasarov, Kh. S., Kaminskii, A. A., Kevorkov, A. M., and Prokhorov, A. M. Rare earth scandium-aluminum garnets with impurity of TR^{3+} ions as active media for solid state lasers, *Sov. Phys. Dokl.* 19, 671 (1975).
280. Hanna, D. C., Jones, J. K., Large, A. C., Shepherd, D. P., Tropper, A, C., Chandler, P. J., Rodman, M. J., Townsend, P. D. and Zhang, L., Quasi-three level 1.03 μm laser operation of a planar ion-implanted Yb:YAG waveguide, *Optics Commun.* 99, 211 (1993).
281. Lacovara, P., Choi, H. K., Wang, C. A., Aggarwal, R. L. and Fan, T. Y., Room-temperature diode-pumped Yb:YAG laser, *Optics. Lett.* 16, 1089 (1991).
282. Sugimoto, N., Ohishi, Y., Katoh, Y., Tate, A., Shimokozono, M. and Sudo, S., A ytterbium- and neodymium-codoped yttrium aluminum garnet-buried channel waveguide laser pumped at 0.81 μm, *Appl. Phys. Lett.* 67, 582 (1995).
283. Sumida, D. S., Fan, T. Y. and Hutcheson, R., Spectroscopy and diode-pumped lasing of Yb^{3+}-doped $Lu_3Al_5O_{12}$ (Yb:LuAG), *OSA Proc. Adv. Solid State Lasers*, Chai, B. H. T. and Payne, S. A., Eds., 24, 348 (1995).
284. Payne, S. A., Smith, L. K., DeLoach, L. D., Kway, W. L., Tassano, J. B. and Krupke, W. F., Laser, optical and thermomechanical properties of Yb-doped fluoroapatite, *IEEE J. Quantum Electron.* 30, 170 (1994).
285. Robinson, M. and Asawa, C. K., Stimulated emission from Nd^{3+} and Yb^{3+} in noncubic sites of neodymium- and ytterbium-doped CaF_2, *J. Appl. Phys.* 38, 4495 (1967).
286. Kaminskii, A. A., Mikaelyan, R. C., and Zigler, I. N., Room-temperature induced emission of CaF_2-SrF_2 crystals containing Nd^{3+}, *Phys. Status Solidi* 31, K85 (1969).
287. Stephens, E., Schearer, L. D. and Verdun, H. R., A tunable Nd:$CaYAlO_4$ laser, *Optics Commun.* 90, 79 (1992).
288. Danielmeyer, H. G., Jeser, J. P., Schonherr, E., and Stetter, W., The growth of laser quality NdP_5O_{14} crystals, *J. Cryst. Growth* 22, 298 (1974).
289. Miyazawa, S. and Kubodera, K., Fabrication of $KNdP_4O_{12}$ laser epitaxial waveguide, *J. Appl. Phys.* 49, 6197 (1978).
290. Johnson, L. F., Optically pumped pulsed crystal lasers other than ruby, in *Lasers, A Series of Advances*, Vol. I, Levine, A. K., Ed., Marcel Dekker, New York (1966), p.137.

291. Johnson, L. F. and Ballman, A. A., Coherent emission from rare-earth ions in electro-optic crystals, *J. Appl. Phys.* 40, 297 (1969).
292. Vylegzhanin, D. N. and Kaminskii, A. A., Study of electron-phonon interaction in $LaF_3:Nd^{3+}$ crystals, *Sov. Phys.-JETP* 35, 361 (1972).
293. Voron'ko, Yu. K., Dmitruk, M. V., Kaminskii. A. A., Osiko, V. V., and Shpakov, V. N., CW stimulated emission in an $LaF_3:Nd^{3+}$ laser at room temperature, *Sov. Phys.-JETP* 27, 400 (1968).
294. O'Connor, J. R. and Hargreaves, W. A., Lattice energy transfer and stimulated emission from $CeF_3:Nd^{3+}$, *Appl. Phys. Lett.* 4, 208 (1964).
295. Dmitruk, M. V., Kaminskii, A. A., and Shcherbakov, I. A., Spectroscopic studies of stimulated emission from a $CeF_3:Nd^{3+}$ laser, *Sov. Phys.-JETP* 27 900 (1968).
296. Hoskins, R. H. and Soffer, B. H., Fluorescence and stimulated emission from $La_2O_3:Nd^{3+}$, *J. Appl. Phys.*, 36, 323 (1965).
297. Evtuhov, V. and Neeland, J. K., A continuously pumped repetitively Q-switched ruby laser and applications to frequency-conversion experiments, *IEEE J. Quantum Electron.* QE-5, 207 (1969).
298. Chou, H., Alpers, P., Cassanho, A. and Jenssen, H.P., CW tunable laser emission of $Nd^{3+}:Na_{0.4}Y_{0.6}F_{2.2}$ in *Tunable Solid-State Lasers II*, Vol. 52, Budgor, A. B., Esterowitz, L., and DeShazer, L. G., Eds., Springer-Verlag, New York (1986), p. 322.
299. Oldberg, L. S. and Bradford, J. N., Passive mode locking and picosecond pulse generation in Nd:lanthanum beryllate, *Appl. Phys. Lett.* 28, 585 (1976).
300. Robinson, M. and Asawa, C. K., Stimulated emission from Nd^{3+} and Yb^{3+} in noncubic sites of neodymium- and ytterbium-doped CaF_2, *J. Appl. Phys.* 38, 4495 (1967).
301. Johnson, L. F., Optical maser characteristics of Nd^{3+} in CaF_2, *J. Appl. Phys.* 33, 756 (1962).
302. Kaminskii, A. A., Korniyenko, L. S., and Prokhorov, A. M., Spectral study of stimulated emission from Nd^{3+} in CaF_2, *Sov. Phys.-JETP* 21, 318 (1965).
303. Moulton, P. F., Fahey, R. E., and Krupke, W. F., Advanced solid-state lasers, in 1981 Laser Program Annual Report, George, E. V., Ed., Lawrence Livermore National Laboratory, Livermore, CA UCRL-50021 (1982), p. 7, available from National Technical Information Service.
304. Moulton, P. F., Advances in tunable transition-metal lasers, *Appl. Phys. B* 28, 233 (1982).
305. Emmett, J. L., Krupke, W. F., and Trenholme, J. B., Future development of high-power solid-state laser systems, *Sov. J. Quantum Electron.* 13, 1 (1983).
306. Telle, H. R., Tunable CW laser oscillation of stoichiometric Nd-materials, in *Proc. Int. Conf. Lasers '85*, Wang. C. P., Ed., STS Press, McLean, VA (1986) 460.
307. Otsuka, K., Li, H., and Telle, H. R., CW Nd-lasers with broad tuning range, *Opt. Commun.* 63, 57 (1987).
308. Bagdasarov, Kh. S., Voron'ko, Yu. K., Kaminskii, A. A., Osiko, V. V., and Prokhorov, A. M., Stimulated emission in yttro-fluorite:Nd^{3+} crystals at room temperature, *Sov. Phys.-Crystallogr.* 10, 626 (1966).
309. Kaminskii, A. A., Osiko, V. V., Prokhorov, A. M., and Voron'ko, Yu. K., Spectral investigation of the stimulated radiation of Nd^{3+} in CaF_2-YF_3, *Phys. Lett.* 22, 419 (1966).
310. Yariv, A., Porto, S. P. S., and Nassau, K., Optical laser emission from trivalent praseodymium in calcium tungstate, *J. Appl. Phys.* 33, 2519 (1962).
311. DeLoach, L. D., Payne, S. A., Smith, L. K., Kway, W. L. and Krupke, W. F., Laser and spectroscopic properties of $Sr_5(PO_4)_3F:Yb$, *J. Opt. Soc. Am. B* 11, 269 (1994).

312. Marshall, C. D., Smith, L. K, Beach, R. J. et al., Diode-pumped ytterbium-doped $Sr_5(PO_4)_3F$ laser performance, *IEEE J. Quantum Electron.* 32, 650 (1996).
313. Collonjgues, R., Lejus, A. M., Thery, J. and Vivien, D., Crystal growth and characterization of new laser materials, *J. Cryst. Growth* 128, 986 (1993).
314. Barnes, N. P., Gettemy, D. J., Esterowitz, L., and Allen, R. E., Comparison of Nd 1.06 and 1.33 μm operation in various hosts, *IEEE J. Quantum Electron.* QE-23, 1434 (1987).
315. Pollak, T. M., Wing, W. F., Grasso, R. J., Chicklis, E. P., and Jenssen, H. P., CW laser operation of Nd:YLF, *IEEE J. Quantum Electron.* QE-18, 159 (1982).
316. Fan, T. Y., Dixon, G. J., and Byer, R. L., Efficient GaAlAs diode-laser pumped operation of Nd:YLF at 1.047 μm with intracavity doubling to 523.6 nm, *Opt. Lett.* 11, 204 (1986).
317. Gverev, G. M., Kuratev, I. I., and Shestakov, A. V., Solid-state microlasers based on crystals with a high concentration of neodymium ions, *Bull. Acad. Sci. USSR Phys. Ser.* 46, 108 (1982).
318. Bagdasarov, Kh. S., Dorozhkin, L. M., Kevorkov, A. M., Krasilov, Yu. I., Potemkin, A. V., Shestakov, A. V., and Kuratev, I. I., Continuous lasing in $La_{1-x}Nd_xMgAl_{11}O_{19}$ crystals, *Sov. J. Quantum Electron.* 13, 639 (1983).
319. Krühler, W. W., Plättner, R. D., and Stetter, W., CW oscillation at 1.05 and 1.32 μm of $LiNd(PO_3)_4$ lasers in external resonator and in resonator with directly applied mirrors, *Appl. Phys.* 20, 329 (1979).
320. Bagdasarov, Kh. S., Kaminskii, A. A., Kevorkov, A. M. et al., Laser properties of Y_2SiO_5-Nd^{3+} crystals irradiated at the $^4F_{3/2} \rightarrow {}^4I_{11/2}$ and $^4F_{3/2} \rightarrow {}^4I_{13/2}$ transitions, *Sov. Phys.-Dokl.* 18, 664 (1974).
321. Harmer, A. L., Linz, A., and Gabbe, D. R., Fluorescence of Nd^{3+} in lithium yttrium fluoride, *J. Phys. Chem. Sol.* 30, 1483 (1969).
322. Le Coff, D., Bettinger, A., and Labadens, A., Etude d'un oscillateur a blocage de modes utilisant un cristal de $LiYF_4$ dope au neodyme, *Opt. Commun.* 26, 108 (1978), in French.
323. Bagdasarov, Kh. S., Kaminskii, A. A., Kevorkov, A. M., and Prokhorov, A. M., Investigation of the stimulated radiation emitted by Nd^{3+} ions in $CaSc_2O_4$ crystals, *Sov. J. Quantum Electron.* 4, 927 (1975).
324. Bagdasarov, Kh. S., Kaminskii, A. A., Kevorkov, A. M. et al., Stimulated emission of Nd^{3+} ions in an $SrAl_2O_4$ crystal at the transitions $^4F_{3/2} \rightarrow {}^4I_{11/2}$ and $^4F_{3/2} \rightarrow {}^4I_{13/2}$, *Sov. Phys.-Dokl.* 19, 350 (1974).
325. Hoskins, R. H. and Soffer, B. H., Stimulated emission from Y_2O_3:Nd^{3+}, *Appl. Phys. Lett.* 4, 22 (1964).
326. Fan, T. Y. and Kokta, M. R., End-pumped Nd:LaF_3 and Nd:$LaMgAl_{11}O_{19}$ lasers, *IEEE J. Quantum Electron.* 25, 1845 (1989).
327. Weston, J., Chiu, P. H., and Aubert, R., Ultrashort pulse active passive mode-locked Nd:YLF laser, *Opt. Commun.* 61, 208 (1987).
328. Kaminskii, A. A., Stimulated-emission spectroscopy of Er^{3+} ions in cubic $(Y,Ln)_3Al_5O_{12}$ and monoclinic $K(Y,Ln)W_2O_8$ single crystals, *Phys. Status Solidi A* 96 K175 (1986).
329. Antipenko, B. M., Rab, O. B., Seiranyan, K. B. and Sukhnreva, L. K., Quasi-continuous lasing of an $LiYF_4$:Er:Pr crystal at 0.85 μm, *Sov. J. Quantum Electron.* 13, 1237 (1983).
330. Kaminskii, A. A., Mill', B. V. and Butashin, A. V., New low-threshold noncentrosymmetric $LaBGeO_5$:Nd^{3+} laser crystal, *Sov. J. Quantum Electron.* 20, 875 (1990).
331. Kaminskii, A.A., Mill', B.V. and Butashin, A.V., Stimulated emission from Nd^{3+} ions in acentric $LaBGeO_5$ crystals, *Phys. Status Solidi A* 115, K59 (1990).

332. Nakano, J., Kubodera, K., Miyazawa, S., Kondo, S., and Koizumi, H., $LiBi_xNd_{1-x}P_4O_{12}$ waveguide laser layer epitaxially grown on $LiNdP_4O_{12}$ substrate, *J. Appl. Phys.* 50, 6546 (1979).
333. Yamada, T., Otsuka, K., and Nakano, J., Fluorescence in lithium neodymium ultraphosphate single crystals, *J. Appl. Phys.* 45, 5096 (1974).
334. Hong, H. Y-P. and Chinn, S. R., Influence of local-site symmetry on fluorescence lifetime in high Nd-concentration laser materials, *Mater. Res. Bull.* 11, 461 (1976).
335. Nakano, J., Kubodera, K., Yarnada, T., and Miyazawa, S., Laser-emission cross sections of $MeNdP_4O_{12}$ (Me = Li,Na,K) crystals, *J. Appl. Phys.* 50, 6492 (1979).
336. Otsuka, K. and Yamada, T., Transversely pumped LNP laser performance, *Appl. Phys. Lett.* 26, 311 (1975).
337. Chinn, S. R. and Hong, H. Y-P., Low-threshold cw $LiNdP_4O_{12}$ laser, *Appl. Phys. Lett.* 26, 649 (1975).
338. Otsuka, K., Yamada, T., Saruwatari, M., and Kimura, T., Spectroscopy and laser oscillation properties of lithium neodymium tetraphosphate, *IEEE J. Quantum Electron.* QE-I I, 330 (1975).
339. Kaminskii, A. A., Pavlyuk, A. A., Butaeva, T. I., Bobovieh, L. I. and Lyuhchenko, V. V., Stimulated emission in the 2.8-μm band by a self-activated crystal of $KEr(WO_4)_2$, *Inorg. Mater. (USSR)* 15, 424 (1979).
340. Hemmati, H., 2.07-μm CW diode-laser-pumped Tm, Ho:$YLiF_4$ room-temperature laser, *Opt. Lett.* 14, 435 (1989).
341. Antipenko, B. M., Glebov, A. S., Krutova, L. I., Solntsev, V. M., and Sukhareva, L. K., Active medium of lasers operating in the 2-μm spectra range and utilizing gadolinium scandium gallium garnet crystals, *Sov. J. Quantum Electron.* 16, 995 (1986).
342. Shcherbakov, I. A., Optically dense active media for solid-state lasers, *IEEE J. Quantum Electron.* 24, 979 (1988).
343. D'yakanov, G. I., Egorov, G. N., Zharikov, E. V., Mlkhailov, V. A., Pak, S. K., Prokhorov, A. M., and Shcherbakov, I. A., Chromium- and neodymium-activated yttrium scandium gallium garnet laser with an efficiency of 3.6% emitting linearly polarized radiation of energy of 0.46 J per single pulse and a pulse repetition frequency 50 Hz, *Sov. J. Quantum Electron.* 18, 43 (1988).
344. Voron'ko, Yu. K., Kaminskii, A. A., Korniyenko, L. S., Osiko, V. V., Prokhorov, A. M., and Udoven'chik, V. T., Investigation of stimulated emission from $CaF_2:Nd^{3+}$- (Type II) crystals at room temperature, *JETP Lett.* 1, 39 (1965).
345. Korzhik, M. V., Livshits, M. G., Bagdasarov, Kh. S., Kevorkov, A. M., Melkonyan, T. A., and Mellman, M. L., Efficient pumping of Nd^{3+} ions via charge transfer bands of Fe^{2+} ions in YAG, *Sov. J. Quantum Electron.*, 19, 344 (1989).
346. Keen, S. J., Maker, G. T., and Ferguson, A. J., Mode-locking of diode laser-pumped Nd:YAG laser at 1.3 μm, *Electron. Lett.* 25, 490 (1989).
347. Khurgin, J. and Zwicker, W. K., High efficiency nanosecond miniature solid-state laser, *Appl. Opt.* 24, 3565 (1985).
348. Kaminow, I. P. and Stulz, L. W., Nd:$LiNbO_3$ laser, *IEEE J. Quantum Electron.* QE-11, 306 (1975).
349. Kubodera, K., Otsuka, K., and Miyazawa, S., Stable $LiNdP_4O_{12}$ miniature laser, *Appl. Opt.* 18, 844 (1979).
350. Kubodera, K. and Otsuka, K., Efficient $LiNdP_4O_{12}$ lasers pumped with a laser diode, *Appl. Opt.* 18, 3882 (1979).
351. Saruwatari, M., Kimura, T., Yamada, T., and Nakano, J., $LiNdP_4O_{12}$ laser pumped with an Al_xGa_{1-x} As electroluminescent diode, *Appl. Phys. Lett.* 27, 682 (1975).
352. Saruwatari, M. and Kimura, T., LED pumped lithium neodymium tetraphosphate lasers, *IEEE J. Quantum Electron.* QE-12, 584 (1976).

353. Kubodera, K. and Otsuka, K., Laser performance of a glass-clad $LiNdP_4O_{12}$ rectangular waveguide, *J. Appl. Phys.* 50, 6707 (1979).
354. Krühler, W. W., Plättner, R. D. and Stetter, W., CW oscillation at 1.05 μm and 1.32 μm of $LiNd(PO_3)_4$ lasers in external resonator and in resonator with directly applied mirrors, *Appl. Phys.* 20, 329 (1979).
355. Chinn, S. R., Zwicker, W. K., and Colak, S., Thermal behavior of NdP_5O_{14} lasers, *J. Appl. Phys.* 53, 5471 (1982).
356. Sarkisov, S. E., Lomonov, V. A., Kaminskii, A. A. et al., Spectroscopic investigation of the stable crystalline mixed system $Bi_4(Ge_{1-x},Si_x)_3O_{12}$, in *Abstracts of Papers of Fifth All-Union Symposium on Spectroscopy of Crystals*, Kazan (1976) p. 195 (in Russian).
357. Lempicki, A., McCollum, B. C., and Chinn, S. R., Spectroscopy and lasing in $K_5NdLi_2F_{10}$ (KNLF), *IEEE J. Quantum Electron.* QE-15, 896 (1979).
358. Danilov, A. A., Zharikov, E. V., Zagumennyl, A. I., Lutts, G. B., Nlkolskii, M. Yu., Tsvetkov, V. B., and Shcherhakov, I. A., Self-Q-switched high-power laser utilizing gadolinium scandium gallium garnet activated with chromium and neodymium, *Sov. J. Quantum Electron.* 19, 315 (1989).
359. Kaminskii, A. A. and Khaidukov, N. M., New low-threshold $LiKYF_5:Nd^{3+}$ laser crystal, *Sov. J. Quantum Electron.* 22, 193 (1992) and *Phys. Status Solidi A* 129, K65 (1992).
360. Comaskey, B., Albrecht, G. F., Beach, R. J., Moran, B. D. and Solarz, R. W., Flash-lamp-pumped laser operation of $Nd^{3+}:Y_2SiO_5$ at 1.074 μm, *Optics Lett.* 18, 2029 (1993).
361. Viscakas, J. and Syrusas, V., Stimulated raman scattering self-conversion of laser radiation and the potential for creating multifrequency solid state lasers, *Sov. Phys. - Collect.* 27, 31 (1987).
362. Dmitruk, M. V., Kaminskii, A. A., Osiko, V. V., and Tevosyan, T. A., Induced emission of hexagonal LaF_3-$SrF_2:Nd^{3+}$ crystals at room-temperature, *Phys. Status Solidi* 25, K75 (1968).
363. Bagdasarov, Kh. S., Kaminskii, A. A., Kevorkov, A. M. et al., Investigation of the stimulated emission of cubic crystals of $YScO_3$ with Nd^{3+} ions, *Sov. Phys.-Dokl.* 20, 681 (1975).
364. Eggleston, J. M., DeShazer, L. G., and Kangas, K. W., Characteristics and kinetics of laser-pumped Ti:Sapphire oscillators, *IEEE J. Quantum Electron.* 24, 1009 (1988).
365. Shand, M. L., Progress in alexandrite lasers, in *Proc. Int. Conf. Lasers 85*, Wang, C. P., Ed., STS Press, McLean, VA (1986), p. 732.
366. Kaminskii, A. A. and Sobolev, B. N., Monoclinic fluoride BaY_2F_8-Nd^{3+}—a new low-threshold inorganic laser materials, *Inorg. Mater. (USSR)* 19, 1718 (1983).
367. Kaminskii, A. A., Sarkisov, S. E, Seiranyan, K. B., and Sobolev, B. P., Study of stimulated emission in $Sr_2Y_5F_{19}$ crystals with Nd^{3+} ions, *Sov. J. Quantum Electron.* 4, 112 (1974).
368. Kaminskii,, A. A., Petrosyran, A. G., Ovanesyan, K. L., Shironyan, G. O., Fedorov, V. A., and Sarkisov, S. E., Concentrational 3 μm stimulated emission tuning in the $(Gd_{1-x}Er_x)_3Al_5O_{12}$ crystal system. *Phys. Status Solidi A* 82, K185 (1984).
369. Kaminskii, A. A., Agamalyan, N. R., Denisenko, G. A., Sarkisov, S. E., and Fedorov, P. P., Spectroscopy and laser emission of disordered GdF_3-$CaF_2:Nd^{3+}$ trigonal crystals, *Phys. Status Solidi A* 70, 397 (1982).
370. Kevorkov, A. M., Kaminskii, A. A., Bagdasarov, Kh. S., Tevosyan, T. A., and Sarkisov, S. E., Spectroscopic properties of $SrAl_2O_4:Nd^{3+}$ crystals, *Inorg. Mater. (USSR)* 9, 1637 (1973).

371. Alves, R. V., Buchanan, R. A., Wickersheim, K. A., and Yates, E. A., Neodymium-activated lanthanum oxysulfide: a new high-gain laser material, *J. Appl. Phys.* 42, 3043 (1971).
372. Soffer, B. H. and Hoskins, R. H., Fluorescence and stimulated emission from $Gd_2O_3:Nd^{3+}$ at room temperature and 77 K, *Appl. Phys. Lett.* 4, 113 (1964).
373. Chinn, S. R. and Hong, H. Y-P., Fluorescence and lasing properties of $NdNa_5(WO_4)_4$, $K_3Nd(PO_4)_2$ and $Na_3Nd(PO_4)_2$, *Opt. Commun.* 18, 87 (1976).
374. Kaminskii, A. A., Zhmurova, Z. I., Lomonov,V. A., and Sarkisov, S.E., Two stimulated emission $^4F_{3/2} \rightarrow {}^4I_{11/2,13/2}$ channels of Nd^{3+} ions in crystals of the CaF_2-ScF_3 system, *Phys. Status Solidi A* 84, K81 (1984).
375. Marling, J. B., 1.05-1.44 μm tunability and performance of the CW Nd^{3+}:YAG laser, *IEEE J. Quantum Electron.* QE-14, 56 (1978).
376. Chinn, S. R. and Zwicker, W. K., A comparison of flash-lamp-excited $Nd_xLa_{1-x}P_5O_{14}$(x = 1.0, 0.75, 0.20) lasers, *J. Appl. Phys.*, 52, 66 (1981).
377. Bagdasarov, Kh. S. and Kaminskii, A. A., RE^{3+}-doped $YAlO_3$ as an active medium for lasers, *JETP Lett.* 9, 303 (1969).
378. Huber, G., Krühler, W. W., Bludau, W., and Danielmeyer, H. G., Anisotropy in the laser performanceof NdP_5O_{14}, *J. Appl. Phys.* 46, 3580 (1975).
379. Bagdasarov, Kh. S., Kaminskii, A. A., and Sobolev, B. P., Laser based on $5NaF{\bullet}9YF_3:Nd^{3+}$ cubic crystals, *Sov. Phys.-Crystallogr.* 13, 779 (1969).
380. Kaminskii, A. A., Markosyan, A. A., Pelevin, A. V., Polyakova, Yu. A., and Uvarova, T. V., Single-crystal $Cd_{1-x}Sc_xF_{2+x}$, with Nd^{3+} ions and its stimulated emission, *Inorg. Mater. (USSR)* 22, 777 (1986).
381. Nakano, J., Otsuka, K., and Yamada, T., Fluorescence and laser-emission cross sections in $NaNdP_4O_{12}$, *J. Appl. Phys.* 47, 2749 (1976).
382. Otsuka, K., Miyazawa, S., Yamada, T., Iwasaki, H., and Nakano, J., CW laser oscillations in $MeNdP_4O_{12}$ (Me = Li,Na,K) at 1.32 μm, *J. Appl. Phys.* 48, 2099 (1977).
383. Belabaev, K. C., Kaminskii, A. A., and Sarkisov, S. E., Stimulated emission from ferroelectric $LiNbO_3$ crystals containing Nd^{3+} and Mg^{2+}- ions, *Phys. Status Solidi* 28a, K17 (1975).
384. Gualtieri, J. G. and Aucoin, T. R., Laser performance of large Nd-pentaphosphate crystals, *Appl. Phys. Lett.* 28, 189 (1976).
385. Szymanski, M., Karolczak, J., and Kaczmarek, F., Temporal studies of intensity dependent laser and spontaneous emission in $NdLaP_5O_{14}$ monocrystals, *Acta Phys. Pol. A* A60 95 (1981).
386. He, N. J., Lu, G. X., Li, Y. C. and Zhao, L. X., A CW $La_{0.1}Nd_{0.9}P_5O_{14}$ laser at 1.051 and 1.32 μm, *Chin. Phys.* 2, 455 (1982).
387. Winzer, G., Mockd, P. G., Oberbacher, R., and Vite, L., Laser emission from polished NdP_5O_{14} crystals with directly applied mirrors, *Appl. Phys.* 11, 121 (1976).
388. Grigor'yanb, V. V., Makovetskil, A. A., and Tishchenko, R. P., Kinetics of emission from a neodymium pentaphosphate microlaser pumped by short pulses, *Sov. J. Quantum Electron.* 10, 1286 (1980).
389. Liu, J., Wang, M., Zhao, X., Liang, Y., Wang, B., and Lu, B., A miniature pulsed NdP_5O_{14} laser, *Chin. Phys. Lasers* 14, 45 1987.
390. Weber, H. P., Damen, T. C., Danielmeyer, H. G., and Tofield, B. C., Nd-ultraphosphate laser, *Appl. Phys. Lett.* 22, 534 (1973).
391. Damen, T. C., Weber, H. P., and Tofield, B. C., NdLa pentaphosphate laser performance, *Appl. Phys. Lett.* 23, 519 (1973).

392. Danielmeyer, H. G., Huber, G., Krühler, W. W., and Jeser, J. P., Continuous oscillation of a (Sc,Nd) pentaphosphate laser with 4 milliwatts pump threshold, *Appl. Phys.* 2, 335 (1973).
393. Chinn, S. R., Pierce, J. W., and Heckscher, H., Low-threshold, transversely excited NdP_5O_{14} laser, *IEEE J. Quantum Electron.* QE-11, 747 (1975).
394. Weber, H. P., Liao, P. F., Tofield, B. C., and Bridenbaugh, P. M., CW fiber laser of NdLa pentaphosphate, *Appl. Phys. Lett.* 26, 692 (1975).
395. Chinn, S. R., Intracavity second-harmonic generation in a Nd pentaphosphate laser, *Appl. Phys. Lett.* 29, 176 (1976).
396. Chinn, S. R. and Zwicker, W. K., FM mode-locked $N_{0.5}$,$La_{0.5}P_5O_{14}$ laser, *Appl. Phys. Lett.* 34, 847 (1979).
397. Budin, J. -P., Neubauer, M., and Rondot, M., Miniature Nd-pentaphosphate laser with bonded mirrors side pumped with low-current-density LED's, *Appl. Phys. Lett.* 33, 309 (1978).
398. Budin, J. -P., Neubauer, M., and Rondot, H., On the design of neodymium miniature lasers, *IEEE J. Quantum Electron.* QE-14, 831 (1978).
399. Krühler, W. W., Huber, G., and Danielmeyer, H. G., Correlations between site geometries and level energies in the laser system $Nd_xY_{1-x}P_5O_{14}$, *Appl. Phys.* 8, 261 (1975).
400. Krühler, W. W. and Plättner, R. D., Laser emission of (Nd,Y)-pentaphosphate at 1.32 μm, *Opt. Commun.* 28, 217 (1979).
401. Krühler, W. W., Jeser, J. P., and Danielmeyer, H. G., Properties and laser oscillation of the (Nd,Y) pentaphosphate system, *Appl. Phys.* 2, 329 (1973).
402. Gualtieri, J. G. and Aucoin, T. R., Laser performance of large Nd-pentaphosphate crystals, *Appl. Phys. Lett.* 28, 189 (1976).
403. Krühler, W. W., Plättner, R. D., Fabian, W., Mockel, P., and Grabmaier, J. G., Laser oscillation of $N_{0.14}Y_{0.86}P_5O_{14}$ layers epilaxially grown on $Gd_{0.33}Y_{0.67}P_5O_{14}$ substrates, *Opt. Commun.* 20, 354 (1977).
404. Gueugnon, C. and Budin, J. P., Determination of fluorescence quantum efficiency and laser emission cross sections of neodymium crystals: application to $KNdP_4O_{12}$, *IEEE J. Quantum Electron.* QE 16, 94 (1980).
405. Chinn, S. R. and Hong, H. Y-P., CW laser action in acentric $NdAl_3(BO_3)_4$ and $KNdP_4O_{12}$, *Opt. Commun.* 15, 345 (1975).
406. Kubodera, K., Miyazawa, S., Nakano, J., and Otsuka, K., Laser performance of an epitaxially grown $KNdP_4O_{12}$ waveguide, *Opt. Commun.* 27, 345 (1978).
407. Badalyan, A. A., Sapondzhyan, S. O., Sarkisyan, D. G., and Torosyan, G. A., Mode-locked Nd:YAG laser with output at 1052, 1061, 1064, and 1074 nm, *Sov. Tech. Phys. Lett.* 11, 513 (1985).
408. DiDomenico, M., Jr., Geusic, J. E., Marcos, H. M., and Smith, R. G., Generation of ultrashort optical pulses by mode locking the YAlG:Nd laser, *Appl. Phys. Lett.* 8, 180 (1966).
409. Osterink, L. M. and Foster, J. D., A mode-locked Nd:YAG laser, *J. Appl. Phys.* 39, 4163 (1968).
410. Clobes, A. R. and Brienza, M. J., Passive mode locking of a pulsed Nd:YAG laser, *Appl. Phys. Lett.* 14, 287 (1969).
411. Dewhurst, R. J. and Jacoby, D., A mode-locked unstable Nd:YAG laser, *Opt. Commun.*, 28, 107 (1979).
412. Reali, G. C., Operation of a mode-locked Nd:YAG oscillator at 1.05 μm, *Appl. Optics* 18, 3975 (1979).
413. Kaminskii, A. A., Sobolev, B. P., Bagdasarov, Kh. S. et al., Investigation of stimulated emission from crystals with Nd^{3+} ions, *Phys. Status Solidi* 23A, K135 (1974).
414. Smith, R. G., New room temperature CW laser transition in YAlG:Nd, *IEEE J. Quantum Electron.*, QE-4, 505 (1968).

415. Dubinskii, M. A., Kazakov, B. N., and Yagudin, Sh. I., Spectroscopy and stimulated emission of Nd^{3+} ions in yttrium trifluoride crystals, *Opt. Spectrosc. (USSR)* 63, 412 (1987).
416. Tucker, A. W., Birnbaurn, M., and Fincher, C. L., Repetitive Q-switched operation of x-axis Nd:$La_2Be_2O_5$, *J. Appl. Phys.* 52, 5434 (1981).
417. Kaminskii, A. A., Kursten, G. D., and Shultze, D., A new laser ferroelectric, $Pb_5Ge_3O_{17}$-Nd^{3+}, *Sov. Phys. Dokl.* 28, 492 (1983).
418. Kaminskii, A. A., Kirsten, H. D., and Schultze, D., Stimulated emission of ferroelectric $Pb_5Ge_3O_{17}$:Nd^{3+}, *Phys. Status Solidi A* 81, K19 (1984).
419. Gualtieri, J. G. and Aucoin, T. R., Laser performance of large Nd-pentaphosphate crystals, *Appl. Phys. Lett.* 28, 189 (1976).
420. Korovkin, A. M., Morozova, L. G., Petrov, M. V., Tkachuk, A. M. and Feofilov, P. P., Spontaneous and induced emission of neodymium in the crystals of yttrium silicates and rare-earth silicates, *Digest of Technical Papers VI All-Union Conf. on Spectroscopy of Crystals*, September 21-25, 1979, Krasnodar, p. 156.
421. Kaminskii, A. A., Sarkisov, S. E., Seiranyan, K. B., and Sobolev, B. P., Stimulated emission from Nd^{3+}- ions in SrF_2-GdF_3 crystals, *Inorg. Mater. (USSR)* 9, 310 (1973).
422. Verdun, H.R. and Thomas, L.M., Nd:$CaYAlO_4$–A new crystal for solid-state lasers emitting at 1.08 μm, *Appl. Phys. Lett.* 56, 608 (1990).
423. Shen, H., Zhou, Y., Yu, G., Huang, X., Wu, C., and Ni, Y., Influences of thermal effects on high power CW outputs of b-axis Nd:YAP lasers, *Chin. Phys.* 3, 45 (1983).
424. Hanson, F., Laser-diode side-pumped Nd:$YAlO_3$ laser at 1.08 and 1.34 μm, *Opt. Lett.* 14, 674 (1989).
425. Kaminskii, A. A., Sobolev, B. R., Zhmurova, Z. R., and Sarkisov, S. E., Generation of stimulated radiation by Nd^{3+} ions in the disordered crystal of a solid solution of the system CaF_2—LuF_3, *Inorg. Mater. (USSR)* 20, 759 (1984).
426. Souriau, J. C., Romero, R., Borel, C. and Wyon, C., Room-temperature diode-pumped continuous-wave $SrY_4(SiO_4)_3O$:Yb^{3+}, Er^{3+} crystal laser at 1554 nm, *Appl. Phys. Lett.* 64, 1189 (1994).
427. Tsuiki, H., Masumoto, T., Kitazawa, K., and Fueki, K., Effect of point defects on laser oscillation properties of Nd-doped Y_2O_3, *Jpn. J. Appl. Phys.* 21, 1017 (1982).
428. Voron'ko, Yu. K., Gessen, S. B., Es'kov, N. A., Osiko, V. V., Sobol', A. A., Tlmoshechkin, M. I., Ushakov, S. N., and Tsymbal, L. I., Spectroscopic and lasing properties of calcium niobium gallium garnet activated with Cr^{3+} and Nd^{3+}, *Sov. J. Quantum Electron.* 18, 198 (1988).
429. Vanherzeele, H., Optimization of a CW mode-locked frequency-doubled Nd:LiYF4 laser, *Appl. Opt.* 27, 3608 (1988).
430. Loth, C. and Bruneau, D., Single-frequency active-passive mode-locked Nd:YLF oscillator at 1.053 μm, *Appl. Opt.* 21, 2091 (1982).
431. Schearer, L. D., Leduc, M., Vlvien, D., Lejus, A. M., and Thery, J., LNA: A new CW Nd laser tunable around 1.05 and 1.08 μm, *IEEE J. Quantum Electron.* QE-22, 713 (1986).
432. Vivien, D., Lejus, A. M., Thery, J., Collongues, R., Aubert, J. J., Moncorge, R., and Auzel, F., Observation of the continuous laser effect in $La_{0.9}Nd_{0.1}MgAl_{11}O_{19}$ aluminate single crystals (LNA) grown by the Czochralski method, *C. R. Seances Acad. Sci. Ser.* 11, 298, 195 (1984).
433. Kaminskii, A. A., Mill, B. V., Belokoneva, E. L. and Butashin, A. V., Structure refinement and laser properties of orthorhombic chromium-containing $LiNbGeO_5$ crystals, *Neorgan. Mater.* 27, 1899 (1991).

434. Kaminskii, A. A., Sobolev, B. P., Bagdasarov, Kh. S. et al., Investigation of stimulated emission of the $^4F_{3/2} \rightarrow {}^4I_{13/2}$ transition of Nd^{3+} ions in crystals, *Phys. Status Sol.* 26A, K63 (1974).
435. Kaminskii, A. A., On the possibility of investigation of the 'Stark' structure of TR^{3+} ion spectra in disordered fluoride crystal systems, *Sov. Phys.-JETP* 31, 216 (1970).
436. Kaminskii, A. A., Klevtsov, P. V., Bagdararov, Kh. S., Maier, A. A., Pavlyuk, A. A., Petrosyan, A. G. and Provotorov, M. V., New CW crystal lasers, *JETP Lett.* 16, 387 (1972).
437. Kvapil, J., Perner, B., Kvapil, J., Manek, B., Hamal, K., Koselja, M., and Kuhecek, V., Laser properties of coactivated YAP:Nd free of colour centers, *Czech. J. Phys. B* 38, 1281 (1988).
438. Kaminskii, A. A., Dsiko, V. V., and Voron'ko, Yu. K., Five-component fluoride: a new laser material, *Sov. Phys.-Crystallogr.* 13, 267 (1968).
439. Kaminskii, A. A., On the laws of crystal-field disorder of Ln^{3+} ions in insulating crystals, *Phys. Status Solidi A* 102, 389 (1987).
440. Kaminskii, A. A., Sobolev, B. P., Bagdararov, Kh. S., Tkachenko, N. L., Sarkisov, S. E. and Seiranyan, K. B., Investigation of stimulated emission from crystals with Nd^{3+} ions, *Phys. Status Solidi* (a) 23, K135 (1974).
441. Portella, M. T., Montelmacher, P., Bourdon, A., Evesque, P., Duran, J., and Boltz, J. C., Characteristics of a Nd-doped yttrium-aluminum-perovskite picosecond laser, *J. Appl. Phys.* 61, 4928 (1987).
442. Chen, L., Chen, S., and Xie, Z., A passively mode-locked Nd:YAP laser, *Laser J. (China)* 8, 4 (1981), (in Chinese).
443. Honda, T., Kuwano, T., Masumoto, T., and Shiroki, K., Laser action of pulse-pumped Nd^{3+}:$Gd_3Ga_5O_{12}$ at 1.054 μm, *J. Appl. Phys.* 51, 896 (1980).
444. Hamel, J. Cassimi, A., Abu-Safia, H., Leduc, M., and Schearer, L. D., Diode pumping of LNA lasers for helium optical pumping, *Opt. Commun.* 63 114 (1987).
445. Kaminskii, A. A. and Lomonov, V. A., Stimulated emission of $M_{1-x}Nd_xF_{2+x}$ solid solutions with the structure of fluorite. *Inorg. Mater. (USSR)* 20, 1799 (1984).
446. Kaminskii, A. A. and Lomonov, V. A., Low-threshold stimulated emission of Nd^{3+} ions in disordered SrF_2-ScF_3 crystals, *Sov. Phys. Dokl.* 30, 388 (1985).
447. Alam, M., Gooen, K. H., DiBartolo, B., Linz, A., Sharp, E., Gillespie, L. F., and Janney, G., Optical spectra and laser action of neodymium in a crystal $Ba_2MgGe_3O_{12}$, *J. Appl. Phys.* 39, 4728 (1968).
448. Horowitz, D. J., Gillespie, L. F., Miller, J. E., and Sharp, E. J., Laser action of Nd^{3+} in a crystal $Ba_2ZnGe_3O_{12}$, *J. Appl. Phys.* 43, 3527 (1972).
449. Dmitriev, V. G., Raevskii, E. V., Rubina, N. M., Rashkovieh, L. N., Silichev, O. O., and Fomichev, A. A., Simultaneous emission at the fundamental frequency and the second harmonic in an active nonlinear medium: neodymium-doped lithium metaniobate, *Sov. Tech. Phys. Lett.* 5, 590 (1979).
450. Bagdasarov, Kh. S., Dorozhkin, L. M., Ermakov, L. A., Kevorkov, A. M., Krasilov, Yu. I., Kuznetsov, N. T., Kurstev, I. I., Potemkin, A. V., Ralskaya, L. N., Tseitlin, P. A., and Shestakov, A. V., Spectroscopic and lasing properties of lanthanum neodymium magnesium hexaaluminate, *Sov. J. Quantum Electron.* 13, 1082 (1983).
451. Demehouk, M. I., Mikhailov, V. P., Gilev, A. K., Zabaznov, A. M., and Shkadarevieh, A. P., Investigation of the passive mode locking in a La-Nd-Mg hexaaluminate-doped laser, *Opt. Commun.* 55, 33 (1985).
452. Dubinskii, M. A., Khaidukov, N. M., Garipov, I. G., Naumov, A. K. and Semashko, V. V., Spectroscopy and stimulated emission of Nd^{3+} in an acentric CsY_2F_7 host, *Appl. Optics* 31, 4158 (1992).

453. Karapetyan, V. E., Korovkin, A. M., Morozova, L. G., Petrov. M. V., and Feofilov, P. P., Luminescence and stimulated emission of neodymium ions in scandium silicate single crystals, *Opt. Spectrosc. (USSR)* 49, 109 (1980).
454. Hong, H. Y-P. and Chinn, S. R., Crystal structure and fluorescence lifetime of potassium neodymium orthophosphate, $K_3Nd(PO_4)_2$, a new laser material, *Mater. Res. Bull.* 11, 421 (1976).
455. Garmash, V. M., Kaminskii, A. A., Polyakov, M. I., Sarkisov, S. E., and Filimonov, A. A., Luminescence and stimulated emission of Nd^{3+} ions in $LaMgAl_{11}O_{19}$ crystals in the $^4F_{3/2} \rightarrow {}^4I_{11/2}$ and $^4F_{3/2} \rightarrow {}^4I_{13/2}$ transitions, *Phys. Status Solidi A* 75, K111 (1983).
456. Morozov, A, M., Morozova, L. G., Fedorov, V. A., and Feofilov, P. P., Spontaneous and stimulated emission of neodymium in lead fluorophosphate crystals, *Opt. Spectrosc.*, 39, 343 (1975).
457. Johnson, L. F. and Ballman, A. A., Coherent emission from rare-earth ions in electro-optic crystals, *J. Appl. Phys.* 40, 297 (1969).
458. Blistanov, A. A., Gabgan, B. I., Denker, B.I., Ivleva, L. I., Osiko, V. V., Polozkov, N. M., and Sverchkov, Yu. E., Spectral and lasing characteristics of $CaMoO_4$:Nd^{3+} single crystals, *Sov. J. Quantum Electron.* 19, 747 (1989).
459. Arsenev, P. A., Bienert, K. E. and Sviridova, R. K., Spectral properties of neodymium ions in the lattice of $GdScO_3$ crystals, *Phys. Status Solidi* (a) 9, K103 (1972).
460. Bagdararov, Kh. S., Kaminskii, A. A., Kevorkov, A. M. and Prokhorov, A. M., Rare earth scandium-aluminum garnets with impurity TR^{3+} ions as active media for solid state lasers, *Sov. Phys. Dokl.* 19, 671 (1975).
461. Abdulsabirov, R. Yu., Dubinskii, M. A., Kazakov, B. N., Silkin, N. I., and Yagudln, Sh. I., New fluoride laser matrix, *Sov. Phys. Crystallogr.* 32, 559 (1987).
462. Kaminskii, A. A., Khadokov, N. M., Joubert, M. F., Boulin, G. and Makou, R., $CsGd_2F_7$:Nd^{3+} - a new laser crystal, *Phys. Status Solidi A* 142, K51 (1994).
463. Johnson, L. F., Optical maser characteristics of rare-earth ions, *J. Appl. Phys.* 34, 897 (1963).
464. Kaminskii, A. A., Sobolev, B. P., Bagdararov, Kh. S., Tkachenko, N. L., Sarkisov, S. E. and Seiranyan, K. B., Investigation of stimulated emission from crystals with Nd^{3+} ions, *Phys. Status Solidi* (a) 23, K135 (1974).
465. Kaminskii, A. A., Sobolev, B. P., Bagdararov, Kh. S., Kevorkov, A. M., Fedorov, P. P. and Sarkisov, S. E., Investigation of stimulated emission in the $^4F_{3/2} \rightarrow {}^4I_{13/2}$ transition of Nd^{3+} ions in crystals (VII), *Phys. Status Solidi* (a) 26, K63 (1974).
466. Tkachuk, A. M., Przhevusskii, A. K., Morozova, L. G., Poletimova, A. V., Petrov, M. V., and Korovkin, A. M., Nd^{3+} optical centers in lutetium, yttrium, and scandium silicate crystals and their spontaneous and stimulated emission, *Opt. Spectrosc. (USSR)* 60, 176 (1986).
467. Kaminskii, A. A., Mill', B. V., Butashin, A. V., Belokoneva, E. L., and Kurbanov, K., Germanates with $NdAlGe_2O_7$-type structure, *Phys. Status Solidi A* 103, 575 1987.
468. Garashina, L. S., Kaminskii, A. A., Li, L., and Sobolev, B. P., Laser based on SrF_2-YF_3:Nd^{3+} cubic cryslals, *Sov. Phys.-Crystallogr.* 14, 799 (1970).
469. Antipenko, B. M., Mak, A. A., Sinitsin, B. E. and Uvarova, T. V., Laser converter on the base of $BaYb_2F_8$:Ho^{3+}, *Digest of Technical Papers VI All-Union Conf. on Spectroscopy of Crystals*, September 21-25 (1979), Krasnodar, p. 30.
470. Kaminskii, A. A., Mill', B. V., Kurbanov, I. and Butashin, A. V., Concentration quenching of luminescence and stimulated emission of Nd^{3+} in a monoclinic $NdGaGe_2O_7$ crystal, *Inorg. Mater. (USSR)* 23, 530 (1987).

471. Cordova-Plaza, A., Fan, T. Y., DiRonnet, M. J. F., Byer, R. L., and Shaw, H. J., Nd:MgO:$LiNbO_3$ continuous-wave laser pumped by a laser diode, *Opt. Lett.* 13, 209 (1988).
472. Kaminskii, A. A., Mill', B. V., Butashin, A. V., and Dosmagambetov, E. S., Stimulated emission of $GdGaGe_2O_7$-Nd^{3+} crystals. *Inorg. Mater. (USSR)* 23, 626 (1987).
473. Kevorkov, A. M., Kaminskii, A. A., Bagdararov, Kh. S., Tevosyan, T. A. and Sarkisov, S. E., Spectroscopic properties of crystals of $SrAl_4O_7$ - Nd^{3+}, *Inorg. Mater. (USSR)* 9, 1637 (1973).
474. Duncan, R. C., Continuous room-temperature Nd^{3+}:$CaMoO_4$ laser, *J. Appl. Phys.* 36, 874 (1965).
475. Kaminskii, A. A., Pavlyuk, A. A., Kurbanov, K., Ivannikova, N. V., and Polyakova, L. A., Growth of $CsLa(WO_4)_2$-Nd^{3+} crystals and study of their spectral-generation properties, *Inorg. Mater. (USSR)* 24, 1144 (1988).
476. Peterson, G. E. and Bridenbaugh, P. M., Laser oscillation at 1.06 μ in the series $Na_{0.5}Gd_{0.5-x}Nd_xWO_4$, *Appl. Phys. Lett.* 4, 173 (1964).
477. Kaminskii, A. A., Bagdasarov, Kh. S., Bogomolova, G. A. et al., Luminescence and stimulated emission ofNd^{3+} ions in $Gd_3Sc_2Ga_5O_{12}$ crystals, *Phys. Status Solidi* 34a, K109 (1976).
478. Kaminskii, A. A., Sarkisov, S. E., and Bagdasarov, Kh. S., Study of stimulated emission from Nd^{3+} ions in crystals at the $^4F_{3/2} \rightarrow {}^4I_{13/2}$ transition. II, *Inorg. Mater. (USSR)* 9, 457 (1973).
479. Flournoy, P. A. and Brixner, L. H., Laser characteristics of niobium compensated $CaMoO_4$ and $SrMoO_4$, *J. Electrochem. Soc.* 112, 779 (1965).
480. Cordova-Plaza, A., Digonnet, M. J. F., and Shaw, H. J., Miniature CW and active internally Q-switched Nd:MgO:$LiNbO_3$ lasers, *IEEE J. Quantum Electron.* QE-23, 262 (1987).
481. Demchuk, M. I., Mikhallov, V. P., Gilev, A. K., Ishchenko, A. A., Kudinova, M. A., Slominskii, Yu. L., and Tolmachev, A. I., Optimization of the passive mode-locking state in an yttrium aluminate laser, *J. Appl. Spectrosc. (USSR)* (Engl. Transl.) 42, 477 (1985).
482. Kaminskii, A. A., Mill', B. V., Tamazyan, S. A., Sarkisov, S. E., and Kurbanov, K., Luminescence and stimulated emission from an acentric crystal of $Ca_3Ga_4O_9$-Nd^{3+}, *Inorg. Mater. (USSR)* 21, 1733 (1986).
483. Johnson, L. F., Boyd, G. D., Nassau, K., and Soden, R. R., Continuous operation of a solid-state optical maser, *Phys. Rev.* 126, 1406 (1962).
484. Johnson, L. F., Characteristics of the $CaWO_4$:Nd^{+3} optical maser, in *Quantum Electronics Proceedings of the Third International Congress,* Grivet, P. and Bloembergen, N., Eds., Columbia University Press, New York (1964), p. 1021.
485. Kaminskii, A. A., Korniyenko, L. S., Maksimova, G. V., Osiko, V. V., Prokhorov, A. M., and Shipulo, G. P., CW $CaWO_4$:Nd^{3+} laser operating at room temperature, *Sov. Phys.-JETP* 22, 22 (1966).
486. Kaminskii, A. A., Spectral composition of stimulated emission from a $CaWO_4$:Nd^{3+} laser, *Inorg. Mater. (USSR)* 6, 347 (1970).
487. Hopkins, R. H., Steinbruegge, K. B., Melamed, N. T. et al., Technical RPT. AFAL-TR-69-239, Air Force Avionics Laboratory (1969).
488. Johnson, L. F. and Guggenheim, H. J., Phonon-terminated coherent emission from V^{2+}- ions in MgF_2, *J. Appl. Phys.* 38, 4837 (1967).
489. Mougel, F., Aka, G., Kahn-Harari, A., Hubert, H., Benitez, J. M. and Vivien, D., Infrared laser performance and self-frequency doubling of Nd^{3+}:$Ca_4GdO(BO_3)_3$ (Nd:GdCOB), *Opt. Mater.* 8, 161 (1997).

490. Bagdasarov, Kh. S., Bogomolova, G. A., Gritsenko, M. M., Kaminskii, A. A., Kevorkov, A. M., Prokhorov, A. M., and Sarkisov, S. E., Spectroscopy of stimulated emission from $Gd_3Ga_5O_{12}:Nd^{3+}$ crystals, *Sov. Phys.-Dokl.* 19, 353 (1974).
491. Karninskii, A. A., *Laser Crystals,* Springer-Verlag, New York (1980).
492. Geusic, J. E., Marcos, H. M., and Van Uitert, L. G., Laser oscillations in Nd-doped yttrium aluminum, yttrium gallium and gadolinium garnets, *Appl. Phys. Lett.* 4, 182 (1964).
493. Danilov, A. A., Zharikov, E. V., Zavartsev, Yu. D., Noginov, M. A., Nikol'skii, M. Yu., Ostroumov, V. G., Smirnov, V. A., Studenikin, P. A., and Shcherbakov, I. A., YSGG:Cr^{3+}:Nd^{3+} as a new effective medium for pulsed solid-state lasers, *Sov. J. Quantum Electron.* 17, 1048 (1987).
494. Tucker, A.W. and Birnbaum, M., Energy levels and laser action in Nd:$CaY_2Mg_2Ge_3O_{12}$ (CAMGAR), Proc. Internat. Conf. Laser '78, Orlando (STS Press, McLean, VA, 1979), p 168.
495. Bagdarasov, Kh. S. and Kaminskii, A. A., $YAlO_3$ with TR^{3+} ion impurity as an active laser medium, *JETP Lett.* 9, 303 (1969).
496. Kaminskii, A. A., Mayer, A. A., Provotorov, M. V., and Sarkisov, S. E., Investigation of stimulated emission from $LiLa(MoO_4)_2:Nd^{3+}$ crystal laser, *Phys. Status Solidi* 17a, K115 (1973).
497. Steinbruegge, K. B., Hennigsen, R. H., Hopkins, R., Mazelsky, R., Melamed, N. T., Riedd, E. P., and Roland, G. W., Laser properties of Nd^{3+} and Ho^{3+} doped crystals with the apatite structure, *Appl. Optics* 11, 999 (1972).
498. Kaminskii, A. A., Klevtsov, P. V., Li, L. et al., Stimulated emission of radiation by crystals of $KLa(MoO_4)_2$ with Nd^{3+} ions, *Inorg. Mater. (USSR)* 9, 1824 (1973).
499. Faure, N., Borel, C., Templier, R., Couchaud, M., Calvat, C. and Wyon, C., Optical properties and laser performance of neodymium doped fluoroapatites $Sr_xCa_{5-x}(PO_4)_3F$ (x= 0,1,2,3,4 and 5), *Opt. Mater.* 6, 293 (1996).
500. Kariss, Ya. E., Tolstoy, M. N., and Feofilov, P. P., Stimulated emission from neodymium in lead molybdate single crystals, *Opt. Spectrosc.* 18, 99 (1965).
501. Kaminskii, A. A., Kozeeva, L. P., and Pavlyuk, A. A., Stimulated emission of Er^{3+} and Ho^{3+} ions in $KLa(Mo_4)_2$ crystals *Phys. Status Solidi A* 83, K65 (1984).
502. Zhang, L., Liu, L., Liu, H., and Lin, C., Growth and investigation of substituted gadolinium gallium garnet laser crystals, *J. Cryst. Growth* 80, 257 (1987).
503. Xun, D., Zhu, H., Jin, F., Liu, H., and Zhang, L., Growth and measurement of GGG(Ca Mg,Zr):(Nd Cr) laser crystals, *Chin. Phys. Lasers* 13, 820 (1986).
504. Bagdasarov, Kh. S., Kaminskii, A. A., Lapsker, Ya. Ye., and Sobolev, B. P., Neodymium-doped α-gagarinite laser, *JETP Lett.* 5, 175 (1967).
505. Bagdasarov, Kh. S., Bogomolova, C. A., Kaminskii, A. A. et al., Study of the stimulated emission of $Lu_3Al_5O_{12}$ crystals containing Nd^{3+} ions at the transitions $^4F_{3/2} \rightarrow {}^4I_{11/2}$ and $^4F_{3/2} \rightarrow {}^4I_{13/2}$, *Sov. Phys.-Dokl.* 19, 584 (1975).
506. Kaminskii, A. A., Mill', B. V., Bulashin, A. V., Sarkisov, S. E., and Nikol'skaya, O. K., Two channels of stimulated emission of Nd^{3+} ions in $Ca_3(Nb, Ga)_2Ga_3O_{12}$ crystal, *Inorg. Mater. (USSR)* 21, 1834 (1985).
507. Seas, A., Petricevic, V. and Alfano, R. R., Continuous-wave mode-locked operation of a chromium-doped forsterite laser, *Optics Lett.* 16, 1668 (1991).
508. Deka, C., Chai, B. H. T., Shimony, Y., Zhang, X. X., Munin, E. and Bass, M., Laser performance of Cr^{4+}:Y_2SiO_5, *Appl. Phys. Lett.* 61, 2141 (1992).
509. Merkle, L. D., Pinto, A., Verdun, H. R. and McIntosh, B., Laser action from Mn^{5+} in $Ba_3(VO_4)_2$, *Appl. Phys. Lett.* 61, 2386 (1992).

510. Kevorkov, A. M., Kaminskii, A. A., Bsgdasarov, Kh. S., Tevosyan, T. A., and Sarkisov, S. E., Spectroscopic properties of $CaAl_4O_7$:Nd^{3+} crystals, *Inorg. Mater. (USSR)* 9, 146 (1973).
511. Sharp, E. J., Mitler, J. E., Horowitz, D. J. et al., Optical spectra and laser action in Nd^{3+}-doped $CaY_2M_2Ge_3O_{12}$, *J. Appl. Phys.* 45, 4974 (1974).
512. Springer, J., Clausen, R., Huber, G., Petermann, K. and Mateika, D., New Nd-doped perovskite for diode-pumped solid-state lasers, *OSA Proc. Adv. Solid State Lasers*, Dube, G. and Chase, L. (eds) 10, 346 (1991).
513. Ryba-Romanowski, W., Jezowska-Trzeblalowska, B., Piekarczyk, W. and Berkowski, M., Optical properties and lasing of $BaLaGa_3O_7$ single crystals doped with neodymium, *J. Phys. Chem. Solids* 49, 199 (1988).
514. Carrig, T. J. and Pollock, C. R., Tunable, cw operation of a multiwatt forsterite laser, *Optics Lett.* 16, 1662 (1991).
515. Zhang, X. X., Hong, P., Loutts, G. B., Lefaucheur, J., Bass, M. and Chai, B. H. T., Efficient laser performance of Nd^{3+}:$Sr_5(PO_4)_3F$ at 1.059 and 1.328 μm, *Appl. Phys. Lett.* 64, 3205 (1994).
516. Zhou, W.-L., Zhang, X. X. and Chai, B. T. H., Laser oscillation at 1059 nm of new laser crystal: Nd^{3+}-doped $NaY(WO_4)_2$, *Proceedings Advanced Solid State Lasers* (1997), to be published.
517. Jansen, M., Alfrey, A., Stafsudd, O. M., Dunn, B., Yang, D. L., and Farrington, G. C., Nd^{3+} beta alumina platelet laser, *Opt. Lett.* 10, 119 (1984).
518. Kaminskii, A. A., Sarkisov, S. E. and Sobolev, B. P., unpublished.
519. Kaminskii, A. A., Mill', B. V., and Butashin, A. V., Growth and stimulated emission spectroscopy of $Ca_3Ga_2Ge_3O_{12}$-Nd^{3+} garnet crystals, *Phys. Status Solidi A*: 78, 723 (1983).
520. Michel, J.-C., Morin, D., and Auzel, F., Intensite' de fluorescence et duree de vie du niveau $^4F_{3/2}$ de Nd^{3+} dans une chlorapatite fortement dopee. Comparison avec d'autres matériaux, *C. R. Acad. Sci. Ser. B* 281,445 (1975).
521. Budin, J.-P., Michel, J.-C., and Auzel, F., Oscillator strengths and laser effect in $Na_2Nd_2Pb_6(PO_4)_6Cl_2$ (chloroapatite), a new high-Nd-concentration laser material, *J. Appl. Phys.* 50, 641 (1979).
522. Bagdararov, Kh. S., Bogomolova, G. A., Grotsenko, M. M., Kaminskii, A. A., Kevorkov, A. M., Prokhorov, A. M. and Sarkisov, S. E., Spectroscopy of the stimulated emission of $Gd_3Al_5O_{12}$-Nd^{3+} crystals, *Sov. Phys. Dokl.* 19, 353 (1974).
523. Kaminskii, A. A., Osiko, V. V., and Voron'ko, Yu. K., Mixed systems on the basis of fluorides as new laser materials for quantum electronics. The optical and emission parameters, *Phys. Status Solidi*, 21, 17 (1967).
524. Kiss, Z. J. and Duncan, R. C., Optical maser action in CaF_2:Tm^{2+}, *Proc. IRE* 50, 1532 (1962).
525. Duncan, R. C. and Kiss, Z. J., Continuously operating CaF_2:Tm^{2+} Optical Maser, *Appl. Phys. Lett.* 3, 23 (1963).
526. Kaminskii, A. A., Mill', B. V., Bebkoneva, E. L., Tomazyan, S. A., Bubshin, A. V., Kurbanov, K., and Dosmagambetov, E. S., Germanates of the $NdAlCe_2O_7$ structure: synthesis, structure of La-$GaGe_2O_7$, absorption-luminscence properties, and stimulated emission with Nd^{3+} activator, *Inorg. Mater.* 22, 1763 (1986).
527. Brandle, C. D. and Vanderleeden, J. C., Growth, optical properties and CW laser action of neodymium-doped gadolinium scandium aluminum garnet, *IEEE J. Quantum Electron.* QE-10, 67 (1974).
528. Kaminskii, A. A., Pavlyuk, A. A., Chan, Ng. et al., 3 μm stimulated emission by Ho^{3+} ions in $KY(WO_4)_2$ crystals at 300 K, *Sov. Phys.-Dokl.* 24, 201 (1979).

529. Bagdararov, Kh. S., Bogomolova, G. A., Kaminskii, A. A., Kevorkov, A. M., Li, L., Prokhorov, A. M. and Sarkisov, S. E., Study of the stimulated emission of $Lu_3Al_5O_{12}$ crystals containing Nd^{3+} ions at the transitions $^4F_{3/2} \rightarrow {}^4I_{11/2}$ and $^4F_{3/2} \rightarrow {}^4I_{13/2}$, *Sov. Phys. Dokl.* 19, 584 (1975).
530. Bagdararov, Kh. S., Kaminskii, A. A. and Sobolev, B. P., Laser action in cubic $5NaF{\bullet}9YF_3$–Nd^{3+} crystals, *Sov. Phys.-Crystallogr.* 13, 900 (1969).
531. Catfey, D. P., Utano, R. A., and Allik, T. H., Diode array side-pumped neodymium-doped gadolinium scandium gallium garnet rod and slab lasers, *Appl. Phys. Lett.* 56, 808 (1990).
532. Kaminskii, A. A., Kolodnyy, G. Ya., and Sergeyeva, N. I., CW $NaLa(MoO_4)_2$:Nd^{3+} crystal laser operating at 300 K, *J. Appl. Spectrosc. (USSR)* 9, 1275 (1968).
533. Morozov, A. M., Tolstoy, M. N., Feofilov, P. P., and Shapovalov, V. N., Fluorescence and stimulated emission in neodymium in lanthanum molybdate-sodium crystals, *Opt. Spectrosc. (USSR)* 22, 224 (1967).
534. Zverev, C. M. and Kolodnyy, G. Ya., Stimulated emission and spectroscopic studies of neodymium-doped binary lanthanum molybdate-sodium single crystals, *Sov. Phys.-JETP* 25, 217 (1967).
535. Johnson, L. F., Guggenheim, H. J., and Thomas, R. A., Phonon-terminated optical masers, *Phys. Rev.* 149, 179 (1966).
536. Zharikov, E. V., Zhitnyuk, V. A., Zverev, G. M., Kalitin, S. P., Kuratev, I. I., Leptev, V. V., Onishchenko, A. M., Osiko, V. V., Pashkov, V. A., Pimenov, A. S., Prokhorov, A. M., Smirnov, V. A., Stel'makh, M. F., Shestakov, A. V., and Shcherbakov, I. A., Active media for high-efficiency neodymium lasers with nonselective pumping, *Sov. J. Quantum Electron.* 12, 1652 (1982).
537. Kaminskii, A. A., Sarkisov, S. E. and Bagdararov, Kh. S., Stimulated emission by Nd^{3+} ions in crystals, due to the $^4F_{3/2} \rightarrow {}^4I_{13/2}$ transition, *Inorg. Mater. (USSR)* 9, 457 (1973).
538. Kaminskii, A. A. and Sarkisov, S. E., Study of stimulated emission from Nd^{3+} ions in crystals emitting at the $^4F_{3/2} \rightarrow {}^4I_{13/2}$ transition. I., *Inorg. Mater. (USSR)* 9, 453 (1973).
539. Kaminskii, A. A., Orthorhombic $NaREGeO_4$ crystals with Nd^{3+} ions; structure and formation. Luminescence properties and stimulated emission, *Appl. Phys. A.* 46, 173 (1988).
540. Kaminskii, A. A., Timofeevs, V. A., Bykov, A. B., and Agsmajyan, N. R., Low-threshold stimulated emission by Nd^{3+} ions in $NaLuGeO_4$. *Sov. Phys. Dokl.* 29, 220 (1984).
541. Kaminskii, A. A. and Lomonov, V. A., Stimulated emission of $M_{1-x}Nd_xF_{2+x}$ solid solutions with the structure of fluorite. *Inorg. Mater. (USSR)* 20, 1799 (1984).
542. Pruss, D., Huber, G., Beimowskl, A., Laptev, V. V., Shcherbakov, I. A., and Zharikov, Y. V., Efficient Cr^{3+} sensitized Nd^{3+}:GdScGa-garnet laser at 1.06 μm, *Appl. Phys. B* 28, 355 1982.
543. Moulton, P. F. and Mooradian, A., Broadly tunable CW operation of Ni:MgF_2 and Co:MgF_2 lasers, *Appl. Phys. Lett.* 35, 838 (1979).
544. Kaminskii, A. A., Belokoneva, E. L., Mill', B. V., Tamazyan, S. A., and Kurbanov, K., Crystal structure, spectral-luminescence properties and stimulated radiation of gallium gehlenite, *Inorg. Mater. (USSR)* 22, 993 (1986).
545. Kaminskii, A. A., Belokoneva, E. L., Mill', B. V., Sarkisov, S. E., and Kurbanov, K., Crystal structure absorption luminescence properties, and stimulated emission of Ga gehlenite ($Ca_{2-x}Nd_xGa_{2+x}Si_{1-x}O_7$), *Phys. Status Solidi A* 97, 279 (1986).
546. Kaminskii, A. A., Laws of crystal-field disorderness of Ln^{3+} ions in insulating laser crystals, *J. Phys. (Paris)*, Colloq. 12, C7–359 (1987).

547. Kaminskii, A. A., Mayer, A. A., Nikonova, N. S., Provotorov, M. V., and Sarkisov, S. E., Stimulated emission from the new $LiGd(MoO_4)_2:Nd^{3+}$ crystal laser, *Phys. Status Solidi* 12a, K73 (1972).
548. Antipenko, B. M., Mak, A. A., Nikolaev, B. V., Raba, O. B., Seiranyan, K. B., and Uvarova, T. V., Analysis of laser situations in $BaYb_2F_8:Er^{3+}$ with stepwise pumping sehemes, *Opt. Spectrsoc. (USSR)* 56, 296 (1984).
549. Kaminskii, A. A., Sarkisov, S. E. and Li, L., Investigation of stimulated emission in the $^4F_{3/2} \rightarrow {}^4I_{13/2}$ transition of Nd^{3+} ions in crystals (III), *Phys. Status Solidi* (a) 15, K141 (1973).
550. Zagumennyi, A. I., Ostroumov, V. G., Shcherbakov, I. A., Jensen, T., Meyen, J. P. and Huber, G., The Nd:$GdVO_4$ crystal: a new material for diode-pumped lasers, *Sov. J. Quantum Electron.* 22, 1071 (1992).
551. Aivea, A. F.,Westinghouse, unpublished.
552. Petricevic, V., Gayen, S. K. and Alfano, R. R., Continuous-wave laser operation of chromium-doped forsterite, *Optics Lett.* 14, 612 (1989).
553. Lu, B., Wang, J., Pan, I., and Jiang, M., Excited emission and self-frequency-doubling effect of $Nd_xY_{1-x}Al_3(BO_3)_4$ crystal, *Chin. Phys. Lett. (China)* 3, 413 (1986).
554. Hayakawa, H., Maeda, K., Ishlkawa, T., Yokoyama, T., and Yoshimasa, F., High average power Nd:$Gd_3Ga_5O_{12}$ slab laser, *Jpn. J. Appl. Phys.* 26, L1623 (1987).
555. Zhang, L., Lin, C., Liu, H., Liu, L., Zhu, H., and Lin, X., Investigation of growth and laser properties of CGG:(Nd,Cr) single crystals, *Chin. J. Phys.* (Engl. Transl.) 5, 136 (1985).
556. Caird, J. A., Shinn, M. D., Kirchoff, T. A., Smith, L. K., and Wilder, R. E., Measurements of losses and lasing efficiency in GSGG:Cr, Nd and YAG:Nd laser rods, *Appl. Opt.* 25, 4294 (1986).
557. Zharikov, E. V., Il'lchev, N. N., Laptev, V. V., Malyutin, A. A., Ostroumov, V. G., Pashlnin, P. P., and Shcherbakov, I. A., Sensitization of neodymium ion luminescence by chromium ions in a $Gd_3Ga_5O_{12}$ crystal, *Sov. J. Quantum Electron.* 12, 338 (1982).
558. Behrens, E.G., Jani, M.G., Powell, R.C., Verdon, H.R. and Pinto, A., Lasing properties of chromium-aluminum-doped forsterite pumped with an alexandrite laser, *IEEE J. Quantum Electron.* 27, 2042 (1991).
559. Allik, T., Merkle, L. D., Utano, R. A., Chai, B.H.T., Lefaucheur, J.-L. V., Voss, H. and Dixon, G, J., Crystal growth, spectroscopy, and laser performance of $Nd^{3+}:KYF_4$, *J. Opt. Soc. Am. B* 10, 633 (1993).
560. Kubodera, K. and Otsuka, K., Efficient $LiNdP_4O_{12}$ lasers pumped with a laser diode, *Appl. Opt.* 18, 3882 (1979).
561. Kubodera, K. and Noda, J., Pure single-mode $LiNdP_4O_{12}$ solid-state laser transmitter for 1.3-μm fiber-optic communications, *Appl. Opt.* 21, 3466 (1982).
562. Berenberg, V. A., Ivanov, A. O., Krutova, L. I., Mochalov, I. V., and Terpugov, V. S., Spectral-luminescent characteristics and stimulated emission of the Nd^{3+} ion in $Gd_{2-x}Nd_x(WO_4)_3$ crystals, *Opt. Spectrosc. (USSR)* 57, 274 (1984).
563. Aleksandrov, V. I., Voron'ko, Yu. K., Mikhalevich, V. G. et al., Spectroscopic properties and emission of Nd^{3+} in ZnO_2 and HfO_2 crystals, *Sov. Phys.-Dokl.* 16, 657 (1972).
564. Aleksandrov, V. I., Kaminskii, A. A., Maksimova, C. V. et al., Stimulated radiation of Nd^{3+} ions in crystals for the $^4F_{3/2} \rightarrow {}^4I_{13/2}$ transition, *Sov. Phys.-Dokl.* 18, 495 (1974).
565. Zharikov, E. V., Zabaznov, A. M., Prokhorov, A. M., Shkadarevich, A. P., and Shcherbakov, 1. A., Use of GSGG:Cr:Nd crystals with photochromic centers as active elements in solid lasers, *Sov. J. Quantum Electron.* 16, 1552 (1986).

566. Telle, H. R., Tunable CW laser oscillation of NdP_5O_{14} at 1.3 μm, *Appl. Phys. B* 35 195 (1984).
567. Avanesov, A. G., Denker, B. I., Galagan, B. I., Osiko, V. V., Shestakov, A. V. and Sverchkov, S. E., Room-temperature stimulated emission from chromium (IV)-activated yttrium orthosilicate, *Quantum Electron.* 24, 198 (1994).
568. Borchardl, H. J. and Bierstedl, P. E., $Gd_2(MoO_4)_3$: A ferro-electric laser host. *Appl. Phys. Lett.* 8, 50 (1966).
569. Kaminskii, A. A., Laser and spectroscopic properties of activated ferroelectrics, *Sov. Phys.-Crystallogr.* 17, 194 (1972).
570. Bethea, C. G., Megawatt power at 1.318 μ in Nd^{3+}:YAG and simultaneous oscillation at both 1.06 and 1.318 μ, *IEEE J. Quantum Electron.* QE-9, 254 (1973).
571. Kaminskii, A. A., Mill, B. V., and Butashin, A. V., New possibilities for exciting stimulated emission in inorganic crystalline materials with the garnet structure, *Inorg. Mater. (USSR)* 19, 1808 (1983).
572. Kück, S., Petermann, K. and Huber, G., Near infrared Cr^{4+}:$Y_3Sc_xAl_{5-x}O_{12}$ lasers, *OSA Proc. Adv. Solid State Lasers*, Fan, T. Y. and Chai, B. H. T. (eds) 20, 180 (1995).
573. Zhang, X. X., Hong, P., Bass, M. and Chai, B. H. T., Multisite nature and efficient lasing at 1041 and 1302 nm in Nd^{3+} doped potassium yttrium fluoride, *Appl. Phys. Lett.* 66, 926 (1995).
574. Blatte, M., Danielmeyer, H. G., and Ulrich R., Energy transfer and the complete level system of NdUP, *Appl. Phys.* 1, 275 (1973).
575. Choy, M. M., Zwicker, W. K., and Chinn, S. R., Emission cross section and flashlamp-excited NdP_5O_{14} laser at 1.32 μm, *Appl. Phys. Lett.* 34, 387 (1979).
576. Saruwatari, M., Otsuka, K., Miyazawa, S., Yamada, T. and Kimura, T., Fluorescence and oscillation characteristics of $LiNdP_4O_{12}$ lasers at 1.317 μm, *IEEE J. Quantum Electron.* QE-13, 836 (1977).
577. Kaminskii, A. G., Timofeevs, V. A., Bykov, A. B., and Sarkisov, S. E., Luminescence and stimulated emission in tne $^4F_{3/2} \rightarrow {}^4I_{11/2}$ and $^4F_{3/2} \rightarrow {}^4I_{13/2}$ channels of Nd^{3+} ions in orthorhombic $NaYGeO_4$ crystals, *Phys. Status Solidi A*: 83, K165 (1984).
578. Morris, P. J., Lüthy, W., Weber, H. P., Zavartsev, Yu. D., Studenikin, P. A., Shcherbakov, I. and Zaguminyi, A. I., Laser operation and spectroscopy of Tm:Ho:$GdVO_4$, *Opt. Commun.* 111, 493 (1994).
579. Beck, R. and Gurs, K., Ho laser with 50-W output and 6.5% slope efficiency, *J. Appl. Phys.* 46, 5224 (1975).
580. Viana, B., Saber, D., Lejus, A. M., Vivien, D., Borel, C., Romero, R. and Wyon, C., Nd^{3+}:$Ca_2Al_2SiO_7$ a new solid-state laser material for diode pumping, in *Advanced Solid-State Lasers*, Pinto, A. A. and Fan, T. Y., Eds., Proceedings Vol. 15, Optical Society of America, Washington, DC (1993), p. 244.
581. Zharikov, E. V., Il'ichev, N. N., Laptev, V. V., Malyutin, A. A., Ostroumov, V. G., Pashinin, P. P., Pimenov, A. S., Smirnov, V. A., and Shcherhakov, I. A., Spectral, luminescence, and lasing properties of gadolinium scandium gallium garnet crystals activated with neodymium and chromium ions, *Sov. J. Quantum Electron.* 13, 82 (1983).
582. Brauch, U. and Dürr, U., Vibronic laser action of V^{2+}:$CsCaF_3$, *Opt. Commun.* 55, 35 (1985).
583. Reed, E., A flashlamp-pumped, Q-switched Cr:Nd:GSGG laser, *IEEE J. Quantum Electron.* QE-21, 1625 (1985).
584. Balakireva, T. P., Briskin, Ch. M., Vakulyuk, V. V., Vasil'ev, E. V., Zolin, V. F., Maier, A. A., Markushev, V. M., Murashov, V. A., and Provotorov, M. V., Luminescence and stimulated emission from $BaGd_{2-x}Nd_x(MoO_4)_4$ single crystals, *Sov. J. Quantum Electron.* 11, 398 (1981).

585. Eilers, H., Dennis, W. M., Yen, W. M., Kück, S., Peterman, K., Huber, G. and Jia, W., Performance of a Cr:YAG laser, *IEEE J. Quantum Electron.* 30, 2925 (1994).
586. Steinbruegge, K. B., High average power characteristics of CaLaSOAP:Nd laser materials, in Digest of Technical Papers CLEA 1973 IEEE/OSA, Washington, DC (1973), p. 49.
587. Antipenko, B. M., Mak, A. A., Sinitsyn, B. V., Raba, O. B., and Uvurova, T. V., New excitation schemes for laser transitions, *Sov. Phys. Tech. Phys.* 27, 333 (1982).
588. Kaminskii, A. A. and Li, L., Spectroscopic and laser studies on crystalline compounds in the system CaO-Nb_2O_5, $Ca(NbO_3)_2$ -Nd^{3+} crystals, *Inorg. Mater. (USSR)* 6, 254 (1970).
589. French, P. M. W., Rizvi, N. H., Taylor, J. R. and Shestakov, A. V., Continuous-wave mode-locked Cr^{4+}:YAG laser, *Optics. Lett.* 18, 39 (1992).
590. Steinbruegge, K. B. and Baldwin, G. D., Evaluation of CaLa SOAP:Nd for high-power flash-pumped Q-switched lasers, *Appl. Phys. Lett.* 25, 220 (1974).
591. Geuie, J. E., Marcos, H. M., and Van Uitert, L. G., Laser oscillations in Nd-doped yttrium aluminum, yttrium gallium and gadolinium garnets, *Appl. Phys. Lett.* 4, 182 (1964).
592. Kiss, Z. I. and Duncan, R. C., Cross-pumped Cr^{3+}-Nd^{3+}:YAG laser system, *Appl. Phys. Lett.* 5, 200 (1964).
593. Kaminskii, A. A., Osiko, V. V., Sarkisov, S. E., Timoshechkin, M. I., Zhekov, E. V., Bohm, J., Reiche, P. and Schulzte, D., Growth, spectroscopic investigations, and some new stimulated emission data on $Gd_3Al_5O_{12}$-Nd^{3+} single crystals, *Phys. Status Solidi* (a) 49, 305 (1978).
593. Stone, I. and Burrus, C. A., Nd:Y_2O_3-single-crystal fiber laser: room-temperature CW operation at 1.07- and 1.35-μm wavelength, *J. Appl. Phys.* 49, 2281 (1978).
594. DeSerno, U., Röss, D. and Zeidler, G., Quasicontinuous giant pulse emission of $^4F_{3/2} \rightarrow {}^4I_{13/2}$ transition at 1.32 μm in YAG-Nd^{3+}, *Phys. Lett.* A 28, 422 (1968).
595. Kaminskii, A. A., Koptsik, V. A., Maskaek, Yu. A. et al., Stimulated emission from Nd^{3+} ions in ferroelectric $Ba_2NaNb_5O_{15}$ crystals (bananas), *Phys. Status Solidi* 28a, K5 (1975).
596. Babushkin, A. V., Vorob'ev, N. S., Zharakov, E. V., Kalitiin, S. P., Osiko, V. V., Prokhorov, A. N., Serdyuchenko, Yu. N., Shchelev, M. Ya., and Shcherbakov, I. A., Picosecond laser made of gadolinium scandium gallium garnet crystal doped with Cr and Nd, *Sov. J. Quantum Electron.* 16, 428 (1986).
597. Voron'ko, Yu. K., Maksimova, G. V., Mikhalevieh, V. G., 0siko, V. V., Sobol', A. A., Timosheekin, M. I., and Shipulo, G. P., Spectroscopic properties of and stimulated emission from yttrium-lutecium-aluminum garnet crystals, *Opt. Spectrosc.* 33, 376 (1972).
598. Kaminskii, A. A. and Li, L., Spectroscopic and stimulated emission studies of crystal compounds in a CaO-Nb_2O_5 system, $Ca(NbO_3)_2$:Nd^{3+} crystals, *Inorg. Mater. (USSR)* 6, 254 (1970).
599. Kaminskii, A. A., Karlov, N. V., Sarkisov, S. E., Stelmakh, O. M., and Tukish, V. E., Precision measurement of the stimulated emission wavelength and continuous tuning of $YAlO_3$:Nd^{3+} laser radiation due to $^4F_{3/2} \rightarrow {}^4I_{13/2}$ transition, *Sov. J. Quantum Electron.* 6, 1371 (1976).
600. Bagdasarov, Kh. S., Gritsenko, M. M., Zubkova, F. M., Kaminskii, A. A., Kevorkov, A. M., and Li, L., CW $Ca(NbO_3)_2$:Nd^{3+} crystal laser, *Sov. Phys.-Crystallogr.* 15, 323 (1970).
601. Krühker, W. W. and Plättner, R. D., Laser emission of (Nd,Y)-pentaphosphate at 1.32 μm, *Optics Commun.* 28, 217 (1979).

602. Kaminskii, A. A. and Sarkisov, S. E., Stimulated emission by Nd^{3+} ions in crystals, due to the $^4F_{3/2} \rightarrow {}^4I_{13/2}$ transition, *Inorg. Mater. (USSR)* 9, 453 (1973).
603. Miller, J. E., Sharp, F. J., and Horowitz, D. J., Optical spectra and laser action of neodymium in a crystal $BaO_{0.25}Mg_{2.75}Y_4Ge_3O_{12}$, *J. Appl. Phys.* 43, 462 (1972).
604. Kaminskii, A. A., Timoreeva, V. A., Agamalyan, N. R., and Bykov, A. B., Infrared laser radiation from $NaCdGeO_4$-Nd^{3+} crystals growth from solution in a melt, *Sov. Phys. Crystallogr.* 27, 316 (1982).
605. Kaminskii, A. A., Timoreeva, V. A., Agamalyan, N. R., and Bykov, A. B., Stimulated emission by Nd^{3+} ions in $NaGdGeO_4$ by the $^4F_{3/2} \rightarrow {}^4I_{11/2}$ and $^4F_{3/2} \rightarrow {}^4I_{13/2}$ transitions at 300 K, *Inorg. Mater. (USSR)* 17, 1703 (1981).
606. Aleksandrov, V. I., Kaminskii, A. A., Maksimova, G. V., Prokhorov, A. M., Sarkisov, S. E., Sobol', A. A., and Tatarintsev, V. M., Study of stimulated emission from Nd^{3+} ions in crystals emitting at the $^4F_{3/2} \rightarrow {}^4I_{13/2}$ transition, *Sov. Phys.-Dokl.*, 18, 495 (1974).
607. Kaminskii, A. A., Sarkisov, S. E., and Li, L., Investigation of stimulated emission in the $^4F_{3/2} \rightarrow {}^4I_{13/2}$ transition of Nd^{3+} ions in crystals (III), *Phys. Status Solidi* 15a, K141 (1973).
608. Steinbruegge, K. B., Hennigsen, R. H., Hopkins, R., Mazelsky, R., Melamed, N. T., Riedd, E. P., and Roland, G. W., Laser properties of Nd^{3+} and Ho^{3+} doped crystals with the apatite structure, *Appl. Optics* 11, 999 (1972).
609. Allik, T. H., Morrison, C. A., Gruber, J. B. and Kokta, M. R., Crystallography, spectroscopic analysis, and lasing properties of Nd^{3+}:$Y_3Sc_2Al_3O_{12}$, *Phys. Rev. B* 41, 21 (1990).
610. Kaminskii, A. A., Bodretsova, A. I., Petrosyan, A. G., and Pavlyuk, A. A., New quasi-CW pyrotechnically pumped crystal lasers, *Sov. J. Quantum Electron.* 13, 975 (1983).
611. Godina, N. A., Tolstoi, M. N. and Feofilov, P. P., Luminescence of neodymium in yttrium and lanthanum niobates and tantalates, *Opt. Spectrosc. (USSR)* 23, 411 (1967).
612. Kutovoi, S. A., Laptev, V. V., Lebedev, V. A., Matsnev, S. Yu., Pisarenko, V. F. and Chuev, Yu. M., Spectral luminescent and lasing properties of the new laser crystals lanthanum scandium borate doped with neodymium and chromium, *Zh. Prikl. Spektr.* 53, 370 (1990).
613. Doroshenko, M.E., Osiko, V.V., Sigachev, V.B. and Timoshechkin, M.I., Stimulated emission from a neodymium-doped gadolinium gallium garnet crystal due to the $^4F_{3/2} - {}^4I_{13/2}$ (λ = 1.33 μm) transition, *Sov. J. Quantum Electron.* 21, 266 (1991).
614. Voron'ko, Yu. K., Zverev, G. M., and Prokhorov, A. M., Stimulated emission from Er^{3+}- ions in CaF_2, *Sov. Phys.-JETP* 21, 1023 (1964).
615. Bagdararov, Kh. S., Kaminskii, A. A., Kevorkov, A. M., Li, L., Prokhorov, A. M., Sarkisov, S. E. and Tevosyan, T. A., Stimulated emission of Nd^{3+} ions in an $SrAl_{12}O_{19}$ crystal at the transitions $^4F_{3/2} \rightarrow {}^4I_{11/2}$ and $^4F_{3/2} \rightarrow {}^4I_{13/2}$, *Sov. Phys. Dokl.* 19, 350 (1974).
616. Chinn, S. R., Pierce, J. W. and Heckscher, H., Low-threshold transversely excited NdP_5O_{14} laser, *Appl. Optics* 15, 1444 (1976).
617. Chinn, S. R., Intracavity second-harmonic generation in a Nd pentaphosphate laser, *Appl. Phys. Lett.* 29, 176 (1976).
618. Trutna, W. R., Jr., Donald, D. K., and Nazarathy, M., Unidirectional diode-laser-pumped Nd:YAG ring laser with a small magnetic field. *Opt. Lett.* 12, 248 (1987).

619. Bakhsheyeva, G. F., Karapetyan, V. Ye., Morozov, A. M., Morozova, L. G., Tolstoy, M. N., and Feofilov, P. P., Optical constants, fluorescence and stimulated emission of neodymium-doped lanthanum niobate single crystals, *Opt. Spectrosc.* 28, 38 (1970).
620. Kaminskii, A. A., Bogomolova, G. A., and Li, L., Absorption, fluorescence, stimulated emission and splitting of the Nd^{3+} levels in a YVO_4 crystal, *Inorg. Mater. (USSR)* 5, 573 (1969).
621. Chinn, S. R. and Zwicker, W. K., Flash-lamp-excited NdP_5O_{14} laser, *Appl. Phys. Lett.* 31, 178 (1977).
622. Singh, S., Miller, D. C., Potopowicz, J. R., and Shick, L. K., Emission cross section and fluorescence quenching of Nd^{3+} lanthanum pentaphosphate, *J. Appl. Phys.* 46, 1191 (1975).
623. Kaminskii, A. A., Kurbanov, K., and Uvarova, T. V., Stimulated radiation from single crystals of $BaYb_2F_8$-Pr^{3+}, *Inorg. Mater. (USSR)* 23, 940 (1987).
624. Sandrock, T., Heumann, E., Huber, G. and Chai, B. H. T., Continuous-wave Pr,Yb:$LiYF_4$ upconversion laser in the red spectral range at room temperature, *OSA Trends in Optics and Photonics on Advanced Solid State Lasers*, Vol. 1, Payne, S. A. and Pollack, C. R., Eds., Optical Society of America, Washington, DC (1996), p. 550.
625. Kaminskii, A. A., Sarkisov, S. E., Maier, A. A. et al., Eulytine with TR^{3+} ions as a laser medium, *Sov. Tech. Phys. Lett.* 2, 59 (1976).
626. Kaminskii, A. A., High-temperature spectroscopic investigation of stimulated emission from lasers based on crystals activated with Nd^{3+} ions, *Phys. Status Sol.* 13a, 573 (1970).
627. Bagdasarov, Kh. S., Iwtova, O. Ye., Kaminskii, A. A., Li, L., and Sobolev, B. P., Optical and laser properties of mixed CdF_2-YF_3:Nd^{3+}- crystals, *Sov. Phys.-Dokl.* 14, 939 (1970).
628. Hong, H. Y-P. and Dwight, K., Crystal structure and fluorescence lifetime of a laser material $NdNa_5(WO_4)_4$, *Mater. Res. Bull.* 9, 775 (1974).
629. Petricevic, V., Bykov, A. B., Evans, J. M. and Alfano, R. R., Room-temperature near-infrared tunable laser operation of Cr^{4+}:Ca_2GeO_4, *Opt. Lett.* 21, 1750 (1996).
630. Hattendorff, H.-D., Huber, G., and Lutz, F., CW laser action in $Nd(Al,Cr)_3(BO_3)_4$, *Appl. Phys. Lett.* 34, 437 (1979).
631. Lutz, F., Ruppel, D., and Leiss, M., Epitaxial layers of the laser material $Nd(Ga,Cr)_3(BO_3)_4$, *J. Cryst. Growth* 48, 41 (1980).
632. Ohlmann, R. C., Steinbruegge, K. B., and Mazelsky, R., Spectroscopic and laser characteristics of neodymium-doped calcium fluorophosphate, *Appl. Optics* 7, 905 (1968).
633. Bruk, Z. M., Voron'ko, Yu. K., Maksimova, G. V., Osiko, V. V., Prokhorov, A. M., Shipilov, K. F., and Shcherbakov, I. A., Optical properties of a stimulated emission from Nd^{3+} in fluorapatite crystals, *JETP Lett.* 8, 221 (1968).
634. Chinn, S. R., Research studies on neodymium pentaphosphate miniature lasers, Final Report ESD TR-78-392, (DDC Number AD-A073140, Lincoln Laboratory, M. I. T., Lexington, MA (1978).
635. Kaczmarek, F. and Szymanski, M., Performance of NdLa pentaphosphate laser pumped by nanosecond pulses, *Appl. Phys.* 13, 55 (1977).
636. Wilson, J., Brown, D. C., and Zwicker, W. K., XeF excimer pumping of NdP_5O_{14}, *Appl. Phys. Lett.* 33, 614 (1978).
637. Gaiduk, M. I., Grigor'yants, V. V., Zhabotinskii, M. E., Makovestskii, A. A., and Tishchenko, R. P., Neodymium pentaphosphate microlaser pumped by the second harmonic of a YAG:Nd^{3+} laser, *Sov. J. Quantum Electron.* 9, 250 (1979).
638. Weber, H. P. and Tofield, B. C., Heating in a cw Nd-pentaphosphate laser, *IEEE J. Quantum Electron.* QE-11, 368 (1975).

639. Chinn, S. R., Pierce, J. W., and Heckscher H., Low-threshold transversely excited NdP_5O_{14} laser, *Appl. Opt.* 15, 1444 (1976).

640. Chinn, S. R., Hong, H. Y-P., and Pierce, J. W., Spiking oscillations in diode-pumped NdP_5O_{14} and $NdAl_3(BO_3)_4$ lasers, *IEEE J. Quantum Electron.* QE-12, 189 (1976).

641. Luo, Z., Jiang, A., Huang, Y., and Qui, M., Laser performance of large neodymium aluminum borate ($NdAl_3(BO_3)_4$) crystals, *Chin. Phys. Lett.* 3, 541 (1986).

642. Huang, Y., Qiu, M., Chen, G., Chen, J., and Luo, Z., Pulsed laser characteristics of neodymium aluminum borate [$NdAl_3(BO_3)_4$] (NAB) crystals, *Chin. Phys. Lasers* 14, 623 (1987).

643. Dianov, E. M., Dmitruk, M. V., Karasik, A. Ya., Kirpickenkova, E. O., Osiko, V. V., Ostrounov, V. G., Timoshechin, M. I., and Shcherbakov, I. A., Synthesis and investigation of spectral, luminescence, and lasing properties of alumoborate crystals activated with chromium and neodymium ions, *Sov. J. Quantum Electron.* 10, 1222 (1980).

644. Ivanov, A. O., Morozova, L. G., Mochalov, I.V. and Fedorov, V. A., Spectra of a neodymium ion in Ca,LaSOAP and Ca,YSOAP crystals and stimulated emission in Ca,LaSOAP-Nd crystals, *Opt. Spectrosc. (USSR)* 42, 556 (1977).

645. Arkhipov, R. N., Evstigneev, V. L., Zharikov, E. V., Pabenichnikov, S. M., Shcherbakov, I. A., and Yumashev, V. E., Optimization of the conditions of utilization of the stored energy by Q switching active elements in the form of gadolinium scandium gallium garnet crystals activated with Cr and Nd, *Sov. J. Quantum Electron.* 16, 688 (1986).

646. Zharikov, E. V., Zhitkova, M. B., Zverev, G. M., Isaev, M. P., Kalitin, S. P., Kurabv, I. I., Kushir, V. R., Laptev, V. V., Osiko, V. V., Pashkov, V. A., Pimenov, A. S., Prokhorov, A. M., Smirnov, V. A., Stel'makh, M. F., Shectakov, A. M., and Shcherbakov, I. A., Output characteristics of a gadolinium scandium gallium garnet laser operating in the pulse-periodic regime, *Sov. J. Quantum Electron.* 13, 1306 (1983).

647. Shestakov, A.V., Borodin, N.I., Zhitnyuk, V.A., Ohrimtchyuk, A.G. and Gapontsev, V.P., Tunable Cr^{4+}:YAG lasers, *CPDP12-1*, 594 (1993).

648. Lutz, P., Leiss, M., and Muller, J., Epitaxy of $NdAl_3(BO_3)_4$ for thin film miniature lasers, *J. Cryst. Growth*, 47, 130 (1979).

649. Lazarev, V. V. and Kandaurov, A. S., Lasing properties of $La_2Be_2O_5$:Nd^{3+} at 1.35-μm wavelength, *Opt. Spectrosc. (USSR)* 63, 519 (1987).

650. Grigor'ev, V. N., Egorov, G. N., Zharikov, E. V., Mlkhailov, V. A., Pak, S. K., Pinskii, Yu. A., Shklovskii, E. I., and Shcherbakov, I. A., Prism-resonator CSGG Cr:Nd laser with polarization coupling out of radiation, *Sov. J. Quantum Electron.* 16, 1554 (1986).

651. Matrosov, V. I., Timosheckin, M. I., Tsvetkov, E. I. et al., Investigation of the conditions of crystallization of lanthanum beryllate in *Abstracts of the Fifth All-Union Conference on Crystal Growth*, Proc. Acad. Sci. Georgian Sov. Soc. Rep., Tiflis (1977), p. 167 (in Russian).

652. Hattendorff, H.-D., Huber, G. and Danielmeyer, H. G., Efficient cross pumping of Nd^{3+} by Cr^{3+} in $Nd(Al, Cr)_3(BO_3)_4$ lasers, *J. Phys. C* 11, 2399 (1978).

653. Gondra, A. D., Gradov, V. M., Danilov, A. A., Dybko, V. V., Zharlkov, E. V., Kondantlnov, B. A., Nlkol'skii, M. Yu., Rogal'skii, Yu. I., Smotryaev, S. A., Terent'ev, Yu. I., Shcherbakov, A. A., and Shcherbakov, I. A., Chromium- and neodymium-activated gadolinium scandium gallium garnet laser with efficient pumping and Q switching, *Sov. J. Quantum Electron.* 17, 582 (1987).

654. Belokrinitskiy, N. S., Belousov, N. D., Bonchkovskiy, V. I., Kobzaraklenko, V. A., Skorobogatov. B. S., and Soskin, M. S., Study of stimulated emission from Nd^{3+}-doped $NaLa(WO_4)_2$ single crystals. *Ukrainskiyl Fizicheskiy Zhurnal* 14, 1400, 1969 (in Russian).

655. Kaminskii, A. A. and Sarkisov, S. E., Study of stimulated emission from Nd^{3+} ions in crystals at the $^4F_{3/2} \rightarrow {}^4I_{13/2}$ transition. 4, *Sov. J. Quantum Electron.* 3, 248 (1973).
656. Huber, G. and Danielmeyer, H. G., NdP_5O_{14} and $NdAl_3(BO_3)_4$ lasers at 1.3 μm, *Appl. Phys.* 18, 77 (1979).
657. Malinowski, M., Joubert, M. F. and Jacquier, B., Simultaneous laser action at blue and orange wavelengths in YAG:Pr^{3+}, *Phys. Status Solidi A* 140, K49 (1993).
658. Winzer G., Mockel, P. G., and Krühler, W., Laser emission from miniaturized $NdAl_3(BO_3)_4$ crystals with directly applied mirrors, *IEEE J. Quantum Electron.* QE-14, 840 (1978).
659. Danilov, A. A., Evstigneev, V. L., Il'lchev, N. N., Mdyutin, A. A., Nikol'skii, M. Yu., Umyhkov, A. F., and Shcherbakov, I. A., Compact GSGG:Cr^{3+}:Nd^{3+} laser with passive Q switching, *Sov. J. Quantum Electron.* 17, 573 (1987).
660. DeSerno, U., Ross, D., and Zeidler, G., Quasicontinuous giant pulse emission of $^4F_{3/2} \rightarrow {}^4I_{13/2}$ transition at 1.32 μm in YAG:Nd^{3+}, *Phys. Lett.* 28A, 422 (1968).
661. Kaminskii, A. A., Schultze, D., Hermoneit, B. et al., Spectroscopic properties and stimulated emission in the $^4F_{3/2} \rightarrow {}^4I_{11/2}$ and $^4F_{3/2} \rightarrow {}^4I_{13/2}$ transitions of Nd^{3+} ions from cubic $Bi_4Ge_3O_{12}$ crystals, *Phys. Status Solidi* 33a, 737 (1976).
662. Moulton, P. F., Mooradian, A., Chen, Y., and Abraham, M. M., unpublished.
663. Richards, J., Fueloep, K., Seymour, R. S., Cashmore, D., Picone, P. J. and Horsburgh, M. A., Nd:BeL laser at 1356 nm, in *Tunable Solid State Lasers*, Shand, M. L. and Jenssen, H. P., Eds., Proceedings Vol. 5, Optical Society of America, Washington, DC (1989), p. 119.
664. French, P. M. W., Rizvi, N. H., Taylor, J. R. and Shestakov, A. V., Continuous-wave mode-locked Cr^{4+}:YAG laser, *Opt. Lett.* 18, 39 (1993).
665. Kaminskii, A. A., Kholov, A., Klevstov, P. V. and Khafizov, S. K., Growth and generation properties of $NaBi(WO_4)_2$–Nd^{3+} single crystals, *Neorgan. Mater.* 25, 1054, 1989
666. Schearer, L. D. and Tin, P., Laser performance and tuning characteristics of a diode pumped Nd:$YAlO_3$ laser at 1083 nm, *Opt. Commun.* 71, 170 (1989).
667. Baturina, O. A., Grechushnikov, B. N., Kaminskii, A. A., Konstantinova, A. F., Markosyan, A., A., Mill', B. V., and Khodzhabagyan, G. G., Crystal-optical investigations of compounds with the structure of trigonal Ca-gallogermanate ($Ca_3Ga_2Ge_4O_{14}$), *Sov. Phys. Crystallogr.* 32, 236 (1987).
668. Kaminskii, A. A., Mill', B. V., Belokonevs, E. L., Sarkisov, S. E., Pastukhova, T. Yu., and Khodazhabagyan, G. G., Crystal structure and stimulated emission of $La_3Ga_{5.5}Nb_{0.5}O_{12}$-$Nd^{3+}$, *Inorg. Mater. (USSR)* 20, 1793 (1984).
669. Gavrilovic, P., O'Neill, M. S., Meehan, K., Zarrabi, J. H., Singh, S. and Grodliewicz, W. H., Temperature-tunable, single frequency microcavity lasers fabricated from flux-grown YCeAG:Nd, *Appl. Phys. Lett.* 60, 1652 (1992).
670. Kvapil, J., Kvapil, Jos., Kubelka, J., and Perner, B., Laser properties of YAG:Nd Ti, *Czech. J. Phys. B* 32, 817 (1982).
671. Kvapil, J., Kvapil, Jos., Perner, B., Kubelka, J., Mamek, B., and Kubecek, V., Laser properties of YAG:Nd Cr,Ce, *Czech. J. Phys. B* 34, 581 (1984).
672. Shi, W. Q., Kurtz, R., Machan, J., Bass, M., Birnbaum, M., and Kokta, M., Simultaneous, multiple wavelength lasing of (Er,Nd):$Y_3Al_5O_{12}$, *Appl. Phys. Lett.* 51, 1218 (1987).
673. Machan, J., Kurtz, R., Bass, M., Birnbaum, M., and Kokta, M., Simultaneous multiple wavelength lasing of (Ho,Nd):$Y_3Al_5O_{12}$, *Appl. Phys. Lett.* 51, 1313 (1987).
674. Forrester, P. A. and Simpson, D. F., A new laser line due to energy transfer from colour centers to erbium ions in CaF_2, *Proc. Phys. Soc.* 88, 199 (1966).

675. Es'kov, N. A., Osiko, V. V., Sobol, A. A. et al., A new laser garnet $Ca_3Ca_2Ge_3O_{12}$-Nd^{3+}, *Inorg. Mater. (USSR)* 14, 1764 (1978).
676. Bezrodnyi, V. I., Tlkhonov, E. A., and Nedbaev, N. Ya., Generation of controlled-duration ultrashort pulses in a passively mode-locked YAG:Nd^{3+} laser, *Sov. J. Quantum Electron.* 16, 796 (1986).
677. Smith, R. J., Rice, R. R., and Aden, L. B., Jr., 100 mW laser diode pumped Nd:YAG laser, in *Proc. Soc. Photo. Opt. Instrum. Eng., Advances in Laser Engineering and Applicatons,* Vol. 247 Stitch, M. L., Ed., SPIE, Bellingham, WA (1980), p. 144.
678. Berger, J., Welch, D. F., Streifer, W., Scifres, D. R., Hoffman, N. J., Smith, J. J., and Radecki, D., Fiber-bundle coupled, diode end-pumped Nd:YAG laser, *Opt. Lett.* 13, 306 (1988).
679. Sipes, D. L., Highly efficient neodymium:yttrium aluminum garnet laser end pumped by a semiconductor laser array *Appl. Phys. Lett.* 47, 74 (1985).
680. Berger, J., Welch, D. F., Scifres, D. R., Streifer, W., and Cross, P. S., 370 mW, 1.06 µm, CW TEM_{00} output from an Nd:YAG laser rod end-pumped by a monolithic diode array, *Electron. Lett.* 23, 669 (1987).
681. Berger, J., Welch, D. F., Scifres, D. R., Strelfer, W., and Cross, P. S., High power, high efficient neodymium:yttrium aluminum garnet laser end pumped by a laser diode array, *Appl. Phys. Lett.* 51, 1212 (1987).
682. Zhou, B., Kane, T. J., Dixon, G. J., and Byer, R. L., Efficient, frequency-stable laser-diode-pumped Nd:YAG laser, *Opt. Lett.* 10, 62 (1985).
683. Kane, T. J., Nilsson, A. C., and Byer, R. L., Frequency stability and offset locking of a laser-diode-pumped Nd:YAG monolithic nonplanar ring oscillator, *Opt. Lett.* 12, 175 1987.
684. Fields, R. A., Birnbaum, M., and Fincher, C. L., Highly efficient Nd:YVO4 diode-laser end-pumped laser, *Appl. Phys. Lett.*, 51, 1885 (1987).
685. Allik, T. H., Hovis, W. W., Caffey, D. P., and King, V., Efficient diode-array-pumped Nd:YAG and Nd:Lu:YAG lasers, *Opt. Lett.* 14, 116 (1989).
686. Reed, M. K., Kozlovsky, W. J., Byer, R. L., Harmgel, G. L., and Cross, P. S., Diode-laser-array- pumped neodymium slab oscillators, *Opt. Lett.* 13, 204 (1988).
687. Berger, J., Harnagel, G., Welch, D. F., Scifres, D. R., and Strelfer, W., Direct modulation of a Nd:YAG laser by combined side and end laser diode pumping, *Appl. Phys. Lett.* 53, 268 (1988).
688. Maker, G. T. and Ferguson, A. I., Single-frequency Q-switched operation of a diode-laser-pumped Nd:YAG laser, *Opt. Lett.* 13, 461 (1988).
689. Denisov, N. N., Manenkov, A. A., and Prokhorov, A. M., Kinetics of generation and amplification of YAG:Nd^{3+} laser radiation in a periodic Q-switched regime with pulsed pumping, *Sov. J. Quantum Electron.* 14, 597 (1984).
690. De Silvestri, S., Laporta, P., and Magni, V., 14-W continuous-wave mode-locked Nd:YAG laser, *Opt. Lett.* 11, 785 (1986).
691. Kuizenga, D. J., Short-pulse oscillator development for the Nd:Glass laser-fusion systems, *IEEE J. Quantum Electron.* QE-17, 1694 (1981).
692. Dawes, J. B. and Sceats, M. G., A high repetition rate pico-synchronous Nd:YAG laser, *Opt. Commun.* 65, 275 (1988).
693. Prokhorenko, V. I., Tikhonov, E. A., Yatskiv, D. Ya., and Bushmakin, E. N., Generation of ultra-short pulses in a YAG:Nd^{3+} laser in a colliding pulse scheme, *Sov. J. Quantum Electron.* 17, 505 (1987).
694. Varnavskii, O. P., Leontovich, A. M., Mozharovskii, A. M., and Solomatin, I. I., Mode-locked YAG:Nd^{3+} laser with a high output energy and brightness, *Sov. J. Quantum Electron.* 13, 1251 (1983).
695. Kaminskii, A. A., Silvestrova, I. M., Sarkisov, S. E., and Denisenko, G. A., Investigation of trigonal $(La_{1-x}Nd_x)_3Ga_5SiO_{14}$ crystals, *Phys. Status Solidi A* 80, 607 (1983).

696. Kaminskii, A. A., Sarkisov, S. E., Mill', B. V., and Khodzhabagyan, G. G., Generation of stimulated emission of Nd^{3+} ions in a trigonal acentric $La_3Ga_5SiO_{14}$ crystal, *Sov. Phys. Dokl.* 27, 403 (1982).
697. Kaminskii, A. A., Mill', B. V., Silvestrova, I. M., and Khodzhabagyan, G. G., The nonlinear active material $(La_{1-x}Nd_x)_3Ga_5SiO_{14}$, *Bull. Acad. Sci. USSR Phys. Ser.* 47, 25 (1983).
698. Kaminskii, A. A., Sarkisov, S. E., Mill', B. V., and Khodzhabagyan, G. G., New inorganic material with a high concentration of Nd^{3+} ions for obtaining stimulated emission at the $^4F_{3/2} \rightarrow {}^4I_{11/2}$ and $^4F_{3/2} \rightarrow {}^4I_{13/2}$ transitions, *Inorg. Mater. (USSR)* 18, 1189 (1982).
699. Kaminskii, A. A., Agamalyan, N. R., Kozeeva, L. P., Nesterenko, V. F., and Pavlyuk, A. A., New data on stimulated emission of Nd^{3+} ions in disordered crystals with scheelite structure, *Phys. Status Solidi A* 75, Kl (1983).
700. Rosenblatt, G. H., Stoneman, R. C. and Esterowitz, L., Diode-pumped room-temperature cw 1.45-μm Tm;Tb:YLF laser, in *Advanced Solid-State Lasers*, Jenssen, H. P. and Dubé, G., Eds., Proceedings Vol. 6, Optical Society of America, Washington, DC (1990), p. 26.
701. Moncorgé, R., Manaa, H, Deghoul, F., Borel, C. and Wyon, Ch., Spectroscopic study and laser operation of Cr^{4+}-doped $(Sr,Ca)Gd_4(SiO_4)_3O$ single crystal, *Opt. Commun.* 116, 393 (1995).
702. Kaminskii, A. A., Temperature pulsations and multi-frequency laser action in $YAlO_3$:Nd^{3+}, *JETP Lett.* 14, 222 (1971).
703. Kaminskii, A. A., Sobolev, B. P., Sarkisov, S. E., Denisenko, G. A., Ryabchenkov, V. V., Fedorov, V. A., and Uvarova, T. V., Physicochemical aspects of the preparation spectroscopy, and stimulated emission of single crystals of $BaLn_2F_8$-Ln^{3+}, *Inorg. Mater. (USSR)* 18, 402 (1982).
704. Moulton, P. F., Pulse-pumped operation of divalent transition-metal lasers, *IEEE J. Quantum Electron.* QE-18, 1185 (1982).
705. Kaminskii, A. A., Kurbanov, K., Markosyan, A. A., Mill', B. V., Sarkisov, S. E., and Khodzhabagyan, G. G., Luminescence-absorption properties and low-threshold stimulated emission of Nd^{3+} ions in $La_3Ga_{5.5}Ta_{0.5}O_{14}$, *Inorg. Mater. (USSR)* 21, 1722 (1985).
706. Petricevic, V., Gayen, S. K., and Alfano, R. R., Laser action in chromium-activated forsterite for near-infrared excitation: is Cr^{3+} the lasing ion?, *Appl. Phys. Lett.* 53, 2590 (1988).
707. Petricevic, V., Gayen, S. K., and Alfano, R. R., Continuous-wave operation of chromium-doped forsterite. *Opt. Lett.* 14, 612 (1989).
708. Verdun, H. R., Thomas, L. M., Andrauskas, D. M., McCollum, T., and Pinto, A., Chromium-doped forsterite laser pumped with 1.06 μm radiation, *Appl. Phys. Lett.* 53, 2593 (1988).
709. Petricevic, V., Gayen, S. K., Alfano, R. R., Yamagishi, K., Anzai, H., and Yamaguchi, Y., Laser action in chromium-doped forsterite, *Appl. Phys. Lett.* 52, 1040 (1988).
710. Antipenko, B. M., Krutova, L. I., and Sukhareva, L. K., Dual-frequency lasing of GSGG-Cr^{3+} Tm crystals, *Opt. Spectrsoc. (USSR)* 60, 252 (1986).
711. Antipenko, B. M., Mak, A. A., Raba, O. B., Seiranyan, K. B., and Uvarova, T. V., New lasing transition in the Tm^{3+} ion, *Sov. J. Quantum Electron.* 13, 558 (1983).
712. Akmanov, A. G., Val'shin, A. M., and Yamaletdinov, A. G., Frequency-tunable $YAlO_3$:Nd^{3+} laser, *Sov. J. Quantum Electron.* 15, 1555 (1985).
713. Kaminskii, A. A., Butaeva, T. I., Kevorkov, A. M., Fedorov, V. A., Petrosyan, A. G. and Gritsenko, M. M., New data on stimulated emission by crystals with high concentrations of Ln^{3+} ions, *Inorg. Mater. (USSR)* 12, 1238 (1976).

714. Antipenko, B. M., Dumbravyanu, R. V., Perlin, Yu. E., Raba, O. B., and Sukhareva, L. K., Spectroscopic aspects of the $BaYb_2F_8$ laser medium, *Opt. Spectrosc. (USSR)* 59, 377 1985.
715. Bass, M. and Weber, M. J., Nd, Cr:$YAlO_3$ laser tailored for high-energy Q-switched operation, *Appl. Phys. Lett.* 17, 395 (1970).
716. Weber, M. J., Bass, M., Andringa, K., Monchamp, R. R., and Comperchio, L., Czochralski growth and properties of $YAlO_3$ laser crystals, *Appl. Phys. Lett.* 15, 342 (1969).
717. Massey, G. A. and Yarborough, J. M., High average power operation and nonlinear optical generation with the Nd:$YAlO_3$ laser, *Appl. Phys. Lett.* 18, 576 (1971).
718. Schearer, L. and Leduc, M., Tuning characteristics and new laser lines in an Nd:YAP CW laser, *IEEE J. Quantum Electron.* QE-22, 756 (1986).
719. Esterowitz, L., Eckardt, R. C., and Allen, R. E., Long-wavelength stimulated emission via cascade laser action in Ho:YLF, *Appl. Phys. Lett.* 35, 236 (1979).
720. Singh, S., Van Uitert, L. G., Potopowicz, I. R., and Grodkiewicz, W. H., Laser emission at 1.065 μm from neodymium-doped anhydrous cerium trichloride at room temperature, *Appl. Phys. Lett.* 24, 10 (1974).
721. Singh, S., Chesler, R. B., Grodkiewicz, W. H. et al., Room temperature CW Nd^{3+}:$CdCl_2$ laser, *J. Appl. Phys.* 46, 436 (1975).
722. Kaminskii, A. A., Fedorov, V. A., Sarkisov, S. E. et al., Stimulated emission of Ho^{3+} and Er^{3+} ions in $Gd_3Ga_5O_{12}$ crystals and cascade laser action of Ho^{3+} ions over the $^5S_2$—$^5I_5$—$^5I_6$—$^5I_8$ scheme, *Phys. Status Solidi* 53a, K219 (1979).
723. Wong, S.K., Mathieu, P. and Pace, P., Eye-safe Nd:YAG laser, *Appl. Phys. Lett.*, 57, 650 (1990).
724. Hodgson, N., Nighan, Jr., W. L., Golding, D. J. and Eisel, D., Efficient 100-watt Nd:YAG laser operating at the wavelength of 1.444 μm, *Optics Lett.* 19, 1328 (1994).
725. Sorokin, E., Sorokina, I., Wintner, E., Zagumennyi, A. I. and Shcherbakov, I. A., CW passive mode-locking of a new Nd^{3+}:$GdVO_4$ crystal laser, in *Advanced Solid-State Lasers*, Pinto, A. A. and Fan, T. Y., Eds., Proceedings Vol. 15, Optical Society of America, Washington, DC (1993), p. 238.
726. Bagdasarov, Kh. S., Kaminskii, A. A., Kevorkov, A. M., Prokhorov, A. M., Sarkisov, S. E., and Tevosyan, T. A., Stimulated emission from RE^{3+} ions in YAG crystals, *Sov. Phys.-Dokl.* 19, 592 (1975).
727. Wang, Q., Zhao, S. and Zhang, X., Laser characterization of low-threshold high-efficiency Nd:$Sr_5(VO4)_3F$ crystal, *Opt. Lett.* 20, 1262 (1995).
728. DeLoach, L., Payne, S. A., Chai, B. H. T. and Loutts, G., Laser demonstration of neodymium-doped strontium chlorovandate, *Appl. Phys. Lett.* 65, 1208 (1994).
729. Bass, M. and Weber, M. J., YALO:Robust at age 2, *Laser Focus*, 34 (1971).
730. Demchouk, M. I., Gilev, A. K., Zabaznov, A. M., Mikhailov, V. P., Stavrov, A. A., and Shkadarevich, A. P., Lasing of ultrashort pulses by a Nd,Cr-doped gadolinium-scandium-gallium garnet laser, *Opt. Commun.* 55, 207 (1985).
731. Bagdasarov, Kh. S., Danilov, V. P., Zhekov, V. I. et al., Pulse-periodic $Y_3Al_5O_{12}$:Er^{3+} laser with high activator concentration, *Sov. J. Quantum Electron.* 8, 83 (1978).
732. Danelyus, R., Kuratev, I., Piskarskas, A., Sirutkaitis, V., Shvom, E., Yuozapavichyus, A., and Yankauskas, A., Generation of picosecond pulses by a gadolinium scandium gallium garnet laser, *Sov. J. Quantum Electron.* 15, 1160 (1985).
733. Simondi-Teisseire, B., Viana, B. and Vivien, D., Near-infrared Er^{3+} laser properties in melilite type crystals $Ca_2Al_2SiO_7$ and $SrLaGa_3O_7$, *Proceedings Advanced Solid State Lasers* (1997), to be published.

734. Heumann, E., Mobert, P. and Huber, G., Room-temperature upconversion-pumped cw Yb,Er:YLF laser at 1.234 μm, *OSA Trends in Optics and Photonics on Advanced Solid State Lasers*, Vol. 1, Payne, S. A. and Pollack, C. R., Eds., Optical Society of America, Washington, DC (1996), p. 288.
735. Kaminskii, A. A., Spectral composition of laser light from neodymium-doped calcium tungstate crystals, *Inorg. Mater. (USSR)* 347 (1970).
736. Kaminskii, A. A., Mill', B. V., Belokoneva, E. L., and Khodzhabagyan, G. G., Growth and crystal structure of a new inorganic lasing material $La_3Ga_5GeO_{14}$-Nd^{3+} *Inorg. Mater. (USSR)* 19, 1559 1983.
737. Paylyuk, A. A., Kozeeva, L. I., Folin, K. G., Gladyshev, V. G., Gulyaev, V. S., Pivbov, V. S., and Kaminskii, A. A., Stimulated emission on the transition $^4F_{3/2} \rightarrow {}^4I_{11/2}$ of Nd^{3+} ions in $RbNd(WO_4)_2$ and $CsNd(MoO_4)_2$, *Inorg. Mater. (USSR)* 19, 767 (1983).
738. White, K. O. and Schlenser, S. A., Coincidence of Er:YAG laser emission with methane absorption at 1645.1 nm, *Appl. Phys. Lett.* 21, 419 (1972).
739. Johnson, L. F., Dietz, R. E., and Guggenheim, H. J., Optical maser oscillation from Ni^{2+} in MgF_2 involving simultaneous emission of phonons, *Phys. Rev. Lett.* ll, 318 (1963).
740. Johnson, L. F., Guggenheim, H. J. and Thomas, R. A., Phonon-terminated optical masers, *Phys. Rev.* 149, 179 (1966).
741. White, K. O. and Schleusener, S. A., Coincidence of Er:YAG laser emission with methane absorption at 1645.1 nm, *Appl. Phys. Lett.* 21, 419 (1972).
742. Kaminskii, A. A., Kurbanov, K., Sattarova, M. A., and Fedorov, P. P., Stimulated IR emission of Nd^{3+} ions in nonstoichiometric cubic fluorides, *Inorg. Mater. (USSR)* 21, 609 (1985).
743. Stoneman, R.C. and Esterowitz, L., Continuous-wave 1.50-μm thulium cascade laser, *Optics Lett.* 16, 232 (1991).
744. Simondi-Teisseire, B., Viana, B., Lejus, A.-M. et al., Room-temperature CW laser operation at ~1.55 μm (eye-safe region) of Yb:Er and Yb:Er:Ce:$Ca_2Al_2SiO_7$ crystals, *IEEE J. Quantum Electron.* 32, 2004 (1996).
745. Kaminskii, A. A., Lapsker, Ya. Ye., and Sobolev, B. P., Induced emission of $NaCaCeF_6$:Nd^{3+} at room temperature, *Phys. Status Solidi* 23, K5 (1967).
746. Quaries, G.J., Rosenbaum, A., Marquardt, C.L. and Esterowitz, L., High-efficiency 2.09 μm flashlamp-pumped laser, *Appl. Phys. Lett.* 55, 1062 (1989).
747. Brinkman, R., Sohler, W. and Suche, H., Continuous-wave erbium-diffused $LiNbO_3$ waveguide laser, *Electron. Lett.* 415 (1991).
748. Klein, P. B., Furneaux, J. E., and Henry, R. L., Laser oscillation at 3.53 μm from Fe^{2+} in n-InP:Fe, *Appl. Phys. Lett.* 42, 638 (1983).
749. Morozov, A. M., Pogkolzina, I. G., Tkachuk, A. M., Fedorov, V. A. and Feofilov, P. P., Luminescence and induced emission of lithium-erbium and lithium-holmium binary fluorides, *Opt. Spectrosc. (USSR)* 39, 605 (1975).
750. Gifeisman, Sh. N., Tkachuk, A. M. and Prizmak, V. V., Optical spectra of Ho^{3+} ion in $LiYF_4$ crystals, *Opt. Spectrosc. (USSR)* 44, 68 (1978).
751. Lenth, W., Hattendorff, H.-D., Huber, G., and Lutz, F., Quasi-cw laser action in $K_5Nd(MoO_4)_4$, *Appl. Phys.* 17, 367 (1978).
752. Becker, P., Brinkmann, R., Dinand, M., Sohler, W. and Suche, H., Er-diffused Ti:$LiNbO_3$ waveguide laser of 1563 and 1576 nm emission wavelengths, *Appl. Phys. Lett.* 61, 1257 (1992).
753. Ballman, A. A., Porto, S. P. S., and Yariv, A., Calcium niobate $Ca(NbO_3)_2$–A new laser host, *J. Appl. Phys.* 34, 3155 (1963).
754. Moulton, P. F. and Mooradian, A., Broadly tunable CW operation of Ni:MgF_2, and Co:MgF_2, lasers, *Appl. Phys. Lett.* 35, 838 (1979). (Unpublished results which improve upon those indicated in this reference have been included.)

755. Schmaul, B., Huber, G., Clausen, R., Chai, B., LiKamWa, P. and Bass, M., Er^{3+}:$YLiF_4$ continuous wave cascade laser operation at 1620 and 2810 nm at room temperature, *Appl. Phys. Lett.* 62, 541 (1993).
756. Camargo, M. B., Stultz, R. D. and Birnbaum, M., Passive Q switching of the Er^{3+}:$Y_3Al_5O_{12}$ laser at 1.64 μm, *Appl. Phys. Lett.* (1995).
757. Kiss, Z. J. and Duncan, R. C., Optical maser action in $CaWO_4$:Er^{3+}, *Proc. IRE* 50, 1531 (1962).
758. Kaminskii, A. A., Sarkisov, S. E., Bohm, J. et al., Growth, spectroscopic and laser properties of crystals in the $K_5Bi_{1-x}Nd_x(MoO_4)_4$ system, *Phys. Status Solidi* 43a, 71 (1977).
759. Antipenko, B. M., Asbkalunin, A. L., Mak, A. A., Sinitsyn, B. V., Tomashevlch, Yu. V., and Shakhkalamyan, G. S., Three-micron laser action in Dy^{3+}, *Sov. J. Quantum Electron.* 10, 560 (1980).
760. Pollack, S. A., Stimulated emission in CaF_2:Er^{3+}, *Proc. IEEE* 51, 1793 (1963).
761. Bowman, S. R., Ganem, J., Feldman, B. J. and Kueny, A. W., Infrared laser characteristics of praseodymium-doped lanthanum trichloride, *IEEE J. Quantum Electron.* 30, 2925 (1994).
762. Schmid, F. and Khattak, C. R., Growth of Co:MgF_2 and Ti:Al_2O_3 crystals for solid state laser applications, in *Tunable Solid State Lasers,* Proc. Int. Conf., Vol. 47, Hammerling, P., Budgor, A. B., and Pinto, A., Eds., Springer-Verlag, New York (1985), p. 22.
763. Kaminskii, A. A., Kurbanov, K., Sarkisov, S. E., Sattarova, M. M., Uvarova, T. V., and Fedorov, P. P., Stimulated emission of Nd^{3+} ions in non-stoichiometric $Cd_{1-x}Ce_xF_{2+x}$ and $Cd_{1-x}Nd_xF_{2+x}$ fluorides with fluorite structure. *Phys. Status Solidi A*: 90, K55 (1985).
764. Aleksandrov, V. I., Murina, T. M., Zhekov, V. K., and Tatarintsev, V. M., Stimulated emission from Tm^{3+} and Ho^{3+} in zirconium dioxide crystals, in Sbornik. Kratkiye Soobshcheniya po Fizike, *An SSSR Fizicheskiy Institut im P. N. Lebedeva* (1973), No. 2, p. 17 (in Russian).
765. Kaminskii, A. A., Klevtsov, P. V., and Pavlyuk, A. A., Stimulated emission from $KY(MoO_4)_2$:Nd^{3+}- crystal laser, *Phys. Status Solidi* la, K91 (1970).
766. Moulton, P. F., Recent advances in transition metal-doped lasers, in *Tunable Solid State Lasers*, Proc. Int. Conf., Vol. 47, Hammerling, P., Budgor, A. B., and Pinto, A., Eds., Springer-Verlag, New York (1985), p. 4.
767. Brixner, L. H. and Flournoy, A. P., Calcium orthovanadate $Ca_3(VO_4)$—a new laser host crystal, *J. Electrochem. Soc.* 112, 303 (1965).
768. Moulton, P. F., An investigation of the Co:MgF_2 laser system. *IEEE J. Quantum Electron.* QE-21, 1582 (1985).
769. Welford, D. and Moulton, P. F., Room-temperature operation of a Co:MgF_2 laser, *Opt. Lett.* 13, 975 (1988).
770. Lovold, S., Moulton, P. F., Kilhinger, D. K., and Menyuk, N., Frequency tuning characteristics of a Q-switched Co:MgF_2 laser, *IEEE J. Quantum Electron.* QE-21, 202 (1985).
771. Johnson, B. C., Moulton, P. F., and Mooradian, A., Mode-locked operation of Co:MgF_2 and Ni:MgF_2 lasers, *Opt. Lett.* 10, 116 (1984).
772. Johnson, B. C., Rosenbluh, M., Moulton, P. F., and Mooradian, A., High average power mode-locked Co:MgF_2 laser, in *Ultrafast Phenomena IV*, Proc., Vol. 38, Auston, D. H. and Eisenthal, K. B., Eds., Springer-Verlag, New York (1984), p. 35.
773. Muciel, A. C., Maly, P., and Ryan, J. F., Simultaneous modelocking and Q-switching of a Co:MgF_2 laser by loss-modulation frequency detuning, *Opt. Commun.* 61, 125 (1987).

774. Johnson, L. F., Guggenheim, H. J., and Bahnck, D., Phonon-terminated laser emission from Ni^{2+} ions in $KMgF_3$, *Opt. Lett.* 8, 371 (1983).
775. Breteau, J. M., Mekhenin, D., and Auzel, F., Study of the Ni^{2+}:MgF_2 tunable laser, *Rev. Phys. Appl.* 22, 1419, 1987 (in French).
776. Huber, G., Duczynski, E. W., and Petermann, K., Laser pumping of Ho-, Tm-, Er-doped garnet lasers at room temperature, *IEEE J. Quantum Electron.* 24, 920 (1988).
777. Dätwyler, M., Lüthy, W., and Weber, H. P., New wavelengths of the $YAlO_3$:Er laser, *IEEE J. Quantum Electron.* QE-23, 158 (1987).
778. Tucker, A. W., Birnbaum, M., Fincher, C. L., and DeShazer, L. G., Continuous-wave operation of Nd:YVO_4, at 1.06 and 1.34 μm, *J. Appl. Phys.*, 47, 232 (1976).
779. Lisitsyn, V. N., Matrosov, V. N., Pestryakov, E. V., and Trunov, V. I., Generation of picosecond pulses in solid-state lasers using new active media, *J. Sov. Laser Res.* 7, 364 (1986).
780. Telle, H. R., Injection locking of a 1.3 μm laser diode to an $LiNdP_4O_{12}$ laser yields narrow linewidth emission, *Electron. Lett.*, 22, 150 (1986).
781. Greskovich, C. and Chernoch, I. P., Improved polycrystalline ceramic lasers, *J. Appl. Phys.*, 45, 4495 (1974).
782. Tocho, J. O., Jaque, F., Sole, J. G., Camarillo, E., Cusso, F. and Munoz Santiuste, J. E., Nd^{3+} active sites in Nd:MgO:$LiNbO_3$ lasers, *Appl. Phys. Lett.* 60, 3206 (1992).
783. Arsen'yev, P. A., Potemkin, A. V., Fenin, V. V. and Senff, I., Investigation of stimulated emission of Er^{3+} ions in mixed crystals with perovskite structure, *Phys. Status Solidi* 43a, K 15 (1977).
784. Arsen'yev, P. A. and Bienert, K. E., Absorption, luminescence, and stimulated emission spectra of Er^{3+} ions in $GdAlO_3$ crystals, *Phys. Status Solidi* 10a, K85 (1972).
785. Klinzel, W., Knierim, W., and Dürr, U., CW infrared laser action of optically pumped Co^{2+}:$KZnF_3$, *Opt. Commun.* 36, 383 (1981).
786. German, K. R., Dürr, U., and Künzel, W., Tunable single-frequency continuous-wave laser action in Co^{2+}:$KZnF_3$, *Opt. Lett.* 11, 12 (1986).
787. Dischler, B. and Wettling, W., Investigation of the laser materials $YAlO_3$:Er and $LiYF_4$:Ho, *J. Phys. D* 17, 1115 (1984).
788. Kaminskii, A. A. and Fedorov, V. A., Cascade stimulated emission in crystals with several metastable states of Ln^{3+} ions, in *Proc. 2d Int. Conf. Trends in Quantum Electron.*, Prokhorov, A. M. and Ursu, I., Eds., Springer-Verlag, New York (1986), p. 69.
789. Kaminskii, A. A., Cascade laser generation by Er^{3+} ions in $YAlO_3$ crystals by the scheme $^4S_{3/2} \rightarrow {}^4I_{11/2}$ and $^4S_{3/2} \rightarrow {}^4I_{13/2}$, *Sov. Phys. Dokl.* 27, 1039 (1982).
790. Kaminskii, A. A. and Osiko, V. V., Sensitization in a CaF_2-ErF_3:Tm^{3+} laser, *Inorg. Mater. (USSR)* 3, 519 (1967).
791. Kaminskii, A. A., Sarkisov, S. E., Rysbcbenkov, V. V., Arakelysn, A. Z., Seiranyan, K. B., and Sharkilatunyan, R. O., Growth of CaF_2-HoF_3 and CaF_2-ErF_3 crystals, and their laser properties, *Sov. Phys. Crystallogr.* 27, 118 (1982).
792. Kaminskii, A. A., Fedorov, V. A., and Mochalov, I. V., New data on the three-micron lasing of Ho^{3+} and Er^{3+} ions in aluminates having the perovskite structure, *Sov. Phys. Dokl.* 25, 744 (1980).
793. Thornton, J. R., Rushworth, P. M., Kelly, E. A., McMillan, R. W., and Harper, L. L., in *Proceedings 4th Conference Laser Technology,* Vol. 11, University of Michigan, Ann Arbor (1970), p. 1249.

794. Kaminskii, A. A., Bagdasarov, Kh. S., Petrosyan, A. G., and Sarkisov, S. E., Investigation of stimulated emission from $Lu_3Al_5O_{12}$ crystal with Ho^{3+}, Er^{3+} and Tm^{3+} ions, *Phys. Status Solidi* 18a, K31 (1973).
795. Ivanov, A. O., Mochalov, I. V., Petrov, M. V. et al., Spectroscopic properties of single crystals of rare-earth aluminum garnet and rare-earth-orthoaluminate activated by the ions Ho^{3+}, Er^{3+} and Tm^{3+}, in *Abstracts of Papers of Fifth All-Union Symposium on Spectroscopy of Crystals*, Kazan (1976), 195 (in Russian).
796. Kaminskii, A. A., Pavlyuk, A. A., Butaeva, T. I. et al., Stimulated emission by subsidiary transitions of Ho^{3+} and Er^{3+}- ions in $KGd(WO_4)_2$ crystals, *Inorg. Mater. (USSR)* 13, 1251 (1977).
797. Andreae, T., Meschede, D. and Hänsch, T.W., New cw laser lines in the Er:$YAlO_3$ crystal, *Optics Comm.* 79, 1062 (1989).
798. Kaminskii, A. A. and Osiko, V. V., Sensitization in optical quantum generators based on CaF_2-ErF_3-Tm^{3+}, *Inorg. Mater. (USSR)* 3, 519 (1967).
799. Rines, D. M., Moulton, P. F., Welford, D. and Rines, G. A., High-energy operation of a Co:MgF_2 laser, *Optics Lett.* 19, 628 (1994).
800. Stoneman, R.C. and Esterowitz, L., Efficient, broadly tunable, laser-pumped Tm:YAG and Tm:YSGG cw lasers, *Optics Lett.* 15, 486 (1990).
801. Johnson, L. F., Dietz, R. E., and Guggenheim, H. J., Spontaneous and stimulated emission from Co^{2+}- ions in MgF_2 and ZnF_2, *Appl. Phys. Lett.* 5, 21 (1964).
802. Zverev, G. M., Garmash, V. M., Onischenko, A. M. et al., Induced emission by trivalent erbium ions in cryslals of yttrium-aluminum garnet, *J. Appl. Spectrosc. (USSR)* 21, 1467 (1974).
803. Ivanov, A. O., Mochalov, I. V., Tkachuk, A. M., Fedorov, V. A., and Feofilov, P. P., Spectral characteristics of the thulium ion and cascade generation of stimulated radiation in a $YAlO_3$:Tm^{3+}:Cr^{3+} crystal, *Sov. J. Quantum Electron.* 5, 117 (1975).
804. Arsenev, P. A. and Bienert, K. E., Absorption, luminescence and stimulated emission spectra of Tm^{3+} in $GdAlO_3$ crystals, *Phys. Status Solidi* 13a, K125 (1972).
805. Bagdasarov, Kh. S., Kaminskii, A. A., and Sobolev, B. P., Stimulated emission in lasers based on α-$NaCaErF_6$:Ho^{3+} and α-$NaCaErF_6$:Tm^{3+} crystals, *Inorg. Mater. (USSR)* 5, 527 (1969).
806. Moulton, P. F., Mooradian, A., and Reed, T. B., Efficient CW optically pumped Ni:MgF_2 laser, *Opt. Lett.* 3, 164 (1978).
807. Wallace, R. W., Oscillation of the 1.833 μm line in Nd^{3+}:YAG, *IEEE J. Quantum Electron.* QE-7, 203 (1971).
808. Barnes, N. P., Allen, R. E., Esterowitz, L., Chickis, E. P., Knights, M. G., and Jenssen, H. R., Operation of an Er YLF laser at 1.73 μm, *IEEE J. Quantum Electron.* QE-22, 337 (1986).
809. Kaminskii,, A. A., Pavlyuk, A. A., Polyakov, A. I., and Lyubchenko, V. V., A new lasing channel in a self-activated erbium crystal $KEr(WO_4)_2$, *Sov. Phys. Dokl.* 28, 154 (1983).
810. Krupke, W. F. and Gruber, J. B., Energy levels of Er^{3+} in LaF_3 and coherent emission at 1.61 μm, *J. Chem. Phys.* 41, 1225 (1964).
811. Esterowitz, L., Allen, R., and Eckardt, R., Cascade laser action in Tm^{3+}:YLF, in *The Rare Earths in Modern Science and Technology*, Vol. 3, McCarthy, G. I., Silber, H. B. and Rhyne, J. J., Eds., Plenum Press, New York (1982), p. 159.
812. Kaminskii, A. A., Two lasing channels of Tm^{3+} ions in lithium-yttrium fluoride *Inorg. Mater. (USSR)* 19, 1247 (1983).

813. Kaminskii, A. A., Sobolev, B., and Uvarova, T. V., Stimulated emission of Ho^{3+} in $BaTm_2F_8$ and Tm^{3+} ions in $BaYb_2F_8:Er^{3+}$ crystals, *Phys. Status Solidi A* 78, K13 (1983).
814. Antipenko, B. M., Glebov, A. S., Kiseleva, T. I., and Plamennyi, V. A., A new spectroscopic scheme of an active medium for the 2-μm band, *Opt. Spectrosc. (USSR)* 60, 95 (1986).
815. Antipenko, B. M., Glebov, A. S., Kiseleva, T. I., and Plamennyi, V. A., Conversion of absorbed energy in YAG:Cr^{3+},Tm^{3+},Ho^{3+} crystals, *Opt. Spectrosc. (USSR)* 64, 221 (1988).
816. Soffer, B. H. and Hoskins, R. H., Energy transfer and CW laser action in Tm^{3+}-Er_2O_3, *Appl. Phys. Lett.* 6, 200 (1968).
817. Johnson, L. F., Geusic, J. E., and Van Uitert, L. G., Efficient, high-power coherent emission from Ho^{3+} ions in yttrium aluminum garnet, assisted by energy transfer, *Appl. Phys. Lett.* 8, 200 (1966).
818. Smirnov, V. A. and Shcherbakov, I. A., Rare-earth scandium chromium garnets as active media for solid-state lasers, *IEEE J. Quantum Electron.* 24, 949 (1988).
819. Giorbachov, V. A., Zhekov, V, I., Murina, T. M. et al., Spectroscopic and growth properties of erbium aluminate with Tr^{3+} impurity ions, *Short Communications in Physics* 4, 16 (1973).
820. Johnson, L. F., Boyd, G. D. and Nassau, K., Optical maser characteristics of Ho^{3+} in $CaWO_4$, *Proc. IRE* 50, 87 (1962).
821. Johnson, L. F., Van Uitert, L. A., Rubin, J. J., and Thomas, R. A., Energy transfer from Er^{3+} to Tm^{3+} and Ho^{3+} ions in crystals, *Phys. Rev.*, 133A, 494 (1964).
822. Johnson, L. F., Boyd, G. D., and Nassau, K., Optical maser characteristics of Tm^{3+} in $CaWO_4$, *Proc. IRE.* 50, 86 (1962).
823. Van Uitert, L. G., Grodkiewicz, W. H., and Dearborn, E. F., Growth of large optical-quality yttrium and rare-earth aluminum garnets, *J. Am. Ceram. Soc.* 48, 105 (1965).
824. Barnes, N. P., Murray, K. E., Jani, M. G. and Kokta, M., Flashlamp pumped Ho:Tm:Cr:LuAG laser, *OSA Proc. Adv. Solid State Lasers*, Chai, B. H. T. and Payne, S. A., Eds., 24, 352 (1995).
825. Arsenyev, P. A. and Bienert, K. E., Spectral properties of Ho^{3+} in $GdAlO_3$ crystals, *Phys. Status Sol.* 13a, K129 (1972).
826. Rosenblatt, G. H., Quarles, G. J., Esterowitz, E., Randles, M., Creamer, J. and Belt, R., Continuous-wave 1.94-μm Tm:$CaY_4(SiO_4)_3O$ laser, *Optics Lett.* 18, 1523 (1993).
827. Saito, H., Chaddha, S., Chang, R. S. F. and Djeu, N., Efficient 1.94-μm Tm^{3+} laser in YVO_4 host, *Opt. Lett.* 17, 189 (1992).
828. Pinto, J. F. and Esterowitz, L., Tunable, flashlamp-pumped operation of a Cr,Tm:YAG laser between 1.945 and 2.014 μm, in *Advanced Solid-State Lasers*, Jenssen, H. P. and Dubé, G., Eds., Proceedings Vol. 6, Optical Society of America, Washington, DC (1990), p. 134.
829. Alpat'ev, A. N., Denisov, A. L., Zharikov, E. V., Zubenko, D. A., Kalitin, S. P., Noginov, M. A., Saidov, Z. S., Smirnov, V.A., Umyskov, A. F. and Shcherbakov, I. A., Crystal YSGG:Cr^{3+}:Tm^{3+} laser emitting in the 2-μm range, *Sov. J. Quantum Electron.* 20, 780 (1990).
830. Voron'ko, Yu. K., Gessen, S. B., Es'kov, N. A., Zverev, A. A., Ryabochkina, P. A., Sobol' A. A., Ushakov, S, N. and Tsymbal, L. I., Spectroscopic and lasing properties of calcium niobium gallium garnet activated with Tm^{3+} and Cr^{3+} ions, *Sov. J. Quantum Electron.* 22, 581 (1992).

831. Jani, M. G., Barnes, N. P., Murray, K. E., Hart, D. W., Quarles, G. J. and Castillo, V. K., Diode-pumped Ho:Tm:$LuLiF_4$ laser at room temperature, *IEEE J. Quantum Electron.* 32, 113 (1997).
832. Rubin, J. J. and Van Uitert, L. G., Growth of large yttrium vanadate single crystals for optical maser studies, *J. Appl. Phys.* 37, 2920 (1966).
833. Wunderlich, J. A., Sliney, J. A., and DeShazer, L. G., Stimulated emission at 2.04 μm in Ho^{3+}- doped $ErVO_4$ and YVO_4, *IEEE J. Quantum Electron.* QE-13, 69 (1977).
834. Dmitruk, M. V. and Kaminskii, A. A., Stimulated emission in a laser based on CaF_2-YF_3 crystals with Ho^{3+} and Er^{3+} ions, *Sov. Phys.-Crystallogr.* 14, 620 (1970).
835. Dmitruk, M. V., Kaminskii, A. A., Osiko, V. V., and Fursikov, M. M., Sensitization in CaF_2- ErF_3:Ho^{3+} lasers, *Inorg. Mater. (USSR)* 3, 516 (1967).
836. Voron'ko, Yu. K., Dmitruk, M. V., Murina, T. M., and Osiko, V. V., CW lasers based on mixed yttrofluorine-type crystals, *Inorg. Mater. (USSR)* 5, 422 (1969).
837. Robinson, M. and Devor, D. P., Thermal switching of laser emission of Er^{3+} at 2.69 μ and Tm^{3+} at 1.861 μ in mixed crystals of CaF_2:ErF_3:TmF_3, *Appl. Phys. Lett.* 10, 167 (1967).
838. Johnson, L. F. and Guggenheim, H. J., Electronic- and phonon-terminated laser emission from Ho^{3+} in BaY_2F_8, *IEEE J. Quantum Electron.* QE-10, 442 (1974).
839. Korovkin, A. M., Morozov, A. M., Tkachuk, A. M., Fedorov, A. A., Fedorov, V. A., and Feofilov, P. P., Spontaneous and stimulated emission from holmium in $NaLa(MoO_4)_2$ and $LaNbO_4$ crystals, in *Sbomik. Spektroskopiya Krisaliov*, Nauka, Moskova (1975).
840. Bagdasarov, Kh. S., Kaminskii, A. A., Kevorkov, A. M., Sarkisov, S. E., and Tevosyan, T. A., Stimulated emission from (Er,Lu)AlO_3 crystals with Ho^{3+} and Tm^{3+} ions, *Sov. Phys.-Crystallogr.* 18, 681 (1974).
841. Kalisky, Y., Kagan, J., Lotem, H., and Sagie, D., Continuous wave operation of multiply doped Ho:YLF and Ho:YAG laser, *Opt. Commun.* 65, 359 (1988).
842. Barnes, N. P., Eye-safe solid-state lasers for LIDAR applications, in Proc. Soc. Photo. Opt. Instrum. Eng., *Laser Radar Technology and Applications,* Vol. 663, Cruickshank, I. M. and Harney R. C., Eds., SPIE, Bellingham, WA (1986), p. 2.
843. Erbil, A. and Jenssen, H. P., Tunable Ho^{3+}:YLF laser at 2.06 μm, *Appl. Opt.* 19 1729 (1980).
844. Petrov, M. V. and Tkachuk, A. M., Delayed stimulated afterglow from holmium ions in crystals with coactivators, *Sov. J. Quantum Electron.* 10, 1478 (1980).
845. Anan'eva, G. V., Baranov, E. N., Zarzhitskaya, M. N., Ivanova, I. A., Koryakina, L. F., Petrova, M. A., Podkolzina, I. G., Semenov, T. S., and Yagmurova, G. P., Growth and physico-chemical investigation of single crystals of tysonite solid solutions $(Y,Ln)_{1-x}Sr_xF_{3-x}$, *Inorg. Mater. (USSR)* 16, 52 (1980).
846. Antipenko, B. M., Vorykhalov, I. V., Sirilitsyn, B. V., and Uvarova, T. V., Laser frequency converter based on a $BaYb_2F_8$:Ho^{3+} crystal stimulated emission at 2 μm, *Sov. J. Quantum Electron.* 10, 114 (1980).
847. Gillespie, P. S., Armstrong, R. L. and White, K. O., Spectral characteristics and atmospheric CO_2 absorption of the Ho^{+3}:YLF laser at 2.05 μm, *Appl. Optics* 15, 865 (1976).
848. Kaminskii, A. A., Petrosyan, A. G., Federov, V. A., Ryabchenkov, V. V., Pavlyuk, A. A., Lyubachenko, V. V., and Lukin, A. V., Two-micron stimulated emission of radiation by Ho^{3+} based on the $^5I_7 \rightarrow {}^5I_8$ transition in sensitized crystals, *Inorg. Mater. (USSR)* 17, 1430 (1981).

849. Cockayne, B., Plant, J. G., and Clay, R. A., The Czochralski growth and laser characteristics of Li(Y,Er,Tm,Ho)F, and Li(Lu,Er,Tm,Ho)F_4 scheelite single crystals, *J. Cryst. Growth* 54, 407 (1981).

850. Antipenko, B. M., Podkolzina, I. G., and Tomashevich, Yu. V., Use of $LiYbF_4$:Ho^{3+} as an active medium in a laser frequency converter, *Sov. J. Quantum Electron.* 10, 370 1980.

851. Kaminskii, A. A., Fedorov, V. A., Ryabchenkov, V., Sarkisov, S. E., Schultze, D., Bohm, J., and Reiche, P., Cascade generation of Ho^{3+} ions in $Gd_3Ga_5O_{12}$ crystal by the scheme $^5I_6$—$^5I_7$—$^5I_8$, *Inorg. Mater. (USSR)* 17, 828 (1981).

852. Podkolizina, I. G., Tkachuk, A. M., Fedorov, V. A. and Feofilov, P. P., Multifrequency generation of stimulated emission of Ho^{3+} ion in $LiYF_4$ crystals, *Opt. Spectrosc. (USSR)* 40, 111 (1976).

853. Chicklis, E. P., Naiman, C. S., Folweiller, R. C., Gabbe, D. R., Jenssen, H. P., and Linz, A., High efficiency room-temperature 2.06-μm laser using sensitized Ho^{3+}:YLF, *Appl. Phys. Lett.* 19, 119 (1971).

854. Remski, R. L., James, L. T., Gooen, K. H., DiBartolo, B., and Linz, A., Pulsed laser action in $LiYF_4$:Er^{3+}, Ho^{3+} at 77°K, *IEEE J. Quantum Electron.* QE-5, 212 (1969).

855. Gong, G. Z., Xu, G., Han, K. and Zhai, G., *Electron. Lett.* 26, 2063 (1990).

856. Morozov, A. M., Petrov, M. V., Startsev, V. R. et al., Luminescence and stimulated emission of holmium in yttrium- and erbium-oxyortho-silicate single crystals, *Opt. Spectrosc.* 41, 641 (1976).

857. Johnson, L. F., Dillon, J. F., and Remeika, J. P., Optical properties of Ho^{3+} ions in yttrium gallium garnet and yttrium iron garnet, *Phys. Rev.*, Bl, 1935 (1970).

858. Jani, M. G., Reeves, R. J. and Powell, R. C., Alexandrite-laser excitation of a Tm:Ho:$Y_3Al_5O_{12}$ laser, *J. Opt. Soc. Am. B* 8, 741, (1991).

859. Cha, S., Sugimoto, N., Chan, K. and Killinger, D. K., Tunable 2.1 μm Ho laser for DIAL remote sensing of atmospheric water vapor, in *Advanced Solid-State Lasers*, Jenssen, H. P. and Dubé, G., Eds., Proceedings Vol. 6, Optical Society of America, Washington, DC (1990), p. 165.

860. Dixon, G. J. and Johnson. L. F., Low-threshold 2-μm holmium laser excited by nonradiative energy transfer from Fe^{3+} in YGG, *Opt. Lett.* 17, 1782 (1992).

861. Alpat'ev, A. N., Zharikov, E. V., Zagumennyi, A. I., Zubenko, D. A., Kalitin, S. P., Lutts, G. B., Noginov, M. A., Smirnov, V. A., Umyskov, A. F. and Shcherbakov, I. A., Holmium GSAG:Cr^{3+}:Tm^{3+}:Ho^{3+} crystal laser (λ=2.09 μm) operating at room temperature, *Sov. J. Quantum Electron.* 19, 1400 (1989).

862. Barnes, N. P., Filer, E. D., Naranjo, F. L., Rodriguez, W. J. and Kokta, M. R., Spectroscopic and lasing properties of Ho:Tm:LuAG, *Optics Lett.* 18, 708 (1993).

863. McFarlane, R. A., Robinson, M., Pollack, S. A., Chang, D. B., and Jenssen, H. P., Visible and infrared laser operation by upconversion pumping of erbium-doped fluorides in *Tunable Solid State Lasers*, Shand, M. L. and Jenssen H. P., Eds., Optical Society of America, Washington, DC (1989), p. 179.

864. Ashurov, M. Kh., Voron'ko, Yu. K., Zharikov, E. V., Kaminskii, A. A., Osiko, V. V., Sobol', A. A., Timoshechkin, M. I., Fedorov, V. A., and Shabaltai, A. A., Structure, spectroscopy, and stimulated emission of crystals of yttrium holmium aluminum garnets, *Inorg. Mater. (USSR)* 15, 979 (1979).

865. Beck, R. and Gurs, K., Ho laser with 50-W output and 6.5% slope efficiency, *J. Appl. Phys.* 46, 5224 (1975).

866. Arsen'yev, P. A., Spectral parameters of trivalent holmium in a YAG lattice, *Ukrainskiy Fizicheskiy Zhurnal* 15, 689, 1970 (in Russian).

867. Diening, A., Dicks, B.-M., Heumann, E., Meyn, J. P., Petermann, K. and Huber, G., Continuous wave lasing near 2 μm in Tm^{3+} doped Y_2O_3, *Proceedings Advanced Solid State Lasers* (1997), to be published.

868. Johnson, L. F., Geusic, J. E., and Van Uitert, L. G., Coherent oscillations from Tm^{3+}, Ho^{3+}, Yb^{3+} and Er^{3+}- ions in yttrium aluminum garnet, *Appl. Phys. Lett.* 7, 127 (1965).
869. Duczynski, E. W., Huber, G., Ostroumov, V. G., and Shcherbakov, I. A., CW double cross pumping of the $^5I_7 - {}^5I_8$ laser transition in Ho^{3+}-doped garnets, *Appl. Phys. Lett.* 48, 1562 (1986).
870. Antipenko, B. M., Mak, A. A., and Sukhareva, L. K., Cross-relaxation $BaEr_2F_8$:Tm + Ho laser, *Sov. Tech. Phys. Lett.* 10, 217 (1984).
871. Antipenko, B. M., Glebov, A. S., Danbravyanu, R. V., Sobolev, B. P., and Uvarova, T. V., Spectroscopy and lasing characteristics of $BaEr_2F_8$:Tm:Ho crystals, *Sov. J. Quantum Electron.* 17, 424 (1987).
872. Antipenko, B. M., Glebov, A. S., and Dumbravyanu, R. V., Physics of energy storage in a $BaEr_2F_8$:Tm:Ho active medium, *Sov. J. Quantum Electron.* 18, 806 (1988).
873. Antipenko, B. M., Krutov, L. I., and Sulbsrevs, L. K., Cascade lasing of GSGG-Cr + Tm + Ho crystals, *Opt. Spectrosc. (USSR)* 61, 414 (1986).
874. Alpat'ev, A. N., Zharikov, E. V., Klifin, S. P., Laptev, V. V., Osiko, V. V., Ostroumov, V. G., Prokhorov, A. M., Salaov, Z. S., Smirnov, V. A., Sorokov, I. T., Umyskov, A. F. and Shcherbakov, I. A., Lasing of holmium ions as a result of the $^5I_7 \rightarrow {}^5I_8$ transition at room temperature in an yttrium scandium gallium garnet crystal activated with chromium, thulium, and holmium ions, *Sov. J. Quantum Electron.* 16, 1404 (1986).
875. Alpat'ev, A. N., Zharikov, E. V., Kalitin, S. P., Umyskov, A. F., and Shcherbakov, I. A., Efficient room-temperature lasing ($\lambda = 2.088$ μ) of yttrium scandium gallium garnet activated with chromium, thulium, and holmium ions, *Sov. J. Quantum Electron.* 17, 587 (1987).
876. Alpat'ev, A. N., Zharikov, E. V., Kalitin, S. P., Smirnov, V. A., Umyskov, A. F., and Shcherbakov, I. A., Q-switched laser utilizing an yttrium scandium gallium garnet crystal activated with holmium ions, *Sov. J. Quantum Electron.* 18, 617 (1988).
877. Kostin, V, V., Kulevsky, L. A., Murina, T. M. et al., $CaF_2:Dy^{2+}$ giant pulse laser with high repetition rate. *IEEE J. Quantum Electron.* QE-2, 611 (1966).
878. Barnes, N. R. and Gettemy, D. J., Pulsed Ho:YAG oscillator and amplifier, *IEEE J. Quantum Electron.* QE-17, 1303 (1981).
879. Fan, T. Y., Huber, G., Byer, R. L., and Mitzscherlich, P., Continuous-wave operation at 2.1 μm of a diode-laser-pumped, Tm-sensitized Ho:$Y_3Al_5O_{12}$ laser at 300 K *Opt. Lett.* 12, 678 (1987).
880. Fan, T. Y., Huber, G., Byer, R. L., and Mitzscherlich, P., Spectroscopy and diode laser-pumped operation of Tm Ho:YAG, *IEEE J. Quantum Electron.* 24, 924 (1988).
881. Kaminskii, A. A., Kurbanov, K., and Petrosyan, A. G., Spectral composition and kinetics of 2 μm stimulated emission of Ho^{3+} ions in sensitized $Y_3Al_5O_{12}$ and $Lu_3Al_5O_{12}$ single crystals, *Phys. Status Solidi A* 98, K57 (1987).
882. Kaminskii, A. A., Petrosyan, A. G., and Fedorov, V. A., Cross-cascade stimulated emission of Er^{3+}, Ho^{3+}, and Tm^{3+} ions in $Er_3Al_5O_{12}$, *Sov. Phys. Dokl.* 26, 309 (1981).
883. Storm, M. E., Laser characteristics of a Q-switched Ho:Tm:Cr:YAG, *Appl. Opt.* 27, 4170 (1988).
884. Hemmati, H., Efficient holmium:yttrium lithium fluoride laser longitudinally pumped by a semiconductor laser array, *Appl. Phys. Lett.* 51, 564 (1987).
885. Allen, R., Esterowitz, L., Goldberg, L., and Weiler, J. F., Diode-pumped 2 μm holmium laser, *Electron. Lett.* 22, 947 (1986).

886. Esterowitz, L., Allen, R., Goldberg, L., Weller, J. F., Sterm, M., and Abella, I., Diode-pumped 2 μm holmium laser, in *Tunable Solid-State Lasers II*, Vol. 52. Budgor, A. B., Esterowitz, L. and DeShazer, L. G., Eds., Springer-Verlag, New York (1986), p. 291.
887. Kaminskii, A. A., Petrosyan, A. G., Fedorov, V. A., Sarkisov, S. E., Ryabchenkov, V. V., Pavlyuk, A. A., Lyubchenko, V. V. and Mechalov, I. V., Two-micron stimulated emission by crystals with Ho^{3+} ions based on the transition $^5I_7$–$^5I_8$, *Sov. Phys. Dokl.* 26, 846 (1981).
888. Kintz, G. J., Esterowitz, L., and Allen, R., CW diode-pumped Tm^{3+}, Ho^{3+}:YAG 2.1 μm room- temperature laser, *Electron. Lett.* 23, 616 (1987).
889. Kaminskii, A. A., Petrosyan, A. G., and Ovanesyan, K. L., Cross-cascade generation of the stimulated emission of Tm^{3+} and Ho^{3+} ions in $Lu_3Al_5O_{12}$:Cr^{3+}-Tm^{3+}, Ho^{3+}, *Inorg. Mater. (USSR)* 19 1098 (1983).
890. Antipenko, B. M., Glebov, A. S., Kiseleva, T. I., and Pismennyi, V. A., 2.12-μm Ho:YAG laser *Sov. Tech. Phys. Lett.* 11, 284 (1985).
891. Antipenko, B. M., Glebov, A. S., Kiseleva, T. I., and Pismennyi, V. A., Interpretation of the temperature dependence of the YAG:Cr $^{3+}$ Tm^{3+} Ho lasing threshold, *Opt. Spectrosc. (USSR)* 63, 230 (1987).
892. Ivanov, A. O., Mochalov, I. V., Tkachuk, A. M., Fedorov, V. A. and Feofilov, P. P., Emission of $\lambda = 2$ μ stimulated radiation by holmium in aluminum holmium garnet crystals, *Sov. J. Quantum Electron.* 5, 115 (1975).
893. Hoskins, R. H. and Soffer, B. H., Energy transfer and CW laser action in Ho^{3+}:Er_2O_3, *IEEE J. Quantum Electron.* QE-2, 253 (1966).
894. Ivanov, A. O., Mochalov, I. V., Tkachuk, A. M., Fedorov, V. A., and Feofilov, P. P., Emission of $\lambda = 2$μ stimulated radiation by holmium in aluminum holmium garnet crystals, *Sov. J. Quantum Electron.* 5, 115 (1975).
895. Bakradze, R. V., Zverev, G. M., Kolodnyi, G. Ya. et al., Sensitized luminescence and stimulated radiation from yttrium-aluminum garnet crystals, *Sov. Phys.-JETP* 26, 323 (1968).
896. Remski, R. L. and Smith, D. J., Temperature dependence of pulsed laser threshold in YAG: Er^{3+}, Tm^{3+}, Ho^3, *IEEE J. Quantum Electron.* QE-6, 750 (1970).
897. Hopkins, R. H., Melamed, N. T., Henningsen, T., and Roland, G. W., Technical Rpt. AFAL-TR-70-103 (1970), Air Force Avionics Laboratory, Dayton, OH.
898. Porto, S. P. S. and Yariv, A., Trigonal sites and 2.24 micron coherent emission of U^{3+} in CaF_2, *J. Appl. Phys.* 33, 1620 (1962).
899. Porto, S. P. S. and Yariv, A., Low lying energy levels and comparison of laser action of U^{3+} in CaF_2 in *Proceedings 3rd International Conference Quantum Electronics*, Grivet, P. and Bloembergen, N., Eds., Columbia University Press, New York (1964), p. 717.
900. Wittke, J. P., Kiss, Z. J., Duncan, R. C., and McCormick, J. J., Uranium-doped calcium fluoride as a laser material, *Proc. IEEE* 51, 56 (1963).
901. Porto, S. P. S. and Yariv, A., Optical maser characteristics BaF_2:U^{3+}, *Proc. IRE* 50, 1542 (1962).
902. Porto, S. P. S. and Yariv, A., Excitation, relaxation and optical maser action at 2.407 microns in SrF_2:U^{3+}, *Proc. IRE* 50, 1543 (1962).
903. Zolotov, Ye. M., Osiko, V. V., Prokhorov, A. M., and Shipulo, O. P., Study of fluorescence and laser properties of SrF_2:Dy^{2+} crystals, *J. Appl. Spectrosc. (USSR)* 8, 627 (1968).
904. Caird, J. A., DeShazer, L. G. and Nella, J., Characteristics of room-temperature 2.3 μm laser emission from Tm^{3+} in YAG and $YAlO_3$, *IEEE J. Quantum Electron.* QE-11, 874 (1975).

905. Hobrock, L. M., DeShazer, L. G., Krupke, W. F., Keig, G. A., and Witter, D. E., Four-level operation of Tm:Cr:$YAlO_3$ laser at 2.35 μm, in *Digest of Technical Papers VII International Quantum Electronics Conference*, Montreal (1972), p. 15.
906. Kiss, Z. J. and Duncan, R. C., Pulsed and continuous optical maser action in $CaF_2:Dy^{2+}$, *Proc. IRE* 50, 1531 (1962).
907. Yariv, A., Continuous operation of a $CaF_2:Dy^{2+}$ optical maser, *Proc. IRE* 50, 1699 (1962).
908. Kiss, Z. J., The CaF_2-Tm^{2+} and the CaF_2-Dy^{2+} optical maser systems, in *Quantum Electronics Proceedings of the Third International Congress*, Grivet, P. and Bloembergen, N., Eds., Columbia University Press, New York (1964), p. 805.
909. Kiss, Z. J., Lewis, H. R., and Duncan, R. C., Sun pumped continuous optical maser, *Appl. Phys. Lett.* 2, 93 (1963).
910. Pressley, R. J. and Wittke, J. P., $CaF_2:Dy^{2+}$-lasers, *IEEE J. Quantum Electron.* QE-3, 116 (1967).
911. Hatch, S. E., Parsons, W. F., and Weagley, J. R., Hot-pressed polycrystalline $CaF_2:Dy^{3+}$ laser, *Appl. Phys. Lett.* 5, 153 (1964).
912. Pinto, J. F., Esterowitz, L. and Rosenblatt, G. H., Tm^{3+}:YLF laser continuously tunable between 2.20 and 2.46 μm, *Optics Lett.* 19, 883 (1994).
913. Stoneman, R. C., Esterowitz, L. and Rosenblatt, G. H., Tunable 2.3 μm $Tm^{3+}:LiYF_4$ laser, in *Tunable Solid State Lasers*, Shand, M. L. and Jenssen, H. P., Eds., Proceedings Vol. 5, Optical Society of America, Washington, DC (1989), p. 154.
914. Page, R. H., Schaffers, K. I., DeLoach, L. D., Wilke, G. D., Patel, F. D., Tassano, J. B., Payne, S. A. and Krupke, W. F., Cr^{2+}-doped chalcogenides as efficient, widely-tunable mid-infrared lasers, *IEEE J. Quantum Electron.* 33, 609 (1997).
915. DeLoach, L. D., Page, R. H., Wilke, G. D., Payne, S. A. and Krupke, W. F., Transition metal-doped zinc chalcogenides: spectroscopy and laser demonstration of a new class of gain media, *IEEE J. Quantum Electron.* 32, 885 (1996).
916. Arutyunyan, S. M., Kostanyan, R. B., Petrosyan, A. G., and Sanamyan, T. V., $YAlO_3:Er^{3+}$ crystal laser, *Sov. J. Quantum Electron.* 17, 1010 (1987).
917. Frauchiger, J., Lüthy, W., Albers, P. and Weber, H. P., Laser properties of selectively excited $YAlO_3$:Er, *Opt. Lett.* 13, 964 (1988).
918. Kaminskii, A. A., Petrosyan, A. G., Denisenko, G. A., Butaeva, T. I., Fedorov, V. A., and Sarkisov, S. E., Spectroscopic properties and 3 μm stimulated emission of Er^{3+} ions in the $(Y^{3+},Er^{3+})_3Al_5O_{12}$ and $(Lu^{3+},Er^{3+})_3Al_5O_{12}$ garnet crystal systems, *Phys. Status Solidi A* 71, 291 (1982).
919. Andriasyan, M. A., Vardanyan, N. V. and Kostanyan, R. B., Influence of the absorption of the excitation energy from the $^4I_{13/2}$ level of erbium ions on the operation of $Lu_3Al_5O_{12}$:Er crystal lasers, *Sov. J. Quantum Electron.* 12, 804 (1982).
920. Kaminskii, A. A. and Petrosyan, A. G., New functional scheme for 3-μ crystal lasers, *Sov. Phys. Dokl.* 24, 363 (1979).
921. Kaminskii, A. A., Seiranyan, K. B., and Arakelysn, A. Z., Peculiarity of the 3-μm stimulated emission of Er^{3+} ions in disordered fluoride crystals, *Inorg. Mater. (USSR)* 18, 446 (1982).
922. Stalder, M., Lüthy, W., and Weber, H. P., Five new 3-μm laser lines in $YAlO_3$:Er, *Opt. Lett.* 12, 602 (1987).
923. Pollack, S. A., Chang, D. B., and Moise, N. L., Continuous wave and Q-switched infrared erbium laser, *Appl. Phys. Lett.* 49 1578 (1986).
924. Pollack, S. A., Chang, D. B., and Moise, N. L., Upconversion-pumped infrared erbium laser, *J. Appl. Phys.* 60, 4077 (1986).

925. Kaminskii, A. A. and Petrosyan, A. G., Stimulated emissions from Er^{3+} ions at 1.7 μm in self-activated oxygen-containing Er crystals, *Inorg. Mater. (USSR)* 18, 1645 (1982).
926. Pollack, S. A. and Chang, D. B., Ion-pair upconversion pumped laser emission in Er^{3+} ions in YAG, YLF, SrF_2, and CaF_2 crystals, *J. Appl. Phys.* 64, 2885 (1988).
927. Kaminskii, A. A., Sobolev, B. P., Sarkisov, S. E., Fedorov, V. A., Ryabchenkov, V. V., and Uvarova, T. V., A new self-activated crystal for producing three-micron stimulated emission, *Inorg. Mater. (USSR)* 17, 829 (1981).
928. Kurtz, R., Fathe, L. and Birnbaum, M., New laser lines of erbium in yttrium aluminum garnet, in *Advanced Solid-State Lasers*, Jenssen, H. P. and Dubé, G., Eds., Proceedings Vol. 6, Optical Society of America, Washington, DC (1990), p. 247.
929. Es'kov, N. A., Kulevskii, L. A., Lukashev, A. V., Pashinin, P. P., Randoshkin, V. V. and Timoshechkin, M. I., Lasing of a calcium niobium gallium garnet crystal activated with chromium and erbium (λ=2.71 μm), *Sov. J. Quantum Electron.* 20, 785 (1990).
930. Kaminskii, A. A., Fedorov, V. A., Ivanov, A. O., Mochalov, I. V., and Krutov, L. I., Three-micron lasers based on $YAlO_3$ crystals with a high concentration of Ho^{3+} and Er^{3+} ions, *Sov. Phys. Dokl.* 27, 725 (1982).
931. Petrov, M. V. and Tkachuk, A. M., Optical spectra and multifrequency generation of induced emission of $LiYF_4$-Er^{3+} crystals, *Opt. Spectrosc. (USSR)* 45, 147 (1978).
932. Sorokin. P. P. and Stevenson, M. J., Stimulated infrared emission from trivalent uranium, *Phys. Rev. Lett.* 5, 557 (1960).
933. Boyd, G. D., Collins, R. J., Porto, S. P. S., Yariv, A., and Hargreaves, W. A., Excitation, relaxation and continuous maser action in 2.613 μm transition of CaF_2:U^{3+}, *Phys. Rev. Lett.* 8, 269 (1962).
934. Antipenko, B. M. and Dolgoborodov, L. E., $Y_3Al_5O_{12}$:Cr,Tm-Er four-level laser medium, in *Advanced Solid-State Lasers*, Jenssen, H. P. and Dubé, G., Eds., Proceedings Vol. 6, Optical Society of America, Washington, DC (1990), p. 244.
935. Rabinovich, W. S., Bowman, S. R., Feldman, B. J. and Winings, M. J., Tunable laser pumped 3 μm Ho:$YAlO_3$ laser, *IEEE J. Quantum Electron.* 27, 895 (1991).
936. Robinson, M. and Devor, D. P., Thermal switching of laser emission of Er^{3+} at 2.69 μ and Tm^{3+}- at 1.86 μ in mixed crystals of CaF_2:ErF_3:TmF_3, *Appl. Phys. Lett.* 10, 167 (1967).
937. Bagdasarov, S. Kh., Kulevskii, L. A., Prokhorov, A. M. et al., Erbium-doped CaF_2, crystal laser operating at room temperature, *Sov. J. Quantum Electron.* 4, 1469 (1975).
938. Kaminskii, A. A., Inorganic materials with Ln^{3+} ions for producing stimulated radiation in the 3 μ band, *Inorg. Mater. (USSR)* 15, 809 (1979).
939. Chicklis, E. P., Esterowitz, L., Allen, R., and Kruer, M., Stimulated emission at 2.81 μm in Er:YLF, in *Proceedings of LASERS '78*, Orlando, FL (1978).
940. Prokhorov, A. M., Kaminskii, A. A., Osiko, V. V. et al., Investigations of the 3 μm stimulated emission from Er^{3+} ions in aluminum garnets at room temperature, *Phys. Status Solidi* 40a, K69 (1977).
941. Kaminskii, A. A. and Petrosyan, A. G., Sensitized stimulated emission from self-saturating $^{3+}$1~m transitions of Ho^{3+} and Er^{3+} ions in $Lu_3Al_5O_{12}$ crystals, *Inorg. Mater. (USSR)* 15, 425 (1979).
942. Dinerman, B. J. and Moulton, P. F., 3-μm cw laser operations in erbium-doped YSGG, GGG, and YAG, *Optics Lett.* 19, 1143 (1994).
943. Hubert, S., Meichenin, D., Zhou, B.W. and Auzel, F., Emission properties, oscillator strengths and laser parameters of Er^{3+} in $LiYF_4$ at 2.7 μm, *J. Lumin.* 50, 7 (1991).

944. Xie, P. and Rand, S.C., Continuous-wave, pair-pumped laser, *Optics Lett.* 15, 848 (1990).
945. Meichenin, D., Auzel, F., Hubert, S., Simoni, E., Louis, M. and Gesland, J. Y., New room temperature CW laser at 2.82 μm:U^{3+}/$LiYF_4$, *Electron. Lett.* 30, 1309 (1994).
946. Kaminskii, A. A., Pavlyuk, A. A., Butaeva, T. I., Bobovich, L. I., and Lyubchenko, V. V., Stimulated emission in the 2.8 μm band by a self-activated crystal of $KEr(WO_4)_2$, *Inorg. Mater. (USSR)* 15, 424 (1979).
947. Zharikov, E. V., Zhekov, V. I., Murina, T. M., Osiko, V. V., Timoshechkin, M. I., and Shcherbakov, I. A., Cross section of the $^4I_{11/2} \rightarrow {}^4I_{13/2}$ laser transition in Er^{3+} ions in yttrium-erbium-aluminum garnet crystals, *Sov. J. Quantum Electron.* 7, 117 (1977).
948. Kaminskii, A. A., *Moderne Problem der Laserkristallphysik*, Physikalische Gesellschaft der DDR, Dresden (1977).
949. Basiev, T. T., Zharikov, E. V., Zhekov, V. I., Murina, T. M., Osiko, V. V., Prokhorov, A. M., Starikov, B. P., Timoshechkin, M. I., and Shcherbakov, I. A., Radiative and nonradiative transitions exhibited by Er^{3+} ions in mixed yttrium-erbium aluminum garnets, *Sov. J. Quantum Electron.* 6, 796 (1976).
950. Bowman, S. R., Shaw, L. B., Feldman, B. J. and Ganem, J., A. seven micron solid-state laser, *OSA Topical Meeting on Advanced Solid State Lasers* (1995).
951. Pinto, J. F., Rosenblatt, G. H. and Esterowitz, L., Continuous-wave laser action in Er^{3+}:YLF at 3.41 μm, *Electron. Lett.* 30, 1596 (1994).
952. Bowman, S. R., Rabinovich, W. S., Feldman, B. J. and Winings, M. J., Tuning the 3-μm Ho:$YAlO_3$ laser, in *Advanced Solid-State Lasers*, Jenssen, H. P. and Dubé, G., Eds., Proceedings Vol. 6, Optical Society of America, Washington, DC (1990), p. 254.
953. Antipenko, B. M., Buehenkov, V. A., Nikitiehev, A. A., Sobolev, B. P., Stepanov, A. I., Sukhareva, L. K., and Uvarova, T. V., Optimization of a $BaYb_2F_8$:Er active medium, *Sov. J. Quantum Electron.* 16, 759 1986.
954. Spariosu, K. and Birnbaum, M., Room-temperature 1.644-micron Er:YAG Lasers, in *Advanced Solid-State Lasers*, Chase, L. L. and Pinto, A. A., Eds., Proceedings Vol. 13, Optical Society of America, Washington, DC (1992), p. 127.
955. Bowman, S. R., Shaw, L. B., Feldman, B. J. and Ganem, J., A 7-μm praseodymium-based solid-state laser, *IEEE J. Quantum Electron.* 32, 646 (1996).
956. Breguet, J., Umyuskov, A. F., Lüthy, W. A. R., Shcherbakov, I. A. and Weber, H. P., Electrooptically q-switched 2.79 μm YSGG:Cr:Er laser with an intracavity polarizer, *IEEE J. Quantum Electron.* 27, 274 (1991).
957. Vodop'y nov, K. L., Kulevskii, L. A., Malyutin, A. A., Pashinin, P. P., and Prokhorov, A. M., Active mode locking in an yttrium erbium aluminum garnet crystal laser (λ = 2.94 μm), *Sov. J. Quantum Electron.* 12, 541 (1982).
958. Kaminskii, A. A., Fedorov, V. A., and Chan, Ng., Three-micron stimulated emission by Ho^{3+} ions in an $LaNbO_4$ crystal. *Inorg. Mater. (USSR)* 14, 1061 (1978).
959. Kaminskii, A. A., Advances in inorganic laser crystals. *Inorg. Mater. (USSR)* 20, 782 (1984).
960. Johnson, L. F. and Guggenheim, H. J., Laser emission at 3 μ from Dy^{3+} in BaY_2F_8, *Appl. Phys. Lett.* 23, 96 (1973).
961. Kintz, G. J., Alban, R. and Esterowitz, L., CW and pulsed 2.8 μm laser emission from diode-pumped Er^{3+}:LiYF, at room temperature, *Appl. Phys. Lett.* 50, 1553 (1987).

962. Vodop'ganov, K. L., Kulevskii, L. A., Pashlnin, P. P., Umyskov, A. F., and Shcherbakov, I. A., Bandwidth-limited picosecond pulses from a YSGG:Cr^{3+}:Er^{3+} laser (λ = 2.79 μm) with active mode locking, *Sov. J. Quantum Electron.* 17, 776 (1987).
963. Chickis, E. P., Esterowitz, L., Allen, R. and Kruer, M., Stimulated emission at 2.8 μm in Er^{3+}:YLF. in *Proc. Int. Conf. Lasers*, Corcoran, V. L., Ed., STS Press, McLean, VA (1979), p. 172.
964. Kaminskii, A., Stimulated emission in the presence of strong nonradiative decay in crystals in *Proc. Conf. Int. Sch. At. Mol. Spectrosc. Radiationless Processes*, Vol. 62, DiBartolo, B., Ed., Plenum Press, New York (1980), p. 499.
965. Kaminskii, A. A., Modern tendencies in the development of the physics and spectroscopy of laser crystals, *Bull. Acad. Sci. USSR Phys. Ser.* 45, 106 (1981).
966. Gilliland, G. D. and Powell, R. C., Spectral and up-conversion dynamics and their relationship to the laser properties of $BaYb_2F_8$:Ho^{3+}, *Phys. Rev. B* 38, 9958 (1988).
967. Eckardt, R. C. and Esterowitz, L., Multiwavelength mid-IR laser emission in Ho:YLF, *Digest for Conf. on Lasers and Electrooptics*, IEEE, Piscataway, NJ (1982), p. 160.
968. Stalder, M. and Lüthy, W., Polarization of 3 μm laser emission in $YAlO_3$:Er, *Opt. Commun.* 61, 274 (1987).
969. Andreeva, L. I., Vodop'yanov, K. L., Kaidalov, S. A., Kalinin, Yu. M., Karasev, M. E., Kulevskii, L. A., and Lukashev, A. V., Picosecond erbium-doped YAG laser (λ = 2.94 μm) with active mode locking, *Sov. J. Quantum Electron.* 16, 326 (1986).
970. Zharikov, E. V., Il'ichev, N. N., Kalitin, S. P., Laptev, V. V., Malyutin, A. A., Osiko, V. V., Pashinin, P. P., Prokhorov, A. M., Saidov, Z. S., Smirnov, V. A., Umyskov, A. P., and Shcherbakov, I. A., Spectral luminescence, and lasing properties of a yttrium scandium gallium garnet crystal activated with chromium and erbium, *Sov. J. Quantum Electron.* 16, 635 (1986).
971. Moulton, P. F., Manni, J. G., and Rlnes, G. A., Spectroscopic and laser characteristics of Er,Cr:YSGG, *IEEE J. Quantum Electron.* 24, 960 (1988).
972. Al'bers, P., Ostroumov, V. G., Umyskov, A. F., Shnell, S., and Sbcherbakov, I. A., Low threshold YSGG:Cr:Er laser for the 3-μm range with a high pulse repetition frequency. *Sov. J. Quantum Electron.* 18, 558 (1988).
973. Bagdasarov, Kh. S., Zhekov, V. I., Lobabev, V. A., Murina, T. M., and Prokhorov, A. M., Steady-state emission from a $Y_3Al_5O_{12}$:Er^{3+} laser (λ = 2.94 μm, T = 300 K), *Sov. J. Quantum Electron.* 13, 262 (1983).
974. Bagdasarov, Kh. S., Zbekov, V. I., Kukvskii, L. A., Murina, T. M., and Prokhorov, A. M., Giant laser radiation pulses from erbium-doped yttrium aluminum garnet crystals, *Sov. J. Quantum Electron.* 10, 1127 (1980).
975. Frauchiger, J. and Lüthy, W., Power limits of a YAG:Er laser, *Opt. Laser Technol.* 19, 312 (1987).
976. Strange, H., Petermann, K., Huber, G., and Duczynski, E. W., Continuous-wave 1.6 μm laser action in Er-doped garnets at room temperature, *Appl. Phys. B* 49, 269 (1989).
977. Becker, T., Clausen, R., Huber, G., Duczynski E. W., and Mitzscherlich, P., Spectroscopic and laser properties of Tm-doped YAG at 2 μm, in *Tunable Solid State Lasers*, Shand, M. L. and Jenssen H. P., Eds., Optical Society of America, Washington, DC (1989), p. 150.
978. Quarles, G. J., Rosenbaum, A., Marquardt, C. L., and Esterowitz, L., Efficient room-temperature operation of a flash-lamp-pumped Cr,Tm:YAG laser at 2.01 μm, *Opt. Lett.* 15, 42 (1990).

979. Duczynski, E. W., Huber, G., and Mitzscberlich, P., Laser action of Cr,Nd Tm,Ho-doped garnets in *Tunable Solid-State Lasers II*, Vol. 52, Budgor, A. B., Esterowitz, L. and DeShazer, L. C., Eds., Springer-Verlag, New York (1986), p. 282.
980. Kaminskii, A. A., Sarkisov, S. E., Klevtsov, P. V., Bagdasarov, Kh. S., Pavlyuk, A. A.. and Petrosyan, A. G., Investigation of stimulated emission in the $^4F_{13/2}$ transition of Nd^{3+} ions in crystals. V, *Phys. Status Solidi* 17a, K75 (1973).
981. Shkadarevich, A. P., Recent advances in tunable solid state lasers in *Tunable Solid State Lasers*, Shand, M. L. and Jenssen, H. P., Eds., Optical Society of America, Washington, DC (1989), p. 66.
982. Zverev, G. M. and Shesttkov, A. V., Tunable near-infrared oxide crystal lasers, *ibid.*
983. Kaminskii, A. A., Ueda, K., Uehara, N. and Verdun, H. R., Room-temperature diode-laser-pumped efficient cw and quasi-cw single-mode lasers based on Nd^{3+}-doped cubic disordered α-$NaCaYF_6$ and tetragonal ordered $LiLuF_4$ crystals, *Phys. Status Solidi A* 140, K45 (1993).
984. Davydov, S.V., Kulak, I.I., Mit'kovets, A.I., Stavrov, A.A. and Sckadarevich, A.P., Lasing in YAG:Nd^{3+} and KGdW:Nd^{3+} crystals pumped with semiconductor lasers, *Sov. J. Quantum Electron.* 21, 16 (1991).
985. Kaminskii, A. A. and Verdun, H. R., New room-temperature diode-laser-pumped cw lasers based on Nd^{3+}-ion doped crystals, *Phys. Status Solidi A* 129, K119 (1992).
986. Dmitruk, M. V, Zhekov, V. I., Prokhorov, A. M., and Timoshechkin, M. I., Spectroscopic properties of $Er_3Al_{5-x}Ga_xO_{12}$ films obtained by liquid-phase epitaxy, *Inorg. Mater. (USSR)* 15, 976 (1979).
987. O'Connor, J. R., Unusual crystal-field energy levels and efficient laser properties of YVO_4:Nd^{3+}, *Appl. Phys. Lett.* 9, 407 (1966).
988. Schearer, L. D., Loduc, M., and Zachorowski, J., CW laser oscillations and tuning characteristics of neodymium-doped lithium niobate crystals, *IEEE J. Quantum Electron.* QE-23, 1996 (1987).
989. Antipenko, B. M., Sinitsyn, B. V., and Uvarova, T. V., Laser converter utilizing $BaYb_2F_8$:Ho^{3+} with a three-micron output, *Sov. J. Quantum Electron.* 10, 1168 (1980).
990. Barnes, N.P. and Allen, R.E., Room temperature Dy:YLF laser operation at 4.34 µm, *IEEE J. Quantum Electron.* 27, 277 (1991).
991. Kaminskii, A. A., New lasing channels of crystals with rare earth ions, in *Laser and Applications, Pt. 1, Proc.*, Ursu, I. and Prokhorov, A. M., Eds., CIP Press, Bucharest, Romania (1982), p. 587.
992. Vodop'pnov, K. L., Vorob'ev, N. S., Kulevskii, L. A., Prokhorov, A. M., and Shchelev, M. Ya., Image-converter recording of picosecond pulses emitted by an erbium laser (A = 2.94 µ) with active mode locking, *Sov. J. Quantum Electron.* 13, 272 (1983).
993. Kaminskii, A. A., Federev, V. A., Petrosyan, A. G., Pavlyuk, A. A., Bohm, I., Reiche, P. and Schulz, D., Stimulated emission by Ho^{3+} ions in oxygen-containing crystals at low temperatures, *Inorg. Mater. (USSR)* 15, 1180 (1979).
994. Kaminskii, A. A., High-temperature spectroscopic investigation of stimulated emission from lasers based on crystals activated with Nd^{3+} ions, *Phys. Status Solidi*, la, 573 (1970).
995. Shen, H. Y., Lin, W. X., Zeng, R. R. et al., 1079.5 and 1341.4 nm: larger energy from a dual-wavelength Nd:$YAlO_3$ pulsed laser, *Appl. Optics* 32, 5952 (1993).
996. Field, S. J., Hanna, D. C., Shepherd, D. P., Tropper, A. C., Chandler, M. J., Townsend, P. D. and Zhang, L. Ion-implanted Nd:MgO:$LiNbO_3$ planar waveguide laser, *Optics Lett.* 16, 481 (1991).
997. Batygov, S. Kh., Kulevskii, L. A., Lavrukhin, S. A. et al., Laser based on CaF_2-ErF_3 crystals, *Kurzfassungen Internat. Tagung Laser und ihre Anwendungen*, Dresden (1973), Teil 2, K97.

998. Batygov, S. Kh., Kulevskii, L. A., Prokhorov, A. M. et al., Erbium-doped CaF_2, crystal laser operating at room temperature, *Sov. J. Quantum Electron.* 4, 1469 (1975).
999. Jones, J. K., de Sandro, J. P., Hempstead, M., Shepherd, D. P., Large, A. C., Tropper, A. C. and Wilkinson, J. S., Channel waveguide laser at 1 μm in Yb-diffused $LiNbO_3$, *Opt. Lett.* 20, 1477 (1995).
1000. Kaminskii, A. A., Verdun, G. R., Mill', B. V. and Butashin, A. V., New diode-laser-pumped continuous lasers based on compounds having the structure of calcium gallogermanate with Nd^{3+} ions, *Neorgan. Mater. (USSR)* 28, 141 (1992).
1001. Koch, G. J., Deyst, J. P. and Storm, M. E., Single-frequency lasing of monolithic Ho,Tm:YLF, *Opt. Lett.* 18, 1235 (1993).
1002. Kaminskii, A. A., Karasev, V. A., Dubrov, V. D., Yakunin, V. P., Mill', B. V. and Butashin, A. V., New disordered $Ca_2Ga_2SiO_7$:Nd^{3+} crystal for high-power solid-state lasers, *Sov. J. Quantum Electron.* 22, 97 (1992).
1003. Manaa, H., Guyot, Y. and Moncorge, R., Spectroscopic and tunable laser properties of Co^{2+}-doped single crystals, *Phys. Rev. B* 48, 3633 (1993).
1004. Antipenko, B. M., Mak, A. A., Raba, O. B., Sukhareva, L. K., and Uvarova, T. V., 2-μm-range rare earth laser, *Sov. Tech. Phys. Lett.* 9, 227 (1983).
1005. Dmitriev, V. G., Raevskii, E. V., Rubina, N. M., Rashkovich, L. N., Silichev, O. O. and Formichev, A. A., 2-μm range rare earth laser, *Sov. Tech. Phys. Lett.* 4, 590 (1979).
1006. Li, M. J., Wang, L., Xie, C., Peng, K. and Xu, G., *Proc. SPIE* 1726, 519 (1992).
1007. Eckhardt, R. C., DeRosa, J. L., and Letellier, J. P., Characteristics of an Nd:CaLaSOAP mode-locked oscillator, *IEEE J. Quantum Electron.* QE-10, 620 (1974).
1008. Cordova-Plaza, A., Fan, T. Y., Digonnet, M. J. F., Byer, R. L. and Shaw, H. J., Nd:MgO:$LiNbO_3$ continuous-wave laser pumped by a laser diode, *Opt. Lett.* 13, 209 (1988).
1009. de Micheli, M. P., in *Guided Wave Nonlinear Optics,* Ostrowsky, D. B. and Renisch, R., Eds., Kluwer Academic Publishers, Dordrecht (1992), p. 147.
1010. Li, M. J., de Micheli, M. P., He, Q. and Ostrowsky, D. B., *IEEE J. Quantum Electron.* QE-26, 1384 (1990).
1011. He, Q., de Micheli, M. P., Ostrowsky, D. B. et al., Self-frequency-doubled high Δn proton exchanged Nd:$LiNbO_3$ waveguide laser, *Opt. Commun.* 89, 54 (1992).
1012. Yamamoto, J. K., Sugimoto, A. and Yamagishi, K., Self-frequency doubling in Nd,Sc2O3:$LiNbO_3$ at room temperature, *Opt. Lett.* 19, 1311 (1994).
1013. Kaminskii, A. A., Mill, B. V. and Butashin, A. V., Stimulated emission from Nd^{3+} ion in acentric $LaBGeO_5$ crystals, *Phys. Status Solidi* (a) 118, K59 (1990).
1014. Kaminskii, A. A., Butashin, A. V., Maslyanitsin, I. A. et al., Pure and Nd^{3+}-,Pr^{3+}-ion doped acentric $LaBGeO_5$ single crystals, *Phys. Status Solidi* (a) 125, 671 (1991).
1015. Kaminskii, A. A., Bagaev, S. N., Mill, B. V. and Butashin, A. V., New inorganic material $LaBGeO_5$-Nd^{3+} for crystalline lasers with self-multiplied generation frequency, *Neorg. Mater. (Russia)* 29, 545 (1993).
1016. Kaminskii, A. A., Shkadarevich, A. P., Mill, B. V., Koptev, V. G., Butashin, A. V. and Demidovich, A. A., *Neorg. Mater. (Russia)* 24, 690 (1986).
1017. Kaminskii, A. A., Butashin, A. V., Demidovich, A. A., Koptev, V. G., Mill, B. V. and Shkadarevich, A. P., *Phys. Status Solidi* (a) 112, 197 (1989).
1018. Knappe, R., Bartschke, J., Becher, C., Beier, B., Scheidt, M., Boller, K. J. and Wallenstein, R., *Conf. Proc. 1994 IEEE Nonlinear Optics* (1994), p. 39.
1019. Lu, B., Wang, J., Pan, H., Jiang, M., Liu, E. and Hou, X., *Chin. Phys. Lett.* 3, 423 (1986).
1020. Osiko, V. V., Sigachev, V. B., Strelov, V. I. and Timoshechkin, M. I., Erbium gadolinium gallium garnet crystal laser, *Sov. J. Quantum Electron.* 21, 159 (1991).

1021. Dorozhkin, L. M., Kuratev, I. I., Leonyuk, N. I., Timochenko, T. I. and Shestakov, A. V., Cerenkov configuration second harmonic generation in proton-exchanged lithium niobate guides, *Sov. Tech. Phys. Lett.* 7, 555 (1981).

1022. Kaminskii, A. A., Butashin, A. V., Demidovich, M. I., Zhavaronkov, N. I., Mikhailov, V. P. and Shkadarevich, A. P., Generation of picosecond pulses by acentric disordered silicates $La_3Ga_5SiO_{14}$ and $Ca_2Ga_2SiO_7$ with Nd^{3+} ions, *Neorg. Mater. (Russia)* 24, 2075 (1986).

1023. Kaminskii, A. A., Butashin, A. V. and Bagaev, S. N., New Nd^{3+}:$BaLu_2F_8$ laser crystal, *Quantum Electron.* 26, 753 (1996).

1024. Kaminskii, A. A., New high-temperature induced transition of an optical quantum generator based on SrF_2-Nd^{3+} crystals (type I), *Inorg. Mater. (USSR)* 5, 525 (1969).

1025. Tin, P. and Schearer, L. D., A high power, tunable, arc-lamp pumped Nd-doped lanthanum-hexaluminate laser, *J. Appl. Phys.* 68, 950 (1990).

1026. Spariosu, K., Birnbaum, M. and Kokta, M., Room-temperature 1.643-μm Er^{3+}:$Y_3Sc_2Ga_3O_{12}$ (Er:YSGG) laser, *Appl. Optics* 34, 8272 (1995).

1027. Loutts, G. B., Bonner, C., Meegoda, C. et al., Crystal growth, spectroscopic characterization, and laser performance of a new efficient laser material Nd:$Ba_5(PO_4)_3F$, *Appl. Phys. Lett.* 71, 303 (1997).

1028. Antipenko, B. M., Mak, A. A., Sinitsyn, B. V., Raba, O. B., and Uvurova, T. V., New excitation schemes for laser transitions, *Sov. Phys. Tech. Phys.* 27, 333 (1982).

1029. Djeu, N., Hartwell, V. E., Kaminskii, A. A. and Butashin, A. V., Room-temperature 3.4-μm Dy:BaY_2F_8 laser, *Optics Lett.* 22, 997 (1997).

1030. Möbert, P. E.-A., Heumann, E., Huber, G. and Chai, B. H. T., Green Er^{3+}:$YLiF_4$ upconversion laser at 551 nm with Yb^{3+} codoping: a novel pumping scheme, *Opt. Lett.* 22, 1412 (1997).

1031. Hömmerich, U., Wu, X. and Davis, V. R., Demonstration of room-temperature laser action at 2.5 μm from Cr^{2+}:$Cd_{0.85}Mn_{0.15}Te$, *Opt. Lett.* 22, 1180 (1997).

1032. Evans, J. M., Petricevic, V., Bykov, A. B., Delgado, A. and Alfano, R. R., Direct diode-pumped continuous-wave near-infrared tunable laser opertion of Cr^{4+}:forsterite and Cr^{4+}:Ca_2GeO_4, *Opt. Lett.* 22, 1171 (1997).

Section 2.2

GLASS LASERS

Introduction to the Tables

Glass lasers have used lanthanide ions almost exclusively as the active ion. The lasing transitions for these ions and approximate wavelengths are given in Table 2.2.1. Although the exact wavelength varies slightly with host glass composition and temperature, this table can generally be used to identify the transition involved in laser action.

A few ion-glass systems in which only gain has been reported are listed in Table 2.2.2.

Glass lasers and amplifiers based on lanthanide ions are arranged in order of increasing wavelength in Table 2.2.3. Lasers that have been tuned over a range of wavelengths are listed by the lowest wavelength reported; the tuning range given is that for the configuration and conditions used and may not represent the extremes possible. The lasing ion is listed in the second column. The host glass type is specified by the glass network former (silicate, phosphate, etc.) and, if known, the principal glass network modifier cation(s). If codopants or sensitizing ions are added, they are listed following the colon. The form of the glass—bulk (rod, disk), fiber, or planar waveguide—is noted in the fourth column.

Further Reading

Davey, S. T., Ainslie, B. J. and Wyatt, R., Waveguide Glasses, in *Handbook of Laser Science and Technology, Suppl. 2: Optical Materials,* CRC Press, Boca Raton, FL (1995), p. 635.

Desurvire, E., *Erbium-Doped Fiber Amplifiers*, John Wiley & Sons, New York, (1994).

Digonnet, M. J. F., Ed., *Selected Papers on Rare-Earth-Doped Fiber Laser Sources and Amplifiers,* SPIE Milestone Series, Vol. MS37, SPIE Optical Engineering Press, Bellingham, WA (1992).

Digonnet, M. J. F., Ed., *Rare Earth Doped Fiber Lasers and Amplifiers*, Marcel Dekker, New York (1993).

France, P. W., Ed., *Optical Fibre Lasers and Amplifiers*, Blackie and Sons, Ltd., Glasgow and London (1991).

Hall, D. W. and Weber, M. J., Glass Lasers, in *Handbook of Laser Science and Technology, Suppl. 1: Lasers*, CRC Press, Boca Raton, FL (1991), p. 137.

Miniscalco, W. J., Erbium-doped glasses for fiber amplifiers at 1500 nm, *J. Lightwave Techn.* 9, 234 (1991).

Rapp, C. F., Laser Glasses: Bulk Glasses, in *Handbook of Laser Science and Technology, Suppl. 2: Optical Materials,* CRC Press, Boca Raton, FL (1995), p. 619.

Stokowski, S. E., Glass Lasers, in *Handbook of Laser Science and Technology, Vol. I: Lasers and Masers*, CRC Press, Boca Raton, FL (1982), p. 215.

Urquhart, P., Review of rare earth doped fibre lasers and amplifiers, *IEE Proc.* 135, 385 (1988).

Weber, M. J., Science and technology of laser glass, *J. Non-Cryst. Solids*, 123, 208 (1991).

Table 2.2.1
Glass Laser Transitions

Wavelength* (µm)	Ion	Transition
0.38	Nd^{3+}	$^4D_{3/2} \rightarrow {}^4I_{11/2}$
0.41	Nd^{3+}	$^2P_{3/2} \rightarrow {}^4I_{11/2}$
0.46	Tm^{3+}	$^1D_2 \rightarrow {}^3F_4$
0.48	Tm^{3+}	$^1G_4 \rightarrow {}^3H_6$
0.49	Pr^{3+}	$^3P_0 \rightarrow {}^3H_4$
0.52	Pr^{3+}	$^3P_1 \rightarrow {}^3H_5$
0.54	Tb^{3+}	$^5D_4 \rightarrow {}^7F_5$
0.55	Er^{3+}	$^4S_{3/2} \rightarrow {}^4I_{15/2}$
0.55	Ho^{3+}	$^5S_2 \rightarrow {}^5I_8$
0.61	Pr^{3+}	$^3P_0 \rightarrow 3H_6$
0.64	Pr^{3+}	$^3P_0 \rightarrow {}^3F_2$
0.65	Sm^{3+}	$^4G_{5/2} \rightarrow {}^6H_{9/2}$
0.70	Pr^{3+}	$^3P_1 \rightarrow {}^3F_4$
0.72	Pr^{3+}	$^3P_1 \rightarrow 3F_4$
0.75	Ho^{3+}	$^5S_2 \rightarrow {}^5I_7$
0.82	Tm^{3+}	$^3H_4 \rightarrow {}^3H_6$
0.85	Er^{3+}	$^4S_{3/2} \rightarrow {}^4I_{13/2}$
0.88	Pr^{3+}	$^3P_1 \rightarrow {}^1G_4$
0.89	Pr^{3+}	$^3P_0 \rightarrow {}^1G_4$
0.93	Nd^{3+}	$^4F_{3/2} \rightarrow {}^4I_{9/2}$
0.93	Pm^{3+}	$^5F_1 \rightarrow {}^5I_5$
0.99	Er^{3+}	$^4I_{11/2} \rightarrow {}^4I_{15/2}$
1.03	Yb^{3+}	$^2F_{5/2} \rightarrow {}^2F_{7/2}$
1.06	Nd^{3+}	$^4F_{3/2} \rightarrow {}^4I_{11/2}$
1.08	Pr^{3+}	$^1D_2 \rightarrow {}^3F_{3,4}$
1.10	Pm^{3+}	$^5F_1 \rightarrow {}^5I_6$
1.2	Ho^{3+}	$^5I_6 \rightarrow {}^5I_8$
1.31	Pr^{3+}	$^1G_4 \rightarrow {}^3H_5$
1.34	Nd^{3+}	$^4F_{3/2} \rightarrow {}^4I_{13/2}$
1.38	Ho^{3+}	$^5S_2, {}^5F_4 \rightarrow {}^5I_5$

Table 2.2.1—*continued*
Glass Laser Transitions

Wavelength* (μm)	Ion	Transition
1.48	Tm^{3+}	$^3H_4 \rightarrow {}^3F_4$
1.51	Tm^{3+}	$^1D_2 \rightarrow {}^1G_4$
1.55	Er^{3+}	$^4I_{13/2} \rightarrow {}^4I_{15/2}$
1.66	Er^{3+}	$^2H_{11/2} \rightarrow {}^4I_{9/2}$
1.72	Er^{3+}	$^4S_{3/2} \rightarrow {}^4I_{9/2}$
1.88	Tm^{3+}	$^3F_4 \rightarrow {}^3H_6$
2.05	Ho^{3+}	$^5I_7 \rightarrow {}^5I_8$
2.3	Tm^{3+}	$^3H_4 \rightarrow {}^3H_5$
2.75	Er^{3+}	$^4I_{11/2} \rightarrow {}^4I_{13/2}$
2.9	Ho^{3+}	$^5I_6 \rightarrow {}^5I_7$
3.5	Er^{3+}	$^4F_{9/2} \rightarrow {}^4I_{9/2}$
3.9	Ho^{3+}	$^5I_5 \rightarrow {}^5I_6$

*Wavelengths of transitions are only approximate; exact wavelengths are dependent on the host and temperature and the specific Stark levels involved.

Table 2.2.2
Ions in Glasses Exhibiting Gain

Wavelength (μm)	Ion	Host glass (sensitizer)	Form	Reference
0.543	Ho^{3+}	fluorozircoaluminate (Yb)	fiber	1
0.546	Er^{3+}	fluorozirconate	fiber	2
0.560–0.585	Cu^{+}	aluminoborosilicate	bulk	3
0.625	CdSSe	silicate	bulk	4
0.633	Cu^{+}	fluorohafnate	bulk	5
1.083	Nd^{3+}	chalcogenide	fiber	6
1.536	Er^{3+}	silica	fiber	7
2.05	Ho^{3+}	alumino-zirco-fluoride(Tm^{3+})	bulk	8

References—Table 2.2.2

1. Shikida, A., Yanagita, H. and Toratani, H., Ho-Yb fluoride glass fiber for green lasers, in *Advanced Solid-State Lasers*, Pinto, A. A. and Fan, T. Y., Eds., Proceedings Vol. 15, Optical Society of America, Washington, DC (1993), p. 261.
2. Ugawa, T. S., Komukai, T. and Miyajuina, Y., Optical amplification in Er^{3+} doped single mode fluoride fiber, *IEEE Phot. Techn. Lett.* 2, 475 (1990).
3. Kruglik, G. S., Skripko, G. A., Shkadarevich, A. P., Ermolenko, N. N., Gorodetskaya, O. G., Belokon, M. V., Shagov, A. A. and Zolotareva, L. E., Amplification of yellow-green light in copper-activated glass, *Opt. Spectrosc. (USSR)* 59, 439 (1985); Copper-doped aluminoborosilicate glass spectroscopic characteristics and stimulated emission, *J. Lumin.* 34, 343 (1986).
4. Zhou, F., Qin, W., Jin, C. et al., Optical gain of CdSSe-doped glass, *J. Lumin.* 60 & 61, 353 (1994).
5. DeShazer, L. G., Cuprous ion doped crystals for tunable lasers, in *Tunable Solid State Lasers,* Hammerling, P., Budgar, A. B., and Pinto, A., Eds., Springer-Verlag, Berlin (1985), p. 91.
6. Mori, A., Ohishi, Y., Kanamori, T. and Sudo, S., Optical amplification with neodymium-doped chalcogenide glass fiber, *Appl. Phys. Lett.* 70, 1230 (1997).
7. Nakazawa, M., Kimura, Y. and Suzuki, K., Efficient Er^{3+}-doped optical fiber amplifier pumped by a 1.48 μm InGaAsP laser diode, *Appl. Phys. Lett.* 54, 295 (1989).
8. Doshida, M., Teraguchi. K. and Obara, M., Gain measurement and upconversion analysis in Tm^{3+}, Ho^{3+} co-doped alumino-zirco-fluoride glass, *IEEE J. Quantum Electron.* 31, 911 (1995).

Table 2.2.3
Glass Lasers Arranged in Order of Increasing Wavelength

Wavelength[a] (μm)	Ion	Host glass[b]	Form	Reference
0.381	Nd^{3+}	fluorozirconate	fiber	1
0.412	Nd^{3+}	fluorozirconate	fiber	2
0.455	Tm^{3+}	fluorozirconate	fiber	3,4
0.480	Tm^{3+}	fluorozirconate	fiber	3,5
0.482	Tm^{3+}	fluorozirconate	fiber	15
0.491	Pr^{3+}	fluorozirconate	fiber	6,7
0.491–0.493	Pr^{3+}	fluorozirconate (Yb)	fiber	8
0.492	Pr^{3+}	fluorozirconate	fiber	9
0.517–0.540	Pr^{3+}	fluorozirconate (Yb)	fiber	8
0.520	Pr^{3+}	fluorozirconate	fiber	6,7
0.521	Pr^{3+}	fluorozirconate	fiber	10
~0.539–550	Ho^{3+}	fluorozirconate	fiber	17
0.54[c]	Tb^{3+}	borate	bulk	11
0.540–0.545[d]	Er^{3+}	fluorozirconate	fiber	12
0.540–0.553	Ho^{3+}	fluorozirconate	fiber	13
0.546	Er^{3+}	fluorozirconate	fiber	12,14
0.548	Er^{3+}	fluorozirconate	fiber	16
0.599–0.611	Pr^{3+}	fluorozirconate	fiber	7
0.601–0.618	Pr^{3+}	fluorozirconate	fiber	18
0.602	Pr^{3+}	fluorozirconate (Yb)	fiber	19
0.605	Pr^{3+}	fluorozirconate	fiber	6
0.605–0.622	Pr^{3+}	fluorozirconate (Yb)	fiber	8
0.631–0.641	Pr^{3+}	fluorozirconate	fiber	7,18
0.6328	Pr^{3+}	fluorozirconate	fiber	20
0.635	Pr^{3+}	fluorozirconate	fiber	6,7,10
0.635	Pr^{3+}	fluorozirconate (Yb)	fiber	19,23
0.635–0.637	Pr^{3+}	fluorozirconate (Yb)	fiber	8
0.651	Sm^{3+}	silica	fiber	24
0.690–0.703	Pr^{3+}	fluorozirconate	fiber	18
0.707–0.725	Pr^{3+}	fluorozirconate	fiber	18

Table 2.2.3—*continued*
Glass Lasers Arranged in Order of Increasing Wavelength

Wavelength[a] (μm)	Ion	Host glass[b]	Form	Reference
0.715	Pr^{3+}	fluorozirconate	fiber	6,7
0.7495–0.7545	Ho^{3+}	fluorozirconate	fiber	13
0.800–0.830	Tm^{3+}	fluorozirconate	fiber	25,26
0.803–0.816	Tm^{3+}	fluorozirconate	fiber	27
0.815–0.825	Tm^{3+}	fluorozirconate	fiber	28
0.85	Er^{3+}	fluorozirconate	fiber	29,30
0.880–0.886	Pr^{3+}	fluorozirconate	fiber	18
0.89	Pr^{3+}	silica (Ge)	fiber	31
0.899–0.951	Nd^{3+}	silica	fiber	32,33
0.902–0.916	Pr^{3+}	fluorozirconate	fiber	18
0.905	Nd^{3+}	phosphate	planar	21
0.918	Nd^{3+}	Na–Ca silicate	bulk	34
0.92	Nd^{3+}	silicate	bulk	35
0.933	Pm^{3+}	Pb–In phosphate	bulk	36
0.938	Nd^{3+}	silica	fiber	37,38
0.974	Yb^{3+}	silica	fiber	39
0.980	Yb^{3+}	silica (Al,P)	fiber	40
0.981–1.004	Er^{3+}	fluorozirconate	fiber	41
1.00	Er^{3+}	fluorozirconate	fiber	42
1.000–1.050	Yb^{3+}	fluorozirconate	fiber	43
1.000–1.085	Pr^{3+}	silica	fiber	44
1.010– 1.162	Yb^{3+}	silica	fiber	39
1.015	Yb^{3+}	Li–Mg–Al silicate	bulk	45,46
1.015–1.140	Yb^{3+}	silica	fiber	47
1.018	Yb^{3+}	Ca–Li borate (Nd)	bulk	48
1.020	Yb^{3+}	fluorozirconate	fiber	43
1.028–1.064	Yb^{3+}	silica	fiber	47
1.047	Nd^{3+}	fluoroberyllate	bulk	49
1.05	Nd^{3+}	phosphate	fiber	50
1.05	Nd^{3+}	phosphate (LG–760)	planar	51

Table 2.2.3—*continued*
Glass Lasers Arranged in Order of Increasing Wavelength

Wavelength[a] (μm)	Ion	Host glass[b]	Form	Reference
1.050	Nd^{3+}	fluorozirconate	fiber	52
1.050–1.075	Nd^{3+}	silicate (P)	fiber	53
1.051	Nd^{3+}	fluorophosphate	bulk	54
1.051	Nd^{3+}	fluorozirconate	microsphere	197
1.0515	Nd^{3+}	silica (P)	planar	55
1.0525	Nd^{3+}	silica (P)	planar	56
1.053	Nd^{3+}	silica	planar	57
1.0535	Nd^{3+}	Na–K–Cd phosphate	bulk	58
1.054	Nd^{3+}	phosphate (LHG–5)	planar	59
1.054	Nd^{3+}	phosphate (LHG–8)	bulk	60
1.054	Nd^{3+}	Zn–Li phosphate	bulk	61,62
1.0546	Nd^{3+}	phosphate (APG-1)	bulk	22
1.055	Nd^{3+}	Li–Nd–La phosphate	bulk	63
1.055	Nd^{3+}	silica (P)	fiber	64
1.057	Nd^{3+}	silicate (LG–660)	planar	65
1.057	Nd^{3+}	phosphate	planar	21
1.058	Nd^{3+}	borosilicate (BK–7)	planar	66
1.06	Nd^{3+}	Ba silicate	planar	67
1.06	Nd^{3+}	borate	bulk	68
1.06	Nd^{3+}	borosilicate	planar	69
1.06	Nd^{3+}	glass ceramic	bulk	70,71
1.06	Nd^{3+}	lead silicate (F7)	fiber	72
1.06	Nd^{3+}	Li germanate	bulk	73
1.06	Nd^{3+}	Li–La phosphate (Cr)	bulk	74
1.06	Nd^{3+}	silicate	bulk	75
1.06	Nd^{3+}	silicate	fiber	76
~1.06	Nd^{3+}	silicate (GLS2,GLS3)	planar	77
1.06	Yb^{3+}	K–Ba silicate	bulk	78
1.060	Nd^{3+}	silica (Al)	fiber	79
1.060	Nd^{3+}	silica (Al)	planar	80
1.0605	Nd^{3+}	Li tellurite	bulk	58,81

Table 2.2.3—*continued*
Glass Lasers Arranged in Order of Increasing Wavelength

Wavelength[(a)] (μm)	Ion	Host glass[(b)]	Form	Reference
1.061	Nd^{3+}	K–Ba silicate	bulk	82
1.061	Nd^{3+}	tellurite	fiber	83
1.062	Nd^{3+}	silicate (buffer)	bulk	84
1.0635	Nd^{3+}	silica (Al)	bulk	85
1.064	Nd^{3+}	Ba silicate	planar	86
1.066	Nd^{3+}	tellurite	bulk	87
1.069–1.144	Nd^{3+}	silica	fiber	33,34,88
1.08	Pr^{3+}	silica	bulk	31
1.080	Nd^{3+}	chalcogenide (Ga:La:S)	bulk, fiber	89
1.088	Nd^{3+}	silica (Al)	fiber	90,91
1.088	Nd^{3+}	silica (Ge)	fiber	92
1.098	Pm^{3+}	Pb–In phosphate	bulk	36
1.12	Yb^{3+}	silica (Ge)	fiber	93
1.128	Yb^{3+}	germanosilicate	fiber	195
1.134	Yb^{3+}	germanosilicate	fiber	195
1.19	Ho^{3+}	fluorozirconate	fiber	127
1.24–1.34	Pr^{3+}	fluorozirconate	fiber	94
1.28–1.34	Pr^{3+}	fluorozirconate	fiber	95
1.290–1.320	Pr^{3+}	fluorozirconate	fiber	96
1.294	Pr^{3+}	fluorozirconate	fiber	97
1.3	Pr^{3+}	fluorozirconate (Yb)	fiber	98
1.31	Pr^{3+}	fluorozirconate	fiber	99
1.310–1.370	Nd^{3+}	fluorozirconate	fiber	100
1.31–1.36	Nd^{3+}	fluorozirconate	fiber	101
1.32	Nd^{3+}	Li–Nd–La phosphate	bulk	63
1.323	Nd^{3+}	fluorophosphate	fiber	102
1.325	Nd^{3+}	phosphate (LHG–5)	planar	103
1.328	Nd^{3+}	fluorophosphate	fiber	102
1.334	Nd^{3+}	fluorozirconate	microsphere	197
1.338	Nd^{3+}	fluorozirconate	fiber	104
1.340	Nd^{3+}	fluorozirconate	fiber	105

Table 2.2.3—*continued*
Glass Lasers Arranged in Order of Increasing Wavelength

Wavelength[a] (μm)	Ion	Host glass[b]	Form	Reference
1.345	Nd^{3+}	fluorozirconate	fiber	106
1.350	Nd^{3+}	fluorozirconate	fiber	107
1.355	Nd^{3+}	fluorophosphate	fiber	102
1.355	Nd^{3+}	phosphate (LHG–5)	planar	103
1.356	Nd^{3+}	phosphate	planar	21
1.36	Nd^{3+}	silica (P)	fiber	4
1.363	Nd^{3+}	phosphate	fiber	108
1.366	Nd^{3+}	phosphate (LHG–8)	fiber	109
1.37	Nd^{3+}	La–Ba–Th borate	bulk	110
1.38	Ho^{3+}	fluorozirconate	fiber	111
1.46–1.51	Tm^{3+}	fluorozirconate	fiber	28
~1.47	Tm^{3+}	fluorozirconate	fiber	112
1.47	Tm^{3+}	fluorozirconate (Tb^{3+})	bulk	113
1.475	Tm^{3+}	fluorozirconate	fiber	114
1.475[e]	Tm^{3+}	fluorozirconate	fiber	115
1.48	Tm^{3+}	fluorozirconate	fiber	28
1.481	Tm^{3+}	fluorozirconate (Tb)	fober	116
1.50–1.70	Er^{3+}	silica	fiber	117
1.51	Tm^{3+}	fluorozirconate	fiber	3
1.52–1.57	Er^{3+}	silica (Ge,Al,P)	fiber	118
1.527–1560	Er^{3+}	aluminosilicate	fiber	119
1.529–1.554	Er^{3+}	silica	fiber	32
1.530–1.565	Er^{3+}	silica (Ge,Ca,Al)	fiber	120
1.530–1.570	Er^{3+}	fluorozirconate	fiber	121
1.531–1.540	Er^{3+}	phosphate (Yb)	bulk	122
1.5321–1.5348	Er^{3+}	phosphate (Yb)	bulk	123
1.533	Er^{3+}	phosphate (Yb)	bulk	124
1.535	Er^{3+}	phosphate	fiber	125
1.535	Er^{3+}	phosphate (Yb)	microchip	126
1.535	Er^{3+}	silica	fiber	127

Table 2.2.3—*continued*
Glass Lasers Arranged in Order of Increasing Wavelength

Wavelength[(a)] (μm)	Ion	Host glass[(b)]	Form	Reference
1.5354	Er^{3+}	silica	fiber	128
1.536	Er^{3+}	silica (Ge)	fiber	129
1.536–1.596	Er^{3+}	fluorophosphate (Cr,Yb)	bulk	130
1.54	Er^{3+}	Al–Zn phosphate (Yb)	bulk	131
1.54	Er^{3+}	fluorophosphate (Yb)	bulk	132
1.54	Er^{3+}	Ba phosphate (Cr,Yb)	bulk	133
1.54	Er^{3+}	phosphate	bulk	134
1.54	Er^{3+}	phosphate	fiber	135
1.54	Er^{3+}	phosphate (Cr,Yb)	bulk	136
1.54	Er^{3+}	phosphate (Yb)	bulk	137-139
1.54	Er^{3+}	silica (Yb)	fiber	140
1.540	Er^{3+}	borosilicate (BK 7)	planar	141
1.540	Er^{3+}	phosphate (Yb)	bulk	142
1.543	Er^{3+}	fluorozirconate	fiber	143
1.543	Er^{3+}	Na–K Ba silicate (Yb)	bulk	144
1.545	Er^{3+}	phosphate (Yb)	bulk	124
1.546	Er^{3+}	silica (P)	planar	145
1.549–1.563	Er^{3+}	phosphate (Yb)	bulk	122
1.55	Er^{3+}	Na–K–Ba silicate	bulk	146
1.55	Er^{3+}	silica	fiber	147
1.55	Er^{3+}	silica	fiber	148
1.55	Er^{3+}	silica (Ge)	fiber	149
1.552	Er^{3+}	silica (Ge)	fiber	150
1.553	Er^{3+}	silica	fiber	127
1.553–1.603	Er^{3+}	silica	fiber	151
1.56	Er^{3+}	silica (Al,P,Yb)	fiber	152,153
1.56	Er^{3+}	silica (Yb)	fiber	154
1.560	Er^{3+}	silica (Al,P)	fiber	155
1.566	Er^{3+}	silica (Ge)	fiber	156
1.57–1.61	Er^{3+}	silica (Al, Ge)	fiber	157

Table 2.2.3—*continued*
Glass Lasers Arranged in Order of Increasing Wavelength

Wavelength[a] (μm)	Ion	Host glass[b]	Form	Reference
1.598	Er^{3+}	silica (P)	planar	158
1.6	Er^{3+}	fluoroaluminate	bulk	159
1.604	Er^{3+}	silica (P)	planar	158
1.65–1.86	Tm^{3+}	silica (Ge)	fiber	160
1.653–1.691	Tm^{3+}	silica	fiber	161
1.660	Er^{3+}	fluorozirconate	fiber	162
1.70–2.00	Tm^{3+}	silica (Al)	fiber	160
1.720	Er^{3+}	fluorozirconate	fiber	162
1.724[f]	Er^{3+}	fluorozirconate	fiber	163
1.780–2.056	Tm^{3+}	silica	fiber	164
1.818–1.858	Tm^{3+}	fluorozirconate	fiber	165
1.84–1.94	Tm^{3+}	fluorozirconate	fiber	28
1.85	Tm^{3+}	Li–Mg–Al silicate	bulk	166
1.870–1.930	Tm^{3+}	fluorozirconate	fiber	165
1.88	Tm^{3+}	fluorozirconate	bulk	167
1.88	Tm^{3+}	lead germanate	fiber	168
1.88–1.96	Tm^{3+}	silica	fiber	169
~1.9	Tm^{3+}	fluorozirconate	fiber	112
1.9	Tm^{3+}	silica	fiber	170
1.905	Tm^{3+}	lead germanate	fiber	168
1.925	Tm^{3+}	fluorozirconate	fiber	165
1.937	Tm^{3+}	silica	fiber	171
1.94–1.96	Tm^{3+}	silica (Ge)	fiber	172
1.960–2.032	Ho^{3+}	silica (Tm)	fiber	173
1.972	Tm^{3+}	fluorozirconate	fiber	174
2.007	Tm^{3+}	silica	fiber	171
2.015	Tm^{3+}	Li–Mg–Al silicate	bulk	166
2.04	Ho^{3+}	fluorozirconate (Tm)	fiber	175
2.04	Ho^{3+}	silica (Ge)	fiber	176
2.049	Tm^{3+}	silica	fiber	171

Table 2.2.3—*continued*
Glass Lasers Arranged in Order of Increasing Wavelength

Wavelength(a) (μm)	Ion	Host glass(b)	Form	Reference
2.054	Ho^{3+}	fluorozirconate (Tm)	fiber	177
2.06–2.10	Ho^{3+}	silicate (Er,Yb)	fiber	178
2.076	Ho^{3+}	fluorozirconate (Tm)	fiber	177
2.08	Ho^{3+}	fluorozirconate	fiber	111
2.08	Ho^{3+}	Li–Mg–Al silicate	bulk	179
2.09	Ho^{3+}	germanate	bulk	196
2.102	Tm^{3+}	silica	fiber	171
2.25	Tm^{3+}	fluorozirconate	bulk	166
2.25–2.50	Tm^{3+}	fluorozirconate	fiber	180
2.27–2.40	Tm^{3+}	fluorozirconate	fiber	28
~2.3	Tm^{3+}	fluorozirconate	fiber	112
2.3	Tm^{3+}	fluorozirconate	fiber	181
2.3	Tm^{3+}	fluorozirconate	fiber	182
2.35	Tm^{3+}	fluorozirconate	fiber	28
2.35	Tm^{3+}	fluorozirconate	fiber	165
2.65–2.77	Er^{3+}	fluorozirconate (Pr)	fiber	183
2.69–2.78	Er^{3+}	fluorozirconate	bulk	184
2.70	Er^{3+}	fluorozirconate	bulk	167
2.702	Er^{3+}	fluorozirconate	fiber	185,186
2.71	Er^{3+}	fluorozirconate	fiber	187
2.714	Er^{3+}	fluorozirconate	fiber	188
2.715(g)	Er^{3+}	fluorozirconate	fiber	163
2.716	Er^{3+}	fluorozirconate	fiber	189
2.75	Er^{3+}	fluorozirconate	fiber	187
2.78	Er^{3+}	fluorozirconate	fiber	187,190
2.83–2.95	Ho^{3+}	fluorozirconate	fiber	191
3.45	Er^{3+}	fluorozirconate	fiber	192
3.483	Er^{3+}	fluorozirconate	fiber	193
3.535	Er^{3+}	fluorozirconate	fiber	193
3.95(h)	Ho^{3+}	fluorozirconate	fiber	194

(a) If a tuning range is reported, the laser is listed by the lowest wavelength cited.
(b) Codopant and fluorescence sensitizing ions are included in parentheses.
(c) Proof of lasing was based on the appearance of emission spikes.
(d) Co-lasing at 1.55 μm.
(e) Co-lasing at 1.88 μm.
(f) Co-lasing at 2.7 μm.
(g) Co-lasing at 1.7 μm.
(h) Operates in cascade with simultaneous laser emission at ~1.2 μm.

References—Table 2.2.4

1. Funk, D. S., Carlson, J.W. and Eden, J. G., Ultraviolet (381 nm) room temperature laser in neodymium-doped fluorozirconate fibre, *Electron. Lett.* 30, 1859 (1994).
2. Funk, D. S., Carlson, J. W. and Eden, J. G., Room-temperature fluorozirconate glass fiber laser in the violet (412 nm), *Optics Lett.* 20, 1474 (1995).
3. Allain, J. Y., Monerie, M., and Poignant, H., Blue upconversion fluorozirconate fiber laser, *Electron. Lett.* 26, 166 (1990).
4. Le Flohic, M. P., Allain, J. Y., Stephan, G. M. and Maze, G., Room-temperature continuous-wave upconversion laser at 455 nm in a Tm^{3+} fluorozirconate fiber, *Optics Lett.* 19 (1982 (1994).
5. Grubb, S. G., Bennett, K. W., Cannon, R. S. and Humer, W. F., CW room-temperature blue upconversion fibre laser, *Electron. Lett.* 28, 1243 (1992).
6. Smart, R. G., Hanna, D. C., Tropper, A. C., Davey, S. T., Carter, S. F., and Szebesta, D., CW room temperature upconversion lasing at blue, green and red wavelengths in infrared-pumped Pr^{3+}-doped fluoride fiber, *Electron. Lett.* 27, 1307 (1991).
7. Smart, R. G., Carter, J. N., Tropper, A. C., Hanna, D. C., Davey, S. T., Carter, S. F. and Szebesta, D., CW room temperature operation of praseodymium-doped fluorozirconate glass fiber lasers in the blue-green, green, and red spectral regions, *Optics Commun.* 86, 337 (1991).
8. Ping, X. and Gosnell, T. R., Room temperature upconversion fiber laser tunable in the red, orange, green and blue spectral regions, *Opt. Lett.* 20, 1014 (1995).
9. Zhao, Y. and Poole, S., Efficient blue Pr^{3+}-doped fluoride fibre upconversion laser, *Electron. Lett.* 30, 967 (1994); Baney, D. M., Rankin, G. and Chang, K.-W., Blue Pr^{3+}-doped ZBLAN fiber upconversion laser, *Opt. Lett.* 21, 1372 (1996).
10. Piehler, D., Carven, D., Kwong, N. and Zarem, H., Laser-diode-pumped red and green upconversion fibre lasers, *Electron. Lett.* 29, 1857 (1993).
11. Andreev, S. I., Bedilov, M. R., Karapetyan, G. O., and Likhachev, V. M., Stimulated radiation of glass activated by terbium, *Sov. J. Opt. Tech.* 34, 819 (1967); *Opt.-Mekh. Promst.* 34, 60 (1967).
12. Allain, J. Y., Monerie, M., and Poignant, H., Tunable green upconversion erbium fibre laser, *Electron. Lett.* 28, 111 (1992).
13. Allain, J. Y., Monerie, M. and Poignant, H., Room temperature cw tunable green upconversion holmium fiber laser, *Electron. Lett.* 26, 261 (1990).
14. Whitley, T. J., Millar, C. A., Wyatt, R., Brierley, M. C. and Szebesta, D., Upconversion pumped green lasing in erbium doped fluorozirconate fibre, *Electron. Lett.* 27, 1786 (1991).
15. Booth, I. J., Mackechnie, C. J. and Ventrudo, B. F., Operation of diode laser pumped Tm^{3+} ZBLAN upconversion fiber laser at 482 nm, *IEEE J. Quantum Electron.* 32, 118 (1996).
16. Hirao, K., Todoroki, S. and Soga, N., CW room temperature upconversion lasing in Er^{3+}-doped fluoride glass fiber, *J. Non-Cryst. Solids* 143, 40 (1992).

17. Funk, D. S., Stevens, S. B., Wu, S. S. and Eden, J. G., Tuning, temporal, spectral characteristics of the green (λ~549 nm), holmium-doped fluorozirconate glass fiber laser, *IEEE J. Quantum Electron.* 32, 638 (1996).
18. Allain, J. Y., Monerie, M. and Poignant, H., Tunable cw lasing around 610, 635, 695, 715, 885, and 910 nm in praseodymium-doped fluorozirconate fibre, *Electron. Lett.* 27, 189. (1991).
19. Baney, D. M., Yang, L., Ratcliff, J. and Chang, K. W., Red and orange Pr^{3+}/Yb^{3+} doped ZBLAN fibre upconversion lasers, *Electron. Lett.* 31, 1842 (1995).
20. Petreski, B. P., Murphy, M. M., Collins, S. F. and Booth, D. J., Amplification in Pr^{3+}-doped fluorozirconate optical fibre at 632.8 nm, *Electron. Lett.* 29, 1421 (1993).
21. Malone, K. J., Sanford, N. A., Hayden, J. S. and Sapak, D. L., Integrated optic laser emitting at 905, 1057, and 1356 nm, in *Advanced Solid-State Lasers*, Pinto, A. A. and Fan, T. Y., Eds., Proceedings Vol. 15, Optical Society of America, Washington, DC (1993), p. 286.
22. Payne, S. A., Marshall, C. D., Bayramian, A. J., Wilke, G. D. and Hayden, J. S., Properties of a new average power Nd-doped phosphate glass, *OSA Proc. Adv. Solid State Lasers*, Chai, B. H. T. and Payne, S. A., Eds., 24, 211(1995).
23. Allain, J. Y., Monerie, M. and Poignant, H., Red upconversion Yb-sensitised Pr doped fluoride fibre laser pumped in the 0.8 μm region, *Electron. Lett.* 27, 1156 (1991).
24. Farries, M. C., Morkel, P. R. and Townsend, J. E., $Samarium^{3+}$-doped glass laser operating at 651 nm, *Electron. Lett.* 24, 709 (1988).
25. Carter, J. N., Smart, R. G., Hanna, D. C. and Tropper, A. C., Lasing and amplification in the 0.8 μm region in thulium doped fluorozirconate fibers, *Electron. Lett.* 26, 1759 (1990).
26. Smart, R. G., Carter, J. N., Tropper, A. C., Hanna, D. C., Carter, S. F. and Szebesta, D., A 20 dB gain thulium-doped fluorozirconate fiber amplifier operating at around 0.8 μm, *Electron. Lett.* 27, 1123 (1991).
27 Dennis, M. L., Dixon, J. W. and Aggarwal, I., High power upconversion lasing at 810 nm in Tm:ZBLAN fibre, *Electron. Lett.* 30, 136 (1994).
28. Allain, J. Y., Monerie, M. and Poignant, H., Tunable cw lasing around 0.82, 1.48, 1.88, and 2.35 μm in a thulium-doped fluorozirconate fiber, *Electron. Lett.* 25, 1660 (1989).
29. Millar, C. A., Brierley, M. C., Hunt, M. H. and Carter, S. F., Efficient up-conversion pumping at 800 nm of an erbium-doped fluoride fiber laser operating at 850 nm, *Electron. Lett.* 26, 1876 (1990).
30. Whitney, T. J., Millar, C. A., Brierley, M. C. and Carter, S. F., 23 dB gain upconversion pumped erbium doped fiber amplifier operating at 850 nm, *Electron. Lett.* 27, 189 (1991).
31. Percival, R. M., Phillips, M. W., Hanna, D. C. and Tropper, A. C., Characterization of spontaneous and stimulated emission from praseodymium ions doped into a silica based monomode optical fiber, *IEEE J. Quantum Electron.* 25, 2119 (1989).
32. Reekie, L., Mears, R. J., Poole, S. B. and Payne, D. N., Tunable single-mode fiber lasers, *J. Lightwave Tech.* LT-4, 956 (1986).
33. Alcock, I. P., Ferguson, A. I., Hanna, D. C. and Tropper, A. C., Tunable, continuous-wave neodymium-doped monomode-fiber laser operating at 0.900-0.945 and 1.070-1.135 μm, *Optics Lett.* 11, 709 (1986).
34. Maurer, R. D., Operation of a Nd^{3+} glass optical maser at 9180 Å, *Appl. Opt.* 2, 87 (1963).
35. Artem'ev, E. P., Murzin, A. G. and Fromzel, V. A., Room-temperature laser action at 0.92 μm in neodymium glasses, *Sov. Phys. Tech. Phys.* 22, 274 (1977); *Zh. Tekh. Fiz.* 47, 456 (1977).
36. Krupke, W. F., Shinn, M. D., Kirchoff, T. A., Finch, D. B. and Boatner, L.A., Promethium-doped phosphate glass laser at 933 and 1098 nm, *Appl. Phys. Lett.* 51, 2186 (1987).

37. Alcock, I. P., Ferguson, A. I., Hanna, D. C. and Tropper, A. C., Continuous-wave oscillation of a monomode neodymium-doped fiber laser at 0.9 μm on the $^4F_{3/2} \rightarrow ^4I_{9/2}$ transition, *Opt. Commun.* 58, 405 (1986).
38. Reekie, L., Jauncey, I. M., Poole, S. B. and Payne, D. N., Diode-laser-pumped Nd^{3+}-doped fiber laser operating at 938 nm, *Electron. Lett.* 23, 884 (1987).
39. Hanna, D. C., Percival, R. M., Perry, I. R., Smart, R. G., Suni, P. J. and Tropper, A. C., An ytterbium-doped monomode fiber laser: broadly tunable operation from 1.010 μm to 1.162 μm and three-level operation at 974 nm, *J. Mod. Opt.* 37, 517 (1990).
40. Armitage, J. R., Wyatt, R., Ainslie, B. J. and Craig-Ryan, S. P., Highly efficient 980 nm operation of an Yb^{3+}-doped silica fiber laser, *Electron. Lett.* 25, 298 (1989).
41. Allain, J. Y., Monerie, M. and Poignant, H., Q-switched 0.98 μm operation of erbium-doped fluorozirconate fiber laser, *Electron. Lett.* 25, 1082 (1989).
42. Allain, J. Y., Monerie, M. and Poignant, H., Lasing at 1.00 μm in erbium-doped fluorozirconate fibers, *Electron. Lett.* 25, 318 (1989).
43. Allain, J. Y., Monerie, M., and Poignant, H., Ytterbium-doped fluoride fibre laser operating at 1.02 μm, *Electron. Lett.* 28, 988 (1992).
44. Shi, Y., Poulsen, C. V., Sejka, M., Ibsen, M. and Poulsen, O., Tunable Pr^{3+}-doped silica-based fibre laser, *Electron. Lett.* 29, 1426 (1993).
45. Etzel, H. W., Gandy, H. W. and Ginther, R. J., Stimulated emission of infrared radiation from ytterbium-activated silicate glass, *Appl. Opt.* 1, 534 (1962).
46. Gandy, H. W. and Ginther, R. J., Simultaneous laser action of neodymium and ytterbium ions in silicate glass, *Proc. IRE* 50, 2114 (1962).
47. Hanna, D. C., Percival, R. M., Perry, I. R., Smart, R. G., Suni, P. J., Townsend, J. E., and Tropper, A. C., Continuous-wave oscillation of a monomode ytterbium-doped fiber laser, *Electron. Lett.* 24, 1111 (1988).
48. Pearson, A. D. and Porto, S.P.S., Non-radiative energy exchange and laser oscillation in Yb^{3+}- Nd^{3+}-doped borate glass, *Appl. Phys. Lett.* 4, 202 (1964).
49. Petrovksii, G. T., Tolstoi, M. N., Feofilov, P. P., Tsurikova, G. A., and Shapovalov, V. N., Luminescence and stimulated emission of neodymium in beryllium fluoride glass, *Opt. Spectrosc. (USSR)* 21, 72 (1966); *Opt. Spektrosk.* 21, 126 (1966).
50. Yamashita, T., Ammano, S., Masuda, I., Izumitani, T., and Ikushima, A. J., Nd and Er-doped phosphate glass fiber lasers, in *Conference on Lasers and Electro-Optics Technical Digest Series,* Opt. Soc Amer., Washington, DC (1988) p. 320.
51. Robertson, G.R.F. and Jessop, P. E., Optical waveguide laser using an rf sputtered Nd:glass film, *Appl. Opt.* 30, 276 (1991).
52. Brierley, M. C. and France, P. W., Neodymium-doped fluorozirconate fiber laser, *Electron. Lett.* 23, 815 (1987).
53. Liu, K., Digonnet, M., Fesler, K., Kim, B. Y., and Shaw, H. J., Broadband diode-pumped fiber laser, *Electron. Lett.* 24, 838 (1988).
54. Stokowski, S. E., Martin, W. E., and Yarema, S. M., Optical and lasing properties of fluorophosphate glass, *J. Non-Cryst. Solids* 40, 48 (1980).
55. Hibino, Y., Kitagawa, T., Shimizu, M., Hanawa, F., and Sugita, A., Neodymium-doped silica optical waveguide laser on silicon substrate, *IEEE Phot. Tech. Lett.* 1, 349 (1990).
56. Kitagawa, T., Hattori, K., Hibino, Y. and Ohmori, Y., Neodymium-doped silica-based planar waveguide lasers, *J. Lightwave Techn.* 12, 436 (1994).
57. Hattori, K., Kitagawa, T., Ohmori, Y. and Kobayashi, M., Laser-diode pumping of waveguide laser based on Nd-doped silica planar lightwave circuit, *IEEE Phot. Techn. Lett.* 3, 882 (1991).
58. Michel, J. C., Morin, D. and Auzel, F., Properietes spectroscopiques et effet laser d'un verre tellurite et d'un verre phosphate fortement dopes en neodyme, *Rev. Phys. Appl.* 11, 859 (1978).

59. Aoki, H., Maruyama, O. and Asahara, Y., Glass waveguide laser, *IEEE Phot. Tech. Lett.* 2, 459 (1990).
60. Kishida, S., Washio, K. and Yoshikawa, S., CW oscillation in a Nd:phosphate glass laser, *Appl. Phys. Lett.* 34, 273 (1979).
61. Deutschbein, O., Pautrat, C. and Svirchevsky, I. M., Phosphate glasses, new laser materials, *Rev. Phys. Appl.* 1, 29 (1967).
62. Deutschbein, O. K. and Pautrat, C. C., CW laser at room temperature using vitreous substances, *IEEE J. Quantum Electron.* QE-4, 48 (1968).
63. Vodop'yanov, K. L., Denker, B. I., Maksimova, G. V., Malyutin, A. A., Osiko, V. V., Pashinin, P. P. and Prokhorov, A. M., Characteristics of simulated emission from LiNdLa phosphate glass, *Sov. J. Quantum Electron.* 8, 403 (1978).
64. Hakimi, F., Po, H., Tumminelli, R., McCollum, B. C., Zenteno, L., Cho, N. M. and Snitzer, E., Glass fiber laser at 1.36 μm from SiO_2:Nd, *Opt. Lett.* 14, 1060 (1989).
65. Sanford, M. A., Malone, K. J. and Larson, D. R., Integrated-optic laser fabricated by field-assisted ion exchange in neodymium-doped soda-lime-silicate glass, *Opt. Lett.* 15, 366 (1990).
66. Mwarania, E. K., Reekie, L., Wang, J. and Wilkinson, J. S., Low-threshold monomode ion-exchanged waveguide lasers in neodymium-doped BK-7 glass, *Electron. Lett.* 26, 1317 (1990).
67. Yajima, H., Kawase, S. and Sakimoto, Y., Amplification at 1.06 μm using a Nd-glass thin-film waveguide, *Appl. Phys. Lett.* 21, 407 (1972).
68. Young, C. G., Continuous glass laser, *Appl. Phys. Lett.* 2, 151 (1963).
69. Saruwatari, M. and Izawa, T., Nd-glass laser with three-dimensional optical waveguide, *Appl. Phys. Lett.* 24, 603 (1974).
70. Rapp, C. F. and Chrysochoos, J., Neodymium-doped glass-ceramic laser material, *J. Mater. Sci.* 7, 1090 (1972).
71. Müller, G. and Neuroth, N., Glass ceramic - a new host material, *J. Appl. Phys.* 44, 2315 (1973).
72. Wang, J., Reekie, L., Brocklesby, W. S., Chow, Y. T. and Payne, D. N., Fabrication, spectroscopy and laser performance of Nd^{3+}-doped lead-silicate glass fibers, *J. Non-Cryst. Solids* 180, 207 (1995).
73. Birnbaum, M., Fincher, C. L., Dugger, C. O., Goodrum, J. and Lipson, H., Laser characteristics of neodymium-doped lithium germanate glass, *J. Appl. Phys.* 41, 2470 (1970).
74. Härig, T., Huber, G. and Shcherbakov, I., Cr^{3+} sensitized Nd^{3+}:Li-La phosphate glass laser, *J. Appl. Phys.* 52, 4450 (1981).
75. Galaktionova, N. M., Garkavi, G. A., Zubkova, V. S., Mak, A. A., Soms, L. N. and Khaleev, M. M., Continuous Nd-glass laser, *Opt. Spectrosc. USSR* 37, 90 (1974); *Opt. Spektrosk.* 37, 162 (1974).
76. Koester, C. J. and Snitzer, E., Amplification in a fiber laser, *Appl. Opt.* 3, 1182 (1964).
77. Babukova, M. V., Berenberg, V. A., Glebov, L. B., Nikonorov, N. V., Petrovskii, G. T. and Terpugov, V. S., Investigation of neodymium silicate glass diffused waveguides, *Sov. J. Quantum Electron.* 15, 1304 (1985).
78. Snitzer, E., Laser emission at 1.06 μ from Nd^{3+}-Yb^{3+} glass, *IEEE J. Quantum Electron.* QE-2, 562 (1966).
79. Stone, J. and Burrus, C. A., Neodymium-doped silica lasers in end-pumped fiber geometry, *Appl. Phys. Lett.* 23, 388 (1973).
80. Tumminelli, R., Hakimi, F. and Haavisto, J., Integrated optic Nd glass laser fabricated by flame hydrolysis deposition using chelates, *Optics Lett.* 16, 1098 (1991).

81. Balashov, I. F., Berezin, B. G., Brachkovskaya, N. B., Volkova, V. V., Ivanov, V. N., Ovcharenko, N. V., Petrov, A. A., Przhevuskii, A. K., and Smirnova, T. V., Experimental study of tellurite laser glass doped with neodymium, *Zh. Prikl. Spekt.* 52, 781 (1989).
82. Snitzer, E., Optical maser action of Nd^{3+} in a barium crown glass, *Phys. Rev. Lett.* 7, 444 (1961).
83. Wang, J. S., Machewirth, D. P., Wu, F., Snitzer, E. and Vogel, E. M., Neodymium-doped tellurite single-mode fiber laser, *Optics Lett.* 19, 1448 (1994).
84. Galant, E. I., Kondrat'ev, Yu.N., Przhevuskii, A. K., Prokhorova, T. I., Tolstoi, M. N., and Shapovalov, V., N., Emission of neodymium ions in quartz glass, *Sov. JETP Lett.* 18, 372 (1973).
85. Thomas, I. M., Payne, S. A. and Wilke, G. D., Optical properties and laser demonstration of Nd-doped sol-gel silica glasses, *J. Non-Cryst. Solids* 151, 183 (1992).
86. Chen, B. -U. and Tang, C. L., Nd-glass thin-film waveguide in an active medium for Nd thin-film laser, *Appl. Phys. Lett.* 28, 435 (1976).
87. Lei, N., Xu, B. and Jiang, Z., Ti:sapphire laser pumped Nd:tellurite glass laser, *Optics Commun.* 127, 263 (1996).
88. Chaoyu, Y., Jiangde, P. and Bingkun, Z., Tunable Nd^{3+}-doped fiber ring laser, *Electron. Lett.* 25, 101 (1989).
89. Schweizer, T., Hewak, D. W., Payne, D. N., Jensen, T. and Huber, G., Rare-earth doped chalcogenide glass laser, *Electron. Lett.* 32, 666 (1996) and Schweizer, T., Samson, B. N., Moore, R. C., Hewak, D. W. and Payne, D. N., Rare-earth doped chalcogenide glass fibre laser, *Electron. Lett.* 33, 414 (1997).
90. Stone, J. and Burrus, C. A., Neodymium-doped fiber lasers: room temperature cw operation with an injection laser pump, *Appl. Opt.* 13, 1256 (1974).
91. Poole, S. B., Payne, D. N. and Fermann, M. E., Fabrication of low-loss optical fibers containing rare-earth ions, *Electron. Lett.* 21, 737 (1985).
92. Mears, R. J., Reekie, L., Poole, S. B. and Payne, D. N., Neodymium-doped silica single-mode fiber lasers, *Electron. Lett.* 21, 738 (1985).
93. Mackechnie, C. J., Barnes, W. L., Hanna, D. C. and Townsend, High power ytterbium (Yb^{3+})-doped fibre lasr operating in the 1.12 μm region, *Electron. Lett.* 29, 52 (1993).
94. Carter, S.F., Szebesta, D., Davey, S.T., Wyatt, R., Brierley, M.C. and France, P.W., Amplification at 1.3 μm in a Pr^{3+}-doped single-mode fluorozirconate fiber, *Electron. Lett.* 27, 628 (1991).
95. Ohishi, Y. Kanamori, T., Kitagawa, T., Snitzer, E. and Sigel, Jr., G.H., Pr^{3+}-doped fluoride fiber amplifier operating at 1.31 μm, *Optics Lett.* 6, 1747 (1991).
96. Whitley, T., Wyatt, R., Szebesta, D., Davey, S. and Williams, J. R., Quarter-watt output at 1.3 μm from a praseodymium-doped fluoride fibre amplifier pumped with a diode-pumped Nd:YLF laser, *IEEE Phot. Tech. Lett.* 4, 399 (1993).
97. Durteste, Y., Monerie, M. Allain, J.Y. and Poignant, H., Amplification and lasing at 1.3 μm in praseodymium-doped fluorozirconate fibers, *Electron. Lett.* 27, 626 (1991).
98. Ohishi, Y., Kanamori, T, Temmyo, J. et al., Laser diode pumped Pr^{3+}-doped and Pr^{3+}-Yb^{3+}-codoped fluoride fibre amplifiers operating at 1.3 μm, *Electron. Lett.* 27,1995 (1991); Ohishi, Y., Kanamori, T, Nishi, T., Takahashi, S. and Snitzer, E., Gain characteristics of Pr^{3+}-Yb^{3+} codoped fluoride fiber for 1.3 μm amplification, *IEEE Phot. Tech. Lett.* 3, 990 (1991).
99. Ohishi, Y., Kanamori, T., Kitagawa, T. Takahashi, S., Snitzer, E. and Sigel, G. H., Pr^{3+}-doped fluoride fiber superfluorescent laser, *Jap. J. Appl. Phys.* 30, L1282 (1991).
100. Brierley, M., Carter, S. and France, P., Amplification in the 1300 nm telecommunications window in a Nd-doped fluoride fiber, *Electron. Lett.* 26, 329 (1990).

101. Miyajima, Y., Komuakai, T. and Sugawa, T., 1.31-1.36 μm optical amplification in Nd^{3+}-doped fluorozirconate fiber, *Electron. Lett.* 26 (194 (1990).
102. Ishikawa, E., Aoki, H., Yamashita, T. and Asahara, Y., Laser emission and amplication at 1.3 μm in neodymium-doped fluorophosphate fibres, *Electron. Lett.* 28, 1497 (1992).
103. Aoki, H., Maruyama, O. and Asahara, Y., Glass waveguide laser operated around 1.3 μm, *Electron. Lett.* 26, 1910 (1990).
104. Miyajima, Y., Sugawa, T. and Komukai, T., Efficient 1.3 μm-band amplification in a Nd^{3+}-doped single-mode fluoride fiber, *Electron. Lett.* 17, 1397 (1990).
105. Miniscalco, W. J., Andrews, L. J., Thompson, B. A., Quimby, R. S., Vaca, L.J.B. and Drexhage, M. G., 1.3 μm fluoride fiber laser, *Electron. Lett.* 24, 28 (1988).
106. Millar, C. A., Fleming, S. C., Brierley, M. C. and Hunt, M. H., Single transverse mode operation at 1345 nm wavelength of a diode-laser pumped neodymium-ZBLAN multimode fiber laser, *IEEE, Phot. Tech. Lett.* 2, 415 (1990).
107. Brierley, M. C. and Millar, C. A., Amplification and lasing at 1350 nm in a neodymium doped fluorozirconate fiber, *Electron. Lett.* 24, 438 (1988).
108. Grubb, S. G., Barnes, W. L., Taylor, E. R. and Payne, D. N., Diode-pumped 1.36 μm Nd-doped fiber laser, *Electron. Lett.* 26, 121 (1990).
109. Yamashita, T., Nd- and Er-doped phosphate glass for fiber laser, *SPIE Fiber Laser Sources and Amplifiers,* vol.1171, 291 (1989).
110. Mauer, P. B., Laser action in neodymium-doped glass at 1.37 microns, *Appl. Opt.* 3, 153 (1964).
111. Brierley, M. C., France, P. W. and Millar, C. A., Lasing at 2.08 μm and 1.38 μm in a holmium doped fluorozirconate fiber laser, *Electron. Lett.* 24, 539 (1988).
112. Smart, R.G., Carter, J.N., Tropper, A.C. and Hanna, D.C., Continuous-wave oscillation of Tm^{3+}-doped fluorozirconate fibre lasers at around 1.47 μm, 1.9 μm and 2.3 μm when pumped at 790 nm, *Optics Commun.* 82, 563 (1991).
113. Rosenblatt, G. H., Ginther, R. J., Stoneman, R. C. and Esterowitz, L., Laser emission at 1.47 μm from fluorozirconate glass doped with Tm^{3+} and Tb^{3+}, *Technical Digest - Tunable Solid State Laser Conference*, Opt. Soc. Amer. (1989), paper WE2-1.
114. Pericival, R. M., Szebesta, D. and Williams, J. R., Highly efficient 1.064 μm upconversion pumped 1.47 μm thulium doped fluoride fibre laser, *Electron. Lett.* 30, 1057 (1994).
115. Percival, R. M., Szebesta, D. and Davey, S. T., Highly efficient cw cascade operation of 1.47 and 1.82 μm transitions in Tm-doped fluoride fibre laser, *Electron. Lett.* 28, 1866 (1992).
116. Percival, R. M., Szebesta, D. and Davey, S. T., Thulium-doped terbium sensitised cw fluoride fibre laser operating on the 1.47 μm transition, *Electron. Lett.* 29, 1054 (1993).
117 Sankawa, I. Izumita, H. Furukawa, S. and Ishihara, An optical fiber amplifier for wide-band wavelength range around 1.65 μm, *IEEE Phot. Tech. Lett.* 2, 422 (1990).
118. Wyatt, R., High-power broadly tunable Er^{3+}-doped silica fiber laser, *Electron. Lett.* 25, 1498 (1989).
119 Yamada, M., Shimizu, M., Horiguchi, M., Okayasu, M. and Sugita, E., Gain characteristics of an Er^{3+}-doped multicomponent glass single-mode optical fiber, *IEEE Phot. Tech. Lett.* 2, 656 (1990).
120. Saifi, M. A., Andrejco, M. J., Way, W. I., Von Lehman, A., Yi-Yan, A., Lin, C., Bilodeau, F. and Hill, K. O., Er^{3+} doped GeO_2-CaO-Al_2O_3 silica core fiber amplifier pumped at 813 nm, *Optical Fiber Commun.* (OFC 91), Optical Soc. Am., San Diego, CA (1991), paper FA6.
121. Spirit, D. M., Walker, G. R., France, P. W., Carter, S. F. and Szebesta, D., Characterization of diode-pumped erbium-doped fluorozirconate fibre optical amplifier, *Electron. Lett.* 26, 1218 (1990).

122. Taccheo, S., Laporta, P. and Svelto, O., Widely tunable single-frequency erbium-ytterbium phosphate glass laser, *Appl. Phys. Lett.* 68 2621 (1996)
123. Laporta, P., Taccheo, S.and Svelto, O., High-power and high-efficiency diode-pumped Er:Yb:glass laser, *Electron. Lett.* 28, 490 (1992).
124. Hutchinson, J. A. and Allik, T. H., Diode array-pumped Er,Yb:phosphate glass laser, *Appl. Phys. Lett.* 60, 1424 (1992).
125. Yamashita, Y., Nd and Er doped phosphate glass for fiber laser, *Fiber Laser Sources and Amplifiers, SPIE* vol. 1171, 291 (1989).
126. Laporta, P., Taccheo, S., Longhi, S. and Svelto, O., Diode-pumped microchip Er-Yb:laser, *Optics Lett.* 18, 1232 (1993).
127 Többen, H. and Wetenkamp, L., High-efficiency cw Ho-doped fluorozirconate fiber laser at 1.19 μm, in *Advanced Solid-State Lasers*, Chase, L. L. and Pinto, A. A., Eds., Proceedings Vol. 13, Optical Society of America, Washington, DC (1992), p. 119.
128. Nakazawa, M., Kimura, Y. and Suzuki, K., High gain erbium fiber amplifier pumped by 800 nm band, *Electron. Lett.* 26, 548 (1990).
129. Horiguchi, H., Shimizu, M., Yamada, M., Yoshino, K. and Hanafusa, H., Highly efficient optical fiber amplifier pumped by a 0.8 nm band laser diode, *Electron. Lett.* 26, 1758 (1990).
130. Ledig, M., Heumann, E., Ehrt, D. and Seeber, W., Spectroscopic and laser properties of Cr^{3+}:Yb^{3+}:Er^{3+} fluoride phosphate glass, *Opt. Quantum Electron.* 22, S107 (1990).
131. Snitzer, E., Woodcock, R. F. and Segre, J., Phosphate glass Er^{3+} laser, *IEEE J. Quantum Electron.* QE-4, 360 (1968).
132. Auzel, F., Stimulated emission of Er^{3+} in fluorophosphate glass, *C. R. Acad. Sci. B* 263, 765 (1966).
133. Lunter, S. G., Murzin, A. G., Tolstoi, M. N., Fedorov, Yu.K. and Fromzel, V. A., Possibility of improving the efficiency of lamp pumping of erbium-glass lasers, *Opt. Spectrosc. USSR* 55, 345 (1983; Energy parameters of lasers utilizing erbium glasses sensitized with ytterbium and chromium, *Sov. J. Quantum Electron.* 14, 66 (1984).
134. Maksimova, G. V., Sverchkov, S. E. and Sverchkov, Yu. E., Lasing tests on new ytterbium-erbium laser glass pumped by neodymium lasers, *Sov. J. Quantum Electron.* 21, 1324 (1991).
135. Astakhov, A. V., Butusov, M. M., Galkin, S. L., Ermakova, N. V. and Fedorov, Yu.K., Fiber laser with 1.54 μm radiation wavelength, *Opt. Spectrosc. USSR,* 62, 140 (1987).
136. Gapontsev, V. P., Gromov, A. K., Izyneev, A. A., Sadouskii, P. I., Stavrov, A. A., Tipenko, Yu.S. and Shkadarevich, A. P., Low-threshold erbium glass minilaser, *Sov. J. Quantum Electron.* 18, 447 (1989).
137. Laporta, P., De Silvestri, S. Magni, V. and Svelto, O., Diode-pumped bulk Er:Yb:glass laser, *Optics Lett.* 16, 1952 (1991).
138. Hanna, D. C., Kazer, A. and Shepherd, D. P., A 1.54 μm Er glass laser pumped by a 1.064 μm Nd:YAG laser, *Opt. Commun.* 63, 417 (1987).
139. Estie, D., Hanna, D. C., Kazer, A. and Shepherd, D. P., CW operation of Nd:YAG pumped Er:Yb phosphate glass laser at 1.54 μm, *Opt. Commun.* 69, 153 (1988).
140. Townsend, J. E., Barnes, W. L., Jedrzejewski, K. P. and Grubb, S. G., Yb^{3+} sensitised Er^{3+} doped silica optical fibre with ultrahigh transfer efficiency and gain, *Electron. Lett.* 27, 1958 (1991).
141. Feuchter, T., Mwarania, E. K., Wang, J., Reekie, L. and Wilkinson, J. S., Erbium-doped ion-exchanged waveguide lasers in BK-7 glass, *IEEE Phot. Tech. Lett.* 4, 542 (1992).
142. Laporta, P., Longhi, S., Taccheo, S., Svelto, O.and Sacchi, G., Single-mode cw erbium-ytterbium glass laser at 1.5 μm, *Optics Lett.* 18, 31 (1993).

143. Ronarc'h, D., Guibert, M., Ibrahim, H., Monerie, M., Poignant, H. and Tromeur, A., 30 dB optical net gain at 1.543 μm in Er^{3+} doped fluoride fibre pumped around 1.48 μm, *Electron. Lett.* 27, 908 (1991)
144. Snitzer, E. and Woodcock, R., Yb^{3+}-Er^{3+} glass laser, *Appl. Phys. Lett.* 6, 45 (1965).
145. Kitagawa, T., Bilodeau, F. Malo, B., Theriault, S., Albert, J., Jihnson, D. C., Hill, K. O., Hattori, K. and Hibino, Y., Single-frequency Er^{3+}-doped silica based planar waveguide laser with integrated photo-imprinted Bragg reflectors, *Electron. Lett. 30*, 1311 (1994).
146. Gandy, H. W., Ginther, R. J. and Weller, J. F., Laser oscillations in erbium activated silicate glass, *Phys. Lett.* 16, 266 (1965).
147. Mears, R. J., Reekie, L., Poole, S. B. and Payne, D. N., Low-threshold tunable CW and Q-switched fiber laser operating at 1.55 μm, *Electron. Lett.* 22, 159 (1986).
148. Reekie, L., Jauncey, I. M., Poole, S. B., and Payne, D. M., Diode-laser-pumped operations of an Er^{3+}-doped single mode fiber laser, *Electron. Lett.* 23, 1076 (1987).
149. Millar, C. A., Miller, I. D., Ainslie, B. J., Craig, S. P. and Armitage, J. R., Low-threshold cw operation of an erbium-doped fiber laser pumped at 807 nm wavelength, *Electron. Lett.* 23, 865 (1987).
150. Petersen, B., Zemon, S. and Miniscalco, W.J., Erbium doped fiber amplifiers pumped in the 800 nm band., *Electron. Lett.* 27, 1295 (1991).
151. Kimura, Y. and Nakazawa, M., Lasing characteristics of Er^{3+}-doped silica fibers from 1553 up to 1603 nm, *J. Appl. Phys.* 64, 516 (1988).
152. Hanna, D. C., Percival, R. M., Perry, I. R., Smart, R. G. and Tropper, A. C., Efficient operation of an Yb-sensitized Er fiber laser pumped in 0.8 μm region, *Electron. Lett.* 24, 1068 (1988).
153. Fermann, M. E., Hanna, D. C., Shepherd, D. P., Suni, P. J. and Townsend, J. E., Efficient operation of an Yb-sensitized Er fiber laser at 1.56 μm, *Electron. Lett.* 24, 1135 (1988).
154. Maker, G. T. and Ferguson, A. I., 1.56 μm Yb-sensitized Er fiber laser pumped by diode-pumped Nd:YAG and Nd:YLF lasers, *Electron. Lett.* 24, 1160 (1988).
155. Wyatt, R., Ainslie, B.J., and Craig, S.P., Efficient Operation of an Array Pumped Er^{3+}-Doped Silica Fiber Laser at 1.5 μm, *Electron. Lett.* 24, 1362 (1988).
156. O'Sullivan, M.S., Chrostowski, J., Desurvire, E. and Simpson, J.R., High-power, narrow-linewidth Er^{3+}-Doped Fiber laser, *Optics Lett.*, 14, 438 (1989).
157. Massicott, J.F., Armitage, J.R., Wyatt, R., Ainslie, B.J. and Craig-Ryan, S.P., High gain, broad bandwidth 1.6 μm Er^{3+} doped silica fiber amplifier , *Electron. Lett.* 26, 1645 (1990).
158. Kitagawa, T., Hattori, K., Shimizu, M., Ohmori and Y., Kobayashi, M., Guided-wave laser based on erbium-doped silica planar lightwave circuit , *Electron. Lett.* 27, 334 (1991).
159. Heumann, E., Ledig, M.,Ehrt, D. and Seeber, W., CW laser action of Er^{3+} in double sensitized fluoroaluminate glass at room temperature, *Appl. Phys. Lett.* 52, 255 (1988).
160. Barnes, W. L. and Townsend, J. E., Highly tunable and efficient diode pumped operation of Tm^{3+} doped fiber lasers, *Electron. Lett.* 26, 726 (1990).
161. Sankawa, I., Izumita, H., Furukawa, S. and Ishihara, K., An optical fiber amplifier for wide-band wavelength range around 1.65 μm, *IEEE Phot. Tech. Lett.* 2, 422 (1990).
162 Smart, R. G., Carter, J. N., Hanna, D. C., and Tropper, A. C., Erbium doped fluorozirconate fiber laser operating at 1.66 and 1.72 μm, *Electron. Lett.* 26, 649 (1990).
163. Ghisler, C., Pollnau, M., Bunea, G., Bunea, M., Lüthy, W. and Weber, H. P., Up-conversion cascade laser at 1.7 μm with simultaneous 2.7 μm lasing in erbium ZBLAN fibre, *Electron. Lett.* 31, 373 (1995).

164. Hanna, C., Percival, R. M., Smart, R. G. and Tropper, A. C., Efficient and tunable operation of a Tm-doped fiber laser, *Opt. Commun.* 75, 283 (1990).
165. Percival, R. M., Szebesta, D. and Davey, S. T., Highly efficient and tunable operation of two colour Tm-doped fluoride fibre laser, *Electron. Lett.* 28, 671 (1992).
166. Gandy, H. W., Ginther, R. J. and Weller, J. F., Stimulated emission of Tm^{3+} radiation in silicate glass, *J. Appl. Phys.* 38, 3030 (1967).
167 Esterowitz, L., Allen, R., Kintz, G., Aggarwal, I. and Ginther, R. J., Laser emission in Tm^{3+} and Er^{3+}-doped fluorozirconate glass at 2.25, 1.88, and 2.70 μm, in *Conference on Lasers and Electro-Optics Technical Digest Series,* 7, Optical Society of America, Washington, DC (1988), p. 318).
168. Lincoln, J. R., Mackechnie, C. J., Wang, J., Brocklesby, W. S., Deol, R. S., Person, A., Hanna, D. C. and Payne, D. N., New class of fibre laser based on lead-germanate glass, *Electron. Lett.* 28, 1021 (1992).
169. Hanna, D. C., Percival, R. M., Perry, I. R., Smart, R. G., Suni, P. J. and Tropper, A. C., Continuous-wave oscillation of a monomode thulium-doped silica fiber laser, *Technical Digest - Tunable solid State Laser Conference,* Optical Society of America (1989), p.WE3-1.
170. Yamamoto, T. Miyajima, Y and Komukai, T., 1.9 μm Tm-doped silica fibre laser pumped at 1.57 μm, *Electron. Lett.* 30, 220 (1994).
171. Boj, S., Delavaque, E., Allain, J. Y., Bayon, J. F., Niay, P. and Bernage, P., High efficiency diode pumped thulium-doped silica fibre lasers with intracore Bragg gratings in the 1.9-2.1 μm band, *Electron. Lett.* 30, 1019 (1994)
172. Hanna, D. C., Jauncey, I. M., Percival, R. M., Perry, I. R., Smart, R. G., Suni, P. J., Townsend, J. E. and Tropper, A. C., Continuous-wave oscillation of a monomode thulium-doped fluorozirconate fiber, *Electron. Lett.* 24, 935 (1988).
173. Ghisler, C., Luethy, W. and Weber, H. P., Tuning of a Tm^{3+}:Ho^{3+}:silica fiber laser at 2 μm, *IEEE J. Quant. Electron.* 31 1877 (1995)
174. Carter, J.N., Smart, R.G., Hanna, D.C. and Tropper, A.C., CW diode-pumped operation of 1.97 μm thulium-doped fluorozirconate fiber laser, *Electron. Lett.* 26, 599 (1990).
175. Allain, J. Y., Monerie, M. and Poignant, H., High-efficiency cw thulium-sensitised holmium-doped fluoride fibre laser operating at 2.04 μm, *Electron. Lett.* 27, 1513, (1991)
176. Hanna, D. C., Percival, R. M., Smart, R. G., Townsend, J. E., and Tropper, A. C., Continuous wave oscillation of a holmium-doped silica fiber laser, *Electron. Lett.* 25, 593 (1989).
177. Percival, R. M., Szebesta, D., Davey, S. T. Swain, N. A. and King, T. A., High efficiency operation of 890-pumped holmium fluoride fibre laser, *Electron. Lett.* 28, 2064, (1992)
178. Veinberg, T. I., Zhmyreva, I. A., Kolobkov, V. P. and Kudryashov, P. I., Laser action of Ho^{3+} ions in silicate glasses coactivated by holmium, erbium, and ytterbium, *Opt. Spectrosc. USSR* 24, 441 (1968); *Opt. Specktrosk.* 24, 823 (1968).
179. Gandy, H. W. and Ginther, R. J., Stimulated emission from holmium activated silicate glass, *Proc. IRE* 50, 2113 (1962).
180. Percival, R. M., Carter, S. F., Szebesta, D., Davey and Stallard, W. A., Thulium-doped monomode fluoride fibre laser broadly tunable from 2.25 to 2.5 μm, S. T., Highly efficient and tunable operation of two colour Tm-doped fluoride fibre laser, *Electron. Lett.* 27, 1912 (1991).
181. Allen, R. and Esterowitz, L., CW diode pumped 2.3 μm fiber laser, *Appl. Phys. Lett.* 55, 721 (1989).
182. Esterowitz, L., Allen, R. and Aggarwal, I., Pulsed laser emission at 2.3 μm in a thulium-doped fluorozirconate fiber, *Electron. Lett.* 24, 1104 (1988).
183. Allain, J. Y., Monerie, M. and Poignant, H., Energy transfer in Er^{3+}/Pr^{3+}-doped fluoride glass fibers and application to lasing at 2.7 μm, *Electron. Lett.* 27, 445 (1991).

184. Auzel, F., Meichenin, D. and Poignant, H., Tunable continuous-wave room-temperature Er^{3+}-doped ZrF_4-based glass laser between 2.69 and 2.78 μm, *Electron. Lett.* 24, 1463 (1988)
185. Brierley, M. C. and France, P.W., Continuous wave lasing at 2.7 μm in an erbium doped fluorozirconate fiber, *Electron. Lett.* 24, 935 (1988).
186. O'Sullivan, M. S., Chrostowski, J., Desurvire, E. and Simpson, J.R., High-power, narrow-linewidth Er^{3+}-doped fiber laser, *Optics Lett.* 14, 438 (1989).
187 Allen, R. and Esterowitz, L., Diode-pumped single-mode fluorozirconate fiber laser from the $^4I_{11/2} \rightarrow {}^4I_{13/2}$ transition in erbium, *Appl. Phys. Lett.* 56, 1635 (1990).
188. Allain, J. Y., Monerie, M. and Poignant, H., Erbium-doped fluorozirconate single-mode fiber lasing at 2.71 μm, *Electron. Lett.* 25, 28 (1989).
189. Ronarch, D., Guibert, M., Auzel, F., Mechenin, D., Allain, J. Y. and Poignant, H., 35 dB optical gain at 2.716 μm in erbium doped ZBLAN fiber pumped at 0.642 μm, *Electron. Lett.* 27, 511 (1991).
190. Pollack, S. A. and Robinson, M., Laser emission of Er^{3+} in ZrF_4-based fluoride glass, *Electron. Lett.* 24, 320 (1988).
191. Wetenkamp, L., Efficient cw operation of a 2.9 μm Ho^{3+}-doped fluorozirconate fiber laser pumped at 640 nm, *Electron. Lett.* 26, 883 (1990).
192. Többen, H., CW lasing at 3.45 μm in erbium-doped fluorozirconate fibres, *Frequenz* 45, 250 (1991).
193. Többen, H., Room temperature cw fibre laser at 3.5 μm in Er^{3+}-doped ZBLAN glass, *Electron. Lett.* 28, 1361 (1992).
194. Schneider, J., Fluoride fibre laser operating at 3.9 μm, *Electron. Lett.* 31, 1250 (1995).
195. Mackechnie, C. J., Barnes, W. L., Carman, R. J., Townsend, J. E. and Hanna, D. C., High power ytterbium doped fiber laser operating at around 1.2 mm, in *Advanced Solid-State Lasers*, Pinto, A. A. and Fan, T. Y., Eds., Proceedings Vol. 15, Optical Society of America, Washington, DC (1993), p. 192.
196. Jiang, S., Myers, J., Belford, R., Rhonehouse, D., Myers, M. and Hamlin, S., Flashlamp pumped lasing of Ho:germanate glass at room temperature, *OSA Proc. Adv. Solid State Lasers*, Fan, T. Y. and Chai, B. H. T. (Eds.)20, 116 (1995).
197. Miura, K., Tanaka, K. and Hirao, K., CW laser oscillation on both the $^4F_{3/2}–{}^4I_{11/2}$ and $^4F_{3/2}–{}^4I_{13/2}$ transitions of Nd^{3+} ions using a fluoride glass microsphere, *J. Non-Cryst. Solids* 213&214, 276 (1997).

Section 2.3

SOLID STATE DYE LASERS

Introduction to the Table

Solid-state lasers based on organic dye molecules in various host materials are listed in order of increasing wavelength in Table 2.3.1. Lasers that have been tuned over a range of wavelengths are listed by the lowest wavelength reported; the tuning range given is that for the experimental configuration and conditions used and may not represent the extremes possible. The dye molecule and host material are listed together with the primary reference to laser action. The references should be consulted for details of the chemical composition and molecular structure of the dyes and host compounds.

The lasing wavelength and output of dye lasers depend on the characteristics of the optical cavity, the dye concentration, the optical pumping source and rate, and other operating conditions. The original references should therefore also be consulted for this information and its effect on the lasing wavelength.

Further Reading

Bezrodnyi, V. I., Bondar, M. V., Kozak, G. Yu., Przhonskaya, O. V. and Tikhonov, E. A., Dye-activated polymer media for frequency-tunable lasers (review), *Zh. Prikl. Spektrosk. (USSR)* 50, 711 (1989).

Bezrodnyi, V. I., Przhonskaya, O. V., Tikhonov, E. A., Bondar, M. V. and Shpak, M. T., Polymer active and passive laser elements made of organic dyes, *Sov. J. Quantum Electron.* 12, 1602 (1982).

Dodabalapur, A., Chanddross, E. A., Berggren, M. and Slusher, R. L. Organic solid-state lasers: past and future, *Science* 277, 1787 (1997).

Dyumaev, K. M., Manenkov, A. A., Maslyukov, A. P., Matyushin, G. A., Nechitailo, V. S. and Prokhorov, A. M., Dyes in modified polymers: problems of photostability and conversion efficiency at high intensities, *J. Opt. Soc. Am. B* 9, 143 (1992).

Maeda, M., *Laser Dyes*, Academic Press, New York (1984).

O'Connell, R. M. and Saito, T. T., Plastics for high-power laser applications: a review, *Opt. Engin.* 22, 393 (1983).

Rahn, M. D. and King, T. A., Comparison of laser performance of dye molecules in sol-gel, polycom, ormosil, and poly(methyl methacryalte) host media, *Appl. Optics* 34, 8260 (1995).

Schäfer, F. P., Ed., *Dye Lasers*, 3rd edition, Springer-Verlag, Berlin (1990).

Tagaya, A., Teramoto, S., Nihei, E., Sasake, K. and Koike, Y., High-power and high-gain organic dye-doped polymer optical fiber amplifiers: novel techniques for preparation and spectral investigation, *Appl. Optics* 36, 572 (1997).

Zink, J. I. and Dunn, B. S., Photonic materials by the sol-gel process, *J. Cer. Soc. Jpn.* 99, 878 (1991).

Table 2.3.1
Solid State Dye Lasers Arranged in Order of Increasing Wavelength

Wavelength (nm)	Dye (a)	Host (b)	Reference
376	4-phenylstilbene	naphthalene (4.2 K)	1
376	Exalite 377E	silica	2
381–394	BBQ	polystyrene	3
383	diphenylbutadiene (DPB)	para-terphenyl	4
385	1-(2-naphthyl)-2-phenylethylene	naphthalene (4.2 K)	1
~395	α-NPO	polyisobutylmethacrylate	5
395	ββ-dinaphthylethylene (ββ-DNE)	naphthalene	6,7
395	β-naphthyl-p-biphenylethylene (β-BNE)	naphthalene	7
396	α-NPO	PMMA	8
396	diphenylbutadiene (DPB)	para-terphenyl	7
397	1,2-di(2-naphthyl)ethylene	naphthalene (4.2 K)	1
397	1-(1-naphthyl)-2-(2-naphthyl)ethylene	naphthalene (4.2 K)	1
397	1-(4-biphenyl)-2-(2-naphthyl)ethylene	naphthalene (4.2 K)	1
398	azulene	naphthalene (4.2 K)	1
400	2-(4-biphenylyl)-5-(p-styrylphenyl)-1,3,4-oxadiazole	PMMA	8
403	LD 390	silica	2
408–410	anthracene	2,3-dimethylnaphthalene	9
408–410	anthracene	dibenzofurane	9
408–410	anthracene	fluorene	1,9,10
408–410	anthracene	sym-octahydroanthracene	9
410	2-(4-biphenylyl)-5-(1-naphthyl)oxazole	PMMA	8

~410	POPOP	polyisobutylmethacrylate	5
411	a-NPO	polystyrene	8
414	5-phenyl-2-(p-styrylphenyl)oxazole	PMMA	8
414	dimethyl-POPOP	polystyrene	8
415	1-styryl-4-[l-(2-naphthyl)]vinylbenzene	naphthalene (4.2 K)	1
415	POPOP	PMMA	8
416	2-(1-naphthyl)-5-styryl-1,3,4-oxadiazole	PMMA	8
416–426	POPOP	polystyrene	3
418	1,2-di-4-biphenylylethylene	diphenyl (4.2 K)	1
418	2-phenyl-5-[p-(4-phenyl-1,3-buladrenyl)phenyl]-1,3,4-oxadiazole	PMMA	8
~420	dimethyl-POPOP	polyisobutylmethacrylate	5
420	POPOP	polystyrene	8
422	1,4-bis(2-naphthyl)styrylbenzene	naphthalene (4.2 K)	1
423	dimethyl-POPOP	PMMA	8
424	dimethyl-POPOP	polystyrene	8
425	2-[p-[2-(2-naphthyl)vinyl]phenyl]-5-phenyloxazole	PMMA	8
425	dimethyl-POPOP	Polyvinylxylene	8
426–445	dimethyl-POPOP	polystyrene	3
427	dimethyl-POPOP	polystyrene	8
428	5-phenyl-2-[p-(-phenylstyryl)phenyl]oxazole	PMMA	8
428–438	dimethyl-POPOP	polystyrene	11
~430	BBOT	polyisobutylmethacrylate	5
433–457	coumarin 1	silica gel	12
437	β-naphthyl-p-biphenylethylene (β-BNE)	dibenzyl	7
440	stilbene	silica	80
440–472	BBOT	polystyrene	3
441	pyrene	K_2SO_4 crystal	13

Table 2.3.1—*continued*
Solid State Dye Lasers Arranged in Order of Increasing Wavelength

Wavelength (nm)	Dye [(a)]	Host [(b)]	Reference
442	1-styryl-4-[l-(2-naphthyl)vinylbenzene	naphthalene (4.2 K)	1
443	5-(4-biphenylyl)-2-[p-(4-phenyl-1,3-baladienyl)phenyl]oxazole	PMMA	8
443.9	coronene	methyl-cyclohexane/isopentane	14
444	1,2-bis(5-phenyloxazolyl)ethylene	PMMA	5
449	1,4-bis(2-naphthyl)styrylbenzene	naphthalene (4.2 K)	1
450	1,5-diphenyl-3-styryl-2-pyrazoline	PMMA	8
450–540[(c)]	methylumbelliferone	Knox[(d)] gelatin	15
~453[(c)]	7-diethylamino-4-methylcoumarin	Knox[(d)] gelatin	15
455	1,4-bis[4-[5-(4-biphenylyl)-2-oxazolyl]styryl benzene	PMMA	8
455–475	perylene	PMMA(BBOT)[(e)]	16
457	3-p-chlorostyryl-1,5-diphenyl-2-pyrazoline	PMMA	8
~460	coumarin 460	PBD	90
461	β-naphthyl-p-biphenylethylene (β-BNE)	dibenzyl	7
464–480	BBOT + perylene	polystyrene	3
468–494	coumarin 460	silica	17
469	POPOP	durol	7
475	coumarin 460	silica	80
478	(see Ref. 8)	PMMA	8
480	2-[p-[2-(9-anthryl)vinyl]phenyl]-5-phenyloxazole	PMMA	8
484–529	CF_3-coumarin	PMMA	18
487–495	coumarin 102	silica gel	12

490–500	1,3-dimethylisobenzofuran	polystyrene	3
490–500	acridine yellow	PMMA(coumarin 1)$^{(f)}$	16
497	COP-2$^{(g)}$	MMA	20
493	2-(2'-hydroxy-5'-fluorophenyl) benzimidazole	PMMA	19
494	tetracene	dibenzyl (4.2 K)	1
494	5(6)-methoxycarbonyl-2(2'-hydroxphenyl)benzimidazole	PMMA	20
498–574	coumarin 153	silica gel (ORMOSIL)	21
500–520	acriflavine	PMMA(coumarin 1)$^{(f)}$	16
508	5(6)-methoxycarbonyl-12-(5'-fluoro-2'-hydroxyphenyl)benzimidazole	PMMA	20
509–514	fluoran	PMMA	8
511	COP-3$^{(g)}$	MMA	20
515	coumarin 540A	PMMA	76
521	tetracene	dibenzyl (4.2 K)	1
522	pyrromethene 567	HTP	22
525	coumarin 540A	silica gel (ORMOSIL)	23
525–650	rhodamine 6G	polyacrylamide gel	24
528	pyrromethene 597	HTP	22
528	naphthacene	para-terphenyl	7
530	tetracene	para-terphenyl	9
530–560	uranine	PMMA(coumarin 1)$^{(f)}$	16
530–630	perylimide (BASF 241)	composite glass	25
535	coumarin 540A	P(HEMA:MMA 1:1)	76
535	coumarin 481	silica	79
541	pyrene	K_2SO_4 crystal	14
542–606	pyrromethene 567	ORMOSIL (VTEOS, MTEOS)	26
543–603	pyrromethene 567	ORMOSIL (VTEOS, MTEOS)	27
545–572	coumarin 153	silica gel	12

Table 2.3.1—*continued*
Solid State Dye Lasers Arranged in Order of Increasing Wavelength

Wavelength (nm)	Dye (a)	Host (b)	Reference
545–630	rhodamine 6G	silica	28–30
549	pyrromethene 567	silica	31,83
550	rhodamine 6G	PMMA	8
550–570	pyrromethene 580	HTP	33
550–570(c)	Na fluorescein	Knox(d) gelatin	15
552–595	rhodamine 6G	PMMA	33,34
553–564	fluorescein (sodium salt)	polystyrene	11
554--584	pyrromethene 580	ORMOSIL (VTEOS, MTEOS)	26
555–565	rhodamine 6G	PMMA	36
556	pyrromethene 567	sol-gel glass	83
557	coumarin 540A	aluminosilicate	23,35
557–598	rhodamine RG6	silica gel (ORMOSIL)	21
558	coumarin 521	silica	80
559–587	rhodamine RG6	silica gel (ORMOSIL)	21
560	rhodamine 6G	silica	37
~560(h)	rhodamine 6G	Al_2O_3	38
560–570	rhodamine 6G	mPMMA	39
560–570	rhodamine 6G chloride	mPMMA	40
560–570	rhodamine 6G percholate	mPMMA	40
560–570	rhodamine III	mPMMA	40
560–570	dye "II B"	mPMMA	40

560–610	rhodamine 6G	silica film (ORMOSIL)	18
562–590	rhodamine 6G	silica (ORMOSIL)	41
564	pyrromethene 567	acrylic copolymer	94
564	rhodamine 590	silica	31,83
565–590	pyrromethene 597	HTP	32
565.1	rhodamine 6G	PSI-gel	42
565.5	rhodamine 6G	PMMA	42
566.3	pyrromethene 570	HTP	42
567–594	peryline orange	ORMOSIL (VTEOS, MTEOS)	27
567–605	rhodamine 6G	PMMA	43
568	rhodamine 6G	silica (ORMOSIL)	23
568	rhodamine 6G	P(HEMA:MMA 1:1)	78
568	rhodamine 590	sol-gel glass	83
568–583	Lumogen LFO240	silica film (ORMOSIL)	17
569.6	pyrromethene 570	PMMA	42
569.7	1,3,5,7,8-pentamethyl-2,6-di-n-butylpyrromethene-BF_2	MMA(i)	44
570	DCM	PMMA	45
570	rhodamine 6G	aluminosilicate	23,35
570	pyrromethene 580	acrylic copolymer	46
570–610	rhodamine 6G	silica gel	47,86
570–620(c)	rhodamine 6G	Knox(d) gelatin	16
570.9	rhodamine 6G	PMMA	42
571	pyrromethene 580	acrylic copolymer	95
571.0	1,3,5,7,8-pentamethyl-2,6-di-n-butylpyrromethene-BF_2	MMA(i)	44
571.4	pyrromethene 567	MMA(i)	44
572	pyrromethene 567	ORMOSIL (silica/PMMA)	31
572	pyrromethene 567	polycom glass(m)	83

Table 2.3.1—*continued*
Solid State Dye Lasers Arranged in Order of Increasing Wavelength

Wavelength (nm)	Dye (a)	Host (b)	Reference
572(563)	rhodamine 590	polycom glass(m)	83
572	rhodamine 590	ORMOSIL (silica/PMMA)	31
573	rhodamine 590	PMMA	31,83
573	rhodamine 590 chloride	polycom glass(m)	83
574–606	pyrromethene 597	ORMOSIL (VTEOS, MTEOS)	26
575–590	BASF-241	ORMOSIL (silica/PMMA)	48
575–590	PPV7	PBD	91
577	rhodamine 6G	silica	37
577–590	rhodamine 6G	PMMA	49
578	peryline orange	PMMA	31
578	peryline orange	ORMOSIL (silica/PMMA)	31,50
578	peryline orange (KF 241)	ORMOSIL (silica/PMMA)	31,83
578	peryline orange (KF 241)	polycom glass(m)	83
580–620	rhodamine 6G	PHEMA+DEGMA	51
582–592	peryline orange	ORMOSIL (VTEOS, MTEOS)	26
584	rhodamine 6G	P(HEMA:MMA 3:7)	78
584	pyrromethene 567	ORMOSIL	46
585	peryline orange	silica	31,83
585	rhodamine 590	PMMA	52
585–606	ASPI	silica-MMA	85
585–635	rhodamine 610	silica film (ORMOSIL)	17

587	pyrromethene 597	acrylic copolymer	46,94
587	pyrromethene 597	high temperature plastic	93
587	rhodamine-Bz-MA	P(HEMA:MMA 1:1)	77
587	rhodamine 6G	P(HEMA:MMA 7:3)	78
587.4	rhodamine B	PMMA	53
589	diphenylbutadiene (DPB)	para-terphenyl	7
589	POPOP	para-terphenyl	7
589	Cl-POPOP	para-terphenyl	7
589	rhodamine-Al	P(HEMA:MMA 1:1)	77
589–635$^{(n)}$	DCM	Alq_3	96
~590$^{(h)}$	rhodamine B	Al_2O_3	38
590.7–654.3	DCM	ORMOSIL (TiO_2)	82
593	PM-HMC	high temperature plastic	22
593	rhodamine-Bz-MA	P(HEMA:MMA 7:3)	77
593	rhodamine 6G	P(HEMA:MMA 1:1)	78
594.0	rhodamine B	PMMA	53
595	rhodamine C	PMMA	8
595–620	rhodamine	K_2SO_4 crystal	13
595–640	peryline red	ORMOSIL (VTEOS, MTEOS)	26
595–644	peryline red	ORMOSIL (VTEOS, MTEOS)	27
595–650	DCM	ORMOSIL (film)	54
598	PM-TEDC	HTP	21
599	rhodamine 6G	P(HEMA:EDGMA 9:1)	78
~600	sulforhodamine B	acrylic monomers	55
~600	rhodamine 640	polystyrene	57
~600	ASPT	HEMA	87
~600	DCM II	PBD	90

Table 2.3.1—*continued*
Solid State Dye Lasers Arranged in Order of Increasing Wavelength

Wavelength (nm)	Dye [a]	Host [b]	Reference
~600–620[c]	rhodamine B	Knox[d] gelatin	15
600–620	rhodamine B	PMMA(rhodamine 6G)[f]	16
600–625	rhodamine B	silica (ORMOSIL)	41
600–650[j]	Wratten[d] 22 filter	———	15
600–650	sulforhodamine 640	silica gel	57
601.0	rhodamine 6G	PMMA	58
604	peryline red	ORMOSIL (silica/PMMA)	31
604	perylene red (KF 856)	polycom glass [m]	83
~605	MEH-PPV	PS:TiO_2	92
605	rhodamine 6G	ORMOSIL (silica/GPTA)	53
605–630	Lumogen LFR300	silica film (ORMOSIL)	17
605–630	red perylimide dye	ORMOSIL (silica/PMMA)	49
605–648	sulforhodamine 640	silica gel	59
~606	DHASI	HEMA	98
609	rhodamine 640	pig fat	60
~610	ASPI[k]	PHEMA	81
610–620	rhodamine B	Al_2O_3 (rhodamine 6G)[f]	61
610–620	rhodamine B	silica gel	12
610–635	rhodamine 6G	polyurethane	62–64
613	peryline red	PMMA	31,83
613	rhodamine 640	chicken tissue	60

613	DCM II	Alq_3	88
614	peryline red	ORMOSIL (silica/PMMA)	50,83
614–624	rhodamine 6G	polystyrene	11
~615	DCM II	NAPOXA	90
615–629	Lumogen LFR300	silica film (ORMOSIL)	9
615–635	PM-TEDC	HTP	32
615–635	PM-HMC	HTP	32
617.9	1,3,5,7-tetramethyl-8-cyanopyrromethene-2,6-dicarboxylate-BF_2	MMA(i)	44
618.2	sulforhodamine B	MMA(i)	44
620–640	oxazine-17	mPMMA	39,40
620–640	cresyl violet	PMMA (rhodamine 6G)(f)	16
620–670	sulforhodamine 101	PMMA (rhodamine 6G)(f)	16
624	PM650	HTP	46
625	coumarin 560	silica	80
~628	rhodamine B	silica (ORMOSIL-γ-GLYMO	66
~630	rhodamine 6G	gelatin	67
632.4	rhodamine B	PMMA	58
632.8	rhodamine B	polyurethane (film)	68
~640(h)	oxazine 4	Al_2O_3	38
643	rhodamine 110	polyurethane	69
645	DCM	Alq_3	89
654	rhodamine 640	HEMA/MMA copolymer	70
~655	DCM	Alq_3	97
655	coumarin 640	silica	80
657	Wratten(d) 29 filter	———	15
670	cresyl violet 670	polyimide (Probimide 414)(l)	84
680	indodicarbocyanine (PK 643)	polyurethane	71

Table 2.3.1—*continued*
Solid State Dye Lasers Arranged in Order of Increasing Wavelength

Wavelength (nm)	Dye (a)	Host (b)	Reference
680–746	Nile Blue	silica (ORMOSIL)	72
695–720	Nile Blue	PMMA (rhodamine 6G)(f)	16
727–747	LD800	ORMOSIL	73
~805	LDS821	PBD	90
819–844	HITC	ORMOSIL	73
845–865	HITC	PMMA (rhodamine 6G/nile blue)(f)	16

Footnotes – Table 2.3.1

(a) See M. Maeda, *Laser Dyes,* Academic Press, New York (1984), Steppel, R. N., Organic Dye Lasers, in *Handbook of Laser Science and Technology, Vol. I: Lasers and Masers*, CRC Press, Boca Raton, FL (1982), p. 299, and Steppel, R. N., Organic Dye Lasers, in *Handbook of Laser Science and Technology, Suppl. 1: Lasers*, CRC Press, Boca Raton, FL (1991), p. 219 for complete compositions and chemical structures of the dyes.

(b) Host materials: Alq_3 — tris-(8-hydroxyquinoline) aluminum, DEGMA — ethylene glycol dimethacrylate, EGDMA—ethylene glycol dimethacrylate, GLYMO — γ glycidyloxyropyl trimethoxy silan, GPTA — glycerol propoxy triacrylate, HEMA — hydroxy ethyl methacrylate, HTP — high temperature plastic (Korry Electronics), MMA — methylmethacrylate, mPMMA — modified polymethylmethacrylate, MTEOS — methyl-triethoxysilane, NAPOXA — 2-napthyl-4,5-bis(4-methoxyphenyl)-1,3-oxazole, ORMOSIL - organically modified silicate, PBD — 2-(4-biphenylyl)-5-(4-*t*-butylphenyl)-1,3,4-oxadiazole, PHEMA — poly (2-hydroxyethyl methacrylate), PMMA — polymethylmethacrylate, PS — polystyrene, PSI — proprietary polymer, VTEOS — vinyltriethoxysilane.

(c) Lasing wavelengths were not given but were stated to be close to those produced by the same dyes in liquid solutions.

(d) Knox and Wratten are commercial product names.

(e) Optical gain was also reported in Ref. 16 for 3-carboethoxy-7-hydroxycoumarin, 3-n-butyl-4-methyl-7-hydroxycoumarin, 4,6-dimethyl-7- ethylaminocoumarin, 4-methyl-7-dimethylaminocoumarin, 4-MU, and DAM-coumarin dyes in PMMA but no wavelengths were given.

(f) Host materials were polymethylmethacrylate and polyvinylalcohol. Dyes in parentheses serve as a donor dye in an energy transfer laser.

(g) See Ref. 20 for molecular structure of copolymer.

(h) This is the wavelength of the fluorescence peak; lasing wavelengths were not cited.

(i) 16% hydroxypropyl acrylate/methyl methacrylate.

(j) Output was stated to be in the orange-red region.

(k) ASPI—trans-4-[p-(N-hydroxyethyl-N-methylamino)styryl]-N-methylpyridinium iodide.

(l) A photosensitive benzophenone tetracarboxylic diahydride-alkylated diamine polyimide.

(m) Sol-gel glass–PMMA composite.

(n) Laser emission varied by changing the thickness of the active organic layer.

(o) Tb^{3+} laser action (~545 nm) from terbium thenoyltrifluoroacetonate in PMMA was reported in Ref. 74, but no wavelengths were given.

(p) Eu^{3+} laser action (~613 nm) from europium tris[4,4,4-trifluoro-1-(2-thienyl-1,3 butanedione in PMMA was reported in Ref. 75, but no wavelengths were given.

References

1. Naboikin, Yu. V., Ogurtsova, L. A., Podgornyi, A. P. and Maikes, L. Ya., Stimulated emission of light from doped molecular crystals at 4.2 K, *Sov. J. Quantum Electron.* 8, 457 (1978).
2. Lam, K. S., Lo, D. and Wong, K. H., Observations of near-UV superradiance emission from dye-doped sol-gel silica, *Optics Comm.* 121, 121 (1995).
3. Muto, S., Ando, A., Yoda, O., Hanawa, T. and Ito, H., Dye laser by sheet of plastic fibers with wide tuning range, *Trans. IECE (Jpn)* E 70, 317 (1987).
4. Budakovskii, S. V., Gruzinskii, V. V., Davydov, S. V., Kulak, I. I. and Kolesnik, E. E., Stimulated emission from electron-beam-pumped diphenylbutadiene molecules imbedded in a crystalline matrix, *Sov. J. Quantum Electron.* 22, 16 (1992).
5. Muto, S., Ichikawa, A., Ando, A., Ito, C. and Inaba, H., Trial to plastic fiber dye laser, *Trans. IECE (Japan)* E 69, 374 (1986).
6. Bokhonov, A. F., Davydov, S. V., and Kulam, I. I. et al., Spontaneous and stimulated-radiation of impurity molecular crystals with optical and electron pumping, *Zh. Prinkl. Spektrosk.* 50, 966 (1989).
7. Budakovskii, S. V., Gruzinskii, V. V., Davydov, S. V. and Kolesnik, E. E., Laser media on impurity organic crystals with pumping into an absorption band of the matrix, *Zh. Prinkl. Spektrosk.* 54, 79 (1991).
8. Naboikin, Yu. V., Ogurtsova, L. A., Podgornyi, A. P. et al., Spectral and energy characteristics of lasers based on organic molecules in polymers and toluol, *Opt. Spektrosk.* 28, 974 (1970); *Sov. Opt. Spectrosc.* 28, 528 (1970).
9. Karl, N., Laser emission from organic molecular crystals, mode selection and tunability, *J. Lumin.* 12-13, 851 (1976).
10. Karl, N., Laser emission from an organic molecular crystal, *Phys. Stat. Sol. A* 13, 651 (1972).
11. Tanuguchi, H., Fujiwara, T., Yamada, H., Tanosake, S. and Baba, M., Whispering-galley-mode dye lasers in blue, green, and orange regions using dye-doped, solid, small spheres, *Appl. Phys. Lett.* 62, 2155 (1993; see also, *J. Appl. Phys.* 73, 7957 (1993).
12. McKiernan, J. M., Yamanaka, S. A., Knoble, E. T., Pouxviel, J. C., Parvench, D. E., Dunn, S. B. and Zink, J. I., *J. Inorg. and Organomet. Polymers* 1, 87 (1991).
13. Rifani, M., Yin, Y.-Y., Elliott, D. S. et al., Solid state dye lasers from stereospecific host-guest interactions, *J. Am. Chem. Soc.* 117, 7572 (1995).
14. Kohlmannsperger, J., Ein organischer laser: coronen in MCH/IP bei 100 K, *Z. Naturf.* 24a, 1547 (1969).
15. Hänsch, T. W., Pernier, M. and Schawlow, A. L., Laser action of dyes in gelatin, *IEEE J. Quantum Electron.* QE-7, 45 (1971).
16. Muto, S., Shiba, T., Iijima, Y., Hattori, K. and Ito, C., Solid thin-film energy transfer dye lasers, *Trans. IECE (Japan)* J69-C, 25 (1986); [*Electron and Commun. Japan*, part 2, 70, 21 (1987)].
17. Lam, K.-S., Lo, D. and Wong, K.-H., Sol-gel silica laser tunable in the blue, *Appl. Optics* 34, 3380 (1995).
18. Itoh, U., Takakusa, M., Moriya, T. and Saito, S., Optical gain of coumarin dye-doped thin film lasers, *Jpn. J. Appl. Phys.* 16, 1059 (1977).
19. Acuña, A. U., Amat-Guerri, F., Costela, A., Douhal, A., Figuera, J. M., Florido, F. and Sastre, R., Proton-transfer lasing from solid organic matrices, *Chem. Phys. Lett.* 187, 98 (1991).
20. Ferrer, M. L., Acuña, A. U., Amat-Guerri, F., Costela, A., Figuera, J. M., Florido, F. and Sastre, R., Proton-transfer lasers from solid polymeric shains with covalently bound 2-(2'-hydroxyphenyl)benzimidazole groups, *Appl. Optics* 33, 2266 (1994).
21. Knobbe, E. T., Dunn, B., Fuqua, P. D. and Nishida, F., Laser behavior and photostability characteristics of organic dye doped silicate gel materials, *Appl. Optics* 29, 2729 (1990).

22. Allik, T., Chandra, S., Robinson, T. R., Hutchinson, J. A., Sathyamoorthi, G. and Boyer, J. H., Laser performance and material properties of a high temperature plastic doped with pyrromethene-BF_2 dyes, Mat. Res. Soc. Proc. Vol. 329, *New Materials for Solid State Lasers* (1994), p. 291.
23. Dunn, B., Mackenzie, J. D., Zink, J. I. and Stafsudd, O. M., Solid-state tunable lasers based on dye-doped sol-gel materials, SPIE Vol. 1328, *Sol-Gel Optics* (1990), p. 174.
24. Kessler, W. J. and Davis, S. J., Novel solid state dye laser host, *Proceedings of the Solid State Dye Laser Workshop* (1994), p. 216.
25. Reisfeld, R., Brusilovsky, D., Eyal, M., Miron, E., Burstein, Z. and Ivri, J., A new solid-state tunable laser in the visible, *Chem. Phys. Lett.* 160, 43 (1989).
26. Canva, M., Dubois, A., Georges, P., Brum, A., Chaput, F., Ranger, A. and Boilot, J.-P., Perylene, pyrromethene and grafted rhodamine doped xerogels for tunable solid state laser, SPIE Vol. 2288, *Sol-Gel Optics III* (1994), p. 298.
27. Canva, M., Georges, P., Perelgritz, J.-F., Brum, A., Chaput, F. and Boilot, J.-P., Perylene- and pyrromethene-doped xerogel for a pulsed laser, *Appl. Optics* 34, 428 (1995).
28. Dul'nev, G. N., Zemskin, V. I., Krynetshi, B. B., Meshkovskii, I. K., Prokhorov, A. M. and Stelmakl, O. M., Tunable solid-state laser with a microcomposition matrix active medium, *Sov. Tech. Phys. Lett.* 4, 420 (1978).
29. Altshuler, G. B., Dulneva, E. G., Meshkovskii, I. K. and Krylov, K. I., Solid state active media based on dyes, *(Zh. Prikl. Spektrosk.), J. Appl. Spectrosc.* 36, 415 (1981).
30. Altshuler, G. B., Dulneva, E. G., Krylov, K. I., Meshkovskii, I. K. and Urbanovich, V. S., Output characteristics of a laser utilizing rhodamine 6G in microporous glass. *Sov. J. Quantum Electron.* 13, 784 (1983).
31. Rahn, M. D. and King, T. A., Lasers based on dye doped sol-gel composite glasses, SPIE Vol. 2288, *Sol-Gel Optics III* (1994), p. 382.
32. Chandra, S. and Allik, T., Compact high-brightness solid state dye laser, *Proceedings of the Solid State Dye Laser Workshop* (1994), p. 29.
33. Soffer, B. H. and McFarland, B. B., Continuously tunable, narrow band organic dye lasers, *Appl. Phys. Lett.* 10, 266 (1967).
34. Kaminow, I. P., Weber, H. P., and Chandross, E. A., Poly(methyl methacrylate) dye laser with internal diffraction grating resonator, *Appl. Phys. Lett.* 18, 497 (1971).
35. McKiernan, J. M., Yamanaka, S. A., Dunn, B. and Zink, J. I., *J. Phys. Chem.* 94, 5652, (1990).
36. Onstott, J. R., Short cavity dye laser excited by an electron beam-pumped semiconductor laser, *Appl. Phys. Lett.* 31, 818 (1977).
37. Whitehurst, C., Shaw, D. J. and King, T. A., Sol-gel glass solid state lasers doped with organic molecules, *Sol-Gel Optics* (1990), p. 183.
38. Kobayashi, Y., Kurokawa, Y. and Imai, Y., A transparent alumina film doped with laser dye and its emission properties, *J. Non-Cryst. Solids* 105, 198 (1988).
39. Gromov, D. A., Dyumaev, K. M., Manenkov, A. A., Maslyukov, A. P., Matyushin, G. A., Nechitailo, V. S. and Prokhorov, A. M., Efficient plastic-host dye lasers, *J. Opt. Soc. Am. B* 2, 1028 (1985).
40. Manenkov, A. A., Maslyukov, A. P., Matyushin, G. A. and Nechitailo, V. S., Modified polymers - effective host materials for solid state dye lasers and laser beam control elements: a review, SPIE Vol. 2115, *Visible and UV Lasers* (1994), p. 136.
41. Altman, J. C., Stone, R. E., Dunn, B. and Nishida, F., Solid state laser using a rhodamine-doped silica gel compound, *IEEE Phot. Techn. Lett.* 3, 189 (1991).
42. Ewanizky, T. F. and Pearce, C. K., Solid state dye lasers in an unstable resonator configuration, *Proceedings of the Solid State Dye Laser Workshop* (1994), p. 154.
43. Fork, R. L., German, I. R. and Chandron, E. A., Photodimer distributed feedback laser, *Appl. Phys. Lett.* 20, 139 (1972).

44. Hermes, R. E, Allik, T. H., Chandra, S. and Hutchison, J. A., High-efficiency pyrromethene doped solid-state dye lasers, *Appl. Phys. Lett.* 63, 877 (1993).
45. Mukherjee, A., Two-photon pumped upconverted lasing in dye doped polymer waveguides, *Appl. Phys. Lett.* 62, 3423 (1993).
46. Boyer, J., Pyrromethene-BF_2 complexes (P-BF_2), *Proceedings of the Solid State Dye Laser Workshop* (1994), p. 66.
47. Altshuler, G. B., Bakhanov, V. A., Dulneva, E. G., Erofeev, A. V., Mazurin, O. V., Roskova, G. P. and Tsekhomskaya, T. S., Laser based on dye-activated silica gel, *Opt. Spectrosc. (USSR)* 62, 709 (1987).
48. Reisfeld, R., Film and bulk tunable lasers in the visible, SPIE Vol. 2288, *Sol-Gel Optics III* (1994), p. 563.
49. Schinke, D. P., Smith, R. G., Spencer, E. G. and Galvin, M. F., Thin-film distributed feedback laser fabricated by ion milling, *Appl. Phys. Lett.* 21, 494 (1972).
50. Rhan, M. D., King, T. A., Capozzi, C. A. and Seldon, A. B., Characteristics of dye doped ormosil lasers, SPIE Vol. 2288, *Sol-Gel Optics III* (1994), p. 364.
51. Amat-Guerri, F., Costela, A., Figuera, J. M., Florido, F. and Sastre, R., Laser action from rhodamine 6G-doped poly(2-hydroxyethyl methacrylate) matrices with different crosslinking degrees, *Chem. Phys. Lett.* 207, 352 (1993).
52. Mandl, A. and Klimek, D. E., 400 mJ long pulse (>1μs) solid state dye laser, *Proceedings of the Solid State Dye Laser Workshop* (1994), p. 192.
53. Wojcik, A. B. and Klein, L. C., Rhodamine 6G-doped inorganic/organic gels for laser and sensor applications, SPIE Vol. 2288, *Sol-Gel Optics III* (1994), p. 392.
54. Reisfeld, R., Film and bulk tunable lasers in the visible, *Proceedings of the Solid State Dye Laser Workshop* (1994), p. 56.
55. Hermes, R. E., McGrew, J. D., Wiswall, C. E., Monroe, S. and Kushina, M., A diode laser-pumped Nd:YAG-pumped polymeric host solid-state dye laser, *Appl. Phys. Commun.* 11, 1 (1992).
56. Misawa, H., Fujisawa, R., Sasaki, K., Kitamura, N. and Masuhara, H., Simultaneous manipulation and lasing of a polymer microparticle using a cw 1064 nm laser beam, *Jpn. J. Appl. Phys.* 32, L788 (1993).
57. Canva, M., Georges, P., Brun, A., Larrue, D. and Zarzycki, J., Impregnated SiO_2 gels used as dye laser matrix hosts, *J. Non-Cryst. Solids* 147&148, 636 (1992).
58. Peterson, O. G. and Snavely, B. B., Stimulated emission from flashlamp excited organic dyes in polymethyl methacrylate, *Appl. Phys. Lett.* 12, 238 (1968).
59. Salin, F., Le Saux, G., Georges, P., Brun, A., Bagnall, C. and Zarzycki, J., Efficient tunable solid-state laser near 630 nm using sulfo rhodamine 640-doped silica gel, *Opt. Lett.* 14, 785 (1989).
60. Siddique, M., Yang, L., Wang, Q. Z. and Alfano, R. R., Mirrorless laser action form optically pumped dye-treated animal tissues, *Optics Commun.* 117, 475 (1995).
61. Sasaki, H., Kobayashi, Y., Muto, S. and Kurokawa, Y., Preparation and photoproperties of a transparent alumina film doped with energy-transfer-type laser dye pairs, *J. Am. Ceram. Soc.* 73, 453 (1990).
62. Weber, H. P. and Ulrich, R., Unidirectional thin-film ring lasers, *Appl. Phys. Lett.*, 20, 38 (1972); Ulrich, R. and Weber, H. P., *Appl. Optics* 11, 428 (1972).
63. Masuda, A. and Iijima, S., Preferential doping of rhodamine 6G in polyurethane optical circuit, *Appl. Phys. Lett.* 30, 571 (1977).
64. Sasaki, K., Fukao, T., Saito, T. and Hamano, O., Thin film waveguide evanescent dye laser and its gain measurement, *J. Appl. Phys.* 51, 3090 (1980).
65. Shamrakov, D. and Reisfeld, R., Super radiant film laser operation of perylimide dyes doped silica-polymethylmethacrylate composite, *Opt. Mater.* 4, 103 (1994).
66. Reisfeld, R., Shamrakov, D. and Sorek, Y., Spectroscopic properties of thin glass films doped by laser dyes prepared by sol-gel method, *J. de Phys. IIII*, Vol. 4, Colloque C4, C4-487 (1994); Sorek, Y. and Reisfeld, R., Light amplification in a dye-doped glass planar waveguide, *Appl. Phys. Lett.* 66, 1169 (1995).

67. Kogelnik, H. and Shank, C. V., Stimulated emission in a periodic structure, *Appl. Phys. Lett.* 18, 152 (1971).
68. Chang, M. S., Burlamacchi, P., Hu, C. and Whinnery, J.R., *Appl. Phys. Lett.* 20, 313 (1972).
69. Sriram, S., Jackson, H. E. and Boyd, J. T., Distributed-feedback dye laser integrated with a channel waveguide formed in silica, *Appl. Phys. Lett.* 36, 721 (1980).
70. Amat-Guerri, F., Costel, A., Figuera, J. M., Florido, F., Garcia-Moreno, I. and Sastre, R., Laser action from a rhodamine 640-doped copolymer of 2-hydroxyethyl methacrylate and methyl methacrylate, *Optics Comm.* 114, 442 (1995).
71. Bondar, M. V., Przhonskaya, O. V. and Tikhonov, E. A., Amplification of light by dyed polymers as the laser pumping frequency changes, *Opt. Spectrosc.* 74, 215 (1993).
72. Dunn, B., Nishida, F., Altman, J. C. and Stone, R. E., Spectroscopy and laser behavior of rhodamine-doped ORMOSILS, in *Chemical Processing of Advanced Materials*, Hench, L. L and West, J. K., Eds., Wiley, New York (1992), p. 941.
73. Dunn, B., Nishida, F., Toda, R., Zink, J. I., Allik, T., Chandra, S. and Hutchinson, J. A., Advances in dye-doped sol gel lasers, *Mat. Res. Soc. Proc.* Vol. 329, New Materials for Solid State Lasers (1994), p. 267.
74. Huffman, E. H., Stimulated optical emission of a terbium ion chelate in a vinylic resin matrix, *Nature* 200, 158 (1963; Stimulated optical emission of a Tb^{3+} chelate in a vinylic resin matrix, *Phys. Lett.* 7, 237 (1963). For additional observations of probable stimulated emision of a terbium ion chlelate in a vinylic resin matrix, see *Nature* 203, 1373 (1964).
75. Wolff, N. E. and Pressley, R. J., Optical laser action in an Eu^{+3}-containing organic matrix, *Appl. Phys. Lett.* 2, 152 (1963).
76. Costela, A., Garcia-Moreno, I., Figuera, J. M., Amat-Guerri, F., Barroso, J. and Sastre, R., Solid-state dye laser based on coumarin 540A-doped polymeric matrices, *Opt. Commun.* 130, 44 (1996).
77. Costela, A., Garcia-Moreno, I., Figuera, J. M., Amat-Guerri, F. and Sastre, R., Solid-state dye lasers based on polymers incorporating covalently bonded modified rhodamine 6G, *Appl. Phys. Lett.* 68, 593 (1996).
78. Costela, A., Florido, F., Garcia-Moreno, I., Duchowicz, R., Amat-Guerri, F., Figuera, J. M. and Sastre, R., Solid-state dye lasers based on copolymers of 2-hydroxyethyl methacrylate and methyl methacrylate doped with rhodamine 6G, *Appl. Phys. B* 60, 383 (1995).
79. Lo. D., Parris, J. E. and Lawless, J. L., Multi-megawatt superradiant emissions from coumarin-doped sol-gel derived silica, *Appl. Phys. B* 55, 365 (1992).
80. Lo. D., Parris, J. E. and Lawless, J. L., Laser and fluorescence properties of dye-doped sol-gel silica from 400 nm to 800 nm, *Appl. Phys. B* 56, 385 (1993).
81. He, G. S., Bhawalkar, J. D., Zhao, C. F. and Park, C. K., Upconversion dye-doped polymer fiber laser, *Appl. Phys. Lett.* 68, 3549 (1996).
82. Hu, W., Chuangdong, H. Y., Jiang, Z. and Zhou, F., All-solid-state tunable dye laser pumped by a diode-pumped Nd:YAG laser, *Appl. Optics* 36, 579 (1997).
83. Rahn, M. D. and King, T. A., Comparison of laser performance of dye molecules in sol-gel, polycom, ormosil, and poly(methyl methancrylate) host media, *Appl. Optics* 34, 8260 (1995).
84. Weiss, M. N., Srivatava, R., Correia, R. R. B., Martins-Filho, J. F. and de Araujo, C. B., Measurement of optical gain at 670 nm in an oxazine-doped polyimide planar waveguide, *Appl. Phys. Lett.* 69, 3653 (1996).
85. Gvishi, R., Ruland, G. and Prasad, P. N., The influence of structure and environment on spectroscopic and lasing properties on dye-doped glasses, *Opt. Mater.* 8, 43 (1997).
86. Finkelstein, I., Ruschin, S., Sorek, Y. and Reisfeld, R., Waveguided visible lasing effects in a dye-doped sol-gel glass film, *Opt. Mater.* 7, 9 (1997).

87. He, G. S., Zhao, C. F., Bhawalkar, J. D. and Prasad, P. N., Two-photon pumped cavity lasing in novel dye doped bulk matrix rods, *Appl. Phys. Lett.* 67, 3703 (1995).
88. Berggren, M., Dodabalapur, A. and Slusher, R. E., Stimulated emission and lasing in dye-doped organic thin films with Forster transfer, *Appl. Phys. Lett.* 71, 2230 (1997).
89. Kozlov, V. G., Bulovic´, V. and Fcrrest, S. R., Temperature independent performance of organic semiconductor lasers, *Appl. Phys. Lett.* 71, 2575 (1997).
90. Berggren, M., Dodabalapur, A., Slusher, R. L. and Bao, Z., Light amplication in organic thin films using cascade energy transfer, *Nature* 389, 466 (1997).
91. Berggren, M., Dodabalapur, A., Bao, Z. and Slusher, R. E., Solid-state droplet laser made from an organic blend with a conjugated polymer emitter, *Adv. Mater.* 9, 968 (1997).
92. Hide, F., Schwartz, B. J., Diaz-Garcia, M. A. and Heeger, A. J., Laser emission from solutions and films containing semiconducting polymer and titanium dioxide nanocrystals, *Chem. Phys. Lett.* 256, 424 (1996).
93. Allik, T. H., Chandra, S., Robinson, T. R., Hutchinson, J. A., Sathyamoorthi, G. and Boyer, J. H., Laser performance and material properties of a high temperature plastic doped with pyrromethene-BF_2 dyes, *Mat. Res. Soc. Symp. Proc.* 329, 291 (1994).
94. Hermes, R.E., Lasing performance of pyrromethene-BF_2 laser dyes in a solid polymer host, *SPIE Proceedings: Visible and UV Lasers*, 2115 (1994).
95. Allik, T. H., Chandra, S., Hermes, R. E., Hutchinson, J. A., Soong, M. L. and Boyer, J. H., Efficient and robust solid-state dye laser, OSA *Proc. on Adv. Solid-State Lasers* 15, 271 (1993).
96. Bulovic´, V., Kozlov, V. G., Khalfin, V. B. and Forrest, S. R., Transform-limited, narrow-linewidth lasing action in organic semiconductor microcavities, *Science* 279, 553 (1998).
97. Berggren, M., Dodabalapur, A., Slusher, R. E., Timko, A. and Nalamasu, O., Organic solid-state lasers with imprinted gratings on plastic substrates, *Appl. Phys. Lett.* 72, 410 (1998).
98. He, G. S., Kim, K.-S., Yuan, L., Cheng, N. and Prasad, P. N., Two-photon pumped partially cross-linked polymer laser, *Appl. Phys. Lett.* 71, 1619 (1997).

Section 2.4
COLOR CENTER LASERS

Introduction to the Table

Color center lasers are listed in order of increasing wavelength in Table 2.4.1. The host crystal and the active center are given in the next two columns. If the host contained additives, they are listed following the colon. Lasers that have been tuned over a range of wavelengths are listed in order of the lowest wavelength reported; the tuning range given is that for the configuration and conditions used and may not represent the extremes possible.

The lasing wavelength and output power of color center lasers depend on the characteristics of the optical cavity, the temperature, the optical pump source, and other operating conditions. The original references should therefore be consulted for this information and its effect on the lasing wavelength.

Further Reading

Basiev, T. T. and Mirov, S. B., *Room Temperature Tunable Color Center Lasers*, Vol. 16 of Laser Science and Technology Series, Gordon & Breach, New York (1994), p. 1.

Basiev, T. T., Mirov, S. B. and Osiko, V. V., Room-temperature color center lasers, *IEEE J. Quantum Electron.* 24, 1052 (1988).

Gellermann, W., Color center lasers, *J. Phys. Chem. Solids* 52, 249 (1991).

German, K. R., Color Center Laser Technology, in *Handbook of Solid-State Lasers*, Cheo, P. K., Ed., Marcel Dekker, New York (1989), p. 457.

Mirov, S. B. and Basiev, T., Progress in color center lasers, in *Semiconductor Lasers, Selected Topics in Quantum Electronics* 1 (June 1995).

Mollenauer, L. F., Color Center Lasers, in *Handbook of Laser Science and Technology, Vol. I: Lasers and Masers*, CRC Press, Boca Raton, FL (1982), p. 171 and *Supplement 1: Lasers* (1991), p. 101.

Mollenauer, L. F., Color Center Lasers, in *Tunable Lasers*, 2nd edition, Mollenauer, L. F., White, J. C. and Pollock, C. R., Eds., Springer-Verlag, Berlin (1992).

Pollock, C. R., Optical properties of laser-active color centers, *J. Lumin.* 35, 65 (1986).

Pollock, C. R., Color Center Lasers, in *Encyclopedia of Lasers and Optical Technology*, Meyers, R. A., Ed., Academic Press, San Diego (1991), p. 9.

Table 2.4.1
Color Center Lasers Arranged in Order of Increasing Wavelength

Wavelength (µm)	Active center	Host crystal	Temperature (K)	Reference
0.357–0.420[(a)]	F^+	CaO	77	1
0.51–0.57	F_3^+	LiF	300	2
0.519–0.722	F_3^+, F_2, unknown	LiF	300	58, 60
0.52–0.56	F_3^+	LiF	300	50
0.53[(a)]	H_3	C(diamond)	300	3
0.54–0.62	unknown	Al_2O_3	300	4–7
0.543	F_3^+	LiF	300	56
0.66	$(F_2)_A$	$MgF_2:Na^+$	300	8–10
0.67	F_2	LiF	300	11
0.67–0.71	F_2	LiF	300	55
0.76	$(F_2)_A$	$CaF_2:Na^+$	300	8–10
0.77–0.93	unknown	Al_2O_3	300	4–7
0.82–1.05	F_2^+	LiF	77	12–14
0.830–1.060	F_2^+	LiF	300	57
0.84–1.10	F_2–F_2^+	LiF	300	15
0.84–1.13	F_2^+	$LiF:OH^-$	300	52
0.85–0.96	unknown	$KMgF_3:Pb^{2+}$	77	16,17
0.85–1.040	F_2^+	$LiF:OH^-,Mg^{2+}$	300	51
0.86–1.02	F_3^-	LiF	300	18
0.89	$(F_2)_A$	$SrF_2:Na^+$	300	8–10
0.94–1.06	unknown	$KMgF_3:Cu^+$	77	17
0.95–1.15	unknown	Al_2O_3	300	4–7
0.98–1.3	$(F_2^+)_A$	$NaF:Li^+$	300	19
0.99–1.22	F_2^+	NaF	77	20
1.03–1.12	$(F_2^+)^*$	$NaF:Mg^{2+}$	77	21, 22
1.09–1.24	F_2^-	LiF	300	15, 53
1.10–1.30	F_2^+	NaF	77	23
1.150–1.172	F_2^-	LiF	300	59
1.23–1.35	N_2	KCl	77	24
1.24–1.45	F_2^+	KF	77	25

Table 2.4.1—*continued*
Color Center Lasers Arranged in Order of Increasing Wavelength

Wavelength (μm)	Active center	Host crystal	Temperature (K)	Reference
1.37–1.77	F_2^+	NaCl:OH^-	300	26, 61
1.40–1.56	F_2^+	NaCl	77	27
1.40–1.60	F_A:$Tl^o(1)$	KCl:Tl^+	77	28, 29
1.41–1.81	F_2^+:O^{2-}	NaCl:OH^-	77	30, 31
1.42–1.76	F_2^+::O^{2-}	NaCl:K^+	77	32
1.43–2.00	F_2^+:S^{2-}	NaCl	77	33
1.450–1.600(b)	$(F_2^+)_H$	NaCl:OH^-	30	34
1.464–1.590	$Tl^o(1)$	NaCl:Tl^+	77	35
1.479–1.705	$(F_2^+)_H$	NaCl:OH^-	<185	35
1.488–1.538	$Tl^o(1)$	KCl:Tl^+	84	36
1.493–1.540	$Tl^o(1)$	NaCl:Tl^+	77	35
~1.575	$(F_2^+)_H$	NaCl:OH^-	77	54
1.61–1.77	F_2^+	KCl	77	27
1.62–1.91	$(F_2^+)_A$	KCl:Na^+	77	37
1.66–1.97	F_2^+:O^{2-}	KCl	77	38
~1.71–2.15	F_2^+:O^{2-}	KCl:Na^+	77	39
1.73–2.10	$(F_2^+)_{AH}$	KCl:Na^+	77	40
1.80–2.16	F_2^+	KBr:O_2^-	77	41
1.86–2.16	F_2^+	KBr:O_2^-	77	38
1.86–2.10	$(F_2^+)_H$	KBr:O_2^-	77	42
1.96—2.35	$(F_2^+)_{AH}$	KBr:Na^+:O_2^-	77	42
2.00–2.50	$(F_2^+)_A$	KCl:Li^+	77	43
2.22–2.75	$F_B(II)$	KCl:Na^+	77	44, 45
2.30–3.10	$F_A(II)$	KCl:Li^+	77	45
2.38–3.99	$(F_2^+)_A$	KI:Li^+	77	46
2.48–3.64	$F_A(II)$	RbCl:Li^+	77	45
4.86	CN^-	KBr	1.7	47
4.86	CN^-	KBr	1.7	48
4.88–5.00	$F_H(CN^-)$	CsCl	77	49

(a) Laser action requires further verification.

(b) Emission of the (a) variety of $(F_2^+)_H$ center.

References

1. Henderson, B., Tunable visible laser using F^+ centers in oxides, *Opt. Lett.* 6, 437 (1981).
2. Voytovich, A. P., Kalinov, V. S., Michonov, S. A. and Ovseichuk, S. I., Investigation of spectral and energy characteristics of green radiation generated in LiF with radiation color centers, *Kvant. Elektron.* 14, 1225 (1987); *Sov. J. Quantum Electron.* 17, 780 (1987).
3. Rand, S. C. and DeShazer, L. G., Visible color-center laser in diamond, *Opt. Lett.* 10, 481 (1985).
4. Martynovich, E. F., Baryshnikov, V. I. and Grigorov, V. A., Lasing in Al_2O_3 color centers at room temperature in thc visible, *Opt. Comm.* 53, 257 (1985); *Sov. Tech. Phys. Lett.* 11, 81 (1985).
5. Martynovich, E. F., Tokarev, A. G., and Grigorov, V. A., Al_2O_3 color center lasing in near infrared at 300 K. *Opt. Commun.* 53, 254 (1985); *Sov. Phys. Tech. Phys.* 30, 243 (1985).
6. Voytovich, A. P., Grinkevich, V. A., Kalinov, V. S., Kononov, V. A. and Michonov, S. A., Spectroscopic and oscillation characteristics of color centers in sapphire crystals in the range of 1.0 μm, *Kvant. Elektron.* 15, 1225 (1987); *Sov. J. Quantum Electron.* 18, 780 (1988).
7. Boiko, B. B., Shakadarevlch, A. P., Zdanov, E. A., Kalosha, I.I., Koptev, V. G. and Demidovich, A. A., Laser action of color centers in the Al_2O_3:Mg crystal, *Kvant. Electron.* 14, 914 (1987); *Sov. J. Quantum Electron.* 17, 581 (1987).
8. Arkhangel'skaya, V. A., Fedorov, A. A. and Feofilov, P. P., Spontaneous and stimulated emission of color centers in MeF_2-Na crystals, *Optica i Spectroskopiya* 44, 409 (1978); *Sov. Opt. Spectroscopy* 44, 240 (1978).
9. Arkhangel'skaya, V. A., Fedorov, A. A. and Feofilov, P. P., Luminescence and stimulated emission of M color centers in fluoride type crystals, *Izv. Akad. Nauk SSR, Ser Fiz.* 43, 1119 (1979); *Bull. Acad. Sci. USSR. Phys. Ser.* 43, 14 (1979).
10. Shkadarevich, A. P. and Yarmolkeich, A. P., New laser media in color centers of compound fluorides, *Inst. Phys. An BSSR, Minsk, USSR* (1985).
11. Gusev, Yu. L., Konoplin, S. N. and Marennikov, S. I., Generation of coherent radiation in F_2 color centers in an LiF single crystal, *Sov. J. Quantum Electron.* 7, 1157 (1977).
12. Mollenauer, L. F., Color center lasers, in *CRC Handbook of Laser Science and Technology,* Vol. 1, Weber, M. J., Ed., CRC Press, Boca Raton, FL (1982), p. 171.
13. Mollenauer, L. F., Color center lasers, in *Laser Handbook,* Stitch, M. and Bass, M., Eds., North-Holland, Amsterdam (1985), chapter 3.
14. Mollenauer, L. F., Color center lasers, in *Tunable Lasers*, Mollenauer, L. F. and White, J. C., Eds., Springer-Verlag, Berlin (1987), chapter 6.
15. Basiev, T. T., Karpushko, F. V., Kulaschik, S. M., Mirov, S. B., Morozov, V. P., Motkin, V. S., Saskevich, N. A. and Sinitsin, G. V., Automatic tunable MASLAN-201 laser, *Kvant. Elektron.* 14, 1726 (1987); *Sov. J. Quantum Electron.* 17, 1102 (1987).
16. Horsch, G. and Paus, H. J., A new color center laser on the basis of lead-doped $KMgF_3$, *Opt. Comm.* 60, 69 (1986).
17. Flassak, W., Goth, A., Horsch, G. and Paus, H. J., Tunable color center lasers with lead- and copper-doped $KMgF_3$, *IEEE J. Quantum Electron.* QE-24, 1070 (1988).
18. Shkadarevich, A. P., Demidovich, A. A., and Protassenya, A. L., Tunable lasers on F_3^--colour centers in LiF Crystals, *OSA Proc. Adv. Solid State Lasers*, Dubé, G. and Chase, L. (Eds) 10, 153 (1991).
19. Gusev, Y. L., Kirpichnikov, S. N., Konoplin, S. N. and Marennikov, S. I., Stimulated emission from $(F_2^+)_A$ color centers in NaF crystal, *Sov. J. Quantum Electron.* 8, 1376 (1981).

20. Mollenauer, L. F., Room-temperature-stable, F_2-like center yields cw laser tunable over the 0.9-1.22 μm range, *Opt. Lett.* 5, 188 (1980).
21. Doualan, J. L., Colour centre laser pumped by a laser diode, *Opt. Commun.* 70, 232 (1991).
22. Mazighi, K., Doualan, J. L., Hamel, J., Margerie, J., Mounier, D. and Ostrovsky, A., Active mode-locked operation of a diode pumped colour-centre laser, *Opt. Commun.* 85, 234 (1997).
23. Mollenauer, L. F., Laser-active, defect stabilized F_2^+ center in NaF:OH and dynamics of defect-stabilized center formation, *Opt. Lett.* 4, 390 (1980).
24. Georgiou, E., Carrig, T. J. and Pollock, C. R., Stable, pulsed, color-center laser in pure KCl tunable from 1.23 to 1.35 μm, *Opt. Lett.* 13, 978 (1988).
25. Mollenauer, L. F. and Bloom, D. M., Color center laser generates picosecond pulses and several watts cw over the 1.24–1.45 μm range, *Opt. Lett.* 4, 247 (1979).
26. Culpepper, C. F., Carrig, T. J., Pinto, J. F., Georgiou, E. and Pollock, C. R., Pulsed, room-temperature operation of a tunable NaCl color-center laser, *Opt. Lett.* 12, 882 (1987).
27. Gellermann, W., Lutz, F., Koch, K. P. and Litfin, G., F_2^+ center stabilization and tunable laser operation in OH^- doped alkali halide, *Phys. Stat. Sol.* (a) 57, 411 (1980).
28. Gellerman, W., Luty, F., and Pollock, C. R., Optical properties and stable broadly tunable cw laser operation of new F_A-type centers in Tl^+-doped alkali-halides, *Opt. Commun.* 39, 391 (1981).
29. Mollenauer, L. F., Vieira, N. D. and Szeto, L., Mode locking by synchronous pumping using a gain medium with microsecond decay times, *Opt. Lett.* 7, 414 (1982).
30. Pinto, J. F., Georgiou, E. and Pollock, C. R., Stable color-center in OH^-doped NaCl operating in the 1.41–1.81-μm region, *Opt. Lett.* 11, 519 (1986).
31. Georgiou, E., Pinto, J. F. and Pollock, C. R., Optical properties and formation of oxygen-perturbed F_2^+ color center in NaCl, *Phys. Rev.* B 35, 7636 (1987).
32. Pinto, J. F., Stratton, L. W. and Pollock, C. R, Stable color-center laser in K-doped NaCl tunable from 1.42 to 1.76 μm, *Opt. Lett.* 10, 384 (1985).
33. Möllmann, K. and Gellermann, W., Optical and laser properties of $(F_2^+)_H$ centers in sulfur-doped NaCl, *Opt. Lett.* 19, 804 (1994).
34. Konaté, A., Doualan, J. L., Girard, S. and Margerie, J., Tunable cw laser emission of the (a) variety of $(F_2^+)_H$ centres in NaCl:OH^-, *Opt. Commun.* 133, 234 (1997).
35. Konaté, A., Doualan, J. L., Girard, S., Margerie, J. and Vicquelin, R., Diode-pumped colour-centre lasers tunable in the 1.5 μm range, *Appl. Phys. B* 62, 437 (1996).
36. Konaté, A., Doualan, J. L. and Margerie, J., Laser diode pumping of a colour centre laser with emission in the 1.5 μm wavelength domain, *Rad. Effects Def. Solids* 136, 61 (1995).
37. Schneider, I. and Marrone, M. J., Continuous-wave laser action of $(F_2^+)_A$ centers in sodium-doped KCl crystals, *Opt. Lett.* 4, 390 (1979).
38. Wandt, D., Gellerman, W., Luty, F. and Welling, H., Tunable cw laser action in the 1.45–2.16 μm range based on F_2^+-like center in O_2^-- doped NaCl, KCl, and KBr crystals, *J. Appl. Phys.* 61, 864 (1987).
39. Wandt, D. and Gellerman, W., Efficient cw color center laser operation in the 1.7 to 2.2 μm range based on F_2^+-like centers in KCl:Na^+:O_2^- crystals, *Opt. Commun.* 61, 405 (1987).
40. Möllmann, K., Mitachke, F. and Gellermann, W., Optical properties and synchronously pumped mode locked 1.73–2.10 μm tunable laser operation of $(F_2^+)_{AH}$ centers in KCl:Na^+:O_2^-, *Opt. Commun.* 83, 177 (1991).
41. Doualan, J. L. and Gellerman, W., 4-W continuous-wave color-center laser pumps a KBr:$O_2^-(F_A^+)_H$ center laser, in *Advanced Solid-State Lasers*, Jenssen, H. P. and Dubé, G., Eds., Proceedings Vol. 6 (Optical Society of America, Washington, DC (1990), p. 276.

42. Möllmann, K. Schrempel, M., Yu, B-K. and Gellermann, W., Subpicosecond and continuous-wave laser operation of $(F_A{}^+)_H$ and $(F_A{}^+)_{AH}$ color-center lasers in the 2-µm range, *Opt. Lett.* 19, 960 (1994).
43. Schneider, I. and Marquardt, C. L., Tunable, cw laser action using $(F_2{}^+)$ centers in Li-doped KCl, *Opt. Lett.* 5, 214 (1980).
44. Litfin, G., Beigang, R. and Welling, H., Tunable cw laser operation in F_B(II) type color center crystals, *Appl. Phys. Lett.* 31, 381 (1977).
45. German, K., Optimization of F_A (II) and F_B (II) color-center lasers, *J. Opt. Soc. Am.* B3, 149 (1986).
46. Schneider, I., Continuous tuning of a color-center laser between 2 and 4 µm, *Opt. Lett.* 7, 271 (1982).
47. Tkach, R. W., Gosnell, T. R., and Sievers, A. J., Solid-state vibrational laser—KBr:CN^-, *Opt. Lett.*. 9, 122, 1984).
48. Gosnell, T. R., Sievers, A. J., and Pollock, C. R., Continuous-wave operation of the KBr:CN^- solid-state vibration laser in the 5-µm region, *Opt. Lett.* 10, 125 (1985).
49. Gellerman, W., Yang, Y. and Luty, F., Laser operation near 5-µm of vibrationally excited F-center CN molecule defect pairs in CsCl crystals, pumped in the visible, *Opt. Commun.* 57, 196 (1986).
50. Tsuboi, T. and Ter-Mikirtychau, V. V., Characteristics of the LiF:$F_3{}^+$ color center laser, *Opt. Commun.* 116, 389 (1995).
51. Ter-Mikirtychau, V. V. and Tsuboi, T., Ultrabroadband LiF:$F_2{}^+$ color center laser using two-prism spatially-dispersive resonator, *Opt. Commun.* 137, 74 (1997).
52. Khulugurov, V. M. and Lobanov, B. D., Color-center lasing at 0.84–1.13 µm in a LiF–OH crystal at 300 K, *Sov. Tech. Phys. Lett.* 4, 595 (1978).
53. Basiev, T. T., Zverov, P. G., Fedorov, V. V. and Mirov, S. B., Multiline, superbroadband and sun-color oscillation of a LiF:$F_2{}^-$ color-center laser, *Appl. Opt.* 36, 2515 (1997).
54. Kennedy, G. T., Grant, R. S. and Sibbett, W., Self-mode-locked NaCl:OH^- color-center laser, *Opt. Lett.* 18, 1736 (1993).
55. Basiev, T. T. and Mirov, S. B., Room-temperature color center lasers, *IEEE J. Quantum Electron.* 24, 1052 (1988).
56. Tsuboi, T. and Gu, H. E., Room-temperature-stable LiF:$F_3{}^+$ color-center laser with a two-mirror cavity, *Appl. Opt.* 33, 982 (1994).
57. Ter-Mikirtychau, V. V., Stable room-temperature LiF:$F_2{}^{+*}$ tunable color-center laser for the 830-1060-nm spectral range pumped by second-harmonic radiation from a neodymium laser, *Appl. Opt.* 34, 6114 (1995).
58. Gu, H.-E., Qi, L. and Wan, L.-F., Broadly tunable laser using some mixed centers in an LiF crystal for the 520-720 band, *Opt. Commun.* 67, 237 (1988).
59. Ter-Mikirtychau, V. V., Arestova, E. L. and Tsuboi, T., Tunable LiF:$F_2{}^-$ color center laser with an intracavity integrated-optic output coupler, *J. Lightwave Technol.* 14, 2353 (1996).
60. Gu, H.-E., Qi, L., Guo, S. and Wan, L.-F., A LiF crystal $F_3{}^+$–F_2 mixed color-center laser, *Chin. Phys.* 11, 148 (1991).
61. Matts, R. É, Stable laser based on color centers in the OH:NaCl crystal tunable in the range 1.4 to 1.7 µm and operating at room temperature, *Quantum Electron.* 23, 44 (1993).

Section 2.5

SEMICONDUCTOR LASERS

Introduction to the Tables

Semiconductor lasers are listed in order of increasing wavelength in Table 2.5.1. Lasers that have been tuned over a range of wavelengths are listed by the lowest lasing wavelength reported; the tuning range given is that for the configuration and conditions used and may not represent the extremes possible. The lasing material, structure and operating configuration, method of excitation, and temperature are also given. If the operation was both pulsed (p) and continuous wave (cw), then the first temperature given is for pulsed operation and the second for continuous wave operation. Only inorganic semiconducting materials are listed in Table 2.5.1. Dye-doped organic semiconductor lasers are included in Table 2.3.1; semiconducting polymer lasers are covered in Section 2.6.

The lasing wavelength and output of semiconductor lasers depend on the chemical composition of the material, structural configuration, optical cavity, temperature, excitation rate, and other operating conditions. The original references should therefore be consulted for this information and its effect on the lasing wavelength.

Abbreviations used in the table to describe the laser structure and operation:

BGSL—broken-gap superlattice
cw—continuous wave
DH—double heterostructure
GRIN—graded index
J—*p-n* junction
MQW—multiple quantum well
QB—quantum box
QD—quantum dot
RW—ridge waveguide
SB-BGSL — strain-balanced BGSL
SCBH—separate confinement buried heterostructure
SCH—separate confinement heterostructure
T2QWL — type II quantum well laser
VCSEL—vertical cavity surface emitting laser
BH—buried heterostructure
DFB—distributed feedback
DQW—double quantum well
H—heterostructure
ML—monolayer
p—pulsed
QC—quantum cascade
QW—quantum well
SL—superlattice
SL-MQW—strained-layer MQW
SLS—strained-layer superlattice
SQW—single quantum well
SSQW—strained SQW
VC—vertical cavity

Further Reading

Agrawal, G. P., Ed., *Semiconductor Lasers, Past, Present and Future*, AIP Press, Woodbury, NY (1995).

Agrawal, G. P. and Dutta, N. K., *Long-Wavelength Semiconductor Lasers*, Van Nostrand Reinhold, New York (1986).

Botez, D. and Scifres, D. R., *Diode Laser Arrays*, Cambridge University Press, Cambridge (1994).

Casey, H. C., Jr. and Parish, M. B., *Heterostucture Lasers, Part A: Fundamental Principles*, Academic Press, Orlando (1978).

Casey, H. C., Jr. and Parish, M. B., *Heterostucture Lasers, Part B: Materials and Operating Characteristics*, Academic Press, Orlando (1978).

Chang-Hasnain, C. J., Ed., Advances of VCSELs, *Optical Society of America Trends in Optics and Photonics Series*, Washington, DC (1997).

Chow, W. W., Koch, S. W. and Sargent, III, M., *Semiconductor Laser Physics*, Springer-Verlag, Berlin (1994).

Coleman, J. J., Ed., *Selected Papers on Semiconductor Diode Lasers*, SPIE Milestone Series, Vol. MS50, SPIE Optical Engineering Press, Bellingham, WA (1992).

Delfyett, P. J. and Lee, C. H., Semiconductor injection lasers, in *Encyclopedia of Lasers and Optical Technology*, Academic Press, New York (1991).

Derry, P., Figueroa, L. and Hong, C. S., Semiconductor Lasers, in *Handbook of Optics*, Vol. 1, 2nd edition, McGraw-Hill, New York (1995), chapter 13.

Kressel, H. and Butler, J. K., *Semiconductor Lasers and Heterojunction LEDs*, Academic Press, New York (1977).

Manasreh, M.O., Ed., *Antimonide Related Heterostructures and Their Applications*, Gordon and Breach, New York (1997).

Nakamura, S., III-V nitride based light-emitting devices, *Solid State Commun.* 102, 237 (1997).

Nakamura, S. and Fasol, G., *The Blue Laser Diode: GaN Based Light Emitters and Lasers*, Springer-Verlag, Heidelburg (1997).

Nurmikko, A. V. and Gunshor, R. L., Physics and device science in II-VI semiconductor visible light emitters, in *Solid State Physics* 49, 205 (1995).

Ohtsu, M., *Highly Coherent Semiconductor Lasers*, Artech House, Boston (1992).

Partin, D. L., Lead salt quantum effect structures, *IEEE J. Quantum Electron.* 24, 1716 (1988).

Sale, T. E., *Vertical Cavity Surface Emitting Lasers*, Wiley, New York (1995).

Sun, G. and Khurgin, J. B., Optically pumped four-level infrared laser based on intersubband transitions in multiple quantum wells: feasibility study, *IEEE J. Quantum Electron.* 29, 1104 (1993).

Thompson, G. H. B., *Physics of Semiconductor Laser Devices*, Wiley, New York (1980).

Zory, Jr., P. S., Ed., *Quantum Well Lasers*, Academic Press, San Diego, CA (1993).

See, also, Far Infrared Semiconductor Lasers, special edition of the *Journal of Optical and Quantum Electronic* 23 (1991); Special Issue on Semiconductor Lasers, *IEEE Journal of Quantum Electronics*, (June 1993); Semiconductor Lasers, *Selected Topics in Quantum Electronics* 1 (June 1995); and Semiconductor Lasers, *Selected Topics in Quantum Electronics* 3 (April 1997).

MRS Internet Journal of Nitride Semiconductor Research (www.mrs.org)

Table 2.5.1
Semiconductor Lasers Arranged in Order of Increasing Wavelength

Wavelength (μm)	Material	Structure	Excitation	Temp. (K)	Ref.
0.3245–0.3300	ZnS	crystal	electron beam (p)	4.2, 77	1
0.33	ZnS	crystal	electron beam (p)	80	2
0.333	CdZnS/ZnS	QW	optical (p)	8	274
0.3497	ZnS	crystal	optical (p)	300	3
~0.356	GaN	SCH QD	optical (p)	20	4
0.357–0.390	CdZnS/ZnS	SLS MQW	optical (p)	300	273
~359	GaN/AlGaN	DH	optical (p)	77	243
0.359	GaN	crystal	optical (p)	2	5
~0.3615	GaN/AlGaN	SCH	optical (p)	300	6
0.362–0.381	GaN	crystal	optical (p)	10–375	7
0.3635	GaN/AlGaN	VCSEL	optical (p)	300	8
~365	GaN/AlGaN	DH	optical (p)	295	243
0.3696	GaN	layer	optical (p)	300	10
0.3749	CdZnS/ZnS	SLS	optical (p)	300	271
0.375–0.400	ZnO	crystal	optical (p)	80–300	11
0.3755	CdZnS/ZnS	SLS MQW	injection (p)	30	273
0.3757	ZnO	crystal	electron beam (p)	77	12
0.376	GaInN	SQW SCH	injection (p)	300	199
0.376–0.378	GaN	layer VC	optical (p)	300	13,14
0.378	GaN	film	optical (p)	300	15

~0.38	ZnO	film	optical (p)	300	16
~0.385	GaN	ML	optical (p)	300	293
0.385	InGaN/GaN	MQW	optical (p)	≤ 220	17
0.387	GaN/AlGaN	H	optical (p)	34	294
0.389–0.399	AlGaN/GaInN	DFB DH	optical (p)	300	295
0.3914	CuCl	crystal QD	optical (p)	77–108	18
0.399–0.402	InGaN	MQW	injection (p, cw)	300	291
0.4025	AlGaN/GaInN	DH	optical (p)	300	19
0.40583	InGaN	MQW	injection (cw)	300	20
0.406	InGaN	ML VC	optical (p)	300	296
0.407–0.411	InGaN	MQW	injection (p)	300	297
0.410	InGaN/InGaN	MQW	optical (p)	300	292
0.415	InGaN/GaN	VC	optical (p)	300	21
0.417	InGaN	MQW	injection (p)	300	22
0.419	InGaN	MQW	injection (p)	300	44
0.427–0.437	InGaN/GaN	MQW	optical (p)	175–575	50
0.430	ZnSe	crystal	optical (p)	≤ 200	23
~0.438	ZnSSe	crystal	opt., elect. beam (p)	≤ 200	23
~0.445–0.455	ZnSe/ZnSSe	SL	optical (p)	14–180	24
0.445	ZnSe/ZnSSe	H	optical (p)	< 400	272
0.4496–0.624	CdSSe	crystal	optical (p)	77	25
~0.453	ZnSe/ZnMnSe	MQW	optical (p)	5.5–80	26
0.454–0.474	ZnSe/ZnSSe	SL	electron beam (p)	100	9
0.460	ZnSe	crystal	electron beam (p)	100	27
0.462	ZnSe/ZnSSe	H	optical (p)	≤ 260	28

Table 2.5.1—*continued*
Semiconductor Lasers Arranged in Order of Increasing Wavelength

Wavelength (μm)	Material	Structure	Excitation	Temp. (K)	Ref.
0.463	ZnSe/ZnMgSSe	SCH	injection (p)	300	267
0.4647–0.4663	CdSe/ZnSe	SL	optical (p)	80	29
0.468	(Zn,Cd)Se/ZnSe	MQW	optical (p)	300	30
0.469	ZnSe	crystal	optical (p)	300	31
0.469–0.475	ZnSe	crystal	optical (p)	300	32
0.474	ZnSe/ZnSSE	SL	electron beam (p)	100	9
0.4784	ZnCdSe/ZnSe	GRINSCH	electron beam (p)	83–225	33
0.480–0.500	(Zn,Cd)Se/ZnSe	MQW	injection (p)	≤ 250	34
0.485	CdZnSe	SQW	electron beam (p)	~20	290
~0.49	CdS	crystal	electron beam (p)	90, 300	35
0.490	CdZnSe	SQW	injection (p)	77	38
0.490–0.512	CdZnSe/ZnSe	QW	injection (cw)	80	39
0.49–0.56	ZnCdTe/ZnSe	SQW	optical (p)	300	36
0.49–0.69	CdSSe	crystal	electron beam (p)	4.2, 77	37
0.491	CdS	crystal	electron beam (p)	4.2, 77	40
0.492	(Zn,Cd)Se/ZnSe	SQW	optical (p)	10	41
~0.494	(Zn,Cd)Se/Zn(S,Se)	MQW	injection (p)	≤ 300	43
0.4943	CdS	crystal	optical (p)	88	43
0.495–0.520	CdS	crystal	optical (p)	90–300	11
0.496	(Zn,Cd)Se	SCH VCSEL	optical (p)	300	45

0.496	ZnCdSe/ZnSSe/ZnMgSSe	SCH	injection (cw)	85	269
0.496, 0.502	CdS	crystal	optical (p)	77	46
0.4963	(Zn,Cd)Se/ZnSe	SQW	optical (p)	150	41
0.4966	CdS	crystal	electron beam (p)	4	48
0.498–0.517	(Zn,Cd)Se/ZnSe	MQW	optical (p, cw)	300–10	49
0.50	(Zn,Cd)Se/ZnSe	SQW	optical (p)	250	41
0.504	ZnCdSe/ZnSSe	MQW	optical (p)	300	51
0.507	BeZnCdSe/BeZnSe/BeMgZnSe	SCH	injection (p)	77	268
0.508–0.535	CdZnSe/ZnSe	QW	injection (cw)	300	39
0.512	ZnCdSe/ZnCdMgSe	GRINSCH SQW	optical (p)	300	53
0.516	ZnCdSe/ZnSSe/ZnMgSSe	SCH	injection (p)	300–394	269
0.520	ZnCdSe/ZnSSe/ZnMgSSe	SCH	injection (p)	300	54
0.5235	ZnCdSe/ZnSSe/ZnMgSSe	SCH	injection (p)	300	270
0.528	ZnTe	crystal	electron beam (p)	4	55
0.533	ZnTe	crystal	electron beam (p)	110	56
0.549–0.562	InGaP	crystal	electron beam (p)	10–150	57
0.5520	InGaP:N	crystal	injection (cw)	77	58
0.575–0.602	CdZnTe/ZnTe	SLS QW	optical (cw)	8–310	59
0.5762–0.5845	GaInP/AlInP	MQW	injection (p)	109–165	60
0.580–0.705	CdSSe	crystal	optical (p)	90–300	11
0.5836	AlGaInP	DH	injection (cw)	77	62
0.585–0.620	ZnCdSe	crystal	electron beam (p)	10–310	64
0.59–0.60	GaSe	crystal	electron beam (p)	77	65
0.59–0.60	GaSe	crystal	optical (p)	77	66
0.6010	GaSe	crystal	optical (p)	2	67,68

Table 2.5.1—*continued*
Semiconductor Lasers Arranged in Order of Increasing Wavelength

Wavelength (μm)	Material	Structure	Excitation	Temp. (K)	Ref.
0.602–0.604	GaSe	crystal	optical (p)	5	69
0.607	GaInP/AlInP	SSQW DFB	injection (p)	140	70
0.615	AlGaAs/InGaP	H	injection (cw)	300	71
0.615	AlGaAs/InGaP	H	injection (cw)	300	71
0.621	AlGaInP	mesa stripe	injection (cw)	273	72
0.625	InGaAlP	MQW	injection (cw)	300	73
0.6262	AlGaInP/AlGaInP	DH	injection (p)	300	74
0.627–0.640	GaInP/GaAlInP	SQW GRINSCH	injection (p)	300	75
0.63–0.65	InGaAlP	H	injection (cw)	295–323	76
~0.636	AlGaAs/AlGaAs	H	optical (p)	300	77
0.6378	InGaAlP	SBR	injection (cw)	298	78
0.640	AlGaInP	H	injection (cw)	293	79
~0.640	In(Al,Ga)P	MQW H	optical/injection (cw)	300	80
0.670–0.690	AlGaInP/AlGaAs	VCSEL QW	injection (cw)	300	81
0.671	GaInP/AlGaInP	DH	injection (cw)	293	82
0.674–0.681	AlGaAs/GaAs	SL QW RW	injection (cw)	300	83
0.680-0.700	AlGaAs	DH	injection (cw)	300	84
0.680–0.785	GaAs/GaAlAs	SL SQW GRINSCH	injection (p, cw)	300	85,86
~0.685	CdSe	crystal	electron beam (p)	4.2	40
0.688–0.729	AlGaAs	H	injection (p,cw)	100	88

0.6897	AlGaInP/GaInP/AlGaInP	DH	injection (cw)	300	89
0.69–0.79	GaAs	SL QW	injection (cw)	300	83
0.6917	CdSe	crystal	optical (p)	77	90
0.695	InGaP	crystal	optical (p)	77	91
0.696–0.760	AlGaAs	layer	electron beam (p)	81	92
0.697	CdSe	crystal	optical (p)	77	93
0.698–0.752	CdSe	crystal	optical (p)	90–300	11
0.7010	GaAsP/InGaP	H	injection (cw)	283	95
0.704	GaAsP	crystal	electron beam (p)	77	96
0.707	InAlAs	QD H	injection (cw)	77	97
0.710	GaAsP	J	injection (p)	77	98
0.750	AlGaAs	H	injection (p)	77, 273	99
0.750–0.855	AlGaAs/GaAs	GRINSCH SQW	injection (cw)	300	100
0.761, 0.763	InGaP	J	injection (p)	4.2, 77	101
0.765	$CdIn_2S_4$	crystal	optical (p)	100–300	102
0.77	$CdSiAs_2$	crystal	electron beam (p)	77	103
0.770	AlGaAs	SL VCSEL	injection (cw)	300	104
0.7812	GaAs/AlGaAs	DH	injection (p)	77	105
~0.785	InGaPAs/GaAs	H	injection (cw)	300	108
0.785	CdTe	crystal	electron beam (p)	10–15	106,107
~0.800–0.845	AlGaAsP	DH	injection (p)	77–300	109
0.808	InGaAsP/GaAs	SQW	injection (cw)	300	110
0.825	GaAs	SQW	injection (cw)	300	83
0.83	GaAs	crystal	optical (p)	300	111
0.83	InGaAsP/InGaP	QW H	injection (cw)	300	112

Table 2.5.1—*continued*
Semiconductor Lasers Arranged in Order of Increasing Wavelength

Wavelength (μm)	Material	Structure	Excitation	Temp. (K)	Ref.
0.8365	GaAs	crystal	optical (p)	77	113
0.837–0.843	GaAs	H	injection (p)	4.2, 77	114
~0.84	GaAs	crystal	electron beam (p)	4	115–117
~0.84–0.86	GaAs/InAs	SCH SSQW	optical (p, cw)	77	118
0.8404	GaAs/AlGaAs	SCH	injection (cw)	300	94
0.842	GaAs	J	injection (p)	77	120
0.843	GaAs	H	injection (p)	77	121
0.843–0.8562	GaAlAs/GaAs	SH	injection (cw)	77	122
0.844–0.852	AlGaAs	H	injection (p)	77, 300	123
0.845	AlGaAs	H	injection (cw)	100	88
0.845	GaAs	crystal VCSEL	injection (cw)	300	125
0.850	InP	crystal	optical (p)	77	126
0.87	GaAs	DH	injection (cw)	300	83
~0.900–0.962	GaAs	VCSEL	injection (cw)	300	127
0.906–0.908	InP	crystal	injection (p)	4, 77	128,129
0.911	InGaAs	QD	injection (p)	80	130
0.915	InP	crystal	optical (p)	300	131
0.935	InGaAs	crystal	optical (p)	300	131
0.942, 0.945	InSe	crystal	optical (p)	5	69
~0.945–0.985	InSe	crystal	optical (p)	20, 90	133

0.9604	InAlGaAs	QD VCSEL	injection (cw)	300	134
0.97	InSe	crystal	electron beam (p)	90	135
~0.978–0.984	InGaAs/GaAs/InGaP	RW QW	injection (cw)	300	136
0.979	InGaAs	SQW	injection (cw)	300	137
0.980	GaAsSb/AlGaAsSb	DH	injection (p)	300	138
0.982–0.992	InAs	GRINSCH QB	injection (cw)	79	139
0.99	InGaAs/GaAs	SL QW	injection (cw)	300	140
1.00–1.05	InAs/GaAs	SCH QD	injection (cw)	100–300	142
1.008	AlGaInAs/AlInAs	DH	injection (p)	300	143
1.011	$CdSnP_2$	crystal	electron beam (p)	80	145
1.028	InGaAs/GaAs	Sl SCH QD	injection (cw)	300	146
1.074	InGa/GaAs	SL QW BH	injection (p)	300	144
1.10–1.60	AlGaSb	crystal	electron beam (p)	83	147
1.17–1.26	InGaNAs/GaInP	DH	optical (p)	300	282
1.17–1.26	InGaNAs/GaInP	DH	optical (p)	300	178
1.27	InGaAsP/InP	DH	injection (p)	300	148
1.3	AlGaInAs	MQW VCSEL	injection (cw)	300	149
1.31	InGaAsP/InP	BH	injection (p)	30	150
1.440–1.640	InGaAs/InP	MQW GRINSCH	injection (cw)	300	151
1.5	InGaAs/InGaAsP	SL-MQW	injection (cw)	283–373	47
1.5–1.6	InGaAs/InAlAs	MQW	injection,optical (p)	300	152
1.51–1.53	GaSb	crystal	electron beam (p)	4, 20	153
1.541	GaSb	crystal	optical (p)	4	154
1.55	InGaAsP	QW VC	injection (p)	300	155
1.55–1.60	GaSb	J	injection (p)	78	141

Table 2.5.1—*continued*
Semiconductor Lasers Arranged in Order of Increasing Wavelength

Wavelength (μm)	Material	Structure	Excitation	Temp. (K)	Ref.
1.57	GaSb	J	injection (p)	77	156
1.60	In_2Se	crystal	electron beam (p)	90	157
1.602	InPAs	crystal	optical (p)	77	158
1.62	InGaAs/InP	SQW GRINSCH	injection (cw)	300	159
1.77, 2.07	InGaAs	J	injection (cw)	1.9	160
1.89–1.97	HgCdTe	MQW SCH	optical (p)	300–10	277
1.9–2.5	HgCdTe	MQW GRINSCH	optical (p)	300	262
2	AlGaAsSb/InGaAsSb	SQW	injection (cw)	283–288	289
~2.0	GaInAsSb/GaSb	DH	injection (cw)	80	161
2.0–2.2	HgCdTe	SCH	optical (p)	150–300	276
2.023	InGaAsSb/AlGaSb	DH	injection (p)	140, 300	162
2.06–2.3	HgCdTe	MQW GRINSCH	optical (p)	300–10	277
2.1	GaInAsSb/AlGaAsSb	DH	injection (p,cw)	300, 190	163
~2.12	Cd_3P_2	crystal	optical (p)	4.2	164
2.18–2.5	HgCdTe	MQW SCH	optical (p)	300–10	277
~2.2	GaInAsSb/AlGaSb	DH	injection (p, cw)	303	165
2.2	InGaAsSb/AlGaSb	DH	injection (p)	300	162
2.36	HgCdTe/CdTe	MQW SCH	optical (p)	12, 77	167
2.5	HgCdTe	DH	optical (p)	110	263
2.6–6.6	PbEuSeTe	DH	injection (p)	≤ 190	193

2.7	InAs/GaInSb/InAs/AlGaInAsSb	SB-BGSL	injection (p)	180	279
2.7	InGaAsSb/AlGaAsSb	MQW	injection (cw)	170–234	283
2.7–3.2	InAs	film	electron beam (p)	80–220	168
2.7–3.9	InAsSb(P)/InAsSbP	DH	injection (cw)	80	169
2.7–6.6	PbEuSeTe	DH	injection (cw)	≤ 147	163,170
2.77–3.14	PbSrS/PbSrS	DH	injection (p)	180–90	179
2.79	HgCdTe	layer	optical (p)	~12	171
2.79–3.44	PbSrS/PbS	MQW	injection (p)	255–90	179
~2.8	HgCdTe	QW DH	optical (p,cw)	>60	173
2.8	GaInSb/InAs	MQW BGSL	injection (p)	225	172
2.86	HgCdTe	DH	injection (p)	77	174
2.88	PbSe/PbEuSe	DH	injection (cw)	100	175
2.9	HgCdTe	DH	injection (p)	40–90	176,177
2.94	InAs	crystal	optical (p)	20	154
2.95-3.84	PbSrS/PbS	DH	injection (p)	250–90	179
2.97	PbSrS/PbS	DH	injection (p,cw)	245, 174	179
~3.0	InAs	crystal	optical (p)	4	180
3–3.3	HgCdTe	MQW GRINSCH	optical (p)	300–10	277
3.06	InGaAsSb/InPSb	DH	injection (p)	35	181
3.06–3.29	InAsSbP/InGaAsSb	DH	optical, (p, cw)	77	181
3.08–3.30	GaInAsSb/GaSb	DH	optical (p)	82–210	284
3.1	GaInSb/InAs	MQW	injection (p)	220	172
3.1	HgCdTe	DH	optical (p)	120	266
3.1	InAs/GaInSb	MQW	injection (p)	190	172
3.1	InAs/InAsSbP	DH	optical (p)	77–100	184

Table 2.5.1—*continued*
Semiconductor Lasers Arranged in Order of Increasing Wavelength

Wavelength (μm)	Material	Structure	Excitation	Temp. (K)	Ref.
3.1	InAs/InAsSbP	DH	optical (p)	77–100	304
3.112	InAs	J	injection (p)	4, 77	185
~3.17	InAsSb	crystal	injection (p)	77	186
~3.2	InAsSbP/InAsSb/InAs	DH	injection (p, cw)	220, 77	288
3.2	GaInSb/InAs	MQW	injection (p)	255	172
3.2	HgCdTe/CdZnTe	QW	optical (p)	80–154	275
3.2	InAs/GaInSb/InAs/AlSb	T2QWL	optical (p)	350	285
3.28–3.90	GaInSb/InAs	MQW	injection (p)	170–84	187
3.3–3.4	InAs/InAsSb	type-II SL	optical (cw)	95	188
3.39–3.58	HgCdTe	DH	optical (p)	78	278
3.4	GaInAs/AlInAs	MQW RW	injection (p, cw)	10–280,10–50	63
3.4	GaInSb/InAs	MQW	injection (p)	195	172
3.4	HgCdTe	DH	injection (p)	40–90	176,177
3.4	HgCdTe	DH	injection (p)	78	264
3.4	HgCdTe	SCH	optical (p)	90	265
3.4	InAs/GaInSb/InAs/AlSb	T2QWL	optical (p)	310	285
3.4–6.5	PbEuTe	DH	injection (p)	> 200	189
3.5	PbCdS	J	injection (cw)	10–20	190
3.5–3.6	InAsSb/	MQW DH	injection (p)	77–135	287
3.57–3.86	InAsSb/InAsP	SLS	optical (p)	80–240	191

~3.6	InAsSb	layer	optical (p)	77	184
~3.6	InAsSb/InAs	J	optical (p)	77	304
3.6	HgCdTe	layer	injection (p)	12–90	192
3.6	InAsSb/InAsSbP	DH	injection (cw)	77–100	286
3.7	InAs/GaInSb/InAs/AlGaInAsSb	SB-BGSL	optical (p)	300	279
3.79–3.8	InAs/(In)GaSb/(In)AlSb	QC type-II MQW	injection (p)	170	194
3.8–3.9	InAsSb/InAs	MQW H	injection (p)	210	195
3.8–6.6	PbEuSeTe	DH	injection (cw)	≤ 147	193
3.86–3.97	InAsSb/GaSb	DH	optical (p)	80–150	284
~3.9	InAs(In)/GaSb/(In)AlSb	QC type-II MQW	injection (p)	40–170	124
3.9	AlGaSb/InAsSb/AlGaSb	DH	optical (p)	80-135	196
3.9	HgCdTe	DH	injection (p)	40–90	176,177
3.9	InAsSb/InAlAsSb	SQW	injection (p, cw)	165, 123	197
3.9	InAsSb/InAs	DH	optical (p)	77–125	184
3.9	InAsSb/InAs	DH	optical (p)	77–125	304
3.9–4.1	InAs/GaInSb/InAs/AlSb	MQW	optical (p)	80–285	198
3.9–8.6	PbSSe	crystal	optical (p)	2	18
3.97	PbEuSeTe/PbTe	SQW	injection (p)	260	170,200
3.97–3.985	InAsSb/AlAsSb	DH	injection (p, cw)	155, 80	202
4.1	GaInSb/InAs	MQW	injection (p)	135	172
4.1	InAs/GaInSb/InAs/AlSb	T2QWL	optical (p)	285	198
4.19–6.49	PbEuSeTe/PbTe	SCBH	injection (cw)	215–20	280,281
4.2–4.5	InAs/GaInSb/InAs/AlSb	QC type-II QW	optical (p)	100–310	203
4.2–6.4	PbEuSeTe/PbTe	BH	injection (cw)	90–203	204
4.26	AlInAs/GaInAs	QC MQW	injection (p)	10–90	205

Table 2.5.1—*continued*
Semiconductor Lasers Arranged in Order of Increasing Wavelength

Wavelength (μm)	Material	Structure	Excitation	Temp. (K)	Ref.
4.3	GaInSb/InAs	MQW	injection (p)	110	172
~4.3	PbS	crystal	electron beam (p)	4.2	206
4.32	PbS	J	injection (cw)	4.2	207,208
4.40–6.50	PbGeTe	J	injection (cw)	4	209
4.4–8.0	PbSe/PbSrSe	H	injection (p, cw)	290, 169	210
4.41–6.45	PbEuSeTe/PbTe	SQW	injection (cw)	174–13	170,200
4.5362–5.7026	PbEuSeTe/PbTe	DH BH	injection (cw)	120–180	211
4.6	GaInAs/AlInAs	QC MQW RW	injection (p, cw)	10–200, 50–85	212
4.77–7.18	PbSnTe/PbEuSeTe	BHG	injection (cw)	20–175	213
~5	GaInAs/AlInAs	MQW RW	injection (p, cw)	10–320, 10–140	87,217-8
~5	GaInAs/AlInAs	QC MQW microdisk	injection (p)	10–150	52
5.085–5.28	InSb	J	injection (p)	10	215
5.1	InAlSb/InSb	H	injection (p)	≤ 90	216
5.16–5.32	InSb	crystal	optical (cw)	20	201
5.2	InAs/GaInSb/InAs/AlGaInAsSb	SB-BGSL	optical (p)	185	279
5.258	InSb	crystal	optical (p)	4	219
5.3	HgMnTe	H	injection (p)	77	220
5.31–5.38	GaInAs/AlInAs	MQW DFB	injection (p)	110–315	61
5.37–5.44	AlInAs/GaInAs	QC MQW DFB	injection (p)	170–300	183
~5.4	HgZnTe	layer	optical (p)	50–70	222

5.55–7.81	PbEuSe/PbSe/PbEuSe	BH	injection (p, cw)	30–160	223
5.7–7.8	PbSe/PbEuSe	DH	injection (p, cw)	220, 174	175
5.90–8.55	PbSnTe/PbTeSe	MQW	injection (p, cw)	~10–204	228
6	PbSnTe	MQW	injection (p, cw)	204, 130	229
6.1	PbSSe	SH DH	injection (cw)	12	230
6.2–6.6	AlInAs/GaInAs	MQW RW	injection (p)	10–280, 10–80	231
6.41	PbTe	crystal	electron beam (p)	4.2	206
6.5	PbTe	crystal	injection (p)	12	232
7.4–8.6	GaInAs/AlInAs	QC MQW RW	injection (p, cw)	210, 110	234
7.7	GaInAs/AlInAs	QC SL RW	injection (p)	10–240	235
7.78–7.93	GaInAs/AlInAs	MQW DFB	injection (p)	80–310	61
8.2–18.5	PbSnTe/PbTe	H	injection (cw)	12–80	225,236
~8.5	GaInAs/AlInAs	MQW RW	injection (p, cw)	10–320, 10–110	87,238
8.5	GaInAs/AlInAs	MQW RW	injection (p)	80–270	132
8.5	PbSe	crystal	electron beam (p)	4	206
8.5	PbSe	J	injection (p)	4	240
~9	PbSnTe/PbTeSe	BH	injection (cw)	80	239
~9.3	GaInAs/AlInAs	QC MQW RW	injection (p)	10–220, 1035	237
9.4	PbSnTe	J	injection (p,cw)	12	241
~9.5	GaInAs/AlInAs	QC microdisk	injection (p)	<140	242
10.2	PbSnSe	J	injection (p, cw)	77	241
~11.1	AlInAs/GaInAs	MQW RW	injection (p, cw)	10–200, 10–30	87,238
11.5	GaInAs/AlInAs	QC MQW microdisk	injection (p)	10	242
~12.5	GaAs/AlGaAs	QW	optical (p)	77	244
12.7	PbSnTe	J	injection (p, cw)	12	241

Table 2.5.1—*continued*
Semiconductor Lasers Arranged in Order of Increasing Wavelength

Wavelength (μm)	Material	Structure	Excitation	Temp. (K)	Ref.
12.8	PbSnTe	DH	injection (p)	77–188	189
12.9	GaInAs/AlInAs	QC MQW RW	injection (p)	10–170	166
13.7	PbSnTe	J	injection (p, cw)	12	241
75–130	Ge(Ga),Ge(Al)	crystal (parallelpiped)	elect./mag. fields (p)	4.2–18	245-256
75–250	Ge(Be),Ge(Zn)	crystal (parallelpiped)	elect./mag. fields (p)	4.2	224,226
80–150	Ge(Cu)	crystal (parallelpiped)	elect./mag. fields (p)	4.2	224,226-27
85–110	Ge(Tl)	crystal (parallelpiped)	elect./mag. fields (p)	4.2	257
~100	Bi_xSb_{1-x}	crystal (plate)	injection (cw)	4.2	258
110–360	Ge(Ga)	crystal (parallelpiped)	elect./mag. fields (p)	4.2	259-261
120–165	Ge(Tl)	crystal (parallelpiped)	elect./mag. fields (p)	4.2	257
170–250	Ge(Ga),Ge(Al)	crystal (parallelpiped)	elect./mag. fields (p)	4.2–18	245-256

References

1. Hurwitz, C. E., Efficient ultraviolet laser emission in electron-beam-excited ZnS, *Appl. Phys. Lett.* 9, 116 (1966
2. Bogdankevich, O. V., Zverev, M. M., Pechenov, A. N., and Sysoev, L. A., Recombination radiation of ZnS single crystals excited by a beam of fast electrons, *Sov. Phys. Sol. Stat.* 8, 2039 (1967).
3. Wang, S. and Chang, C. C., Coherent fluorescence from zinc sulphide excited by two-photon absorption, *Appl. Phys. Lett.* 12, 193 (1968).
4. Tanaka, S., Hirayama, H., Aoyagi, Y., Narukawa, Y., Kawakami, Y., Fujita, S. and Fujita, S., Stimulated emission form optically pumped GaN quantum dots, *Appl. Phys. Lett.* 71, 1299 (1997).
5. Dingle, R., Shaklee, K. L., Leheny, R. F., and Zenerstrom, R. B., Stimulated emission and laser action in gallium nitride, *Appl. Phys. Lett.* 19, 5 (1971).
6. Schmidt, T. J., Yang, X. H., Shan, W., Song, J. J., Salvador, A., Kim, W., Altas, Ö., Botchkarev, A. and Morkoc, H., *Appl. Phys. Lett.* 68, 1820 (1996).
7. Yang, X. H., Schmidt, T. J., Shan, W. and Song, J. J., Above room temperature near ultraviolet lasing from an optically pumped GaN film grown on sapphire, *Appl. Phys. Lett.* 66, 1 (1995).
8. Redwing, J. M., Loeber, D. A. S., Anderson, N. G., Tischler, M. A. and Flynn, J. S., An optically pumped GaN-AlGaN vertical cavity surface emitting laser, *Appl. Phys. Lett.* 69, 1 (1996).
9. Cammack, D. A., Dalby, R. J., Cornelissen, H. J. and Khurgin, J., *J. Appl. Phys.* 62, 3071 (1987).
10. Kurai, S., Naoi, Y., Abe, T., Ohmi, S. and Sakai, S., Photopumped stimulated emission from homoepitaxial GaN grown on bulk GaN prepared by sublimation method, *Jpn. J. Appl. Phys.* 35, L77 (1996).
11. Johnston, Jr., W. D., Characteristics of optically pumped platelet lasers of ZnO, CdS, CdSe and $CdS_{0.6}Se_{0.4}$ between 300 and 80 K, *J. Appl. Phys.* 42, 2731 (1971).
12. Nicoll, F. H., Ultraviolet ZnO laser pumped by an electron beam, *Appl. Phys. Lett.* 9, 13 (1966).
13. Khan. M. A., Olson, D. T., Van Hove, J. M. and Kuznia, J. N., Vertical-cavity, room-temperature stimulated emission from photopumped GaN films deposited over sapphire substrates using low-pressure metalorganic vapor deposition, *Appl. Phys. Lett.* 58, 1515 (1991).
14. Khan, M. A., Kuznia, J. N., Van Hove, J. M., Olson, D. T., Krishnankutty, S. and Kolbas, R. M., Growth of high optical and electrical quality GaN layers using low pressure metalorganic chemical vapor deposition, *Appl. Phys. Lett.* 58, 526 (1991).
15. Amano, H., Asahi, T. and Akasaki, I., Stimulated emission near ultraviolet at room temperature from a GaN film grown on sapphire by MOVPE using an AlN buffer layer, *Jpn. J. Appl. Phys.* 29, L205 (1990).
16. Segawa, Y., Ohtomo, A., Kawasaki, M., Koinuma, H., Tang, Z. K., Yu, P. and Wong, G. K. L., Growth of ZnO thin film by laser MBE: lasing of exciton at room temperature, *Phys. Stat. Sol. (b)* 202, 669 (1997).
17. Khan, M. A., Sun, C. J., Yang, J. W., Chen, Q., Lim, B. W., Anwar, M. Z., Osinsky, A. and Temkin, H., cleaved cavity optically pumped InGaN-GaN laser grown on spinel substrates, *Appl. Phys. Lett.* 69, 2418 (1996).
18. Mooradian, A., Strauss, A. J., and Rossi, J. A., Broad band laser emission from optically pumped $PbS_{1-x}Se_x$, *IEEE J. Quantum Electron.* QE-9, 347 (1973).
19. Amano, H., Tanaka, T., Kunii, Y., Kim, S. T. and Akasaki, I., *Appl. Phys. Lett.* 64, 1377 (1994).
20. Nakamura, S., Senoh, M., Nagahame, S., Iwasa, N., Yamada, T., Matsushita, T., Sugimoto, Y. and Kiyoku. H., Room-temperature continuous-wave operation of InGaN multi-quantum-well structure diodes with a lifetime of 27 hours, *Appl. Phys. Lett.* 70, 1417 (1997).

21. Khan. M. A., Krishnankutty, S., Skogman, R. A., Kuznia, J. N., Olson, D. T. and George, T., Vertical-cavity stimulated emission from photopumped InGaN/GaN heterojunctions at room temperature, *Appl. Phys. Lett.* 65, 520 (1994).
22. Nakamura, S., Senoh, M., Nagahame, S., Iwasa, N., Yamada, T., Matsushita, T., Kiyoku. H. and Sugimoto, Y., InGaN-based multi-quantum-well-structure laser diodes, *Jpn. J. Appl. Phys.* 35, L74 (1996).
23. Yang, X. H., Hays, J. M., Shan, W., Song, J. J. and Cantwell, E., Two-photon pumped blue lasing in bulk ZnSe and ZnSSe, *Appl. Phys. Lett.* 62, 1071 (1993).
24. Suemune, I., Yamada, K., Masato, H., Kan, Y. and Yamanishi, M., *Appl. Phys. Lett.* 54, 981 (1989).
25. Brodin, M. S., VUrikhovskii, N. I., Zakrevskii, S. V., and Reznichenko, V. Ya., Generation in mixed CdS_x—$CdSe_{1-x}$ crystals excited with ruby laser radiation, *Sov. Phys. Sol. Stat.* 8, 2461 (1967).
26. Bylsma, R., Becker, W. M., Bonsett, T. C., Kolodziejski, L. A., Gunshor, R. L., Yamanishi, M. and Datta, S., *Appl. Phys. Lett.* 47, 1039 (1985).
27. Bogdankevich, O. V., Zverev, M. M., Krasiinikov, A. I., and Pechenov, A. N., Laser emission in electron-beam excited ZnSe, *Phys. Stat. Solid* 19, K56 (1967).
28. Nakanishi, K., Suemune, I., Masato, H., Kuroda, Y. and Yamanishi, M., Near-room-temperature photopumped blue lasers in ZnS_xSe_{1-x}/ZnSe multilayer structures, *Jpn. J. Appl. Phys.* 29, L2420 (1990).
29. Ledentsov, N. N., Krestnikov, I. L., Maximov, M. V. et al., Ground state exciton lasing in CdSe submonolayers inserted in a ZnSe matrix, *Appl. Phys. Lett.* 69, 1343 (1996).
30. Jeon, H., Ding, J., Nurmikko, A. V., Luo, H., Samarth, N., Furdyna, J. K., Bonner, W. A. and Nahory, R. E., Room-temperature blue lasing action in (Zn,Cd)Se/ZnSe optically pumped multiple quantum well structures on lattice-matched (Ga,In)As substrates, *Appl. Phys. Lett.* 57, 2413 (1990).
31. Zmudzinski, C. A., Guan, Y. amd Zory, P. S., *IEEE Phot. Technol. Lett. 2, 94 (1990).*
32. Yang, X. H., Hays, J., Shan, W., Song, J. J., Cantwell, E. and Aldridge, J., Optically pumped lasing of ZnSe at room temperature, *Appl. Phys. Lett.* 59, 1681 (1991).
33. Hervé, D., Accomo, R., Molva, E., Vanzetti, L., Paggel, J. J., Sorba, L. and Franciose, A., Microgun-pumped blue lasers, *Appl. Phys. Lett.* 67, 2144 (1995).
34. Jeon, H. Ding, J., Patterson, W., Nurmikko, A. V., Xie, W., Grillo, D. C., Kobayashi, M. and Gunshor, R. L., Blue-green injection laser diodes in (Zn,Cd)Se/ZnSe quantum wells, *Appl. Phys. Lett.* 59, 3619 (1991).
35. Kurbatov, L. N., Mashchenko, V. E., and Mochalkin, N. N., Coherent radiation from cadmium sulfide single crystals excited by an electron beam, *Opt. Spectrosc.* 22, 232 (1967).
36. Ishihara, T., Ikemoto, Y., Goto, T., Tsujimura, A., Ohkawa, K. and Mitsuyu, T., Optical gain in an inhomogeneously broadened exciton system, *J. Lumin.* 58, 241 (1994).
37. Hurwitz, C. E., Efficient visible lasers of CdS_xSe_{1-x} by electron-beam excitation, *Appl. Phys. Lett.* 8, 243 (1966).
38. Haase, M. A., Qiu, J., DePuydt, J. M. and Cheng, H., Blue-green laser diodes, *Appl. Phys. Lett.* 59, 1272 (1991).
39. Walker, C. T., DePuydt, J. M., Haase, M. A., Qiu, J. and Cheng, H., Blue-green II-VI laser diodes, *Physica B* 185, 27 (1993).
40. Hurwitz, C. E., Electron beam pumped lasers of CdSe and CdS, *Appl. Phys. Lett.* 8, 121 (1966).
41. Ding, J., Jeon, H., Nurmikko, A. V., Luo, H., Samarth, N. and Furdyna, J. K., Laser action in the blue-green from optically pumped (Zn,Cd)Se/ZnSe single quantum well structures, *Appl. Phys. Lett.* 57, 2756 (1990).

42. Jeon, H., Ding, J., Nurmikko, A. V., Xie, W., Grillo, D. C., Kobayashi, M. Gunshor, R. L., Hua, G. C. and Otsuka, N., Blue and green diode lasers in ZnSe-based quantum wells, *Appl. Phys. Lett.* 60, 2045 (1992).
43. Konyukhov, V. K., Kulevskii, L. A., and Prokhorov, A. M., Optical oscillation in CdS under the action of two-photon excitation by a ruby laser, *Sov. Phys. Dokl.* 10, 943 (1966).
44. Nakamura, S., Senoh, M., Nagahame, S., Iwasa, N., Yamada, T., Matsushita, T., Kiyoku. H. and Sugimoto, Y., Characteristics of InGaN multi-quantum-well-structure laser diodes, *Appl. Phys. Lett.* 68, 3269 (1996).
45. Jeon, H., Kozlov, V., Kelkar, P. et al., Room-temperature optically pumped blue-green vertical cavity surface emitting laser, *Appl. Phys. Lett.* 67, 1668 (1995).
46. Basov, N. G., Grasyuk, A. Z., Zubarev, I. G., and Katuiin, V. A., Laser action in CdS induced by two-photon optical excitation from a ruby laser, *Sov. Phys. Sol. Stat.* 7, 2932 (1966).
47. Thijs, P. J. A., Binsma, J. J. M., Tiemeijer, L. F., Slootweg, R. W. M., van Roijen, R. and van Dongen, T., Sub-mA threshold operation of λ=1.5 µm strained InGaAs multiple quantum well lasers grown on (311) B InP substrates, *Appl. Phys. Lett.* 60, 3217 (1992).
48. Basov, N. G., Bogdankevid, O. V., and Devyatkov, A. G., Exciting a semiconductor quantum generator (laser) with a fast electron beam, *Sov. Phys. Dokl.* 9, 288 (1964).
49. Jeon, H., Ding, J., Nurmikko, A. V., Luo, H., Samarth, N., and Furdyna, J. K., Low threshold pulsed and continuous-wave laser action in optically pumped (Zn,Cd)Se/ZnSe multiple quantum well lasers in the blue-green, *Appl. Phys. Lett.* 59, 1293 (1991).
50. Bidnyk, S., Schmidt, T. J., Cho, Y. H., Gainer, G. H. and Song, J. J., High temperature stimulated emission in optically pumped InGaN/GaN multi-quantum wells (to be published).
51. Kawakami, Y., Yamaguchi, S., Wu, Y., Ichino, K. Fujita, S. and Fujita, S., Optically pumped blue-green laser operation above room-temperature in $Zn_{0.80}Cd_{0.20}Se$-Zn $S_{0.08}Se_{0.92}$ multiple quantum well structures grown by metalorganic molecular beam epitaxy, *Jpn. J. Appl. Phys.* 30, L605 (1991).
52. Faist, J., Gmachl, C., Striccoli, M., Sirtori, C., Capasso, F., Sivco, D. L. and Cho, A. Y., Quantum cascade disk lasers, *Appl. Phys. Lett.* 69, 2456 (1996).
53. Guo, Y. Aizin, G., Chen, Y. C., Zeng, L., Cavus, A. and Tamargo, M., Photo-pumped ZnCdSe/ZnCdMgSe blue-green quantum well lasers grown on InP substrates, *Appl. Phys. Lett.* 70, 1351 (1997).
54. Yokogawa, T., Kamiyama, S., Yoshii, S., Ohkawa, K., Tsujimura, A. and Sasai, Y., Real-index guided blue-green laser diode with small beam astigmatism fabricated using ZnO buried structure, *Jpn. J. Appl. Phys.* 35, L314 (1996).
55. Hurwitz, C. E., Laser emission from electron beam excited ZnTe, *IEEE J. Quantum Electron.* QE-3, 333 (1967).
56. Vlasov, A. N., Kozina, G. S., and Fedorova, O. B., Stimulated emission from zinc telluride single crystals excited by fast electrons, *Sov. Phys. JETP* 25, 283 (1967).
57. Ermanov, O. N., Garba, L. S., Golvanov, Y. A., Sushov, V. P., and Chukichev, M. V., Yellow-green InGaP and InGaPAs LEDs and electron-beam-pumped lasers prepared by LPE and VPE, *IEEE Trans. Elect. Devices* ED-26, 1190 (1979).
58. Macksey, H. M., Holonyak, Jr., N., Dupuis, R. D., Campbell, J. C. and Zack, G. W., Crystal synthesis, electrical properties, and spontaneous and stimulated photoluminescence of $In_{1-x}Ga_xP$:N grown from solution, *J. Appl. Phys.* 44, 1333 (1973).
59. Glass, A. M., Tai, K., Bylsma, R B., Feldman, R. D., Olson, D. H. and Austin, R. F., *Appl. Phys. Lett.* 53, 834 (1988).
60. Kaneko, Y., Kikuchi, A., Nomura, I. and Kishino, K., Yellow light (576 nm) lasing emission of GaInP/AlInP multiple quantum well lasers prepared by gas-source-molecular-beam-epitaxy, *Electron. Lett.* 26, 657(1990).

61. Faist, J., Gmachl, C., Capasso, F., Sirtori, C., Sivco, D. L., Baillargeon, J. N. and Cho, A. Y., Distributed feedback quantum cascade lasers, *Appl. Phys. Lett.* 70, 2670 (1997).
62. Hino, I., Kawat, S., Gomo, A., Kobayashi, K. and Suzuki, S., Continuous wave operation (77 K) of yellow (583.6 nm) emitting AlGaInP double heterostructure laser diodes, *Appl. Phys. Lett.* 48, 557 (1986).
63. Faist, J., Capasso, F., Sivco, D. L., Hutchinson. A. L., Chu, S. N. G. and Cho, A. Y., Short wavelength (λ~3.4 μm) quantum cascade laser based on strain compensated InGaAs/AlInAs, *Appl. Phys. Lett.* (1998) (in press).
64. Khurgin, J., Fitzpatrick, B. J. and Seemungai, W., Cathodoluminiscence, gain, and stimulated emission in electron-beam-pumped ZnCdSe, *J. Appl. Phys.* 61, 1606 (1987).
65. Basov, N. G., Bogdankevich, O. V., and Abdullaev, A. N., Radiation in GaSe single crystals induced by excitation with fast electrons. *Sov. Phys. Dokl.* 10, 329 (1965).
66. Abdullaev, G. B., Aliev, M. Kh., and Mirzoev, B. R., Laser emission by GaSe under two-photon optical excitation conditions, *Sov. Phys. Semiconductor* 4, 1189 (1971).
67. Nahory, R. E., Shaklee, K. L., Leheny, R. F. and DeWinter, J. C., Stimulated emission and the type of bandgap in GaSe, *Solid State. Commun.* 9, 1107 (1971).
68. Catalano, I. M., Cingolani, A., Ferrara, M. and Minafra, A., Luminescence by exciton-exciton collision in GaSe, *Phys. Stat. Sol. (b)* 68, 341 (1975).
69. Abdullaev, G. B., Godhaev, I. O., Kakhramanov, N. B. and Suleimanov, R.A., Stimulated emission from indium selenide and gallium selenide layer semiconductors, *Sov. Phys. Solid State* 34, 39 (1992).
70. Jang, D.-H., Kaneko, Y. and Kishino, K., Shortest wavelength (607 nm) operations of GaInP/AlInP distributed Bragg reflector lasers, *Electron. Lett.* 28, 428 (1992).
71. Chang, L. B. and Shia, L.Z., Room-temperature continuous wave operation of a visible AlGaAs/InGaP transverse junction stripe laser grown by liquid phase epitaxy, *Appl. Phys. Lett.* 60, 1090 (1992).
72. Kawata, S., Kobayashi, K., Gomyo, A., Hino, I. and Suzuki, T., 621 nm cw operation (0°C) of AlGaInP visible semiconductor lasers, *Electron. Lett.* 22, 1265 (1986).
73. Rennie, J., Okajima, M., Watanabe, M. and Hatakoshi, G., Room temperature cw operation of orange light (625 nm) emitting InGaAlP laser, *Electron. Lett.* 28, 1950(1992).
74. Kobayashi, K., Hino, I. and Suzuki, T., 626.2-nm pulsed operation (300 K) of an AlGaInP double heterostructure laser grown by metalorganic chemical vapor deposition, *Appl. Phys. Lett.* 46, 7 (1985).
75. Ou, S. S., Yang, J. J., Fu, R. J. and Hwang, C. J., High-power 630-640 nm GaInP/GaAlInP laser diodes, *Appl. Phys. Lett.* 61, 842 (1992).
76. Hatakoshi, G., Itaya, K., Ishikawa, M., Okajima, M. and Uematsu, Y., Short-wavelength InGaAlP visible laser diodes, *IEEE J. Quantum Electron.* 27, 1476 (1991).
77. Rinker, M., Kalt, H. and Köhler, K., Indirect stimulated emission at room temperature, *Appl. Phys. Lett.* 57, 584 (1990).
78. Ishikawa, M., Shiozawa, H., Tsuburai, Y. and Uematsu, Y., Short-wavelength (638 nm) room-temperature cw operation of InGaAlP laser diodes with quaternary active layer, *Electron. Lett.* 26, 211 (1990).
79. Kawata, S., Fujii, H., Kobayashi, K., Gomyo, A., Hino, I. and Suzuki, T., Room-temperature continuous-wave operation of a 640 nm AlGaINP visible-light semiconductor laser, *Electron. Lett.* 23, 1328 (1987).
80. Dallesasse, J. M., Nam, D. W., Deppe, D. G. et al., Short-wavelength (<6400Å) room-temperature continuous operation of *p-n* $In_{0.5}(Al_xGa_{1-x})_{0.5}P$ quantum well lasers, *Appl. Phys. Lett.* 53, 1826 (1988).

81. Schneider, Jr., R. P., Hagerott, M., Choquette, K. D., Lear, K. L., Kilcoyne, S. P. and Figiel, J. J., Improved AlGaInP-based red (670-690 nm) surface-emitting lasers with novel C-doped short-cavity epitaxial design, *Appl. Phys. Lett.* 67, 329 (1995).
82. Ikeda, M., Mori, Y., Sato, H., Kaneko, K. and Watanabe, N., Room-temperature continuous-wave operation of an AlGaInP double heterostructure laser grown by atmosphere pressure metalorganic chemical vapor deposition, *Appl. Phys. Lett.* 47, 1027 (1985).
83. Hayakawa, T., Suyama, T., Takahashi, K., Kondo, M., Yamamoto, S. and Hijikata, T., Low-threshold room-temperature cw operation of $(AlGaAs)_m(GaAs)_n$ superlattice quantum well lasers emitting at ~680 nm, *Appl. Phys. Lett.* 51, 70 (1987).
84. Yamamoto, S., Hayashi, H., Hayakawa, T., Miyauchi, N., Yano, S. and Hijikata, T., Room-temperature cw operation in the visible spectral range of 680-700 nm by AlGaAs double heterojunction lasers, *Appl. Phys. Lett.* 41, 796 (1982).
85. Hayakawa, T., Suyama, T., Takahashl, K., Kondo, M., Yamamoto, T. and Hijlkata, T., Low current and threshold AlGaAs visible laser diode with a $(AlGaAs)_m(GaAs)_n$ superlattice quantum well, *Appl. Phys. Lett.* 49(11), 637 (1986).
86. Derry, P. L., Yariv, A., Lau, K. Y., Bar-Chaim, N., Lee, K. and Rosenberg, J., Ultra-low-threshold graded-index separate-confinement single quantum well buried heterostructure (AlGa)As laser with high reflectivity coating, *Appl. Phys. Lett.* 50, 1773 (1987).
87. Capasso, F., Faist, J., Sirtori, C. and Cho, A. Y., Infrared (4-11 μm) quantum cascade laser, *Solid State Commun.* 102, 231 (1997).
88. Kressel, H. and Hawrylo, F. Z., Stimulated emission at 300 K and simultaneous lasing at two wavelengths in epitaxial $Al_xGa_{1-x}As$ injection lasers, *Proc. IEEE* 56, 1598 (1968).
89. Kobayashi, K., Kawata, S., Gomyo, A., Hino, I. and Suzuki, T., Room-temperature cw operation of AlGaInP double-heterostructure visible lasers, *Electron. Lett.* 23, 931 (1985).
90. Holonyak, Jr., N., Sirkis, H. D., Stillman, G. E., and Johnson, M. R., Laser operation of CdSe pumped with a Ga(AsP) laser diode. *Proc. IEEE* 54, 1068 (1966).
91. Burnham, R. D., Holonyak, Jr., N., Keune, D. L., Scrifres, D. R. and Dapkus, P. D., Stimulated emission in $In_{1-x}Ga_xP$, *Appl. Phys. Lett.* 17, 430 (1970).
92. Dolginov, L. M., Druzhinina, L. V. and Kryukova, I. V., Parameters of electron-beam-pumped $Al_xGa_{1-x}As$ lasers in the visible part of the spectrum, *Sov. J. Quantum Electron.* 4, 104 (1975).
93. Grasyuk, A. Z., Eiimkov, V. F., and Zubarev, I. G., Semiconductor CdSe laser with two-photon optical excitation, *Sov. Phys. Sol. Stat.* 8, 1548 (1966).
94. van der Ziel, J. P., Dupuis, R. D., Logan, R. A., Mikulyak, R. M., Pinzone, C. J. and Savage, A., Low threshold pulse and continuous laser oscillation from AlGaAs/GaAs double heterostructure grown by metalorganic chemical vapor deposition, *Appl. Phys. Lett.* 50, 454 (1987).
95. Kressel, H., Olssen, G. H. and Nuese, C. J., Visible $GaAs_{0.7}P_{0.3}$ cw heterostructure laser, *Appl. Phys. Lett.* 30, 249 (1977).
96. Basov, N. G., Bogdankevich, O. V., Eliseev, P. G., and Lavrushin, B. M., Electron beam excited lasers made from solid solutions of GaP_xAs_{1-x}, *Sov. Phys. Sol. Stat.* 8, 1073 (1966).
97. Fafard, S., Hinzer, K., Raymond, S., Dion, M., McCaffrey, J., Feng, Y. and Charbonneau, S., Red-emitting semiconductor quantum dot lasers, *Science* 274, 1350 (1996).
98. Holonyak, Jr., N. and Bevacqua, S. F., Coherent (visible) light emission from $Ga(As_{1-x}P_x)$ junctions, *Appl. Phys. Lett.* 1, 82 (1962).
99. Rupprecht, H., Woodall, J. M. and Pettit, G. D., Stimulated emission from Ga,Al,As diodes at 70 K, *IEEE J. Quantum Electron.* QE-4, 35 (1968).
100. Mehuys, D., Mittelstein, M. and Yariv, A., Optimised Fabry-Perot (AlGa)As quantum-well lasers tunable over 105 nm, *Electron. Lett.* 25, 143 (1989).

101. Macksey, H. M., Holonyak, Jr., N., Scifres, D. R., Dupuis, R. D. and Zack, G. W., $In_{1-x}Ga_xP$ p-n junction lasers, *Appl. Phys. Lett.* 19, 271 (1971).
102. Beauvais, J. and Fortin, E., Optical gain in $CdIn_2S_4$, *J. Appl. Phys.* 62, 1349 (1987).
103. Averkleva, G. K., Goryunova, N. A., and Prochukhan, V. D., Stimulated recombination radiation emitted by $CdSiAs_2$, *Sov. Phys. Semiconductor* 5, 151 (1971).
104. Lee, Y. H., Tell, B., Brown-Goebeler, K. F., Leibenguth, R. E. and Mattera, V. D., Deep-red continuous wave top-surface-emitting vertical-cavity AlGaAs superlattice lasers, *IEEE Phot. Technol. Lett.* 3, 108 (1991).
105. Windhorn, T. H., Metze, G. M., Tsaur, B.-Y. and Fan, J. C. C., *Appl. Phys. Lett.* 45, 309 (1984).
106. Vavilov, V. S. and Nolle, E. L., Cadmium telluride laser with electron excitation, *Sov. Phys. Dokl.* 10, 827 (1966).
107. Vavilov, V. S., Nolle, E. L., and Egorov, V. D., New data on the electron-excited recombination radiation spectrum of cadmium telluride, *Sov. Phys. Sol. Stat.* 7, 749 (1965).
108. Mukai, S., Yajima, H., Mitsuhashi, Y., Shimada, J. and Kutsuwada, N., Continuous operating visible-light-emitting lasers using liquid-phase-epitaxial InGaPAs grown on GaAs substrates, *Appl. Phys. Lett.* 43, 24 (1983).
109. Burnham, R. D., Holonyak, N., Jr., Korb, H. W., Macksey, H. M., Scifres, D. R., and Woodhouse, J. B., Double heterostructure AlGaAsP quaternary laser, *Appl. Phys. Lett.* 19, 25 (1971).
110. Diaz, J., Eliashevich, I., He., X., Yi, H., Wang, L., Kolev, E., Garguzov, D. and Razeghi, M., High-power InGaAsP/GaAs 0.8-μm laser diodes and peculiarities of operational characteristics, *Appl. Phys. Lett.* 65, 1004(1994).
111. Nakamura, M., Yariv, A., Yen, H. W., Somekh, S. and Garvin, H. L., Optically pumped GaAs surface laser with corrugatin feedback, *Appl. Phys. Lett.* 22, (1973).
112. Wade, J. K., Mawst, L. J., Botez, D., Jansen, M., Fang, F. and Nabiev, R. F., High continuous wave power, 0.8 μm-band, Al-free active-region diode lasers, *Appl. Phys. Lett.* 70, 149 (1997).
113. Basov, N. G., Grasyuk, A. Z. and Katulin, V. A., Induced radiation in optically excited gallium arsenide, *Sov. Phys. Dokl.* 10, 343 (1965.
114. Quist, T. M., Rediker, R. H. and Keyes, R., Semiconductor maser of GaAs, *Appl. Phys. Lett.*1, 91 (1962).
115. Hurwitz, C. E. and Keyes, R. J., Electron beam pumped GaAs laser, *Appl. Phys. Lett.* 5, 139 (1964).
116. Kurbatov, L, N., Kabanov, A. N. and Sigriyanskii, B. B., Generation of coherent radiation in gallium arsenide by electron excitation, *Sov. Phys. Dokl.* 10, 1059 (1966).
117. Cusano, D. A., Radiative recombination from GaAs directly excited by electron beams, *Solid State Commun.* 2, 353 (1964).
118. Lee, J. H., Hsieh, K. Y. and Kolbas, R. M., Photoluminescence and stimulated emission from monolayer-thick pseudomorphic InAs single-quantum-well heterostructures, *Phys. Rev. B* 41, 7678 (1990).
119. Chen, H. Z., Ghaflari, A., Wang, H., Morkoc, H. and Yariv, A., *Optics Lett.* 12, 812 (1987).
120. Hall, R. N., Fenner, G. E., Kingley, J. D., Solbs, T. J. and Carlson, R. O., Coherent light emission from GaAs junctions, *Phys. Rev. Lett.* 9, 366 (1962).
121. Nathan, M. E., Dumke, W. P., Burns, C., Dill, Jr., F. H. and Lasher, G. J., Stimulated emission of radiation from GaAs p-n junction, *Appl. Phys. Lett.* 1, 62 (1962).
122. Scifres, D. R., Burnham, R. D. and Streifer, W., Distributed-feedback single heterojunction GaAs diode laser, *Appl. Phys. Lett.* 25, 203 (1974).

123. Susaki, W., Sogo, T. and Oku, T., Optical losses and efficiency in GaAs laser diodes, *IEEE J. Quantum Electron.* QE-4, 122 (1968).
124. Yang, R. Q., Yang, H. B., Lin, C.-H., Zhang, D., Murry, S. J., Wu, H. and Pei, S. S., High power mid-infrared interband cascade lasers based on type-II quantum wells, *Appl. Phys. Lett.* 71, 2409 (1997).
125. Tell, B., Lee, Y. H., Brown-Goebeler, K. F. et al., High-power cw vertical-cavity top surface-emitting GaAS quantum well lasers, *Appl. Phys. Lett.* 57, 1855 (1990).
126. Eliseev, P. G., Ismailo, I., and Mikhaillna, L. I., Coherent emission of InP optically excited by an injection laser, *JETP Lett.* 6, 15 (1967).
127. Yuen, W., Li, G. S. and Chang-Hasnain, C. J., Multiple-wavelength vertical-cavity surface-emitting laser arrays with a record wavelength span, *IEEE Phot. Technol. Lett.* 8, 4 (1996).
128. Weiser, K. and Levitt, R. S., Stimulated light emission from indium phosphide, *Appl. Phys. Lett.* 2, 178 (1963).
129. Basov, N. G., Eliseev, P. G. and Ismailov, I., Some properties of semiconductor lasers based on indium phosphide, *Sov. Phys. Solid State* 8, 2087 (1967).
130. Shoji, H., Ohtsuka, N., Sugawara, M., Uchida, T. and Ishikawa, H., Lasing at three-dimensionally quantum-confined sublevel of self-organized $In_{0.5}Ga_{0.5}As$ quantum dots by current injection, *IEEE Phot. Tech. Lett.* 7, 1385 (1995).
131. Rossi, J. A. and Chinn, S. R., Efficient optically pumped InP and $In_xGa_{1-x}As$ lasers, *J. Appl. Phys.* 43, 4806 (1972).
132. Slivken, S., Jelen, C., Rybaltowski, A., Diaz, J. and Razeghi, M., Gas-source molecular beam epitaxy growth of an 8.5 μm quantum cascade laser, *Appl. Phys. Lett.* 71, 2593 (1997).
133. Cingolani, A., Ferrara, M., Lugara, M. and Lévy, F., Stimulated photoluminescence in indium selenide, *Phys. Rev. B* 25, 1174 (1982).
134. Saito, H., Nishi, K., Ogura, I., Sugou, S. and Sugimoto, Y., Room-temperature lasing operation of a quantum-dot vertical-cavity surface-emitting laser, *Appl. Phys. Lett.* 69, 3140 (1996).
135. Kurbatov, L, N., Dirochka, A. I., and Britov, A. D., Stimulated emission of indium monoselenide subjected by electron bombardment, *Sov. Phys. Semiconductor* 5, 494 (1971).
136. Dutta, N. K., Hobson, W. S., Lopata, J. and Zydzik, G., Tunable InGaAs/GaAS/InGaP laser, *Appl. Phys. Lett.* 70, 1219 (1997).
137. Geels, R. S. and Coldren, L. A., Submilliamp threshold vertical-cavity laser diodes, *Appl. Phys. Lett.* 57, 1605 (1990).
138. Sugiyama K. and Saito, H., GaAsSb-AlGaAsSb double heterostructure laser, *Jpn. J. Appl. Phys.* 11, 1057 (1972).
139. Xie, Q., Kalburge, A., Chen, P. and Madhukar, A., Observation of lasing from vertically self-organized InAs three-dimensional island quantum boxes on GaAs (001), *IEEE Phot. Technol. Lett.* 8, 965 (1996).
140. Feketa, D., Chan, K. I., Ballantyne, J. M., and Eastman, L. F., *Appl. Phys. Lett.* 49, 1659 (1986).
141. Chipaux, C. and Eymard, R., Study of the laser effect in GaSb alloys, *Phys. Stat. Sol. (b)*10, 165 (1965).
142. Heinrichsdorff, F., Mao, M.-H., Kirstaedter, N., Krost, A., Bimberg, D., Kosogov, A. O. and Werner, P., Room-temperture continuous-wave lasing from stacked InAs/GaAs quantum dots grown by metalorganic chemical vapor deposition, *Appl. Phys. Lett.* 71, 22 (1997).
143. Chang-Hasnain, C. J., Bhat, R., Zah, C. E., Koza, M. A., Favire, F. and Lee, T. P., Novel AlGaInAs/AlInAs lasers emitting at 1 μm, *Appl. Phys. Lett.* 57, 2638 (1990).
144. York, P. K., Berenik, K., Fernandez, G. E. and Coleman, J. J., *Appl. Phys. Lett.* 54, 499 (1989).
145. Berkovskil, F. M., Goryunova, N. A., and Ordov, V. M., $CdSnP_2$ laser excited with an electron beam, *Sov. Phys. Semiconductor* 2, 1027 (1969).

146. Kamath, K., Bhattacharya, P., Sosnowski, T., Norris, T. and Phillips, J., Room-temperature operation of $In_{0.4}Ga_{0.6}As/GaAs$ self-organised quantum dot lasers, *Electron. Lett.* 32, 1374 (1996).
147. Akimov, Yu. A., Burov, A. A., and Zagarinskii, E. A., Electron-beam-pumped $AI_xGa_{1-x}Sb$ semiconductor laser, *Sov. J. Quantum Electron.* 5, 37 (1975).
148. Razeghi, M., Defour, M., Omnes, F., Maurel, Ph., Chazelas, J. and Brillouet, F., *Appl. Phys. Lett.* 53, 725 (1988).
149. Qiun, Y., Zhu, Z. H., Lo, Y. H. et al., Long wavelength (1.3 μm) vertical-cavity surface-emitting lasers with a wafer-bonded mirror and an oxygen-implanted confinement region, *Appl. Phys. Lett.* 71, 25 (1997).
150. Wakao, K., Nakai, K., Sanada, T. et al., InGaAsP/InP planar buried heterostructure laser with semi-insulating InP current blocking layer grown by MOCVD, *IEEE J. Quantum Electron.* 23, 943 (1987).
151. Lidgard, A., Tnabun-Ek, T., Logan, R. A., Temkin, H., Wicht, K. W. and Olsson, N. A., External-cavity InGaAs/InP graded index multiquantum well laser with a 200 nm tuning range, *Appl. Phys. Lett.* 56, 816 (1990).
152. Temkin, H., Alavi, K., Wagner, W. R., Pearsall, T. P. and Cho, A. Y., *Appl. Phys. Lett.* 42, 845 (1983).
153. Benoit-a-la Guillaume, C. and Debever, J. M., Laser effect in gallium antimonide by electron bombardment, *Compt. Rend.* 259, 2200 (1964).
154. Benoit-a-la Guillaume, C. and Laurant, J. M., Laser effect in InAs and GaSb by optical excitation, *Compt. Rend.* 262, 275 (1966).
155. Babic, D. I., Dudley, J. J., Strubel, K., Mirin, R. P., Bowers, J. E. and Hu, E. L., Double-fused 1.52-μm vertical-cavity lasers, *Appl. Phys. Lett.* 66, 1030 (1995).
156. Kryukova, I. V., Karnaukhov, V. G. and Paduchikh, L. I., Stimulated radiation from diffused p-n junction in gallium antimonide, *Sov. Phys. Sol. Stat.* 7, 2757 (1966).
157. Kurbatov, D. N., Dirochka, A. I., and Ogorodnik, A. D., Recombination radiation of In2Se, *Sov. Phys. Semiconductor* 4, 1195 (1971).
158. Alexander, F. B., Bird, V. R. and Carpenter, D. B., Spontaneous and stimulated infrared emission from indium phosphide-arsenide diodes, *Appl. Phys. Lett.* 4, 13 (1964).
159. Zah, C.E., Bhat, R., Cheung, K. W. et al., Low-threshold ($\leq$ 92 A/cm^2) 1.6 μm strained-layer single quantum well laser diodes optically pumped by a 0.8 μm laser diode, *Appl. Phys. Lett.* 57, 1608 (1990).
160. Melngailis, I., Strauss, A. J. and Rediker, R. H., Semiconductor diode masers of $In_xGa_{1-x}As$, *Proc. IEEE* 51, 1154 (1963).
161. Kano, H. and Sugiyama, K., 2.0 μm C.W. operation of GaInAsSb/GaSb D.H. lasers at 80 K, *Electron. Lett.* 16, 146 (1980).
162. Chiu, T. H., Tsang, W. T., Ditzenberger, J. A. and van der Ziel, J. P., *Appl. Phys. Lett.* 49, 1051 (1986).
163. Caneau, C., Srivastava, A. K., Zyskind, J. L., Sulhoff, J. W., Dentai, A. G. and Pollack, M. A., CW operation on GaInAsSb/AlGaAsSb laser up to 190 K, *Appl. Phys. Lett.* 49, 55 (1986).
164. Bishop, S. G., Moore, W. J., and Swiggard, E. M., Optically pumped Cd_3P_2 laser, *Appl. Phys. Lett.* 16, 459 (1970).
165. Choi, H. K. and Eglash, S. J., Room-temperature cw operation at 2.2 μm of GaInAsSb/AlGaAsSb diode lasers grown by molecular beam epitaxy, *Appl. Phys. Lett.* 59, 1165 (1991).
166. Capasso, F. and Gmachl, C. (private communication and to be published).
167. Mahavadi, K. K., Bleuse, J., Sivananthan, S. and Faurie, J. P., Stimulated emission from a CdTe/HgCdTe separate confinement heterostructure grown by molecular beam epitaxy, *J. Appl. Phys.* 56, 2077 (1990).
168. Kryukona, I. V., Leskovich, V. I. and Matveenko, E. V., Mechanisms of laser action in epitaxial InAs subjected to electron beam excitation, *Sov. J. Quantum Electron.* 9, 823 (1979).

169. Baranov, A. N., Imenkov, A. N., Sherstnev, V. V. and Yakovlev, Yu. P., 2.7-3.9 μm InAsSb(P)/InAsSbP low threshold diode lasers, *Appl. Phys. Lett.* 64, 2480 (1994).
170. Partin, D. L., Majkowski, R. F. and Swets, D. E., Quantum well diode lasers of lead-europium-selenide-telluride, *J. Vac. Sci. Technol. B* 3, 576 (1985).
171. Harman, T., Optically pumped LPE-grown $Hg_{1-x}Cd_xTe$ lasers, *J. Electron. Mater.* 8, 191 (1979).
172. Hasenberg, T. C., Miles, R. H., Kost, A. R. and West, L., Recent advances in Sb-based midwave-infrared lasers, *IEEE J. Quantum Electron.* 33, 1403 (1997).
173. Giles, N. C., Han, J. W., Cook, J. W. and Schetzina, J. F., Stimulated emission at 2.8 μm from Hg-based quantum well structures grown by photoassisted molecular beam epitaxy, *J. Appl. Phys.* 55, 2026 (1989).
174. Zandian, M., Arias, J. M., Zucca, R., Gil, R. V. and Shin, S. H., HgCdTe double heterostructure injection lasers grown by molecular beam epitaxy, *J. Appl. Phys.* 59, 1022 (1991).
175. Tacke, M., Spanger, B., Lambrecht, A., Norton, P. R., and Böttner, H., *Appl. Phys. Lett.* 53, 2260 (1988).
176. Zucca, R., Zandian, M., Arias, J. M. and Gil, R. V., Mid-IR HgCdTe double heterostructure lasers, *SPIE* Vol. 1634, 161 (1992).
177. Zucca, R., Zandian, M., Arias, J. M. and Gil, R. V., HgCdTe double heterostructure diode lasers grown by molecular-beam epitaxy, *J. Vac. Sci. Technol. B* 10, 1587 (1992).
178. Sato, S., Osawa, Y. and Saitoh, Y., Room-temperature operation of GaInNAs/GaInP DH laser diodes grown by MOCVD, *Jpn. J. Appl. Phys.* 36(5A), 2671 (1997).
179. Ishida, A., Muramatsu, K., Takashiba, H. and Fuyiyasu, H., $Pb_{1-x}Sr_xS$/PbS double-heterostructure lasers prepared by hot-wall expitaxy, *Appl. Phys. Lett.* 55, 430 (1989).
180. Melngailis, I., Optically pumped InAs laser, *IEEE J. Quantum Electron.* QE-1, 104 (1965).
181. Menna, R. J., Capewell, D. R., Martinelli, R. U., York, P. K. and Enstrom, R. E., 3 μm InGaAsSb/InPSb diode lasers grown by organometallic vapor-phase epitaxy, *SPIE* Vol. 1634, 174 (1992).
182. Aidaraliev, M., Zotova, N. V., Karndachev, S.A., Matseev, B. A., Stus', N. M. and Talalakin, G. N., Low-threshold lasers for the interval 3-3.5 μm based on InAsSbP/$In_{1-x}Ga_xAs_{1-y}Sb_y$ double heterostructures, *Sov. Tech. Phys. Lett.* 15, 600 (1989).
183. Gmachl, C., Faist, J., Baillargeon, J. N., Capasso, F., Sirtori, C., Sivco, D. L., Chu, S. N. G. and Cho, A. Y., Complex-coupled quantum cascade distributed-feedback laser, *IEEE Phot. Tech. Lett.* 9, 1090 (1997).
184. van der Ziel, J. P., Logan, R. A., Mikulyak, R. M. and Ballman, A. A., Laser oscillation at 3-4 μm from optically pumped $InAs_{1-x-y}Sb_xP_y$, *IEEE J. Quantum Electron.* QE-21, 1827 (1985).
185. Melngailis, I., Maser action in InAs diodes, *Appl. Phys. Lett.* 2, 176 (1963).
186. Basov, N. G., Dudenkova, A. V., and Krasilnikov, A. I., Semiconductor p-n junction lasers in the $InAs_{1-x}Sb_x$ system, *Sov. Phys. Sol. Stat.* 8, 847 (1966).
187. Chow, D. H., Miles, R. H., Hasenberg, T. C., Kost, A. R., Zhang, Y.-H., Dunlap, H. L. and West, L., Mid-wave infrared diode lasers based on GaInSb/InAs and In As/AlSb superlattices, *Appl. Phys. Lett.* 67, 3700 (1995).
188. Zhang, Y.-H., Continuous wave operation of InAs/$InAs_xSb_{1-x}$ midinfrared lasers, *Appl. Phys. Lett.* 66, 118 (1995).
189. Nishijima, Y., PbSnTe double-heterostructure lasers and PbEuTe double-heterostructure lasers by hot-wall epitaxy, *J. Appl. Phys.* 65, 935 (1989).
190. Nill, K. W., Strauss, A. J., and Blum, F. A., Tunable cw $Pb_{0.98}Cd_{0.02}S$ diode laser emitting at 3.5 μm, *Appl. Phys. Lett.* 22, 677 (1973).

191. Kurtz, S. R., Allerman, A. A. and Biefeld, R. M., Midinfrared lasers and light-emitting diodes with InAsSb/InAsP strained-layer superlattice active regions, *Appl. Phys. Lett.* 70, 3188 (1997).
192. Ravid, A. and Zussman, A., Laser action and photoluminescence in an indium-doped *n*-type $Hg_{1-x}Cd_xTe$ (x=0.375) layer grown by liquid phase epitaxy, *J. Appl. Phys.* 73, 3979 (1993).
193. Partin, D. L. and Thrush, C. M., Wavelength coverage of lead-europium-selenide-telluride diode lasers, *Appl. Phys. Lett.* 45, 193 (1984).
194. Lin, C.-H., Yang, R. Q., Zhang, D., Murry, S. J., Pei, S. S., Allerman, A. A. and Kurtz, S. R., Type-II interband quantum cascade laser at 3.8 μm, *Electron. Lett.* 33, 598 (1997).
195. Allerman, A. A., Biefeld, R. M. and Kurtz, S. R., InAsSb-based mid-infrared lasers (3.8-3.9 μm) and light-emitting diodes with AlAsSb claddings and semimetal electron injection, grown by metalorganic chemical vapor depostion, *Appl. Phys. Lett.* 69, 465 (1996).
196. van der Ziel, J. P., Chui, T. H. and Tsang, W. T., Optically pumped laser oscillation at 3.9 μm from $Al_{0.5}Ga_{0.5}Sb/InAs_{0.91}Sb_{0.09}/Al_{0.5}Ga_{0.5}Sb$ double heterostructure grown by molecular beam epitaxy on GaSb, *Appl. Phys. Lett.* 48, 315 (1986).
197. Choi, H. K. and Turner, G. W., InAsSb/InAlAsSb strained quantum-well diode lasers emitting at 3.9 μm, *Appl. Phys. Lett.* 67, 332 (1995).
198. Malin, J. I., Meyer, J. R., Felix, C. L. et al., Type II mid-infrared quantum well laser, *Appl. Phys. Lett.* 68, 2976 (1996).
199. Akasaki, I., Sota, S., Sakai, H., Tanaka, T., Koike, M., and Amano, H., Shortest wavelength semiconductor laser diode, *Electron. Lett.* 32(12), 1105 (1996).
200. Partin, D. L., Single quantum well lead-europium-selenide-telluride diode lasers, *Appl. Phys. Lett.* 45, 487 (1984).
201. Yoshida, T., Miyazaki, K. and Fujisawa, K., Emission properties of two-photon pumped InSb laser under magnetic field, *Jpn. J. Appl. Phys.* 14, 1987 (1975).
202. Eglash, S. J. and Choi, H. K., InAsSb/AlAsSb double-heterostructure diode lasers emitting at 4 μm, *Appl. Phys. Lett.* 64, 833 (1994).
203. Felix, C. L., Meyer, J. R., Vurgaftman, I., Lin, C.-H., Murry, S. J., Zhang, D. and Pei, S.-S., High-temperature 4.5 μm type-II quantum-well laser with Auger suppression, *IEEE Phot. Technol. Lett.* 9, 734 (1997).
204. Feit, Z., Kostyk, D., Woods, R. J. and Mak, P., Single-mode molecular beam epitaxy grown PbEuSeTe/PbTe buried-heterostructure diode lasers for CO_2 high-resolution spectroscopy, *Appl. Phys. Lett.* 58, 343 (1991).
205. Faist, J., Capasso, F., Sivco, D. L., Sirtori, C., Hutchinson. A. L. and Cho, A. Y., Quantum cascade laser, *Science* 264, 553 (1994).
206. Hurwitz, C. E., Calawa, A. R., and Rediker, R. H., Electron beam pumped laser of PbS, PbSe and PbTe, *IEEE J. Quantum Electron.* QE-1, 102 (1965).
207. Butler, J. F. and Calawa, A. R., PbS diode laser, *J. Electrochem. Soc.* 112, 1056 (1965).
208. Ralston, R. W., Waipole, J. M., and Calawa, A. R., High CW output power in stripe-geometry PbS diode lasers, *J. Appl. Phys.* 45, 1323 (1974).
209. Anticliffe, G. A., Parker, S. G., and Bate, R. T., CW operation and nitric oxide spectroscopy using diode laser of $Pb_{1-x}Ge_xTe$, *Appl. Phys. Lett.* 21, 505 (1972).
210. Spanger, B., Schiesse, U., Lambrecht, A., Böttner, H., and Tacke, M., *Appl. Phys. Lett.* 53, 2583 (1988).
211. Feit, Z., Kostyk, D., Woods, R. J. and Mak, P., PbEuSeTe buried heterostructure lasers grown by molecular-beam epitaxy, *J. Vac. Sci. Tech. B* 8, 200 (1990).
212. Faist, J., Capasso, F., Sirtori, C., Sivco, D. L., Hutchinson. A. L., and Cho, A. Y., Continuous wave operation of a vertical transition quantum cascade laser above T=80 K, *Appl. Phys. Lett.* 67, 3057 (1995).

213. Feit, Z., Kostyk, D., Woods, R. J. and Mak, P., Molecular beam epitaxy-grown PbSnTe-PbEuSeTe buried heterostructure diode lasers, *IEEE Phot. Technol. Lett.* 2, 860 (1990).
214. Faist, J., Capasso, F., Sirtori, C., Sivco, D. L., Hutchinson. A. L., and Cho, A. Y., Room temperature mid-infrared quantum cascade lasers, *Electron. Lett.* 32, 560 (1996).
215. Melngailis, I., Phelan, R. J. and Rediker, R. H., Luminescence and coherent emission in large-volume injection plasma in InSb, *Appl. Phys. Lett.* 5, 99 (1964).
216. Ashley, T., Elliott, C. T., Jefferies, R., Johnson, A. D., Pryce, G. J. and White, A. M., Mid-infrared $In_{1-x}Al_xSb$/InSb heterostructure diode lasers, *Appl. Phys. Lett.* 70, 931 (1997).
217. Faist, J., Capasso, F., Sirtori, C., Sivco, D. L., Baillargeon, J. N., Hutchinson. A. L., Chu, S. N. G. and Cho, A. Y., High power mid-infrared (λ~5 μm) quantum cascade lasers operating above room temperature, *Appl. Phys. Lett.* 68, 3680 (1996).
218. Faist, J., Capasso, F., Sirtori, C., Sivco, D. L., Baillargeon, J. N., Hutchinson. A. L., Chu, S. N. G. and Cho, A. Y., High power mid-infrared quantum cascade lasers with a molecular beam epitaxy grown InP cladding operating above room temperature, *J. Cryst. Growth* 175/176, 22 (1997).
219. Phelan, R. J. and Redlker, R. H., Optically pumped semiconductor lasers, *Appl. Phys. Lett.* 6, 70 (1965).
220. Becla, P., HgMnTe light emitting diodes and laser heterostructures, *J. Vac. Sci. Technol. A* 6, 2725 (1988).
221. Kurbatov, L. N., Britov, A. D., and Dirochka, A. I., Stimulated radiation from solid solutions of chalcogenide of lead and tin in the range of 10 μm, *Quantum Electron.* Basov, N. G., Ed., *Soviet Radio* (1972), p. 97.
222. Ravid, A., Zussman, A. and Sher, A., Optically pumped $Hg_{1-x}Zn_xTe$ lasers grown by liquid phase epitaxy, *J. Appl. Phys.* 58, 337 (1991).
223. Schlereth, K.-H., Spanger, B., Böttner, H., Lambrecht, A. and Tacke, M., Buried waveguide double-heterostructure PbEuSe-lasers grown by MBE, *Infrared Phys.* 30, 449 (1990).
224. Bründermann, E., Linhart, A.M., Reichertz, L., Röser, H.P., Dubon, O.D., Hansen, W.L., Sirmain, G. and Haller, E.E., Double acceptor doped Ge: a new medium for inter-valence-band lasers, *Appl. Phys. Lett.* 68, 3075 (1996).
225. Groves, S. H., Nill, K. W., and Strauss, A. J., Double heterostructure $Pb_{1-x}Sn_xTe$-PbTe lasers with cw operation at 77 K, *Appl. Phys. Lett.* 25, 331 (1974).
226. Reichertz, L. A., Dubon, O. D., Sirmain, G., Bründermann, E., Hansen, W. L., Chamberlin, D. R., Linhart, A. M., Röser, H. P. and Haller, E. E., Stimulated far-infrared emission from combined cyclotron resonances in germanium, *Phys. Rev.* B 56, 12069 (1997).
227. Sirmain, G., Reichertz, L. A., Dubon, O. D., Haller, E. E., Hansen, W. L., Bründermann, E., Linhart, A. M. and Röser, H. P., Stimulated far-infrared emission from copper-doped germanium crystals, *Appl. Phys. Lett.* 70, 1659 (1997).
228. Ishida, A., Fujiyasu, H., Ebe, H. and Shinohara, K., Lasing mechanism of type-I' PbSnTe-PbTeSe multiquantum well laser with doping structure, *J. Appl. Phys.* 59, 3023 (1986).
229. Shinohara, K., Nishijima, Y., Ebe, H., Ishida, A. and Fujiyasu, H., PbSnTe multiple quantum well lasers for pulsed operation at 6 μm up to 204 K, *Appl. Phys. Lett.* 47, 1184 (1985).
230. McLane, G. F. and Sleger, K. J., Vacuum deposited epitaxial layers of $PbS_{1-x}Se_x$ for laser devices. *J. Electron. Mat.* 4, 465 (1975).
231. Faist, J., Capasso, F., Sirtori, C., Sivco, D. L., Hutchinson. A. L., and Cho, A. Y., Laser action by tuning the oscillator strength, *Nature* 387, 777 (1997).
232. Butler, J. F., Calawa, R. A., Phelan, R. J., Jarman, T. C., Strauss, A. J. and Rediker, R. H., PbTe diode laser, *Appl. Phys. Lett.* 5, 75 (1964).

233. Sirtori, C., Faist, J., Capasso, F., Sivco, D. L., Hutchinson, A. L. and Cho, A. Y., Long wavelength infrared ($\lambda \approx 11$ μm) quantum cascade lasers, *Appl. Phys. Lett.* 69, 2810 (1996).
234. Sirtori, C., Faist, J., Capasso, F., Sivco, D. L., Hutchinson, A. L., Chu, S. N. G. and Cho, A. Y., Continuous wave operation of midinfrared (7.4–8.6 μm) quantum cascade lasers up to 110 K temperature, *Appl. Phys. Lett.* 68, 1745 (1996).
235. Scamarcio, G., Capasso, F., Sirtori, C., Faist, J., Hutchinson, A. L., Sivco, D. L. and Cho, A. Y., High-power infrared (8-micrometer wavelength) superlattice lasers, *Science* 276, 773 (1997).
236. Zussman, A., Felt, Z., Eger, D., and Shahar, A., Long wavelength $Pb_{1-x}Sn_xTe$ homostructure diode lasers having a gallium doped cladding layer, *Appl. Phys. Lett.* 42, 344 (1983).
237. Sirtori, C., Faist, J., Capasso, F., Sivco, D. L., Hutchinson, A. L. and Cho, A. Y., Pulsed and continuous-wave operation of long wavelength infrared (λ=9.3 μm) quantum cascade lasers, *IEEE J. Quantum Electron.* 33, 89 (1997).
238. Sirtori, C., Faist, J., Capasso, F., Sivco, D. L., Hutchinson, A. L. and Cho, A. Y., Mid-infrared (8.5 μm) semiconductor lasers operating at room temperature, *IEEE Phot. Tech. Lett.* 9, 294 (1997).
239. Kasemset, D., Rotter, S. and Fonstad, C. G., $Pb_{1-x}Sn_xTe/PbTe_{1-y}Se_y$ lattice-matched buried heterostructure lasers with cw single mode output, *IEEE Electron. Device Lett.* EDL-1, 75 (1980).
240. Butler, J. F., Calawa, A. R., Phelan, R. J., Strauss, A. J., and Rediker, R. H., PbSe diode laser, *Solid State Commun.* 2, 303 (1964).
241. Butler, J. F., Calawa, A. R., and Harman, T. C., Diode lasers of $Pb_{1-x}Sn_xSe$ and $Pb_{1-x}Sn_xTe$, *Appl. Phys. Lett.* 9, 427 (1966).
242. Gmachl, C., Faist, J., Capasso, F., Sirtori, C., Sivco, D. L. and Cho, A. Y., Long-wavelength (9.5-11.5 μm) microdisk quantum-cascade lasers, *IEEE J. Quantum Electron.* 33, 1567 (1997).
243. Aggarwal, R. L., Maki, P. A., Molnar, R. J., Liau, Z.-L. and Melngailis, I., Optically pumped $GaN/Al_{0.1}Ga_{0.9}N$ double-heterostructure ultraviolet laser, *J. Appl. Phys.* 79, 2148 (1996).
244. Gauthier-Lafaye, O., Sauvage, S., Boucaud, P. et al., Intersubband stimulated emission in GaAs/AlGaAs quantum wells: pump-probe experiments using a two-color free-electron laser, *Appl. Phys. Lett.* 70, 3097 (1997).
245. Andronov, A. A., Zverev, I. V., Kozlov, V. A., Nozdrin, Yu. N., Pavlov, S. A. and Shastin, V. N., Stimulated emission in the long-wave IR region from hot holes in Ge in crossed electric and magnetic fields, *Sov. Phys.-JETP Lett.* 40, 804 (1984).
246. Komiyama, S., Iizuka, N., and Akasaka, Y., Evidence for induced far-infrared emission from p-Ge in crossed electric and magnetic fields, *Appl. Phys. Lett.* 47, 958 (1985).
247. Andronov, A. A., Nozdrin, Yu. N. and Shastin, V. N.,Tunable FIR lasers in semiconductors using hot holes, *Infrared Phys.* 27, 31 (1987).
248. Kuroda, S. and Komiyama, S., Far-infrared laser oscillation in p-type Ge: remarkably improved operation under uniaxial stress, *Infrared Phys.* 29, 361 (1989).
249. Kremser, C., Heiss, W., Unterrainer, K., Gornik, E., Haller, E. E. and Hansen, W.L., Stimulated emission from p-Ge due to transitions between light-hole Landau levels and excited states of shallow impurities, *Appl. Phys. Lett.* 60, 1785 (1992).
250. Komiyama, S., Morita, H. and Hosako, I., Continuous wavelength tuning of intervalence-band laser oscillation in p-type germanium over range of 80-120 μm, *Jpn. J. Appl. Phys.* 32, 4987 (1993).
251. Heiss, W., Kremser, C., Unterrainer, K., Strasser, G., Gornik, E., Meny, C. and Leotin, J., Influence of impurities on broadband p-type-Ge laser spectra under uniaxial stress, *Phys. Rev.* B47, 16586 (1993).
252. Bründermann, E., Röser, H. P., Muravjov, A. V., Pavlov, S. G. and Shastin, V. N., Mode fine structure of the FIR p-Ge intervalence band laser measured by heterodyne

mixing spectroscopy with an optically pumped ring gas laser, *Infrared Phys. Technol.* 1, 59 (1995).
253. Bründermann, E., Röser, H. P., Heiss, W., Gornik, E. and Haller, E.E., Miniaturization of p-Ge lasers: progress toward continuous wave operation, *Appl. Phys. Lett.* 68, 1359 (1996).
254. Park, K., Peale, R. E., Weidner, H. and Kim, J. J., Submillimeter p-Ge laser using a Voigt-configured permanent magnet, *IEEE J. Quantum Electron.* 32, 1203 (1996).
255. Hovenier, J. N., Muravjov, A. V., Pavlov, S. G., Shastin, V.N., Strijbos, R. C. and Wenckebach, W. Th., Active mode locking of a p-Ge hot hole laser, *Appl. Phys. Lett.* 71, 443 (1997).
256. Bründermann, E. and Röser, H. P., First operation of a far-infrared p-Germanium laser in a standard closed-cycle machine at 15 Kelvin, *Infrared Phys. Technol.* 38, 201 (1997).
257. Heiss, W., Unterrainer, K., Gornik, E., Hansen, W. L. and Haller, E. E., Influence of impurity absorption on germanium hot-hole laser spectra, *Semicond. Sci. Technol.* 9, B638 (1994).
258. Aleksanyan, A. G., Kazaryan, R. K. and Khachatryan, A. M., Semiconductor laser made of Bi_{1-x} Sb_x, *Sov. J Quantum Electron.* 14, 336 (1984).
259. Ivanov, Yu. L. and Vasil'ev, Yu. B., Submillimeter emission from hot holes in germanium in a transverse magnetic field, *Pis'ma v Zhurnal Tekhnicheskoi Fizika* 9, 613 (1983), Translation: *Sov. Tech. Phys. Lett.* 9, 264 (1983).
260. Mityagin, Yu. A., Murzin, V. N., Stepanov, O. N. and Stoklitsky, S. A., Cyclotron resonance submillimeter laser emission in hot hole Landau level system in uniaxially stressed p-germanium, *Physica Scripta* 49, 699 (1994).
261. Pfeffer, P., Zawadzki, W., Unterrainer, K., Kremser, C., Wurzer, C., Gornik, E., Murdin, B. and Pidgeon, C.R., p-type Ge cyclotron-resonance laser: Theory and experiment, *Phys. Rev.* B 47, 4522 (1993).
262. Bonnet-Gemard, J., Bleuse, J., Magnea, N. and Pautrat, J. L., Optical gain and laser emission in HgCdTe heterostructures, *J. Appl. Phys.* 78, 6908 (1995).
263. Ravid, A., Cinader, G. and Zussman, A., Optically pumped laser action in double-heterostructure HgCdTe grown by metalorganic chemical deposition on a CdTe substrate, *J. Appl. Phys.* 74, 15 (1993).
264. Million, A., Colin, T., Ferret, P., Zanattan, J. P., Bouchot, P., Destéfanis, G. L. and Bablet, J., HgCdTe double heterostructure for infrared injection laser, *J. Cryst. Growth* 127, 291 (1993).
265. Bleuse, J., Magnea, N., Pautrat, J. L. and Mriette, H., Cavity structure effects on CdHgTe photopumped heterostructure lasers, *Semicond. Sci. Technol.* 8, SC266 (1993).
266. Ravid, A., Sher, A., Cinader, G. and Zussman, A., Optically pumped laser action and photoluminescence in HgCdTe layer grown on (211) CdTe by metalorganic chemical vapor deposition, *J. Appl. Phys.* 73, 7102 (1993).
267. Grillo, D. C., Han, J., Ringle, M., Hua, G., Gunshor, R. L., Kelkar, P., Kozlov, V., Jeon, H. and Nurmikko, A. V., Blue ZnSe quantum-well diode laser, *Electron. Lett.* 30, 2131 (1994).
268. Waag, A., Fischer, F., Schüll, K. et al., Laser diodes based on beryllium-chalcogenides, *Appl. Phys. Lett.* 70, 280 (1997).
269. Gaines, J. M., Crenten, R. R., Haberern, K. W., Marshall, T., Mensz, P. and Petruzzello, J., Blue-green injection lasers containing pseudomorphic $Zn_{1-x}Mg_x$-S_ySe_{1-y} cladding layers and operating up to 394 K, *Appl. Phys. Lett.* 62, 2462 (1993).
270. Nakayama, N., Itoh, S., Ohata, T., Nakano, K., Okuyama, H., Ozawa, M., Ishibashi, A., Ikeda, M. and Mori, Y., Room temperature continuous operation of blue-green laser diodes, *Electron. Lett.* 29, 1488 (1993).

271. Yamada, Y., Masumoto, Y., Mullins, J. T. and Taguchi, T., Ultraviolet stimulated emission and optical gain spectra in $Cd_xZn_{1-x}S$-ZnS strained-layer superlattices, *Appl. Phys. Lett.* 61, 2190 (1992).
272. Nakanichi, K., Suemune, I., Fujii, Y., Kuroda, Y. and Yamanishi, M., Extremely-low-threshold and high-temperature operation in a photopumped ZnSe/ZnSSe blue laser, *Appl. Phys. Lett.* 59, 1401 (1991).
273. Taguchi, T., Onodera, C., Yamada, Y. and Masumoto, Y., Band offsets in CdZnS/ZnS strained-layer quantum well and its application to UV laser diode, *Jpn. J. Appl. Phys.* 32, L1308 (1993).
274. Ozanyan, K. B., Nicholls, J. E., O'Neill, M., May, L., Hogg, J. H. C., Hagstom, W. E., Lunn, B. and Ashenford, D. E., Spectroscopic evidence for the excitonic lasing mechanism in ultraviolet ZnS/ZnCdS multiple quantum well lasers, *Appl. Phys. Lett.* 69, 4230 (1996).
275. Le, H. Q., Arias, J. M., Zandian, M., Zucca, R. and Liu, Y.-Z., High-power diode-laser-pumped midwave infrared HgCdTe/CdZnTe quantum-well lasers, *J. Appl. Phys.* 65, 810 (1994).
276. Bleuse, J., Magnea, N., Ulmer, L., Pautrat, J. L. and Mriette, H., Room-temperature laser emission near 2 μm from an optically pumped HgCdTe separate-confinement heterostructure, *J. Cryst. Growth* 117, 1046 (1992).
277. Bonnet-Gemard, J., Bleuse, J., Magnea, N. and Pautrat, J. L., Emission wavelength and cavity design dependence of laser behaviour in HgCdTe heterostructures, *J. Cryst. Growth* 159, 613 (1996).
278. Bouchut, P., Destefanis, G., Bablet, J., Million, A., Colin, T. and Ravetto, M., Mesa stripe transverse injection laser in HgCdTe, *J. Appl. Phys.* 61, 1561 (1992).
279. Flatte', M. E., Hasenberg, T. C., Olesberg, J. T., Anson, S. A., Boggess, T. F., Yan, C. and McDaniel, D. L. Jr., III-V interband 5.2 μm laser operating at 185 K, *Appl. Phys. Lett.* 71, 3764 (1997).
280. Feit, Z., Kostyk, D., Woods, R. J. and Mak, P., Molecular beam epitaxy-grown separate confinement buried heterostructure PbEuSeTe-PbTe diode lasers, *IEEE Phot. Technol. Lett.* 7, 1403 (1995).
281. Feit, Z., McDonald, M., Woods, R. J., Archambault, V. and Mak, P., Low-threshold PbEuSeTe/PbTe separate confinement buried heterostructure diode lasers, *Appl. Phys. Lett.* 68, 738 (1996).
282. Sato, S., Osawa, Y. and Saitoh, Y., Room-temperature operation of GaInNAs/GaInP DH laser diodes grown by MOCVD, *Jpn. J. Appl. Phys.* 36(5A), 2671 (1997).
283. Garbuzov, D. Z., Martinelli, R. U., Menna, R. J., York, P. K., Lee, H., Narayan, S. Y. and Connolly, J. C., 2.7-μm InGaAsSb/AlGaAsSb laser diodes with continuous-wave operation up to -39°C, *Appl. Phys. Lett.* 67, 1346 (1995).
284. Le, H. Q., Turner, G. W., Eglash, S. J., Choi, H. K. and Coppeta, D. A., High-power diode-laser-pumped InAsSb/GaSb and GaInAsSb/GaSb lasers emitting from 3 to 4 μm, *Appl. Phys. Lett.* 64, 152 (1994).
285. Malin, J. I., Felix, C. L., Meyer, J. R., Hoffman, C. A., Pinto, J. F., Lin, C.-H., Chang, P. C., Murry, S. J., and Pei, S. S., Type II mid-IR lasers operating above room temperature, *Electron. Lett.* 32, 1593 (1996).
286. Popov, A., Sherstnev, V., Yakovlev, Y., Mücke, R. and Werle, P., High power InAsSb/InAsSbP double heterostructure laser for continuous wave operation at 3.6 μm, *Appl. Phys. Lett.* 68, 2790 (1996).
287. Kurtz, S. R., Biefeld, R. M., Allerman, A. A., Howard, A. J., Crawford, M. H. and Pelczynski, M. W., Pseudomorphic InAsSb multiple quantum well injection laser emitting at 3.5 μm, *Appl. Phys. Lett.* 68, 1332 (1996).
288. Diaz, J., Yi, H., Rybaltowski, A., Lane, B., Lukas, G., Wu, D., Kim, S., Erdtmann, M., Kaas, E. and Razeghi, M., InAsSbP/InAsSb/InAs laser diodes (λ=3.2 μm) grown by low-pressure metal-organic chemical-vapor deposition, *Appl. Phys. Lett.* 70, 40 (1997).

289. Garbuzov, D. Z., Martinelli, R. U., Lee, H., Menna, R. J., York, P. K., DiMarco, L. A., Harvey, M. G., Matarese, R. J., Narayan, S. Y. and Connolly, J. C., 4 W quasi-continuous-wave output power from 2 μm AlGaAsSb/InGaAsSb single-quantum-well broadened waveguide laser diodes, *Appl. Phys. Lett.* 70, 2931 (1997).
290. Trager-Cowan, C., Bagnall, D. M., McGow, F. et al., Electron beam pumping of CdZnSe quantum well laser structures using a variable energy electron beam, *J. Cryst. Growth* 159, 618 (1996).
291. Nakamura, S., Senoh, M., Nagahama, S., Iwasa, N., Yamada, T., Matsusita, T., Sugimoto, Y. and Kiyoku, H., Subband emissions from InGaN multi-quantum-well laser diodes under room-temperature continuous wave operation, *Appl. Phys. Lett.* 70(20), 2753 (1997).
292. Hofstetter, D., Bour, D. P., Thorton, R. L., and Johnson, N. M., Excitation of a higher order transverse mode in an optically pumped $In_{0.15}Ga_{0.85}N/In_{0.05}Ga_{0.95}N$ multiquantum well laser structure, *Appl. Phys. Lett.* 70(13), 1650 (1997).
293. Gluschenkov, O., Myoung, J. M., Shim, K. H., Kim, K., Gigen, Z. G., Gao, J. and Eden, J. G., Stimulated emission at 300 K from photopumped GaN grown by plasma-assisted molecular beam epitaxy with an inductively coupled plasma source, *Appl. Phys. Lett.* 70, 811 (1997).
294. Nakadaira, A. and Tanaka, H., Stimulated emission at 34 K from an optically pumped cubic GaN/AlGaN heterostructure grown by metalorganic vapor-phase epitaxy, *Appl. Phys. Lett.* 71, 811 (1997).
295. Hofmann, R., Gauggel, H.-P., Griesinger, U. A., Gräbeldinger, H., Adler, F., Ernst, P., Bola, H., Härle, V., Scholz, F., Schweizer, H., and Pilkuhn, M. H., Realization of optically pumped second-order GaInN-distributed-feedback lasers, *Appl. Phys. Lett.* 69, 2068 (1996).
296. Kim, S. T., Amano, H. and Akasaki, I., Surface-mode stimulated emission from optically pumped GaInN at room temperature, *Appl. Phys. Lett.* 67, 267 (1996).
297. Nakamura, S., Senoh, M., Nagahame, S., Iwasa, N., Yamada, T., Matsushita, T., Sugimoto, Y. and Kiyoku. H., Ridge-geometry InGaN multi-quantum-well-structure laser diodes, *Appl. Phys. Lett.* 69, 1477 (1996).

Section 2.6

POLYMER LASERS

Introduction to the Table

Lasers based on neat and dilute blends of conjugated polymers are listed in order of increasing wavelength in Table 2.6.1 (dye-doped polymer lasers are included in Table 2.3.1). Lasers that have been tuned over a range of wavelengths are listed by the lowest wavelength reported; the tuning range given is that for the experimental configuration and conditions used and may not represent the extremes possible. The lasing material, mode of photon confinement, and optical pumping wavelength are also listed together with the primary reference. The references should be consulted for details of the chemical composition and molecular structure of the lasing compounds. All experiments used pulsed excitation and were performed at room temperature. The reported observations may be indicative of lasing or amplified spontaneous emission.

Further Reading

Dodabalapur, A., Chanddross, E. A., Berggren, M. and Slusher, R. L. Organic solid-state lasers: past and future, *Science* 277, 1787 (1997).

Friend, R. H., Denton, G. J., Halls, J. J. M. et al., Electronic excitations in luminescent conjugated polymers, *Solid State Commun.* 102, 249 (1997).

Hide, F., Diaz-Garcia, M. A., Schwartz, B. J., Andersson, M. R., Pei, Q. and Heeger, A. J., Semiconducting polymers: a new class of solid-state laser materials, *Science* 273, 1833 (1996).

Jenekje, S. A. and Wynne, K. J., Eds., *Photonic and Optoelectronic Polymers*, American Chemical Society, Washington, DC (1997).

Abbreviations for the materials in Table 2.6.1:

BCHA	poly 2,5-bis(cholestanoxy)
BDOO-PF	poly(9,9-bis(3,6-dioxaoctyl)-fluorene-2,7-diyl)
BEH	poly 2,5-bis(2'-ethylhexyloxy)
BuEH	poly 2-butyl-5-2'-ethylhexyl
CB	chlorobenzene
CN-PPP	poly(2-(6'-mehylheptyloxy)-1,4-phenylene)
DCM/PS	4-(dicyanomethylene)-2-methyl-6-(4-dimethylaminostyry)-4*H*-pyran
DOO-PPV	2,5-dioctyloxy *p*-phenylene vinylene
HEH-PF	poly(9-hexyl-9-2'-ethylhexyl)-fluorene-2,7-diyl)
m-LPPP	methyl-substituted conjugated laddertype poly(paraphenylene)
M3O	poly 2-methoxy-5-3'-octyloxy
MEH	poly 2-methoxy-5-2'-ethylhexloxy

Abbreviations for the materials in Table 2.6.1—*continued:*

NAPOXA	2-napthyl-4,5-bis(4-methoxyphenyl)-1,3-oxazole
PBD	2-(4-biphenylyl)-5-(4-*t*-butylphenyl)-1,3,4-oxadiazole
PMMA	polymethylmethacrylate
PPPV	phenyl-substituted poly(p-phenylene vinylene)
PPV	poly(1,4-phenylene vinylene)
PS	polystyrene
Si-PPV	poly(dimethylsilylene-*p*-phene-vinylene -(2,5-di-*n*-octyl-*p*-phenylene)-PPV
THF	tetrahydrofuran

Table 2.6.1
Polymer Lasers Arranged in Order of Increasing Wavelength

Wavelength (nm)	Material	Photon confinement	Pump (nm)	Ref.
392	PBD	waveguide	337	1
420*	CN–PPP	waveguide	355	2
425, 445*	HEH–PF	waveguide	355	2
430, 450, 540	BDOO–PF	waveguide	355	2
452	Si–PPV	waveguide	355	10
480–545	PPPV–PMMA	resonator	450	11
~483–492	m-LPPP	waveguide	444	3
520–620	BuEH–PPV	waveguide	310	4
530, 620*	M3O–PPV	waveguide	532	2
540–583	BuEH–PPV	waveguide (DFB)	435	5
540, 570 (Sh)*	BuEH–MEH(97.5:2.5)	waveguide	435	2
540, 630*	BCHA–PPV	waveguide	532	2
~545	PPV	microcavity	335	6
~545	PPV	microcavity	355	7
545, 580(Sh)*	BuEH–MEH(95:5)	waveguide	435	2
~550	BuEH-PPV	waveguide/microcavity	435	8
~550	BuEH–PPV	waveguide	435	2
550, 580(S2)*	BuEH–MEH(90:10)	waveguide	435	2
565, 600*	BuEH–MEH(70:30)	waveguide	532	2
580, 625*	BEH–PPV	waveguide	532	2
580, 625*	BuEH–MEH(10:90)	waveguide	532	2
585, 625*	MEH–PPV	waveguide	532	2
625	DOO-PPV	waveguide	532	9
640*	DCM/PS	waveguide	532	2

* Peak(s) of photoluminescence; lasing wavelength was not reported.
DFB – distributed feedback
Sh – shoulder

References

1. Berggren, M., Dodabalapur, A., Slusher, R. L. and Bao, Z., Light amplication in organic thin films using cascade energy transfer, *Nature* 389, 466 (1997).
2. Hide, F., Diaz-Garcia, M. A., Schwartz, B. J., Andersson, M. R., Pei, Q. and Heeger, A. J., Semiconducting polymers: a new class of solid-state laser materials, *Science* 273, 1833 (1996).
3. Zenz, C., Graupner, W., Tasch, S., Leising, G., Müllen, K. and Scherf, U., Blue green stimulated emission from a high gain conjugated polymer, *Appl. Phys. Lett.* 71, 2566 (1997).
4. Schwartz, B. J., Hide, F., Andersson, M. R. and Heeger, A. J., Ultrafast studies of stimulated emission and gain in solid films of conjugated polymers, *Chem. Phys. Lett.* 265, 327 (1997).
5. McGehee, M. D., Diaz-Garcia, M. A., Hide, F., Gupta, R., Miller, E. K., Moses, D. and Heeger, A. J., Semiconducting polymer distributed feedback lasers, *Appl. Phys. Lett.* (submitted for publication).
6. Tessler, N. Denton, G. J. and Friend, R. H., Lasing from conjugated-polymer microcavities, *Nature* 382, 695 (1996).
7. Friend, R. H., Denton, G. J., Halls, J. J. M. et al., Electronic excitations in luminescent conjugated polymers, *Solid State Commun.* 102, 249 (1997).
8. Diaz-Garcia, M. A., Hide, F., Schwartz, B. J., McGehee, M. D., Andersson, M. R. and Heeger, A. J., "Plastic" lasers: Comparison of gain narrowing with a soluble semiconducting polymer in waveguides and microcavities, *Appl. Phys. Lett.* 70, 3191 (1997).
9. Frolov, S. V., Gellermann, W., Ozaki, M., Yoshino, K. and Vardeny, Z. V., Cooperative emission in π-conjugated polymer thin films, *Phys. Rev. Lett.* 78, 729 (1997). See, also, Frolov, S. V., Shkunov, M., Vardeny, Z. V., and Yoshino, K., *Phys. Rev. B* 656, R4363 (1997).
10. Brouwer, H. J., Krasnikov, V., Hilberer, A. and Hadziioannou, G., Blue superradiance from neat semiconducting alternating copolymer films, *Adv. Mater.* 8, 935 (1996).
11. Wegmann, G., Giessen, H., Hertel, D. and Mahrt, R. F., Blue-green laser emission from a solid conjugated polymer, *Solid State Commun.* 104, 759 (1997).

Section 3: Liquid Lasers

Section 3.1

ORGANIC DYE LASERS

Introduction to the Table

Organic dye lasers are tabulated in order of increasing wavelength of the gain maximum in Table 3.1.1. Lasers that have been tuned over a range of wavelengths are listed by the lowest wavelength reported or the range is listed in parentheses after the gain maximum. The tuning range given is that for the experimental configuration and conditions used and may not represent the extremes possible. In the following columns the dye and solvent are given together with the pump source and primary reference to laser action.

The wavelength, tuning range, and output of dye lasers depend on the characteristics of the dye, the dye concentration, solvent, optical pumping source and rate, optical cavity, and other operating conditions. The multiple listings of a given dye at several wavelengths reflect these differences. The original references should be consulted for the experimental conditions used and their effects on the lasing wavelength.

Chemical nomenclature

The names of various dyes found in the literature have in many instances been reduced to relatively simple pseudo acronyms by reducing the important parts of the name to a letter or by the assignment of a number to more complex molecules. The reader is referred to Steppel, R. N., Organic Dye Lasers, in *Handbook of Laser Science and Technology, Vol. I: Lasers and Masers*, CRC Press, Boca Raton, FL (1982), p. 299; Steppel, R. N., Organic Dye Lasers, in *Handbook of Laser Science and Technology, Suppl. 1: Lasers*, CRC Press, Boca Raton, FL (1991), p. 219; and Maeda, M., *Laser Dyes*, Academic, New York (1984) for common names of the dyes and details of the dye structures. The chemical names have been abbreviated according to the nomenclature system previously used by Fletcher. The shorthand notation for the parent compounds and substituents are as follows with numbers referring to the skeletal position (see numbered structure): AQ, azaquinolone; Q, quinolone; AC, azacoumarin; C, coumarin; A, amino; M, methyl; DM, dimethyl; DMA, dimethylamino; DEA, diethylamino; MO, methoxy; MOR, morpholino; TFM, trifluoromethyl; and OH, hydroxy. Thus 7A-M-AQ designates 7-amino-4-methylazaquinolone. Literature descriptions such as AQIF, QIF and C2H are not necessarily systematic. The first letter(s) refer to the family whereas the H and F refer to hydrogen and fluorine substitution.

Commercial dyes may have different nomenclature depending upon the producer. Some synonyms for laser dyes are given below:

BBQ	BiBuQ
BPBD-365	Butyl PBD
Coumarin 440	Coumarin 120
Coumarin 450	Coumarin 2
Coumarin 456	Coumarin 4

Coumarin 460	Coumarin 1, Coumarin 47
Coumarin 461	Coumarin 311
Coumarin 478	Coumarin 106
Coumarin 480	Coumarin 102
Coumarin 481	Coumarin 152A, Coumarin 1F
Coumarin 485	Coumarin 152 Coumarin 2F
Coumarin 490	Coumarin 151 Coumarin 3F
Coumarin 500	Coumarin SA-28
Coumarin 503	Coumarin 307
Coumarin 504	Coumarin 314
Coumarin 515	Coumarin 30
Coumarin 519	Coumarin 343
Coumarin 521	Coumarin 334
Coumarin 522	Coumarin 8F
Coumarin 523	Coumarin 337
Coumarin 535	Coumarin 7
Coumarin 540	Coumarin 6
Coumarin 540A	Coumarin 153
Cresyl Violet 670 Perchlorate	Kresylviolett
Disodium Fluorescein	Uranin
DMOTC	Methyl DOTC (DmOTC-1)
DMT	BM-Terphenyl
Fluorescein 548	Fluorescein 27, Fluorescein
Fluorol 555	Fluorol 7GA
HIDC lodide	Hexacyanin 2
HITC lodide	Hexacyanin 3
IR-5	Q-switch 5
IR-26	Dye 26
Kiton Red 620	Sulforhodamine B
LD 390	Quinolon 390
LD 466	Coumarin 466
LD 490	Coumarin 6H
LD 690 Perchlorate	Oxazine 4 Perchlorate
LD 700 Perchlorate	Rhodamine 700
LD 800	Rhodamine 800
LDS 698	Pyridine 1
LDS 722	Pyridine 2
LDS 730 Perchlorate	Styryl 6
LDS 750	Styryl 7*
LDS 751	Styryl 8*
LDS 798	Styryl 11
LDS 821	Styryl 9/9M
LDS 925	Styryl 13
LDS 950	Styryl 14
Nile Blue 690 Perchlorate	Nileblau

Oxazine 725 Perchlorate	Oxazine 1 Perchlorate
Oxazine720 Perchlorate	Oxazine 170 Perchlorate
p-Quaterphenyl	PQP
p-Terphenyl	PTP
Phenoxazone 660	Phenoxazone 9
Rhodamine 560 Chloride	Rhodamine 110
Rhodamine 590 Chloride	Rhodamine 6G
Rhodamine 610 Chloride	Rhodamine B
Rhodamine 640 Perchlorate	Rhodamine 101
Stilbene 420	Stilbene 3
Sulforhodamine 640	Sulforhodamine 101

* These two dyes are the same as originally reported by Kato. Dyes supplied by various sources should be consulted to verify that their products correspond to these dyes.

Abbreviations used in Table 3.1.1:

Excitation sources:

Ar (argon-ion laser), bb (broad band), coax (coaxial flashlamp), Cu (copper vapor laser), cw (continuous wave), FL (flashlamp), Kr (krypton-ion laser), KrF (krypton fluoride excimer laser), L (laser), N_2 (nitrogen laser), Nd:glass (neodymium:glass laser), Nd:YAG (neodymium:yttrium aluminum garnet laser), triax (triaxial flashlamp), XeCl (xenon chloride excimer laser).

Solvents:

Ar (argon), BuOAc (butylacetate), BzOH (benzyl alcohol), CCl_4 (carbon tetrachloride), CH_3CN (acetonitrile), CH_3COOH (acetic acid), COT (cyclooctetraene), DB (o-dichlorobenzene), DCE (1,2-dichloroethane), DEC (diethylcarbonate), DEE (diethyl ether), DMA (N,N-dimethylacetamide), DMA/EtOH (dimethylacetamide/ethanol), DMF (N,N-dimethyl-formamide), DMF/EtOH (dimethylformamide/ethanol), DMSO (dimethyl-sulfoxide), DPA (N,N-dipropylacetamide), DX (purified dioxane), EG (ethylene glycol), EtOH (ethanol), EtOH/H_2O (ethanol/water), G (glycerol), glyme (1,2-dimethoxyethane), H_2O (water), HCl (hydrochloric acid), HFIP (hexafluoroisopropanol), HFIP/H_2O (hexafluoroisopropanol/water), LO (ammonyx LO), MC (methylene chloride), MCH (methylcyclohexane), MeOH (methanol), MeOH/H_2O (methanol/water), NB (nitrobenzene), NMP (N-methylpyrrolidone), PPH (1-phenoxy-2-propanol), 2-PrOH (2-propanol), TEA (triethylamine), TFE (trifluoroethanol), THF (tetrahydrofuran), a (acidic), c (cyclohexane), d (DMF), e (ethanol), g (glycerol), m (DMA), p (p-dioxane), t (toluene), w (water).

Further Reading

Duarte, F. J., Ed., *High Power Dye Lasers*, Springer-Verlag, Berlin (1991).

Duarte, F. J., Ed., *Selected Papers on Dye Lasers*, SPIE Milestone Series, Vol. MS45, SPIE Optical Engineering Press, Bellingham, WA (1992).

Duarte, F. J. and Hillman, L. W., Eds., *Dye Laser Principles*, Academic Press, New York (1990).

Duarte, F. J., Paisner, J. A. and Penzkofer, A., Eds., Dye lasers, special issue of *Applied Optics* 31, 6977 (1992).

Maeda, M., *Laser Dyes*, Academic, New York (1984).

Schäfer, S. P., Ed., *Dye Lasers*, 3rd Edition, Springer-Verlag, Berlin (1990).

Steppel, R. N., Organic Dye Lasers, in *Handbook of Laser Science and Technology, Vol. I: Lasers and Masers*, CRC Press, Boca Raton, FL (1982), p. 299.

Steppel, R. N., Organic Dye Lasers, in *Handbook of Laser Science and Technology, Suppl. 1: Lasers*, CRC Press, Boca Raton, FL (1991), p. 219.

Stuke, M., *Dye Lasers: 25 Years*, Springer-Verlag, Berlin (1992).

Table 3.1.1
Organic Dye Lasers Arranged in Order of Increasing Wavelength

Wavelength (nm)	Laser dye*	Solvent	Pump	Ref.
332 (311–360)	terphenyl	cyclohexane	KrF(248)	1
336 (331–342)	oligophenylene 2-1-7	ethanol	KrF(249)	2
336 (331–342)	oligophenylene 2-1-9	ethanol	KrF(249)	2
337	p-terphenyl	cyclohexane	KrF(248)	3
338	p-terphenyl	ethanol	KrF(248)	3
338 (326–358)	p-terphenyl	—	KrF(248)	4
339 (322–366)	p-terphenyl	cyclohexane	KrF(248)	1
339 (332–346)	oligophenylene 2-1-10	ethanol	KrF(249)	2
340	p-terphenyl	p-dioxane	KrF(248)	5,6
340 (323–364)	p-terphenyl	cyclohexane	KrF(248)	7
341	p-terphenyl	DMF	FL	8
347	oxadiazole	cyclohexane	KrF(249)	9
347	oxadiazole	ethanol	KrF(249)	9
353	oxadiazole	ethanol	KrF(249)	9
357	oxadiazole	ethanol	KrF(249)	9
359	oxadiazole	ethanol	KrF(249)	9
362	p-quaterphenyl	cyclohexane	Nd:YAG(266)	10
362–390	p-quaterphenyl	DMF	FL	8
365 (359–391)	oxazole:PPO	toluene	N_2(337)	11
372	oxadiazole	toluene	N_2(337)	12
372	oxazole:PPO	cyclohexane	KrF(248)	3
373	oxadiazole	toluene	N_2(337)	12
374	p-quaterphenyl	toluene	N_2(337)	3
375	oxadiazole	toluene	N_2(337)	12
375–380	pteridine	methanol	N_2(337)	10
377	oxadiazole	toluene	N_2(337)	12
377 (365–400)	Exalite 337E	EG	Ar	13
379	oxadiazole	toluene	N_2(337)	12
380	p-quaterphenyl	cyclohexane	KrF(248)	5
380 (372–406)	oxadiazole	toluene	N_2(337)	11
381	oxazole:POPOP	vapor(Ar+N_2)	e-beam	14
381	oxazole:PPO	cyclohexane	FL	8
381–389	p-quaterphenyl	butanol	FL	15
382 (373–391)	p-quaterphenyl	cyclohexane	Nd:YAG(266)	10
385–415	oxazole:α-NPO	cyclohexane	N_2(337)	16
386	oligophenylene 2-4-7	DMA/EtOH:4/1	FL	17
386	oligophenylene 2-4-7	ethanol	FL	17
386	oligophenylene 2-5-5	DMA/EtOH:1/4	FL	18

Table 3.1.1—*continued*
Organic Dye Lasers Arranged in Order of Increasing Wavelength

Wavelength (nm)	Laser dye*	Solvent	Pump	Ref.
386	p-quaterphenyl	—	KrF(248)	6
386 (373–399)	p-quaterphenyl	toluene/ethanol	N_2(337)	20
386 (380–391)	oligophenylene 2-3-15	p-dioxane	XeCl(308)	21
388	oxazole 4-4-1	p-dioxane	XeCl	22
389	oligophenylene 2-5-4	ethanol	FL	18
332 (311–360)	terphenyl	cyclohexane	KrF(248)	1
336 (331–342)	oligophenylene 2-1-7	ethanol	KrF(249)	2
336 (331–342)	oligophenylene 2-1-9	ethanol	KrF(249)	2
337	p-terphenyl	cyclohexane	KrF(248)	3
338	p-terphenyl	ethanol	KrF(248)	3
338 (326–358)	p-terphenyl	—	KrF(248)	4
339 (322–366)	p-terphenyl	cyclohexane	KrF(248)	1
339 (332–346)	oligophenylene 2-1-10	ethanol	KrF(249)	2
340	p-terphenyl	p-dioxane	KrF(248)	5,6
340 (323–364)	p-terphenyl	cyclohexane	KrF(248)	7
341	p-terphenyl	DMF	FL	8
347	oxadiazole	cyclohexane	KrF(249)	9
347	oxadiazole	ethanol	KrF(249)	9
353	oxadiazole	ethanol	KrF(249)	9
357	oxadiazole	ethanol	KrF(249)	9
359	oxadiazole	ethanol	KrF(249)	9
362	p-quaterphenyl	cyclohexane	Nd:YAG(266)	10
362–390	p-quaterphenyl	DMF	FL	8
365 (359–391)	oxazole:PPO	toluene	N_2(337)	11
372	oxadiazole	toluene	N_2(337)	12
372	oxazole:PPO	cyclohexane	KrF(248)	3
373	oxadiazole	toluene	N_2(337)	12
374	p-quaterphenyl	toluene	N_2(337)	3
375	oxadiazole	toluene	N_2(337)	12
375–380	pteridine	methanol	N_2(337)	10
377	oxadiazole	toluene	N_2(337)	12
377 (365–400)	Exalite 337E	EG	Ar	13
379	oxadiazole	toluene	N_2(337)	12
380	p-quaterphenyl	cyclohexane	KrF(248)	5
380 (372–406)	oxadiazole	toluene	N_2(337)	11
381	oxazole:POPOP	vapor(Ar+N_2)	e-beam	14
381	oxazole:PPO	cyclohexane	FL	8
381–389	p-quaterphenyl	butanol	FL	15

Table 3.1.1—*continued*
Organic Dye Lasers Arranged in Order of Increasing Wavelength

Wavelength (nm)	Laser dye*	Solvent	Pump	Ref.
382 (373–391)	p-quaterphenyl	cyclohexane	Nd:YAG(266)	10
385–415	oxazole:α-NPO	cyclohexane	N_2(337)	16
386	oligophenylene 2-4-7	DMA/EtOH:4/1	FL	17
386	oligophenylene 2-4-7	ethanol	FL	17
386	oligophenylene 2-5-5	DMA/EtOH:1/4	FL	18
386	p-quaterphenyl	—	KrF(248)	6
386 (373–399)	p-quaterphenyl	toluene/ethanol	N_2(337)	20
386 (380–391)	oligophenylene 2-3-15	p-dioxane	XeCl(308)	21
388	oxazole 4-4-1	p-dioxane	XeCl	22
389	oligophenylene 2-5-4	ethanol	FL	18
389–395	oxazole 4-6-1	ethanol	ruby(316)	28
389–395	p-quaterphenyl	DMF	FL	15
390	oxazole 4-4-2	p-dioxane	XeCl	22
390 (370–410)	p-quaterphenyl	DMF	FL	25
390 (385–417)	oxadiazole	toluene	N_2(337)	11
390–395	α-NPO	ethanol, w, t	N_2(337)	27
390–445	7-diethylamino	ethanol, w, t	N_2(337)	27
391	oligophenylene 2-4-8	DMF/EtOH:4/1	FL	17
391	oligophenylene 2-4-8	ethanol	FL	17
391	oligophenylene 2-5-6	DMA/EtOH:1/4	FL	18
391 (380–410)	p-quaterphenyl	EtOH/toluene:1/1	Nd:YAG(355)	28
393	oxazole:POPOP	vapor	N_2(337)	29
393 (375–410)	Exalite 392E	EG	Ar	30,31
395	oligophenylene 2-5-7	DMA/EtOH:1/4	FL	18
395–402	oxazole:4PyPO	ethanol	N_2(337)	32
400	oxazole:α-NPO	ethanol	FL	7
400 (385–425)	Exalite 400E	EG	Ar	30,59
400–420	diphenyl-stilbene	ethanol, w, t	N_2(337)	27
406 (396–416)	stilbene:DPS	p-dioxane	N_2(337)	20
407	furan 1-1-3	toluene	N_2(337)	32
409	stilbene:DPS	DMF	FL	8
411	sodium salicylate	DMF, t, w	XeCl(308)	33
413 (408–422)	styrybenzene	p-dioxane, THF	N_2(337)	34
414	9-methylanthracene	EtOH/MeOH:4/1	N_2(337)	35
414	9,10-dimethylanthracene	MCH/toluene:2/1	N_2(337)	35
414	coumarin;3,7-substituted	benzene	N_2(337)	36
414 (408–423)	styrybenzene	p-dioxane, THF	N_2(337)	36
415	furan 1-1-1	toluene	N_2(337)	32

Table 3.1.1—*continued*
Organic Dye Lasers Arranged in Order of Increasing Wavelength

Wavelength (nm)	Laser dye*	Solvent	Pump	Ref.
415	p-terphenyl	DMF	FL	25
415	Stilbene 1	EG	Kr(UV)	39
415–430	oxazole:POPOP	THF	N_2(337)	16
416	9-dichloroanthracene	ethanol	N_2(337)	35
416	Blankophor R	methanol	N_2(337)	40
416	miscellaneous 13-1-1	DMSO	N_2(337)	41
416 (403–437)	oxazole	toluene	N_2(337)	42
417	9-phenylanthracene	ethanol	N_2(337)	35
417	coumarin:7A-4MO-C	ethanol	FL	43
417–427	p-terphenyl	ethanol (sat.)	N_2(337)	36
418	4-phenylpyridine	ethanol	N_2(337)	44
418	furan 1-1-4	toluene	N_2(337)	32
419	oxazole:POPOP	toluene	FL	46
419 (410–439)	styrybenzene	toluene	N_2(337)	34
419–424	oxazole:POPOP	ethanol	FL	186
420	furan 1-1-5	toluene	N_2(337)	32
420 (413–431)	styrybenzene	methanol	N_2(337)	34
420–470	Stilbene 420; Stilbene 3	EG	Ar(UV)	49
421	coumarin;3,7-substituted	benzene	N_2(337)	36
422 (412–432)	triazolstilbene	methanol	N_2(337)	40
422–430	oxazole 4-6-3	H_2O/HOAc:95/5	ruby(316)	28
423	coumarin;3,7-substituted	benzene	N_2(337)	36
423 (413–431)	styrybenzene	methanol	N_2(337)	34
423 (414–442)	oxazole:POPOP	toluene	N_2(337)	42
424 (411–436)	Stilbene 420; Stilbene 3	methanol	Nd:YAG(355)	51
425	furan 1-1-2	toluene	N_2(337)	32
425	oxadiazole	methyl chloride	N_2(337)	36
425 (400–480)	Stilbene 420; Stilbene 3	—	Kr(UV)	167
425 (408–453)	Stilbene 420; Stilbene 3	methanol	N_2(337)	34
425 (414–438)	triazolstilbene	methanol	N_2(337)	52
425 (416–437)	styrybenzene	p-dioxane, THF	N_2(337)	34
425 (418–443)	triazinylstilbene	methanol	N_2(337)	52
426	coumarin;3,7-substituted	benzene	N_2(337)	36
427 (410–459)	triazolstilbene	methanol	N_2(337)	40
427 (411–465)	oxazole:POPOP	p-dioxane	N_2(337)	40
427 (414–451)	Tinopal PCRP	p-dioxane	N_2(337)	40
429 (404–460)	stilbene 420; stilbene 3	methanol	Nd:YAG(355)	53
429 (406–465)	triazinylstilbene	methanol	N_2(337)	40

Table 3.1.1
Organic Dye Lasers Arranged in Order of Increasing Wavelength

Wavelength (nm)	Laser dye*	Solvent	Pump	Ref.
429 (406–465)	triazinylstilbene	methanol	N_2(337)	40
~430	oxazole:dimethylPOPOP	p-dioxane	FL	32
430	9,10-diphenylanthracene	ethanol	N_2(337)	35
430	coumarin;3,7-substituted	dichloromethane	N_2(337)	36
430	triazolstilbene	benzene	N_2(337)	36
430 (412–462)	triazinylstilbene	methanol	N_2(337)	40
430 (420–438)	triazinylstilbene	methanol	N_2(337)	52
430 (420–445)	triazinylstilbene	methanol	N_2(337)	52
430–445	POPOP	ethanol, w, t	N_2(337)	27
430(418–465)	oxazole:dimethylPOPOP	p-dioxane	N_2(337)	11
431 (415–458)	stilbene 420; stilbene 3	H_2O+NP-10	N_2(337)	34
431 (419–448)	triazinylstilbene	methanol	N_2(337)	52
432	9,10-diphenylanthracene	MCH/toluene:2/1	N_2(337)	35
432 (406–448)	stilbene 420; stilbene 3	EG,methanol:9/1	Ar(UV)	162
432 (407–460)	triazolstilbene	methanol	N_2(337)	40
432 (412–464)	triazolstilbene	methanol	N_2(337)	40
432.6	9,10-diphenylanthracene	cyclohexane	ruby(347)	194
433 (418–448)	triazinylstilbene	methanol	N_2(337)	52
433 (418–449)	triazinylstilbene	methanol	N_2(337)	52
433 (418–461)	triazinylstilbene	methanol	N_2(337)	52
433 (420–447)	triazinylstilbene	methanol	N_2(337)	52
434	coumarin;3,7-substituted	benzene	N_2(337)	36
434	Delft Weiss BSW	methanol	N_2(337)	40
434 (414–472)	Tinopal RBS	MeOH/p–dioxane	N_2(337)	40
435	bimane 5–1–5	p–dioxane	FL	50
435-450	diphenyl–anthracene	ethanol, w, t	N_2(337)	27
436	coumarin:CSA–1	ethanol	FL	43,44
436 (410-462)	triazinylstilbene	methanol	N_2(337)	40
436 (417–455)	triazolstilbene	H_2O, NP10	N_2(337)	52
437	coumarin;3,7-substituted	benzene	N_2(337)	36
437 (420–457)	coumarin:CSA-1	ethanol	N_2(337)	20
438	coumarin;3,7-substituted	benzene	N_2(337)	36
438 (419–466)	coumarin:CSA-1	ethanol	N_2(337)	56
439	coumarin;3,7-substituted	benzene	N_2(337)	36
439	salicylamide	DMF	XeCl(308)	57
440	coumarin:CSA-10	ethanol	N_2(337)	44
440	coumarin:CSA-8	ethanol	N_2(337)	44
440	coumarin:CSA-9	ethanol	N_2(337)	44

Table 3.1.1—*continued*
Organic Dye Lasers Arranged in Order of Increasing Wavelength

Wavelength (nm)	Laser dye*	Solvent	Pump	Ref.
440 (424–475)	Tinopal GS	methanol	N_2(337)	40
441	coumarin:CSA-1	ethanol	Nd:YAG(355)	60
442	coumarin:CSA-1	ethanol	FL	61
442	pyrylium salt	dichloromethane	N_2(337)	36
442 (425–443)	coumarin:CSA-1	ethanol	FL	25
442 (426–458)	coumarin:CSA-1	methanol	Nd:YAG(355)	51
442 (435–455)	coumarin:CSA-1	MeOH/H_2O:1/1	FL	19
445	coumarin:CSA-6	ethanol	N_2(337)	44
445	coumarin:CSA-6	MeOH/H_2O:1/1	FL	26
445 (421–468)	stilbene 420; stilbene 3	H_2O	N_2(337)	34
445–490	4-methylcoumarin	ethanol, w, t	N_2(337)	27
446	coumarin 450; 2	ethanol	FL	37
446	coumarin;3,7-substituted	methanol	N_2(337)	36
446 (428–465)	coumarin 450; 2	ethanol	N_2(337)	20
447	miscellaneous 13–1–2	DMSO	N_2(337)	41
447 (423-461)	triazinylstilbene	methanol	N_2(337)	52
448	pyrylium salt	dichloromethane	N_2(337)	36
449 (435–479)	coumarin 450; 2	ethanol/1.5%LO	FL	25
449 (436–493)	stilbene 420; stilbene 3	EG,methanol:9/1	Ar(UV)	162
450	coumarin:C3H	ethanol	FL	191
450	coumarin:CSA-4	ethanol	N_2(337)	44
450	coumarin:CSA-5	ethanol	N_2(337)	44
450	coumarin;3,7-substituted	water	N_2(337)	36
450 (420–470)	coumarin:CSA-1	DPA,COT	Ar(351/364)	37
450 (427–477)	coumarin:CSA-1	EG	Ar(cw)	47
450 (427–488)	coumarin 450; 2	methanol	FL	59
450 (435–485)	coumarin 450; 2	EG	Ar or Kr(uv)	164
450–511	coumarin 6-4-2	ethanol	XeCl	24
451–465	coumarin 6-1-2	hexane(Ar)	FL(triaxial)	38
452	oxazole 4-2-7	H_2O	ruby(347)	45
452 (430–492)	coumarin 450; 2	EG	Ar(cw)	47
452–480	coumarin 6-1-2	toluene(Ar)	FL(triaxial)	38
453	coumarin 311	ethanol	FL	54
453	coumarin:CSA-1	water	FL	55
453–514	coumarin 6-4-1	ethanol	XeCl	24
454	coumarin 450; 2	methanol	FL	61
454	coumarin 450; 2	methanol	Nd:YAG(355)	60
455	coumarin C3F:CSA-29	ethanol	N_2(337)	44

Table 3.1.1—*continued*
Organic Dye Lasers Arranged in Order of Increasing Wavelength

Wavelength (nm)	Laser dye*	Solvent	Pump	Ref.
455 (433–474)	coumarin 450; 2	methanol	Nd:YAG(355)	51
456	coumarin CSA-11	ethanol	N_2(337)	44
457	coumarin 175	water	FL	55
457	coumarin 311	ethanol	FL	58
457	coumarin;3,7-substituted	water	N_2(337)	36
457 (440–478)	coumarin 460; 1	ethanol	N_2(337)	20
457 (450–484)	coumarin 460; 1	ethanol	FL	25
458	coumarin 450; 2	methanol	FL	191
458	coumarin;3,7-substituted	methanol	N_2(337)	36
458–474	coumarin 6-1-2	toluene(air)	FL(triaxial)	38
458–503	pyrylium dye 14	CH_3CN	N_2(337)	190
~460	coumarin 138	ethanol	N_2(337)	120
460	coumarin 460; 1	ethanol	Nd:YAG(355)	96
460	coumarin 460; 1	ethanol	FL	43,61
460	coumarin 460; 1	ethanol	FL	62
460	coumarin:CSA-6	EG	Ar(cw)	63
460	pteridine	MeOH, alkaline	N_2(337)	10
460 (430–480)	coumarin 450; 2	20%aq. DPA	Ar(351/364)	192
460 (442–490)	coumarin 460; 1	—	Nd:YAG(355)	53
460 (445–482)	coumarin 450; 2	MeOH/H_2O:4/6	FL	19
460–479	coumarin 6-1-2	benzene(air)	FL(triaxial)	38
460–483	coumarin 6–1–2	benzene(Ar)	FL(triaxial)	38
461 (448-489)	coumarin 460; 1	methanol	FL	25
461–479	coumarin 6-1-2	DEC(air)	FL(triaxial)	38
462	oxazole 4-2-7	H_2O	FL	45
462–483	coumarin 6-1-2	BuOAC(air)	FL(triaxial)	38
462–485	coumarin 6-1-2	DEC(Ar)	FL(triaxial)	38
462–489	oxazole 4-6-1	H_2O/HOAc:95/5	ruby(316)	28
463 (450–480)	benzimidazole 3-1-2	methanol	XeCl	128
463–488	coumarin 6-1-2	BuOAC(Ar)	FL(triaxial)	38
464 (446–492)	coumarin C1H: LD 466	EtOH/p-dioxane	N_2(337)	64
464 (447–510)	triazolstilbene	methanol	N_2(337)	40
464–514	pyrylium dye 13	CH_3CN	N_2(337)	190
465	coumarin C2H	ethanol	FL	191
465 (452–480)	coumarin C1H: LD 466	ethanol	N_2(337)	1
465 (454–480)	benzimidazole 3-1-2	ethanol	XeCl	128
467	coumarin;3,7-substituted	water	N_2(337)	36
467 (459–477)	coumarin C1H: LD 466	ethanol	Nd:YAG(355)	65

Table 3.1.1—*continued*
Organic Dye Lasers Arranged in Order of Increasing Wavelength

Wavelength (nm)	Laser dye*	Solvent	Pump	Ref.
467–490	coumarin 6-1-2	glyme(air)	FL(triaxial)	38
467–491	coumarin 6-1-2	THF(Ar)	FL(triaxial)	38
467–497	coumarin 6-1-2	glyme(Ar)	FL(triaxial)	38
468	coumarin 378	water	FL	55
468	pyrylium salt	dichloromethane	N_2(337)	36
468–491	coumarin 6-1-2	THF(air)	FL(triaxial)	38
470	coumarin 360	water	FL	55
470	coumarin 380	water	FL	55
470 (450–495)	coumarin 460; 1	EG	Ar or Kr(uv)	164
470 (460–495)	2-(*o*-hydroxyphenyl/ benzimidazole	dioxane	XeCl(308)	33
470–517	pyrylium dye 15	CH_3CN	N_2(337)	190
471 (448–505)	coumarin 460; 1	—	Kr(uv)	52
472 (446–506)	coumarin 460; 1	EG	Ar(cw)	47
473	coumarin 379	water	FL	55
473 (460–490)	benzimidazole 3-1-2	p-dioxane	XeCl	128
474 (459–489)	benzimidazole 3-1-2	acetonitrile	XeCl	128
474 (462–490)	coumarin C1H: LD 466	methanol+LO	FL	25
475	coumarin 522; C8H	ethanol	FL	37
475	pyrylium salt	dichloromethane	N_2(337)	36
475–500	coumarin 6-1-2	acetone(air)	FL(triaxial)	38
475–504	coumarin 6-1-2	acetone(Ar)	FL(triaxial)	38
477	coumarin C4H	ethanol	FL	191
477 (463–495)	benzimidazole 3-1-2	DMF	XeCl	128
477–526	coumarin 6-4-4	ethanol	XeCl	24
478	coumarin 381	water	FL	55
478	coumarin 478; 106	ethanol	FL	66
478	pyrylium salt	dichloromethane	N_2(337)	36
478–525	coumarin 6-4-6	ethanol	XeCl	24
479	coumarin C2F; 485	p-dioxane	FL	66
479–506	coumarin 6-1-2	CH_3CN(Ar)	FL(triaxial)	38
480	coumarin 480; 102	ethanol	FL	61
480	coumarin C1F:CSA-27	ethanol	N_2(337)	44
480–497	oxazole 4-1-8	ethanol(air)	FL	129
480–528	coumarin 6-4-3	ethanol	XeCl	24
481	coumarin C1F:CSA-27	ethanol	FL	61,66
481	coumarin C1F:CSA-27	p-dioxane	FL	66,67
481	coumarin C1F:CSA-27	p-dioxane	KrF(248)	6

Table 3.1.1—*continued*
Organic Dye Lasers Arranged in Order of Increasing Wavelength

Wavelength (nm)	Laser dye*	Solvent	Pump	Ref.
481 (460–518)	coumarin C1F:CSA-27	p-dioxane	N_2(337)	56
481 (475–490)	coumarin C1F:CSA-27	p-dioxane	FL	25
482–507	coumarin 515; 30	ethanol	N_2He(428)	68
482–509	oxazole 4-1-8	MeOH(air)	FL	129
482–526	coumarin 6-4-5	ethanol	XeCl	24
483	pyrylium salt	dichloromethane	N_2(337)	36
483 (463–516)	coumarin C1F:CSA-27	p-dioxane	N_2(337)	69
483 (477–489)	coumarin 6-3-1	ethanol(air)	FL(triaxial)	130
483 (478–489)	coumarin 6-3-1	ethanol(Ar)	FL(triaxial)	130
483 (560–517)	coumarin C1F:CSA-27	p-dioxane	N_2(337)	20
484	coumarin C3F:CSA-29	ethanol	FL	66
484–507	coumarin 6-1-2	DMA(Ar)	FL(triaxial)	38
484–512	oxazole 4-1-8	MeOH(Ar)	FL	129
485	coumarin C2F; 485	ethanol	N_2(337)	44
485	coumarin:CSA-10	ethanol+HCl	N_2(337)	44
485	coumarin:CSA-6	ethanol+HCl	N_2(337)	44
485	coumarin:CSA-8	ethanol+HCl	N_2(337)	44
485	coumarin:CSA-9	ethanol+HCl	N_2(337)	44
486	coumarin 6-1-2	2-PrOH(air)	FL(triaxial)	38
486	coumarin 486	water	FL	55
486–508	coumarin 6-1-2	DMF(air)	FL(triaxial)	38
487 (465–510)	benzimidazole 3-1-1	acetonitrile	N_2	128
487–503	coumarin 6-1-2	CH_3CN(air)	FL(triaxial)	38
487–506	coumarin 6-1-2	DCE(air)	FL(triaxial)	38
487–511	coumarin 6-1-2	DMA(air)	FL(triaxial)	38
487–512	coumarin 6-1-2	DMF(Ar)	FL(triaxial)	38
488	coumarin C3F:CSA-29	ethanol	FL	70
488	pyrylium salt	dichloromethane	N_2(337)	36
488–511	coumarin 6-1-2	DCE(Ar)	FL(triaxial)	38
489	coumarin 6-1-1	ethanol(air)	FL(triaxial)	38
489	coumarin 6-1-3	ethanol(air)	FL(triaxial)	38
489	coumarin 488	water	FL	55
489	coumarin C1F:CSA-27	p-dioxane	Nd:YAG(355)	60
489 (460–540)	quinoxalinone	ethanol	N_2(337)	71
489 (467–510)	coumarin C3F:CSA-29	methanol	FL	72
489–513	coumarin 6-1-2	NMP(air)	FL(triaxial)	38
489–541	pyrylium dye 44	CH_3CN	N_2(337)	48
490	coumarin C3F:CSA-29	ethanol	FL	61

Table 3.1.1—*continued*
Organic Dye Lasers Arranged in Order of Increasing Wavelength

Wavelength (nm)	Laser dye*	Solvent	Pump	Ref.
490	coumarin C6H; LD 490	—	FL	191
490	oxazole 4-2-1	H_2O	FL	45
490–514	oxazole 4-1-8	EtOH/H_2O(air)	FL	129
490–546	pyrylium dye 36	CH_3CN	N_2(337)	48
491	coumarin 6-1-2	ethanol(air)	FL(triaxial)	38
491 (482–509)	imitrine 7-1-131	ethanol	Nd:YAG(355)	133
491–513	coumarin 6-1-2	NMP(Ar)	FL(triaxial)	38
492	coumarin 6-1-3	ethanol(Ar)	FL(triaxial)	38
492	oxazole 4-3-1	MeOH(air)	FL	131
492	pyrylium salt	dichloromethane	N_2(337)	36
492–504	oxazole 4-1-9	ethanol(air)	FL	129
492–507	oxazole 4-1-2	ethanol(air)	FL	129
492–511	oxazole 4-6-5	H_2O/HOAc:95/5	ruby(316)	28
493	coumarin 6-1-1	ethanol(Ar)	FL(triaxial)	38
493	coumarin 6-1-1	methanol(air)	FL(triaxial)	38
493	coumarin 6-1-3	methanol(air)	FL(triaxial)	38
493	oxazole 4-3-1	ethanol(Ar)	FL	131
493 (475–518)	coumarin 6-5-9	ethanol	XeCl	132
493–508	oxazole 4-1-2	ethanol(Ar)	FL	129
493–512	oxazole:4PyPO	water+HCl(pH2)	N_2(337)	32
494	coumarin 6-1-2	methanol(air)	FL(triaxial)	38
494	coumarin 316	water	FL	55
494	oxazole 4-3-1	H_2O(air)	FL	131
494	oxazole 4-3-1	H_2O(Ar)	FL	131
494	oxazole 4-3-15	ethanol(air)	FL	131
494	oxazole 4-3-4	methanol(air)	FL	131
494	oxazole 4-3-4	methanol(Ar)	FL	131
494–504	oxazole 4-6-5	ethanol	ruby(316)	28
494–512	oxazole 4-1-1	H_2O(Ar)	FL	129
494–512	oxazole 4-1-2	H_2O(air)	FL	129
494–546	pyrylium dye 46	CH_3CN	N_2(337)	190
495	benzimidazole 3-1-4	p-dioxane	N_2	128
495	coumarin 6-1-2	ethanol(Ar)	FL(triaxial)	38
495	oxazole 4-3-1	EtOH/H_2O(Ar)	FL	131
495	oxazole 4-3-15	ethanol(Ar)	FL	131
495	oxazole 4-3-4	ethanol(Ar)	FL	131
495	oxazole 4-3-4	EtOH/H_2O(Ar)	FL	131
495	oxazole 4-3-4	H_2O(air)	FL	131

Table 3.1.1—*continued*
Organic Dye Lasers Arranged in Order of Increasing Wavelength

Wavelength (nm)	Laser dye*	Solvent	Pump	Ref.
495	oxazole 4-3-4	H_2O(Ar)	FL	131
495 (477–515)	coumarin C3F:CSA-29	methanol+LO	FL	25
495–511	oxazole 4-1-2	EtOH/H_2O(Ar)	FL	129
495–514	oxazole:4PyPO-TS	water	N_2(337)	32
496	coumarin C3F:CSA-29	EtOH/H_2O	FL	66
496	oxazole 4-2-1	H_2O	ruby(347)	45
496	oxazole 4-3-1	EtOH/H_2O(air)	FL	131
496	oxazole 4-3-4	EG/H_2O(Ar)	FL	131
496	oxazole 4-3-4	MeOH/H_2O(air)	FL	131
496	oxazole 4-3-8	ethanol(air)	FL	131
496 (484–508)	coumarin 6-3-1	EtOH/H_2O(Ar)	FL(triaxial)	130
496–507	oxazole 4-1-2	EtOH/H_2O(air)	FL	129
496–508	oxazole 4-1-9	MeOH(air)	FL	129
497	coumarin 6-1-1	EG/EtOH(air)	FL(triaxial)	38
497	coumarin 6-1-1	MeOH/H_2O(air)	FL(triaxial)	38
497	coumarin 6-1-3	MeOH(Ar)	FL(triaxial)	38
497	pyrylium salt	dichloromethane	N_2(337)	36
497 (485–508)	coumarin 6-3-1	EtOH/H_2O(air)	FL(triaxial)	130
498	coumarin 6-1-3	EG/EtOH(air)	FL(triaxial)	38
498	coumarin 6-1-3	EG/EtOH(Ar)	FL(triaxial)	38
498	coumarin 6-1-3	MeOH/H_2O(air)	FL(triaxial)	38
498	oxazole 4-3-4	EG/H_2O(Ar)	FL	131
498	oxazole 4-3-4	MeOH/H_2O(Ar)	FL	131
498 (477–531)	coumarin 503; 307	ethanol	FL	25
498–518	oxazole 4-6-6	H_2O	ruby(316)	28
499	coumarin 6-1-2	EG/EtOH(air)	FL(triaxial)	38
499	coumarin 6-1-2	MeOH/H_2O(air)	FL(triaxial)	38
499	oxazole 4-3-8	ethanol(Ar)	FL	131
499 (484–537)	coumarin 504; 314	methanol+LO	FL	25
499–514	oxazole 4-6-6	ethanol	ruby(316)	28
499–547	pyrylium dye 45	CH_3CN	N_2(337)	190
499 (492–507)	coumarin 6-3-5	Ethanol(Ar)	FL(triaxial)	130
500	coumarin 6-1-2	EG/EtOH(Ar)	FL(triaxial)	38
500	coumarin CSA-28; 500	ethanol	KrF(248)	3
500	coumarin CSA-28; 500	ethanol	N_2(337)	44
500 (473–547)	coumarin CSA-28; 500	ethanol	N_2(337)	20
500 (482–517)	coumarin C2F; 485	p-dioxane	N_2(337)	73
500 (492–508)	coumarin 6-3-5	ethanol(air)	FL(triaxial)	130

Table 3.1.1—*continued*
Organic Dye Lasers Arranged in Order of Increasing Wavelength

Wavelength (nm)	Laser dye*	Solvent	Pump	Ref.
500 (494–504)	coumarin CSA-28; 500	ethanol	Nd:YAG(355)	65
500–546	coumarin 6-4-8	ethanol	XeCl	24
500–562	pyrylium dye 18	CH_3CN	N_2(337)	190
501	bimane 5-1-1	HFIP	FL	134
501	coumarin 6-1-1	EtOH/H_2O(air)	FL(triaxial)	38
501	coumarin 6-1-1	EtOH/H_2O(Ar)	FL(triaxial)	38
501	coumarin 6-1-3	EtOH/H_2O(air)	FL(triaxial)	38
501	oxazole 4-2-3	H_2O	FL	45
501	oxazole 4-3-20	ethanol(Ar)	FL	131
501	pyrylium salt	dichloromethane	N_2(337)	36
501 (490–513)	coumarin 519; 334	methanol	FL	25
501 (493–508)	coumarin 6-3-7	ethanol(air)	FL(triaxial)	130
501–550	coumarin 6-4-7	ethanol	XeCl	24
501–563	pyrylium dye 17	CH_3CN	N_2(337)	190
502	coumarin 6-1-2	EtOH/H_2O(air)	FL(triaxial)	38
502	coumarin 6-1-2	MeOH(Ar)	FL(triaxial)	38
502	coumarin 6-1-3	EtOH/H_2O(Ar)	FL(triaxial)	38
502	coumarin 6-1-4	MeOH/H_2O	F	135
502	coumarin 503; 307	ethanol	FL	61
502	coumarin;3,7-substituted	methanol	N_2(337)	36
502	oxazole 4-3-20	ethanol(air)	FL	131
502	oxazole 4-3-21	ethanol(air)	FL	131
502	oxazole 4-3-21	ethanol(Ar)	FL	131
502	phosphorine	benzene	N_2(337)	36
502 (474–556)	quinoxalinone	ethanol	N_2(337)	71
502 (494–510)	coumarin 6-3-7	ethanol(Ar)	FL(triaxial)	130
503	coumarin 6-1-2	EtOH/H_2O(Ar)	FL(triaxial)	38
503	coumarin 6-1-4	MeOH(air)	FL(triaxial)	38
503	oxazole 4-2-3	H_2O	ruby(347)	45
503	oxazole 4-2-5	H_2O	ruby(347)	45
503	oxazole 4-2-7	H_2O	ruby(347)	45
503	oxazole 4-3-21	EtOH/H_2O(air)	FL	131
503	oxazole 4-3-21	EtOH/H_2O(Ar)	FL	131
503	oxazole 4-3-21	H_2O(air)	FL	131
503	oxazole 4-3-21	H_2O(Ar)	FL	131
503–521	oxazole 4-6-2	ethanol	ruby(316)	28
503–557	pyrylium dye 34	CH_3CN	N_2(337)	190
504	coumarin 6-1-3	MeOH/H_2O(Ar)	FL(triaxial)	38

Table 3.1.1—*continued*
Organic Dye Lasers Arranged in Order of Increasing Wavelength

Wavelength (nm)	Laser dye*	Solvent	Pump	Ref.
504	coumarin 504; 314	ethanol	FL	61
504	oxazole 4-2-2	H_2O	FL	45
504	oxazole 4-2-4	H_2O	ruby(347)	45
504	oxazole:4PyPO	water+HCl(pH2)	FL	32
504 (481–530)	coumarin 503; 307	MeOH+LO	FL	25
504 (498–511)	coumarin 6-3-6	ethanol(air)	FL(triaxial)	130
505	coumarin 6-1-2	MeOH/H_2O(Ar)	FL(triaxial)	38
505	pyrylium salt	dichloromethane	N_2(337)	36
505 (495–515)	coumarin 515; 30	aq.,DPA,COT	Ar(cw,458)	192
505 (495–517)	coumarin 504; 314	MeOH/H_2O:1/1	FL	26
505 (498–512)	coumarin 6-3-6	ethanol(Ar)	FL(triaxial)	130
506	oxazole 4-2-6	H_2O	ruby(347)	45
506	oxazole:4PyPO-TS	water	FL	32
507	bimane 5-1-3	HFIP/H_2O:4/1	FL	50
507	coumarin 540A; 153	p-dioxane	N_2(337)	69
507 (481–540)	coumarin 481; CSA-27	dioxane/ethanol	N_2(337)	69
507 (490–537)	imitrine 7-1-124	ethanol	Nd:YAG(355)	133
507–529	coumarin 540; 6	ethanol	N_2He(428)	68
508	coumarin;3,4,7-subst.	methanol	N_2(337)	36
508	oxazole 4-2-2	H_2O	ruby(347)	45
508	oxazole 4-3-5	EtOH(Ar)	FL	131
508	pyrylium salt	dichloromethane	N_2(337)	36
508 (477–548)	coumarin 515; 30	—	Kr(violet)	52
508 (481–573)	coumarin CSA-28; 500	methanol	Nd:YAG(355)	53
508 (490–521)	imitrine 7-1-133	ethanol	Nd:YAG(355)	133
508–588	coumarin 6-4-10	ethanol	XeCl	24
509–531	oxazole 4-6-8	H_2O/HOAc:95/5	ruby(316)	28
510	oxazole 4-3-5	H_2O(Ar)	FL	131
510 (478–570)	quinoxalinone	ethanol	N_2(337)	71
510 (492–550)	coumarin 515; 30	EG	Ar(458)	48
510–522	oxazole 4-6-7	ethanol	ruby(316)	28
510–527	oxazole 4-6-7	H_2O	ruby(316)	28
510–535	coumarin 6-8-1	p-dioxane	Nd:YAG	136
511 (483–556)	quinoxalinone	ethanol	N_2(337)	71
511 (504–518)	coumarin 6-3-9	ethanol(air)	FL(triaxial)	130
512	coumarin 6-1-4	EtOH/H_2O(air)	FL(triaxial)	38
512	pyrylium salt	dichloromethane	N_2(337)	36
512 (504–519)	coumarin 6-3-9	ethanol(Ar)	FL(triaxial)	130

Table 3.1.1—*continued*
Organic Dye Lasers Arranged in Order of Increasing Wavelength

Wavelength (nm)	Laser dye*	Solvent	Pump	Ref.
512 (505–518)	coumarin 6-3-8	ethanol(air)	FL(triaxial)	130
512 (507–517)	coumarin 6-3-8	ethanol(Ar)	FL(triaxial)	130
512–585	coumarin 6-4-9	ethanol	XeCl	24
513	coumarin C4F; 340	ethanol	FL	61
513	oxazole 4-3-14	H_2O(air)	FL	131
513	thiapyrylium salt	dichloromethane	N_2(337)	36
513 (504–523)	coumarin 6-3-5	EtOH/H_2O(Ar)	FL(triaxial)	130
513 (505–520)	coumarin 6-3-6	EtOH/H_2O(air)	FL(triaxial)	130
514	bimane 5-1-4	H_2O	FL	50
514	coumarin 217	water	FL	55
514	oxazole 4-3-14	EtOH/H_2O(air)	FL	131
514	oxazole 4-3-14	H_2O(Ar)	FL	131
514 (482–552)	coumarin CSA-28; 500	methanol	Nd:YAG(355)	51
514 (505–523)	coumarin 6-3-6	EtOH/H_2O(Ar)	FL(triaxial)	130
514 (505–524)	coumarin 6-3-5	EtOH/H_2O(air)	FL(triaxial)	130
514 (505–524)	coumarin 6-3-7	EtOH/H_2O(air)	FL(triaxial)	130
515	coumarin 6-1-4	MeOH/H_2O(air)	FL(triaxial)	38
515	oxazole 4-3-14	ethanol(air)	FL	131
515	pyrylium salt	dichloromethane	N_2(337)	36
515 (492–545)	coumarin C1F:CSA-27	ethanol	N_2(337)	69
515 (495–545)	coumarin 515; 30	EG	Kr(400-420)	164
515 (502–531)	imitrine 7-1-138	ethanol	Nd:YAG(355)	133
515 (505–535)	imitrine 7-1-132	ethanol	Nd:YAG(355)	133
516	2-pyrazoline	methanol	N_2(337)	36
516	bimane 5-1-2	TFE	FL	137
516	oxazole 4-3-14	EtOH/H_2O(Ar)	FL	131
516 (490–566)	coumarin C1F:CSA-27	ethanol	N_2(337)	56
516 (511–522)	coumarin 6-3-10	ethanol(air)	FL(triaxial)	130
516–550	pyrylium dye 7	CH_3CN	N_2(337)	190
516–550	pyrylium dye 37	CH_3CN	N_2(337)	190
517 (512–521)	coumarin 6-3-10	ethanol(Ar)	FL(triaxial)	130
518	coumarin 521; 334	water	FL	55
518 (497–547)	coumarin 6-5-1	1,2-DCE	XeCl	132
519	coumarin 217523; 337	ethanol	FL	75
519	coumarin C2F; 485	ethanol	FL	66,70
519	oxazole 4-3-13	EtOH/H_2O(Ar)	FL	131
520	benzoxazinone 8-1-3	ethanol	N_2(337)	68
520	coumarin C2F; 485	ethanol	FL	61

Table 3.1.1—*continued*
Organic Dye Lasers Arranged in Order of Increasing Wavelength

Wavelength (nm)	Laser dye*	Solvent	Pump	Ref.
520	oxazole 4-3-13	ethanol(Ar)	FL	131
520	oxazole 4-3-13	H_2O(air)	FL	131
520 (490–562)	coumarin C2F; 485	ethanol	N_2(337)	44
520 (498–556)	coumarin C8F; 522	ethanol	FL	25
520 (506–544)	coumarin 504; 314	MeOH/H_2O:1/1	FL	19
520 (509–537)	benzimidazole 3-1-5	cyclohexane/p	N_2	128
520 (513–527)	coumarin 6-3-9	EtOH/H_2O(air)	FL(triaxial)	130
520 (513–528)	coumarin 6-3-9	EtOH/H_2O(Ar)	FL(triaxial)	130
520–600	fluorescein	ethanol, w, t	N_2(337)	27
521	coumarin 540A, 153	ethanol	FL	61
522	coumarin 355	ethanol	FL	61
522	coumarin 540A, 153	ethanol	FL	61
522	coumarin C340	ethanol	FL	66
522	oxazole 4-3-13	EtOH/H_2O(air)	FL	131
522	oxazole 4-3-14	ethanol(Ar)	FL	131
522	phosphorine	benzene	N_2(337)	36
522	pteridine	water, alkaline	N_2(337)	10
522	pyrylium salt	dichloromethane	N_2(337)	36
522 (500–548)	coumarin CSA-28; 500	MeOH/H_2O:1/1	FL	26
522 (500–572)	coumarin C8F; 522	ethanol	FL	66
522 (518–525)	coumarin 6-3-8	EtOH/H_2O(Ar)	FL(triaxial)	130
522 (537–580)	disodium fluorescein	EG,COT	Ar	47
523	coumarin C2F; 485	methanol	FL	25
523	oxazole 4-3-13	H_2O(Ar)	FL	131
524	oxazole 4-3-13	ethanol(Ar)	FL	131
524	pyrylium salt	dichloromethane	N_2(337)	36
524 (475–542)	benzimidazole 3-1-3	DMF	N_2	128
524 (507–540)	benzimidazole 3-1-3	acetonitrile	N_2	128
524 (509–542)	benzimidazole 3-1-3	cyclohexane/p	N_2	128
524 (515–549)	benzimidazole 3-1-3	p-dioxane	N_2	128
525	bimane 5-1-5	H_2O	FL	50
525 (500–575)	coumarin 535; 7	DPA+LO,COT	Ar(cw,477)	192
525 (502–573)	coumarin C2F; 485	methanol	Nd:YAG(355)	53
526	pyrylium salt	dichloromethane	N_2(337)	36
526 (501–568)	coumarin C8F; 522	methanol+LO	FL	25
526 (520–530)	coumarin 6-3-10	EtOH/H_2O(Ar)	FL(triaxial)	130
526–596	pyrylium dye 22	CH_3CN	N_2(337)	190
527	coumarin CSA-25	ethanol	N_2(337)	44

Table 3.1.1—*continued*
Organic Dye Lasers Arranged in Order of Increasing Wavelength

Wavelength (nm)	Laser dye*	Solvent	Pump	Ref.
527	oxazole 4-3-16	EtOH(Ar)	FL	131
527	pyrylium salt	dichloromethane	N_2(337)	36
528	coumarin 6-7-5	p-dioxane	N_2	139
528 (520–536)	coumarin 6-3-8	EtOH/H_2O(air)	FL(triaxial)	130
529	phosphorine	benzene	N_2(337)	36
529 (511–559)	coumarin 6-5-3	1,2-DCE	XeCl	132
529–536	oxazole 4-6-8	ethanol	ruby(316)	28
530 (514–552)	benzimidazole 3-1-5	p-dioxane	N_2	128
530–570	coumarin 6-8-1	EtOH+HOAc	N_2	136
531	coumarin 531	ethanol	N_2(337)	44
531 (510–556)	coumarin 540; 6	methanol	FL	25
532	coumarin CSA-24	ethanol	N_2(337)	44
532	pyrylium salt	dichloromethane	N_2(337)	36
532–550	coumarin 6–7–5	DMF+HOAc	N_2	139
532-609	pyrylium dye 16	CH_3CN	N_2(337)	190
532–622	pyrylium dye 20	CH_3CN	N_2(337)	190
533	pyrromethene-BF_2 9-1-1	MeOH	FL	141,142
533 (510–558)	coumarin 6-5-2	1,2-DCE	XeCl	132
533 (515–570)	coumarin C8F; 522	DMF+MeOH	FL	19
533 (525–541)	coumarin 6-3-4	ethanol(air)	FL(triaxial)	130
534	pyrylium salt	dichloromethane	N_2(337)	36
534 (525–543)	coumarin 6-3-4	ethanol(Ar)	FL(triaxial)	130
534 (526–541)	coumarin 6-3-2	ethanol(air)	FL(triaxial)	130
535 (500–565)	coumarin 535; 7	EG	Ar(477)	164
535 (505–565)	coumarin 535; 7	EG	Kr(400-420)	164
535 (512–562)	coumarin 6-5-10	ethanol	XeCl	132
536	phosphorine	benzene	N_2(337)	36
536 (515–557)	imitrine 7-1-142	ethanol	Nd:YAG(355)	133
536 (515–583)	coumarin 540A; 153	ethanol	FL	25
536 (517–576)	coumarin 540A; 153	ethanol	N_2(337)	20
536 (522–546)	imitrine 7-1-130	ethanol	Nd:YAG(355)	133
536 (522–548)	imitrine 7-1-125	ethanol	Nd:YAG(355)	133
536 (526–546)	imitrine 7-1-136	ethanol	Nd:YAG(355)	133
537	oxazole 4-3-17	ethanol(Ar)	FL	131
537	oxazole 4-3-17	EtOH/H_2O(Ar)	FL	131
537 (531–544)	coumarin 6-3-2	ethanol(Ar)	FL(triaxial)	130
537–623	pyrylium dye 32	CH_3CN	N_2(337)	190
538	coumarin 540A; 153	ethanol	FL	66

Table 3.1.1—*continued*
Organic Dye Lasers Arranged in Order of Increasing Wavelength

Wavelength (nm)	Laser dye*	Solvent	Pump	Ref.
538 (521–551)	coumarin 540; 6	methanol	FL	59
538–620	pyrylium dye 21	CH_3CN	N_2(337)	190
539	oxazole 4-3-17	ethanol(Ar)	FL	131
539 (530–548)	coumarin 6-3-2	EtOH/H_2O(Ar)	FL(triaxial)	130
540	bimane 5-1-1	H_2O	FL	134
540	coumarin 540A; 153	ethanol	FL	61
540 (515–566)	coumarin 540; 6	EG	Ar(cw)	47
540 (515–585)	coumarin 540; 6	DPA+LO,COT	Ar(488)	76
540 (516–590)	coumarin 540A; 153	methanol	Nd:YAG(355)	51
540 (521–559)	imitrine 7-1-141	ethanol	Nd:YAG(355)	133
540 (537–560)	pyrromethene 567	DMA/MeOH	FL (coaxial)	142
540 (523–557)	imitrine 7-1-135	ethanol	Nd:YAG(355)	133
541	oxazole 4-3-17	ethanol(Ar)	FL	131
541 (520–586)	coumarin 540A; 153	methanol	FL	25
541 (532–549)	coumarin 6-3-10	EtOH/H_2O(air)	FL(triaxial)	130
542 (520–570)	coumarin 6-5-4	1,2-DCE	XeCl	132
542 (523–580)	pyrromethene 546	methanol	FL (triaxial)	143,144
542 (530–554)	imitrine 7-1-123	ethanol	Nd:YAG(355)	133
542 (532–565)	pyrromethene 546	DMA/MeOH	FL (coaxial)	142
542–630	pyrylium dye 19	CH_3CN	N_2(337)	190
543	benzoxazinone 8-1-1	ethanol	N_2(337)	138
543	coumarin 540A; 153	ethanol	FL	70
543–550	coumarin 6-7-5	p-dioxane+HOAc	N_2	139
544	2-pyrazoline	methanol alkaline	N_2(337)	36
544	oxazole 4-3-3	ethanol(air)	FL	131
544 (526–570)	coumarin 540; 6	methanol	FL	26
544 (536–552)	imitrine 7-1-139	ethanol	Nd:YAG(355)	133
545	oxazole 4-3-3	ethanol(Ar)	FL	131
545	pyrylium salt	dichloromethane	N_2(337)	36
545 (537–553)	coumarin 6-3-4	EtOH/H_2O(air)	FL(triaxial)	130
545 (537–553)	coumarin 6-3-4	EtOH/H_2O(Ar)	FL(triaxial)	130
546	pyrromethene-BF_2	ethanol	FL	141
546 (527–583)	pyrromethene 556	EG	Ar (488)	145
546 (536–556)	imitrine 7-1-137	ethanol	Nd:YAG(355)	133
546 (bb)	pyrromethene 546	ethanol	FL	189
547 (523–582)	pyrromethene 556	EG	Ar (699-1488)	23
547 (540–555)	coumarin 6-3-3	ethanol(Ar)	FL(triaxial)	130
547 (541–554)	coumarin 6-3-3	ethanol(air)	FL(triaxial)	130

Table 3.1.1—*continued*
Organic Dye Lasers Arranged in Order of Increasing Wavelength

Wavelength (nm)	Laser dye*	Solvent	Pump	Ref.
548 (537–605)	pyrromethene 556	methanol	FL (triaxial)	143
548 (542–554)	coumarin 6-3-2	EtOH/H_2O(air)	FL(triaxial)	130
548–652	dibenzoxanthylium	TFE	Cu	121
549–630	pyrylium dye 39	CH_3CN	N_2(337)	190
550	pyrylium salt	dichloromethane	N_2(337)	36
550	Rhodamine 6G; 590	methanol	Nd:YAG(532)	65
550 (527–584)	pyrromethene 556	EG	Ar (514.5)	145
550–580	coumarin 6-8-2	p-dioxane	Nd:YAG	136
550–580	fluorol 555; 7GA	—	FL	77
551	pyrromethene-BF_2 9-1-3	methanol	FL	141
551 (545–559)	xanthene 10-1-2	H_2OMeOH:1/2	Nd:YAG(532)	147
552	fluorol 555; 7GA	MeOH/LO	FL	78
552 (536–602)	rhod. 560 chloride; 110	EG	Ar(cw)	47
552 (538–573)	disodium fluorescein	EG	Ar(458,514)	164
552 (543–562)	coumarin 6-3-3	EtOH/H_2O(Ar)	FL(triaxial)	130
552 (545–586)	pyrromethene 580	methanol	Nd:YAG(532)	122
552 (545–585)	pyrromethene 580	ethanol	Nd:YAG (532)	148
553 (530–624)	pyrromethene 556	EG	Ar (458-514)	149
553–655	pyrylium dye 47	CH_3CN	N_2(337)	190
554	benzoxazinone 8-1-6	ethanol	N_2(337)	138
555	fluorol 555; 7GA	alcohol or water	FL	69
555 (545–585)	pyrromethene 556	DMA/MeOH	FL (coaxial)	142
555 (547–564)	coumarin 6-3-3	EtOH/H_2O(air)	FL(triaxial)	130
556	pyrromethene-BF_2(5)	methanol	FL	123
556–623	pyrylium dye 41	CH_3CN	N_2(337)	190
556–629	pyrylium dye 40	CH_3CN	N_2(337)	190
558–603	pyrylium dye 43	CH_3CN	N_2(337)	190
558–632	pyrylium dye 35	CH_3CN	N_2(337)	190
559	oxazole 4-3-11	ethanol(air)	FL	131
559	oxazole 4-3-2	2-PrOH	FL	131
559	pyrromethene-BF_2(6)	ethanol	FL	123
559 (548–588)	xanthene 10-1-1	H_2OMeOH:1/2	Nd:YAG(532)	147
559–582	oxazole 4-1-3	2-PrOH(Ar)	FL	129
560	coumarin 6-7-5	DMF+HOAc	N_2	139
560	pyrylium salt	dichloromethane	N_2(337)	36
560	rhodamine 560 chloride	ethanol	FL	80
560 (510–570)	coumarin 540; 6	EG/bz alcohol	Ar(488)	76
560 (530–623)	pyrromethene 556(9-1-3)	EG	Ar(458-524,cw)	124

Table 3.1.1—*continued*
Organic Dye Lasers Arranged in Order of Increasing Wavelength

Wavelength (nm)	Laser dye*	Solvent	Pump	Ref.
560 (543–584)	pyrromethene 567	PPH	Ar (514.5)	145
560 (548–580)	rhodamine 6G; 590	methanol	Nd:YAG(532)	53
560 (551–570)	xanthene 10-1-3	H_2OMeOH:1/2	Nd:YAG(532)	147
560–583	oxazole 4-1-3	ethanol	FL	129
560–598	oxazole 4-6-12	ethanol	ruby(316)	28
561 (540–580)	pyrromethene 556	methanol	FL	144-46
562 (508–573	brilliant sulfaflavine	COT	FL	81,82
562 (520–575)	coumarin 540A; 153	ethanol	Ar(476)	76
562 (522–618)	brilliant sulfaflavine	methanol	FL	26
562 (546–592)	rhodamine 6G; 590	methanol	Nd:YAG(532)	83
562–595	oxazole 4-6-4	H_2O/HOAc:4/6	ruby(316)	28
563	benzoxazinone 8-1-2	ethanol	N_2(337)	138
563 (541–583)	rhodamine 560 chloride	methanol	FL	26
563 (550–590)	rhodamine 6G; 590	methanol	Nd:YAG(532)	84
563–580	xanthene 10-1-6	H_2OMeOH:1/2	Nd:YAG(532)	147
563–600	oxazole 4-6-12	H_2O	ruby(316)	28
564	rhodamine 6G; 590	methanol	Nd:YAG(532)	51
564–633	pyrylium dye 48	CH_3CN	N_2(337)	190
565	oxazole 4-5-9	methanol	FL(coaxial)	150
565 (544–589)	rhodamine 560 chloride	methanol	FL	25
565 (550–595)	Pyrromethene 567	PPH	Nd:YAG(532)	122
565 (551–579)	imitrine 7-1-127	DEE	Nd:YAG(355)	133
565–610	coumarin 6-8-2	EtOH+HOAc	N_2	136
565–620	Rhodamine 6G	ethanol, w, t	N_2(337)	27
566	benzoxazinone 8-1-5	ethanol	N_2(337)	138
566	oxazole 4-3-11	ethanol(Ar)	FL	131
566	pyrromethene-BF_2(3)	ethanol	FL	123
566 (549–592)	pyrromethene 567	PPH	Nd:YAG (532)	122
567	oxazole 4-3-2	ethanol(Ar)	FL	131
567	pyrromethene 567	ethanol	FL	123
567	pyrromethene-BF_2(4)	methanol	FL	123
567 (546–587)	rhodamine 560 chloride	ethanol	FL	25
567–587	oxazole 4-1-3	ethanol(Ar)	FL	129
568	oxazole 4-5-8	MeOH	FL(coaxial)	150
568 (550–608)	pyrromethene 567	PPH	Ar(458-524,cw)	124
568 (559–580)	xanthene 10-1-4	H_2O/MeOH:1/2	Nd:YAG(532)	70
568–598	oxazole 4-6-11	H_2O	ruby(316)	28
568–598	oxazole 4-6-11	H_2O/HOAc:1/1	ruby(316)	28

Table 3.1.1—*continued*
Organic Dye Lasers Arranged in Order of Increasing Wavelength

Wavelength (nm)	Laser dye*	Solvent	Pump	Ref.
568–598	oxazole 4-6-11	H_2O/HOAc:95/5	ruby(316)	28
569	oxazole 4-5-1	methanol	FL(coaxial)	150
569	oxazole 4-5-2	methanol	FL(coaxial)	150
569	oxazole 4-5-3	methanol	FL(coaxial)	150
569	oxazole 4-5-4	methanol	FL(coaxial)	150
569	oxazole 4-5-5	methanol	FL(coaxial)	150
569 (533–600)	rhodamine 560 chloride	—	Kr(blue/green)	52
569 (545–583)	pyrromethene 580	ethanol	Nd:YAG (532)	148
570	benzoxazinone 8-1-4	ethanol	N_2(337)	68
570	oxazole 4-3-2	EtOH/H_2O(air)	FL	131
570	pyrromethene 567	methanol	FL	144
570	rhodamine 560 chloride	ethanol	FL	80
570 (540–590)	fluorol 555; 7GA	ethanol	FL	25
570 (540–600)	rhodamine 560 chloride	EG	Ar(458,514)	164
570 (553–587)	imitrine 7-1-127	ethanol	Nd:YAG(355)	133
570 (bb)	pyrromethene 580	ethanol	Nd:YAG (532)	153
570–600	oxazole 4-6-13	H_2O	ruby(316)	28
571	oxazole 4-3-2	ethanol(Ar)	FL	131
571	pyrromethene 597	ethanol	Nd:YAG (532)	148
571	pyrromethene-BF_2(2)	ethanol	FL	125
571 (552–608)	pyrromethene 567	NMP/PPH	Ar (all lines)	149
571 (560–600)	pyrromethene 597	ethanol	Nd:YAG(532)	122
571–588	oxazole 4-1-3	methanol(Ar)	FL	129
571–591	oxazole 4-1-3	H_2O(Ar)	FL	129
572	oxazole 4-5-7	MeOH	FL(coaxial)	150
572 (560–584)	imitrine 7-1-134	ethanol	Nd:YAG(355)	133
572 (564–600)	rhodamine 6G; 590	ethanol	Cu(511,578)	85
573	oxazole 4-3-22	ethanol(air)	FL	131
573	oxazole 4-5-6	methanol	FL(coaxial)	150
573–682	dibenzoxanthylium	CH_3CN	Cu	121
574	oxazole 4-3-22	ethanol(Ar)	FL	131
574	xanthene 10-1-5	H_2OMeOH:1/2	Nd:YAG(532)	147
574 (535–590)	fluorol 555; 7GA	methanol	FL	77
574 (542–592)	fluorol 555; 7GA	methanol	FL	25
575	rhodamine 575	ethanol	FL	80
575 (555–592)	pyrromethene 580	PPH	Ar (all lines)	154
575 (569–585)	perylene 11-1-1	DMF	Nd:YAG(532)	155
576 (550–592)	imitrine 7-1-129	ethanol	Nd:YAG(355)	133

Table 3.1.1—*continued*
Organic Dye Lasers Arranged in Order of Increasing Wavelength

Wavelength (nm)	Laser dye*	Solvent	Pump	Ref.
576 (555–618)	rhodamine 6G; 590	ethanol	N_2(337)	86
577 (563–602)	rhodamine 575	ethanol	FL	25
577 (566–600)	perylene BASF-241	chloroform	Nd:YAG(532)	156
578	oxazole 4-5-9	H_2O	FL(coaxial)	150
578 (565–612)	rhodamine 6G; 590	methanol	FL	25
579	oxazole 4-3-12	ethanol(Ar)	FL	131
579 (568–605)	rhodamine 6G; 590	ethanol	N_2(337)	20
579 (570–596)	rhodamine B; 610	—	Nd:YAG(532)	83
580	oxazole 4-5-8	H_2O	FL(coaxial)	150
580	rhodamine 6G; 590	ethanol	KrF(248)	3
580 (bb)	pyrromethene 580	ethanol	FL	157
581	oxazole 4-5-1	methanol	FL(coaxial)	150
581	oxazole 4-5-7	H_2O	FL(coaxial)	150
581–598	oxazole 4-6-13	ethanol	ruby(316)	28
582	oxazole 4-3-12	EtOH/H_2O(Ar)	FL	131
582	oxazole 4-3-12	H_2O(Ar)	FL	131
582	oxazole 4-5-6	H_2O	FL(coaxial)	150
582 (561–593)	imitrine 7-1-128	ethanol	Nd:YAG(355)	133
583 (570–604)	Kiton Red 620; sulforhodamine B	methanol	Nd:YAG(532)	53
584 (570–618)	rhodamine 6G; 590	ethanol	FL	25
585	rhodamine 575	ethanol	FL	80
586 (563–625)	rhodamine 6G; 590	methanol	FL	59
586 (570–606)	rhodamine B; 610	—	Nd:YAG(532)	53
587 (579–601)	rhodamine B; 610	methanol	Nd:YAG(532)	65
587 (bb)	pyrromethene 597	ethanol	Nd:YAG (532)	153
590	rhodamine 6G; 590	methanol	FL	62
590	rhodamine 6G; 590	p-dioxane	KrF(248)	36
590 (566–610)	rhodamine 575	MeOH/H_2O	FL	19
590 (570–650)	rhodamine 6G; 590	EG	Ar(458,514)	164
590 (578–610)	rhodamine B; 610	—	Nd:YAG(532)	84
590–614	oxazole 4-6-9	H_2O/HOAc:1/1	ruby(316)	28
590–614	oxazole 4-6-9	H_2O/HOAc:95/5	ruby(316)	28
590–620	oxazole 4-6-10	H_2O/HOAc:1/1	ruby(316)	28
590–620	oxazole 4-6-10	H_2O/HOAc:95/5	ruby(316)	28
590–625	oxazole 4-6-3	ethanol	ruby(316)	28
590–640	sulfohodamine 640; 101	ethanol	Nd:YAG(532)	29
590–616	coumarin 6-9-1	ethanol	L(520)	158

Table 3.1.1—*continued*
Organic Dye Lasers Arranged in Order of Increasing Wavelength

Wavelength (nm)	Laser dye*	Solvent	Pump	Ref.
591 (582–618)	rhodamine B; 610	TFE	Cu(511,578)	85
591–614	oxazole 4-6-10	H_2O	ruby(316)	28
592 (578–629)	rhodamine B; 610	methanol	Nd:YAG(532)	51
593	oxazole 4-3-7	ethanol(Ar)	FL	131
593	pyrromethene-BF_2(3)	ethanol	FL	125
593 (bb)	pyrromethene 597	p-dioxane	FL	159
594–612	oxazole 4-6-9	H_2O	ruby(316)	28
595–617	coumarin 6-9-1	ethanol	L(520)	158
596 (577–614)	rhodamine 6G; 590	MeOH/H_2O:1/3	FL	26
597	oxazole 4-3-9	ethanol(Ar)	FL	131
597 (bb)	pyrromethene 597	ethanol	FL	157
~598	MEH-PPV	xyl./chloroform	Nd:YAG(532)	126
598 (577–625)	rhodamine 6G; 590	MeOH/H_2O:1/1	FL	19
599–635	ASPI	ethanol	Nd:YAG(532)	33
600	rhodamine 6G; 590	4%LO/H_2O	FL	62
600 (567–657)	rhodamine 6G; 590	EG	Ar(cw)	47
600–700	DCM-OH	methanol	FL	127
601	thiapyrylium salt	dichloromethane	N_2(337)	36
602	pyrromethene-BF_2(4)	ethanol	FL	125
602 (560–654)	rhodamine 6G; 590	—	Kr(blue/green)	53
602 (589–623)	rh. 640 perchlorate: 101	—	Nd:YAG(532)	83
602 (592–624)	rh. 640 perchlorate: 101	methanol	Nd:YAG(532)	53
603 (580–670)	coumarin 6-5-5	1,2-DCE	XeCl	132
605	coumarin 6-9-2	methanol	L(520)	158
605 (bb)	pyrromethene 605	ethanol	FL	157
607	coumarin 6-9-2	CH_3CN	L(520)	158
608	coumarin 6-9-2	ethanol	L(520)	158
609	coumarin 6-9-1	benzene	L(520)	158
609 (594–643)	rhodamine B; 610	—	N_2(337)	20
610 (585–633)	rhodamine 6G; 590	4%LO/H_2O	FL	19
610 (596–632)	coumarin 6-5-7	1,2-DCE	XeCl	132
610–695	DCM-OH	methanol/H_2O	FL	127
611	rh. 640 perchlorate; 101	methanol	Nd:YAG(532)	65
612	coumarin 6-9-1	CH_3CN	L(520)	158
612	pyrromethene-BF_2(5)	ethanol	FL	125
612	thiapyrylium salt	dichloromethane	N_2(337)	36
612 (598–640)	rh. 640 perchlorate; 101	—	Nd:YAG(532)	84
612 (604–630)	pyrromethene 650	—	Nd:YAG (532)	160

Table 3.1.1—*continued*
Organic Dye Lasers Arranged in Order of Increasing Wavelength

Wavelength (nm)	Laser dye*	Solvent	Pump	Ref.
612 (606–618)	imitrine 7-1-140	ethanol	Nd:YAG(355)	133
613 (596–645)	rhodamine B; 610	methanol	FL	25
613 (602–657)	rh. 640 perchlorate; 101	methanol	Nd:YAG(532)	51
615–628	coumarin 6-9-1	ethanol	FL	158
617 (595–639)	Kiton Red 620; sulforhodamine B	TFE	Cu(511,578)	85
617 (598–647)	rhodamine B; 610	ethanol	FL	87
618	thiapyrylium salt	dichloromethane	N_2(337)	36
620	coumarin 6-9-1	EtOH+HCl	L(520)	158
620	Kiton Red 620; sulforhodamine B	methanol	FL	88
620 (580–630)	Kiton Red 620; sulforhodamine B	ethanol	FL	87
620 (596–647)	rhodamine B; 610	ethanol	FL	25
621 (608–634)	Kiton Red 620; sulforhodamine B	methanol+COT	FL	26
623	pyrylium salt	dichloromethane	N_2(337)	36
623 (598–649)	Kiton Red 620; sulforhodamine B	ethanol+COT	FL	25
626	coumarin 6-9-2	benzene	L(520)	158
627	pyrylium salt	dichloromethane	N_2(337)	36
627 (595–629)	Kiton Red 620; sulforhodamine B	methanol+COT	FL	25
628 (603–647)	Kiton Red 620; sulforhodamine B	TFE	N_2(337)	86
630	rhodamine 640 perchlorate; 101	ethanol	FL	80
630 (601–675)	rhodamine B; 610	EG	Ar(458,514)	164
631	pyrylium salt	dichloromethane	N_2(337)	36
631 (600–660)	Kiton Red 620; sulforhodamine B	methanol	FL	59,87
631 (bb)	pyrromethene 650	xylene	Nd:YAG (532)	153
633	DODC-I	methanol	Nd:YAG(532)	89
633 (615–655)	cresyl violet 670		Nd:YAG(532)	83
635	oxazole 4-3-18	EtOH(Ar)	FL	131
636 (603–670)	Kiton Red 620; sulforhodamine B	EG	FL	16

Table 3.1.1—*continued*
Organic Dye Lasers Arranged in Order of Increasing Wavelength

Wavelength (nm)	Laser dye*	Solvent	Pump	Ref.
637	Kiton Red 620; sulforhodamine B	DMSO	FL	86
637 (608–682)	rhodamine B; 610	EG	Ar(cw)	47
637 (620–660)	cresyl violet 670	methanol	Nd:YAG(532)	53
637–650	coumarin 6-9-2	ethanol	FL	158
638 (610–670)	Kiton Red 620; sulforhodamine B	EG	Ar(cw)	47
639 (620–670)	cresyl violet 670	—	Nd:YAG(532)	84
639 (645–705)	cresyl violet 670	methanol	FL	81
640	rh. 640 perchlorate; 101	ethanol	FL	5
640 (610–680)	DCM	methanol	Nd:YAG(532)	53
640 (605–680)	DCM	BzOH/EG	Ar(514)	90
640 (620–670)	cresyl violet 670	MeOH/H_2O	Nd:YAG(532)	91
640 (620–680)	rh. 640 perchlorate; 101	ethanol	N_2(337)	92
642 (622–665)	Kiton Red 620; sulforhodamine B	4%LO/H_2O	FL	19
642 (627–657)	rh. 640 perchlorate; 101	methanol	FL	26
643 (623–657)	rh. 640 perchlorate; 101	ethanol	FL	25
644 (620–673)	rh. 640 perchlorate; 101	ethanol	N_2(337)	93
645 (620–690)	rh. 640 perchlorate; 101	EG	Ar(458514)	164
645–705	cresyl violet 670	methanol	FL	81
646 (6250660)	cresyl violet 670	methanol	Nd:YAG(532)	91
647	cresyl violet 670	—	Nd:YAG(532)	91
648 (608–710)	rh. 640 perchlorate; 101	—	Kr(568)	52
649	DCM	DMF	FL	86
649 (615–688)	DCM	DMSO	Cu(511,578)	85
650	rh. 640 perchlorate; 101	methanol	FL	62
652 (620–687)	rh. 640 perchlorate; 101	MeOH/H_2O:1/1	FL	19
654 (601–716)	DCM	DMSO	N_2(337)	86
654–669	coumarin 6-9-2	EtOH+HCl	FL	158
655	DCM	DMSO	FL	86
655 (646–697)	cresyl violet 670	methanol	FL	25
656	coumarin 6-9-2	EtOH+HCl	L(520)	158
656	sulfohodamine 640; 101	MeOH/H_2O:1/1	cw	71
659 (650–695)	cresyl violet 670	ethanol	FL	25
660	LD 690 perchlorate	methanol	Nd:YAG(532)	94
660 (641–687)	cresyl violet 670	ethanol	N_2(337)	20
662 (600–706)	coumarin 6-5-6	1,2-DCE	XeCl	132

Table 3.1.1—*continued*
Organic Dye Lasers Arranged in Order of Increasing Wavelength

Wavelength (nm)	Laser dye*	Solvent	Pump	Ref.
662 (648–682)	sulfohodamine 640; 101	MeOH/H_2O:1/1	FL	26
664 (631–705)	cresyl violet 670	methanol	FL	59
668	benzoxazinone 8-1-7	ethanol	N_2(337)	138
668 (646–680)	sulfohodamine 640; 101	EG	Ar	90
668 (649–700)	oxazine 720; 170	methanol	Nd:YAG(532)	53
668 (655–705)	LD 690 perchlorate	ethanol	Nd:YAG(532)	93
670	pyrylium salt	dichloromethane	N_2(337)	36
670 (660–716)	LD 690 perchlorate	DMSO/EtOH:2/1	N_2(337)	93
671 (613–708)	oxazine 720; 170	methanol	Nd:YAG(532)	51
671 (634–704)	rh. 640 perchlorate: 101	DMSO+HCl	N_2(337)	86
672	oxazine 720; 170	ethanol	Nd:YAG(532)	81
673 (650–696)	cresyl violet 670	EG	Ar(cw)	47
674 (641–712)	coumarin 6-5-7	1,2-DCE	XeCl	132
675	HIDC-I	methanol	Nd:YAG→585	95
675 (628–712)	coumarin 6-5-8	1,2-DCE	XeCl	132
681	oxazine 725; 1	CH_2Cl_2	FL→R610	96
681 (662–710)	Nile Blue 690; A	—	Nd:YAG(532)	84
683	Nile Blue 690; A	methanol	Nd:YAG(532)	97
684 (661–724)	LDS 698/pyridine 1	methanol	Cu(511,578)	161
687–826	oxazine 725; 1	DMSO/EG/COT	Kr(647)	98
690	LD 700 perchlorate	alcohol	Nd:YAG→585	96
690	Nile Blue 690; A	ethanol	laser	80
690	oxazine 725; 1	CH_2Cl_2	Nd:YAG(532)	97
690 (684–736)	carbazine 720; 122	ethanol	Nd:YAG(532)	53
692 (676–698)	oxazine 720; 170	methanol	FL	25
695	oxazine 725; 1	DMSO	FL→R610	96
695 (658–738)	LDS 698/pyridine 1	methanol	Nd:YAG(532)	26
695 (675–708)	cresyl violet 670	EG	Ar(458-514)	164
696 (683–710)	Nile Blue 690; A	ethanol	N_2(337)	20
696–780	LD 690 perchlorate	EG	Kr(cw)	99
698 (682–720)	oxazine 720; 170	methanol	FL	26
699 (675–711)	oxazine 720; 170	ethanol	FL	25
699 (680–740)	carbazine 720; 122	ethanol	Nd:YAG(532)	84
700	carbazine 720; 122	ethanol	Nd:YAG(532)	97
700 (680–738)	carbazine 720; 122	methanol,TEA	FL	73,100
705	Nile Blue 690; A	methanol	FL	25
705 (675–730)	oxazine 720; 170	MeOH/H_2O	FL	19
705–735	DTDC-1	DMSO	ruby(694.3)	89

Table 3.1.1—*continued*
Organic Dye Lasers Arranged in Order of Increasing Wavelength

Wavelength (nm)	Laser dye*	Solvent	Pump	Ref.
706 (692–752)	LD 700 perchlorate	ethanol	N_2(337)	93
710 (670–760)	LDS 698/pyridine 1	DMSO	XeCl(308)	163
710 (690–740)	oxazine 720; 170	methanol	FL	70
710–755	Dmo-DTDC-1	DMSO	ruby(694.3)	89
711	DmOTC-1	acetone	ruby(694.3)	101
711	DTDC-1	acetone	ruby(694.3)	101
714	10-Cl-5,6-Dbz-DTDC-1	acetone	ruby(694.3)	101
715	oxazine 725; 1	ethanol	FL	80
716 (692–743)	LDS 730/Styryl 6	methanol	Nd:YAG(532)	164
717 (689–750)	Nile Blue 690; A	methanol	FL	59
718	carbazine 720; 122	DMSO	Nd:YAG(532)	102
718 (675–750)	LDS 698/pyridine 1	DMSO	N_2(337)	165
720	carbazine 720; 122	ethanol	FL	102
720	HIDC-1	methanol	Nd:YAG→585	95
720 (680–760)	carbazine 720; 122	ammonyx TEA	Nd:YAG(532)	91
720 (698–758)	LD 700 perchlorate	ethanol	N_2(337)	93
720–775	DTDC-1	DMSO	Kr	103
722	Nile Blue 690; A	methanol	FL	87
722 (685–760)	LDS 722/pyridine 2	methanol	Nd:YAG(532)	164
722 (687–755)	LDS 722/pyridine 2	methanol	Cu(511,578)	161
722 (698–743)	LDS 750/styryl 7	methanol	Nd:YAG(532)	164
722 (704–786)	oxazine 750 perchlorate	methanol	Nd:YAG(532)	51
723 (688–800	oxazine 725; 1	—	Kr(red)	52
724 (695–761)	oxazine 725; 1	methanol	Nd:YAG(532)	51
725 (705–745)	oxazine 725; 1	MeOH/R590	FL	59
725 (705–750)	oxazine 725; 1	ethanol	N_2(337)	20
725–765	DOTC-I (DEOTC)	DMSO	ruby(694.3)	96
725–780	DmOTC-I	DMSO	ruby(694.3)	89
726 (688–808)	LDS 698/pyridine 1	PC/EG:15/85	Ar(458–514)	164
730 (692–782)	Nile Blue 690; A	EG	Kr(cw)	47
732	DOTC-1 (DEOTC)	MeOH/N_2/COT	FL→R610	95
734 (683–776)	carbazine 720; 122	EG(ethanol)	Kr(SF)	102
735 (700–780)	LDS 722/pyridine 2	DMSO	N_2(337)	166
737 (700–810)	LD 700 perchlorate	EG	Kr(SF)	13
740	carbazine 720; 122	DMSO	FL	73,100
740	DOTC-I (DEOTC)	DMSO	Nd:YAG→585	95
740 (700–820)	LD 700 perchlorate	EG	Kr(647,676)	13
740 (720–758)	oxazine 725; 1	CH_2Cl_2	FL	25

Table 3.1.1—*continued*
Organic Dye Lasers Arranged in Order of Increasing Wavelength

Wavelength (nm)	Laser dye*	Solvent	Pump	Ref.
740–770	D-2-QDC-1	EG	ruby(694.3)	89
742	DOTC-I (DEOTC)	acetone	ruby(694.3)	101
742–874	DOTC-I (DEOTC)	EG/DMSO:3/1	Kr(647)	98
744	DTDC-I	DMSO	N_2(337)	104
745	11-Br-Dm-2-QDC-I	glycerin	ruby(694.3)	101
745	DOTC-I (DEOTC)	DMSO	FL→R610	96
745 (645–810)	oxazine 725; 1	DMSO/EGor G	Kr(647,676)	103
745 (700–785)	oxazine 750 perchlorate	ethanol	FL→R640	25
746	DTDC-P	DMSO	N_2(337)	104
747 (682–810)	LDS 722/pyridine 2	PC/EG	Ar(458-514)	167
747 (687–811)	carbazine 720; 122	EG(ethanol)	Kr(647,676)	102
749	Dm-4-QC-I	glycerin	ruby(694.3)	101
750 (695–801)	oxazine 725; 1	DMSO/EG	Kr(647,676)	164
750 (710–790)	Nile Blue 690; A	EG	R590(Ar)	164
750 (714–790)	LDS 751/styryl 8	methanol	Nd:YAG(532)	164
750–810	DmOTC-I	DMSO	ruby(694.3)	48
750–825	DOTC-I (DEOTC)	methanol	FL→R640	25
751	Dm-4-QC-I-Br	glycerin	ruby(694.3)	101
754	Dm-4-QC-I-Br	glycerin	ruby(694.3)	101
754	DOTC-I(DEOTC)	DMSO	Nd:YAG(532)	105
756 (736–793)	DOTC-I(DEOTC)	DMSO	Nd:YAG(532)	51
760–775	oxazine 750 perchlorate	DMSO	Nd:YAG→585	95
764 (738–800)	LDS 765	methanol	Nd:YAG(532)	164
765 (715–840)	LDS 751/styryl 8	PC/EG	Ar(458–514)	164
767 (743–787)	LDS 765	methanol	Cu(511,578)	161
770	DOTC-I (DEOTC)	DMSO	Nd:YAG→585	95
770 (750–835)	oxazine 750 perchlorate	EG/DMSO:4/1	Kr(647)	106
770–830	HITC-I	DMSO	FL→R620	88
774	DTTC-I	acetone	ruby(694.3)	101
775 (747–885)	oxazine 750 perchlorate	EG/DMSO:84/16	Kr(647,676)	164
776 (747–801)	oxazine 750 perchlorate	EG/DMSO:2/1	Kr(647,676)	107
780 (749–825)	oxazine 750 perchlorate	EG	Kr(647,676)	107
780–883	HITC-I	DMSO	ruby(694.3)	28
782	DOTC-I (DEOTC)	DMSO	N_2(337)	104
783 (750–833)	DOTC-I (DEOTC)	EG	Kr(647,676)	108
785	DTTC-I	DMSO	N_2(337)	104
788	DmOTC-I	DMSO	N_2(337)	104
790–840	HITC-I	DMSO	ruby(694.3)	89

Table 3.1.1—*continued*
Organic Dye Lasers Arranged in Order of Increasing Wavelength

Wavelength (nm)	Laser dye*	Solvent	Pump	Ref.
790–871	DTTC-I	DMSO	ruby(694.3)	28
795 (765–875)	DOTC-P (DEOTC)	DMSO+EG	Kr(647,676)	164
795 (768–850)	LDS 798/styryl 11	PC/EG:15/85	Ar	168
795–815	IR-123-I (DTTC)-I	DMSO	ruby(694.3)	89
798 (765–845)	LDS 798/styryl 11	methanol	Nd:YAG(532)	164
~800 (750–864)	DmOTC-I	DMSO+EG	Kr(647,676)	103
800 (755–870)	DOTC-I (DEOTC)	DMSO+EG	Kr(647,676)	103
800–870	IR-144 (HITC type)	EG	FL→R640	109
800–882	HITC-I	EG	FL→R640	25
805–872	IR-140-P(DTTC)	ethanol	FL→Ox720	25
806 (788–832)	HITC-I	DMSO	Nd:YAG(532)	51
807 (784–830)	HITCI	ethanol	Rh6G(587)	188
808 (756–871)	DOTC-I (DEOTC)	EG	Kr(red)	52
810–830	DTTC-Br	DMSO	ruby(694.3)	89
815	11-Br-D-2-QDC-I	glycerin	ruby(694.3)	101
815 (793–845)	LDS 821/styryl 9**	methanol	Cu(511,578)	161
815–870	DTTC-I	EG	FL→R640	25
818 (785–850)	LDS 821/styryl 9**	methanol	XeCI(308)	169
818 (785–851)	LDS 821/styryl 9**	methanol	Nd:YAG(532)	164
819	HITC-I	acetone	ruby(694.3)	101
820–875	5-Temo-DTTC-I	DMSO	ruby(694.3)	89
820–900	DTTC-I	DMSO	FL→KR620	88
821 (802–852)	LDS 821/styryl 9**	PC	N_2(337)	170
822	HITC-I	DMSO	Nd:YAG→700	110
828 (813–859)	DTTC-I	DMSO	Nd:YAG(532)	51
829	DTTC-I	acetone	ruby(694.3)	101
830	11-Br-Dm-4-QDC-I	methanol	ruby(694.3)	101
830	IR-123-I(DTTC)-I	DMSO	FL	109
832–911	HITC-I	EG/DMSO:3/1	Kr(647)	98
834	DTTC-I	ethanol	Nd:YAG→700	110
834 (817–842)	LDS 821/styryl 9**	methanol	FL	26
834–892	IR-144 (HITC type)	EG/DMSO	Kr(752,799)	111
834–900	4,5-Dbz-DTTC-I	DMSO	ruby(694.3)	28
835-890	IR-144 (HITC type)	DMSO	ruby(694.3)	98
836	HITC-I	EG	ruby(694.3)	112
839 (826–850)	HIDC-I	DMSO	N_2(337)	20
840–870	DTTC-I	DMSO	Kr	59
840–920	IR-125-I(HITC type)	DMSO	ruby(694.3)	89

Table 3.1.1—*continued*
Organic Dye Lasers Arranged in Order of Increasing Wavelength

Wavelength (nm)	Laser dye*	Solvent	Pump	Ref.
845 (780–960)	LDS 821/styryl 9**	PC/EG	Ar(458–514)	164
845–920	D-4-QDC-I	DMSO	ruby(694.3)	89
849	HITC-I	DMSO	Nd:YAG(532)	111
850–930	IR-140-P(DTTC)	DMSO	ruby(694.3)	89
853	5,6-Temo-DTTC-I	acetone	ruby(694.3)	101
855-885	5,6-Temo-DTTC-I	DMSO	ruby(694.3)	89
860	4,5-Dbz-DTTC-I	acetone	ruby(694.3)	101
862	HITC-I	DMSO	N_2(337)	104
862 (851–890)	LDS 867	methanol	Cu(511,578)	161
863 (816–855)	DTTC-I	DMSO	ruby(694.3)	89
863 (844–885)	IR-144 (HITC type)	DMSO	Nd:YAG→700	25,110
863 (846–907)	IR-125-I(HITC type)	DMSO	Nd:YAG	51
863–1048	IR-132-P(DTTC)	EG/DMSO:3/1	Kr(752)	113
865 (825–912)	HITC-I	—	Kr(red)	52
865–920	D-2-QTC-I-I	DMSO	ruby(694.3)	28
866 (830–910)	LDS 867	methanol	Nd:YAG(532)	164
869	IR-144 (HITC type)	DMSO	Nd:YAG(532)	105
869 (832–888)	HITC-I	EG	Kr(752,799)	108
870 (812–929)	HITC-P	DMSO+EG	Kr(752,799)	103
870 (828–909)	HITC-P	DMSO+EG	Kr(647,676)	103
873 (819–937)	HITC-I	EG/DMSO:3/1	Kr(752)	113
874	IR-144 (HITC type)	DMSO	Nd:YAG(532)	114
875	IR-109-I(DTTC)	DMSO	FL	109
875 (840–940)	HITC-P	EG/DMSO:84/16	Kr(647,676)	164
875–916	IR-140-P(DTTC)	DMSO	FL→Ox720	25
875–920	IR-132-P(DTTC)	DMSO	ruby(694.3)	89
876	DTTC-I	DMSO	N_2(337)	104
880 (793–923)	LDS 821/styryl 9**	PC/EG	Kr(647)	171-2
883	IR-139-P(DTTC)	DMSO	FL	109
885	IR-116-I(DTTC)	DMSO	FL	109
887–986	IR-140-P(DTTC)	EG/DMSO:3/1	Kr(752,799)	98
888	IR-134-P(DTTC)	DMSO	FL	109
889	DTTC-I	DMSO	FL	151,152
893 (882–913)	IR-140-P(DTTC)	DMSO	Nd:YAG(532)	51
898	D-2-QTC-I-I	acetone	ruby(694.3)	101
898	IR-140-P(DTTC)	DMSO	Nd:YAG→700	101
903	IR-125-I(HITC type)	DMSO	Nd:YAG→700	101
908	IR-140-P(DTTC)	DMSO	Nd:YAG(532)	105

Table 3.1.1—*continued*
Organic Dye Lasers Arranged in Order of Increasing Wavelength

Wavelength (nm)	Laser dye*	Solvent	Pump	Ref.
910	IR-132-P(DTTC)	DMSO	Nd:YAG(532)	111
910	IR-140-P(DTTC)	DMSO	Nd:YAG(532)	111
913	IR-125-I(HITC type)	DMSO	Nd:YAG(532)	111
916–984	IR-132-P(DTTC)	EG/DMSO:3/1	Kr(752,799)	98
920 (880–965)	6,7-DBZ-HITC-P	DMSO+EG	Kr(752,799)	103
920–950	12A-DTTeC-P	DMSO	ruby(694.3)	89
925 (875–1050)	LDS 925/styryl 13	PC/EG	Ar	173
927 (855–1032)	IR-137-P(DTTC)	EG/DMSO:3/1	Kr(752)	113
927 (858–1030)	IR-140-P(DTTC)	EG/DMSO:3/1	Kr(752)	113
928	4,5-Dbz-DTTC-I	DMSO	FL	104
935–1019	12A-DTTeC-P	EG/DMSO:1/1	Kr(752,799)	115
940	IR-125-I(HITC type)	DMSO	FL	109
946	IR-141-I(DTTC)	DMSO	FL	109
946 (922–963)	LDS 867	DMSO	FL	26
949–880	IR-144 (HITC type)	DMSO	FL	89
950	IR-137-P(DTTC)	DMSO	FL	109
950	IR-140-P(DTTC)	DMSO	FL	109
950 (862–1013)	IR-140-P(DTTC)	EG/DMSO:1/1	Kr(752,799)	115
960 (902–1023)	LDS 925/Styryl 13	DMSO	XeCI(308)	174
960 (913–1020)	IR-143-P(DTTC)	EG/DMSO:1/1	Kr(752,799)	115
962	IR-140-P(DTTC)		Kr(IR)	52
970 (894–1095)	IR-143-P(DTTC)	EG/DMSO:3/1	Kr(752)	113
970 (915–1058)	12A-DTTeC-P	EG/DMSO:3/1	Kr(752)	113
972	IR-132-P(DTTC)	DMSO	FL	109
972	IR-143-P(DTTC)	DMSO	FL	109
975 (930–1040)	LDS 925/styryl 13	PC/EG	Nd:YAG(532)	175
980 (928–1084)	styryl 14	PC/EG	Ar	173
983–1081	D-4-QTC-I-I	DMSO	ruby(694.3)	89
1000	D-4-QTC-I-I	acetone	ruby(694.3)	101
1084–1125	DNDTPC-P	DMSO	Nd:YAG(1064)	116-7
1124 (1102–1148)	9,11,15,17-Dnp-DTPC-p	DMSO	Nd:YAG(1064)	118
1140	Dye 9860, Q-switch II	DCE	Nd:YAG(1.06)	176,180
1140 (1107–1187)	9,11,15,17-Dnp-5,6-Temo-DTPC-P	—	Nd:YAG(1064)	119
1172 (1151–1198)	9,11,15,17-Dnp-6,7-DBZ-DTPC-P	DMSO	Nd:YAG(1064)	118
1180–1530	Dye 5, Q-switch	—	Nd:YAG(1064)	183-4

Table 3.1.1—*continued*
Organic Dye Lasers Arranged in Order of Increasing Wavelength

Wavelength (nm)	Laser dye*	Solvent	Pump	Ref.
1190 (1150–1240)	Dye 26, IR-26	DCE	Nd:YAG	117,182
1231 (1192–1285)	DNDTPC-P	DMSO	Nd:YAG(1064)	119
1270 (1200–1320)	Dye 26, IR-26	BzOH	Nd:YAG	179
1320 (1180–1400)	Dye 5, Q-switch 5	DCE	Nd:glass	176,181
1550	S401	DCE	Nd:glass	178
1600	S301	DCE	Nd:glass	178
1800	S501	DB	Nd:glass	178

* For common names of the dyes and details of the dye structures, see Steppel, R. N., Organic Dye Lasers, in *Handbook of Laser Science and Technology, Vol. I: Lasers and Masers*, CRC Press, Boca Raton, FL (1982), p. 299, and Steppel, R. N., Organic Dye Lasers, in *Handbook of Laser Science and Technology, Suppl. 1: Lasers*, CRC Press, Boca Raton, FL (1991), p. 219.
** Rigidized

References

1. Zapka, W. and Brickmann, U., Shorter dye laser wavelengths from substituted p-terphenyl, *Appl. Phys.* 20, 283 (1979).
2. Gusten. H. and Rinke, M., Photophysical properties and laser performance of photostable uv laser dyes III. Sterically hindered p-quarterphenyls, *Appl. Phys.* B 45, 279 (1988).
3. Tomin, V. I., Alcock, A. J., Sarjeant, W. J., and Leopold, K. E., Some characteristics of efficient dye laser emission obtained by pumping at 248 nm with a high-power KrF* discharge laser, *Opt. Commun.* 26(3), 396 (1978).
4. Tomin, V. I., Alcock, A. J., Sarjeant, W. J., and Leopold, K. E., Tunable, narrow bandwidth, 2 MW dye laser pumped by a KrF* discharge laser, *Opt. Commun.* 28(3), 336 (1979).
5. McKee, T. J., Stoicheff, B. P., and Wallace, S. C., Tunable, coherent radiation in the Lyman-region (1210-1290 Å) using magnesium vapor, *Opt. Lett.* 3(6), 207 (1978).
6. McKee, T. J. and James, D. J., Characterization of dye laser pumping using a high-power KrF excimer laser at 248 nm, *Can. J. Phys.* (Sept. 1979).
7. Godard, B. and de Witte, O., Efficient laser emission in para-terphenyl tunable between 323 and 364 nm, *Opt. Commun.* 19(3), 325 (1976).
8. Furumoto, H. W. and Ceccon, H. L., Ultraviolet organic liquid lasers, *IEEE J. Quantum Electron.* QE-6, 262 (1970).
9. Rulliere, C., Morand, J. P., and de Witte, O., KrF laser pumps new dyes in the 3500 Å spectral range, *Opt. Commun.* 20(3), 339 (1977).
10. Ziegler, L. D. and Hudson, B. S., Tuning ranges of 266 nm pumped dyes in the near uv, *Opt. Commun.* 32(1), 119 (1980).
11. Dunning, F. B. and Stebbings, R. F., The efficient generation of tunable near UV radiation using an N_2 pumped dye laser, *Opt. Commun.* 11, 112 (1974).
12. Ducasse, L., Rayez, J. C., and Rulliere, C., Substitution effects enhancing the lasing ability of organic compounds, *Chem. Phys. Lett.* 57(4), 547 (1978).
13. Profitt, W., Coherent, Inc. (private communication, Steppel, R. N. 1980).
14. Marowsky, G., Cordray, R., Tittel, F. K., Wilson, W. L., and Collins, C. B., Intense laser emission from electron-beam-pumped ternary mixtures of Ar, N_2, and POPOP vapor, *Appl. Phys. Lett.* 33(1), 59 (1978).
15. Rulliere, C. and Joussat-Dubien, J., Dye laser action at 330 nm using benzoxazole: a new class of lasing dyes, *Opt. Commun.* 24(1), 38 (1978).
16. Myer, J. A., Itzkan, I., and Kierstead, E., Dye lasers in the ultraviolet, *Nature (London)* 225, 544, (1970).
17. Kelley, C. J., Ghiorghis, A., Neister, E., Armstrong, L. and Prause, P. R., Bridged quarterphenyls as flashlamp-pumpable laser dyes, *Laser Chem.* 7, 343 (1987).
18. Kauffman, J. M., Kelley, C. J., Ghiorghis, A., Neister, E. and Armstrong, L., Cyclic ether auxofluors on oligophenylene laser dyes, *Laser Chem.* 8, 335 (1988).
19. Chromatix, Inc., Sunnyvale, CA 94086.
20. Molectron Corporation, Sunnyvale, CA 94086.
21. Rinke, M., Gusten, H. and Ache, H. J., Photophysical properties and laser performance of photostable uv laser dyes II. Ring-bridged p-quarterphenyls, *J. Phys. Chem.* 90, 2666 (1986).
22. Gusten, H., Rinke, M., Kao, C., Zhou, Y., Wang, M. and Pan, J., New efficient laser dyes for operation in the near UV range, *Opt. Commun.* 59(5,6), 379 (1986).
23. Michelson, M. M. (private communication, Steppel, R. N., 1993).
24. Chen, C. H., Fox, J. L., Duarte, F. J. and Ehrlich, J. J., Lasing characteristics of new coumarin-analog dyes: broadband and narrow-linewidth performance, *Appl. Optics* 27(3), 443 (1988).
25. Phase-R Company, New Durham, NH.
26. Candela Corporation, Natick, MA.

27. Myer, J. A., Johnson, C. L., Kierstead, E., Sharma, R. D. and Itzkan, I., Dye laser stimulation with a pulsed N_2 laser line at 3371 Å, *Appl. Phys. Lett.* 16, 3 (1970).
28. Alekseeva, V. I., Afanasiadi, L. S., Volkov, V. M., Krasovitskii, B. M., Vernigor, E. M., Lebedev, S. A., Savvina, L. P. and Tur, I. N., Spectral luminescent and lasing properties of pyridylaryloxazoles, *J. Appl. Spectrosc. (USSR)* 44, 244 (1986): *Zh. Prikl. Spectrosc.* 44, 403 (1986).
29. Smith, P. W., Liao, P. F., Schank, C. V., Gustafson, T. K., Lin, C., and Maloney, P. J., Optically excited organic dye vapor laser, *Appl. Phys. Lett.* 25, 144 (1974).
30. Guggenheimer, S. C., Peterson, A. B., Knaak, L. E. and Steppel, R. N. (to be published).
31. Tully, F. P. and Durant, Jr., J. L., Exalite 392E: A new laser dye for efficient cw operation between 373 and 408 nm, *Appl. Optics* 27(11), 2096 (1988).
32. Gomez, M. S. and Guerra Perez, J. M., The ACDF. A new class of lasing dyes, *Optics Commun.* 40(2), 144 (1981).
33. Acuña, A. U., Amat, F., Catalan, J., Costel, A., Figuera, J. M. and Muñoz, J. M., Pulsed liquid lasers from proton transfer in the excited state, *Chem. Phys. Lett.* 132, 567 (1986).
34. Telle, H., Brinkmann, U., and Raue, R., Laser properties of bis-styryle compounds, *Opt. Commun.* 24(3), 248 (1978).
35. Ferguson, J. and Mau, A. W. H., Laser emission from meso-substituted anthracenes at low temperatures, *Chem. Phys. Lett.* 14, 245 (1972).
36. Basting, D., Schafer, F. P., and Steyer, B., New laser dyes, *Appl. Phys.* 3, 81 (1974).
37. Srinivasan, R., New materials for flash-pumped organic lasers, *IEEE J. Quantum Electron.* QE-5, 552 (1969).
38. Fletcher, A. N., Pietrak, M. E. and Bliss, D. E., Laser dye stability. Part II. The fluorinated azacoumarin dyes, *Appl. Phys.* B 42, 79 (1987).
39. Huffer, W., Schieder, R., Telle, H., Raue, R., and Brinkwerth, W., CW dye laser emission down to the near uv, *Opt. Commun.* 28(3), 353 (1979).
40. Majewski, W. and Krasinski, J., Laser properties of fluorescent brightening agents, *Opt. Commun.*, 18(3), 255 (1976).
41. Ebeid, E. M., Sabry, M. F. and El-Daly, S. A., 1,4-bis (β-pyridyl-2-vinyl)benzene (p2vb) and 2,5-distyryl-pyrazine (dsp) as blue laser dyes, *Laser Chem.* 5, 223 (1985).
42. Beterov, I. M., Ishchenko, V. N., Kogan, B. Ya., Krasovitskii, B. M., and Chernenko, A. A., Stimulated emission from 2-phenyl-5(4-difluoromethylsulfonyl-phenyl)oxazole pumped with nitrogen laser radiation, *Sov. J. Quantum Electron.* 7(2), 246 (1977).
43. Hammond, P. R., Fletcher, A. N., Henry, R. A., and Atkins, R. L., Search for efficient, near uv lasing dyes, *Appl. Phys.* 8, 311 (1975).
44. Srinivasan, R., von Gutfield, R. J., Angadiyavar, C. S., and Tynan, E. E., Photochemical studies on organic lasers, Air Force Materials Laboratory, Wright-Patterson Air Force Base, Dayton, Ohio, AFML-TR-74-110 (1974). a. With Rhodamine 6G.
45. Dzyubenko, M. I., Krainov, I. P. and Maslov, V. V., Lasing properties of water-soluble dyes in the blue-green region, *Optics Spectrosc. (USSR)* 57(1), 58 (1984).
46. Furumoto, H. W. and Ceccon, H. L., Flashlamp pumped organic scintillator lasers, *J Appl. Phys.* 40, 4204 (1969).
47. Yarborough, J. M., CW laser emission spanning the visible spectrum, *Appl. Phys. Lett.* 24(12), 629 (1974).
48. Decker, C. D. and Tittel, P. K., Broadly tunable, narrow linewidth dye laser emission in the near infrared, *Opt. Commun.* 7(2), 155 (1974).
49. Eckstein, N. J., Ferguson, A. I., Hansch, T. W., Minard, C. A., and Chan, C. K., Production of deep blue tunable picosecond light pulses by synchronous pumping of a dye laser, *Opt. Commun.* 27(3), 466 (1978).

50. Pavlopoulos, T. G., Boyer, J. H., Politzer, I. R. and Lau, C. M., Syn-dioxabimanes as laser dyes, *Optics Commun.* 64(4), 367 (1987).
51. Quantel International, 928 Benecia Avenue, Sunnyvale, CA.
52. Telle, H., Brinkmann, U., and Raue, R., Laser properties of triazinyl stilbene compounds, *Opt. Commun.* 24, 33 (1978).
53. Quanta-Ray, 1250 Charleston Rd., Mountain View, CA.
54. Huth, B. G. and Farmer, G. I., Laser action in 9,10 diphenylanthracene, *IEEE J. Quantum Electron.* QE-4, 427 (1968).
55. Drexhage, K. H., Erickson, O. R., Hawks, G. H., and Reynolds, G. A., Water soluble coumarin dyes for flashlamp-pumped dye lasers, *Opt. Commun.* 15, 399 (1975).
56. Kittrell, C. (private communication, R. N. Steppel, 1977).
57. Acuña, A. U., Costel, A. and Muñoz, J. M., A proton-transfer laser, *J. Chem. Phys.* 90, 2807 (1986).
58. Fletcher, A. N. and Bliss, D. E., Laser dye stability. V. Efforts of chemical substituents of bicyclic dyes upon photodegradation parameters, *Appl. Phys.* 16, 289 (1978).
59. Marling, J. B., Hawley, J. H., Liston, E. M., and Grant, W. B., Lasing characteristics of seventeen visible-wavelength dyes using a coaxial-flashlamp-pumped laser, *Appl. Optics*13(10), 2317 (1974). [a. With rhodamine 6G.]
60. Kato, K., 3547-Å pumped high power dye laser in the blue and violet, *IEEE J. Quantum Electron.* QE-II, 373 (1975).
61. Reynolds, G. A. and Drexhage, K. H., New coumarin dyes with rigidized structures for flashlamp pumped dye lasers, *Opt. Commun.* 13(3), 222 (1975).
62. Allain, J. Y., High energy pulsed dye lasers for atmospheric sounding, *Appl. Optics* 18(3), 287 (1979).
63. Roullard, F. P. (private communication, R. N. Steppel, 1976).
64. Williamson, A. (private communication, R. N. Steppel, 1977).
65. Green, W. R. (private communication, R. N. Steppel, 1977).
66. Schimitschek, E. J., Trias, J. A., Hammond, P. R., and Atkins, R. L., Laser performance and stability of fluorinated coumarin dyes, *Opt. Commun.* 1 l, 352 (1974).
67. Schimitschek, E. J., Trias, J. A., Taylor, M., and Celto, J. E., New improved laser dyes for the blue-green spectral region, *IEEE J. Quantum Electron.* QE-9, 781 (1974).
68. Collins, C. H., Taylor, K. N., and Lee, F. W., Dyes pumped by the nitrogen ion laser, *Opt. Commun.* 26(1), 101 (1978).
69. Halstead, J. A. and Reeves, R. R., Mixed solvent systems for optimizing output from a pulsed dye laser, *Opt. Commun.* 27(2), 273 (1978).
70. Drexhage, K. H. and Reynolds, G. A., New highly efficient laser dyes, VII Int. Quantum Electronics Conf., Paper F. 1, San Francisco, Calif. (1974); see also References 80 and 100.
71. Petty, B. W. and Morris, K., *Opt. Quantum Electron.* 8(4), 371 (1976).
72. Morton, R. O., Mack, M. E. and Itzkan, I., Efficient cavity dumped dye laser, *Appl. Opt.* 17, 3268 (1978).
73. Ledbetter, J. W. (private communication, R. N. Steppel, 1977).
74. Gacoin, P., Bokobza, A., Bos, F., LeBris, M. T., and Hayat, G., New class of high-efficiency laser dyes: the quinoxalinones, Conference on Laser and Electrooptical Systems, San Diego, (1978), S6, WEE6.
75. Fletcher, A. N., Laser dye stability. III. Bicyclic dyes in ethanol, *Appl. Phys.* 14, 295 (1977).
76. Blazy, J. (private communication, R. N. Steppel, 1978).
77. Lill, E., Schneider, S., and Dorr, F., Passive mode-locking of a flashlamp-pumped fluorol 7GA dye laser in the green spectral region, *Opt. Commun.* 20(2), 223 (1977).
78. Blazej, D. (private communication, R. N. Steppel, 1977).

79. Lambropoulos, M., Fluorol 7CA: An efficient yellow-green dye for flashlamp-pumped lasers, *Opt. Commun.* 15, 35 (1975).
80. Drexhage, K. H.,What's ahead in laser dyes, *Laser Focus* 9, 35 (1974).
81. Marling, J. B., Wood, L. L., and Gregg, D. W., Long pulse dye laser across the visible spectrum, *IEEE J. Quantum Electron.* QE-7,498 (1971).
82. Marling, J. B., Gregg, D. W. and Thomas, S. J., Effect of oxygen on flashlamp-pumped organic-dye lasers, *IEEE J. Quantum Electron.* QE-6, 570 (1970).
83. Hartig, W., A high power dye-laser pumped by the second harmonic of a Nd-YAG laser, *Opt. Commun.* 27(3), 447 (1978).
84. J. K. Lasers Ltd., Somers Road, Rugby, Warwickshire, U.K.
85. Hargrove, R. S. and Kan, T., Efficient, high average power dye amplifiers pumped by copper vapor lasers, *IEEE J. Quantum Electron.* QE-13, 28D (1977).
86. Hammond, P. R., Laser dye DCM, spectral properties, synthesis and comparison with other dyes in the red, *Opt. Commun.*, Preprint.
87. Drake, J. M., Steppel, R. N., and Young, D., Kiton red s and rhodamine b. The spectroscopy and laser performance of red laser dyes, *Chem. Phys. Lett.* 35(2), 181 (1975).
88. Passner, A. and Venkatesan, T., Inexpensive, pulsed, tunable ir dye laser pumped by a driven dye laser, *Rev. Sci. Instrum.* 49(10), 1413 (1978).
89. Oettinger, P. E. and Dewey, C. F., Lasing efficiency and photochemical stability of infrared laser dyes in the 710-1080 nm region, *IEEE J. Quantum Electron.* QE-12(2), 95 (1976).
90. Wayashita, M., Kasarnatsu, M., Kashiwagi, H., and Machida, K., The selective excitation of lithium isotopes by intracavity nonlinear absorption in a cw dye laser, *Opt. Commun.* 26(3), 343 (1978).
91. McDonald, J. (private communication, R. N. Steppel, 1974).
92. Woodruff, S. and Ahlgren, D. (private communication, R. N. Steppel, 1977).
93. Holton, G. (private communication, R. N. Steppel, 1978).
94. (a) Shirley, J., private communication (1977); (b) Hall, R. J., Shirley, J. A., and Eckbreth, A. C., Coherent anti-Stokes Raman spectroscopy: Spectra of water vapor in flames, *Opt. Lett.* 4, 87 (1979).
95. Drell, P. (private communication, R. N. Steppel, 1978).
96. Mahon, R., McIlrath, T. J., and Koopman, D. W., High-power TEM_{00} tunable laser system, *Appl. Optics* 18(6), 891 (1979).
97. Kato, K., A high-power dye laser at 6700-7700 Å, *Opt. Commun.* 19(1), 18 (1976).
98. Kuhl, J., Lambrich, R., and von der Linde, D., Generation of near-infrared picosecond pulses by mode locked synchronous pumping of a jet-stream dye laser, *Appl. Phys. Lett.* 31(10), 657 (1977).
99. Jarett, S., Spectra Physics (private communication, Steppel, R. N., 1980).
100. Drexhage, K. H., Structure and properties of laser dyes, in *Dye Lasers*, Vol. 1, Schafer, F. P., Ed., Springer-Verlag, Berlin (1973), p. 44 and references therein.
101. Miyazoe, Y. and Mitsuo, M., Stimulated emission from 19 polymethine dyes-laser action over the continuous range 710-1060 μm, *Appl. Phys. Lett.* 12, 206 (1968).
102. Szabo, A., National Research Council of Canada, private communication (1980), Jessop, P. E. and Scabo, A., Single frequency cw dye laser operation in the 690-700 nm gap, *IEEE J. Quantum Electron.* 16, 812 (1980).
103. Romanek, K. M., Hildebrand, O., and Gobel, E., High power CW dye laser emission in the near IR from 685 nm to 965 nm, *Opt. Commun.* 21(1), 16 (1977); *Spectra-Physics Laser Review* 4(1), April (1977).
104. Hildebrand, O., Nitrogen laser excitation of polymethine dyes for emission wavelength up to 9500 Å, *Opt. Commun.* 10(4), 310 (1974).
105. Moore, C. A. and Decker, C. D., Power-scaling effects in dye lasers under high power laser excitation, *J. Appl. Phys.* 49, 47, (1978).

106. Fehrenback, G. W., Oruntz, K. J., and Ulbrich, R. G., Subpicosecond light pulses from a synchronously mode-locked dye laser with composite gain and absorber medium, *Appl. Phys. Lett.* 33(2), 159 (1978).
107. Bryon, D, A., McDonnell Douglas Astronautics Company, (private communication, R. N. Steppel, 1979).
108. Donzel, A. and Weisbach, C., CW dye laser emission in the range 7540-8880 Å, *Opt. Commun.*, 17(2), 153 (1976).
109. Webb, J. P., Webster, F. G., and Plourde, B. E., Sixteen new infrared laser dyes excited by a simple, linear flashlamp, *IEEE J. Quantum Electron.* QE-11, 114 (1975).
110. Kato, K., Near infrared dye laser pumped by a carbazine 122 dye laser, *IEEE J. Quantum Electron.* QE-12, 442 (1976).
111. Decker, C. D., Excited state absorption and laser emission from infrared dyes optically pumped at 532 nm, *Appl. Phys. Lett.* 27(11), 607 (1975).
112. Miyazoe, Y. and Maeda, M., Polymethine dye lasers, *Opto Electronics* 2(4), 227 (1970).
113. Leduc, M., Synchronous pumping of dye lasers up to 1095 nm, *Opt. Commun.* 31, 66 (1979).
114. Ammann, E. O., Decker, C. D., and Falk, J., High-peak-power 532 nm pumped dye laser, *IEEE J. Quantum Electron.* QE-10, 463 (1974).
115. Leduc, M. and Weisbach, C., CW dye laser emission beyond 1000 nm, *Opt. Commun.* 26, (11), 78 (1978).
116. Ferrario, A., A., 13 MW peak power dye laser tunable in the 1.1 μm range, *Opt. Commun.* 30(1), 83 (1979).
117. Ferrario, A., A picosecond dye laser tunable in the 1.1 μm region, *Opt. Commun.* 30(1), 85 (1979).
118. Kato, K., Nd:YAG laser pumped infrared dye laser, *IEEE J. Quantum Electron.* QE-14, 7 (1978).
119. Kato, K., Broadly tunable dye laser emission to 12850 Å, *Appl. Phys. Lett.* 33(16), 509 (1978).
120. Yenagi, J. V., Gorbal, M. R., Savadatti, M. I. and Naik, D. B., A new laser dye, *Opt. Commun.* 85, 223 (1991).
121. Doizi, D., Jaraudias, J. and Salvetat, G., Laser performance of dibenzoxanthylium salts, *Opt. Commun.* 99, 207 (1993).
122. O'Neil, M. P., Synchronously pumped visible laser dye with twice the efficiency of rhodamine 6G, *Opt. Lett.* 18, 37 (1993).
123. Pavlopoulos, T. G., Boyer, J. H., Shah, M., Thangaraj, K. and Soong, M.-L., Laser action from 2,6,8-position trisubstituted 1,3,5,7-tetramethylpyrromethene-BF_2, *Appl. Optics* 29, 3885 (1990).
124. Guggenheimer, S. C., Boyer, J. H., Thangaraj, K., Shah, M., Soong, M.-L. and Pavlopoulos, T. G., Efficient laser action from two cw laser-pumped pyrromethene-BF_2 complexes, *Appl. Optics* 32, 3942 (1993).
125. Boyer, J. H., Haag, A., Soong, M.-L., Thangaraj, K. and Pavlopoulos, T. G., Laser action from 2,6,8-position trisubstituted 1,3,5,7-tetramethyl-pyrromethene-BF_2 complexes: part 2, *Appl. Optics* 30, 3788 (1991).
126. Moses, D., High quantum efficiency luminescence from a conducting polymer in solution: a novel polymer laser dye, *Appl. Phys. Lett.* 60, 3215 (1992).
127. Said, J. and Boquillon, Lasing characteristics of a new DCM derivative under flash-lamp pumping, *Opt. Commun.* 82, 51 (1991).
128. Costela, A., Amat, F., Catalan, J., Douhal, A., Figuera, J. M., Munoz, J. M. and Acuna, A. U., Phenylbenzimidazole proton-transfer laser dyes: spectral and operational properties, *Optics Commun.* 64(5), 457 (1987).
129. Fletcher, A. N., Henry, R. A., Kubin, R. F. and Hollins, R. A., Fluorescence and lasing characteristics of some long-lived flashlamp-pumpable, oxazole dyes, *Optics Commun.* 48(5), 352 (1984).

130. Fletcher, A. N., Bliss D. E. and Kauffman, J. M., Lasing and fluorescent characteristics of nine, new, flashlamp-pumpable, coumarin dyes in ethanol and ethanol:water, *Optics Commun.* 47(1), 57 (1983).
131. Fletcher, A. N., Henry, R. A., Pietrak, M. E. and Bliss D. E., Laser dye stability, part 12. The pyridinium salts, *Appl. Phys.* B 43, 155 (1987).
132. Raue, R., Harnisch, H. and Drexhage, K. H., Dyestuff lasers and light collectors-two new fields of application for fluorescent heterocyclic compounds, *Heterocycles* 21(1), 167 (1984).
133. Komel'kova, L. A., Kruglenko, V. P., Logunov, O. A., Povstyanoi, M. V., Startsev, A. V., Stoilov, Yu. and Timoshin, A. A., Imitrines. IV. New laser compounds in the imitrine class operating in the 482-618 nm range, *Sov. J. Quantum Electron.* 13(4), 549 (1983).
134. Pavlopoulos, T. G., Boyer, J. H., Politzer, I. R. and Lau, C. M., Laser action from syn-(methyl, methyl)bimane, *J. Appl. Phys.* 60(11), 4028 (1986).
135. Everett, P. N., Aldag, H. R., Ehrlich, J. J., Janes, G. S., Klimek, D. E., Landers, F. M. and Pacheco, D. P., Efficient 7-1 flashlamp-pumped dye laser at 500-nm wavelength, *Appl. Optics*, 25(13), 2142 (1986).
136. Asimov, M. M., Katarkevich, V. M., Kovalenko, A. N., Nikitchenko, V. M., Novikov, A. I., Rubinov, A. N. and Efendiev, T. Sh., Spectroluminescence and lasing characteristics of a new series of bifluorophoric laser dyes, *Opt. Spectrosc. (USSR)* 63(3), 356 (1987).
137. Pavlopoulos, T. G., McBee, C. J., Boyer, J. H., Politzer, I. R. and Lau, C. M., Laser action from syn-(methyl,chloro)bimane, *J. Appl. Phys.* 62(1), 36 (1987).
138. Dupuy, F., Rulliere, C., Le Bris, M. T. and Valeur, B., A new class of laser dyes: benzoxazinone derivatives, *Optics Commun.* 51(1), 36 (1984).
139. Asimov, M. M., Nikitchenko, V. M., Novikov, A. I., Rubinov, A. N., Bor, Zs. and Gaty, L., New high-efficiency biscoumarin laser dyes, *Chemical Phys. Lett.* 149(2), 140 (1988).
140. Pavlopoulos, T. G., Shah, M. and Boyer, J. H., Efficient laser action from 1,3,5,7,8-dentamethylpyrromethene-BF_2 complex and its disodium 2,6-di-sulfonate derivative, *Optics Commun.* 70(5), 425 (1989).
141. Pavlopoulos, T. G., Shah, M. and Boyer, J. H., Laser action from a tetramethyl-pyrromethene-BF_2 complex, *Appl. Optics* 27(24), 4998 (1988).
142. Neister, S.E. (private communication, Steppel, R. N.).
143. Davenport, W. E., Ehrlich, J. J. and Neister, S. E., Characterization of pyrromethene-BF_2 complexes as laser dyes, *Proceedings of the International Conferences on Lasers '89*, New Orleans, LA, (1989), p. 408.
144. Shah, M., Thangaraj, K., Soong, M. L., Wolford, L. T., Boyer, J. H., Politzer, I. R. and Pavlopoulos, T. G., Pyrromethene-BF_2 complexes as laser dyes: 1, *Heteroatom. Chem.* 1 (5), 389 (1990).
145. Benson, M., Coherent Laser Group (private communication, R. N. Steppel, 1994).
146. Hsia, J., Candela Laser Corporation (private communication, R. N. Steppel, 1989).
147. Piechowski, A. P. and Bird, G. R., A new family of lasing dyes from an old family of fluors, *Optics Commun.* 50(6), 386 (1984).
148. Partridge, Jr., W. P., Laurendeau, N. M., Johnson, C.C. and Steppel, R. N., Performance of pyrromethene 580 and 597 in a commercial Nd:YAG-pumped dye laser system, *Optics Lett.* 19(20), 1(1994).
149. Guggenheimer, S. G., Boyer, J. H., Thangaraj, K., Shah, M., Soong, M. L. and Pavlopoulos, T. G., Efficient laser action from two cw laser pumped pyrromethene-BF_2 complexes, *Appl. Optics* 32(21), 3942 (1993).
150. Kauffman, J. M. and Bently, J. H., Effect of various anions and zwittenons on the lasing properties of a photostable cationic laser dye, *Laser Chem.* 8 (1988).
151. Maeda, M. and Miyazoe, Y., Flashlamp-excited organic liquid laser in the range from 342 to 889 nm, *Jpn. J. Appl. Phys.* 11(5), 692 (1972).

152. Loth, C. and Gacoin, P., Improvement of infrared flashlamp-pumped dye laser solution with a double effect additive, *Opt. Commun.* 15(2), 179 (1975).
153. Allik, T. H., Hermes, R. E., Sathyamoorthi, G. and Boyer, J. H., Spectroscopy and laser performance of new BF_2-complex dyes in solution, *SPIE Proceedings: Visible and UV Lasers* 2115, 240 (1994).
154. Shinn, M. D., Bryn Mawr College (private communication, R. N. Steppel, 1994).
155. Sadrai, M. and Bird, G. R., A new laser dye with potential for high stability and a broad band of lasing action: perylene-3,4,9,10 tetracarboxylic acid-bis-n,n' (2',6'xylidyl) diimide, *Optics Commun.* 51(1), 62 (1984) .
156. Ivri, J. Burshtein, Z., Miron, E., Reisfeld, R. and Eyal, M., The perylene derivative BASF-241 solution as a new tunable dye laser in the visible, *IEEE J. Quantum Electron.* 26, 1516 (1990).
157. Boyer, J. H., Haag, A. M., Sathyamoorthi, G., Soong, M. L. and Thangara, K., Pyrromethene-BF_2 complexes as laser dyes: 2, *Heteroatom. Chem.* 4(1), 39 (1993).
158. Maslov, V. V., Dzyubenko, M. I., Kovalenko, S. N., Nikitchenko, V. M. and Nivikov, A. I., New efficient dyes for the red part of the lasing spectrum, *Sov. J. Quantum Electron.* 17(8), 998 (1987).
159. Boyer, J. H., Haag, A., Shah, M., Soong, M. L., Thangaraj, K. and Pavlopoulos, T. G., Laser action from 2,6,8-trisubstituted-1,3,5,7-tetramethyl-pyrromethene-BF_2 complexes: Part 2, *Appl. Optics* 30(27), 3788 (1991).
160. Richter, D. (private communication, Steppel, R. N., 1994).
161. Broyer, M., Chevaleyre, J., Delacretaz, G. and Woste, L., CVL-pumped dye laser for spectroscopic application, *Appl. Phys.* B 35, 31 (1984).
162. Kuhl, J., Telle, H., Scheider, R., and Brinkmann, U., New efficient and stable laser dyes for cw operation in the blue and violet spectral range, *Opt. Commun.* 24, 251 (1978).
163. Antonov, V. S. and Hohla, K. L., Dye stability under excimer-laser pumping II. visible and UV dyes, *Appl. Phys.* B. 32, 9 (1983).
164. Spectra-Physics, 1250 W. Middlefield Road, Mountain View, CA 94039.
165. Friedrich, D. M., Nitrogen pumped LDS 698, (private communication, Steppel, R. N., 1985).
166. Jasny, J., Novel method for wavelength tuning of distributed feedback dye lasers, *Optics Commun.* 53, 238 (1985).
167. Coherent Inc., 3210 Porter Dr., Palo Alto, CA 94304.
168. Hoffnagle, J., Roesch, L. Ph., Schlumpf, N. and Weis, A., CW operation of laser dyes Styryl-9 and Styryl-11, *Optics Commun.* 42, 267 (1982); K. Kato, see Reference 5 therein.
169. Lumonics Inc., 105 Schneider Road, Kanata (Ottawa), Ontario, Canada K2K IY3.
170. Klein, P. (private communication, Steppel, R. N., 1983).
171. Giberson, K. W., Jeys, T. H. and Dunning, F. B., Generation of tunable cw radiation near 875 nm, *Appl. Optics* 22(18), 2768 (1983).
172. Schellenberg, F. (private communication, Steppel, R. N., 1982).
173. Kato, K., Ar-ion-laser-pumped infrared dye laser at 875-1084 nm, *Optics Lett.* 9(12), 544 (1984).
174. Bloomfield, L. A., Excimer-laser pumped infrared dye laser at 907-1023 nm, *Optics Commun.* 70(3), 223 (1989).
175. Stark, T. S., Dawson, M. D. and Smirl, A. L., Synchronous and hybrid mode-locking of a Styryl 13 dye laser, *Optics Commun.* 68(5), 361 (1988).
176. Seilmeier, A., Kopainsky, B. and Kaiser, W., Infrared fluorescence and laser action of fast mode-locking dyes, *Appl. Phys.* 22, 355 (1980).
177. Reynolds, G. A. and Drexhage, K. H., Stable heptamethine pyrylium dyes that absorb in the infrared, *J. Org. Chem.* 42(5), 885 (1977).
178. Seilmeier, A., Kaiser, W., Kussler, M., Marx, N. J., Sens, B. and Drexhage, K. H., Picosecond dye laser emission in the infrared between 1.4 and 1.81 μm, *Appl. Phys.* B32, 53 (1983).

179. Seilmeier, A., Kaiser, W., Sens, B. and Drexhage, K. H., Tunable picosecond pulses around 1.3 μm generated by a synchronously pumped infrared dye laser, *Optics Lett.* 8(4), 205 (1983).
180. Kopainsky, B., Kaiser, W. and Drexhage, K. H., New ultrafast saturable absorbers for Nd:lasers, *Optics Commun.* 32(3), 451 (1980).
181. Elsaesser, T., Polland, H. J., Seilmeier, A. and Kaiser, W., Narrow-band infrared picosecond pulses tunable between 1.2 and 1.4 μm generated by a traveling-wave dye laser, *IEEE J. Quantum Electron.* QE-20(3). 191 (1984).
182. Kopainsky, B., Qiu, P., Kaiser, W., Sens, B. and Drexhage, K. H., Lifetime, photostability, and chemical structure of ir heptamethine cyanine dyes absorbing beyond 1 μm, *Appl. Phys.* B 29, 15 (1982).
183. Looentanzer, H. and Polland, H. J., Generation of tunable picosecond pulses between 1.18 μm and 1.53 μm in a ring laser configuration using dye no. 5., *Optics Commun.* 62(1), 35 (1987).
184. Alfano, R. R., Schiller, N. H. and Reynolds, G. A., Production of picosecond pulses by mode locking an nd:glass laser with dye #5, *IEEE J. Quantum Electron.* QE-17(3), 290 (1981).
185. Rinke, M., Gusten, H. and Ache, H. J., Photophysical properties and laser performance of photostable uv laser dyes I. Substituted p-quarterphenyls, *J. Phys. Chem.* 90, 2661 (1986).
186. Schafer, F. P., Bor, Zs., Luttke, W., and Liphardt, B., Bifluorophoric laser dyes with intramolecular energy transfer, *Chem. Phys. Lett.* 56, 455 (1978).
187. Coherent Inc., 3210 Porter Dr., Palo Alto, CA.
188. Ivri, J., Burshtein, Z. and Miron, E., Characteristics of 1,1',3,3,3',3'-hexa-methyl-indotricarbocyanine iodide as a tunable dye laser in the near infrared, *Appl. Optics* 30, 2484 (1991).
189. Pavlopoulos, T. G., Shah, M. and Boyer, J. H., Efficient laser action from 1,3,5,7,8-pentamethylpyrromethene-BF_2 complex and its disodium 2,6- disulfonate derivative, *Opt. Commun.*, 70(5), 425 (1989).
190. Valat, P., Tascano, V., Kossanyi, J. and Bos, F., Laser effect of a series of variously substituted pyrylium and thiopyrylium salts, *J. Lumin.* 37, 149 (1987).
191. SchimitscheK, E. J., Trias, J. A., Hammond, P. R., Henry, R. A., and Atkins, R. L., New laser dyes with blue-green emissions, *Opt. Commun.* 16, 313 (1976).
192. Tuccio, S. A., Drexhage, K. H., and Reynolds, G. A., CW laser emission from coumarin dyes in the blue and green, *Opt. Commun.* 7(3), 248 (1974).
193. Hoffnagle, J. (private communication, Steppel, R. N., 1987).
194. Hammond, P. R., Fletcher, A. N., Henry, R. A., and Atkins, R. L., Search for efficient, near uv lasing dyes. II. Aza substitution in bicyclic dyes, *Appl. Phys.* 8, 315 (1975).

Section 3.2

RARE EARTH LIQUID LASERS

Introduction to the Table

Rare Earth Chelate Lasers

Rare earth chelate lasers are listed in order of increasing wavelength in Table 3.2.1 together with the lasing ion, chelating agent, and references. The lasing wavelength and output of these lasers depend on the characteristics of the optical cavity and the rare earth concentration. The original references should therefore be consulted for this information and its effect on the lasing wavelength.

Rare Earth Aprotic Lasers

Rare earth aprotic lasers are listed in order of increasing wavelength in Table 3.2.2 together with the lasing ion, solvent, and references. Rare earth aprotic laser amplifiers are listed separately in Table 3.2.3. The wavelength and output of these lasers depend on the characteristics of the optical cavity and the rare earth concentration. The original references should therefore be consulted for this information and its effect on the lasing wavelength.

Further Reading

Lempicki, A. and Samelson, H., Organic laser systems, in *Lasers*, Levine, A. D., Ed., Marcel Dekker, New York (1966), p. 181.

Lempicki, A., Samelson, H. and Brecher, C., Laser action in rare earth chelates, in *Applied Optics, Suppl. 2 of Chemical Lasers* (1965), p. 205.

Samelson, H., Inorganic Liquid Lasers, in *Handbook of Laser Science and Technology, Vol. I: Lasers and Masers*, CRC Press, Boca Raton, FL (1982), p. 397.

Samelson, H., Inorganic Liquid Lasers, in *Handbook of Laser Science and Technology, Suppl. 1: Lasers*, CRC Press, Boca Raton, FL (1991), p. 319.

Table 3.2.1
Rare Earth Chelate Lasers Arranged in Order of Increasing Wavelength

Wavelength (μm)	Ion	Ligand	Solvent	Reference
0.547	Tb^{3+}	trifluoroacetylacetone	acetonitrile	1
0.547	Tb^{3+}	trifuoroacetylacetone	dibenzoylmethide	1
0.611	Eu^{3+}	benzoylacetonate	ethanol-methanol (3:1)	2,3
0.611	Eu^{3+}	benzoylacetonate	dimethyl formamide	3
0.6117	Eu^{3+}	m-chlorobenzoyl-trifluoroacetonate	acetonitrile	4
0.6117	Eu^{3+}	p-chlorobenzoyl-trifluoroacetonate	acetonitrile	4
0.6117	Eu^{3+}	o-fluorobenzoyl-trifluoroacetonate	acetronitrile	4
0.6117	Eu^{3+}	p-fluorobenzoyl-trifluoroacetonate	acetonitrile	4
0.6117	Eu^{3+}	o-bromobenzoyl-trifluoroacetonate	acetonitrile	4
0.6117	Eu^{3+}	m-bromobenzoyl-trifluoroacetonate	acetonitrile	4
0.6117	Eu^{3+}	p-bromobenzoyl-trifluoroacetonate	acetonitrile	4
0.6118	Eu^{3+}	m-fluorobenzoyl-trifluoroacetonate	acetonitrile	4
0.6118	Eu^{3+}	α-naphthoyl-trifluoroacetonate	acetonitrile	5
0.6118	Eu^{3+}	o-chlorobenzoyl-trifluoroacetonate	acetonitrile	4,6
0.6118	Eu^{3+}	benzoyltrifluoroacetonate	acetonitrile	7
0.6118	Eu^{3+}	benzoyltrifluoroacetonate	acetonitrile	8
0.6119	Eu^{3+}	benzoyltrifluoroacetonate	ethanol-methanol dimethyl formade (7.5:2.5:1)	9
0.6119	Eu^{3+}	benzoyltrifluoroacetonate	acetonitrile	10
0.6119	Eu^{3+}	trifluoroacetylacetonate	acetonitrile	9

0.612	Eu^{3+}	dibenzoylmethide	ethanol-methanol-dimethyl formade (9:3:2)	11
0.612[a]	Eu^{3+}	4,4,4-trifluoro-phenyl-1,3-butanedione	acetonitrile	12
0.6122	Eu^{3+}	trifluoroacetylacetonate	ethanol-methanol (3:1)	13
0.6123	Eu^{3+}	benzoylacetonate	acetonitrile	14
0.6123	Eu^{3+}	thenoyltrifluoroacetonate	acetonitrile	14
0.6125	Eu^{3+}	thenoyltrifluoroacetonate	acetonitrile	8
0.6129	Eu^{3+}	benzoylacetonate	ethanol-methanol (3:1)	14–23
0.613	Eu^{3+}	benzoylacetonate	ethanol	22,24
0.613	Eu^{3+}	benzoylacetonate	ethanol-methanol (3:1)	13
0.613[b]	Eu^{3+}	dibenzoylmethane-phenantroline	ethanol-glycerol (3:1)	25
0.6131	Eu^{3+}	benzoylacetonate	ethyl/methyl alcohol (3:1)	26
1.054	Nd^{3+}	deuterotributyl phosphate	hexafluorobenzene	27
1.054	Nd^{3+}	deuterotributyl phosphate	carbon tetrachloride	27
1.057	Nd^{3+}	pentafluoropropionate + 1.10 phenanthroline	dimethyl sulfoxide	28
(c)	Nd^{3+}	$(TTA)_4Py$	acrylonitrile/carbon tetrachloride (1:9)	29

Note: Eu^{3+} laser action was also reported in Refs. 30–32 but no wavelengths were given.

(a) Stimulated emission in a planar microcavity.
(b) Microdroplet; lasing due to morphology-dependent resonances.
(c) Nd^{3+} laser action was reported in Ref. 29, but no wavelengths were given.

Chelate Laser References

1. Bjorklund, S., Kellermeyer, G., Hurt, C. R., McAvoy, N., and Filipescu, N., Laser action from terbium trifluoroacetylacetonate in p-dioxane and acetonitrile at room temperature, *Appl. Phys. Lett.* 10, 160 (1967).
2. Samelson, H., Brophy, V. A., Brecher, C., and Lempicki, A., Shift of laser emission of europium benzoylacetonate by inorganic ions, *J. Chem. Phys.* 41, 3998 (1964).
3. Meyer, Y., Astier, R., and Simon, J., Emission stimulee a 6111Å dans le benzoylacetonate d'europium active au sodium, *Compt. Rend.* 259, 4604 (1964).
4. Schimitschek, E. J., Nehrich, R. B., and Trais, J.A., Fluorescence properties and stimulated emission in substituted europium chelates, *J. Chem. Phys.* 64, 173 (1967).
5. Riedel, E. P. and Charles, R. G., Spectroscopic and laser properties of europium naphthoyl-trifluoroacetonate in solution, *J. Chem. Phys.* 42 (1908 1966).
6. Schimitschek, E. J., Nehrich, R. B., and Trias, J. A., Recirculating liquid laser, *Appl. Phys. Lett.* 9, 103 (1966).
7. Schimitschek, E. J., Nehrich, R. B., and Trais, J. A., Laser action in fluorinated europium chelates in acetonitrile, *J. Chem. Phys.* 42, 788 (1965).
8. Schimitschek, E. J., Trais, J. A., and Nehrich, R. B., Stimulated emission in an europium chelate solution at room temperature, *J. Appl. Phys.* 36, 867 (1965).
9. Brecher, C., Samelson, H., and Lempicki, A., Laser phenomena in europium chelates, III: spectroscopic effects of chemical composition and molecular structure, *J. Chem. Phys.* 42, 1081 (1965).
10. Samelson, H., Lempicki, A., Brecher, C., and Brophy, V., Room temperature operation of a europium chelate liquid laser, *Appl. Phys. Lett.* 5, 173 1964).
11. Schimitschek, E. J. and Nehrich, R. B., Laser action in europium dibenzoylmethide, *J. Appl. Phys.* 35, 2786 (1964).
12. Ebina, K., Okadam Y., Yamasaki, A. and Ujihara, K., Spontaneous and stimulated emission by Eu-chelate in a planar microcavity, *Appl. Phys. Lett.* 66, 2783 (1995).
13. Nehrich, R. B., Schimitschek, E. J., and Tras, J. A., Laser action in europium chelates prepared with NH_3, *Phys. Lett.* 12, 198 (1964).
14. Malashkevich, G. E. and Kuznetsova, V. V., Laser excited lasing in solutions of some europium chelates, *J. Appl. Spectr.* 22, 170 (1975).
15. Bykov, V. P., Intramolecular energy transfer and quantum generators, *J. Exptl. Theor. Phys. (U.S.S.R.)* 43, 1634 (1962).
16. Charles, R. G. and Ohlmann, R. C., Europium thenoyl trifluoro acetonate, *J. Inorg. Nucl. Chem.* 27, 255 (1965).
17. Metlay, M., Fluorescence lifetime of the europium dibenzoylmethides, *J. Phys. Chem.* 39, 491 (1963).
18. Bhaumik, M. L., Fletcher, P. C., Nugent, L. J., Lee, S. M., Higa, S., Telk, C. L., and Weinberg, M., Laser emission from a europium benzoylacetonate alcohol solution, *J. Phys. Chem.* 68, 1490 (1964).
19. Ohlmann, R. C. and Charles, R. G., Fluorescence properties of europium dibenzoylmethide and its complexes with Lewis bases, *J. Chem. Phys.* 41, 3131 (1964).
20. Aristov, A. V., Maslyukov, Yu. S., and Reznikova, I. I., Luminescence of a europium chelate solution under intense pulsed excitation, *Opt. Spectr.* 21, 286 (1966).
21. Lempicki, A., Samelson, H., and Brecher, C., Laser phenomena in europium chelates, IV. Characteristics of the europium benzoylacetonate laser, *J. Chem. Phys.* 41, 1214 (1964).
22. Lempicki, A., Samelson, H., and Brecher, C., Laser action in rare earth chelates, in *Applied Optics Supplement 2 of Chemical Lasers*, 205 (1965).
23. Aristov, A. V. and Maslyukov, Yu. S., Stimulated emission in europium benzoylacetonate solutions, *J. Appl. Spectr.* 8, 431 (1968).

24. Schimitschek, E. J., Stimulated emission in rare earth chelate (europium benzoylacetonate) in a capillary tube, *Appl. Phys. Lett.* 3, 117 (1963).
25. Taniguchi, H., Tomisawa, H. and Kido, J., Ultra-low-threshold europium chelate laser in morphology-dependent resonances, *Appl. Phys. Lett.* 66, 1578 (1995).
26. Lempicki, A. and Samelson, H., Optical maser action in europium-benzoylacetone, *Phys. Lett.* 4, 133 (1963).
27. Goryaeva, E. M., Shablya, A. V., and Serov, A. P., Luminescence and stimulated emission for solutions of complexes of neodymium nitrate with perdeutero-tributylphosphate, *J. Appl. Spectr.* 28, 55 (1976).
28. Heller, A., Fluorescence and room temperature laser action of trivalent neodymium in an organic liquid solution, *J. Am. Chem. Soc.* 89, 167 (1967).
29. Whittaker, B., Low threshold laser action of a rare earth chelate in liquid and solid host media, *Nature* 228, 157 (1970).
30. Samelson, H., Brecher, C. and Lempicki, A., Europium chelate lasers, *J. Chem. Phys.* 64, 165 (1967).
31. Ross, D. L., Blanc, J. and Pressley, R. J., Deuterium isotope effect on the performance of europium chelate lasers, *Appl. Phys. Lett.* 8, 101 (1966).
32. Ross, D. L. and Blanc, J., Europium chelates as laser materials, in *Advances in Chemistry Series*, No. 71, American Chemical Society, Washington, DC (1967), chapter 12.

Table 3.2.2
Neodymium Aprotic Liquid Lasers

Wavelength (μm)	Solvent	Operation	Reference
1.050	$POCl_3$–$SnCl_4$–UO_2^{2+}	long-pulse	1
1.052	$POCl_3$–$SnCl_4$	long-pulse	2
1.053	$POCl_3$–$ZrCl_4$	long-pulse	3
1.053	$POCl_3$–$AlCl_3$	long-pulse	3
1.054	$POCl_3$–$AlCl_3$	long-pulse	3
1.054	$POCl_3$–BBr_3	long-pulse	3
1.054	$POCl_3$–$AlCl_3$	long-pulse	4
1.054	$POCl_3$–$SOCl_2$–$SnCl_4$	long-pulse	5
1.055	$SeOCl_2$–$SnCl_4$	long-pulse	6
1.055	$POCl_3$–$SnCl_4$	long-pulse	7,8
1.056	$SeOCl_2$–$SnCl_4$	long-pulse	4,9–12
1.056	$SeOCl_2$–$SnCl_4$	Q-switched	13–15
1.056	$SeOCl_2$–$SbCl_5$	long-pulse	4,11
1.056	$SeOCl_2$–$SbCl_5$	Q-switched	16
1.056	$POCl_3$–$SnCl_4$	long-pulse	12,17
1.056	$POCl_3$–$ZrCl_4$	long-pulse	3,4,12,18,19
1.056	$POCl_3$–$ZrCl_4$	Q-switched	20–25
1.056	$POCl_3$–$AlCl_3$	long-pulse	4
1.058	$GaCl_3$–$SOCl_2$	long-pulse	26,27
1.058	$SeOCl_2$–$SnCl_4$	long-pulse	7,12
1.060	$POCl_3$–$SnCl_4$	long-pulse	28
1.061	$GaCl_3$–CCl_3	long-pulse	29
1.066	PBr_3–$SbBr_3$–$AlBr_3$	long-pulse	30
1.330	$SeOCl_2$–$SnCl_4$	long-pulse	31

Table 3.2.3
Neodymium Liquid Aprotic Single-Pass Laser Amplifiers

Wavelength (μm)	Oscillator	Solvent	Reference
1.056	Nd:silicate glass	$SeOCl_2$–$SnCl_4$	12,15,32
1.056	Nd:$SeOCl_2$–$SnCl_4$	$SeOCl_2$–$SnCl_4$	15
1.052	Nd:$POCl_3$–$ZrCl_4$	$POCl_3$–$ZrCl_4$	19
1.062	Nd:$POCl_3$	$POCl_3$–$ZrCl_4$	18,25,33,35
1.058	Nd:glass	$POCl_3$–$ZrCl_4$	35-37
1.053	Nd:ethylene glyol	ethylene glyol	38

Aprotic Laser References

1. Dvachenko, P. P., Kalinin, V. V., Seregina, E. A. et al., Inorganic liquid laser doped with neodymium and uranyl, *Laser and Particle Beams* 11, 493 (1993).
2. Collier, F., Michon, M. and LeSergent, C., Parametres laser du systeme liquide Nd^{+3}-$POCl_3$-$SnCl_4(H_2O)$ compares a ceux du YAG et du verre dope au neodyme, *Compt. Rend.* 272, 945 (1971).
3. Schimitschek, E. J., Laser emission of a neodymium salt dissolved in $POCl_3$, *J. Appl. Phys.* 39, 6120 (1968).
4. Weichselgartner, H. and Perchermeier, J., Anorganischer flüssigkeits laser, *Z. Naturforsch.* 25a, 1244 (1970).
5. Alekseev, N. E., Zhabotinski, M. E., Ivanova, E. B., Malashko, Ya. I. and Rudnitskii, Y. P., Effect of thionyl chloride on the laser characteristics of the liquid phosphor $POCl_3$ - $SnCl_4$ - Nd^{+3}, *Inorg. Mater.* 9, 215 (1973).
6. Samelson, H., Lempicki, A. and Brophy, V., Output properties of the Nd^{+3}:$SeOCl_2$ liquid lasers, *IEEE J. Quantum Electron.* QE-4, 849 (1968). See, also, Watson, W., Reich, S., Lempicki, A. and Lech, J., A circulating liquid laser system, *ibid.*, p. 842.
7. LeSergent, C., Michon, M., Rousseau, S., Collier, F., Dubost, H. and Raoult, G., Characteristics of the laser emission obtained with the solution $POCl_3$, $SnCl_4$, Nd_2O_3, *Compt. Rend.* 268, 1501 (1969).
8. Voronko, Yu. K., Krotova, L. V., Sychugov, V. A. and Shipulo, G. P., Lasers with liquid active materials based on $POCl_3$:Nd^{+3}, *J. Appl. Spect.* 10, 168 (1969).
9. Lempicki, A. and Heller, A., Characteristics of the Nd^{+3}:$SeOCl_y$ liquid laser, *Appl. Phys. Lett.* 9, 108 (1966).
10. Kato, D. and Shimoda, K., Liquid $SeOCl_2$:Nd^{+3} laser of high quality, *Jpn. J. Appl. Phys.* 7, 548 (1968).
11. Heller, A., A high gain, room-temperature liquid laser: trivalent neodymium in selenium oxychloride, *Appl. Phys. Lett.* 9, 106 (1966), and Liquid lasers - design of neodymium based inorganic systems, *J. Mol. Spectrosc.* 28, 101 (1968).
12. Samelson, H., Kocher, R., Waszak, T. and Kellner, S., Oscillator and amplifier characteristics of lasers based on Nd^{+3} dissolved in aprotic solvents, *J. Appl. Phys.* 41, 2459 (1970).
13. Yamaguchi, G., Endo, F., Murakawa, S., Okamura, S. and Yamanaka, C., Room temperature, Q-switched liquid laser ($SeOCl_2$-Nd^{+3}), *Jpn. J. Appl. Phys.* 7, 179 (1968).
14. Samelson, H. and Lempicki, A., Q switching and mode locking of Nd^{+3}:$SeOCl_2$ liquid laser, *J. Appl. Phys.* 39, 6115 (1968).
15. Yamanaka, C., Yamanaka, T., Yamaguchi, G., Sasaki, T. and Nakai, S., Tandem amplifier systems of glass and $SeOCl_2$ liquid lasers doped with neodymium, *Nachrichten Tech. Fachberichte* 35, 791 (1968).
16. Lang, R. S., Die erzeugung von reisen impulsen durch einen aktiv und passiv geschalteten anorganischen neodym-flüssigkeits laser, *Z. Naturforsch.* 25a, 1354 (1970).
17. Zaretskii, A. I., Vladimirova, S. I., Kirillov, G. A., Kormes, S. B., Negiva, V. R. and Sukharov, S. A., Some characteristics of a $POCl_3$ + $SnCl_4$ + Nd^{+3} inorganic liquid laser, *Sov. J. Quantum Electron.* 4, 646 (1974).
18. Samelson, H. and Kocher, R., Final Technical Report, High Energy Liquid Lasers, Contract N0001468-C-0110 (1974).
19. Green, M., Reou, D., Little, V. I. and Selden, A. C., A multigigawatt liquid laser amplifier, *J. Phys.* D 9, 701 (1976).
20. Ueda, K., Hongyo, M., Sasaki, T. and Yamanaka, C., High power Nd^{+3} $POCl_3$ liquid laser system, *IEEE J. Quantum Electron.* QE-7, 291 (1971).
21. Brinkschulte, H., Fill, E. and Lang, R., Spectral output properties of an inorganic liquid laser, *J. Appl. Phys.* 43, 1807 (1972).

22. Brinkschulte, H., Perchermeier, J. and Schimitschek, E. J., A repetitively pulsed, Q-switched, inorganic liquid laser, *J. Phys.* D-7, 1361 (1974).
23. Andreou, D., Little, V., Selden, A. C. and Katzenstein, J., Output characteristics of a Q-switched laser system, $Nd^{+3}:POCl_3:ZrCl_4$, *J. Phys.* D-5, 59 (1972).
24. Fahlen, T. S., High average power Q-switched liquid laser, *IEEE J. Quantum Electron.* QE-9, 493 (1973).
25. Hongyo, M., Sasaki, T., Ngao, Y., Ueda, K. and Yamanaka, C., High power $Nd^{+3}:POCl_3$ liquid laser system, *IEEE J. Quantum Electron.* QE-8,192 (1972).
26. Mochalov, I. V., Bondareva, N. P., Bondareva, A. S. and Markosov, S. A., Spectral, luminescence and lasing properties of Nd^{3+} ions in systems utilizing $GaCl_3$-$SOCl_2$ and $AlCl_3$-$SOCl_2$ inorganic liquid media, *Sov. J. Quantum Electron.* 12, 647 (1982); Mokhova E. A. and Sviridov, V. V., Luminescence and lasing properties of $SOCl_2$-$GaCl_3$-Nd^{3+} inorganic laser liquids, *Zh. Prinkl. Spectrosk.* 50, 609 (1989).
27. Batyaev, I. M., Kabatskii, Yu. A. and Shilov, S. M., Luminescence spectrum and lasing parameters for Nd^{3+} in the $SOCl_2$-$GaCl_3$-$NdCl_3$ system, *Inorg. Mater.* 27, 1633 (1991).
28. Blumenthal, N., Ellis, C. B. and Grafstein, D., New room temperature liquid laser: Nd(III) in $POCl_3$-$SnCl_4$, *J. Chem. Phys.* 48, 5726 (1968).
29. Batyaev, I. M. and Kabatskii, Yu. A., Luminescence spectrum and lasing parameters for the CCl_3-$GaCl_3$-Nd^{3+} system, *Inorg. Mater.* 27, 1630 (1991).
30. Bondarev, A. S., Buchenkov, V. A., Volyukin, V. M., Mak, A. A., Pogodaev, A. K., Przhevaskii, A. K., Sidorenko, Yu. K., Soms, L. N. and Stepanov, A. I., New low toxicity inorganic Nd^{+3}-activated liquid medium for lasers, *Sov. J. Quantum Electron.* 6, 202 (1976).
31. Heller, A. and Brophy, V., Liquid lasers: stimulated emission of Nd^{+3} in selenium oxychloride solutions in the $^4F_{3/2}$ to $^4I_{13/2}$ transition, *J. Appl. Phys.* 39, 6120 (1968).
32. Sasaki, T., Yamanaka, T., Yamaguchi, G. and Yamanaka, C., A construction of the high power laser amplifier using glass and selenium oxychloride doped with Nd^{+3}, *Jpn. J. Appl. Phys.* 8, 1037-1045 (1969).
33. Andreou, D., A high power liquid laser amplifier, *J. Phys.* D-7, 1073 (1974).
34. Andreou, D., On the growth of stimulated Raman scattering in amplifying media, *Phys. Lett.* 57A, 250 (1976).
35. Fill, E. E., Ein Nd-$POCl_3$ laser verstärker, *Z. Angew. Phys.* 32. 356 (1972).
36. Andreou, D., Selden, A. C. and Little, V. I., Amplification of mode locked trains with a liquid laser amplifier, $Nd^{+3}:POCl_3:ZrCl_4$, *J. Phys.* D-5, 1405 (1972).
37. Andreou, D. and Little, V. I., The effect of frequency shifts on the power gain of a laser amplifier, *Opt. Commun.* 6, 180 (1972).
38. Han, K. G., Kong, H. J., Kim, H. S. and Um, G. Y., Nd^{3+}:ethylene glyol amplifier and its stimulated emission cross section, *Appl. Phys. Lett.* 67, 1501 (1995).

Section 4: Gas Lasers

Section 4.1
NEUTRAL ATOM, ION, AND MOLECULAR GAS LASERS

Introduction to the Table

Lasers involving neutral atoms, ions, and molecules are listed in order of increasing wavelength in Table 4.1.1. The uncertainty in the wavelength determination and whether the wavelength value is for air or vacuum, if given, are noted in columns two and three. The lasing species and charge are given in the following two columns. The references generally include the original report of lasing plus other reports relevant to the identification of the lasing transition and operation. The references with titles or descriptions of the contents are given in Section 4.3.

Further Reading

Bennett, W. R., Jr., *Atomic Gas Laser Transition Data,* Plenum, New York (1979).

Bridges, W. B., Ionized Gas Lasers, in *Handbook of Laser Science and Technology, Vol. II: Gas Lasers*, CRC Press, Boca Raton, FL (1982), p. 171.

Chang, T.-Y., Vibrational Transition Lasers, in *Handbook of Laser Science and Technology, Vol. II: Gas Lasers*, CRC Press, Boca Raton, FL (1982), p. 313 and *Suppl. 1: Lasers,* CRC Press, Boca Raton, FL (1991), p. 387.

Cheo, P. K., Ed., *Handbook of Molecule Lasers*, Marcel Dekker Inc., New York (1987).

Davis, C. C., Neutral Gas Lasers, in *Handbook of Laser Science and Technology, Vol. II: Gas Lasers*, CRC Press, Boca Raton, FL (1982), p. 3.

Davis, R. S. and Rhodes, C. K., Electronic Transition Lasers, in *Handbook of Laser Science and Technology, Vol. II: Gas Lasers*, CRC Press, Boca Raton, FL (1982), p. 273.

Douglas, N. G., *Millimetre and Submillimetre Wavelength Lasers: A Handbook of cw Measurements*, Springer-Verlag, Berlin (1987).

Eden, J. G., Electronic Transition Lasers, in *Handbook of Laser Science and Technology, Suppl. 1: Lasers,* CRC Press, Boca Raton, FL (1991), p. 341.

Eden, J. G., Ed., *Selected Papers on Gas Lasers*, SPIE Milestone Series, SPIE Optical Engineering Press, Bellingham, WA (in press).

Evans, J. D., Ed., *Selected Papers on CO_2 Lasers*, SPIE Milestone Series, Vol. MS 24, SPIE Optical Engineering Press, Bellingham, WA (1990).

Goldhar, J., Ionized Gas Lasers, in *Handbook of Laser Science and Technology, Suppl. 1: Lasers,* CRC Press, Boca Raton, FL (1991), p. 325.

Hooker, S. M. and Webb, C. E., Progress in vacuum ultraviolet lasers, *Progress in Quantum Electronics* 18, 227 (1994).

King, D., Photoionization-Pumped Short Wavelength Lasers, in *Handbook of Laser Science and Technology, Suppl. 1: Lasers,* CRC Press, Boca Raton, FL (1991), p. 531.

Knight, D. J. E., Far-Infrared CW Gas Lasers, in *Handbook of Laser Science and Technology, Vol. II: Gas Lasers*, CRC Press, Boca Raton, FL (1982), p. 411 and *Suppl. 1: Lasers,* CRC Press, Boca Raton, FL (1991), p. 415.

Peterson, A. B., Ionized Gas Lasers in *Handbook of Laser Science and Technology, Suppl. 1: Lasers,* CRC Press, Boca Raton, FL (1991), p. 335.

Rhodes, C. K. (Ed.), *Excimer Lasers*, 2nd edition, Springer-Verlag, Berlin (1984).

Waynant, R. W. and Ediger, M. N., Eds., *Selected Papers on UV, VUV, and X-Ray Lasers*, SPIE Milestone Series, Vol. MS71, SPIE Optical Engineering Press, Bellingham, WA (1993).

Witteman, W. J., *The CO_2 Laser*, Springer Verlag, Berlin (1987).

Table 4.1.1
Neutral Atom, Ion, and Molecular Gas Lasers Arranged in Order of Increasing Wavelength

Wavelength (μm)	Uncertainty	Medium	Species	Charge	Reference
0.0907		vacuum	Kr	2	892
0.0969		vacuum	Cs	0	903
0.1089		vacuum	Xe	2	892,896
0.10982		air	para-H_2	0	66
0.11020		air	H_2	0	66-72,74
0.11134		air	D_2	0	66
0.11152		air	para-H_2	0	66
0.11189		air	H_2	0	66-72,74
0.11377		air	D_2	0	66
0.11386		air	HD	0	66
0.11415		air	HD	0	66
0.11446		air	para-H_2	0	66
0.11476		air	D_2	0	66
0.11486		air	H_2	0	66-72,74
0.11520		air	HD	0	66
0.11565		air	D_2	0	66
0.11584		air	D_2	0	66
0.11600		air	Para-H_2	0	66
0.11613		air	H_2	0	66-72,74
0.11639		air	H_2	0	66-72,74
0.11662		air	H_2	0	66-72,74
0.11746		air	para-H_2	0	66
0.11758		air	H_2	0	66-72,74
0.11763			H_2	0	1535
0.11777			H_2	0	1535
0.11781		air	HD	0	66
0.11783		air	H_2	0	66-72,74
0.11805		air	H_2	0	66-72,74
0.11881		air	D_2	0	66
0.11893		air	H_2	0	66-72,74
0.11900		air	HD	0	66
0.11901		air	D_2	0	66
0.11928		air	HD	0	66
0.11975		air	D_2	0	66
0.11994		air	D_2	0	66

Table 4.1.1—*continued*
Neutral Atom, Ion, and Molecular Gas Lasers Arranged in Order of Increasing Wavelength

Wavelength (μm)	Uncertainty	Medium	Species	Charge	Reference
0.12010		air	HD	0	66
0.12054		air	para–H_2	0	66
0.12064		air	D_2	0	66
0.12067		air	H_2	0	66–72,74
0.12082		air	D_2	0	66
0.12093		air	H_2	0	66–72,74
0.12113		air	HD	0	66
0.12173		air	H_2	0	66–72,74
0.12177		air	para–H_2	0	66
0.12189		air	H_2	0	66–72,74
0.12214		air	H_2	0	66–72,74
0.12236		air	H_2	0	66–72,74
0.12280		air	D_2	0	66
0.12284		air	HD	0	66
0.12287		air	para–H_2	0	66
0.12299		air	H_2	0	66–72,74
0.12323		air	H_2	0	66–72,74
0.12356		air	D_2	0	66
0.12383		air	para–H_2	0	66
0.12394		air	H_2	0	66–72,74
0.124–0.128			Ar_2	0	1606,1622
0.12417		air	H_2	0	66--72,74
0.12424		air	D_2	0	66
0.12441		air	D_2	0	66
0.12457		air	HD	0	66,68
0.12462		air	para–H_2	0	66
0.12483		air	D_2	0	66
0.12500		air	D_2	0	66
0.12520		air	para–H_2	0	66
0.12528		air	HD	0	66,68
0.12533		air	D_2	0	66
0.1261		air	Ar_2	0	1,51,116
0.1270		vacuum	Zn	2	1817
0.12752			H_2	0	1535
0.12795		air	para–H_2	0	66,68

Table 4.1.1—*continued*
Neutral Atom, Ion, and Molecular Gas Lasers Arranged in Order of Increasing Wavelength

Wavelength (μm)	Uncertainty	Medium	Species	Charge	Reference
0.1280			H_2	0	1535
0.13033		air	HD	0	66,68
0.13036		air	D_2	0	66,68
0.1306		vacuum	Zn	2	1817
0.1319		vacuum	Zn	2	1817
0.13336			H_2	0	1535
0.13386		air	para–H_2	0	66,68
0.1339			H_2	0	1535
0.13423		air	H_2	0	66–72,74
0.13459		air	D_2	0	66,68
0.13551		air	HD	0	66,68
0.13598		air	para–H_2	0	66,68
0.13680		air	para–H_2	0	66,68
0.13888		air	D_2	0	66,68
0.13944			H_2	0	1535
0.13990		air	para–H_2	0	66,68
0.13998			H_2	0	1535
0.14026		air	H_2	0	66–72,74
0.14075		air	para–H_2	0	66,68
0.14077		air	HD	0	66,68
0.14187		air	H_2	0	66–72,74
0.14288			H_2	0	1535
0.14322		air	D_2	0	66,68
0.14326		air	para–H_2	0	66,68
0.14362		air	H_2	0	66–72,74
0.14376		air	para–H_2	0	66,68
0.14406		air	para–H_2	0	66,68
0.14409		air	H_2	0	66–72,74
0.14409			H_2	0	1535
0.145–0.146			Kr_2	0	1606,1622
0.14555			H_2	0	1535
0.1457		air	Kr_2	0	40,73,84,116
0.14602		air	para–H_2	0	66,68
0.14609			H_2	0	1535
0.14638		air	H_2	0	66–72,74

Table 4.1.1—*continued*
Neutral Atom, Ion, and Molecular Gas Lasers Arranged in Order of Increasing Wavelength

Wavelength (μm)	Uncertainty	Medium	Species	Charge	Reference
0.14641		air	para–H_2	0	66,68
0.14670		air	H_2	0	66–72,74
0.14684		air	para–H_2	0	66,68
0.14865		air	H_2	0	66–72,74
0.14876			H_2	0	1535
0.14884		air	HD	0	66,68
0.14917		air	para–H_2	0	66,68
0.14942		air	H_2	0	66–72,74
0.14952		air	H_2	0	66–72,74
0.14996			H_2	0	1535
0.15136		air	HD	0	66,68
0.15157		air	para–H_2	0	66,68
0.15199		air	para–H_2	0	66,68
0.15233		air	H_2	0	66–72,74
0.15299		air	HD	0	66,68
0.15315			H_2	0	1535
0.15349		air	para–H_2	0	66,68
0.15449		air	H_2	0	66–72,74
0.15454			H_2	0	1535
0.154820	±0.000030	vacuum	C	3	1099,1048
0.15501		air	para–H_2	0	66,68
0.155090	±0.000030	vacuum	C	3	1099,1048
0.15534		air	H_2	0	66–72,74
0.15574			H_2	0	1535
0.15620		air	HD	0	66,68
0.15640		air	para–H_2	0	66,68
0.15655		air	H_2	0	66–72,74
0.15663		air	H_2	0	66–72,74
0.15671		air	F_2	0	63–65
0.15673		air	H_2	0	66–72,74
0.15675		air	para–H_2	0	66,68
0.15708			H_2	0	1535
0.15713		air	HD	0	66,68
0.15720		air	H_2	0	66–72,74
0.15727		air	HD	0	66,68

Table 4.1.1—*continued*
Neutral Atom, Ion, and Molecular Gas Lasers Arranged in Order of Increasing Wavelength

Wavelength (μm)	Uncertainty	Medium	Species	Charge	Reference
0.15743		air	HD	0	66,68
0.15743		air	para–H_2	0	66,68
0.15748		air	F_2	0	63–65
0.15758		air	D_2	0	66,68
0.15759		air	F_2	0	63–65
0.15774		air	H_2	0	66–72,74
0.15777		air	H_2	0	66–72,74
0.15777		air	para–H_2	0	66,68
0.15792		air	H_2	0	66–72,74
0.15800		air	H_2	0	66–72,74
0.15800		air	para–H_2	0	66,68
0.15801		air	HD	0	66,68
0.15808		air	H_2	0	66–72,74
0.15809		air	HD	0	66,68
0.15811		air	para–H_2	0	66,68
0.15814		air	para–H_2	0	66,68
0.15819		air	HD	0	66,68
0.15825		air	HD	0	66,68
0.15831		air	HD	0	66,68
0.15849			H_2	0	1535
0.15863		air	D_2	0	66,68
0.15864		air	D_2	0	66,68
0.15867		air	D_2	0	66,68
0.15869		air	D_2	0	66,68
0.15871		air	D_2	0	66,68
0.15872		air	D_2	0	66,68
0.15890		air	H_2	0	66–72,74
0.15890		air	para–H_2	0	66,68
0.15898		air	D_2	0	66,68
0.15913		air	D_2	0	66,68
0.15913		air	H_2	0	66–72,74
0.15914		air	D_2	0	66,68
0.15923		air	D_2	0	66,68
0.15926		air	D_2	0	66,68
0.15934		air	H_2	0	66–72,74

Table 4.1.1—*continued*
Neutral Atom, Ion, and Molecular Gas Lasers Arranged in Order of Increasing Wavelength

Wavelength (μm)	Uncertainty	Medium	Species	Charge	Reference
0.15934		air	para–H_2	0	66,68
0.15938		air	HD	0	66,68
0.15955		air	HD	0	66,68
0.15961		air	H_2	0	66–72,74
0.15974		air	HD	0	66,68
0.15993		air	para–H_2	0	66,68
0.16009		air	D_2	0	66,68
0.16021		air	D_2	0	66,68
0.16023		air	HD	0	66,68
0.16024		air	para–H_2	0	66,68
0.16035		air	D_2	0	66,68
0.16037		air	HD	0	66,68
0.16046		air	HD	0	66,68
0.16049		air	H_2	0	66–72,74
0.16052		air	HD	0	66,68
0.16057		air	HD	0	66,68
0.16058		air	D_2	0	66,68
0.16059		air	H_2	0	66–72,74
0.16059		air	para–H_2	0	66,68
0.16062		air	H_2	0	66–72,74
0.16062		air	para–H_2	0	66,68
0.16065		air	D_2	0	66,68
0.16065		air	HD	0	66,68
0.16067		air	HD	0	66,68
0.16068		air	D_2	0	66,68
0.16068		air	HD	0	66,68
0.16069		air	HD	0	66,68
0.16075		air	H_2	0	66–72,74
0.16075		air	HD	0	66,68
0.16077		air	D_2	0	66,68
0.16079		air	HD	0	66,68
0.16083		air	H_2	0	66–72,74
0.16083		air	HD	0	66,68
0.16083		air	para–H_2	0	66,68
0.16084		air	H_2	0	66–72,74

Table 4.1.1—*continued*
Neutral Atom, Ion, and Molecular Gas Lasers Arranged in Order of Increasing Wavelength

Wavelength (μm)	Uncertainty	Medium	Species	Charge	Reference
0.16085		air	D_2	0	66,68
0.16090		air	H_2	0	66–72,74
0.16091		air	HD	0	66,68
0.16096		air	D_2	0	66,68
0.16096		air	para–H_2	0	66,68
0.16103		air	H_2	0	66–72,74
0.16103		air	HD	0	66,68
0.16103		air	para–H_2	0	66,68
0.16107		air	D_2	0	66,68
0.16108		air	D_2	0	66,68
0.16109		air	para–H_2	0	66,68
0.16113		air	HD	0	66,68
0.16115		air	D_2	0	66,68
0.16117		air	D_2	0	66,68
0.16117		air	H_2	0	66–72,74
0.16117		air	para–H_2	0	66,68
0.16120		air	D_2	0	66,68
0.16124		air	D_2	0	66,68
0.16126		air	D_2	0	66,68
0.16132		air	D_2	0	66,68
0.16132		air	H_2	0	66–72,74
0.16132		air	para–H_2	0	66,68
0.16141		air	D_2	0	66,68
0.16148		air	H_2	0	66–72,74
0.16149		air	para–H_2	0	66,68
0.16165		air	H_2	0	66–72,74
0.16166		air	D_2	0	66,68
0.16395		air	H_2	0	66–72,74
0.164		vacuum	He	1	1813
0.16415		air	H_2	0	66–72,74
0.16429		air	H_2	0	66–72,74
0.16429		air	para–H_2	0	66,68
0.16444		air	H_2	0	66–72,74
0.16460		air	para–H_2	0	66,68
0.1690		air	ArCl	0	23,35,36

Table 4.1.1—*continued*
Neutral Atom, Ion, and Molecular Gas Lasers Arranged in Order of Increasing Wavelength

Wavelength (μm)	Uncertainty	Medium	Species	Charge	Reference
0.170–0.176			Xe_2	0	1606,1622
0.1716		air	Xe_2	0	35,15–20
0.1750		air	ArCl	0	23,35,36
0.175641	±0.000003	air	Kr	3	990
0.181085		vacuum	CO	0	55,56
0.183243	±0.000003	air	Kr	4	990
0.184343	±0.000003	vacuum	Ar	4	990
0.1850		vacuum	In	2	1818
0.187831		vacuum	CO	0	55,56
0.189784		vacuum	CO	0	55,56
0.1933		air	ArF	0	33–35,37,38
0.195006		vacuum	CO	0	55,56
0.196808	±0.000003	air	Kr	3	990
0.197013		vacuum	CO	0	55,56
0.201842	±0.000001	air	Ne	3	989,1081
0.202219	±0.000001	air	Ne	3	989,990,1081
0.205108	±0.000001	air	Kr	3	989,990
0.2055		air	Cd	1	1816
0.206530	±0.000001	air	Ne	2	989,990
0.206530	±0.000001	air	Ne	3	989,990
0.211398	±0.000001	air	Ar	3	989
0.2163			C	2	1834
0.2177			C	2	1834
0.217770	±0.000001	air	Ne	2	989
0.218086	±0.000001	air	Ne	2	989
0.2181			NO	0	1843,1844
0.21822	±0.00008	air	NO	0	1808
0.219192	±0.000001	air	Kr	3	989,990
0.21920			Kr	3	926
0.222		air	KrCl	0	23,33,36,93,94
0.223244	±0.000001	air	Xe	2	989
0.223244	±0.000001	air	Xe	3	989
0.2235	±.0005		KrCl	0	1572
0.224340	±0.000020	air	Ag	1	1001,937
0.224884	±0.000001	air	Ar	3	989

Table 4.1.1—*continued*
Neutral Atom, Ion, and Molecular Gas Lasers Arranged in Order of Increasing Wavelength

Wavelength (μm)	Uncertainty	Medium	Species	Charge	Reference
0.225464	±0.000001	air	Kr	3	989
0.22613	±0.00008	air	NO	0	1808
0.226400		air	Au	1	938
0.226570		air	Ne	4	989
0.227760	±0.000020	air	Ag	1	1001
0.228579	±0.000001	air	Ne	3	989
0.2312		air	Cd	1	1815
0.23152			Xe	3	1487
0.231536	±0.000001	air	Xe	2	989
0.231536	±0.000001	air	Xe	3	989
0.233848	±0.000001	air	Kr	3	989
0.235255	±0.000001	air	Ne	3	989
0.235798	±0.000001	air	Ne	3	653,989
0.236246	±0.000001	air	Br	3	991
0.237			NO	0	1583
0.237320	±0.000001	air	Ne	3	989
0.23869			Kr	3	926
0.24071			Kr	3	926
0.24177			Kr	3	926
0.241784	±0.000001	air	Kr	2	989
0.241784	±0.000001	air	Kr	3	989
0.24317			Kr	3	926
0.24433			Xe	3	1487
0.24504			Kr	3	926
0.2470		air	CO	1	61
0.247340	±0.000001	air	Ne	2	653,989
0.247739	±0.000003	air	Xe	0	653,874,983
0.248			NO	0	1583
0.2484		air	KrF	0	37,38,96–102
0.248580	±0.000010	air	Cu	1	937,999
0.24891			Kr	3	926
0.2491		air	KrF	0	37,38,96–102
0.250650	±0.000010	air	Cu	1	999
0.251330	±0.000002	air	Ar	3	989
0.252666			Xe	3	1487

Table 4.1.1—*continued*
Neutral Atom, Ion, and Molecular Gas Lasers Arranged in Order of Increasing Wavelength

Wavelength (μm)	Uncertainty	Medium	Species	Charge	Reference
0.252666	±0.000001	air	Xe	3	989
0.252920	±0.000010	air	Cu	1	937,1000
0.253350	±0.000020	air	Au	1	673
0.25563			Xe	3	1487
0.2573		air	Cd	1	1815
0.2580		air	Cl_2	0	55,56
0.258125	±0.000001	air	Br	3	991
0.259060	±0.000010	air	Cu	1	918,937,999
0.25973			Kr	3	926
0.259900	±0.000010	air	Cu	1	691,937,999
0.25993			Ar	3	1825
0.260030	±0.000010	air	Cu	1	691,937,1000
0.26048			Kr	3	926
0.260998	±0.000001	air	Ne	2	989
0.261340	±0.000010	air	Ne	2	989
0.261640	±0.000020	air	Au	1	673
0.262138	±0.000001	air	Ar	3	989
0.262488	±0.000001	air	Ar	3	653,989
0.263269	±0.000001	air	Cl	2	220,991,1020
0.263896	±0.000001	air	S	2	991,1489
0.263896	±0.000001	air	S	3	991,1489
0.264936	±0.000001	air	Kr	3	653,989,1125
0.26494			Kr	3	926
0.26644			Kr	3	926
0.266440	±0.000001	air	Kr	3	653,989,1125
0.2677			Ne	2	450,907,989
0.267792	±0.000001	air	Ne	2	450,907,989
0.267869	±0.000001	air	Ne	2	450,907,989
0.26919			Xe	3	1487
0.269194	±0.000001	air	Xe	3	653,874,983
0.270070	±0.000030	air	Cu	1	1161
0.270310	±0.000010	air	Cu	1	918,937,1000
0.2722		air	Cu	1	937,938
0.274138	±0.000001	air	Kr	3	653,989
0.27414			Kr	3	926

Table 4.1.1—*continued*
Neutral Atom, Ion, and Molecular Gas Lasers Arranged in Order of Increasing Wavelength

Wavelength (μm)	Uncertainty	Medium	Species	Charge	Reference
0.2748		air	Cd	1	1815
0.275388	±0.000001	air	Ar	2	450,985
0.275959	±0.000006	air	F	2	220,1022
0.27620			Ar	2	1825
0.27680			Xe	2	1487
0.277763	±0.000001	air	Ne	2	450,989
0.278150	±0.000050	air	O	4	908
0.278762	±0.000001	air	Br	2	991
0.2818		air	XeBr	0	123–126
0.28199			Xe	2	1487
0.282220	±0.000020	air	Au	1	673,691,937
0.28227			Xe	2	1487
0.2823–0.2832			ClF	0	1534
0.282608	±0.000006	air	F	3	220
0.2840			ClF	0	1527
0.2844			ClF	0	1527,1534
0.284720	±0.000020	air	Au	1	673,691,937
0.2849			ClF	0	1534
0.2850		air	ClF	0	56
0.285537	±0.000001	air	Ar	2	989
0.2860			ClF	0	1534
0.286673	±0.000001	air	Ne	2	450,653,989
0.28741			Xe	2	1487
0.288422	±0.000001	air	Ar	2	653,989,1153
0.289350	±0.000020	air	Au	1	673,691,937
0.29085			Kr	3	926
0.291292	±0.000001	air	Ar	3	450,907,989
0.2915		air	Br_2	0	50,52–54
0.29152			Kr	3	926
0.291810	±0.000020	air	Au	1	673,691,937
0.2925			Ta	0	1849
0.292623	±0.000001	air	Ar	3	450,653,989
0.29310			Kr	3	926
0.29547			Xe	2	1487
0.29707			Xe	2	1487

Table 4.1.1—*continued*
Neutral Atom, Ion, and Molecular Gas Lasers Arranged in Order of Increasing Wavelength

Wavelength (μm)	Uncertainty	Medium	Species	Charge	Reference
0.298389	±0.000006	air	O	2	450,908,989
0.29858			Kr	3	926
0.299951		air	Fe	0	415
0.300–0.350			SO	0	1848
0.300264	±0.000001	air	Ar	2	450,907,989
0.301618		air	Fe	0	415
0.302400	±0.000050	air	Ar	2	450,985
0.303164		air	Fe	0	415
0.304043		air	Fe	0	415
0.30438			Xe	0	1487
0.30438			Xe	2	1487
0.304715	±0.000006	air	O	2	450,653,908
0.30497			Kr	2	926
0.304970	±0.000002	air	Kr	2	450,653,989
0.3051		air	Tm	0	1493
0.305480	±0.000050	air	Ar	2	450,985
0.30633			Xe	0	1487
0.306346	±0.000006	air	O	3	653,908,989
0.30649			Ar	2	1825
0.3070		air	XeCl	0	3–6,33–36,126
0.3073		air	XeCl	0	3–6,33–36,126
0.30765		air	XeCl	0	3–6,33–36,126
0.30783			Ar	2	1825
0.3079		air	Tm	0	1493
0.30792		air	XeCl	0	3–6,33–36,126
0.307974	±0.000002	air	Xe	2	450,653,989
0.30817		air	XeCl	0	3–6,33–36,126
0.30843		air	XeCl	0	3–6,33–36,126
0.31049			Xe	0	1487
0.31089			Xe	3	1487
0.31095			Xe	0	1487
0.31104			Ar	2	1825
0.312150	±0.000001	air	F	2	220,991,1022
0.3122784		air	Au	0	132,134,135
0.312436	±0.000001	air	Kr	2	653,985,989

Table 4.1.1—*continued*
Neutral Atom, Ion, and Molecular Gas Lasers Arranged in Order of Increasing Wavelength

Wavelength (μm)	Uncertainty	Medium	Species	Charge	Reference
0.31260			Ar	2	1825
0.31260			Xe	2	1487
0.31379			Xe	0	1487
0.31722			Kr	2	926
0.317418	±0.000006	air	F	2	220,1022
0.318060	±0.000020	air	Ag	1	691,1001
0.31839			Xe	0	1487
0.319142	±0.000001	air	Cl	2	220,991
0.320274	±0.000006	air	F	1	220,1021
0.3227			Ta	0	1849
0.323951	±0.000001	air	Kr	2	450,653,989
0.324692	±0.000001	air	Xe	3	653,983,1024
0.3250		air	Cd	1	676–678,1034
0.3250		air	Cd	1	1812,1814
0.32596			Xe	0	1487
0.32676		air	Sb	0	1491
0.3269		air	Ge	0	1490
0.3281			Ta	0	1849
0.3304			Ta	0	1849
0.330596	±0.000001	air	Xe	3	450,653,923
0.33060			Xe	3	1487
0.331975	±0.000001	air	Ne	1	907,1030,1153
0.332375	±0.000001	air	Ne	1	907,987,1169
0.332486	±0.000001	air	S	2	991,1489
0.332750	±0.000050	air	Ne	1	450,1030
0.332902	±0.000010	air	Ne	1	1030,1153
0.333087	±0.000002	air	Xe	3	450,983,1122
0.333107	±0.000010	air	Ne	2	1153
0.33330			Xe	0	1487
0.333621	±0.000006	air	Ar	2	450,985,1169
0.334479	±0.000006	air	Ar	2	450,985,1169
0.334545	±0.000002	air	Ne	1	454,653,1153
0.334776	±0.000006	air	P	3	220
0.334974	±0.000006	air	Xe	2	653,874,983
0.335852	±0.000006	air	Ar	2	450,653,985

Table 4.1.1—*continued*
Neutral Atom, Ion, and Molecular Gas Lasers Arranged in Order of Increasing Wavelength

Wavelength (μm)	Uncertainty	Medium	Species	Charge	Reference
0.3364903		air	N_2	0	104,105,109–11
0.3365474		air	N_2	0	104,105,109–11
0.3365537		air	N_2	0	104,105,109–11
0.3366156		air	N_2	0	104,105,109–11
0.3366211		air	N_2	0	104,105,109–11
0.3366682		air	N_2	0	104,105,109–11
0.3366911		air	N_2	0	104,105,109–11
0.3367218		air	N_2	0	104,105,109–11
0.336732	±0.000006	air	N	2	653
0.3368432		air	N_2	0	104,105,109–11
0.3368917		air	N_2	0	104,105,109–11
0.3369250		air	N_2	0	104,105,109–11
0.3369361		air	N_2	0	104,105,109–11
0.3369502		air	N_2	0	104,105,109–11
0.3369542		air	N_2	0	104,105,109–11
0.3369555		air	N_2	0	104,105,109–11
0.3369575		air	N_2	0	104,105,109–11
0.3369760		air	N_2	0	104,105,109–11
0.3369838		air	N_2	0	104,105,109–11
0.3369852		air	N_2	0	104,105,109–11
0.3370081		air	N_2	0	104,105,109–11
0.3370121		air	N_2	0	104,105,109–11
0.3370138		air	N_2	0	104,105,109–11
0.3370161		air	N_2	0	104,105,109–11
0.3370169		air	N_2	0	104,105,109–11
0.3370297		air	N_2	0	104,105,109–11
0.3370316		air	N_2	0	104,105,109–11
0.3370360		air	N_2	0	104,105,109–11
0.3370374		air	N_2	0	104,105,109–11
0.3370434		air	N_2	0	104,105,109–11
0.3370472		air	N_2	0	104,105,109–11
0.3370529		air	N_2	0	104,105,109–11
0.3370551		air	N_2	0	104,105,109–11
0.3370559		air	N_2	0	104,105,109–11
0.3370614		air	N_2	0	104,105,109–11

Table 4.1.1—*continued*
Neutral Atom, Ion, and Molecular Gas Lasers Arranged in Order of Increasing Wavelength

Wavelength (μm)	Uncertainty	Medium	Species	Charge	Reference
0.3370623		air	N_2	0	104,105,109–11
0.3370663		air	N_2	0	104,105,109–11
0.3370682		air	N_2	0	104,105,109–11
0.3370716		air	N_2	0	104,105,109–11
0.3370731		air	N_2	0	104,105,109–11
0.3370757		air	N_2	0	104,105,109–11
0.3370762		air	N_2	0	104,105,109–11
0.3370787		air	N_2	0	104,105,109–11
0.3370803		air	N_2	0	104,105,109–11
0.3370821		air	N_2	0	104,105,109–11
0.3370843		air	N_2	0	104,105,109–11
0.3370924		air	N_2	0	104,105,109–11
0.3370941		air	N_2	0	104,105,109–11
0.3370990		air	N_2	0	104,105,109–11
0.3371042		air	N_2	0	104,105,109–11
0.3371082		air	N_2	0	104,105,109–11
0.3371120		air	N_2	0	104,105,109–11
0.3371129		air	N_2	0	104,105,109–11
0.3371141		air	N_2	0	104,105,109–11
0.3371147		air	N_2	0	104,105,109–11
0.3371179		air	N_2	0	104,105,109–11
0.3371271		air	N_2	0	104,105,109–11
0.3371312		air	N_2	0	104,105,109–11
0.3371371		air	N_2	0	104,105,109–11
0.3371398		air	N_2	0	104,105,109–11
0.3371427		air	N_2	0	104,105,109–11
0.3371433		air	N_2	0	104,105,109–11
0.337500	±0.000050	air	Kr	2	450,985,1169
0.337826	±0.000001	air	Ne	1	450,454,907
0.3379898		air	N_2	0	104,105,109–11
0.338134	±0.000006	air	O	3	653,1130
0.338134	±0.000006	air	O	3	653,1130
0.338554	±0.000006	air	O	3	653,908,1130
0.339280	±0.000001	air	Ne	1	450,1018,1169
0.339286	±0.000001	air	Cl	2	220,991

Table 4.1.1—*continued*
Neutral Atom, Ion, and Molecular Gas Lasers Arranged in Order of Increasing Wavelength

Wavelength (μm)	Uncertainty	Medium	Species	Charge	Reference
0.339340	±0.000010	air	Ne	1	1030,1153
0.339344	±0.000001	air	Cl	2	220,991
0.3400–0.3600			Na_2	0	1584
0.3420		air	I_2	0	85–87
0.3420–0.3428			I_2	0	1548–1553
0.3423		air	I_2	0	85–87
0.3424		air	I_2	0	85–87
0.3428		air	I_2	0	85–87
0.345132		air	B	1	200
0.345425	±0.000001	air	Xe	2	653,1024,1169
0.347			Ni	0	1845
0.347876	±0.000005	air	N	3	664,907,989
0.348302	±0.000006	air	N	3	653,902
0.348331	±0.000003	air	Xe	3	450,653,874,
0.3488		air	XeF	0	7–9,12,37
0.3497		air	Tm	0	1493
0.349733	±0.000001	air	S	2	991,1489
0.34995			Ar	2	1825
0.35035			Ar	2	1825
0.350742	±0.000001	air	Kr	2	450,1092,1169
0.35092			Ar	2	1825
0.351			XeF	0	1607
0.3511		air	XeF	0	7–9,12,37
0.351112	±0.000006	air	Ar	2	450,454,997
0.3514			Ta	0	1849
0.351415	±0.000006	air	Ar	2	450,997,1123
0.353			XeF	0	1608,1609
0.353002	±0.000001	air	Cl	2	220,991
0.3531		air	XeF	0	7–9,12,37
0.3540		air	XeF	0	7–9,12,37
0.3542			BrF	0	1527
0.354231	±0.000005	air	Xe	2	454
0.3545			BrF	0	1527
0.356063	±0.000001	air	Cl	2	220,991
0.356420	±0.000006	air	Kr	2	653,989,1088

Table 4.1.1—*continued*
Neutral Atom, Ion, and Molecular Gas Lasers Arranged in Order of Increasing Wavelength

Wavelength (μm)	Uncertainty	Medium	Species	Charge	Reference
0.3566		air	Tm	0	1493
0.3568		air	Tm	0	1493
0.3575980		air	N_2	0	104,105,109–11
0.3576194		air	N_2	0	104,105,109–11
0.3576250		air	N_2	0	104,105,109–11
0.3576320		air	N_2	0	104,105,109–11
0.3576571		air	N_2	0	104,105,109–11
0.3576613		air	N_2	0	104,105,109–11
0.3576778		air	N_2	0	104,105,109–11
0.3576899		air	N_2	0	104,105,109–11
0.357690	±0.000050	air	Ar	1	450,1005
0.3576955		air	N_2	0	104,105,109–11
0.358			Ar	0	1807
0.358660	±0.000010	air	Al	1	423
0.358740	±0.000020	air	Al	1	878
0.359600	±0.000100	air	Xe	2	923,1169
0.360210	±0.000006	air	Cl	2	220
0.361210	±0.000006	air	Cl	2	220
0.362–0.570			S_2	0	1598
0.3620			S_2	0	1599
0.362269	±0.000006	air	Cl	2	220
0.363786	±0.000004	air	Ar	2	450,997,1172
0.36395677		air	Pb	0	204,239
0.36395677		air	Pb	0	204,239
0.364548	±0.000001	air	Xe	3	653,923,1153
0.36482			Xe	0	1487
0.36483			Xe	0	1487
0.36492			Kr	3	926
0.365–0.570		air	S_2	0	120,121
0.365015		air	Hg	0	156
0.365483		air	Hg	0	156
0.366288		air	Hg	0	156
0.366328		air	Hg	0	156
0.36667			Xe	0	1487
0.366920	±0.000003	air	Xe	2	653,874,1123

Table 4.1.1—*continued*
Neutral Atom, Ion, and Molecular Gas Lasers Arranged in Order of Increasing Wavelength

Wavelength (μm)	Uncertainty	Medium	Species	Charge	Reference
0.370520	±0.000050	air	Ar	0	450,1117
0.3706		air	Ca	1	976,1109,1166
0.370935	±0.000001	air	S	2	991,1489
0.3712			S_2	0	1599
0.371300	±0.000100	air	Ne	1	1030,1169,1172
0.3718		air	Tm	0	1493
0.372044	±0.000001	air	Cl	2	220,991
0.3722		air	Tm	0	1493
0.37268			Ne	1	1825
0.37270			Ne	1	1825
0.372711	±0.000050	air	O	1	908
0.37303			Xe	0	1487
0.37304			Xe	0	1487
0.3737		air	Ca	1	976,1109,1166
0.374573	±0.000006	air	Xe	2	653,1088,1169
0.374877	±0.000001	air	Cl	2	220,991
0.374947	±0.000004	air	O	1	450,664,964,989
0.375468	±0.000004	air	O	2	450,664,908,964
0.3755			O	2	1846
0.37572		air	O	2	989,1026
0.375989	±0.000005	air	O	2	450,664,908,964
0.375994	±0.000003	air	Xe	3	450,874,923,983
0.3760			O	2	1846
0.37622			Xe	2	1487
0.37688			Xe	0	1487
0.377134	±0.000005	air	Kr	1	454,1007
0.37738		air	O	2	1026
0.3775727		air	Tl	0	180
0.37765			Xe	2	1487
0.378099	±0.000002	air	Xe	2	450,653,1169
0.379528	±0.000006	air	Ar	2	450,1123
0.3801		air	Sn	0	1490
0.380329	±0.000003	air	Xe	3	653,874,1025
0.38035			Xe	3	1487
0.3807			S_2	0	1599

Table 4.1.1—*continued*
Neutral Atom, Ion, and Molecular Gas Lasers Arranged in Order of Increasing Wavelength

Wavelength (μm)	Uncertainty	Medium	Species	Charge	Reference
0.381			Ni	0	1845
0.38281			Xe	0	1487
0.384100	±0.000100	air	Xe	2	1123
0.384100	±0.000100	air	Xe	2	1123
0.385		air	POPOP	0	32
0.385826	±0.000006	air	Ar	2	450,1123
0.3869757			Se_2	0	1601
0.3870360			Se_2	0	1601
0.38742			Ar	2	1825
0.3907			S_2	0	1599
0.39079			Ar	2	1825
0.391			N_2	1	1581,1582
0.3914		air	N_2	1	112–115,127
0.3925932			Se_2	0	1601
0.3926552			Se_2	0	1601
0.3954		air	CO	1	61
0.3962			Al	0	1832
0.397301	±0.000003	air	Xe	3	653,874,1025
0.39732			Xe	3	1487
0.3983450			Se_2	0	1601
0.398399	±0.000002	air	Hg	1	930
0.3984086			Se_2	0	1601
0.399300	±0.000100	air	Xe	2	1016,1123
0.399300	±0.000100	air	Xe	2	1016,1123
0.399499	±0.000001	air	N	1	1092,1170,1173
0.4011			S_2	0	1599
0.402478	±0.000006	air	F	1	220,1021
0.4032987		air	Ga	0	177
0.404			XeF	0	1610
0.404		air	Hg	0	1504
0.4042357			Se_2	0	1601
0.4043010			Se_2	0	1601
0.4044136		air	K	0	350
0.4044136		air	K	0	350
0.40472602		air	K	0	350

Table 4.1.1—*continued*
Neutral Atom, Ion, and Molecular Gas Lasers Arranged in Order of Increasing Wavelength

Wavelength (μm)	Uncertainty	Medium	Species	Charge	Reference
0.404990	±0.000020	air	Xe	2	1015,1092
0.40578067		air	Pb	0	204,207,239
0.4059			N_2	0	1578–1580
0.406048	±0.000006	air	Xe	2	450,989,1169
0.4062			$PbBr_2$	0	1805
0.40621360		air	Pb	0	204,207,239
0.40621360		air	Pb	0	204,207,239
0.406736	±0.000006	air	Kr	2	450,1092,1172
0.4078		air	Sr	1	895
0.408620	±0.000020	air	Ag	1	691,948
0.408860	±0.000020	air	Ar	0	1123,1125
0.408890	±0.000010	air	Si	3	450,1090,1134
0.4095			V	0	1826
0.409729	±0.000006	air	N	2	664,907,989
0.40982			Ar	2	1825
0.4101745		air	In	0	178
0.410336	±0.000002	air	N	2	664,989,1092
0.4118		air	Tm	0	1493
0.4119			S_2	0	1599
0.4120		air	N	0	212
0.4122		air	Tm	0	1493
0.413138	±0.000006	air	Kr	2	450,974,1092
0.413250	±0.000010	air	Cl	1	1107
0.41349			Kr	3	926
0.41420			Xe	2	1487
0.414530	±0.000060	air	Xe	2	1016,1123
0.414660	±0.000004	air	Ar	2	450,1123
0.41500			Xe	0	1487
0.4151		air	Tm	0	1493
0.415445	±0.000004	air	Kr	2	450,1123
0.41604			Xe	0	1487
0.4162		air	Sr	1	976,1109
0.4164526			Se_2	0	1601
0.4165214			Se_2	0	1601
0.417181	±0.000010	air	Kr	2	450

Table 4.1.1—*continued*
Neutral Atom, Ion, and Molecular Gas Lasers Arranged in Order of Increasing Wavelength

Wavelength (μm)	Uncertainty	Medium	Species	Charge	Reference
0.4172042		air	Ga	0	177
0.418292	±0.000006	air	Ar	2	450,1081,1128,
0.41968			Xe	0	1487
0.420185		air	Rb	0	350
0.4210		air	CO	1	61
0.4210		air	CO	1	61
0.421405	±0.000006	air	Xe	2	450,1169
0.421556		air	Rb	0	350
0.422225	±0.000006	air	P	2	220
0.422651	±0.000006	air	Kr	2	450
0.42266			Kr	2	
0.4227890			Se_2	0	1601
0.4228597			Se_2	0	1601
0.4233			S_2	0	1599
0.42371			Kr	0	926
0.42375			Xe	0	1487
0.424026	±0.000010	air	Xe	2	450,1169
0.427260	±0.000006	air	Xe	2	450,1169
0.42781		air	N_2	1	112–115,127
0.428			N_2	1	1581,1582
0.428592	±0.000006	air	Xe	2	450
0.429633	±0.000005	air	Xe	1	454,1092
0.430	±.25		Kr_2F	0	1573
0.430–0.437			ICl	0	1557
0.430–0.452			Na_2	0	1587
0.4300		air	Kr_2F	0	129
0.4305		air	Sr	1	976,1109
0.430575	±0.000003	air	Xe	3	450,989,1092
0.43148			Ti	0	1850–1855
0.431800	±0.000020	air	Kr	1	978,1019,1151
0.43199			Kr	0	926
0.43283955		air	N	0	213
0.4340465		air	H	0	130
0.434738	±0.000004	air	O	1	450,664,908
0.435		air	$TbAl_3Cl_{12}$	0	26,27

Table 4.1.1—*continued*
Neutral Atom, Ion, and Molecular Gas Lasers Arranged in Order of Increasing Wavelength

Wavelength (μm)	Uncertainty	Medium	Species	Charge	Reference
0.435126	±0.000004	air	O	1	450,664,1092
0.4352			S_2	0	1599
0.435835		air	Hg	0	156
0.4359310			Se_2	0	1601
0.4360			Na_2	0	1586
0.4360		air	Tm	0	1493
0.4360198			Se_2	0	1601
0.437073	±0.000006	air	Ar	1	450,1005,1118
0.43760			Ar	1	1825
0.43760			Ar	1	1825
0.438360	±0.000060	air	Ar	1	1005,1016
0.438610	±0.000020	air	Kr	1	978,107,11517
0.441300	±0.000060	air	Xe	2	923,1016
0.441439	±0.000004	air	O	1	450,664,1092
0.4415		air	Tm	0	1493
0.441560	±0.000070	air	Cd	1	870,1064,1065
0.4416		air	Cd	1	1812,1814
0.441697	±0.000004	air	O	1	450,664,1092
0.442–0.444			HgI	0	1547
0.4427765			Se_2	0	1601
0.4428531			Se_2	0	1601
0.4430		air	HgI	0	78,79
0.443422	±0.000010	air	Xe	2	450,921
0.444328	±0.000004	air	Kr	2	450
0.4450		air	HgI	0	78,79
0.446800	±0.000050	air	Se	1	963
0.4472			CS_2	0	1804
0.4477			S_2	0	1599
0.448200	±0.000100	air	Ar	1	857,973,1005
0.448850	±0.000020	air	I	1	993,1037
0.450	±.01		Kr_2F	0	1574
0.450350	±0.000060	air	Xe	2	1016
0.450660	±0.000030	air	Cu	1	1174
0.451089	±0.000002	air	N	2	664,989,1092
0.4511			In	0	1832

Table 4.1.1—*continued*
Neutral Atom, Ion, and Molecular Gas Lasers Arranged in Order of Increasing Wavelength

Wavelength (μm)	Uncertainty	Medium	Species	Charge	Reference
0.4511299		air	In	0	178
0.451486	±0.000003	air	N	2	664,989,1092
0.4525		air	N	0	212
0.452862		air	Fe	0	416
0.453379	±0.000003	air	I	2	968
0.453379	±0.000003	air	I	3	968
0.4538			S_2	0	1524
0.4544			S_2	0	1524
0.454504	±0.000010	air	Ar	1	450,888,1096
0.455259	±0.000006	air	Si	2	220,1020,1091
0.4555276		air	Cs	0	350
0.455630	±0.000030	air	Cu	1	1174
0.4557			S_2	0	1524
0.455874	±0.000006	air	Xe	3	922,923,1092
0.456070	±0.000010	air	Bi	2	957
0.4563			S_2	0	1524
0.456784	±0.000006	air	Si	2	220,1020,1091
0.4569772			Se_2	0	1601
0.4570582			Se_2	0	1601
0.457720	±0.000010	air	Kr	1	450,861,1116
0.457936	±0.000016	air	Ar	1	880,888,1005
0.458300	±0.000100	air	Kr	1	941,978,1151
0.460302	±0.000004	air	Xe	1	450,884,1116–7
0.460460	±0.000050	air	Se	1	915,1034,1069
0.460552	±0.000009	air	O	0	664
0.46075			$^{130}Te_2$	0	1602
0.4608			S_2	0	1599
0.46087			$^{130}Te_2$	0	1602
0.460957	±0.000010	air	Ar	1	450,1005
0.461520	±0.000010	air	Kr	1	1007,1014
0.461910	±0.000050	air	Se	1	963
0.461917	±0.000010	air	Kr	1	450,1007,1116–7
0.462100	±0.000080	air	N	1	1016,1092,1170
0.463051	±0.000002	air	N	1	664,989,1170
0.463392	±0.000006	air	Kr	1	450,1116,1118

Table 4.1.1—*continued*
Neutral Atom, Ion, and Molecular Gas Lasers Arranged in Order of Increasing Wavelength

Wavelength (μm)	Uncertainty	Medium	Species	Charge	Reference
0.46392			Kr	0	926
0.4643593			Se_2	0	1601
0.464390	±0.000080	air	N	1	1016,1092,1170
0.4644427			Se_2	0	1601
0.464740	±0.000004	air	C	2	450,664,874
0.464740	±0.000004	air	Xe	3	450,874,923
0.464860	±0.000050	air	Se	1	915,1069
0.464908	±0.000010	air	O	1	212,450
0.465016	±0.000010	air	Kr	1	450,1007
0.465021	±0.000004	air	C	2	450,664,874
0.465025	±0.000001	air	Xe	2	450,983,1117
0.465025	±0.000001	air	Xe	3	450,983,1117
0.4654			CS_2	0	1804
0.46576			$^{130}Te_2$	0	1602
0.465795	±0.000002	air	Ar	1	888,965,1005
0.46588			$^{130}Te_2$	0	1602
0.4662		air	Tm	0	1493
0.467			IF	0	1558
0.4671			S_2	0	1524
0.467320	±0.000030	air	Cu	1	1174
0.467373	±0.000006	air	Xe	2	450,1092,1118
0.467440	±0.000003	air	I	2	968
0.467440	±0.000003	air	I	3	968
0.4675		air	Be	1	1108
0.467560	±0.000020	air	I	1	993,1037
0.4677			S_2	0	1524
0.468045	±0.000006	air	Kr	1	450,1116,1117
0.468050	±0.000070	air	In	1	1064
0.468290	±0.000030	air	Cu	1	1174
0.468357	±0.000006	air	Xe	2	450,1122
0.4690			S_2	0	1524
0.469410	±0.000020	air	Kr	1	978,1151
0.4697			S_2	0	1524
0.47085			$^{130}Te_2$	0	1602
0.4709		air	N_2	1	112–115,127

Table 4.1.1—*continued*
Neutral Atom, Ion, and Molecular Gas Lasers Arranged in Order of Increasing Wavelength

Wavelength (μm)	Uncertainty	Medium	Species	Charge	Reference
0.47098			$^{130}Te_2$	0	1602
0.471030	±0.000060	air	Kr	2	1016
0.47226			Ti	0	1850–1855
0.4712			Ti	0	1850–1855
0.471850	±0.000050	air	Se	1	963
0.4719389			Se_2	0	1601
0.4720249			Se_2	0	1601
0.4722			Bi	0	1806
0.4722		air	Zn	0	1503
0.472252		air	Bi	0	223,386,769
0.472357	±0.000005	air	Xe	2	923,983,1014
0.4725			IF	0	1558
0.472689	±0.000004	air	Ar	1	888,1005,1096
0.4737		air	Tl	1	931
0.474040	±0.000010	air	Cl	1	1107
0.474060	±0.000050	air	Se	1	963
0.474266	±0.000003	air	Br	1	955,1043
0.4745			S_2	0	1599
0.474894	±0.000001	air	Xe	2	874,983,1118
0.475			CdI	0	1530
0.4750295		air	N	0	212
0.475450	±0.000030	air	Kr	2	1016
0.47603			$^{130}Te_2$	0	1602
0.47616			$^{130}Te_2$	0	1602
0.476244	±0.000006	air	Kr	1	450,861
0.476410	±0.000050	air	Se	1	1069
0.476488	±0.000004	air	Ar	1	929,965,1096
0.4765		air	Ar	1	1812,1814
0.476510	±0.000050	air	Se	1	915,963
0.476571	±0.000010	air	Kr	1	450,1116,1117
0.476874	±0.000006	air	Cl	1	220,1107
0.478134	±0.000003	air	Cl	1	857,927,997
0.4787			IF	0	1559
0.478840	±0.000020	air	Ag	1	691,948
0.479450	±0.000060	air	Xe	2	1016

Table 4.1.1—*continued*
Neutral Atom, Ion, and Molecular Gas Lasers
Arranged in Order of Increasing Wavelength

Wavelength (μm)	Uncertainty	Medium	Species	Charge	Reference
0.479630	±0.000060	air	Kr	1	1007,1016
0.479700	±0.000010	air	Hg	2	876,996,1131
0.4797236			Se_2	0	1601
0.4798120			Se_2	0	1601
0.4800		air	Cd	0	1503
0.4810			S_2	0	1524
0.4811		air	Zn	0	1503
0.48130			$^{130}Te_2$	0	1602
0.48143			$^{130}Te_2$	0	1602
0.4816			S_2	0	1524
0.482518	±0.000006	air	Kr	1	450,861,1116
0.4830			S_2	0	1524
0.4830		air	XeF	0	10,11
0.4838			S_2	0	1524
0.484060	±0.000050	air	Se	1	1069
0.484330	±0.000040	air	Te	1	904,1073
0.484666	±0.000006	air	Kr	1	450,1007,1117
0.4847			IF	0	1559
0.485			XeF	0	1609,1611–16
0.485–0.491			IF	0	1565,1566
0.485–0.507			Hg_3	0	1536
0.4852			CS_2	0	1804
0.485500	±0.000050	air	Se	1	915,958,1069
0.485580	±0.000030	air	Cu	1	1174
0.4860		air	XeF	0	10,11
0.4861324		air	H	0	130
0.486200	±0.000100	air	Xe	1	979,1049,1151
0.48667			$^{130}Te_2$	0	1602
0.48680			$^{130}Te_2$	0	1602
0.486948	±0.000006	air	Xe	2	450,1118
0.4877210			Se_2	0	1601
0.4878120			Se_2	0	1601
0.487986	±0.000004	air	Ar	1	888,965,1096
0.4882		air	Cd	1	1027
0.488700	±0.000100	air	Xe	1	857

Table 4.1.1—*continued*
Neutral Atom, Ion, and Molecular Gas Lasers Arranged in Order of Increasing Wavelength

Wavelength (μm)	Uncertainty	Medium	Species	Charge	Reference
0.48884122		vacuum	Se	0	245,390
0.488906	±0.000006	air	Ar	1	450,1005,1118
0.489688	±0.000003	air	Cl	1	857,873,997
0.490473	±0.000003	air	Cl	1	857,927,997
0.4907			IF	0	1527,1559–61
0.490970	±0.000030	air	Cu	1	1174
0.49116		air	Zn	1	945,1059,1085,
0.491766	±0.000003	air	Cl	1	857,927,997
0.4920			S_2	0	1524
0.49212			$^{130}Te_2$	0	1602
0.49226			$^{130}Te_2$	0	1602
0.4925		air	Zn	1	1027,1039,1085
0.492560	±0.000006	air	S	1	1489
0.493170	±0.000030	air	Cu	1	975,1174
0.493467	±0.000003	air	I	2	968
0.493467	±0.000003	air	I	3	968
0.495–0.512			I_2	0	1554,1555
0.4950			S_2	0	1524
0.495418	±0.000003	air	Xe	3	450,923,1092
0.4959397			Se_2	0	1601
0.4960333			Se_2	0	1601
0.4965			IF	0	1559
0.496508	±0.000006	air	Xe	1	450,857
0.496509	±0.000002	air	Ar	1	888,965,1096
0.497271	±0.000005	air	Xe	1	454,911,1150
0.497610	±0.000050	air	Se	1	915,958,1069
0.49768			$^{130}Te_2$	0	1602
0.49782			$^{130}Te_2$	0	1602
0.4981		air	Tl	1	931
0.498670	±0.000020	air	I	1	266,993,1042
0.4990		air	HgBr	0	75–78
0.499255	±0.000005	air	Ar	0	877
0.499290	±0.000050	air	Se	1	915,1034,1069
0.500780	±0.000003	air	Xe	3	450,923,1122
0.501330	±0.000030	air	Cu	1	1174

Table 4.1.1—*continued*
Neutral Atom, Ion, and Molecular Gas Lasers Arranged in Order of Increasing Wavelength

Wavelength (μm)	Uncertainty	Medium	Species	Charge	Reference
0.501424	±0.000006	air	S	1	1489
0.501639		air	N	1	872,1170
0.501640	±0.000010	air	Kr	2	1016
0.5018		air	HgBr	0	75–78
0.5020–0.5026			HgBr	0	1537–1545
0.502000	±0.000040	air	Te	1	904,1073
0.502190	±0.000030	air	Cu	1	1174
0.502200	±0.000100	air	Kr	1	857,1118
0.5023		air	HgBr	0	75–78
0.50259		air	Cd	1	1027
0.5026		air	HgBr	0	75–78
0.502720	±0.000020	air	Ag	1	948
0.503262	±0.000006	air	S	1	1489
0.50333			$^{130}Te_2$	0	1602
0.50348			$^{130}Te_2$	0	1602
0.503750	±0.000060	air	Kr	1	1016,1157
0.5039		air	HgBr	0	75–78
0.5042		air	HgBr	0	75–78
0.5043880			Se_2	0	1601
0.5044847			Se_2	0	1601
0.504489	±0.000006	air	Xe	1	450,884,1116–7
0.5046		air	HgBr	0	75–78
0.505210	±0.000030	air	Cu	0	1174
0.505463	±0.000005	air	Br	1	955,1043
0.506050	±0.000030	air	Cu	1	1174
0.5061		air	Tm	0	1493
0.506210	±0.000025	air	Ar	1	975,1005,1111
0.506870	±0.000050	air	Se	1	915,963,1069
0.5072			CS_2	0	1804
0.507830	±0.000003	air	Cl	1	857,927,997
0.5079		air	Tl	1	931
0.5086		air	Cd	0	1503
0.509610	±0.000050	air	Se	1	915,943,1069
0.5100			Na_2	0	1524,1588
0.510310	±0.000010	air	Cl	1	1107

Table 4.1.1—*continued*
Neutral Atom, Ion, and Molecular Gas Lasers Arranged in Order of Increasing Wavelength

Wavelength (μm)	Uncertainty	Medium	Species	Charge	Reference
0.510554		air	Cu	0	336,338,340,805
0.5115		air	Tm	0	1493
0.512600	±0.000010	air	Kr	1	880,978,1151
0.5130754			Se_2	0	1601
0.513150	±0.000070	air	Ge	1	1063,1064
0.5131750			Se_2	0	1601
0.5137			Na_2	0	1524,1588
0.514180	±0.000005	air	Ar	1	450,1005,1118
0.514190	±0.000050	air	Se	1	958,1069
0.5144			CS_2	0	1804
0.514533	±0.000002	air	Ar	1	888,965,1096
0.514570	±0.000050	air	C	1	1031
0.51495			$^{130}Te_2$	0	1602
0.51510			$^{130}Te_2$	0	1602
0.5152		air	Tl	1	931,977
0.515704	±0.000006	air	Xe	3	922,923,983
0.515908	±0.000003	air	Xe	3	450,923,1122
0.516032	±0.000006	air	S	0	1489
0.517600	±0.000050	air	Se	1	915,963,1069
0.517840	±0.000070	air	Ge	1	1063,1064
0.5180		air	Xe_2Cl	0	21,22,34
0.518238	±0.000002	air	Br	1	857,943,955
0.520–0.530			Xe_2Cl	0	1604,1605
0.520–0.565			NaK	0	1842
0.520832	±0.000004	air	Kr	1	450,1116,1117
0.52092			$^{130}Te_2$	0	1602
0.5210		air	Hg	2	1131
0.52107			$^{130}Te_2$	0	1602
0.521430	±0.000020	air	I	1	993,1037
0.521630	±0.000020	air	I	1	266,993,1042
0.521790	±0.000003	air	Cl	1	873,927,997
0.521820	±0.000040	air	Kr	1	1007,1016
0.521962	±0.000006	air	S	0	1489
0.5220119			Se_2	0	1601
0.5221138			Se_2	0	1601

Table 4.1.1—*continued*
Neutral Atom, Ion, and Molecular Gas Lasers Arranged in Order of Increasing Wavelength

Wavelength (μm)	Uncertainty	Medium	Species	Charge	Reference
0.522130	±0.000003	air	Cl	1	857,927,997
0.522340	±0.000060	air	Xe	2	1016
0.522760	±0.000050	air	Se	1	963,1034,1069
0.5228		air	N_2	1	112,114,115,127
0.5237			$^6Li_2$	0	1576
0.523826	±0.000004	air	Br	1	857,955,1043
0.523889	±0.000006	air	Xe	2	450,1074,1118
0.5245		air	Na_2	0	117–119
0.52504			$^{130}Te_2$	0	1602
0.5251			Na_2	0	1524,1588
0.525260	±0.000050	air	Se	1	958,1034,1069
0.525640	±0.000040	air	Te	1	904,1073
0.525650	±0.000060	air	Xe	3	923,1016,1487
0.52575			$^{130}Te_2$	0	1602
0.526017	±0.000003	air	Xe	3	450,923,1117
0.526043	±0.000003	air	Xe	1	923,1117,1118
0.526150	±0.000100	air	Xe	1	450,1077,1116
0.5263			$^6Li_2$	0	1576
0.52633		air	Na_2	0	117–119
0.5269			$^6Li_2$	0	1576
0.527130	±0.000050	air	Se	1	958,1069
0.5272		air	Be	1	1108
0.5274		air	Na_2	0	117–119
0.528700	±0.000100	air	Ar	1	888,1005,1096
0.5289			Na_2	0	1524,1588
0.52982		air	Na_2	0	117,118
0.52995		air	Na_2	0	117,118
0.530550	±0.000050	air	Se	1	915,958,1069
0.5307		air	Tm	0	1493
0.530868	±0.000004	air	Kr	1	450,1116,1117
0.5310			Na_2	0	1524,1588
0.5312057			Se_2	0	1601
0.5313116			Se_2	0	1601
0.531400	±0.000100	air	Xe	1	979,1077,1151
0.5319			Na_2	0	1524,1588

Table 4.1.1—*continued*
Neutral Atom, Ion, and Molecular Gas Lasers Arranged in Order of Increasing Wavelength

Wavelength (μm)	Uncertainty	Medium	Species	Charge	Reference
0.532088	±0.000006	air	S	1	857,869,1489
0.5321		air	Na_2	0	117–119
0.5326			Na_2	0	1524,1588
0.53318			$^{130}Te_2$	0	1602
0.533203	±0.000003	air	Br	1	857,943,955
0.5333		air	Na_2	0	117–119
0.53334			$^{130}Te_2$	0	1602
0.5337		air	Cd	1	868,953,1085
0.5338		air	Cd	1	868,953,1085
0.5339			Na_2	0	1524,1588
0.5340		air	Na_2	0	117–119
0.5341065		air	Mn	0	406,407,409–414
0.53415		air	Na_2	0	117,118
0.53428		air	Na_2	0	117,118
0.534334	±0.000005	air	Xe	3	923,983,1074
0.534583	±0.000006	air	S	1	857,869,1489
0.5346			Na_2	0	1524,1588
0.5347			Na_2	0	1524,1588
0.53490		air	Na_2	0	117–119
0.5349472		air	Ca	0	137
0.535065		air	Tl	0	181,182,187–189
0.5352			Na_2	0	1524,1588
0.535290	±0.000003	air	Xe	3	450,983,1122
0.5355			Na_2	0	1524,1588
0.5358			$^6Li_2$	0	1576
0.5362			$^6Li_2$	0	1576
0.5363			Na_2	0	1524,1588
0.536700	±0.000060	air	Xe	2	1016
0.53690		air	Na_2	0	117–119
0.5370		air	Na_2	0	117–119
0.5371		air	Na_2	0	117–119
0.537210	±0.000070	air	Pb	1	1064
0.5374			Na_2	0	1524,1588
0.5376		air	Na_2	0	117–119
0.5376		air	XeO	0	14,39,41

Table 4.1.1—*continued*
Neutral Atom, Ion, and Molecular Gas Lasers Arranged in Order of Increasing Wavelength

Wavelength (μm)	Uncertainty	Medium	Species	Charge	Reference
0.53781		air	Na_2	0	117–119
0.5382			Na_2	0	1524,1588
0.5383			Na_2	0	1524,1588
0.5384			CS_2	0	1804
0.53850		air	Na_2	0	117,118
0.538510	±0.000040	air	As	1	981,1033,1041
0.53863		air	Na_2	0	117,118
0.5387			Se_2	0	1601
0.5388			Na_2	0	1524,1588
0.5389		air	Na_2	0	117–119
0.5391			Na_2	0	1524,1588
0.539215	±0.000003	air	Cl	1	857,873,927,997
0.539460	±0.000003	air	Xe	3	450,923,1122
0.53949			$^{130}Te_2$	0	1602
0.53965			$^{130}Te_2$	0	1602
0.54005616		air	Ne	0	188,336,446,448
0.540090	±0.000030	air	Xe	2	1012
0.54024		air	Na_2	0	117–119
0.5405			$^6Li_2$	0	1576
0.5406689			Se_2	0	1601
0.540750	±0.000020	air	I	1	266,866,993
0.5407782			Se_2	0	1601
0.5409			Na_2	0	1524,1588
0.54131		air	Na_2	0	117–119
0.541350	±0.000060	air	Xe	2	1016
0.5416		air	Na_2	0	117–119
0.5417			$^6Li_2$	0	1576
0.5417			$^6Li_2$	0	1576
0.5418			Na_2	0	1524,1588
0.541916	±0.000006	air	Xe	1	450,884,1116
0.5420368		air	Mn	0	406–7,411,413
0.5424			$^6Li_2$	0	1576
0.5424			Na_2	0	1524,1588
0.542874	±0.000006	air	S	1	869,1489
0.543287	±0.000006	air	S	1	857,869,1489

Table 4.1.1—*continued*
Neutral Atom, Ion, and Molecular Gas Lasers
Arranged in Order of Increasing Wavelength

Wavelength (μm)	Uncertainty	Medium	Species	Charge	Reference
0.5435161		vacuum	Ne	0	449
0.54353			$^{130}Te^{128}Te$	0	1602
0.54355			$^{130}Te_2$	0	1602
0.5440		air	N	0	212
0.54402			$^{130}Te^{128}Te$	0	1602
0.5442		air	XeO	0	14,39,41
0.5443		air	I_2	0	88–92
0.54430			$^{130}Te_2$	0	1602
0.5446			$^6Li_2$	0	1576
0.54469		air	Na_2	0	117–119
0.5448		air	Na_2	0	117–119
0.544980	±0.000040	air	Te	1	904,1073
0.545		air	$TbAl_3Cl_{12}$	0	26,27
0.5451			$^6Li_2$	0	1576
0.54529		air	Eu	0	421
0.545388	±0.000006	air	S	1	857,869,1489
0.5454		air	Na_2	0	117–119
0.5454		air	Te	0	443
0.545400	±0.000050	air	Te	1	443,904
0.545460	±0.000060	air	Xe	2	1016,1101
0.5459			$^6Li_2$	0	1576
0.546074		air	Hg	0	156,159
0.5467		air	Na_2	0	117–119
0.5469		air	Na_2	0	117–119
0.547			XeO	0	1519
0.5470640		air	Mn	0	406,407,413
0.547374	±0.000006	air	S	1	869,1489
0.54742			Ti	0	1850–1855
0.547930	±0.000040	air	Te	1	904,1073
0.5480		air	Na_2	0	117–119
0.5481345		air	Mn	0	336,407
0.5485		air	Na_2	0	117–119
0.54916		air	Na_2	0	117–119
0.5496			$^6Li_2$	0	1576
0.549680	±0.000040	air	As	1	981,1033,1041

Table 4.1.1—*continued*
Neutral Atom, Ion, and Molecular Gas Lasers Arranged in Order of Increasing Wavelength

Wavelength (μm)	Uncertainty	Medium	Species	Charge	Reference
0.549760	±0.000040	air	As	1	943,1033,1041
0.5499		air	Na_2	0	117–119
0.549931	±0.000004	air	Xe	3	922,923,1074
0.5500428		air	N	0	212
0.550150	±0.000050	air	Kr	2	1016
0.550220	±0.000050	air	Ar	2	450
0.5504125			Se_2	0	1601
0.5505254			Se_2	0	1601
0.55053			Br_2	0	1526
0.5508			$^6Li_2$	0	1576
0.550990	±0.000006	air	S	1	1489
0.55125			Ti	0	1850–1855
0.55144			Ti	0	1850–1855
0.55145			Ti	0	1850–1855
0.5516		air	HgCl	0	75,78–83
0.5516777		air	Mn	0	336,406,407,413
0.552280	±0.000050	air	Se	1	915,1034,1069
0.5523		air	HgCl	0	75,78–83
0.55245			$^{130}Te_2$	0	1602
0.552450	±0.000050	air	Xe	1	1118
0.55262			$^{130}Te_2$	0	1602
0.5528			$^6Li_2$	0	1576
0.55374			$^{128}Te_2$	0	1602
0.5537749		air	Mn	0	336,406,407,413
0.55400			$^{128}Te_2$	0	1602
0.5540307		air	N	0	212
0.5545			C_2	0	1528
0.5550		air	HgCl	0	75,78–83
0.5550		air	I_2	0	88–92
0.555820	±0.000040	air	As	1	943,1033,1041
0.55645			$^{130}Te_2$	0	1602
0.556511	±0.000006	air	S	1	1489
0.5567		air	I_2	0	88–92
0.556710	±0.000050	air	Se	1	958,963
0.5571		air	Te_2	0	48

Table 4.1.1—*continued*
Neutral Atom, Ion, and Molecular Gas Lasers Arranged in Order of Increasing Wavelength

Wavelength (μm)	Uncertainty	Medium	Species	Charge	Reference
0.55723			$^{130}Te_2$	0	1602
0.5575		air	Te_2	0	48
0.5576		air	HgCl	0	75,78–83
0.557650	±0.000040	air	Te	1	442–3,904,1073
0.557714		air	Eu	0	421
0.5578		air	Te_2	0	48
0.55781		air	KrO	0	2,39,41,42,44,47
0.55788939		vacuum	O	0	224,225,387,388
0.5579		air	Te_2	0	48
0.558			HgCl	0	1546
0.5580		air	ArO	0	2,39,41–47
0.5581		air	Na_2	0	117–119
0.5584		air	HgCl	0	75,78–83
0.5588			$^6Li_2$	0	1576
0.5589		air	Sn	1	202
0.559			HgCl	0	1546
0.5590		air	HgCl	0	75,78–83
0.5591		air	Na_2	0	117–119
0.55912			$^{130}Te_2$	0	1602
0.559160	±0.000050	air	Se	1	915,1069
0.5592			O	2	1846
0.55921		vacuum	CO	0	59,60
0.559235	±0.000005	air	Xe	3	904,923,967,983
0.559237	±0.000006	air	O	2	450,664,1026
0.55929			$^{130}Te_2$	0	1602
0.5593			$^6Li_2$	0	1576
0.559310	±0.000020	air	I	1	993,1037
0.55949		vacuum	CO	0	59,60
0.5597		air	Na_2	0	117–119
0.55975		vacuum	CO	0	59,60
0.559770	±0.000100	air	Kr	2	1016
0.5599			$^6Li_2$	0	1576
0.55998		vacuum	CO	0	59,60
0.56019		vacuum	CO	0	59,60
0.56040		vacuum	CO	0	59,60

Table 4.1.1—*continued*
Neutral Atom, Ion, and Molecular Gas Lasers Arranged in Order of Increasing Wavelength

Wavelength (μm)	Uncertainty	Medium	Species	Charge	Reference
0.5604483			Se_2	0	1601
0.56053		vacuum	CO	0	59,60
0.5605648			Se_2	0	1601
0.560860	±0.000050	air	Pb	1	1072
0.562280	±0.000050	air	Se	1	963
0.5625		air	I	1	266,993
0.5626		air	Te_2	0	48
0.56308			$^{130}Te_2$	0	1602
0.56353			$^{130}Te^{128}Te$	0	1602
0.5638		air	Te_2	0	48
0.56388			$^{130}Te_2$	0	1602
0.5640		air	Te	0	443
0.564012	±0.000006	air	S	1	857,869,1489
0.56405			$^{130}Te^{128}Te$	0	1602
0.564050	±0.000050	air	Te	1	443,904
0.5642		air	Te_2	0	48
0.5643		air	Te_2	0	48
0.5646		air	Te_2	0	48
0.5647		air	Te_2	0	48
0.564716	±0.000006	air	S	1	869,1489
0.56486			$^{128}Te^{126}Te$	0	1602
0.5649		air	Te_2	0	48
0.5650		air	Te_2	0	48
0.565200	±0.000100	air	As	1	943,981
0.56541			$^{128}Te^{126}Te$	0	1602
0.565900	±0.000100	air	Xe	1	857,923,1117
0.566610	±0.000040	air	Te	1	904,1073
0.566662	±0.000003	air	N	1	872,927,1170
0.5670			$^6Li_2$	0	1576
0.5676		air	Tm	0	1493
0.567603	±0.000003	air	N	1	927,1170
0.5678		air	Hg	1	924,933
0.567820	±0.000020	air	I	1	266,866,1042
0.567953	±0.000003	air	N	1	212,450,872
0.5680		air	I_2	0	88–92

Table 4.1.1—*continued*
Neutral Atom, Ion, and Molecular Gas Lasers Arranged in Order of Increasing Wavelength

Wavelength (μm)	Uncertainty	Medium	Species	Charge	Reference
0.568192	±0.000004	air	Kr	1	450,861,1116–17
0.5685			Se_2	0	1601
0.568690	±0.000080	air	N	1	1016,1170
0.5689			$^{6}Li_2$	0	1576
0.5696		air	Te_2	0	48
0.5697		air	I_2	0	88–92
0.569790	±0.000050	air	Se	1	915,1069
0.570024		air	Cu	0	349
0.5701		air	Te_2	0	48
0.5707882			Se_2	0	1601
0.570850	±0.000050	air	Te	1	442,904,1098
0.5709091			Se_2	0	1601
0.5711		air	Te_2	0	48
0.5714		air	Te_2	0	48
0.5715		air	Te_2	0	48
0.5719		air	Te_2	0	48
0.571920	±0.000010	air	Bi	1	957
0.5720		air	Te_2	0	48
0.5721		air	Te_2	0	48
0.5721		air	Te_2	0	48
0.5724		air	Te_2	0	48
0.572700	±0.000100	air	Xe	1	857,1117,1151
0.57284			$^{130}Te_2$	0	1602
0.57284			Te_2	0	1602
0.57302			$^{130}Te_2$	0	1602
0.574			K_2	0	1858
0.574150	±0.000040	air	Te	1	904,1073
0.57434			$^{128}Te_2$	0	1602
0.5745		air	I_2	0	88–92
0.57462			$^{128}Te_2$	0	1602
0.574790	±0.000050	air	Se	1	1069
0.575100	±0.000100	air	Xe	1	857,1117
0.575340	±0.000050	air	Kr	1	857,921,1007
0.5755			$^{6}Li_2$	0	1576
0.575570	±0.000040	air	Te	1	904,1073

Table 4.1.1—*continued*
Neutral Atom, Ion, and Molecular Gas Lasers Arranged in Order of Increasing Wavelength

Wavelength (μm)	Uncertainty	Medium	Species	Charge	Reference
0.576070	±0.000020	air	I	1	1032,1042
0.5764		air	I_2	0	88–92
0.576490	±0.000040	air	Te	1	904,1073
0.5766		air	Te_2	0	48
0.5766		air	Tm	0	1493
0.5767		air	Te_2	0	48
0.57671			$^{130}Te_2$	0	1602
0.576959		air	Hg	0	156
0.5773			$^6Li_2$	0	1576
0.5773		air	Te_2	0	48
0.5774		air	Te_2	0	48
0.57754			$^{130}Te_2$	0	1602
0.5780		air	Te_2	0	48
0.5783		air	Te_2	0	48
0.5784		air	Te_2	0	48
0.5785		air	Te_2	0	48
0.5787			$^6Li_2$	0	1576
0.5787		air	Te_2	0	48
0.5788			Se_2	0	1601
0.5789		air	Te_2	0	48
0.5790		air	Te_2	0	48
0.5793		air	Te_2	0	48
0.5794		air	Te_2	0	48
0.5797		air	Te_2	0	48
0.5798		air	Te_2	0	48
0.579870	±0.000070	air	Sn	1	202,1064
0.57989			Te_2	0	1602
0.57989			$^{130}Te_2$	0	1602
0.580–0.635			Xe_2F	0	1518
0.58008			Te_2	0	1602
0.58008			$Te_2{}^{130}$	0	1602
0.58048			Br_2	0	1526
0.5807			$^6Li_2$	0	1576
0.5809			$^6Li_2$	0	1576
0.5809		air	Tm	0	1493

Table 4.1.1—*continued*
Neutral Atom, Ion, and Molecular Gas Lasers Arranged in Order of Increasing Wavelength

Wavelength (μm)	Uncertainty	Medium	Species	Charge	Reference
0.58090			Br_2	0	1526
0.5814462			Se_2	0	1601
0.5815		air	I_2	0	88–92
0.5815369			Se_2	0	1601
0.581935	±0.000006	air	S	1	1489
0.5830		air	I_2	0	88–92
0.58372			$^{130}Te_2$	0	1602
0.583800	±0.000040	air	As	1	981,1033,1041
0.5841		air	Te_2	0	48
0.584280	±0.000050	air	Se	1	963
0.58457			$^{130}Te_2$	0	1602
0.58471			$^{130}Te^{128}Te$	0	1602
0.5849		air	Te_2	0	48
0.585			IF	0	1567
0.585100	±0.000040	air	Te	1	904,1073
0.58525		air	Ne	0	1495–1497
0.58526			$^{130}Te^{128}Te$	0	1602
0.5857		air	Te_2	0	48
0.5857452		air	Ca	0	137
0.5859		air	Te_2	0	48
0.58627			$^{128}Te^{126}Te$	0	1602
0.5865		air	Te_2	0	48
0.586670	±0.000050	air	Se	1	915,963
0.5867			$^6Li_2$	0	1576
0.58686			$^{128}Te^{126}Te$	0	1602
0.5869		air	Te_2	0	48
0.5870		air	Te_2	0	48
0.5874		air	Te_2	0	48
0.5880		air	I_2	0	88–92
0.5889			$^6Li_2$	0	1576
0.589330	±0.000003	air	Xe	1	454,874
0.5894		air	Zn	1	1034,1039
0.58959236		air	Na	0	350,463
0.58995		air	Tm	0	1494
0.5905		air	I_2	0	88–92

Table 4.1.1—*continued*
Neutral Atom, Ion, and Molecular Gas Lasers Arranged in Order of Increasing Wavelength

Wavelength (μm)	Uncertainty	Medium	Species	Charge	Reference
0.59205			$^{130}Te^{128}Te$	0	1602
0.5924		air	Te_2	0	48
0.5924350			Se_2	0	1601
0.5925637			Se_2	0	1601
0.59261			$^{130}Te^{128}Te$	0	1602
0.5927		air	Te_2	0	48
0.5929		air	Bi_2	0	48,49
0.59293			Bi_2	0	1524,1525
0.5934		air	Te_2	0	48
0.593530	±0.000060	air	Kr	1	1007,1016
0.593530	±0.000060	air	Kr	2	1016
0.5936		air	Te_2	0	48
0.593650	±0.000050	air	Te	1	443,904,1073
0.59369			$^{129}Te^{126}Te$	0	1602
0.59409633		vacuum	Ne	0	185,212,451
0.59442			$^{130}Te_2$	0	1602
0.59448342		air	Ne	0	444,453–55
0.59461			$^{130}Te_2$	0	1602
0.5949		air	Tl	1	897–8,977
0.595567	±0.000003	air	Xe	3	923,1024,1074
0.5968		air	I_2	0	88–92
0.5971		air	Tm	0	1493
0.597112	±0.000006	air	Xe	1	450,1116–17
0.597430	±0.000040	air	Te	1	904,1073
0.59813			$^{130}Te_2$	0	1602
0.59902			$^{130}Te_2$	0	1602
0.6002		air	Te_2	0	48
0.6004		air	Te_2	0	48
0.6005		air	Te_2	0	48
0.6008		air	Te_2	0	48
0.6009		air	Te_2	0	48
0.6011			ZnI	0	1520,1541
0.601470	±0.000040	air	Te	1	904,1073
0.6018			ZnI	0	1520,1541
0.60190			Te_2	0	1602

Table 4.1.1—*continued*
Neutral Atom, Ion, and Molecular Gas Lasers Arranged in Order of Increasing Wavelength

Wavelength (μm)	Uncertainty	Medium	Species	Charge	Reference
0.60190			$^{130}Te_2$	0	1602
0.6021		air	Zn	1	946,1145
0.60210			Te_2	0	1602
0.60210			$^{130}Te_2$	0	1602
0.602427	±0.000006	air	P	1	220,869,992
0.6025			ZnI	0	1520,1541
0.6025		air	I_2	0	88–92
0.6031			ZnI	0	1520,1541
0.60311			IF	0	1567,1839
0.603419	±0.000006	air	P	1	220,992
0.60372			$Te_2{}^{128}$	0	1602
0.603760	±0.000080	air	Kr	1	1007,1015
0.603760	±0.000080	air	Kr	2	1015
0.6037695			Se_2	0	1601
0.6039			ZnI	0	1520,1541
0.6039026			Se_2	0	1601
0.604			K_2	0	1858
0.60403			$Te_2{}^{128}$	0	1602
0.604322	±0.000006	air	P	1	220,869,992
0.60461348		air	Ne	0	185,451
0.6048		air	I_2	0	88–92
0.60555			$^{130}Te_2$	0	1602
0.605630	±0.000050	air	Se	1	915,1034,1069
0.605736		air	Eu	0	420
0.60645			$^{130}Te_2$	0	1602
0.60646		vacuum	CO	0	59,60
0.606610	±0.000050	air	Se	1	963
0.60674		vacuum	CO	0	59,60
0.606900	±0.000020	air	I	1	993,1037,1104
0.60699		vacuum	CO	0	59,60
0.607200	±0.000100	air	Kr	0	656
0.60722		vacuum	CO	0	59,60
0.60742		vacuum	CO	0	59,60
0.6075			C_2	0	1528
0.60759		vacuum	CO	0	59,60

Table 4.1.1—*continued*
Neutral Atom, Ion, and Molecular Gas Lasers Arranged in Order of Increasing Wavelength

Wavelength (μm)	Uncertainty	Medium	Species	Charge	Reference
0.6082		air	Te_2	0	48
0.608240	±0.000040	air	Te	1	904,1073
0.6083		air	Te_2	0	48
0.6085		air	Te_2	0	48
0.6087		air	Te_2	0	48
0.608804	±0.000006	air	P	1	220,992
0.6089		air	Te_2	0	48
0.609400	±0.000100	air	Xe	1	1077,1151
0.609474	±0.000003	air	Cl	1	857,927,997
0.610210	±0.000050	air	Se	1	963
0.610280	±0.000070	air	Zn	1	1064,1083–85
0.6111		air	I_2	0	88–92
0.611756	±0.000006	air	Br	1	200,956,1043
0.6119		air	Tm	0	1493
0.61197087		vacuum	Ne	0	212,451,456
0.612		air	K	1	1810
0.6124			Se_2	0	1601
0.612740	±0.000020	air	I	1	993,1032,1037
0.613000	±0.000100	air	Sb	1	442
0.61316			Br_2	0	1526
0.61368			Br_2	0	1526
0.614170	±0.000010	air	Ba	1	145,149
0.61430623		air	Ne	0	447–8,453–461
0.61495			$^{130}Te^{128}Te$	0	1602
0.6150		air	Hg	1	924,951,1132
0.61546			Bi_2	0	1524,1525
0.6154650			Se_2	0	1601
0.61555			$^{130}Te^{128}Te$	0	1602
0.6156027			Se_2	0	1601
0.6160		air	Bi_2	0	48,49
0.6162		air	Te_2	0	48
0.61629			Bi_2	0	1524,1525
0.6165		air	Te_2	0	48
0.616574	±0.000006	air	P	1	220,992
0.6168		air	Te_2	0	48

Table 4.1.1—*continued*
Neutral Atom, Ion, and Molecular Gas Lasers Arranged in Order of Increasing Wavelength

Wavelength (μm)	Uncertainty	Medium	Species	Charge	Reference
0.61685			$^{128}Te^{126}Te$	0	1602
0.616878	±0.000006	air	Br	1	956,1043
0.616880	±0.000050	air	Kr	1	1007,1118
0.6170		air	Te_2	0	48
0.6170		air	Tm	0	1493
0.617020	±0.000040	air	As	1	943,1033,1041
0.61749			$^{128}Te^{126}Te$	0	1602
0.6175		air	I_2	0	88–92
0.617520		air	I_2	0	88–92
0.617619	±0.000003	air	Xe	2	921,923,983
0.617730		air	I_2	0	88–92
0.617900		air	I_2	0	88–92
0.617970		air	I_2	0	88–92
0.617990		air	I_2	0	88–92
0.618245		air	I_2	0	88–92
0.618325		air	I_2	0	88–92
0.618490		air	I_2	0	88–92
0.618580		air	I_2	0	88–92
0.6198		air	I_2	0	88–92
0.6204		air	Te_2	0	48
0.620490	±0.000020	air	I	1	993,1037
0.6214		air	Zn	1	1036,1145
0.62290			$^{130}Te^{128}Te$	0	1602
0.62293			IF	0	1568
0.62297			IF	0	1568
0.623040	±0.000040	air	Te	1	904,1073
0.62351			$^{130}Te^{128}Te$	0	1602
0.62369			IF	0	1568
0.62378			IF	0	1568
0.623890	±0.000080	air	Xe	2	921,1015
0.6239		air	Bi_2	0	48,49
0.6239651		air	F	0	248,249
0.623987			Bi_2	0	1524,1525
0.6241			Se_2	0	1601
0.62436			IF	0	1568

Table 4.1.1—*continued*
Neutral Atom, Ion, and Molecular Gas Lasers Arranged in Order of Increasing Wavelength

Wavelength (μm)	Uncertainty	Medium	Species	Charge	Reference
0.624550	±0.000050	air	Te	1	443,904,1073
0.62493			IF	0	1567,1839
0.625		air	K	1	1810
0.62508			IF	0	1568
0.62522			IF	0	1569
0.62524			Te_2	0	1602
0.62524			$^{130}Te_2$	0	1602
0.62541			IF	0	1569
0.62545			Te_2	0	1602
0.62545			$^{130}Te_2$	0	1602
0.62546			IF	0	1569
0.62552			IF	0	1569
0.6258		air	I_2	0	88–92
0.6258			Ti	0	1850–1855
0.62592			IF	0	1568
0.6260		air	I_2	0	88–92
0.62614			IF	0	1569
0.62626			IF	0	1569
0.62634			IF	0	1569
0.62639			IF	0	1569
0.62645			IF	0	1569
0.62651			IF	0	1569
0.62657			IF	0	1569
0.62662			IF	0	1569
0.62664950		air	Ne	0	462
0.627090	±0.000010	air	Xe	1	450,1116,1117
0.62733			$^{128}Te_2$	0	1602
0.6275378			Se_2	0	1601
0.62766			$^{128}Te_2$	0	1602
0.6276804			Se_2	0	1601
0.6278		air	Te_2	0	48
0.6278170		air	Au	0	336,341–2,132
0.62866			$^{130}Te_2$	0	1602
0.628660	±0.000060	air	Xe	3	923,1016,1101
0.6287		air	Te_2	0	48

Table 4.1.1—*continued*
Neutral Atom, Ion, and Molecular Gas Lasers Arranged in Order of Increasing Wavelength

Wavelength (μm)	Uncertainty	Medium	Species	Charge	Reference
0.6288		air	Te_2	0	48
0.62937447		air	Ne	0	451,456
0.6295		air	Te_2	0	48
0.62962			$^{130}Te_2$	0	1602
0.629951			Bi_2	0	1524,1525
0.6300		air	Bi_2	0	48,49
0.631		air	K	1	1810
0.631030	±0.000080	air	Kr	2	1015
0.631260	±0.000080	air	Kr	0	921,926,1015
0.63281646		air	Ne	0	459,462,465
0.6330		air	I_2	0	88–92
0.63308			Bi_2	0	1524,1525
0.63334			Te_2	0	1602
0.63334			$^{130}Te_2$	0	1602
0.63355			Te_2	0	1602
0.63355			$^{130}Te_2$	0	1602
0.6339		air	Bi_2	0	48,49
0.63396			Bi_2	0	1524,1525
0.633990	±0.000020	air	I	1	993,1037
0.634343	±0.000005	air	Xe	2	923,983,1101
0.6345			K_2	0	1841
0.634724	±0.000006	air	Si	1	220,1020,1062
0.6348508		air	F	0	248–250
0.63497		air	Te	0	442
0.63518618		air	Ne	0	451
0.6352		air	I_2	0	88–92
0.63548		air	Cd	1	1057–58,1085
0.63552			$^{128}Te_2$	0	1602
0.635737		air	I	1	873,993
0.63586			$^{128}Te_2$	0	1602
0.63601		air	Cd	1	871,1057,1058
0.63654			Br_2	0	1526
0.63666			$^{130}Te_2$	0	1602
0.63705			Br_2	0	1526
0.6371		air	Te_2	0	48

Table 4.1.1—*continued*
Neutral Atom, Ion, and Molecular Gas Lasers
Arranged in Order of Increasing Wavelength

Wavelength (μm)	Uncertainty	Medium	Species	Charge	Reference
0.637148	±0.000006	air	Si	1	220,1020,1062
0.63735			$^{130}Te_2$	0	1602
0.6379		air	Te_2	0	48
0.638075		air	Sr	0	138
0.6381		air	Te_2	0	48
0.6381			K_2	0	1841
0.6388		air	Te_2	0	48
0.6400050			Se_2	0	1601
0.6401526			Se_2	0	1601
0.6404		air	Tm	0	1493
0.640430	±0.000020	air	Ag	1	1044
0.6413651		air	F	0	248,249
0.6414		air	Bi_2	0	48,49
0.641445			Bi_2	0	1524,1525
0.641700	±0.000100	air	Kr	1	656,1007
0.642134			Bi_2	0	1524,1525
0.6422		air	Bi_2	0	48,49
0.64418			$^{130}Te_2$	0	1602
0.644390	±0.000050	air	Se	1	915,1034,1069
0.644981		air	Ca	0	138
0.645300	±0.000070	air	Sn	1	676,1064
0.6455			K_2	0	1841
0.6461		air	Tm	0	1493
0.647100	±0.000050	air	Kr	1	861,1018,1116
0.6477		air	Te_2	0	48
0.64771			$^{130}Te^{128}Te$	0	1602
0.64812			IF	0	1567,1839
0.648260	±0.000060	air	N	1	986,1016,1170
0.648280	±0.000020	air	Ar	1	941,1005,1077
0.64837			$^{130}Te^{128}Te$	0	1602
0.6484		air	Te_2	0	48
0.648897		air	I	1	873,993
0.6490		air	I_2	0	88–92
0.649010	±0.000050	air	Se	1	915,1034,1069
0.649690	±0.000010	air	Ba	1	145,149

Table 4.1.1—*continued*
Neutral Atom, Ion, and Molecular Gas Lasers Arranged in Order of Increasing Wavelength

Wavelength (μm)	Uncertainty	Medium	Species	Charge	Reference
0.65000			$^{128}Te^{126}Te$	0	1602
0.6501		air	Hg	2	1131
0.65071			$^{128}Te^{126}Te$	0	1602
0.651000	±0.000010	air	Kr	1	880,1007,1077
0.6511		air	I_2	0	88–92
0.651180	±0.000040	air	As	1	981,1033,1041
0.651620	±0.000020	air	I	1	993,1032,1037
0.65183			Bi_2	0	1524,1525
0.6519		air	Tm	0	1493
0.65246			IF	0	1569
0.65264			Bi_2	0	1524,1525
0.652850	±0.000050	air	Xe	1	884,1118
0.6528847			Se_2	0	1601
0.6530380			Se_2	0	1601
0.65315			IF	0	1569
0.65315			$^{130}Te_2$	0	1602
0.653460	±0.000050	air	Se	1	1069
0.6537			IF	0	1570
0.6538			CdI	0	1529,1530
0.65501			IF	0	1569
0.6553			CdI	0	1529–1531
0.65535			IF	0	1569
0.65592			IF	0	1569
0.656			H	0	1836,1837
0.6561		air	Te_2	0	48
0.65632			IF	0	1569
0.65632			$^{130}Te^{128}Te$	0	1602
0.6568			CdI	0	1531
0.65700			$^{130}Te^{128}Te$	0	1602
0.657000	±0.000050	air	Kr	1	1007,1116–18
0.6571			CdI	0	1529,1531
0.6574			CdI	0	1531
0.6574		air	Te_2	0	48
0.6576		air	Bi_2	0	48,49
0.65765			Bi_2	0	1524,1525

Table 4.1.1—*continued*
Neutral Atom, Ion, and Molecular Gas Lasers Arranged in Order of Increasing Wavelength

Wavelength (μm)	Uncertainty	Medium	Species	Charge	Reference
0.657800	±0.000050	air	C	1	1031
0.657903		air	Sn	0	200–202
0.657903	±0.000006	air	Sn	1	200–202
0.6581		air	Te_2	0	48
0.65817			Bi_2	0	1524,1525
0.6582		air	Bi_2	0	48,49
0.65847			Bi_2	0	1524,1525
0.658500	±0.000040	air	Te	1	904,1073
0.658530	±0.000020	air	I	1	910,1032,1037
0.65862			$^{130}Te_2$	0	1602
0.65885			$^{130}Te_2$	0	1602
0.6589		air	Tm	0	1493
0.6592		air	I_2	0	88–92
0.65923			Bi_2	0	1524,1525
0.6593			CdI	0	1531
0.65973		vacuum	CO	0	59,60
0.6599		air	Ne	0	1496
0.660		air	K	1	1810
0.66013		vacuum	CO	0	59,60
0.660280	±0.000080	air	Kr	1	921,1007,1015
0.660280	±0.000080	air	Kr	2	921,1015
0.6603		air	Bi_2	0	48,49
0.660386			Bi_2	0	1524,1525
0.66049		vacuum	CO	0	59,60
0.66082		vacuum	CO	0	59,60
0.66109		vacuum	CO	0	59,60
0.661110			Bi_2	0	1524,1525
0.66133		vacuum	CO	0	59,60
0.66153		vacuum	CO	0	59,60
0.66163			$^{130}Te_2$	0	1602
0.6620		air	Tm	0	1493
0.662250	±0.000020	air	I	1	993,1037
0.66269			$^{130}Te_2$	0	1602
0.664020	±0.000100	air	O	1	986
0.66405			Bi_2	0	1524,1525

Table 4.1.1—*continued*
Neutral Atom, Ion, and Molecular Gas Lasers Arranged in Order of Increasing Wavelength

Wavelength (μm)	Uncertainty	Medium	Species	Charge	Reference
0.66406			Bi_2	0	1524,1525
0.6645		air	Eu	1	1054,1087
0.6645		air	I_2	0	88–92
0.66457			Bi_2	0	1524,1525
0.664820	±0.000040	air	Te	1	904,1073
0.6650		air	Bi_2	0	48,49
0.66500			Bi_2	0	1524,1525
0.66510			Bi_2	0	1524,1525
0.66512			$^{130}Te^{128}Te$	0	1602
0.6658		air	Tm	0	1493
0.66581			$^{130}Te^{128}Te$	0	1602
0.66594			Bi_2	0	1524,1525
0.666010	±0.000050	air	Pb	1	1072
0.6661968			Se_2	0	1601
0.6663315			Se_2	0	1601
0.666545			NaRb	0	1597
0.66661			Bi_2	0	1524,1525
0.66665			Bi_2	0	1524,1525
0.666694		air	O	1	872,1156
0.66710			Bi_2	0	1524,1525
0.667193		air	P	0	220
0.667193	±0.000006	air	Si	1	220,1020,1062
0.667227		air	I	1	873,993
0.66740			Te_2	0	1602
0.66740			$^{130}Te_2$	0	1602
0.66764			Te_2	0	1602
0.66764			$^{130}Te_2$	0	1602
0.667650	±0.000040	air	Te	1	904,1073
0.6678			He	0	1838
0.668681			NaRb	0	1597
0.66943		air	Xe	1	973
0.669950	±0.000030	air	Xe	3	922,923
0.67001			$^{128}Te_2$	0	1602
0.670200	±0.000100	air	Xe	1	884,1106
0.67029			$^{130}Te_2$	0	1602

Table 4.1.1—*continued*
Neutral Atom, Ion, and Molecular Gas Lasers Arranged in Order of Increasing Wavelength

Wavelength (μm)	Uncertainty	Medium	Species	Charge	Reference
0.67038			$^{128}Te_2$	0	1602
0.6707		air	Li	0	1490
0.67137			$^{130}Te_2$	0	1602
0.6717			K_2	0	1841
0.67191			Bi_2	0	1524,1525
0.672138	±0.000004	air	O	1	664,872
0.6723		air	Tm	0	1493
0.67272			Bi_2	0	1524,1525
0.673000	±0.000050	air	Ar	0	921
0.673448			Bi_2	0	1524,1525
0.674241			Bi_2	0	1524,1525
0.67455			Br_2	0	1526
0.6748596			Se_2	0	1601
0.675			IF	0	1567
0.67506			Br_2	0	1526
0.6756099			Se_2	0	1601
0.6762			IF	0	1570
0.6763		air	I_2	0	88–92
0.67637			$^{130}Te_2$	0	1602
0.676457	±0.000010	air	Kr	1	450,861,1116
0.67661			$^{130}Te_2$	0	1602
0.67686			IF	0	1569
0.67704			Bi_2	0	1524,1525
0.67793			IF	0	1569
0.67804			Bi_2	0	1524,1525
0.6782		air	Tm	0	1493
0.678360	±0.000050	air	C	1	1031
0.6799589			Se_2	0	1601
0.6801233			Se_2	0	1601
0.680156			Bi_2	0	1524,1525
0.68036			IF	0	1569
0.68081			IF	0	1569
0.6809		air	Bi_2	0	48,49
0.680917			Bi_2	0	1524,1525
0.68137			IF	0	1569

Table 4.1.1—*continued*
Neutral Atom, Ion, and Molecular Gas Lasers Arranged in Order of Increasing Wavelength

Wavelength (μm)	Uncertainty	Medium	Species	Charge	Reference
0.68210			IF	0	1569
0.682520	±0.000020	air	I	1	993,1037,1104
0.6831			K_2	0	1841
0.68366			Bi_2	0	1524,1525
0.68420			IF	0	1569
0.684400	±0.000070	air	Sn	1	676,1064
0.6845		air	Tm	0	1493
0.68465			Bi_2	0	1524,1525
0.6848		air	Tm	0	1493
0.68510			IF	0	1569
0.686110	±0.000020	air	Ar	1	941,1005,1077
0.68651			Bi_2	0	1524,1525
0.68656			Bi_2	0	1524,1525
0.68687			Bi_2	0	1524,1525
0.6869			K_2	0	1841
0.687096	±0.000010	air	Kr	1	450,1116,1118
0.68710			Bi_2	0	1524,1525
0.68813			$^{130}Te_2$	0	1602
0.688530	±0.000040	air	Te	1	904,1073
0.6887340			Se_2	0	1601
0.68926			$^{130}Te_2$	0	1602
0.6895119			Se_2	0	1601
0.69010			Bi_2	0	1524,1525
0.6904		air	I	1	266,873,993
0.6908			K_2	0	1841
0.69084			Bi_2	0	1524,1525
0.6911			K	0	1840
0.6914		air	Tm	0	1493
0.691998	±0.000010	air	Al	1	423,878,1050
0.69263			$^{130}Te^{128}Te$	0	1602
0.69337			$^{130}Te^{128}Te$	0	1602
0.6936		air	I_2	0	88–92
0.693677			Bi_2	0	1524,1525
0.6939			K	0	1840
0.6941961			Se_2	0	1601

Table 4.1.1—*continued*
Neutral Atom, Ion, and Molecular Gas Lasers Arranged in Order of Increasing Wavelength

Wavelength (μm)	Uncertainty	Medium	Species	Charge	Reference
0.6943668			Se_2	0	1601
0.694482			Bi_2	0	1524,1525
0.6951		air	Tl	1	898,931,977
0.69549			$^{128}Te^{126}Te$	0	1602
0.6956		air	Tm	0	1493
0.6961			K_2	0	1524
0.69629			$^{128}Te^{126}Te$	0	1602
0.6966349		air	F	0	250–52,440
0.69667			Bi_2	0	1524,1525
0.69732			$^{130}Te_2$	0	1602
0.69742			Bi_2	0	1524,1525
0.6975			K_2	0	1524
0.69848			$^{130}Te_2$	0	1602
0.6985			K_2	0	1524
0.6988		air	Tm	0	1493
0.6995			K_2	0	1524
0.6996		air	Sb	1	932
0.6996		air	Tm	0	1493
0.7006		air	Bi_2	0	48,49
0.700636			Bi_2	0	1524,1525
0.7013		air	Bi_2	0	48,49
0.701423			Bi_2	0	1524,1525
0.70220			$^{130}Te^{128}Te$	0	1602
0.70295			$^{130}Te^{128}Te$	0	1602
0.70324		air	Ne	0	1495,1496
0.703300	±0.000020	air	I	1	993,1032,1037
0.7037			K_2	0	1524
0.7037469		air	F	0	197,251–255
0.7038		air	Tm	0	1493
0.703920	±0.000060	air	Te	1	443,904,1073
0.70403			Bi_2	0	1524,1525
0.704209	±0.000010	air	Al	1	423,878,1050
0.70440			$^{130}Te_2$	0	1602
0.70466			$^{130}Te_2$	0	1602
0.70507			Bi_2	0	1524,1525

Table 4.1.1—*continued*
Neutral Atom, Ion, and Molecular Gas Lasers Arranged in Order of Increasing Wavelength

Wavelength (μm)	Uncertainty	Medium	Species	Charge	Reference
0.70518			$^{128}Te^{126}Te$	0	1602
0.705640	±0.000020	air	Al	1	878
0.70600			$^{128}Te^{126}Te$	0	1602
0.706420	±0.000050	air	Se	1	963
0.70670			$^{130}Te_2$	0	1602
0.7067124		vacuum	He	0	422,423
0.7067162		vacuum	He	0	422,423
0.70691661		vacuum	Ar	0	454
0.70723		air	Xe	1	973
0.70725			Bi_2	0	1524,1525
0.70730			Bi_2	0	1524,1525
0.70730			IF	0	1569
0.70759			Bi_2	0	1524,1525
0.707677			Bi_2	0	1524,1525
0.70784			Bi_2	0	1524,1525
0.70789			$^{130}Te_2$	0	1602
0.70840			IF	0	1569
0.708506			Bi_2	0	1524,1525
0.7089			K_2	0	1524
0.7089295			Se_2	0	1601
0.710250	±0.000040	air	As	1	981,1033,1041
0.71065			Bi_2	0	1524,1525
0.71096			Bi_2	0	1524,1525
0.7114		air	I_2	0	88–92
0.71141			Bi_2	0	1524,1525
0.71197			$^{130}Te^{128}Te$	0	1602
0.71199			Bi_2	0	1524,1525
0.7120329		air	Ba	0	138,145
0.71274			$^{130}Te^{128}Te$	0	1602
0.7127890		air	F	0	197,249,253
0.7133974		vacuum	Cd	0	152
0.713900	±0.000020	air	I	1	1032,1104
0.71406			IF	0	1569
0.7141			IF	0	1570
0.71414			$^{130}Te_2$	0	1602

Table 4.1.1—*continued*
Neutral Atom, Ion, and Molecular Gas Lasers Arranged in Order of Increasing Wavelength

Wavelength (μm)	Uncertainty	Medium	Species	Charge	Reference
0.71416			IF	0	1569
0.71441			Te_2	0	1602
0.71466			IF	0	1569
0.71482			IF	0	1569
0.714894	±0.000060	air	Xe	1	884,923,973
0.71508			$^{128}Te^{126}Te$	0	1602
0.71515			IF	0	1569
0.71592			$^{128}Te^{126}Te$	0	1602
0.71630			IF	0	1569
0.71714			Bi_2	0	1524,1525
0.71737			$^{128}Te_2$	0	1602
0.71748			$^{130}Te_2$	0	1602
0.71756			Bi_2	0	1524,1525
0.71770			Bi_2	0	1524,1525
0.71775			Bi_2	0	1524,1525
0.71780			$^{128}Te_2$	0	1602
0.71825			Bi_2	0	1524,1525
0.71848			Bi_2	0	1524,1525
0.7202360		air	F	0	250–253,440
0.7215			IF	0	1570
0.72156			IF	0	1569
0.721908			Bi_2	0	1524,1525
0.72194			IF	0	1569
0.72195			$^{130}Te^{128}Te$	0	1602
0.72205			IF	0	1569
0.72233			IF	0	1569
0.72275			$^{130}Te^{128}Te$	0	1602
0.722799			Bi_2	0	1524,1525
0.7229			Pb	0	1847
0.7229			$PbBr_2$	0	1805
0.72294			IF	0	1569
0.72321			IF	0	1569
0.72369		air	Cd	1	871,1057,1085
0.72410			Te_2	0	1602
0.72410			$^{130}Te_2$	0	1602

Table 4.1.1—*continued*
Neutral Atom, Ion, and Molecular Gas Lasers Arranged in Order of Increasing Wavelength

Wavelength (μm)	Uncertainty	Medium	Species	Charge	Reference
0.72437			Te_2	0	1602
0.72437			$^{130}Te_2$	0	1602
0.72452		air	Ne	0	1495,1496
0.725600	±0.000020	air	Cu	1	1077
0.72746			$^{128}Te_2$	0	1602
0.72790			$^{128}Te_2$	0	1602
0.7281			He	0	1838
0.7284			Na_2	0	1589,1590
0.72843		air	Cd	1	871,1057,1085
0.72915			Bi_2	0	1524,1525
0.7292		air	Bi_2	0	48,49
0.729288			Bi_2	0	1524,1525
0.72940			Bi_2	0	1524,1525
0.7301		air	Bi_2	0	48,49
0.730173			Bi_2	0	1524,1525
0.73068569		vacuum	Ne	0	451
0.7309033		air	F	0	249
0.7311019		air	F	0	250–2,255,440
0.73236			Bi_2	0	1524,1525
0.73238			Bi_2	0	1524,1525
0.73349			Bi_2	0	1524,1525
0.7335		air	Bi_2	0	48,49
0.73355			Bi_2	0	1524,1525
0.7339		air	Tm	0	1493
0.7342		air	Tm	0	1493
0.73427			Te_2	0	1602
0.73427			$^{130}Te_2$	0	1602
0.73454			Te_2	0	1602
0.73454			$^{130}Te_2$	0	1602
0.7346		air	Hg	1	924
0.734804	±0.000005	air	Ar	1	454,1005
0.73601			$^{130}Te_2$	0	1602
0.73615			Bi_2	0	1524,1525
0.73623			Bi_2	0	1524,1525
0.73629			Bi_2	0	1524,1525

Table 4.1.1—*continued*
Neutral Atom, Ion, and Molecular Gas Lasers Arranged in Order of Increasing Wavelength

Wavelength (μm)	Uncertainty	Medium	Species	Charge	Reference
0.7364		air	Bi_2	0	48,49
0.73641			Bi_2	0	1524,1525
0.7366		air	Bi_2	0	48,49
0.73667			Bi_2	0	1524,1525
0.736693			Bi_2	0	1524,1525
0.73682			Bi_2	0	1524,1525
0.73728			$^{130}Te_2$	0	1602
0.737584			Bi_2	0	1524,1525
0.7376		air	Bi_2	0	48,49
0.73777			$^{128}Te_2$	0	1602
0.73822			$^{128}Te_2$	0	1602
0.73881			Bi_2	0	1524,1525
0.7392		air	Tm	0	1493
0.73923			Bi_2	0	1524,1525
0.739240	±0.000050	air	Se	1	963
0.73925			Bi_2	0	1524,1525
0.73937			Bi_2	0	1524,1525
0.73974			Bi_2	0	1524,1525
0.73978			Bi_2	0	1524,1525
0.7398		air	Bi_2	0	48,49
0.7398688		air	F	0	248,250
0.740020	±0.000020	air	Cu	1	1077
0.74014			Bi_2	0	1524,1525
0.740450	±0.000030	air	Cu	1	1051,1174
0.7406		air	Tm	0	1493
0.7408		air	Bi_2	0	48,49
0.74084			Bi_2	0	1524,1525
0.7409		air	Tm	0	1493
0.74181		air	Hg	1	886
0.7425645		air	F	0	256
0.743576	±0.000001	air	Kr	1	989,1007,1089
0.74367			Bi_2	0	1524,1525
0.74372			Bi_2	0	1524,1525
0.74377			Bi_2	0	1524,1525
0.74387			Bi_2	0	1524,1525

Table 4.1.1—*continued*
Neutral Atom, Ion, and Molecular Gas Lasers Arranged in Order of Increasing Wavelength

Wavelength (μm)	Uncertainty	Medium	Species	Charge	Reference
0.7439		air	Bi_2	0	48,49
0.7439		air	Tm	0	1493
0.743900	±0.000020	air	Cu	1	1077
0.74413			Bi_2	0	1524,1525
0.744169			Bi_2	0	1524,1525
0.74438			Bi_2	0	1524,1525
0.74466			Te_2	0	1602
0.74466			$^{130}Te_2$	0	1602
0.7448		air	Tm	0	1493
0.74494			Te_2	0	1602
0.74494			$^{130}Te_2$	0	1602
0.745046			Bi_2	0	1524,1525
0.74615			Bi_2	0	1524,1525
0.74619			$^{130}Te_2$	0	1602
0.74638			Br_2	0	1526
0.74660			Bi_2	0	1524,1525
0.74674			Bi_2	0	1524,1525
0.74675			Bi_2	0	1524,1525
0.7468		air	Bi_2	0	48,49
0.74704			Br_2	0	1526
0.7471		air	Bi_2	0	48,49
0.74713			Bi_2	0	1524,1525
0.747149	±0.000010	air	Al	1	423,878,1050
0.74731			Bi_2	0	1524,1525
0.74749			$^{130}Te_2$	0	1602
0.7475		air	Bi_2	0	48,49
0.74754			Bi_2	0	1524,1525
0.7478		air	Zn	1	1809
0.747830	±0.000160	air	Zn	1	945,1064
0.7481		air	Tm	0	1493
0.7482		air	Bi_2	0	48,49
0.7482187		air	N_2	0	105
0.7482723		air	F	0	256
0.74829			Bi_2	0	1524,1525
0.7485941		air	N_2	0	105

Table 4.1.1—*continued*
Neutral Atom, Ion, and Molecular Gas Lasers Arranged in Order of Increasing Wavelength

Wavelength (μm)	Uncertainty	Medium	Species	Charge	Reference
0.7486135		air	N_2	0	105
0.7486253		air	N_2	0	105
0.7486413		air	N_2	0	105
0.7487409		air	N_2	0	105
0.7488046		air	N_2	0	105
0.7488246		air	N_2	0	105
0.7489107		air	N_2	0	105
0.7489155		air	F	0	250,252,440
0.74893			Bi_2	0	1524,1525
0.7489626		air	N_2	0	105
0.7489809		air	N_2	0	105
0.7490096		air	N_2	0	105
0.7490317		air	N_2	0	105
0.7491510		air	N_2	0	105
0.7491705		air	N_2	0	105
0.7492379		air	N_2	0	105
0.7493082		air	N_2	0	105
0.7493716		air	N_2	0	105
0.7493910		air	N_2	0	105
0.74946			Bi_2	0	1524,1525
0.7495086		air	N_2	0	105
0.7495465		air	N_2	0	105
0.7495660		air	N_2	0	105
0.7496024		air	N_2	0	105
0.7497256		air	N_2	0	105
0.7497524		air	N_2	0	105
0.7497728		air	N_2	0	105
0.7498898		air	N_2	0	105
0.7499013		air	N_2	0	105
0.7499327		air	N_2	0	105
0.7499593		air	N_2	0	105
0.7499825		air	N_2	0	105
0.7500071		air	N_2	0	105
0.7500646		air	N_2	0	105
0.7500734		air	N_2	0	105

Table 4.1.1—*continued*
Neutral Atom, Ion, and Molecular Gas Lasers Arranged in Order of Increasing Wavelength

Wavelength (μm)	Uncertainty	Medium	Species	Charge	Reference
0.7501056		air	N_2	0	105
0.7501295		air	N_2	0	105
0.7501404		air	N_2	0	105
0.7501553		air	N_2	0	105
0.7502139		air	N_2	0	105
0.7502729		air	N_2	0	105
0.7502768		air	N_2	0	105
0.7503035		air	N_2	0	105
0.7503371		air	N_2	0	105
0.7503418		air	N_2	0	105
0.7503642		air	N_2	0	105
0.7503669		air	N_2	0	105
0.7503697		air	N_2	0	105
0.7503838		air	N_2	0	105
0.7503960		air	N_2	0	105
0.7503994		air	N_2	0	105
0.7504106		air	N_2	0	105
0.7504160		air	N_2	0	105
0.7504184		air	N_2	0	105
0.7504274		air	N_2	0	105
0.7504598		air	N_2	0	105
0.7504768		air	N_2	0	105
0.750508	±0.000005	air	Ar	1	454,1005
0.7505113		air	N_2	0	105
0.7505710		air	N_2	0	105
0.7505903		air	N_2	0	105
0.75059341		vacuum	Ar	0	450
0.7506063		air	N_2	0	105
0.7506356		air	N_2	0	105
0.7508145		air	N_2	0	105
0.7509890		air	N_2	0	105
0.7510133		air	N_2	0	105
0.7510923		air	N_2	0	105
0.7511592		air	N_2	0	105
0.7512799		air	N_2	0	105

Table 4.1.1—*continued*
Neutral Atom, Ion, and Molecular Gas Lasers Arranged in Order of Increasing Wavelength

Wavelength (μm)	Uncertainty	Medium	Species	Charge	Reference
0.75130			Bi_2	0	1524,1525
0.7513003		air	N_2	0	105
0.7513569		air	N_2	0	105
0.75137			Bi_2	0	1524,1525
0.75139			Bi_2	0	1524,1525
0.7514357		air	N_2	0	105
0.7514919		air	F	0	256
0.7515079		air	N_2	0	105
0.7515446		air	N_2	0	105
0.7515650		air	N_2	0	105
0.75166			Bi_2	0	1524,1525
0.751735			Bi_2	0	1524,1525
0.7517728		air	N_2	0	105
0.7518013		air	N_2	0	105
0.7522		air	Tm	0	1493
0.752550	±0.000010	air	Kr	1	947,1007
0.7526		air	Tm	0	1493
0.752653			Bi_2	0	1524,1525
0.7532		air	Tm	0	1493
0.75363			Bi_2	0	1524,1525
0.75408			Bi_2	0	1524,1525
0.75409			Bi_2	0	1524,1525
0.75419			Bi_2	0	1524,1525
0.7543		air	Bi_2	0	48,49
0.75441			H_2	0	1535
0.75463			Bi_2	0	1524,1525
0.75466			Bi_2	0	1524,1525
0.75496			Bi_2	0	1524,1525
0.7551		air	Bi_2	0	48,49
0.7552235		air	F	0	248–250
0.75527			$^{130}Te_2$	0	1602
0.75557			$^{130}Te_2$	0	1602
0.755600	±0.000020	air	Au	1	673
0.75580			Bi_2	0	1524,1525
0.7559		air	Tm	0	1493

Table 4.1.1—*continued*
Neutral Atom, Ion, and Molecular Gas Lasers Arranged in Order of Increasing Wavelength

Wavelength (μm)	Uncertainty	Medium	Species	Charge	Reference
0.75622			Bi_2	0	1524,1525
0.7563			Na_2	0	1589,1590
0.75657			$^{130}Te_2$	0	1602
0.75683			Bi_2	0	1524,1525
0.7577		air	Tm	0	1493
0.75791			$^{130}Te_2$	0	1602
0.7586439		air	N_2	0	105
0.758750	±0.000160	air	Zn	1	676,945,1064
0.7587693		air	N_2	0	105
0.759870	±0.000050	air	Bi	2	957
0.760040	±0.000020	air	Au	1	673
0.760154393		air	Kr	0	604
0.7603477		air	N_2	0	105
0.76048			Bi_2	0	1524,1525
0.7606374		air	N_2	0	105
0.76068			Bi_2	0	1524,1525
0.7607626		air	N_2	0	105
0.7608801		air	N_2	0	105
0.7609853		air	N_2	0	105
0.7610759		air	N_2	0	105
0.7611082		air	N_2	0	105
0.761118	±0.000160	air	Zn	1	1064,1145
0.7611514		air	N_2	0	105
0.7612105		air	N_2	0	105
0.76122			Bi_2	0	1524,1525
0.7612528		air	N_2	0	105
0.7613260		air	N_2	0	105
0.7615347		air	N_2	0	105
0.76165			Bi_2	0	1524,1525
0.76169			Bi_2	0	1524,1525
0.7616994		air	N_2	0	105
0.7617357		air	N_2	0	105
0.76177			Bi_2	0	1524,1525
0.761850	±0.000020	air	I	1	993,1037
0.761900	±0.000020	air	Xe	1	1077

Table 4.1.1—*continued*
Neutral Atom, Ion, and Molecular Gas Lasers Arranged in Order of Increasing Wavelength

Wavelength (μm)	Uncertainty	Medium	Species	Charge	Reference
0.7619288		air	N_2	0	105
0.76204			Bi_2	0	1524,1525
0.7620844		air	N_2	0	105
0.7620943		air	N_2	0	105
0.7621029		vacuum	Rb	0	350
0.7621161		air	N_2	0	105
0.76215			Bi_2	0	1524,1525
0.7622235		air	N_2	0	105
0.76223			Bi_2	0	1524,1525
0.76224			Bi_2	0	1524,1525
0.7622565		air	N_2	0	105
0.7622959		air	N_2	0	105
0.7623256		air	N_2	0	105
0.7623311		air	N_2	0	105
0.76235			Bi_2	0	1524,1525
0.7623582		air	N_2	0	105
0.7623686		air	N_2	0	105
0.7623918		air	N_2	0	105
0.7624220		air	N_2	0	105
0.7624690		air	N_2	0	105
0.7624924		air	N_2	0	105
0.7625115		air	N_2	0	105
0.7625445		air	N_2	0	105
0.76256			Bi_2	0	1524,1525
0.7625709		air	N_2	0	105
0.7625770		air	N_2	0	105
0.7625812		air	N_2	0	105
0.7625906		air	N_2	0	105
0.7626007		air	N_2	0	105
0.7626044		air	N_2	0	105
0.7626114		air	N_2	0	105
0.7626180		air	N_2	0	105
0.7626207		air	N_2	0	105
0.7626360		air	N_2	0	105
0.7626560		air	N_2	0	105

Table 4.1.1—*continued*
Neutral Atom, Ion, and Molecular Gas Lasers
Arranged in Order of Increasing Wavelength

Wavelength (μm)	Uncertainty	Medium	Species	Charge	Reference
0.7626700		air	N_2	0	105
0.7626749		air	N_2	0	105
0.7626826		air	N_2	0	105
0.7628854		air	N_2	0	105
0.7629102		air	N_2	0	105
0.7630305		air	N_2	0	105
0.7631880		air	N_2	0	105
0.7632446		air	N_2	0	105
0.7633348		air	N_2	0	105
0.7633985		air	N_2	0	105
0.7634546		air	N_2	0	105
0.7634779		air	N_2	0	105
0.7635474		air	N_2	0	105
0.7636126		air	N_2	0	105
0.7636904		air	N_2	0	105
0.7637586		air	N_2	0	105
0.7638274		air	N_2	0	105
0.7639571		air	N_2	0	105
0.7639715		air	N_2	0	105
0.7640383		air	N_2	0	105
0.7640794		air	N_2	0	105
0.7641929		air	N_2	0	105
0.7642478		air	N_2	0	105
0.7644612		air	N_2	0	105
0.7647			Na_2	0	1589,1590
0.764852			Na_2	0	1589,1590
0.765418			Na_2	0	1589,1590
0.76613			$^{130}Te_2$	0	1602
0.76642			$^{130}Te_2$	0	1602
0.766470	±0.000030	air	Cu	1	1051,1174
0.7664899		air	K	0	350
0.7669		air	Tm	0	1493
0.76718			$^{130}Te_2$	0	1602
0.767324			Na_2	0	1589,1590
0.767490	±0.000050	air	Se	1	963

Table 4.1.1—*continued*
Neutral Atom, Ion, and Molecular Gas Lasers Arranged in Order of Increasing Wavelength

Wavelength (μm)	Uncertainty	Medium	Species	Charge	Reference
0.7676			Na_2	0	1589,1590
0.767606			Na_2	0	1589,1590
0.76763			Bi_2	0	1524,1525
0.76784			Bi_2	0	1524,1525
0.768487			Na_2	0	1589,1590
0.76855			$^{130}Te_2$	0	1602
0.7687			Na_2	0	1589,1590
0.76904			Bi_2	0	1524,1525
0.76916			Bi_2	0	1524,1525
0.76930			Bi_2	0	1524,1525
0.76941			Bi_2	0	1524,1525
0.772360	±0.000050	air	Se	1	963
0.7726542		vacuum	S	0	238
0.773250	±0.000050	air	Zn	1	928,1145
0.773580	±0.000020	air	I	1	993,1037
0.773800			Na_2	0	1589,1590
0.773870	±0.000030	air	Cu	1	1051,1174
0.7743859		air	N_2	0	105
0.77486			Bi_2	0	1524,1525
0.77506			Bi_2	0	1524,1525
0.7752354		air	N_2	0	105
0.775265			Na_2	0	1589,1590
0.7753652		air	N_2	0	105
0.775423			Na_2	0	1589,1590
0.7754696		air	F	0	250,252,440
0.7757		air	Zn	1	868,1139
0.77612			Bi_2	0	1524,1525
0.7761570		vacuum	Rb	0	350
0.77624			Bi_2	0	1524,1525
0.77635			Bi_2	0	1524,1525
0.77647			Bi_2	0	1524,1525
0.7768		air	Tm	0	1493
0.77700379		vacuum	Se	0	245,390
0.77722			$^{130}Te_2$	0	1602
0.77752			$^{130}Te_2$	0	1602

Table 4.1.1—*continued*
Neutral Atom, Ion, and Molecular Gas Lasers Arranged in Order of Increasing Wavelength

Wavelength (μm)	Uncertainty	Medium	Species	Charge	Reference
0.777890	±0.000030	air	Cu	1	1174
0.77801			$^{130}Te_2$	0	1602
0.77941			$^{130}Te_2$	0	1602
0.779620	±0.000050	air	Se	1	963
0.7800212		air	F	0	250,252,254
0.7800268		air	Rb	0	350
0.780160	±0.000060	air	Te	1	904,1073
0.7804		air	Tm	0	1493
0.780530	±0.000030	air	Cu	1	934,1174
0.780780	±0.000030	air	Cu	1	691,1051,1149
0.78218			Bi_2	0	1524,1525
0.78237			Bi_2	0	1524,1525
0.782600	±0.000030	air	Cu	1	934,1174
0.782800	±0.000300	air	Xe	1	1075
0.78330			Bi_2	0	1524,1525
0.78342			Bi_2	0	1524,1525
0.783930	±0.000050	air	Se	1	963
0.784292			Na_2	0	1589,1590
0.7845		air	Tm	0	1493
0.784530	±0.000030	air	Cu	1	934,1051,1174
0.7846		air	P	1	869,992
0.7849		air	Tm	0	1493
0.78493		air	Na_2	0	117,118
0.7851			Na_2	0	1589,1590
0.78569		air	Na_2	0	117,118
0.7879			Na_2	0	1589,1590
0.7888			Na_2	0	1589,1590
0.789204			Na_2	0	1589,1590
0.789600	±0.000030	air	Cu	1	934,1174
0.78974		air	Na_2	0	117,118
0.78979		air	Na_2	0	117,118
0.790270	±0.000030	air	Cu	1	935,1174
0.790912			Na_2	0	1589,1590
0.791003			Na_2	0	1589,1590
0.7913			Li_2	0	1577,1619

Table 4.1.1—*continued*
Neutral Atom, Ion, and Molecular Gas Lasers
Arranged in Order of Increasing Wavelength

Wavelength (μm)	Uncertainty	Medium	Species	Charge	Reference
0.79154			IF	0	1839
0.79178		air	Na_2	0	117,118
0.792140	±0.000060	air	Te	1	904,1073
0.792346			Na_2	0	1589,1590
0.792346			Na_2	0	1589,1590
0.792517			Na_2	0	1589,1590
0.7929		air	Tm	0	1493
0.79295		air	Na_2	0	117,118
0.7931		air	Tm	0	1493
0.79314		air	Kr	1	973,1007
0.793253			Na_2	0	1589,1590
0.79370		air	Na_2	0	117,118
0.793903			Na_2	0	1589,1590
0.794072			Na_2	0	1589,1590
0.794480	±0.000030	air	Cu	1	935,1174
0.7945		air	Hg	1	951,1035,1060
0.7947603		air	Rb	0	350
0.7948			Ar	0	1833
0.7957			Na_2	0	1589,1590
0.795889			Na_2	0	1589,1590
0.796070	±0.000250	air	Ni	1	936
0.796550			Na_2	0	1589,1590
0.7968			Na_2	0	1589,1590
0.79747		air	Na_2	0	117,118
0.797480	±0.000250	air	Ni	1	936
0.7975			Na_2	0	1589,1590
0.797559			Na_2	0	1589,1590
0.79766		air	Na_2	0	117,118
0.7983		air	Tm	0	1493
0.798523			IF	0	1567,1839
0.798553			Na_2	0	1589,1590
0.798820	±0.000030	air	Cu	1	934,1174
0.7989			Na_2	0	1589,1590
0.798900	±0.000300	air	Xe	1	1075
0.79909		air	Na_2	0	117,118

Table 4.1.1—*continued*
Neutral Atom, Ion, and Molecular Gas Lasers Arranged in Order of Increasing Wavelength

Wavelength (μm)	Uncertainty	Medium	Species	Charge	Reference
0.799154			IF	0	1567
0.799300	±0.000050	air	Kr	1	973,1007,1116
0.79966		air	Na_2	0	117,118
0.799752			IF	0	1567,1839
0.800540	±0.000010	air	Ag	1	691,954,1045
0.80084		air	Na_2	0	117,118
0.800991			IF	0	1567,1839
0.801329			Na_2	0	1589,1590
0.8020		air	Tm	0	1493
0.802136			Na_2	0	1589,1590
0.802345			Na_2	0	1589,1590
0.803391			Na_2	0	1589,1590
0.80365		air	Na_2	0	117,118
0.80393		air	Na_2	0	117,118
0.80445		air	Na_2	0	117,118
0.804472			Na_2	0	1589,1590
0.8050			Na_2	0	1589,1590
0.80537		air	Na_2	0	117,118
0.80561		air	Na_2	0	117,118
0.80669		air	Cd	1	871,953,1059
0.806920	±0.000050	air	Bi	2	957
0.80694		air	Na_2	0	117,118
0.80805		air	Na_2	0	117,118
0.808858		air	Cu	1	935
0.8096		air	Cu	1	934
0.810436392		air	Kr	0	454,455,538
0.811–0.816			CdBr	0	1529
0.8144		air	I_2	0	88–92
0.8170		air	Tm	0	1493
0.817020	±0.000020	air	I	1	993,1037
0.8175			Na_2	0	1589,1590
0.8180		air	Tm	0	1493
0.8186		air	Tm	0	1493
0.819228		air	Cu	1	934
0.819228		air	Cu	1	934

Table 4.1.1—*continued*
Neutral Atom, Ion, and Molecular Gas Lasers Arranged in Order of Increasing Wavelength

Wavelength (μm)	Uncertainty	Medium	Species	Charge	Reference
0.8215		air	Tm	0	1493
0.82316376		air	Xe	0	248,588,589
0.8233		air	Tm	0	1493
0.825390	±0.000020	air	I	1	993,1032,1037
0.825450	±0.000010	air	Ag	1	691,935,1044
0.826300	±0.000200	air	Ag	1	806
0.8264			Li_2	0	1577,1619
0.827–0.832			Na_2	0	1591
0.827290		air	Au	1	935
0.827700	±0.000200	air	Cu	1	806
0.82798		air	D_2	0	71,72,74
0.828030	±0.000010	air	Kr	1	947,973,1007
0.828321		air	Cu	1	935
0.830890	±0.000050	air	Se	1	963
0.832206			IF	0	1839
0.832480	±0.000010	air	Ag	1	935,1044,1045
0.832819			IF	0	1839
0.833000	±0.000300	air	Xe	1	1075
0.8334		air	Kr	1	1007,1152
0.8335149		air	C	0	190
0.833532			IF	0	1839
0.834767			IF	0	1839
0.83519		air	H_2	0	71,72,74
0.8358		air	I_2	0	88–92
0.8370			H_2	0	1535
0.837950		air	Ag	1	806,935
0.8389		air	Cd	1	153,152
0.840350	±0.000010	air	Ag	1	691,935,1044
0.84039			Ag	1	1831
0.84091940		air	Xe	0	538,609
0.844300	±0.000300	air	Xe	1	1075
0.844628		air	O	0	226–240,1698
0.844638		air	O	0	226–240,1698
0.844672		air	O	0	226–240,1698
0.844672		air	O	0	226–240,1698

Table 4.1.1—*continued*
Neutral Atom, Ion, and Molecular Gas Lasers Arranged in Order of Increasing Wavelength

Wavelength (μm)	Uncertainty	Medium	Species	Charge	Reference
0.844680		air	O	0	226–240,1698
0.844680		air	O	0	226–240,1698
0.844680		air	O	0	226–240,1698
0.8454			Li_2	0	1577,1619
0.84633569		air	Ne	0	526
0.8473		air	Kr	1	1007,1089
0.8480		air	Tm	0	1493
0.8480		air	Tm	0	1493
0.8482		air	Tm	0	1493
0.851104		air	Cu	1	935
0.8521133		air	Cs	0	350
0.85309		air	Cd	1	871,1034,1059
0.8533		air	Tm	0	1493
0.854180	±0.000060	air	Ca	1	336,1166
0.8547		air	Hg	1	160
0.856900	±0.000300	air	Xe	2	923,983,1075
0.8578		air	I_2	0	88–92
0.8579		air	I_2	0	88–92
0.858200	±0.000300	air	Xe	1	1075
0.858900	±0.000300	air	Kr	2	1075
0.8594005		air	N	0	190,214
0.859900		air	Au	1	938
0.860440	±0.000060	air	Te	1	904,1073
0.8628		air	Hg	1	160
0.8629238		air	N	0	190–92,214–16
0.86376895		vacuum	Ne	0	526
0.8652		air	Cd	1	938
0.866200	±0.000060	air	Ca	1	336,1166
0.8669223		air	N_2	0	105
0.8671332		air	N_2	0	105
0.8672			Li_2	0	1577,1619
0.8677		air	Hg	0	160
0.8677		air	Hg	1	160
0.8680		air	N	0	1505
0.86819216		air	Ne	0	462

Table 4.1.1—*continued*
Neutral Atom, Ion, and Molecular Gas Lasers Arranged in Order of Increasing Wavelength

Wavelength (μm)	Uncertainty	Medium	Species	Charge	Reference
0.8682			Li_2	0	1577,1619
0.8683		air	N	0	1505
0.868400	±0.000100	air	Cr	1	939
0.8686		air	N	0	1505
0.86901		air	Kr	1	973,1007
0.8692580		air	N_2	0	105
0.8696366		air	N_2	0	105
0.8697945		air	N_2	0	105
0.8698263		air	N_2	0	105
0.8699397		air	N_2	0	105
0.8700		air	Sr	1	976
0.8700670		air	N_2	0	105
0.8700684		air	N_2	0	105
0.8701481		air	N_2	0	105
0.8701718		air	N_2	0	105
0.8702451		air	N_2	0	105
0.8702681		air	N_2	0	105
0.8703		air	N	0	1505
0.8703093		air	N_2	0	105
0.8703457		air	N_2	0	105
0.8704549		air	N_2	0	105
0.8707478		air	N_2	0	105
0.8710118		air	N_2	0	105
0.8710273		air	N_2	0	105
0.8712956		air	N_2	0	105
0.8713533		air	N_2	0	105
0.871400	±0.000300	air	Xe	1	973,1075
0.8715519		air	N_2	0	105
0.8716718		air	N_2	0	105
0.8717377		air	N_2	0	105
0.8717970		air	N_2	0	105
0.8718571		air	N_2	0	105
0.8718654		air	N_2	0	105
0.8719537		air	N_2	0	105
0.8719562		air	N_2	0	105

Table 4.1.1—*continued*
Neutral Atom, Ion, and Molecular Gas Lasers Arranged in Order of Increasing Wavelength

Wavelength (μm)	Uncertainty	Medium	Species	Charge	Reference
0.8719791		air	N_2	0	105
0.8720251		air	N_2	0	105
0.8720284		air	N_2	0	105
0.8720308		air	N_2	0	105
0.8720419		air	N_2	0	105
0.8720848		air	N_2	0	105
0.8721155		air	N_2	0	105
0.8721327		air	N_2	0	105
0.8721718		air	N_2	0	105
0.8721971		air	N_2	0	105
0.8722007		air	N_2	0	105
0.8722220		air	N_2	0	105
0.8722341		air	N_2	0	105
0.8722569		air	N_2	0	105
0.8722836		air	N_2	0	105
0.8723057		air	N_2	0	105
0.8726333		air	N_2	0	105
0.8728430		air	N_2	0	105
0.8730		air	Tm	0	1493
0.8730453		air	N_2	0	105
0.8732394		air	N_2	0	105
0.8734247		air	N_2	0	105
0.873430	±0.000060	air	Te	1	904,1073
0.8735995		air	N_2	0	105
0.8737644		air	N_2	0	105
0.8739162		air	N_2	0	105
0.8740559		air	N_2	0	105
0.8742917		air	N_2	0	105
0.8747			Li_2	0	1577,1619
0.874760		air	Ag	1	806,935
0.8757			Li_2	0	1577,1619
0.87614150		air	Cs	0	350
0.87614150		air	Cs	0	350
0.877300	±0.000200	air	Ag	1	806
0.87740648		vacuum	Ne	0	526

Table 4.1.1—*continued*
Neutral Atom, Ion, and Molecular Gas Lasers Arranged in Order of Increasing Wavelength

Wavelength (μm)	Uncertainty	Medium	Species	Charge	Reference
0.878000	±0.000300	air	Ar	1	1075,1077
0.8803		air	Tm	0	1493
0.8804		air	I_2	0	88–92
0.880428	±0.000020	air	I	1	266,993,1037
0.8806		air	I_2	0	88–92
0.8813		air	I_2	0	88–92
0.88228702		vacuum	O	0	190
0.8836		air	Tm	0	1493
0.8845349		air	N_2	0	105
0.8846			Li_2	0	1577,1619
0.8856271		air	N_2	0	105
0.8858470		air	N_2	0	105
0.886347			Li_2	0	1577,1619
0.8865			Li_2	0	1577,1619
0.88653057		air	Ne	0	490,526,527
0.886760		air	Au	1	935
0.887376			Li_2	0	1577,1619
0.887740	±0.000020	air	I	1	993,1037
0.88778		air	Cd	1	871,1039,1085
0.88787		air	H_2	0	71,72,74
0.8880521		air	N_2	0	105
0.8884527		air	N_2	0	105
0.8886204		air	N_2	0	105
0.8886378		air	N_2	0	105
0.8887756		air	N_2	0	105
0.8889111		air	N_2	0	105
0.8889738		air	N_2	0	105
0.8890243		air	N_2	0	105
0.8891133		air	N_2	0	105
0.8891769		air	N_2	0	105
0.8892149		air	N_2	0	105
0.8892940		air	N_2	0	105
0.8896001		air	N_2	0	105
0.8898930		air	N_2	0	105
0.8899078		air	N_2	0	105

Table 4.1.1—*continued*
Neutral Atom, Ion, and Molecular Gas Lasers Arranged in Order of Increasing Wavelength

Wavelength (μm)	Uncertainty	Medium	Species	Charge	Reference
0.89013		air	H_2	0	71,72,74
0.8901733		air	N_2	0	105
0.8902711		air	N_2	0	105
0.8904419		air	N_2	0	105
0.8906097		air	N_2	0	105
0.8906649		air	N_2	0	105
0.8906994		air	N_2	0	105
0.8907920		air	N_2	0	105
0.8908808		air	N_2	0	105
0.8908878		air	N_2	0	105
0.8909451		air	N_2	0	105
0.8909527		air	N_2	0	105
0.8909750		air	N_2	0	105
0.8910132		air	N_2	0	105
0.8910480		air	N_2	0	105
0.8910612		air	N_2	0	105
0.8911001		air	N_2	0	105
0.8911063		air	N_2	0	105
0.8911280		air	N_2	0	105
0.8911502		air	N_2	0	105
0.8911538		air	N_2	0	105
0.8911608		air	N_2	0	105
0.8911898		air	N_2	0	105
0.8912139		air	N_2	0	105
0.8918033		air	N_2	0	105
0.8920			Na_2	0	1592
0.8920184		air	N_2	0	105
0.8922249		air	N_2	0	105
0.8924223		air	N_2	0	105
0.8926099		air	N_2	0	105
0.8927865		air	N_2	0	105
0.892869155		air	Kr	0	588,589
0.8929509		air	N_2	0	105
0.893			$^{130}Te_2$	0	1603
0.8931019		air	N_2	0	105

Table 4.1.1—*continued*
Neutral Atom, Ion, and Molecular Gas Lasers Arranged in Order of Increasing Wavelength

Wavelength (μm)	Uncertainty	Medium	Species	Charge	Reference
0.8933580		air	N_2	0	105
0.8943468		air	Cs	0	350
0.895			$^{130}Te_2$	0	1603
0.8963			Li_2	0	1577,1619
0.8964		air	Br	0	1505
0.8969			Li_2	0	1577,1619
0.897190	±0.000060	air	Te	1	904,1073
0.8972			Li_2	0	1577,1619
0.8974			Li_2	0	1577,1619
0.8978			Li_2	0	1577,1619
0.89784		air	Kr	1	1007,1089
0.8979			Li_2	0	1577,1619
0.8982			Li_2	0	1577,1619
0.8984			Li_2	0	1577,1619
0.8989			Li_2	0	1577,1619
0.8991			Li_2	0	1577,1619
0.89910237		vacuum	Ne	0	490,526
0.899820	±0.000060	air	Te	1	904,1073
0.900			Na_2	0	1593
0.9037			Li_2	0	1577,1619
0.9037		air	I_2	0	88–92
0.9038		air	I_2	0	88–92
0.904			Na_2	0	1593
0.90454514		air	Xe	0	455,538
0.9045878		air	N	0	214
0.9047			Li_2	0	1577,1619
0.9047		air	I_2	0	88–92
0.905			Na_2	0	1593
0.9060		air	I_2	0	88–92
0.906300	±0.000400	air	Xe	1	1075
0.9064			Li_2	0	1577,1619
0.9068			Li_2	0	1577,1619
0.907			Na_2	0	1593
0.9074			Li_2	0	1577,1619
0.910			Na_2	0	1593

Table 4.1.1—*continued*
Neutral Atom, Ion, and Molecular Gas Lasers Arranged in Order of Increasing Wavelength

Wavelength (μm)	Uncertainty	Medium	Species	Charge	Reference
0.9122			Li_2	0	1577,1619
0.912297		air	Ar	0	248,588–9,722
0.9163		air	HD	0	71,72,74
0.91723217		air	Cs	0	350
0.9173		air	Br	0	1505
0.9178		air	Br	0	1505
0.9187449		air	N	0	190
0.918784		air	N	0	190
0.921800	±0.000150	air	Mg	1	920,976,1046
0.9222			H_2	0	1535
0.924400	±0.000150	air	Mg	1	920,976,1046
0.924930	±0.000100	air	Se	1	963
0.926500	±0.000400	air	Xe	1	1075
0.9274		air	I_2	0	88–92
0.9276		air	I_2	0	88–92
0.928700	±0.000400	air	Xe	1	1075
0.9288		air	I_2	0	88–92
0.9295		air	I_2	0	88–92
0.9305		air	I_2	0	88–92
0.9350		air	Tl	1	931
0.937790	±0.000060	air	Te	1	904,1073
0.9386805		air	N	0	190–2,215–218
0.9392789		air	N	0	190–2,215–218
0.9396		air	Hg	1	160
0.9405729		air	C	0	190–192,383
0.9452098		air	Cl	0	258,260
0.94892838		vacuum	Ne	0	528
0.9518		air	I_2	0	88–92
0.9520		air	I_2	0	88–92
0.95326		air	D_2	0	71,72,74
0.9545		air	I_2	0	88–92
0.9555		air	I_2	0	88–92
0.961			Na_2	0	1593
0.962		air	C	0	1505
0.965389		air	N_2	0	105

Table 4.1.1—*continued*
Neutral Atom, Ion, and Molecular Gas Lasers Arranged in Order of Increasing Wavelength

Wavelength (μm)	Uncertainty	Medium	Species	Charge	Reference
0.965779		air	Ar	0	588–9,593,722
0.9658		air	C	0	1505,1506
0.965846		air	N_2	0	105
0.966599		air	N_2	0	105
0.967270		air	N_2	0	105
0.967758		air	N_2	0	105
0.967943		air	N_2	0	105
0.968061		air	N_2	0	105
0.969552		air	N_2	0	105
0.969700	±0.000200	air	Xe	1	1075,1171
0.969879		air	N_2	0	105
0.9766		air	I_2	0	88–92
0.9767		air	I_2	0	88–92
0.9793		air	Br	0	1505
0.97997039		air	Xe	0	455,588,589
0.98		air	I	0	266
0.9838743		vacuum	Cd	0	152
0.9898		air	Eu	1	1054,1087
0.9940		air	Ca	1	976,1166
0.995470	±0.000100	air	Se	1	963
0.9963		air	I_2	0	88–92
0.9973		air	I_2	0	88–92
1.0019		air	I_2	0	88–92
1.0020		air	Eu	1	1078,1087
1.0053		air	I_2	0	88–92
1.008422		air	P	0	221
1.01		air	I	0	266
1.01236025		air	Cs	0	350
1.016–1.340			I_2	0	1556
1.0166		air	Eu	1	1078,1087
1.0225		air	I_2	0	88–92
1.0245		air	I_2	0	88–92
1.0255		air	I_2	0	88–92
1.0274		air	I_2	0	88–92
1.0298238		vacuum	Ne	0	490,526,527

Table 4.1.1—*continued*
Neutral Atom, Ion, and Molecular Gas Lasers
Arranged in Order of Increasing Wavelength

Wavelength (μm)	Uncertainty	Medium	Species	Charge	Reference
1.03		air	I	0	266
1.0324559		vacuum	Yb	0	247
1.033050	±0.000050	air	Sr	1	139,881
1.040940	±0.000100	air	Se	1	963
1.041720	±0.000060	air	I	3	968
1.042			$^{130}Te_2$	0	1603
1.0455451		air	S	0	195,196,239
1.0455985		vacuum	As	0	222
1.04701		air	Ar	0	588
1.0534		air	I_2	0	88–92
1.0563328		air	N	0	219
1.0586		air	Hg	1	160,924
1.059			$^{130}Te_2$	0	1603
1.0600		air	Nd(thd)$_3$	0	25
1.0612556		vacuum	Sn	0	384
1.0617063		vacuum	As	0	222
1.062		air	Sn	1	202,384
1.0623177		air	N	0	219
1.0623574		vacuum	Ne	0	185,490,531–3
1.0634		air	Xe	0	609
1.063400	±0.000600	air	Xe	1	1075
1.0635993		air	S	0	195,196
1.0643981		air	N	0	219
1.06596		air	Kr	1	1007,1089
1.0683082		air	C	0	192–193
1.0685345		air	C	0	190,193
1.0691250		air	C	0	190–196
1.06939			Tm	0	1856,1857
1.0707333		air	C	0	193
1.0730		air	C	0	1505
1.074		air	Sn	1	202
1.0775		air	I_2	0	88–92
1.0788		air	I_2	0	88–92
1.0801000		vacuum	Ne	0	389,526,534
1.0847447		vacuum	Ne	0	389,526,534

Table 4.1.1—*continued*
Neutral Atom, Ion, and Molecular Gas Lasers Arranged in Order of Increasing Wavelength

Wavelength (μm)	Uncertainty	Medium	Species	Charge	Reference
1.0860		air	S_2	0	122
1.0867911		vacuum	Cd	0	152
1.088			$^{130}Te_2$	0	1603
1.089			$^{130}Te_2$	0	1603
1.091450	±0.000050	air	Sr	1	139,881
1.0915		air	S_2	0	122
1.091500	±0.000150	air	Mg	1	920,1046
1.0917		air	S_2	0	122
1.0920		air	S_2	0	122
1.0923		air	S_2	0	122
1.092300	±0.000100	air	Ar	1	594,1117,1143
1.0941		air	S_2	0	122
1.0946		air	S_2	0	122
1.0950		air	Xe	0	609
1.095000	±0.000600	air	Xe	0	1075
1.095200	±0.000150	air	Mg	1	920,1046
1.0990		air	S_2	0	122
1.09963		air	CN	0	57,58
1.09965		air	CN	0	57,58
1.09966		air	CN	0	57,58
1.09974		air	CN	0	57,58
1.09974		air	CN	0	57,58
1.09987		air	CN	0	57,58
1.1000		air	S_2	0	122
1.10007		air	CN	0	57,58
1.10031		air	CN	0	57,58
1.10061		air	CN	0	57,58
1.10096		air	CN	0	57,58
1.10136		air	CN	0	57,58
1.10182		air	CN	0	57,58
1.10232		air	CN	0	57,58
1.10288		air	CN	0	57,58
1.10348		air	CN	0	57,58
1.10414		air	CN	0	57,58
1.10445		air	CN	0	57,58

Table 4.1.1—*continued*
Neutral Atom, Ion, and Molecular Gas Lasers Arranged in Order of Increasing Wavelength

Wavelength (μm)	Uncertainty	Medium	Species	Charge	Reference
1.10485		air	CN	0	57,58
1.10521		air	CN	0	57,58
1.10560		air	CN	0	57,58
1.10603		air	CN	0	57,58
1.10641		air	CN	0	57,58
1.1066		air	I_2	0	88–92
1.10689		air	CN	0	57,58
1.10726		air	CN	0	57,58
1.1073		air	I_2	0	88–92
1.10782		air	CN	0	57,58
1.10879		air	CN	0	57,58
1.10981		air	CN	0	57,58
1.1101		air	Tm	0	1494
1.11090		air	CN	0	57,58
1.11200		air	CN	0	57,58
1.113			$^{130}Te_2$	0	1603
1.11321		air	CN	0	57,58
1.1146071		vacuum	Ne	0	472,526,534
1.1163455		vacuum	P	0	221
1.1165		air	H_2	0	71,72,74
1.1179812		vacuum	Hg	0	160,161
1.1179812		vacuum	Hg	0	160,161
1.1180588		vacuum	Ne	0	529,537,539
1.1186470		vacuum	P	0	221
1.1206		air	I_2	0	88–92
1.1210			H_2	0	1535
1.1214		air	I_2	0	88–92
1.1216		air	I_2	0	88–92
1.1225		air	H_2	0	71,72,74
1.1226		air	I_2	0	88–92
1.1230		air	Sr	1	976
1.1247708		vacuum	As	0	222
1.1255		air	I_2	0	88–92
1.126		air	Al	0	1499
1.1290435		vacuum	Hg	0	156

Table 4.1.1—*continued*
Neutral Atom, Ion, and Molecular Gas Lasers Arranged in Order of Increasing Wavelength

Wavelength (μm)	Uncertainty	Medium	Species	Charge	Reference
1.130304		air	Ba	0	146–148
1.1328		air	I_2	0	88–92
1.1334		air	I_2	0	88–92
1.134			$^{130}Te_2$	0	1603
1.1347		air	I_2	0	88–92
1.1348		air	I_2	0	88–92
1.1350		air	I_2	0	88–92
1.1350		air	Tl	1	931
1.138145		air	Na	0	572–3,679,785
1.1393552		vacuum	Ne	0	528,529,534
1.1403784		air	Na	0	572–3,679,785
1.1412258		vacuum	Ne	0	389,466,526–9
1.144			$^{130}Te_2$	0	1603
1.1453		air	I_2	0	88–92
1.14574813		air	Kr	0	538
1.1464		air	I_2	0	88–92
1.1485		air	Cd	0	152
1.14881		air	Ar	0	588
1.1502		air	I_2	0	88–92
1.1510		air	I_2	0	88–92
1.1515		air	I_2	0	88–92
1.1522		air	I_2	0	88–92
1.1522595		vacuum	As	0	175,222
1.1524056		vacuum	As	0	175,222
1.1525900		vacuum	Ne	0	529,535–40
1.1528174		vacuum	Ne	0	535,537,544
1.1529		air	I_2	0	88–92
1.1547277		vacuum	P	0	221
1.1554		air	Cd	0	152
1.1587		air	S_2	0	122
1.159		air	Pb	1	1086
1.1604712		vacuum	Ne	0	389,466,540
1.161			$^{130}Te_2$	0	1603
1.1617260		vacuum	Ne	0	529,536,540
1.1663677		vacuum	Cd	0	152

Table 4.1.1—*continued*
Neutral Atom, Ion, and Molecular Gas Lasers Arranged in Order of Increasing Wavelength

Wavelength (μm)	Uncertainty	Medium	Species	Charge	Reference
1.1698		air	I_2	0	88–92
1.1703		air	I_2	0	88–92
1.171			$^{130}Te_2$	0	1603
1.1711		air	Bi	0	1492
1.1711		air	I_2	0	88–92
1.1718		air	I_2	0	88–92
1.1740		air	I_2	0	88–92
1.1743		air	Br	0	1505
1.1745636		vacuum	Cd	0	152
1.1750		air	I_2	0	88–92
1.1750		air	Tl	1	931
1.1770013		vacuum	Ne	0	466,526–529
1.177283		air	K	0	350
1.1787698		vacuum	P	0	221
1.1792270		vacuum	Ne	0	490,526–529
1.1874246		vacuum	Cd	0	152,153
1.191			$^{130}Te_2$	0	1603
1.1984187		air	Si	0	199
1.1988192		vacuum	Ne	0	526,529,536
1.2031507		air	Si	0	199
1.2068179		vacuum	Ne	0	535,536,546
1.2096		air	Be	1	1108
1.2115637		vacuum	Ar	0	532
1.21397378		air	Ar	0	588
1.2170		air	I_2	0	88–92
1.2222		air	Hg	0	162,163
1.224			$^{130}Te_2$	0	1603
1.2246		air	Hg	0	162,163
1.240282693		air	Ar	0	588,590
1.243224		air	K	0	772
1.2462797		vacuum	Ne	0	490,526–529
1.2478		air	Ba	1	976
1.252211		air	K	0	350,772
1.2545		air	Hg	0	160
1.2545		air	Hg	1	160

Table 4.1.1—*continued*
Neutral Atom, Ion, and Molecular Gas Lasers Arranged in Order of Increasing Wavelength

Wavelength (μm)	Uncertainty	Medium	Species	Charge	Reference
1.255136		vacuum	Yb	0	247
1.2561370		air	Pb	0	211
1.258790	±0.000100	air	Se	1	963
1.258790	±0.000100	air	Se	1	963
1.2588072		vacuum	Ne	0	553
1.2588088		vacuum	Ne	0	553
1.2598449		vacuum	Ne	0	528,529
1.2692672		vacuum	Ne	0	426,542,553
1.27022810		air	Ar	0	590–593,806
1.271400	±0.000100	air	Yb	1	404,1002
1.2740		air	I_2	0	88–92
1.2760		air	Hg	0	162,163
1.2773017		vacuum	Ne	0	526,528
1.279		air	Li	0	1499
1.279		air	Li	0	1499
1.28027391?		air	Ar	0	532
1.2870		air	I_2	0	88–92
1.2890684		vacuum	Ne	0	490
1.2899		air	Mn	0	406–7,411–13
1.2915545		vacuum	Ne	0	490,553,528
1.2925		air	I_2	0	88–92
1.2940		air	I_2	0	88–92
1.2945989		vacuum	As	0	222
1.2981		air	Hg	0	160
1.2981		air	Hg	1	160
1.3010		air	I_2	0	88–92
1.3020		air	I_2	0	88–92
1.304		air	Ca	0	1499
1.3040		air	I_2	0	88–92
1.305363		air	Zn	0	383
1.3058983		vacuum	Tm	0	246
1.3061		air	H_2	0	71,72,74
1.3069		air	I_2	0	88–92
1.3080		air	I_2	0	88–92
1.3103722		vacuum	Pb	0	383

Table 4.1.1—*continued*
Neutral Atom, Ion, and Molecular Gas Lasers Arranged in Order of Increasing Wavelength

Wavelength (μm)	Uncertainty	Medium	Species	Charge	Reference
1.3104227		vacuum	Tm	0	246
1.31258		vacuum	HF	0	1624
1.3152443		vacuum	I	0	267–283,2
1.3152769		vacuum	Pb	0	383
1.3153		air	I_2	0	88–92
1.3166		air	H_2	0	71,72,74
1.317		air	Ca	0	1499
1.31774118		air	Kr	0	538
1.3192		air	I_2	0	88–92
1.32		air	Rb	0	1501
1.3200		air	I_2	0	88–92
1.32125		vacuum	HF	0	1624
1.3282		air	I_2	0	88–92
1.3291		air	I_2	0	88–92
1.32941		air	Mn	0	336,406–7,411
1.3295		air	Kr	1	1007,1089
1.33053		vacuum	HF	0	1624
1.3310		air	I_2	0	88–92
1.33179		air	Mn	0	336,406–7,411
1.3324		air	I_2	0	88–92
1.3333		air	I_2	0	88–92
1.3349		air	I_2	0	88–92
1.3380		air	I_2	0	88–92
1.3383700		vacuum	Tm	0	246
1.34043		vacuum	HF	0	1624
1.3406		air	I_2	0	88–92
1.3418		air	I_2	0	88–92
1.3421		air	I_2	0	88–92
1.3429		air	I_2	0	88–92
1.342961		air	N	0	215
1.342996		air	In	0	383
1.345300	±0.000100	air	Yb	1	404,1002
1.3476544		vacuum	Ar	0	594
1.35099		vacuum	HF	0	1624
1.3574217		vacuum	Hg	0	156

Table 4.1.1—*continued*
Neutral Atom, Ion, and Molecular Gas Lasers Arranged in Order of Increasing Wavelength

Wavelength (μm)	Uncertainty	Medium	Species	Charge	Reference
1.358133		air	N	0	191,195–6,215
1.358831		air	Cs	0	350
1.360257		air	Cs	0	796
1.3610		air	Eu	1	1078,1087
1.3612294		vacuum	Sn	0	383
1.36219		vacuum	HF	0	1624
1.36224153		air	Kr	0	538
1.3625		air	Mn	0	406,407,413
1.3655		air	Hg	0	160
1.36570559		air	Xe	0	538
1.366501		air	Rb	0	350
1.3677207		vacuum	Hg	0	156,161
1.37282		vacuum	HF	0	1638,1639
1.37406		vacuum	HF	0	1638,1639
1.375883		air	Cs	0	350,796
1.375900	±0.000200	air	Ag	1	806
1.38196		vacuum	HF	0	1638,1639
1.3863		air	Mn	0	406,407,413
1.38633		air	Cl	0	260
1.38931		air	Cl	0	260
1.39175		vacuum	HF	0	1638,1639
1.3954389		vacuum	Hg	0	156
1.396710		air	Ni	0	417
1.3982714		vacuum	Cd	0	152,154,383
1.39970		air	Mn	0	336,406–7,411
1.4019620		air	S	0	240
1.40936399		air	Ar	0	538
1.4124892		vacuum	As	0	222
1.41830		air	CN	0	57,58
1.41849		air	CN	0	57,58
1.41876		air	CN	0	57,58
1.41911		air	CN	0	57,58
1.41954		air	CN	0	57,58
1.42005		air	CN	0	57,58
1.42065		air	CN	0	57,58

Table 4.1.1—*continued*
Neutral Atom, Ion, and Molecular Gas Lasers Arranged in Order of Increasing Wavelength

Wavelength (μm)	Uncertainty	Medium	Species	Charge	Reference
1.42132		air	CN	0	57,58
1.42207		air	CN	0	57,58
1.42289		air	CN	0	57,58
1.42380		air	CN	0	57,58
1.42478		air	CN	0	57,58
1.425		air	Ca	0	1499
1.42583		air	CN	0	57,58
1.42586227712		vacuum	As	0	175
1.4259232		vacuum	As	0	175
1.42696		air	CN	0	57,58
1.42808		air	CN	0	57,58
1.42945		air	CN	0	57,58
1.43081		air	CN	0	57,58
1.4321	±.00051	air	Ne	0	554
1.4330	±.00051	air	Ne	0	554
1.4331602		vacuum	Cd	0	152–154,383
1.4343722		vacuum	Tm	0	246
1.4346		air	Ne	0	554
1.4368	±.00051	air	Ne	0	554
1.441920		air	In	0	383
1.44267933		air	Kr	0	538
1.44693		vacuum	HF	0	1638,1639
1.4478302		vacuum	Cd	0	154,383,775
1.4489080		vacuum	Tm	0	246
1.45304			Tm	0	1856,1857
1.45423		air	N	0	215
1.454250		air	C	0	191,194–198
1.4545941		vacuum	I	0	260
1.4553011		vacuum	N	0	195,196
1.455371		vacuum	Ni	0	155
1.45730		vacuum	HF	0	1638,1639
1.4629079		vacuum	As	0	222
1.469493		air	Cs	0	350
1.47			Cs	0	1835
1.475241		air	Rb	0	350

Table 4.1.1—*continued*
Neutral Atom, Ion, and Molecular Gas Lasers Arranged in Order of Increasing Wavelength

Wavelength (μm)	Uncertainty	Medium	Species	Charge	Reference
1.47654720		air	Kr	0	538
1.4770		air	Eu	1	1054,1087
1.4793059		vacuum	Yb	0	404
1.4848636		vacuum	Ne	0	555
1.4873294		vacuum	Ne	0	555
1.4876248		vacuum	Ne	0	555
1.4892012		vacuum	Ne	0	555
1.4903576		vacuum	Ne	0	555
1.4940304		vacuum	Ne	0	555
1.49578			Tm	0	1856,1857
1.49618939		air	Kr	0	605
1.4998810		vacuum	Tm	0	246
1.50004		air	Ba	0	146–151,708
1.502499		air	Mg	0	146
1.5036834		vacuum	Tm	0	246
1.504605		air	Ar	0	806
1.5234875		vacuum	Ne	0	526,529,550
1.528843		air	Rb	0	350
1.528948		air	Rb	0	350
1.529954		vacuum	Hg	0	160–61,164–74
1.53264796		air	Kr	0	605
1.5335134		vacuum	Pb	0	383
1.5422255		air	S	0	241
1.5533401		vacuum	I	0	400
1.5550		air	Hg	1	160,1102
1.5716351		air	P	0	221
1.5730		air	Cl	0	1505
1.58697		air	Cl	0	197,261,391
1.588441		air	Si	0	199,403
1.5970		air	Cl	0	1505
1.598200	±0.000200	air	Ag	1	806
1.6057666		vacuum	Xe	0	538
1.6180021		air	Ar	0	389,429
1.6383650		vacuum	Tm	0	246
1.6404449		vacuum	Cd	0	152,154,383

Table 4.1.1—*continued*
Neutral Atom, Ion, and Molecular Gas Lasers Arranged in Order of Increasing Wavelength

Wavelength (μm)	Uncertainty	Medium	Species	Charge	Reference
1.6407031		vacuum	Ne	0	551
1.6437081		vacuum	Cd	0	152
1.646400	±0.000200	air	Ag	1	806
1.6486189		vacuum	Cd	0	152,153
1.648791		vacuum	P	0	221
1.649800	±0.000200	air	Yb	1	247,1002,1133
1.65199		air	Ar	0	593
1.6542665		air	S	0	241
1.6758663		vacuum	Tm	0	246
1.68534881		air	Kr	0	538
1.68967525		air	Kr	0	389,429,538
1.6924775		vacuum	Hg	0	160,161
1.69358061		air	Kr	0	389,429
1.6940584		air	Ar	0	389,429,538
1.6946636		vacuum	Hg	0	160,161
1.7077438		vacuum	Hg	0	160,161
1.7114554		vacuum	Hg	0	160,161
1.7166616		vacuum	Ne	0	490,526
1.720200	±0.000200	air	Ag	1	806
1.7323684		vacuum	Tm	0	247
1.7330499		vacuum	Xe	0	588–89,610–13
1.7334185		vacuum	Hg	0	161,167
1.734600	±0.000200	air	Ag	1	806
1.7367231		vacuum	Ga	0	175
1.7459155		vacuum	Yb	0	247
1.748000	±0.000200	air	Ag	1	806
1.7600985		vacuum	Eu	0	247,418
1.771000	±0.000200	air	Cu	1	806
1.78427374		air	Kr	0	389,429
1.79			Ar	0	1807
1.7919615		vacuum	Ar	0	389,429
1.798400		vacuum	Yb	0	404
1.801			Na_2	0	1593
1.804			Na_2	0	1593
1.8053474		vacuum	As	0	175

Table 4.1.1—*continued*
Neutral Atom, Ion, and Molecular Gas Lasers Arranged in Order of Increasing Wavelength

Wavelength (μm)	Uncertainty	Medium	Species	Charge	Reference
1.8057		air	Yb	1	404,1002
1.8068806		vacuum	As	0	222
1.8135329		vacuum	Hg	0	160–61,164–69
1.81673150		air	Kr	0	427
1.81850539		air	Kr	0	389,429
1.8199686		vacuum	Cu	0	806
1.820416		air	Ba	0	147
1.8215302		vacuum	Ne	0	526
1.82340574		vacuum	Cu	0	806
1.8258313		vacuum	Ne	0	185
1.8258357		vacuum	Ne	0	185
1.827659		air	Ne	0	551,557–559
1.828258		air	Ne	0	551,557–559
1.830400		air	Ne	0	551,557,559
1.831000	±0.000100	air	Zn	1	155,1145
1.836		vacuum	DF	0	1281
1.8380629		vacuum	Ag	0	383,806
1.840316		air	Ne	0	550–1,557,559
1.841000	±0.000200	air	Ag	1	806
1.844			Na_2	0	1593
1.844		vacuum	DF	0	1281
1.846400	±0.000200	air	Ag	1	806
1.854		vacuum	DF	0	1281
1.859112		air	Ne	0	551,557,559
1.859730		air	Ne	0	556,557,559
1.868596		air	He	0	424,425
1.870		air	Li	0	1499
1.872400	±0.000200	air	Ag	1	806
1.8736732		vacuum	In	0	175
1.87510		air	H	0	130,339
1.876			Na_2	0	1593
1.879600	±0.000200	air	Ag	1	806
1.882			Na_2	0	1593
1.888			Na_2	0	1593
1.8943842		vacuum	P	0	221

Table 4.1.1—*continued*
Neutral Atom, Ion, and Molecular Gas Lasers Arranged in Order of Increasing Wavelength

Wavelength (μm)	Uncertainty	Medium	Species	Charge	Reference
1.897		air	Ca	0	1499
1.898000	±0.000200	air	Ag	1	806
1.9022415		vacuum	Ba	0	146
1.905		air	Ca	0	1499
1.9123124		vacuum	Cd	0	152
1.91240055d		vacuum	Sm	0	247
1.915600	±0.000200	air	Cu	1	806
1.9216572		vacuum	Kr	0	389,429
1.925			Na_2	0	1593
1.926		air	Eu	0	420
1.931			Na_2	0	1593
1.9371923		vacuum	Ag	0	806
1.948000	±0.000200	air	Cu	1	806
1.954313		air	He	0	424–427
1.95740		air	Ne	0	526,542,553
1.958248		vacuum	Ne	0	434,526,561
1.958985		vacuum	Tm	0	246
1.961		air	Eu	0	420
1.968			Na_2	0	1593
1.971000	±0.000200	air	Cu	1	806
1.971600	±0.000200	air	Ag	1	806
1.9722834		vacuum	Tm	0	246
1.973362		air	Br	0	261,403
1.97546477557		air	Cl	0	199,260–64
1.97546477557		vacuum	As	0	175
1.9814602		vacuum	Ge	0	175
1.982400	±0.000200	air	Ag	1	806
1.9835173		vacuum	Yb	0	246,247
1.98616		air	CN	0	57,58
1.98629		air	CN	0	57,58
1.98658		air	CN	0	57,58
1.98701		air	CN	0	57,58
1.98759		air	CN	0	57,58
1.98831		air	CN	0	57,58
1.98918		air	CN	0	57,58

Table 4.1.1—*continued*
Neutral Atom, Ion, and Molecular Gas Lasers Arranged in Order of Increasing Wavelength

Wavelength (μm)	Uncertainty	Medium	Species	Charge	Reference
1.99020		air	CN	0	57,58
1.99135		air	CN	0	57,58
1.99263		air	CN	0	57,58
1.99406		air	CN	0	57,58
1.9947227		vacuum	Tm	0	
1.99563		air	CN	0	57,58
1.99733		air	CN	0	57,58
1.9988		air	CN	0	57,58
2.000400	±0.000200	air	Cu	1	806
2.0009		air	CN	0	57,58
2.0031		air	CN	0	57,58
2.0041827		vacuum	Yb	0	246,247
2.0055		air	CN	0	57,58
2.0080		air	CN	0	57,58
2.0107		air	CN	0	57,58
2.0135		air	CN	0	57,58
2.0164		air	CN	0	57,58
2.0196		air	CN	0	57,58
2.01994		air	Cl	0	199,260–64
2.020052		vacuum	V	0	155
2.020602		vacuum	Ge	0	175
2.02622395		air	Xe	0	389,538,588–9
2.0282741		vacuum	As	0	175
2.0355792		vacuum	Ne	0	526,537,550
2.0359432		vacuum	Ne	0	537,542,564
2.0482		air	Sm	0	247
2.0581302		air	He	0	181,428
2.060755		vacuum	He	0	185,429,431
2.0616228		air	Ar	0	389,596–8,806
2.0655993		vacuum	C	0	195
2.079400	±0.000200	air	Ag	1	806
2.0962339		vacuum	P	0	221
2.0986110		air	Ar	0	427,565,806
2.1023345		vacuum	Ne	0	185,435,526
2.10409		air	Ne	0	434,537,542

Table 4.1.1—*continued*
Neutral Atom, Ion, and Molecular Gas Lasers Arranged in Order of Increasing Wavelength

Wavelength (μm)	Uncertainty	Medium	Species	Charge	Reference
2.1059135		vacuum	Tm	0	246
2.1129476		vacuum	Tm	0	246
2.11654709		air	Kr	0	389,429,570
2.1186997		vacuum	Yb	0	246,247
2.1332885		air	Ar	0	195
2.1480		air	Yb	1	404,1002
2.1534205		air	Ar	0	427,565,593
2.1573497		vacuum	Ba	0	146,147
2.17074		air	Ne	0	434,537,553
2.19025126		air	Kr	0	389,429,570
2.206		air	Na	0	1500
2.2077181		air	Ar	0	195,264,564
2.24857754		air	Kr	0	602
2.252965		air	Rb	0	183,796
2.2792		vacuum	H_2O	0	1329
2.2801247		vacuum	S	0	242,403
2.286565		air	Br	0	265
2.293247		air	Rb	0	183,350,796
2.3133204		air	Ar	0	538,564,593
2.31933328		air	Xe	0	559,582,607
2.32553		air	Ba	0	146,147
2.3266649		vacuum	Ne	0	526
2.351215		air	Br	0	265
2.3785794		vacuum	In	0	175
2.3851957		vacuum	Tm	0	246
2.3957953		vacuum	Ne	0	542,556,560
2.3962995		vacuum	Ne	0	185,542,567–8
2.3966520		air	Ar	0	428,564,588
2.404160	±0.000300	air	Mg	1	136,1046
2.412520	±0.000300	air	Mg	1	136,1046
2.41381		vacuum	HF	0	1638,1639
2.4162547		vacuum	Ne	0	569
2.4225538		vacuum	Ne	0	185,570
2.4256255		vacuum	Ne	0	434,526,542
2.42605059		air	Kr	0	427,565

Table 4.1.1—*continued*
Neutral Atom, Ion, and Molecular Gas Lasers Arranged in Order of Increasing Wavelength

Wavelength (μm)	Uncertainty	Medium	Species	Charge	Reference
2.43312		vacuum	HF	0	1638,1639
2.436331		air	S	0	241
2.437700	±0.000200	air	Yb	1	247,1002,1133
2.4466627		vacuum	As	0	175
2.44700		air	Cl	0	197,260,264
2.448099		vacuum	V	0	155
2.45381		vacuum	HF	0	1638,1639
2.458			Na_2	0	1593
2.47588		vacuum	HF	0	1638,1639
2.4764593		vacuum	Ba	0	146
2.48247157		air	Xe	0	610
2.498			Na_2	0	1593
2.5014408		vacuum	Ar	0	195,564
2.503			Na_2	0	1593
2.510			Na_2	0	1593
2.5152702		vacuum	Xe	0	602
2.517			Na_2	0	1593
2.520			Na_2	0	1593
2.52338198		air	Kr	0	538,570,608
2.524			Na_2	0	1593
2.537			Na_2	0	1593
2.538			Na_2	0	1593
2.5400115		vacuum	Ne	0	195,564
2.54946		vacuum	Ar	0	195,564
2.55083		vacuum	HF	0	1638,1639
2.5512187		vacuum	Ar	0	195,564
2.55157		air	Ba	0	146,147
2.5531329		vacuum	Ne	0	526
2.557			Na_2	0	1593
2.561			Na_2	0	1593
2.5634025		vacuum	Ar	0	195
2.5668023		vacuum	Ar	0	564
2.57885		vacuum	HF	0	1638,1639
2.5818111		vacuum	Eu	0	247
2.592300	±0.000150	air	Ba	1	146,1133

Table 4.1.1—*continued*
Neutral Atom, Ion, and Molecular Gas Lasers Arranged in Order of Increasing Wavelength

Wavelength (μm)	Uncertainty	Medium	Species	Charge	Reference
2.598577		air	I	0	260
2.60848		vacuum	HF	0	1638,1639
2.6266703		vacuum	Kr	0	195,564
2.62690832		air	Xe	0	559,582,607
2.6288137		vacuum	Kr	0	195,564
2.63976		vacuum	HF	0	1638,1639
2.65108645		air	Xe	0	550,582,610–3
2.6513946		vacuum	O	0	236
2.6550282		vacuum	Ar	0	601
2.6608397		vacuum	Xe	0	559,607
2.66679		vacuum	HF	0	1638,1639
2.6672615		vacuum	Xe	0	601
2.67274		vacuum	HF	0	1638,1639
2.6843026		vacuum	Ar	0	195,564
2.6886382		vacuum	$^{12}C^{16}O$	0	1293,1324
2.689		air	Li	0	1499
2.6914188		vacuum	$^{12}C^{16}O$	0	1293,1324
2.69625		vacuum	HF	0	1638,1639
2.7006079		vacuum	Sm	0	247
2.70752377		vacuum	HF	0	1638,1639
2.71			Br	0	1859
2.7135274		vacuum	Br	0	437–439
2.7152859		vacuum	Ar	0	602
2.7181668		vacuum	Eu	0	247
2.7261939		vacuum	$^{12}C^{16}O$	0	1293,1324
2.72749		vacuum	HF	0	1638,1639
2.7290266		vacuum	$^{12}C^{16}O$	0	1293,1324
2.73		air	Rb	0	1501
2.7319168		vacuum	$^{12}C^{16}O$	0	1293,1324
2.7348646		vacuum	$^{12}C^{16}O$	0	1293,1324
2.7363805		vacuum	Ar	0	195,564
2.7378703		vacuum	$^{12}C^{16}O$	0	1293,1324
2.7409341		vacuum	$^{12}C^{16}O$	0	1293,1324
2.7440564		vacuum	$^{12}C^{16}O$	0	1293,1324
2.74412		vacuum	HF	0	1638,1639

Table 4.1.1—*continued*
Neutral Atom, Ion, and Molecular Gas Lasers Arranged in Order of Increasing Wavelength

Wavelength (μm)	Uncertainty	Medium	Species	Charge	Reference
2.7472374		vacuum	$^{12}C^{16}O$	0	1293,1324
2.7504772		vacuum	$^{12}C^{16}O$	0	1293,1324
2.7537763		vacuum	$^{12}C^{16}O$	0	1638,1639
2.7571350		vacuum	$^{12}C^{16}O$	0	1293,1324
2.757298		air	I	0	265,401
2.7580982		vacuum	Ne	0	195,261
2.76036		vacuum	HF	0	1638,1639
2.7605534		vacuum	$^{12}C^{16}O$	0	1293,1324
2.7640319		vacuum	$^{12}C^{16}O$	0	1638,1639
2.7646881		vacuum	$^{12}C^{16}O$	0	1293,1324
2.7675709		vacuum	$^{12}C^{16}O$	0	1293,1324
2.7675747		vacuum	$^{12}C^{16}O$	0	1638,1639
2.7705202		vacuum	$^{12}C^{16}O$	0	1293,1324
2.7711706		vacuum	$^{12}C^{16}O$	0	1293,1324
2.7765892		vacuum	$^{12}C^{16}O$	0	1638,1639
2.7797133		vacuum	$^{12}C^{16}O$	0	1293,1324
2.78		air	Ba	0	1502
2.78257		vacuum	HF	0	1638,1639
2.7826380		vacuum	Ne	0	185,564,195
2.7828975		vacuum	$^{12}C^{16}O$	0	1293,1324
2.7861419		vacuum	$^{12}C^{16}O$	0	1638,1639
2.7894469		vacuum	$^{12}C^{16}O$	0	1293,1324
2.79023		vacuum	HF	0	1638,1639
2.790537		air	Rb	0	350
2.7928128		vacuum	$^{12}C^{16}O$	0	1293,1324
2.79527		vacuum	HF	0	1638,1639
2.7962400		vacuum	$^{12}C^{16}O$	0	1293,1324
2.7997285		vacuum	$^{12}C^{16}O$	0	1638,1639
2.8032790		vacuum	$^{12}C^{16}O$	0	1293,1324
2.8041545		vacuum	$^{12}C^{16}O$	0	1293,1324
2.8068915		vacuum	$^{12}C^{16}O$	0	1638,1639
2.8070965		vacuum	$^{12}C^{16}O$	0	1293,1324
2.8100991		vacuum	$^{12}C^{16}O$	0	1293,1324
2.8105665		vacuum	$^{12}C^{16}O$	0	1638,1639
2.8143044		vacuum	$^{12}C^{16}O$	0	1293,1324

Table 4.1.1—*continued*
Neutral Atom, Ion, and Molecular Gas Lasers Arranged in Order of Increasing Wavelength

Wavelength (μm)	Uncertainty	Medium	Species	Charge	Reference
2.8181054		vacuum	$^{12}C^{16}O$	0	1293,1324
2.8202417		vacuum	Ar	0	195,564
2.82126		vacuum	HF	0	1638,1639
2.8219699		vacuum	$^{12}C^{16}O$	0	1293,1324
2.8227216		vacuum	$^{12}C^{16}O$	0	1293,1324
2.82310		vacuum	HF	0	1638,1639
2.8245953		vacuum	Ar	0	195,564
2.8258983		vacuum	$^{12}C^{16}O$	0	1293,1324
2.8260316		vacuum	$^{12}C^{16}O$	0	1293,1324
2.8294039		vacuum	$^{12}C^{16}O$	0	1293,1324
2.8298909		vacuum	$^{12}C^{16}O$	0	1293,1324
2.83181		vacuum	HF	0	1638,1639
2.8328388		vacuum	$^{12}C^{16}O$	0	1293,1324
2.8363366		vacuum	$^{12}C^{16}O$	0	1293,1324
2.837716		air	Br	0	265
2.8398977		vacuum	$^{12}C^{16}O$	0	1293,1324
2.8435224		vacuum	$^{12}C^{16}O$	0	1293,1324
2.8446284		vacuum	$^{12}C^{16}O$	0	1293,1324
2.8472110		vacuum	$^{12}C^{16}O$	0	1293,1324
2.8476276		vacuum	$^{12}C^{16}O$	0	1293,1324
2.8506891		vacuum	$^{12}C^{16}O$	0	1293,1324
2.8509638		vacuum	$^{12}C^{16}O$	0	1293,1324
2.85404		vacuum	HF	0	1638,1639
2.8547812		vacuum	$^{12}C^{16}O$	0	1293,1324
2.8586637		vacuum	$^{12}C^{16}O$	0	1293,1324
2.8590043		vacuum	Xe	0	602
2.8610550		air	Kr	0	195,538,559
2.8620231		vacuum	Ar	0	598,603
2.8626114		vacuum	$^{12}C^{16}O$	0	1293,1324
2.8633965		vacuum	Ne	0	195
2.8655717		air	Kr	0	195,538,559
2.86567		vacuum	HF	0	1638,1639
2.8666249		vacuum	$^{12}C^{16}O$	0	1293,1324
2.8703841		vacuum	$^{12}C^{16}O$	0	1293,1324
2.87057		vacuum	HF	0	1638,1639

Table 4.1.1—*continued*
Neutral Atom, Ion, and Molecular Gas Lasers Arranged in Order of Increasing Wavelength

Wavelength (μm)	Uncertainty	Medium	Species	Charge	Reference
2.8707045		vacuum	$^{12}C^{16}O$	0	1293,1324
2.8738903		vacuum	$^{12}C^{16}O$	0	1293,1324
2.8774613		vacuum	$^{12}C^{16}O$	0	1293,1324
2.8782932		vacuum	Ar	0	195,564
2.8810972		vacuum	$^{12}C^{16}O$	0	1293,1324
2.8843088		vacuum	Ar	0	195,564
2.8847986		vacuum	$^{12}C^{16}O$	0	1293,1324
2.8885657		vacuum	$^{12}C^{16}O$	0	1293,1324
2.88889		vacuum	HF	0	1638,1639
2.8892054		vacuum	$^{12}C^{16}O$	0	1293,1324
2.8923278		vacuum	$^{12}C^{16}O$	0	1293,1324
2.8923991		vacuum	$^{12}C^{16}O$	0	1293,1324
2.8944397		vacuum	O	0	235,1698
2.8962987		vacuum	$^{12}C^{16}O$	0	1293,1324
2.9002654		vacuum	$^{12}C^{16}O$	0	1293,1324
2.9042993		vacuum	$^{12}C^{16}O$	0	1293,1324
2.905900	±0.000200	air	Ba	1	146,1133
2.9084008		vacuum	$^{12}C^{16}O$	0	1293,1324
2.91026		vacuum	HF	0	1638,1639
2.91106		vacuum	HF	0	1638,1639
2.9134037		vacuum	Ar	0	603
2.9196521		vacuum	$^{12}C^{16}O$	0	1293,1324
2.92208		vacuum	HF	0	1638,1639
2.9230381		vacuum	Ba	0	146,147
2.9233655		vacuum	$^{12}C^{16}O$	0	1293,1324
2.92555		vacuum	HF	0	1638,1639
2.9271462		vacuum	$^{12}C^{16}O$	0	1293,1324
2.9280662		vacuum	Ar	0	195,564
2.9287507		vacuum	$^{12}C^{16}O$	0	1293,1324
2.9309946		vacuum	$^{12}C^{16}O$	0	1293,1324
2.9317981		vacuum	Cs	0	350
2.9318698		vacuum	$^{12}C^{16}O$	0	1293,1324
2.9349111		vacuum	$^{12}C^{16}O$	0	1293,1324
2.935		vacuum	OH	0	1288
2.9350549		vacuum	$^{12}C^{16}O$	0	1293,1324

Table 4.1.1—*continued*
Neutral Atom, Ion, and Molecular Gas Lasers Arranged in Order of Increasing Wavelength

Wavelength (μm)	Uncertainty	Medium	Species	Charge	Reference
2.9388960		vacuum	$^{12}C^{16}O$	0	1293,1324
2.9429497		vacuum	$^{12}C^{16}O$	0	1293,1324
2.9455858		vacuum	Ne	0	185,195,564
2.9470727		vacuum	$^{12}C^{16}O$	0	1293,1324
2.9512653		vacuum	$^{12}C^{16}O$	0	1293,1324
2.95391		vacuum	HF	0	1638,1639
2.95487		vacuum	HF	0	1638,1639
2.9555279		vacuum	$^{12}C^{16}O$	0	1293,1324
2.95727		vacuum	HF	0	1638,1639
2.9598610		vacuum	$^{12}C^{16}O$	0	1293,1324
2.96432		vacuum	HF	0	1638,1639
2.9663		air	Sm	0	247
2.9676035		vacuum	Ne	0	195,564
2.969		vacuum	OH	0	1288
2.9706062		vacuum	$^{12}C^{16}O$	0	1293,1324
2.9724809		vacuum	$^{12}C^{16}O$	0	1293,1324
2.9745386		vacuum	$^{12}C^{16}O$	0	1293,1324
2.9756629		vacuum	$^{12}C^{16}O$	0	1293,1324
2.9785412		vacuum	$^{12}C^{16}O$	0	1293,1324
2.9789128		vacuum	$^{12}C^{16}O$	0	1293,1324
2.9796792		vacuum	Ar	0	195,564
2.98		air	Ba	0	1502
2.9812503		vacuum	Ne	0	195,564
2.9812503		vacuum	Ne	0	195,564
2.9813368		vacuum	As	0	175
2.9826142		vacuum	$^{12}C^{16}O$	0	1293,1324
2.9844656		vacuum	Kr	0	195,564
2.9867581		vacuum	$^{12}C^{16}O$	0	1293,1324
2.9878091		vacuum	Kr	0	195,564
2.98961		vacuum	HF	0	1638,1639
2.9909732		vacuum	$^{12}C^{16}O$	0	1293,1324
2.9952601		vacuum	$^{12}C^{16}O$	0	1293,1324
2.99891		vacuum	HF	0	1638,1639
2.9996191		vacuum	$^{12}C^{16}O$	0	1293,1324
3.0040507		vacuum	$^{12}C^{16}O$	0	1293,1324

Table 4.1.1—*continued*
Neutral Atom, Ion, and Molecular Gas Lasers Arranged in Order of Increasing Wavelength

Wavelength (μm)	Uncertainty	Medium	Species	Charge	Reference
3.00505		vacuum	HF	0	1638,1639
3.00642		vacuum	HF	0	1638,1639
3.0085555		vacuum	$^{12}C^{16}O$	0	1293,1324
3.01			Cs	0	1835
3.01033		air	Cs	0	350
3.0111339		vacuum	Cs	0	796
3.0118377		vacuum	Sr	0	143,144
3.0131337		vacuum	$^{12}C^{16}O$	0	1293,1324
3.0173828		vacuum	$^{12}C^{16}O$	0	1293,1324
3.0206297		vacuum	$^{12}C^{16}O$	0	1293,1324
3.02635		vacuum	HF	0	1638,1639
3.0267787		vacuum	Ne	0	185,195,564
3.0274972		vacuum	$^{12}C^{16}O$	0	1293,1324
3.0275836		vacuum	Ne	0	185,195,564
3.0317345		vacuum	$^{12}C^{16}O$	0	1293,1324
3.0360451		vacuum	$^{12}C^{16}O$	0	1293,1324
3.036119		air	I	0	401,402
3.0404296		vacuum	$^{12}C^{16}O$	0	1293,1324
3.0448885		vacuum	$^{12}C^{16}O$	0	1293,1324
3.04612		vacuum	HF	0	1638,1639
3.046207		vacuum	Ar	0	195,564
3.04819		vacuum	HF	0	1638,1639
3.0494222		vacuum	$^{12}C^{16}O$	0	1293,1324
3.05		air	Ba	0	1502
3.0536574		vacuum	Kr	0	195,564
3.0540311		vacuum	$^{12}C^{16}O$	0	1293,1324
3.05820		vacuum	HF	0	1638,1639
3.0587159		vacuum	$^{12}C^{16}O$	0	1293,1324
3.0634769		vacuum	$^{12}C^{16}O$	0	1293,1324
3.06517		vacuum	HF	0	1638,1639
3.066080		air	Cl	0	264
3.0663542		air	Kr	0	428,559,564
3.0668182		vacuum	$^{12}C^{16}O$	0	1293,1324
3.0670208		vacuum	Sr	0	143,144
3.0683147		vacuum	$^{12}C^{16}O$	0	1293,1324

Table 4.1.1—*continued*
Neutral Atom, Ion, and Molecular Gas Lasers Arranged in Order of Increasing Wavelength

Wavelength (μm)	Uncertainty	Medium	Species	Charge	Reference
3.0720016		vacuum	Ne	0	571
3.0732299		vacuum	$^{12}C^{16}O$	0	1293,1324
3.0779257		vacuum	$^{12}C^{16}O$	0	1293,1324
3.078		vacuum	OH	0	1288
3.0823354		vacuum	$^{12}C^{16}O$	0	1293,1324
3.0868211		vacuum	$^{12}C^{16}O$	0	1293,1324
3.0913834		vacuum	$^{12}C^{16}O$	0	1293,1324
3.09350		vacuum	HF	0	1638,1639
3.09580		vacuum	HF	0	1638,1639
3.0960229		vacuum	$^{12}C^{16}O$	0	1293,1324
3.0961401		vacuum	Cs	0	183,796
3.09821		vacuum	HF	0	1638,1639
3.0996226		vacuum	Ar	0	195,564
3.1007399		vacuum	$^{12}C^{16}O$	0	1293,1324
3.1055351		vacuum	$^{12}C^{16}O$	0	1293,1324
3.10616		vacuum	HF	0	1638,1639
3.10692302		air	Xe	0	550,559,582
3.1104089		vacuum	$^{12}C^{16}O$	0	1293,1324
3.11255		vacuum	HF	0	1638,1639
3.115		vacuum	OH	0	1288
3.1153618		vacuum	$^{12}C^{16}O$	0	1293,1324
3.1203946		vacuum	$^{12}C^{16}O$	0	1293,1324
3.1255075		vacuum	$^{12}C^{16}O$	0	1293,1324
3.1333028		vacuum	Ar	0	564
3.1344849		vacuum	$^{12}C^{16}O$	0	1293,1324
3.1345761		vacuum	Ar	0	195
3.13495		vacuum	HF	0	1638,1639
3.1391545		vacuum	$^{12}C^{16}O$	0	1293,1324
3.14110		vacuum	HF	0	1638,1639
3.1415224		vacuum	K	0	350,772
3.1439037		vacuum	$^{12}C^{16}O$	0	1293,1324
3.14802		vacuum	HF	0	1638,1639
3.1487329		vacuum	$^{12}C^{16}O$	0	1293,1324
3.14939		vacuum	HF	0	1638,1639
3.1514572		vacuum	Kr	0	195,564

Table 4.1.1—*continued*
Neutral Atom, Ion, and Molecular Gas Lasers Arranged in Order of Increasing Wavelength

Wavelength (μm)	Uncertainty	Medium	Species	Charge	Reference
3.1536426		vacuum	$^{12}C^{16}O$	0	1293,1324
3.157		vacuum	OH	0	1288
3.1586333		vacuum	$^{12}C^{16}O$	0	1293,1324
3.1601267		vacuum	K	0	350,796
3.1637058		vacuum	$^{12}C^{16}O$	0	1293,1324
3.1653653		vacuum	Te	0	155
3.1688605		vacuum	$^{12}C^{16}O$	0	1293,1324
3.16962		vacuum	HF	0	1638,1639
3.17387		vacuum	HF	0	1638,1639
3.1740979		vacuum	$^{12}C^{16}O$	0	1293,1324
3.1748096		vacuum	Pb	0	175
3.1834750		vacuum	$^{12}C^{16}O$	0	1293,1324
3.1882559		vacuum	$^{12}C^{16}O$	0	1293,1324
3.19117		vacuum	HF	0	1638,1639
3.1931188		vacuum	$^{12}C^{16}O$	0	1293,1324
3.1980643		vacuum	$^{12}C^{16}O$	0	1293,1324
3.20293		vacuum	HF	0	1638,1639
3.2030929		vacuum	$^{12}C^{16}O$	0	1293,1324
3.2050778		vacuum	Cs	0	184–186
3.2082052		vacuum	$^{12}C^{16}O$	0	1293,1324
3.2134017		vacuum	$^{12}C^{16}O$	0	1293,1324
3.21501		vacuum	HF	0	1638,1639
3.2186830		vacuum	$^{12}C^{16}O$	0	1293,1324
3.2240498		vacuum	$^{12}C^{16}O$	0	1293,1324
3.22933		vacuum	HF	0	1638,1639
3.234		vacuum	OH	0	1288
3.236285		air	I	0	164,263,265
3.24381		vacuum	HF	0	1638,1639
3.25849		vacuum	HF	0	1638,1639
3.26028		vacuum	HF	0	1638,1639
3.27392788		air	Xe	0	427,550,632
3.274		vacuum	OH	0	1288
3.289000	±0.000200	air	Cd	1	155
3.29196		vacuum	HF	0	1638,1639
3.29463		air	N_2	0	107,128

Table 4.1.1—*continued*
Neutral Atom, Ion, and Molecular Gas Lasers Arranged in Order of Increasing Wavelength

Wavelength (μm)	Uncertainty	Medium	Species	Charge	Reference
3.29913		vacuum	HF	0	1638,1639
3.30149		air	N_2	0	107,128
3.30438		vacuum	HF	0	1638,1639
3.30734		air	N_2	0	107,128
3.3094055		vacuum	Xe	0	601
3.30989		air	N_2	0	107,128
3.31221		air	N_2	0	107,128
3.31426		air	N_2	0	107,128
3.31607		air	N_2	0	107,128
3.31760		air	N_2	0	107,128
3.3182141		vacuum	Ne	0	559,564,570
3.31889		air	N_2	0	107,128
3.32059		vacuum	HF	0	1638,1639
3.32069		air	N_2	0	107,128
3.33348		vacuum	HF	0	1638,1639
3.3341754		vacuum	Ne	0	559,564,570
3.3361448		vacuum	Ne	0	559,564,570
3.3409635		air	Kr	0	195,564,570
3.3510469		vacuum	Ne	0	432
3.3520466		vacuum	Ne	0	432
3.36666991		air	Xe	0	550,582,623
3.37720		vacuum	HF	0	1638,1639
3.3813942		vacuum	Ne	0	195,564
3.3849653		vacuum	Ne	0	195,564
3.389503		vacuum	S	0	242
3.3912244		vacuum	Ne	0	559,564,578–0
3.3922348		vacuum	Ne	0	456,556,572
3.4023945		vacuum	Xe	0	601
3.407422		vacuum	C	0	195
3.408		air	Na	0	1500
3.41		air	Na	0	1500
3.428		vacuum	^{12}COS	0	1316
3.429573		air	I	0	164,263,265
3.4344638		vacuum	Xe	0	564,627,630
3.4480843		vacuum	Ne	0	525,559,570

Table 4.1.1—*continued*
Neutral Atom, Ion, and Molecular Gas Lasers Arranged in Order of Increasing Wavelength

Wavelength (μm)	Uncertainty	Medium	Species	Charge	Reference
3.45184		air	N_2	0	107,128
3.45832		air	N_2	0	107,128
3.46114		air	N_2	0	107,128
3.46368		air	N_2	0	107,128
3.4654		air	Sm	0	247
3.46596		air	N_2	0	107,128
3.46795		air	N_2	0	107,128
3.4679986		vacuum	Kr	0	564,195
3.46967		air	N_2	0	107,128
3.47109		air	N_2	0	107,128
3.4789495		vacuum	Ne	0	432,525
3.4882957		vacuum	Kr	0	195,564
3.4894892		vacuum	Kr	0	195,564
3.4909363		vacuum	Cs	0	796
3.49327		vacuum	DF	0	1281
3.5117661		vacuum	C	0	195
3.52140		vacuum	DF	0	1281
3.5361		air	Sm	0	247
3.543052		air	Cl	0	197
3.55068		vacuum	DF	0	1281
3.572781		vacuum	$H^{35}Cl$	0	1256
3.58110		vacuum	DF	0	1281
3.5844556		vacuum	Ne	0	559,564,570
3.601		air	B	0	176
3.602617		vacuum	$H^{35}Cl$	0	1256
3.61283		vacuum	DF	0	1281
3.6219081		vacuum	Xe	0	559
3.62349		air	N_2	0	108
3.62614		air	N_2	0	108
3.62910		air	N_2	0	108
3.631236		vacuum	Ar	0	602
3.633673		vacuum	$H^{35}Cl$	0	1256
3.636190		vacuum	$H^{37}Cl$	0	1256
3.63630		vacuum	DF	0	1281
3.64313		air	N_2	0	108

Table 4.1.1—*continued*
Neutral Atom, Ion, and Molecular Gas Lasers Arranged in Order of Increasing Wavelength

Wavelength (μm)	Uncertainty	Medium	Species	Charge	Reference
3.64472		air	N_2	0	108
3.64560		vacuum	DF	0	1281
3.64662		air	N_2	0	108
3.64883		air	N_2	0	108
3.65138		air	N_2	0	108
3.6518315		vacuum	Xe	0	564,611–3,625
3.65424		air	N_2	0	108
3.65745		air	N_2	0	108
3.66095		air	N_2	0	108
3.66483		air	N_2	0	108
3.665989		vacuum	$H^{35}Cl$	0	1256
3.66652		vacuum	DF	0	1281
3.668485		vacuum	$H^{37}Cl$	0	1256
3.66899		air	N_2	0	108
3.67352		air	N_2	0	108
3.67834		air	N_2	0	108
3.6789254		vacuum	Mg	0	136
3.6789565		vacuum	Mg	0	136
3.67980		vacuum	DF	0	1256,1282
3.6798859		vacuum	Xe	0	559,582,632
3.6825364		vacuum	Mg	0	136
3.6858866		vacuum	Xe	0	559,611,623
3.69825		vacuum	DF	0	1281
3.699583		vacuum	$H^{35}Cl$	0	1256
3.7013512		vacuum	Ar	0	602
3.702069		vacuum	$H^{37}Cl$	0	1256
3.70711		vacuum	$H^{35}Cl$	0	1256
3.7086023		vacuum	Ar	0	604
3.70971		vacuum	$H^{37}Cl$	0	1256
3.7143477		vacuum	Ar	0	602
3.71552		vacuum	DF	0	1256,1282
3.726		air	Cu	0	1499
3.7265		air	Xe	0	1498
3.73095		vacuum	DF	0	1281
3.734504		vacuum	$H^{35}Cl$	0	1256

Table 4.1.1—*continued*
Neutral Atom, Ion, and Molecular Gas Lasers Arranged in Order of Increasing Wavelength

Wavelength (μm)	Uncertainty	Medium	Species	Charge	Reference
3.736975		vacuum	$H^{37}Cl$	0	1256
3.73830		vacuum	$H^{35}Cl$	0	1256
3.74079		vacuum	$H^{37}Cl$	0	1256
3.75199		vacuum	DF	0	1256,1282
3.75633		vacuum	DF	0	1256,1282
3.76510		vacuum	DF	0	1256,1282
3.770787		vacuum	$H^{35}Cl$	0	1256
3.77100		vacuum	$H^{35}Cl$	0	1256
3.77347		vacuum	$H^{37}Cl$	0	1256
3.7742128		vacuum	Kr	0	604
3.7746325		vacuum	Ne	0	559,564
3.78782		vacuum	DF	0	1256,1282
3.79018		vacuum	DF	0	1256,1282
3.7942		air	N	0	215
3.796602		air	Cl	0	197
3.80071		vacuum	DF	0	1256,1282
3.80499		vacuum	$H^{35}Cl$	0	1256
3.80742		vacuum	$H^{37}Cl$	0	1256
3.808469		vacuum	$H^{35}Cl$	0	1256
3.8135916		vacuum	Tl	0	175
3.8154		air	N	0	215
3.82057		vacuum	DF	0	1256,1282
3.82980		vacuum	DF	0	1256,1282
3.83749		vacuum	DF	0	1256,1282
3.84011		vacuum	$H^{35}Cl$	0	1256
3.84249		vacuum	$H^{37}Cl$	0	1256
3.85		vacuum	HCN	0	1327
3.8502		vacuum	DF	0	1256,1282
3.85091		vacuum	$H^{35}Cl$	0	1256
3.85361		vacuum	$H^{37}Cl$	0	1256
3.85471		vacuum	DF	0	1256,1282
3.863		air	Al	0	1499
3.86573		air	Mg	0	136
3.8696535		vacuum	Xe	0	559,632
3.8704		vacuum	DF	0	1256,1282

Table 4.1.1—*continued*
Neutral Atom, Ion, and Molecular Gas Lasers Arranged in Order of Increasing Wavelength

Wavelength (μm)	Uncertainty	Medium	Species	Charge	Reference
3.87573		vacuum	DF	0	1256,1282
3.87684		vacuum	$H^{35}Cl$	0	1256
3.8817		vacuum	DF	0	1256,1282
3.88395		vacuum	$H^{35}Cl$	0	1256
3.88641		vacuum	$H^{37}Cl$	0	1256
3.89028		vacuum	DF	0	1256,1282
3.8950221		vacuum	Xe	0	559,582,611
3.9128		vacuum	DF	0	1256,1282
3.9145		vacuum	DF	0	1256,1282
3.91491		vacuum	$H^{35}Cl$	0	1256
3.91547		vacuum	DF	0	1256,1282
3.92049		vacuum	$H^{37}Cl$	0	1256
3.92716		vacuum	DF	0	1256,1282
3.928361		air	Hg	0	161,167
3.94867		vacuum	DF	0	1256,1282
3.95365		vacuum	$H^{35}Cl$	0	1256
3.9557248		air	Kr	0	604
3.95602		vacuum	$H^{37}Cl$	0	1256
3.95653		vacuum	DF	0	1256,1282
3.95717		vacuum	DF	0	1256,1282
3.9589222		vacuum	Ba	0	146
3.96541		vacuum	DF	0	1256,1282
3.980630		air	Ne	0	195,564
3.98429		vacuum	DF	0	1256,1282
3.99093		vacuum	$H^{35}Cl$	0	1256
3.9966035		vacuum	Xe	0	559,611,623–4
3.99949		vacuum	DF	0	1256,1282
4.00317		vacuum	DF	0	1256,1282
4.00543		vacuum	DF	0	1256,1274
4.00590		vacuum	$H^{35}Cl$	0	1256,1259
4.0079678		vacuum	Ba	0	146
4.01703		vacuum	$H^{79}Br$	0	1255–1257
4.01760		vacuum	$H^{81}Br$	0	1255–1257
4.0207278		vacuum	Xe	0	601,651
4.02118		vacuum	DF	0	1256,1274

Table 4.1.1—*continued*
Neutral Atom, Ion, and Molecular Gas Lasers Arranged in Order of Increasing Wavelength

Wavelength (μm)	Uncertainty	Medium	Species	Charge	Reference
4.029		air	Li	0	1499
4.02951		vacuum	$H^{35}Cl$	0	1256,1259
4.04042		vacuum	$H^{35}Cl$	0	1256,1259
4.0433		vacuum	DF	0	1256,1274
4.04639		vacuum	DF	0	1256,1274
4.04699		vacuum	$H^{79}Br$	0	1255–1257
4.04755		vacuum	$H^{81}Br$	0	1255–1257
4.0491		vacuum	DF	0	1256,1274
4.0594		vacuum	DF	0	1256,1274
4.0685162		air	Kr	0	604
4.07644		vacuum	$H^{35}Cl$	0	1256,1259
4.07825		vacuum	$H^{79}Br$	0	1255–1257
4.07884		vacuum	$H^{81}Br$	0	1255–1257
4.0891		vacuum	DF	0	1256,1274
4.08949		vacuum	DF	0	1256,1274
4.11066		vacuum	$H^{79}Br$	0	1255–1257
4.11123		vacuum	$H^{81}Br$	0	1255–1257
4.11399		vacuum	$H^{35}Cl$	0	1256,1259
4.1336		vacuum	DF	0	1256,1274
4.1368		air	Sm	0	247
4.13690		vacuum	DF	0	1256,1274
4.14424		vacuum	$H^{79}Br$	0	1255–1257
4.14477		vacuum	$H^{81}Br$	0	1255–1257
4.1526299		vacuum	Xe	0	559
4.1526711		vacuum	Kr	0	604
4.16531		vacuum	$H^{79}Br$	0	1255–1257
4.16585		vacuum	$H^{81}Br$	0	1255–1257
4.17962		vacuum	$H^{81}Br$	0	1255–1257
4.17980		vacuum	DF	0	1256,1274
4.182		vacuum	Xe	0	601
4.18622		vacuum	DF	0	1256,1274
4.19695		vacuum	$H^{79}Br$	0	1255–1257
4.19754		vacuum	$H^{81}Br$	0	1255–1257
4.2013276		vacuum	Mg	0	136
4.2044098		vacuum	Ar	0	601

Table 4.1.1—*continued*
Neutral Atom, Ion, and Molecular Gas Lasers Arranged in Order of Increasing Wavelength

Wavelength (μm)	Uncertainty	Medium	Species	Charge	Reference
4.2181082		vacuum	Cs	0	350
4.2182950		vacuum	Ne	0	571
4.22947		vacuum	$H^{79}Br$	0	1255–1257
4.26334		vacuum	$H^{79}Br$	0	1255–1257
4.26392		vacuum	$H^{81}Br$	0	1255–1257
4.29880		vacuum	$H^{79}Br$	0	1255–1257
4.29937		vacuum	$H^{81}Br$	0	1255–1257
4.3131086		vacuum	$^{12}C^{16}O_2$	0	1295,1296
4.314		vacuum	$^{12}C^{16}O^{18}O$	0	1312
4.3162577		vacuum	$^{12}C^{16}O_2$	0	1295,1296
4.3178659		vacuum	$^{12}C^{16}O_2$	0	1294,1295
4.32031		vacuum	$^{12}C^{16}O_2$	0	1297
4.3211010		vacuum	$^{12}C^{16}O_2$	0	1294,1295
4.3213904		vacuum	Eu	0	247
4.3227096		vacuum	$^{12}C^{16}O_2$	0	1295,1296
4.3243851		vacuum	$^{12}C^{16}O_2$	0	1297
4.32492		vacuum	$^{12}C^{16}O_2$	0	1294,1295
4.32498		vacuum	$H^{79}Br$	0	1255–1257
4.32554		vacuum	$H^{81}Br$	0	1255–1257
4.3260126		vacuum	$^{12}C^{16}O_2$	0	1295,1296
4.32764		vacuum	$^{12}C^{16}O_2$	0	1297
4.3277182		vacuum	$^{12}C^{16}O_2$	0	1294,1295
4.3285152		vacuum	Ba	0	147
4.3293670		vacuum	$^{12}C^{16}O_2$	0	1295,1296
4.3311007		vacuum	$^{12}C^{16}O_2$	0	1294,1295
4.3321362		vacuum	I	0	400
4.3327731		vacuum	$^{12}C^{16}O_2$	0	1295,1296
4.3345326		vacuum	$^{12}C^{16}O_2$	0	1294,1295
4.33539		vacuum	$H^{79}Br$	0	1255–1257
4.33595		vacuum	$H^{81}Br$	0	1255–1257
4.3362312		vacuum	$^{12}C^{16}O_2$	0	1295,1296
4.3380141		vacuum	$^{12}C^{16}O_2$	0	1294,1295
4.3397413		vacuum	$^{12}C^{16}O_2$	0	1295,1296
4.340		vacuum	$^{12}C^{16}O^{18}O$	0	1312
4.3415456		vacuum	$^{12}C^{16}O_2$	0	1294,1295

Table 4.1.1—*continued*
Neutral Atom, Ion, and Molecular Gas Lasers Arranged in Order of Increasing Wavelength

Wavelength (μm)	Uncertainty	Medium	Species	Charge	Reference
4.3433038		vacuum	$^{12}C^{16}O_2$	0	1295,1296
4.3451268		vacuum	$^{12}C^{16}O_2$	0	1294,1295
4.346		vacuum	$^{12}C^{18}O_2$	0	1312
4.3505861		vacuum	$^{12}C^{16}O_2$	0	1295,1296
4.354		vacuum	$^{12}C^{16}O^{18}O$	0	1312
4.35493		vacuum	$^{12}C^{16}O_2$	0	1297
4.35791		vacuum	$H^{79}Br$	0	1255–1257
4.35802		vacuum	$^{12}C^{16}O_2$	0	1297
4.35846		vacuum	$H^{81}Br$	0	1255–1257
4.36121		vacuum	$^{12}C^{16}O_2$	0	1297
4.3638859		vacuum	Mg	0	136
4.36443		vacuum	$^{12}C^{16}O_2$	0	1297
4.36775		vacuum	$^{12}C^{16}O_2$	0	1297
4.371		vacuum	$^{12}C^{18}O_2$	0	1312
4.37109		vacuum	$^{12}C^{16}O_2$	0	1297
4.37445		vacuum	$^{12}C^{16}O_2$	0	1297
4.3747938		vacuum	Kr	0	564,570,604
4.376		vacuum	$^{12}C^{18}O_2$	0	1312
4.3766712		vacuum	Kr	0	195
4.37790		vacuum	$^{12}C^{16}O_2$	0	1297
4.38139		vacuum	$^{12}C^{16}O_2$	0	1297
4.382		vacuum	$^{12}C^{18}O_2$	0	1312
4.384		vacuum	$^{12}C^{18}O_2$	0	1312
4.38495		vacuum	$^{12}C^{16}O_2$	0	1297
4.39012		vacuum	$^{13}C^{16}O_2$	0	1643,1644
4.392		vacuum	$^{12}C^{18}O_2$	0	1312
4.39250		vacuum	$H^{79}Br$	0	1255–1257
4.39267		vacuum	$^{13}C^{16}O_2$	0	1643,1644
4.39306		vacuum	$H^{81}Br$	0	1255–1257
4.39789		vacuum	$^{13}C^{16}O_2$	0	1643,1644
4.398		vacuum	$^{12}C^{18}O_2$	0	1312
4.40056		vacuum	$^{13}C^{16}O_2$	0	1643,1644
4.42813		vacuum	$H^{79}Br$	0	1255–1257
4.42870		vacuum	$H^{81}Br$	0	1255–1257
4.44858		vacuum	$^{13}C^{16}O_2$	0	1643,1644

Table 4.1.1—*continued*
Neutral Atom, Ion, and Molecular Gas Lasers Arranged in Order of Increasing Wavelength

Wavelength (μm)	Uncertainty	Medium	Species	Charge	Reference
4.45210		vacuum	$^{13}C^{16}O_2$	0	1643,1644
4.45931		vacuum	$^{13}C^{16}O_2$	0	1643,1644
4.46299		vacuum	$^{13}C^{16}O_2$	0	1643,1644
4.46524		vacuum	$H^{79}Br$	0	1255–1257
4.46576		vacuum	$H^{81}Br$	0	1255–1257
4.46672		vacuum	$^{13}C^{16}O_2$	0	1643,1644
4.473		vacuum	$^{13}C^{16}O^{18}O$	0	1626
4.50410		vacuum	$H^{79}Br$	0	1255–1257
4.50467		vacuum	$H^{81}Br$	0	1255–1257
4.528		vacuum	$^{13}C^{16}O^{18}O$	0	1626
4.53295		vacuum	$H^{79}Br$	0	1255–1257
4.53348		vacuum	$H^{81}Br$	0	1255–1257
4.5393674		vacuum	Xe	0	571,601,632
4.5615027		vacuum	O	0	235,237,1698
4.5678667		vacuum	Xe	0	601
4.56911		vacuum	$H^{79}Br$	0	1255–1257
4.56965		vacuum	$H^{81}Br$	0	1255–1257
4.5706441		vacuum	Xe	0	601
4.5851		vacuum	$^{15}N^{14}NO$	0	1335
4.6		vacuum	$N_2{}^{15}O$	0	1335
4.60535		air	He	0	432,433
4.6056		air	He	0	433
4.60700		vacuum	$H^{79}Br$	0	1255–1257
4.60755		vacuum	$H^{81}Br$	0	1255–1257
4.6109078		vacuum	Xe	0	559
4.6157396		vacuum	Sn	0	155
4.6189		vacuum	$^{15}N^{14}NO$	0	1335
4.6204		vacuum	$^{14}N^{15}NO$	0	1335
4.64626		vacuum	$H^{79}Br$	0	1255–1257
4.64675		vacuum	$H^{81}Br$	0	1255–1257
4.6699795		vacuum	Ba	0	146,147
4.6812		vacuum	$^{14}N^{15}NO$	0	1335
4.694835		vacuum	Eu	0	247
4.7151680		vacuum	Ar	0	601
4.7169143		vacuum	Ba	0	146

Table 4.1.1—*continued*
Neutral Atom, Ion, and Molecular Gas Lasers Arranged in Order of Increasing Wavelength

Wavelength (μm)	Uncertainty	Medium	Species	Charge	Reference
4.7184144		vacuum	Ba	0	146,147
4.7451305		vacuum	$^{12}C^{16}O$	0	1179,1187
4.7545011		vacuum	$^{12}C^{16}O$	0	1179,1187
4.7639848		vacuum	$^{12}C^{16}O$	0	1179,1187
4.771		vacuum	H_2O	0	1328
4.7735825		vacuum	$^{12}C^{16}O$	0	1179,1187
4.7768928		vacuum	$^{12}C^{16}O$	0	1179,1187
4.7832950		vacuum	$^{12}C^{16}O$	0	1179,1187
4.7860767		vacuum	$^{12}C^{16}O$	0	1179,1187
4.7931233		vacuum	$^{12}C^{16}O$	0	1179,1187
4.7953740		vacuum	$^{12}C^{16}O$	0	1179,1187
4.8021974		vacuum	Yb	0	247
4.8030680		vacuum	$^{12}C^{16}O$	0	1179,1187
4.8047856		vacuum	$^{12}C^{16}O$	0	1179,1187
4.8131300		vacuum	$^{12}C^{16}O$	0	1179,1187
4.8143122		vacuum	$^{12}C^{16}O$	0	1179,1187
4.8233101		vacuum	$^{12}C^{16}O$	0	1179,1187
4.8239548		vacuum	$^{12}C^{16}O$	0	1179,1187
4.8336095		vacuum	$^{12}C^{16}O$	0	1179,1187
4.8337142		vacuum	$^{12}C^{16}O$	0	1179,1187
4.8435910		vacuum	$^{12}C^{16}O$	0	1179,1187
4.8440288		vacuum	$^{12}C^{16}O$	0	1179,1187
4.8535862		vacuum	$^{12}C^{16}O$	0	1179,1187
4.8545692		vacuum	$^{12}C^{16}O$	0	1179,1187
4.8562339		vacuum	$^{12}C^{16}O$	0	1179,1187
4.858472		vacuum	I	0	401,402
4.862914		vacuum	I	0	401,402
4.8637009		vacuum	$^{12}C^{16}O$	0	1179,1187
4.8652314		vacuum	$^{12}C^{16}O$	0	1179,1187
4.8656		air	Sm	0	247
4.8658033		vacuum	$^{12}C^{16}O$	0	1179,1187
4.8739354		vacuum	$^{12}C^{16}O$	0	1179,1187
4.8754905		vacuum	$^{12}C^{16}O$	0	1179,1187
4.8773393		vacuum	Kr	0	195,564
4.8831334		vacuum	Kr	0	195,564

Table 4.1.1—*continued*
Neutral Atom, Ion, and Molecular Gas Lasers Arranged in Order of Increasing Wavelength

Wavelength (μm)	Uncertainty	Medium	Species	Charge	Reference
4.8842914		vacuum	$^{12}C^{16}O$	0	1179,1187
4.8852962		vacuum	$^{12}C^{16}O$	0	1179,1187
4.8947691		vacuum	$^{12}C^{16}O$	0	1179,1187
4.8952217		vacuum	$^{12}C^{16}O$	0	1179,1187
4.9052675		vacuum	$^{12}C^{16}O$	0	1179,1187
4.9053698		vacuum	$^{12}C^{16}O$	0	1179,1187
4.9088830		vacuum	$^{12}C^{16}O$	0	1179,1187
4.9154344		vacuum	$^{12}C^{16}O$	0	1179,1187
4.9160256		vacuum	Ar	0	195,564
4.9160943		vacuum	$^{12}C^{16}O$	0	1179,1187
4.9184946		vacuum	$^{12}C^{16}O$	0	1179,1187
4.9207077		vacuum	Ar	0	195,564
4.9257235		vacuum	$^{12}C^{16}O$	0	1179,1187
4.9282261		vacuum	$^{12}C^{16}O$	0	1179,1187
4.9361360		vacuum	$^{12}C^{16}O$	0	1179,1187
4.9380782		vacuum	$^{12}C^{16}O$	0	1179,1187
4.9466722		vacuum	$^{12}C^{16}O$	0	1179,1187
4.9480521		vacuum	$^{12}C^{16}O$	0	1179,1187
4.9573335		vacuum	$^{12}C^{16}O$	0	1179,1187
4.9581483		vacuum	$^{12}C^{16}O$	0	1179,1187
4.9627720		vacuum	$^{12}C^{16}O$	0	1179,1187
4.9681208		vacuum	$^{12}C^{16}O$	0	1179,1187
4.9683676		vacuum	$^{12}C^{16}O$	0	1179,1187
4.9724254		vacuum	$^{12}C^{16}O$	0	1179,1187
4.9787115		vacuum	$^{12}C^{16}O$	0	1179,1187
4.9790348		vacuum	$^{12}C^{16}O$	0	1179,1187
4.9822006		vacuum	$^{12}C^{16}O$	0	1179,1187
4.9891802		vacuum	$^{12}C^{16}O$	0	1179,1187
4.9900770		vacuum	$^{12}C^{16}O$	0	1179,1187
4.9920988		vacuum	$^{12}C^{16}O$	0	1179,1187
4.9996952		vacuum	Kr	0	601,605
4.9997753		vacuum	$^{12}C^{16}O$	0	1179,1187
5.0012478		vacuum	$^{12}C^{16}O$	0	1179,1187
5.0021206		vacuum	$^{12}C^{16}O$	0	1179,1187
5.00310		vacuum	$D^{35}Cl$	0	1255–56,1265

Table 4.1.1—*continued*
Neutral Atom, Ion, and Molecular Gas Lasers Arranged in Order of Increasing Wavelength

Wavelength (μm)	Uncertainty	Medium	Species	Charge	Reference
5.00984		vacuum	$D^{37}Cl$	0	1255–56,1265
5.0104972		vacuum	$^{12}C^{16}O$	0	1179,1187
5.0122670		vacuum	$^{12}C^{16}O$	0	1179,1187
5.0125487		vacuum	$^{12}C^{16}O$	0	1179,1187
5.0179406		vacuum	$^{12}C^{16}O$	0	1179,1187
5.0213475		vacuum	$^{12}C^{16}O$	0	1179,1187
5.0225389		vacuum	$^{12}C^{16}O$	0	1179,1187
5.0235584		vacuum	Ar	0	602
5.0239807		vacuum	$^{12}C^{16}O$	0	1179,1187
5.0243255		vacuum	Xe	0	601,602
5.0276354		vacuum	$^{12}C^{16}O$	0	1179,1187
5.0322846		vacuum	Ba	0	146
5.0323267		vacuum	$^{12}C^{16}O$	0	1179,1187
5.0329370		vacuum	$^{12}C^{16}O$	0	1179,1187
5.03438		vacuum	$D^{35}Cl$	0	1255–56,1265
5.0355449		vacuum	$^{12}C^{16}O$	0	1179,1187
5.03723		vacuum	CN	0	58
5.0374540		vacuum	$^{12}C^{16}O$	0	1179,1187
5.04124		vacuum	$D^{37}Cl$	0	1255–56,1265
5.0434358		vacuum	$^{12}C^{16}O$	0	1179,1187
5.0434625		vacuum	$^{12}C^{16}O$	0	1179,1187
5.04452		vacuum	$D^{35}Cl$	0	1255–56,1265
5.0472422		vacuum	$^{12}C^{16}O$	0	1179,1187
5.0473978		vacuum	$^{12}C^{16}O$	0	1179,1187
5.04745		vacuum	CN	0	58
5.05140		vacuum	$D^{37}Cl$	0	1255–56,1265
5.0541162		vacuum	$^{12}C^{16}O$	0	1179,1187
5.0546759		vacuum	$^{12}C^{16}O$	0	1179,1187
5.0574675		vacuum	$^{12}C^{16}O$	0	1179,1187
5.05779		vacuum	CN	0	58
5.0590735		vacuum	$^{12}C^{16}O$	0	1179,1187
5.0648991		vacuum	$^{12}C^{16}O$	0	1179,1187
5.0660484		vacuum	$^{12}C^{16}O$	0	1179,1187
5.0660871		vacuum	Eu	0	247
5.06673		vacuum	$D^{35}Cl$	0	1255–56,1265

Table 4.1.1—*continued*
Neutral Atom, Ion, and Molecular Gas Lasers Arranged in Order of Increasing Wavelength

Wavelength (μm)	Uncertainty	Medium	Species	Charge	Reference
5.0676637		vacuum	$^{12}C^{16}O$	0	1179,1187
5.06827		vacuum	CN	0	58
5.0710407		vacuum	$^{12}C^{16}O$	0	1179,1187
5.07357		vacuum	$D^{37}Cl$	0	1255–56,1265
5.07429		vacuum	$D^{35}Cl$	0	1255–56,1265
5.0758123		vacuum	$^{12}C^{16}O$	0	1179,1187
5.0779877		vacuum	$^{12}C^{16}O$	0	1179,1187
5.07887		vacuum	CN	0	58
5.08109		vacuum	$D^{37}Cl$	0	1255–56,1265
5.0831442		vacuum	$^{12}C^{16}O$	0	1179,1187
5.08375		vacuum	CN	0	58
5.0841666		vacuum	$^{12}C^{16}O$	0	1179,1187
5.086000	±0.000400	air	Zn	1	155,1145
5.0868568		vacuum	$^{12}C^{16}O$	0	1179,1187
5.0884401		vacuum	$^{12}C^{16}O$	0	1179,1187
5.08960		vacuum	CN	0	58
5.0920023		vacuum	$^{12}C^{16}O$	0	1179,1187
5.09391		vacuum	CN	0	58
5.0940281		vacuum	$^{12}C^{16}O$	0	1179,1187
5.0953859		vacuum	$^{12}C^{16}O$	0	1179,1187
5.0968277		vacuum	$^{12}C^{16}O$	0	1179,1187
5.0980334		vacuum	$^{12}C^{16}O$	0	1179,1187
5.0990222		vacuum	$^{12}C^{16}O$	0	1179,1187
5.10001		vacuum	$D^{35}Cl$	0	1255–56,1265
5.1004300		vacuum	$^{12}C^{16}O$	0	1179,1187
5.103		vacuum	Ne	0	432
5.1040171		vacuum	$^{12}C^{16}O$	0	1179,1187
5.10418		vacuum	CN	0	58
5.1044128		vacuum	$^{12}C^{16}O$	0	1179,1187
5.10491		vacuum	$D^{35}Cl$	0	1255–56,1265
5.10673		vacuum	$D^{37}Cl$	0	1255–56,1265
5.1072522		vacuum	Tl	0	175
5.1077662		vacuum	$^{12}C^{16}O$	0	1179,1187
5.1089799		vacuum	$^{12}C^{16}O$	0	1179,1187
5.1093435		vacuum	$^{12}C^{16}O$	0	1179,1187

Table 4.1.1—*continued*
Neutral Atom, Ion, and Molecular Gas Lasers Arranged in Order of Increasing Wavelength

Wavelength (μm)	Uncertainty	Medium	Species	Charge	Reference
5.10973514		vacuum	$^{12}C^{16}O$	0	1179,1187
5.11182		vacuum	$D^{37}Cl$	0	1255–56,1265
5.1121182		vacuum	$^{12}C^{16}O$	0	1179,1187
5.1141339		vacuum	$^{12}C^{16}O$	0	1179,1187
5.11459		vacuum	CN	0	58
5.1199442		vacuum	$^{12}C^{16}O$	0	1179,1187
5.1202868		vacuum	$^{12}C^{16}O$	0	1179,1187
5.12057825		vacuum	$^{12}C^{16}O$	0	1179,1187
5.1207879		vacuum	$^{12}C^{16}O$	0	1179,1187
5.1216467		vacuum	Ar	0	195,564
5.1222		vacuum	$D^{35}Cl$	0	1255–56,1265
5.1227527		vacuum	$^{12}C^{16}O$	0	1179,1187
5.1243800		vacuum	$^{12}C^{16}O$	0	1179,1187
5.12513		vacuum	CN	0	58
5.1278917		vacuum	$^{12}C^{16}O$	0	1179,1187
5.1297256		vacuum	$^{12}C^{16}O$	0	1179,1187
5.1311509		vacuum	Kr	0	601
5.1315551		vacuum	$^{12}C^{16}O$	0	1179,1187
5.1323676		vacuum	$^{12}C^{16}O$	0	1179,1187
5.1329488		vacuum	$^{12}C^{16}O$	0	1179,1187
5.13408		vacuum	$D^{35}Cl$	0	1255–56,1265
5.1347558		vacuum	$^{12}C^{16}O$	0	1179,1187
5.13579		vacuum	CN	0	58
5.1359612		vacuum	$^{12}C^{16}O$	0	1179,1187
5.13627		vacuum	$D^{35}Cl$	0	1255–56,1265
5.1368187		vacuum	$^{12}C^{16}O$	0	1179,1187
5.14070		vacuum	$D^{37}Cl$	0	1255–56,1265
5.14221		vacuum	CN	0	58
5.1426650		vacuum	$^{12}C^{16}O$	0	1179,1187
5.14311		vacuum	$D^{37}Cl$	0	1255–56,1265
5.1440321		vacuum	$^{12}C^{16}O$	0	1179,1187
5.1440840		vacuum	$^{12}C^{16}O$	0	1179,1187
5.1452623		vacuum	$^{12}C^{16}O$	0	1179,1187
5.14660		vacuum	CN	0	58
5.15105		vacuum	$D^{35}Cl$	0	1255–56,1265

Table 4.1.1—*continued*
Neutral Atom, Ion, and Molecular Gas Lasers Arranged in Order of Increasing Wavelength

Wavelength (μm)	Uncertainty	Medium	Species	Charge	Reference
5.1513665		vacuum	$^{12}C^{16}O$	0	1179,1187
5.1519668		vacuum	$^{12}C^{16}O$	0	1179,1187
5.15241		vacuum	CN	0	58
5.1539094		vacuum	$^{12}C^{16}O$	0	1179,1187
5.1559008		vacuum	$^{12}C^{16}O$	0	1179,1187
5.1559383		vacuum	$^{12}C^{16}O$	0	1179,1187
5.15754		vacuum	CN	0	58
5.1578		vacuum	$D^{37}Cl$	0	1255–56,1265
5.1588221		vacuum	$^{12}C^{16}O$	0	1179,1187
5.1620001		vacuum	$^{12}C^{16}O$	0	1179,1187
5.16273		vacuum	CN	0	58
5.1652895		vacuum	$^{12}C^{16}O$	0	1179,1187
5.1663999		vacuum	$^{12}C^{16}O$	0	1179,1187
5.16667260		vacuum	$^{12}C^{16}O$	0	1179,1187
5.1679311		vacuum	$^{12}C^{16}O$	0	1179,1187
5.16863		vacuum	CN	0	58
5.16884		vacuum	$D^{35}Cl$	0	1255–56,1265
5.16908		vacuum	$D^{35}Cl$	0	1255–56,1265
5.1711388		vacuum	Ne	0	432
5.1721640		vacuum	$^{12}C^{16}O$	0	1179,1187
5.17320		vacuum	CN	0	58
5.1741002		vacuum	$^{12}C^{16}O$	0	1179,1187
5.1757		vacuum	$D^{37}Cl$	0	1255–56,1265
5.1768063		vacuum	$^{12}C^{16}O$	0	1179,1187
5.1775769		vacuum	$^{12}C^{16}O$	0	1179,1187
5.1796081		vacuum	$^{12}C^{16}O$	0	1179,1187
5.1800642		vacuum	$^{12}C^{16}O$	0	1179,1187
5.18108		vacuum	$D^{35}Cl$	0	1255–56,1265
5.1819239		vacuum	$^{12}C^{16}O$	0	1179,1187
5.1824591		vacuum	$^{12}C^{16}O$	0	1179,1187
5.18382		vacuum	CN	0	58
5.1864363		vacuum	$^{12}C^{16}O$	0	1179,1187
5.18791		vacuum	$D^{37}Cl$	0	1255–56,1265
5.1884610		vacuum	$^{12}C^{16}O$	0	1179,1187
5.1886169		vacuum	$^{12}C^{16}O$	0	1179,1187

Table 4.1.1—*continued*
Neutral Atom, Ion, and Molecular Gas Lasers Arranged in Order of Increasing Wavelength

Wavelength (μm)	Uncertainty	Medium	Species	Charge	Reference
5.1898712		vacuum	$^{12}C^{16}O$	0	1179,1187
5.1923382		vacuum	$^{12}C^{16}O$	0	1179,1187
5.1928864		vacuum	$^{12}C^{16}O$	0	1179,1187
5.1933867		vacuum	$^{12}C^{16}O$	0	1179,1187
5.19454		vacuum	CN	0	58
5.1997924		vacuum	$^{12}C^{16}O$	0	1179,1187
5.2002543		vacuum	$^{12}C^{16}O$	0	1179,1187
5.2004592		vacuum	$^{12}C^{16}O$	0	1179,1187
5.2021		vacuum	$D^{35}Cl$	0	1255–56,1265
5.2034470		vacuum	$^{12}C^{16}O$	0	1179,1187
5.2047548		vacuum	$^{12}C^{16}O$	0	1179,1187
5.20541		vacuum	CN	0	58
5.2076547		vacuum	$^{12}C^{16}O$	0	1179,1187
5.2089		vacuum	$D^{37}Cl$	0	1255–56,1265
5.2111050		vacuum	$^{12}C^{16}O$	0	1179,1187
5.2113155		vacuum	$^{12}C^{16}O$	0	1179,1187
5.21178		vacuum	$D^{35}Cl$	0	1255–56,1265
5.2121879		vacuum	$^{12}C^{16}O$	0	1179,1187
5.2141417		vacuum	$^{12}C^{16}O$	0	1179,1187
5.2149737		vacuum	$^{12}C^{16}O$	0	1179,1187
5.21643		vacuum	CN	0	58
5.2173153		vacuum	$^{12}C^{16}O$	0	1179,1187
5.21855		vacuum	$D^{37}Cl$	0	1255–56,1265
5.2213928		vacuum	$^{12}C^{16}O$	0	1179,1187
5.2224167		vacuum	$^{12}C^{16}O$	0	1179,1187
5.2225555		vacuum	$^{12}C^{16}O$	0	1179,1187
5.2242624		vacuum	$^{12}C^{16}O$	0	1179,1187
5.2249713		vacuum	$^{12}C^{16}O$	0	1179,1187
5.22756		vacuum	CN	0	58
5.2299844		vacuum	$^{12}C^{16}O$	0	1179,1187
5.2300202		vacuum	$^{12}C^{16}O$	0	1179,1187
5.2316028		vacuum	$^{12}C^{16}O$	0	1179,1187
5.23366		vacuum	CN	0	58
5.2341450		vacuum	$^{12}C^{16}O$	0	1179,1187
5.23593867		vacuum	$^{12}C^{16}O$	0	1179,1187

Table 4.1.1—*continued*
Neutral Atom, Ion, and Molecular Gas Lasers Arranged in Order of Increasing Wavelength

Wavelength (μm)	Uncertainty	Medium	Species	Charge	Reference
5.2364797		vacuum	$^{12}C^{16}O$	0	1179,1187
5.2376769		vacuum	$^{12}C^{16}O$	0	1179,1187
5.23887		vacuum	CN	0	58
5.2419468		vacuum	$^{12}C^{16}O$	0	1179,1187
5.2428716		vacuum	$^{12}C^{16}O$	0	1179,1187
5.24348		vacuum	$D^{35}Cl$	0	1255–56,1265
5.24431		vacuum	CN	0	58
5.2454955		vacuum	$^{12}C^{16}O$	0	1179,1187
5.2458747		vacuum	$^{12}C^{16}O$	0	1179,1187
5.24704097		vacuum	$^{12}C^{16}O$	0	1179,1187
5.2488404		vacuum	$^{12}C^{16}O$	0	1179,1187
5.25025		vacuum	$D^{37}Cl$	0	1255–56,1265
5.25028		vacuum	CN	0	58
5.2524253		vacuum	$^{12}C^{16}O$	0	1179,1187
5.25511		vacuum	CN	0	58
5.2558705		vacuum	$^{12}C^{16}O$	0	1179,1187
5.2577457		vacuum	$^{12}C^{16}O$	0	1179,1187
5.2582821		vacuum	$^{12}C^{16}O$	0	1179,1187
5.2613461		vacuum	$^{12}C^{16}O$	0	1179,1187
5.2630393		vacuum	$^{12}C^{16}O$	0	1179,1187
5.26607		vacuum	CN	0	58
5.2690182		vacuum	$^{12}C^{16}O$	0	1179,1187
5.2696629		vacuum	$^{12}C^{16}O$	0	1179,1187
5.2697592		vacuum	$^{12}C^{16}O$	0	1179,1187
5.2737899		vacuum	$^{12}C^{16}O$	0	1179,1187
5.2739976		vacuum	$^{12}C^{16}O$	0	1179,1187
5.27601		vacuum	$D^{35}Cl$	0	1255–56,1265
5.27713		vacuum	CN	0	58
5.28118318		vacuum	$^{12}C^{16}O$	0	1179,1187
5.2819162		vacuum	$^{12}C^{16}O$	0	1179,1187
5.2822426		vacuum	$^{12}C^{16}O$	0	1179,1187
5.2825643		vacuum	Eu	0	247
5.28287		vacuum	$D^{37}Cl$	0	1255–56,1265
5.28467933		vacuum	$^{12}C^{16}O$	0	1179,1187
5.2867966		vacuum	$^{12}C^{16}O$	0	1179,1187

Table 4.1.1—*continued*
Neutral Atom, Ion, and Molecular Gas Lasers Arranged in Order of Increasing Wavelength

Wavelength (μm)	Uncertainty	Medium	Species	Charge	Reference
5.2879276		vacuum	As	0	175
5.28838		vacuum	CN	0	58
5.2924984		vacuum	$^{12}C^{16}O$	0	1179,1187
5.2928466		vacuum	$^{12}C^{16}O$	0	1179,1187
5.2942182		vacuum	$^{12}C^{16}O$	0	1179,1187
5.29570628		vacuum	$^{12}C^{16}O$	0	1179,1187
5.29973		vacuum	CN	0	58
5.2997443		vacuum	$^{12}C^{16}O$	0	1179,1187
5.2999768		vacuum	Kr	0	195,559,564
5.3019553		vacuum	Kr	0	195,559,564
5.3028905		vacuum	$^{12}C^{16}O$	0	1179,1187
5.3032817		vacuum	$^{13}C^{16}O$	0	1253,1254
5.30465169		vacuum	$^{12}C^{16}O$	0	1179,1187
5.3066664		vacuum	$^{12}C^{16}O$	0	1179,1187
5.3068717		vacuum	$^{12}C^{16}O$	0	1179,1187
5.30969		vacuum	$D^{35}Cl$	0	1255–56,1265
5.31121		vacuum	CN	0	58
5.3128419		vacuum	$^{12}C^{16}O$	0	1179,1187
5.3134195		vacuum	$^{12}C^{16}O$	0	1179,1187
5.3145120		vacuum	$^{13}C^{16}O$	0	1253,1254
5.3163		vacuum	$D^{37}Cl$	0	1255–56,1265
5.31660104		vacuum	$^{12}C^{16}O$	0	1179,1187
5.3181789		vacuum	$^{12}C^{16}O$	0	1179,1187
5.3192618		vacuum	$^{12}C^{16}O$	0	1179,1187
5.32286		vacuum	CN	0	58
5.3240867		vacuum	$^{12}C^{16}O$	0	1179,1187
5.32445		vacuum	$D^{35}Cl$	0	1179,1187
5.3258685		vacuum	Ne	0	432
5.3258763		vacuum	$^{13}C^{16}O$	0	1253,1254
5.3260910		vacuum	$^{12}C^{16}O$	0	1179,1187
5.3264331		vacuum	Ne	0	432
5.3284263		vacuum	$^{13}C^{16}O$	0	1253,1254
5.3286950		vacuum	$^{12}C^{16}O$	0	1179,1187
5.3296279		vacuum	$^{12}C^{16}O$	0	1179,1187
5.3297517		vacuum	Bi	0	175

Table 4.1.1—*continued*
Neutral Atom, Ion, and Molecular Gas Lasers Arranged in Order of Increasing Wavelength

Wavelength (μm)	Uncertainty	Medium	Species	Charge	Reference
5.3320059		vacuum	$^{12}C^{16}O$	0	1179,1187
5.3348931		vacuum	$^{12}C^{16}O$	0	1179,1187
5.3373755		vacuum	$^{13}C^{16}O$	0	1253,1254
5.3393194		vacuum	$^{13}C^{16}O$	0	1253,1254
5.3394927		vacuum	$^{12}C^{16}O$	0	1179,1187
5.3409351		vacuum	$^{12}C^{16}O$	0	1179,1187
5.3412198		vacuum	$^{12}C^{16}O$	0	1179,1187
5.3432570		vacuum	$^{12}C^{16}O$	0	1179,1187
5.34431		vacuum	$D^{35}Cl$	0	1255–56,1265
5.3448997		vacuum	$^{12}C^{16}O$	0	1179,1187
5.34583895		vacuum	$^{12}C^{16}O$	0	1179,1187
5.3517193		vacuum	$^{12}C^{16}O$	0	1179,1187
5.35295470		vacuum	$^{12}C^{16}O$	0	1179,1187
5.3533228		vacuum	$^{12}C^{16}O$	0	1179,1187
5.3546483		vacuum	$^{12}C^{16}O$	0	1179,1187
5.3549012		vacuum	$^{12}C^{16}O$	0	1179,1187
5.35616		vacuum	$D^{35}Cl$	0	1255–56,1265
5.3566973		vacuum	Xe	0	571,601
5.3569272		vacuum	$^{12}C^{16}O$	0	1179,1187
5.3579450		vacuum	$^{12}C^{16}O$	0	1179,1187
5.3603139		vacuum	$^{12}C^{16}O$	0	1179,1187
5.3615063		vacuum	$^{13}C^{16}O$	0	1179,1187
5.3621859		vacuum	$^{12}C^{16}O$	0	1179,1187
5.36294		vacuum	$D^{37}Cl$	0	1255–56,1265
5.3648371		vacuum	$^{12}C^{16}O$	0	1179,1187
5.3653405		vacuum	$^{12}C^{16}O$	0	1179,1187
5.3658590		vacuum	$^{12}C^{16}O$	0	1179,1187
5.36815606		vacuum	$^{12}C^{16}O$	0	1179,1187
5.3690416		vacuum	$^{12}C^{16}O$	0	1179,1187
5.3698544		vacuum	$^{12}C^{16}O$	0	1179,1187
5.3711427		vacuum	$^{12}C^{16}O$	0	1179,1187
5.3728021		vacuum	$^{13}C^{16}O$	0	1253,1254
5.3744795		vacuum	$^{12}C^{16}O$	0	1179,1187
5.3759198		vacuum	$^{12}C^{16}O$	0	1179,1187
5.37686330		vacuum	$^{12}C^{16}O$	0	1179,1187

Table 4.1.1—*continued*
Neutral Atom, Ion, and Molecular Gas Lasers Arranged in Order of Increasing Wavelength

Wavelength (μm)	Uncertainty	Medium	Species	Charge	Reference
5.3770194		vacuum	$^{13}C^{16}O$	0	1179,1187
5.3776541		vacuum	$^{12}C^{16}O$	0	1179,1187
5.3779031		vacuum	$^{12}C^{16}O$	0	1179,1187
5.3785452		vacuum	$^{12}C^{16}O$	0	1179,1187
5.3795302		vacuum	$^{12}C^{16}O$	0	1179,1187
5.37990		vacuum	$D^{35}Cl$	0	1255–56,1265
5.3810786		vacuum	$^{12}C^{16}O$	0	1179,1187
5.3842343		vacuum	$^{13}C^{16}O$	0	1253,1254
5.3844944		vacuum	$^{12}C^{16}O$	0	1179,1187
5.3855853		vacuum	$^{12}C^{16}O$	0	1179,1187
5.3866395		vacuum	$^{12}C^{16}O$	0	1179,1187
5.3868992		vacuum	$^{12}C^{16}O$	0	1179,1187
5.3878074		vacuum	$^{12}C^{16}O$	0	1179,1187
5.3878364		vacuum	$^{13}C^{16}O$	0	1253,1254
5.38892		vacuum	$D^{35}Cl$	0	1255–56,1265
5.3890399		vacuum	$^{12}C^{16}O$	0	1179,1187
5.39104599		vacuum	$^{12}C^{16}O$	0	1179,1187
5.3912471		vacuum	Ar	0	601
5.3913826		vacuum	$^{12}C^{16}O$	0	1179,1187
5.3936490		vacuum	$^{12}C^{16}O$	0	1179,1187
5.3946662		vacuum	$^{12}C^{16}O$	0	1179,1187
5.39561		vacuum	$D^{37}Cl$	0	1255–56,1265
5.3958038		vacuum	$^{13}C^{16}O$	0	1253,1254
5.3960306		vacuum	$^{12}C^{16}O$	0	1179,1187
5.3975011		vacuum	$^{12}C^{16}O$	0	1179,1187
5.3980014		vacuum	$^{12}C^{16}O$	0	1179,1187
5.3987880		vacuum	$^{13}C^{16}O$	0	1253,1254
5.4			NO	0	1859
5.4013637		vacuum	$^{12}C^{16}O$	0	1179,1187
5.4016561		vacuum	$^{12}C^{16}O$	0	1179,1187
5.4018460		vacuum	$^{12}C^{16}O$	0	1179,1187
5.4027105		vacuum	$^{12}C^{16}O$	0	1179,1187
5.4043724		vacuum	$^{12}C^{16}O$	0	1179,1187
5.4048094		vacuum	Ne	0	525,564,570
5.4052982		vacuum	$^{12}C^{16}O$	0	1179,1187

Table 4.1.1—*continued*
Neutral Atom, Ion, and Molecular Gas Lasers Arranged in Order of Increasing Wavelength

Wavelength (μm)	Uncertainty	Medium	Species	Charge	Reference
5.4075114		vacuum	$^{13}C^{16}O$	0	1253,1254
5.4085055		vacuum	$^{12}C^{16}O$	0	1179,1187
5.4087770		vacuum	$^{12}C^{16}O$	0	1179,1187
5.4098751		vacuum	$^{13}C^{16}O$	0	1253,1254
5.4101766		vacuum	$^{12}C^{16}O$	0	1179,1187
5.4138378		vacuum	$^{12}C^{16}O$	0	1179,1187
5.4145202		vacuum	$^{12}C^{16}O$	0	1179,1187
5.4147029		vacuum	$^{12}C^{16}O$	0	1179,1187
5.4160296		vacuum	$^{12}C^{16}O$	0	1179,1187
5.4175163		vacuum	$^{12}C^{16}O$	0	1179,1187
5.4186415		vacuum	$^{12}C^{16}O$	0	1179,1187
5.4196535		vacuum	$^{12}C^{16}O$	0	1179,1187
5.4227		vacuum	$D^{35}Cl$	0	1255–56,1265
5.4234147		vacuum	$^{12}C^{16}O$	0	1179,1187
5.4242451		vacuum	$^{12}C^{16}O$	0	1179,1187
5.4264630		vacuum	$^{12}C^{16}O$	0	1179,1187
5.4264777		vacuum	$^{12}C^{16}O$	0	1179,1187
5.4272417		vacuum	$^{12}C^{16}O$	0	1179,1187
5.42950		vacuum	$D^{37}Cl$	0	1255–56,1265
5.4306800		vacuum	Eu	0	247
5.4308152		vacuum	$^{12}C^{16}O$	0	1179,1187
5.4309325		vacuum	$^{12}C^{16}O$	0	1179,1187
5.4309464		vacuum	$^{12}C^{16}O$	0	1179,1187
5.4324602		vacuum	$^{13}C^{16}O$	0	1253,1254
5.4339262		vacuum	$^{12}C^{16}O$	0	1179,1187
5.4359776		vacuum	$^{12}C^{16}O$	0	1179,1187
5.4377867		vacuum	$^{13}C^{16}O$	0	1253,1254
5.4385837		vacuum	$^{12}C^{16}O$	0	1179,1187
5.4385843		vacuum	$^{12}C^{16}O$	0	1179,1187
5.4392407		vacuum	$^{12}C^{16}O$	0	1179,1187
5.4399784		vacuum	$^{12}C^{16}O$	0	1179,1187
5.44238556		vacuum	$^{12}C^{16}O$	0	1179,1187
5.4439604		vacuum	$^{13}C^{16}O$	0	1253,1254
5.4442707		vacuum	$^{12}C^{16}O$	0	1179,1187
5.4448503		vacuum	$^{12}C^{16}O$	0	1179,1187

Table 4.1.1—*continued*
Neutral Atom, Ion, and Molecular Gas Lasers Arranged in Order of Increasing Wavelength

Wavelength (μm)	Uncertainty	Medium	Species	Charge	Reference
5.4463689		vacuum	$^{12}C^{16}O$	0	1179,1187
5.4486595		vacuum	$^{13}C^{16}O$	0	1253,1254
5.4507500		vacuum	$^{12}C^{16}O$	0	1179,1187
5.4508410		vacuum	$^{12}C^{16}O$	0	1179,1187
5.4521724		vacuum	$^{12}C^{16}O$	0	1179,1187
5.4539711		vacuum	$^{12}C^{16}O$	0	1179,1187
5.4542888		vacuum	$^{12}C^{16}O$	0	1179,1187
5.4556237		vacuum	$^{13}C^{16}O$	0	1253,1254
5.45771		vacuum	$D^{35}Cl$	0	1255–56,1265
5.4578843		vacuum	$^{12}C^{16}O$	0	1179,1187
5.4596692		vacuum	$^{13}C^{16}O$	0	1253,1254
5.460		air	Cu	0	1499
5.4616665		vacuum	$^{12}C^{16}O$	0	1179,1187
5.4623443		vacuum	$^{12}C^{16}O$	0	1179,1187
5.4632491		vacuum	$^{12}C^{16}O$	0	1179,1187
5.4652592		vacuum	$^{12}C^{16}O$	0	1179,1187
5.4657051		vacuum	$^{12}C^{16}O$	0	1179,1187
5.4673804		vacuum	$^{13}C^{16}O$	0	1253,1254
5.4676586		vacuum	Ar	0	559,564
5.4677812		vacuum	Ar	0	559,564
5.4705352		vacuum	$^{12}C^{16}O$	0	1179,1187
5.4708172		vacuum	$^{13}C^{16}O$	0	1253,1254
5.4716574		vacuum	$^{12}C^{16}O$	0	1179,1187
5.4727279		vacuum	$^{12}C^{16}O$	0	1179,1187
5.4749269		vacuum	Xe	0	601
5.4758098		vacuum	$^{12}C^{16}O$	0	1179,1187
5.4775888		vacuum	$^{12}C^{16}O$	0	1179,1187
5.4785025		vacuum	$^{12}C^{16}O$	0	1179,1187
5.4798		air	Ba	0	146
5.4821042		vacuum	$^{13}C^{16}O$	0	1253,1254
5.4839363		vacuum	$^{12}C^{16}O$	0	1179,1187
5.4855914		vacuum	$^{12}C^{16}O$	0	1179,1187
5.4885248		vacuum	$^{12}C^{16}O$	0	1179,1187
5.4896227		vacuum	$^{12}C^{16}O$	0	1179,1187
5.4913685		vacuum	$^{13}C^{16}O$	0	1253,1254

Table 4.1.1—*continued*
Neutral Atom, Ion, and Molecular Gas Lasers Arranged in Order of Increasing Wavelength

Wavelength (μm)	Uncertainty	Medium	Species	Charge	Reference
5.4919041		vacuum	$^{12}C^{16}O$	0	1179,1187
5.49348		vacuum	$D^{35}Cl$	0	1255–56,1265
5.4935308		vacuum	$^{13}C^{16}O$	0	1253,1254
5.4952920		vacuum	$^{12}C^{16}O$	0	1179,1187
5.498705		vacuum	I	0	401,402
5.5000140		vacuum	$^{13}C^{16}O$	0	1253,1254
5.501		vacuum	Xe	0	601
5.5013949		vacuum	$^{12}C^{16}O$	0	1179,1187
5.5018084		vacuum	$^{12}C^{16}O$	0	1179,1187
5.504		vacuum	NO	0	1285–1287
5.5035779		vacuum	$^{13}C^{16}O$	0	1253,1254
5.5050989		vacuum	$^{13}C^{16}O$	0	1253,1254
5.5054647		vacuum	$^{12}C^{16}O$	0	1179,1187
5.5067965		vacuum	$^{12}C^{16}O$	0	1179,1187
5.50849		vacuum	$D^{35}Cl$	0	1255–56,1265
5.5109423		vacuum	$^{13}C^{16}O$	0	1253,1254
5.5141471		vacuum	$^{12}C^{16}O$	0	1179,1187
5.5144216		vacuum	$^{12}C^{16}O$	0	1179,1187
5.515		vacuum	Ne	0	432
5.5164741		vacuum	$^{12}C^{16}O$	0	1179,1187
5.5168092		vacuum	$^{13}C^{16}O$	0	1253,1254
5.5184507		vacuum	$^{12}C^{16}O$	0	1179,1187
5.5191863		vacuum	$^{12}C^{16}O$	0	1179,1187
5.5220103		vacuum	$^{13}C^{16}O$	0	1253,1254
5.5266402		vacuum	$^{12}C^{16}O$	0	1179,1187
5.5274442		vacuum	$^{12}C^{16}O$	0	1179,1187
5.5276064		vacuum	$^{12}C^{16}O$	0	1179,1187
5.5286626		vacuum	$^{13}C^{16}O$	0	1253,1254
5.5302559		vacuum	$^{12}C^{16}O$	0	1179,1187
5.53036		vacuum	$D^{35}Cl$	0	1255–56,1265
5.5330701		vacuum	$^{12}C^{16}O$	0	1179,1187
5.5332186		vacuum	$^{13}C^{16}O$	0	1253,1254
5.5385629		vacuum	$^{12}C^{16}O$	0	1179,1187
5.5392885		vacuum	$^{12}C^{16}O$	0	1179,1187
5.5406607		vacuum	$^{13}C^{16}O$	0	1253,1254

Table 4.1.1—*continued*
Neutral Atom, Ion, and Molecular Gas Lasers Arranged in Order of Increasing Wavelength

Wavelength (μm)	Uncertainty	Medium	Species	Charge	Reference
5.5409505		vacuum	$^{12}C^{16}O$	0	1179,1187
5.54221360		vacuum	$^{12}C^{16}O$	0	1179,1187
5.54228		vacuum	$D^{35}Cl$	0	1255–56,1265
5.5445687		vacuum	$^{13}C^{16}O$	0	1253,1254
5.5471178		vacuum	$^{12}C^{16}O$	0	1179,1187
5.547327		vacuum	Ca	0	138–141
5.5498306		vacuum	$^{12}C^{16}O$	0	1179,1187
5.5520938		vacuum	$^{12}C^{16}O$	0	1179,1187
5.5528041		vacuum	$^{13}C^{16}O$	0	1253,1254
5.55432395		vacuum	$^{12}C^{16}O$	0	1179,1187
5.5544555		vacuum	$^{12}C^{16}O$	0	1220,1230
5.5560612		vacuum	$^{13}C^{16}O$	0	1253,1254
5.5612490		vacuum	$^{12}C^{16}O$	0	1179,1187
5.5613304		vacuum	$^{12}C^{16}O$	0	1179,1187
5.5636		air	Ba	0	146
5.5650570		vacuum	$^{12}C^{16}O$	0	1179,1187
5.5665892		vacuum	$^{12}C^{16}O$	0	1179,1187
5.5676972		vacuum	$^{13}C^{16}O$	0	1253,1254
5.56886		vacuum	$D^{79}Br$	0	1255–56,1258
5.5699805		vacuum	Kr	0	185,195,564
5.57038		vacuum	$D^{81}Br$	0	1255–56,1258
5.5728186		vacuum	$^{12}C^{16}O$	0	1179,1187
5.5747348		vacuum	$^{13}C^{16}O$	0	1253,1254
5.5754726		vacuum	Xe	0	570,582,632
5.57759		vacuum	$D^{35}Cl$	0	1255–56,1265
5.5781797		vacuum	$^{12}C^{16}O$	0	1179,1187
5.5790103		vacuum	$^{12}C^{16}O$	0	1179,1187
5.5794779		vacuum	$^{13}C^{16}O$	0	1253,1254
5.58454162		vacuum	$^{12}C^{16}O$	0	1253,1254
5.5858604		vacuum	$^{13}C^{16}O$	0	1253,1254
5.5862769		vacuum	Kr	0	559,564,570
5.5914045		vacuum	$^{13}C^{16}O$	0	1253,1254
5.5914636		vacuum	$^{12}C^{16}O$	0	1179,1187
5.5915884		vacuum	$^{12}C^{16}O$	0	1179,1187
5.5948932		vacuum	$^{12}C^{16}O$	0	1179,1187

Table 4.1.1—*continued*
Neutral Atom, Ion, and Molecular Gas Lasers Arranged in Order of Increasing Wavelength

Wavelength (μm)	Uncertainty	Medium	Species	Charge	Reference
5.59641727		vacuum	$^{12}C^{16}O$	0	1179,1187
5.5971292		vacuum	$^{13}C^{16}O$	0	1253,1254
5.59785		vacuum	$D^{79}Br$	0	1255–56,1258
5.5983205		vacuum	C	0	185,195
5.59939		vacuum	$D^{81}Br$	0	1255–56,1258
5.6034328		vacuum	Xe	0	601
5.6034784		vacuum	$^{13}C^{16}O$	0	1253,1254
5.6043250		vacuum	$^{12}C^{16}O$	0	1179,1187
5.6049096		vacuum	$^{12}C^{16}O$	0	1179,1187
5.6060681		vacuum	$^{12}C^{16}O$	0	1179,1187
5.60844849		vacuum	$^{12}C^{16}O$	0	1179,1187
5.6085416		vacuum	$^{13}C^{16}O$	0	1253,1254
5.61369		vacuum	$D^{35}Cl$	0	1255–56,1265
5.6172214		vacuum	$^{12}C^{16}O$	0	1179,1187
5.6173950		vacuum	$^{12}C^{16}O$	0	1179,1187
5.6185192		vacuum	$^{12}C^{16}O$	0	1179,1187
5.6200996		vacuum	$^{13}C^{16}O$	0	1253,1254
5.62063454		vacuum	$^{12}C^{16}O$	0	1179,1187
5.62762		vacuum	$D^{79}Br$	0	1255–56,1258
5.6280714		vacuum	$^{13}C^{16}O$	0	1253,1254
5.6288754		vacuum	$^{12}C^{16}O$	0	1179,1187
5.62914		vacuum	$D^{81}Br$	0	1255–56,1258
5.6302787		vacuum	$^{12}C^{16}O$	0	1179,1187
5.6305126		vacuum	Kr	0	559,564,570
5.6322937		vacuum	$^{12}C^{16}O$	0	1179,1187
5.6329815		vacuum	$^{12}C^{16}O$	0	1179,1187
5.6434982		vacuum	$^{12}C^{16}O$	0	1179,1187
5.6436555		vacuum	$^{13}C^{16}O$	0	1253,1254
5.6454853		vacuum	$^{12}C^{16}O$	0	1179,1187
5.6512718		vacuum	$^{13}C^{16}O$	0	1253,1254
5.6523006		vacuum	$^{12}C^{16}O$	0	1179,1187
5.6568816		vacuum	$^{12}C^{16}O$	0	1179,1187
5.6581491		vacuum	$^{12}C^{16}O$	0	1179,1187
5.65822		vacuum	$D^{79}Br$	0	1255–56,1258
5.65973		vacuum	$D^{81}Br$	0	1255–56,1258

Table 4.1.1—*continued*
Neutral Atom, Ion, and Molecular Gas Lasers Arranged in Order of Increasing Wavelength

Wavelength (μm)	Uncertainty	Medium	Species	Charge	Reference
5.6626004		vacuum	$^{13}C^{16}O$	0	1253,1254
5.6642477		vacuum	$^{12}C^{16}O$	0	1179,1187
5.6660928		vacuum	$^{13}C^{16}O$	0	1253,1254
5.6667372		vacuum	Ne	0	195,564,570
5.6678052		vacuum	$^{13}C^{16}O$	0	1253,1254
5.6704306		vacuum	$^{12}C^{16}O$	0	1179,1187
5.6709741		vacuum	$^{12}C^{16}O$	0	1179,1187
5.6740756		vacuum	$^{13}C^{16}O$	0	1253,1254
5.6763531		vacuum	$^{12}C^{16}O$	0	1179,1187
5.6801058		vacuum	$^{13}C^{16}O$	0	1179,1187
5.6839619		vacuum	$^{12}C^{16}O$	0	1179,1187
5.6841461		vacuum	$^{12}C^{16}O$	0	1179,1187
5.6856987		vacuum	$^{13}C^{16}O$	0	1253,1254
5.6866913		vacuum	$^{12}C^{16}O$	0	1179,1187
5.6886174		vacuum	$^{12}C^{16}O$	0	1179,1187
5.68961		vacuum	$D^{79}Br$	0	1255–56,1258
5.69113		vacuum	$D^{81}Br$	0	1255–56,1258
5.692		vacuum	Xe	0	601
5.6971140		vacuum	$^{12}C^{16}O$	0	1179,1187
5.6974710		vacuum	$^{13}C^{16}O$	0	1253,1254
5.6980297		vacuum	$^{12}C^{16}O$	0	1179,1187
5.6982330		vacuum	$^{12}C^{16}O$	0	1179,1187
5.7		vacuum	HCOOH	0	1325,1326
5.7010424		vacuum	$^{12}C^{16}O$	0	1179,1187
5.7051640		vacuum	$^{13}C^{16}O$	0	1253,1254
5.7067951		vacuum	Ne	0	432
5.7104313		vacuum	$^{12}C^{16}O$	0	1179,1187
5.7110		vacuum	$D^{79}Br$	0	1255–56,1258
5.7127		vacuum	$D^{81}Br$	0	1255–56,1258
5.7136291		vacuum	$^{12}C^{16}O$	0	1179,1187
5.7214666		vacuum	$^{13}C^{16}O$	0	1253,1254
5.7217904		vacuum	$^{12}C^{16}O$	0	1179,1187
5.7239156		vacuum	$^{12}C^{16}O$	0	1179,1187
5.7263792		vacuum	$^{12}C^{16}O$	0	1179,1187
5.7296871		vacuum	$^{13}C^{16}O$	0	1253,1254

Table 4.1.1—*continued*
Neutral Atom, Ion, and Molecular Gas Lasers Arranged in Order of Increasing Wavelength

Wavelength (μm)	Uncertainty	Medium	Species	Charge	Reference
5.7308403		vacuum	$^{13}C^{16}O$	0	1253,1254
5.7336925		vacuum	$^{13}C^{16}O$	0	1253,1254
5.7338082		vacuum	$^{12}C^{16}O$	0	1179,1187
5.7375681		vacuum	$^{12}C^{16}O$	0	1179,1187
5.7392933		vacuum	$^{12}C^{16}O$	0	1179,1187
5.7409		vacuum	$D^{79}Br$	0	1253,1254
5.7412249		vacuum	$^{13}C^{16}O$	0	1253,1254
5.7435		vacuum	$D^{81}Br$	0	1253,1254
5.7459875		vacuum	$^{12}C^{16}O$	0	1179,1187
5.7460720		vacuum	$^{13}C^{16}O$	0	1253,1254
5.7513904		vacuum	$^{12}C^{16}O$	0	1179,1187
5.7523733		vacuum	$^{12}C^{16}O$	0	1179,1187
5.7529127		vacuum	$^{13}C^{16}O$	0	1253,1254
5.754965		vacuum	Ga	0	175
5.7577843		vacuum	$^{12}C^{16}O$	0	1198
5.7583287		vacuum	$^{12}C^{16}O$	0	1179,1187
5.7586063		vacuum	$^{13}C^{16}O$	0	1253,1254
5.7653841		vacuum	$^{12}C^{16}O$	0	1179,1187
5.7656205		vacuum	$^{12}C^{16}O$	0	1220,1230
5.7708336		vacuum	$^{12}C^{16}O$	0	1220,1230
5.7712966		vacuum	$^{13}C^{16}O$	0	1253,1254
5.7717		vacuum	$D^{79}Br$	0	1255–56,1258
5.772238		vacuum	Eu	0	247
5.7733		vacuum	$D^{81}Br$	0	1255–56,1258
5.7767459		vacuum	$^{13}C^{16}O$	0	1253,1254
5.7773913		vacuum	Ne	0	432
5.7790363		vacuum	$^{12}C^{16}O$	0	1179,1187
5.7795507		vacuum	$^{12}C^{16}O$	0	1179,1187
5.7811505		vacuum	$^{12}C^{16}O$	0	1179,1187
5.7835037		vacuum	$^{12}C^{16}O$	0	1179,1187
5.7841443		vacuum	$^{13}C^{16}O$	0	1253,1254
5.7888926		vacuum	$^{13}C^{16}O$	0	1253,1254
5.7926222		vacuum	$^{12}C^{16}O$	0	1220,1230
5.7930755		vacuum	$^{12}C^{16}O$	0	1179,1187
5.7938915		vacuum	$^{12}C^{16}O$	0	1179,1187

Table 4.1.1—*continued*
Neutral Atom, Ion, and Molecular Gas Lasers Arranged in Order of Increasing Wavelength

Wavelength (μm)	Uncertainty	Medium	Species	Charge	Reference
5.7963397		vacuum	$^{12}C^{16}O$	0	1179,1187
5.7971505		vacuum	$^{13}C^{16}O$	0	1253,1254
5.7984477		vacuum	$^{13}C^{16}O$	0	1253,1254
5.8011949		vacuum	$^{13}C^{16}O$	0	1253,1254
5.8025		vacuum	$D^{79}Br$	0	1255–56,1258
5.8035		vacuum	$D^{79}Br$	0	1255–56,1258
5.8037601		vacuum	Ar	0	428,601,603
5.8042		vacuum	$D^{81}Br$	0	1255–56,1258
5.8051641		vacuum	$^{12}C^{16}O$	0	1179,1187
5.80528		vacuum	$D^{81}Br$	0	1255–56,1258
5.8063795		vacuum	$^{12}C^{16}O$	0	1220,1230
5.8084085		vacuum	$^{12}C^{16}O$	0	1179,1187
5.8093435		vacuum	$^{12}C^{16}O$	0	1179,1187
5.8100469		vacuum	$^{13}C^{16}O$	0	1253,1254
5.8103166		vacuum	$^{13}C^{16}O$	0	1253,1254
5.8136537		vacuum	$^{13}C^{16}O$	0	1253,1254
5.8174165		vacuum	$^{12}C^{16}O$	0	1179,1187
5.8203098		vacuum	$^{12}C^{16}O$	0	1220,1230
5.8225163		vacuum	$^{12}C^{16}O$	0	1220,1230
5.8262700		vacuum	$^{13}C^{16}O$	0	1253,1254
5.8298347		vacuum	$^{12}C^{16}O$	0	1179,1187
5.8319		vacuum	$D^{79}Br$	0	1255–56,1258
5.8336		vacuum	$D^{81}Br$	0	1255–56,1258
5.8337074		vacuum	$^{13}C^{16}O$	0	1253,1254
5.8344146		vacuum	$^{12}C^{16}O$	0	1220,1230
5.8358593		vacuum	$^{12}C^{16}O$	0	1220,1230
5.8360		vacuum	$D^{79}Br$	0	1255–56,1258
5.8374		vacuum	$D^{81}Br$	0	1255–56,1258
5.8390454		vacuum	$^{13}C^{16}O$	0	1253,1254
5.8424197		vacuum	$^{12}C^{16}O$	0	1179,1187
5.8457708		vacuum	$^{13}C^{16}O$	0	1253,1254
5.84618		vacuum	NO	0	1285–1287
5.8477292		vacuum	Ar	0	559,564
5.8486958		vacuum	$^{12}C^{16}O$	0	1220,1230
5.8493742		vacuum	$^{12}C^{16}O$	0	1220,1230

Table 4.1.1—*continued*
Neutral Atom, Ion, and Molecular Gas Lasers Arranged in Order of Increasing Wavelength

Wavelength (μm)	Uncertainty	Medium	Species	Charge	Reference
5.8519808		vacuum	$^{13}C^{16}O$	0	1253,1254
5.8542337		vacuum	$^{12}C^{16}O$	0	1198
5.85487		vacuum	NO	0	1285–1287
5.8551729		vacuum	$^{12}C^{16}O$	0	1179,1187
5.8579913		vacuum	$^{13}C^{16}O$	0	1253,1254
5.85840		vacuum	NO	0	1285–1287
5.86197		vacuum	$D^{79}Br$	0	1255–56,1258
5.8630624		vacuum	$^{12}C^{16}O$	0	1220,1230
5.86362		vacuum	$D^{81}Br$	0	1255–56,1258
5.8648		vacuum	Hg	0	161,166,167
5.8650779		vacuum	$^{13}C^{16}O$	0	1253,1254
5.8662257		vacuum	$^{12}C^{16}O$	0	1198
5.8665236		vacuum	Ar	0	195
5.8680955		vacuum	$^{12}C^{16}O$	0	1179,1187
5.8703698		vacuum	$^{13}C^{16}O$	0	1253,1254
5.87058		vacuum	NO	0	1285–1287
5.8769258		vacuum	$^{12}C^{16}O$	0	1220,1230
5.8783840		vacuum	$^{12}C^{16}O$	0	1179,1187
5.87893		vacuum	NO	0	1285–1287
5.8858082		vacuum	Ne	0	432
5.8899		air	Ba	0	146,147
5.8907098		vacuum	$^{12}C^{16}O$	0	1179,1187
5.8909648		vacuum	$^{12}C^{16}O$	0	1220,1230
5.8924207		vacuum	$^{13}C^{16}O$	0	1253,1254
5.89282		vacuum	$D^{79}Br$	0	1255–56,1258
5.89442		vacuum	$D^{81}Br$	0	1255–56,1258
5.8944550		vacuum	$^{12}C^{16}O$	0	1179,1187
5.9032044		vacuum	$^{12}C^{16}O$	0	1179,1187
5.90361		vacuum	NO	0	1285–1287
5.9043954		vacuum	$^{13}C^{16}O$	0	1253,1254
5.9051820		vacuum	$^{12}C^{16}O$	0	1220,1230
5.9078949		vacuum	$^{12}C^{16}O$	0	1220,1230
5.90831		vacuum	NO	0	1285–1287
5.9084670		vacuum	$^{13}C^{16}O$	0	1253,1254
5.913		vacuum	Xe	0	601

Table 4.1.1—*continued*
Neutral Atom, Ion, and Molecular Gas Lasers Arranged in Order of Increasing Wavelength

Wavelength (μm)	Uncertainty	Medium	Species	Charge	Reference
5.9158695		vacuum	$^{12}C^{16}O$	0	1179,1187
5.9195787		vacuum	$^{12}C^{16}O$	0	1220,1230
5.9214909		vacuum	$^{13}C^{16}O$	0	1253,1254
5.9215094		vacuum	$^{12}C^{16}O$	0	1253,1254
5.92456		vacuum	$D^{79}Br$	0	1255–56,1258
5.92610		vacuum	$D^{81}Br$	0	1255–56,1258
5.9287060		vacuum	$^{12}C^{16}O$	0	1179,1187
5.9288227		vacuum	$^{13}C^{16}O$	0	1253,1254
5.9353011		vacuum	$^{12}C^{16}O$	0	1220,1230
5.94234		vacuum	NO	0	1285–1287
5.9480336		vacuum	$^{13}C^{16}O$	0	1253,1254
5.9492707		vacuum	$^{12}C^{16}O$	0	1220,1230
5.9495478		vacuum	Eu	0	247
5.9535415		vacuum	$^{12}C^{16}O$	0	1179,1187
5.9538949		vacuum	$^{13}C^{16}O$	0	1253,1254
5.95461		vacuum	NO	0	1285–1287
5.9549		vacuum	NO	0	1285–1287
5.9548990		vacuum	$^{12}C^{16}O$	0	1179,1187
5.95735		vacuum	$D^{79}Br$	0	1255–56,1258
5.9578742		vacuum	Ne	0	432
5.95901		vacuum	$D^{81}Br$	0	1255–56,1258
5.9615550		vacuum	$^{13}C^{16}O$	0	1253,1254
5.96317		vacuum	NO	0	1285–1287
5.9634200		vacuum	$^{12}C^{16}O$	0	1220,1230
5.9648404		vacuum	$^{13}C^{16}O$	0	1179,1187
5.9659403		vacuum	$^{12}C^{16}O$	0	1179,1187
5.9666761		vacuum	$^{13}C^{16}O$	0	1253,1254
5.96726		vacuum	NO	0	1285–1287
5.9682587		vacuum	$^{12}C^{16}O$	0	1179,1187
5.9752452		vacuum	$^{13}C^{16}O$	0	1253,1254
5.97564		vacuum	NO	0	1285–1287
5.9768819		vacuum	$^{13}C^{16}O$	0	1253,1254
5.9777508		vacuum	$^{12}C^{16}O$	0	1220,1230
5.9785113		vacuum	$^{12}C^{16}O$	0	1179,1187
5.9796224		vacuum	$^{13}C^{16}O$	0	1253,1254

Table 4.1.1—*continued*
Neutral Atom, Ion, and Molecular Gas Lasers Arranged in Order of Increasing Wavelength

Wavelength (μm)	Uncertainty	Medium	Species	Charge	Reference
5.97990		vacuum	NO	0	1285–1287
5.9817950		vacuum	$^{12}C^{16}O$	0	1179,1187
5.9830082		vacuum	O	0	235,1698
5.9833948		vacuum	Hg	0	175
5.98817		vacuum	NO	0	1285–1287
5.9909		vacuum	$D^{79}Br$	0	1255–56,1258
5.9927		vacuum	$D^{81}Br$	0	1255–56,1258
5.9927349		vacuum	$^{13}C^{16}O$	0	1253,1254
5.99308		vacuum	NO	0	1285–1287
5.9955104		vacuum	$^{12}C^{16}O$	0	1220,1230
6.00100		vacuum	NO	0	1285–1287
6.0014521		vacuum	$^{13}C^{16}O$	0	1253,1254
6.0041759		vacuum	$^{12}C^{16}O$	0	1179,1187
6.00536		vacuum	NO	0	1285–1287
6.0060150		vacuum	$^{13}C^{16}O$	0	1253,1254
6.0094059		vacuum	$^{12}C^{16}O$	0	1220,1230
6.0172722		vacuum	$^{12}C^{16}O$	0	1179,1187
6.01917		vacuum	NO	0	1285–1287
6.0194643		vacuum	$^{13}C^{16}O$	0	1253,1254
6.02090		vacuum	$D^{79}Br$	0	1255–56,1258
6.02246		vacuum	$D^{81}Br$	0	1255–56,1258
6.0234832		vacuum	$^{12}C^{16}O$	0	1220,1230
6.0255		vacuum	$D^{79}Br$	0	1255–56,1258
6.02667		vacuum	NO	0	1285–1287
6.0266802		vacuum	$^{13}C^{16}O$	0	1253,1254
6.0274		vacuum	$D^{81}Br$	0	1255–56,1258
6.0305462		vacuum	$^{12}C^{16}O$	0	1179,1187
6.03242		vacuum	NO	0	1285–1287
6.0330839		vacuum	$^{13}C^{16}O$	0	1285–1287
6.0377435		vacuum	$^{12}C^{16}O$	0	1220,1230
6.0386		vacuum	NO	0	1285–1287
6.0395438		vacuum	$^{13}C^{16}O$	0	1253,1254
6.04018		vacuum	NO	0	1285–1287
6.04186		vacuum	NO	0	1285–1287
6.0431867		vacuum	$^{12}C^{16}O$	0	1179,1187

Table 4.1.1—*continued*
Neutral Atom, Ion, and Molecular Gas Lasers Arranged in Order of Increasing Wavelength

Wavelength (μm)	Uncertainty	Medium	Species	Charge	Reference
6.05290		vacuum	$D^{79}Br$	0	1255–56,1258
6.0530155		vacuum	Ar	0	195,564
6.05429		vacuum	NO	0	1285–1287
6.05440		vacuum	$D^{81}Br$	0	1255–56,1258
6.0558337		vacuum	$^{12}C^{16}O$	0	1179,1187
6.0576338		vacuum	$^{12}C^{16}O$	0	1179,1187
6.0592473		vacuum	Eu	0	247
6.06281		vacuum	NO	0	1285–1287
6.0635085		vacuum	$^{13}C^{16}O$	0	1253,1254
6.0657774		vacuum	$^{13}C^{16}O$	0	1253,1254
6.06726		vacuum	NO	0	1285–1287
6.0686579		vacuum	$^{12}C^{16}O$	0	1179,1187
6.0714500		vacuum	$^{12}C^{16}O$	0	1179,1187
6.0791497		vacuum	$^{13}C^{16}O$	0	1253,1254
6.08010		vacuum	NO	0	1285–1287
6.0816611		vacuum	$^{12}C^{16}O$	0	1179,1187
6.0854502		vacuum	$^{12}C^{16}O$	0	1220,1230
6.08576		vacuum	$D^{79}Br$	0	1255–56,1258
6.08732		vacuum	$D^{81}Br$	0	1255–56,1258
6.08839		vacuum	NO	0	1285–1287
6.0885551		vacuum	$^{13}C^{16}O$	0	1253,1254
6.0926946		vacuum	$^{13}C^{16}O$	0	1253,1254
6.09336		vacuum	NO	0	1285–1287
6.0948439		vacuum	$^{12}C^{16}O$	0	1179,1187
6.0996356		vacuum	$^{12}C^{16}O$	0	1220,1230
6.1013312		vacuum	$^{13}C^{16}O$	0	1253,1254
6.10147		vacuum	NO	0	1285–1287
6.1082086		vacuum	$^{12}C^{16}O$	0	1179,1187
6.1099892		vacuum	$^{12}C^{16}O$	0	1179,1187
6.1140080		vacuum	$^{12}C^{16}O$	0	1220,1230
6.11995		vacuum	$D^{79}Br$	0	1255–56,1258
6.1203065		vacuum	$^{13}C^{16}O$	0	1179,1187
6.12044		vacuum	NO	0	1285–1287
6.12156		vacuum	$D^{81}Br$	0	1255–56,1258
6.1217561		vacuum	$^{12}C^{16}O$	0	1179,1187

Table 4.1.1—*continued*
Neutral Atom, Ion, and Molecular Gas Lasers Arranged in Order of Increasing Wavelength

Wavelength (μm)	Uncertainty	Medium	Species	Charge	Reference
6.1225323		vacuum	$^{12}C^{16}O$	0	1179,1187
6.1273954		vacuum	$^{13}C^{16}O$	0	1253,1254
6.1285693		vacuum	$^{12}C^{16}O$	0	1220,1230
6.132		vacuum	Xe	0	601
6.1352546		vacuum	$^{12}C^{16}O$	0	1179,1187
6.1354880		vacuum	$^{12}C^{16}O$	0	1179,1187
6.1398711		vacuum	$^{13}C^{16}O$	0	1253,1254
6.1406862		vacuum	$^{13}C^{16}O$	0	1253,1254
6.14168		vacuum	NO	0	1285–1287
6.1477551		vacuum	Ga	0	175
6.1481584		vacuum	$^{12}C^{16}O$	0	1179,1187
6.1494061		vacuum	$^{12}C^{16}O$	0	1179,1187
6.1523828		vacuum	$^{13}C^{16}O$	0	1253,1254
6.15377		vacuum	NO	0	1285–1287
6.1541514		vacuum	$^{13}C^{16}O$	0	1253,1254
6.15460		vacuum	$D^{79}Br$	0	1255–56,1258
6.1546		vacuum	NO	0	1285–1287
6.15616		vacuum	$D^{81}Br$	0	1255–56,1258
6.15764		vacuum	NO	0	1285–1287
6.1612442		vacuum	$^{12}C^{16}O$	0	1179,1187
6.1650658		vacuum	$^{13}C^{16}O$	0	1253,1254
6.16629		vacuum	NO	0	1285–1287
6.1677922		vacuum	$^{13}C^{16}O$	0	1253,1254
6.1745142		vacuum	$^{12}C^{16}O$	0	1179,1187
6.1779214		vacuum	$^{13}C^{16}O$	0	1253,1254
6.17917		vacuum	NO	0	1285–1287
6.1816101		vacuum	$^{13}C^{16}O$	0	1253,1254
6.18383		vacuum	NO	0	1285–1287
6.1879691		vacuum	$^{12}C^{16}O$	0	1179,1187
6.19034		vacuum	$D^{79}Br$	0	1255–56,1258
6.1909503		vacuum	$^{13}C^{16}O$	0	1253,1254
6.19184		vacuum	$D^{81}Br$	0	1255–56,1258
6.19214		vacuum	NO	0	1285–1287
6.1956064		vacuum	$^{13}C^{16}O$	0	1253,1254
6.19721		vacuum	NO	0	1285–1287

Table 4.1.1—*continued*
Neutral Atom, Ion, and Molecular Gas Lasers Arranged in Order of Increasing Wavelength

Wavelength (μm)	Uncertainty	Medium	Species	Charge	Reference
6.1973		vacuum	NO	0	1285–1287
6.2016112		vacuum	$^{12}C^{16}O$	0	1179,1187
6.2040666		vacuum	$^{12}C^{16}O$	0	1179,1187
6.2041540		vacuum	$^{13}C^{16}O$	0	1253,1254
6.20548		vacuum	NO	0	1285–1287
6.2097831		vacuum	$^{13}C^{16}O$	0	1253,1254
6.21095		vacuum	NO	0	1285–1287
6.2154414		vacuum	$^{12}C^{16}O$	0	1179,1187
6.2168641		vacuum	$^{12}C^{16}O$	0	1179,1187
6.2175344		vacuum	$^{13}C^{16}O$	0	1253,1254
6.21914		vacuum	NO	0	1285–1287
6.22374		vacuum	$D^{81}Br$	0	1255–56,1258
6.22487		vacuum	NO	0	1285–1287
6.22723		vacuum	$D^{79}Br$	0	1255–56,1258
6.22886		vacuum	$D^{81}Br$	0	1255–56,1258
6.2294615		vacuum	$^{12}C^{16}O$	0	1179,1187
6.2298468		vacuum	$^{12}C^{16}O$	0	1179,1187
6.2308351		vacuum	$^{13}C^{16}O$	0	1253,1254
6.2310925		vacuum	$^{13}C^{16}O$	0	1253,1254
6.23807		vacuum	NO	0	1285–1287
6.2430152		vacuum	$^{12}C^{16}O$	0	1179,1187
6.2435937		vacuum	$^{13}C^{16}O$	0	1253,1254
6.2436728		vacuum	$^{12}C^{16}O$	0	1179,1187
6.2448297		vacuum	$^{13}C^{16}O$	0	1253,1254
6.25113		vacuum	NO	0	1285–1287
6.2563717		vacuum	$^{12}C^{16}O$	0	1179,1187
6.2565279		vacuum	$^{13}C^{16}O$	0	1253,1254
6.25657		vacuum	$D^{79}Br$	0	1255–56,1258
6.25810		vacuum	$D^{81}Br$	0	1255–56,1258
6.2587474		vacuum	$^{13}C^{16}O$	0	1253,1254
6.26021		vacuum	NO	0	1285–1287
6.26449		vacuum	NO	0	1285–1287
6.270		vacuum	$^{14}NH_3$	0	1203,1204
6.2696393		vacuum	$^{13}C^{16}O$	0	1253,1254
6.2699176		vacuum	$^{12}C^{16}O$	0	1179,1187

Table 4.1.1—*continued*
Neutral Atom, Ion, and Molecular Gas Lasers Arranged in Order of Increasing Wavelength

Wavelength (μm)	Uncertainty	Medium	Species	Charge	Reference
6.2728473		vacuum	$^{13}C^{16}O$	0	1253,1254
6.27782		vacuum	NO	0	1285–1287
6.2829292		vacuum	$^{13}C^{16}O$	0	1253,1254
6.2836541		vacuum	$^{12}C^{16}O$	0	1179,1187
6.28650		vacuum	NO	0	1285–1287
6.2909		vacuum	$D^{79}Br$	0	1255–56,1258
6.29129		vacuum	NO	0	1285–1287
6.29164		vacuum	$D^{79}Br$	0	1255–56,1258
6.2925		vacuum	$D^{81}Br$	0	1255–56,1258
6.29323		vacuum	$D^{81}Br$	0	1255–56,1258
6.2963990		vacuum	$^{13}C^{16}O$	0	1253,1254
6.2975826		vacuum	$^{12}C^{16}O$	0	1179,1187
6.29977		vacuum	NO	0	1285–1287
6.3007581		vacuum	$^{12}C^{16}O$	0	1179,1187
6.3015999		vacuum	$^{13}C^{16}O$	0	1253,1254
6.30509		vacuum	NO	0	1285–1287
6.3100504		vacuum	$^{13}C^{16}O$	0	1253,1254
6.3117052		vacuum	$^{12}C^{16}O$	0	1179,1187
6.3120378		vacuum	Xe	0	601
6.31361		vacuum	NO	0	1285–1287
6.3138189		vacuum	$^{12}C^{16}O$	0	1179,1187
6.3154334		vacuum	Xe	0	601
6.31912		vacuum	NO	0	1285–1287
6.3238843		vacuum	$^{13}C^{16}O$	0	1253,1254
6.3260230		vacuum	$^{12}C^{16}O$	0	1179,1187
6.32699		vacuum	NO	0	1285–1287
6.3270696		vacuum	$^{12}C^{16}O$	0	1179,1187
6.32787		vacuum	$D^{79}Br$	0	1255–56,1258
6.32943		vacuum	$D^{81}Br$	0	1255–56,1258
6.33356		vacuum	NO	0	1285–1287
6.3372336		vacuum	$^{13}C^{16}O$	0	1253,1254
6.3379027		vacuum	$^{13}C^{16}O$	0	1253,1254
6.3405128		vacuum	$^{12}C^{16}O$	0	1179,1187
6.3405381		vacuum	$^{12}C^{16}O$	0	1179,1187
6.3504272		vacuum	$^{13}C^{16}O$	0	1253,1254

Table 4.1.1—*continued*
Neutral Atom, Ion, and Molecular Gas Lasers Arranged in Order of Increasing Wavelength

Wavelength (μm)	Uncertainty	Medium	Species	Charge	Reference
6.3521065		vacuum	$^{13}C^{16}O$	0	1253,1254
6.3541489		vacuum	$^{12}C^{16}O$	0	1179,1187
6.3552521		vacuum	$^{12}C^{16}O$	0	1179,1187
6.3638031		vacuum	$^{13}C^{16}O$	0	1253,1254
6.3664982		vacuum	$^{13}C^{16}O$	0	1253,1254
6.3679796		vacuum	$^{12}C^{16}O$	0	1179,1187
6.3687777		vacuum	Se	0	155
6.3701664		vacuum	$^{12}C^{16}O$	0	1179,1187
6.37637		vacuum	NO	0	1285–1287
6.3773627		vacuum	$^{13}C^{16}O$	0	1253,1254
6.3810785		vacuum	$^{13}C^{16}O$	0	1253,1254
6.3820070		vacuum	$^{12}C^{16}O$	0	1179,1187
6.38941		vacuum	NO	0	1285–1287
6.3962327		vacuum	$^{12}C^{16}O$	0	1179,1187
6.39799		vacuum	NO	0	1285–1287
6.4001778		vacuum	$^{12}C^{16}O$	0	1179,1187
6.40311		vacuum	NO	0	1285–1287
6.4050388		vacuum	$^{13}C^{16}O$	0	1253,1254
6.4135104		vacuum	$^{12}C^{16}O$	0	1179,1187
6.422525		vacuum	K	0	131,142
6.4252842		vacuum	$^{12}C^{16}O$	0	1179,1187
6.42616		vacuum	NO	0	1285–1287
6.4270392		vacuum	$^{12}C^{16}O$	0	1179,1187
6.43207		vacuum	NO	0	1285–1287
6.4401134		vacuum	$^{12}C^{16}O$	0	1179,1187
6.4407658		vacuum	$^{12}C^{16}O$	0	1179,1187
6.4468647		vacuum	$^{13}C^{16}O$	0	1253,1254
6.4479667		vacuum	$^{13}C^{16}O$	0	1253,1254
6.4546		air	Ba	0	146
6.4546912		vacuum	$^{12}C^{16}O$	0	1179,1187
6.4551478		vacuum	$^{12}C^{16}O$	0	1179,1187
6.4566866		vacuum	Sr	0	139,141
6.4575288		vacuum	K	0	131,142
6.4605142		vacuum	$^{13}C^{16}O$	0	1253,1254
6.4626595		vacuum	$^{13}C^{16}O$	0	1253,1254

Table 4.1.1—*continued*
Neutral Atom, Ion, and Molecular Gas Lasers Arranged in Order of Increasing Wavelength

Wavelength (μm)	Uncertainty	Medium	Species	Charge	Reference
6.4688175		vacuum	$^{12}C^{16}O$	0	1179,1187
6.4703883		vacuum	$^{12}C^{16}O$	0	1179,1187
6.4743525		vacuum	$^{13}C^{16}O$	0	1253,1254
6.4774391		vacuum	Hg	0	161,166,169
6.4831461		vacuum	$^{12}C^{16}O$	0	1179,1187
6.4883811		vacuum	$^{13}C^{16}O$	0	1253,1254
6.4887747		vacuum	Hg	0	161,166,169
6.5024478		vacuum	$^{12}C^{16}O$	0	1179,1187
6.5026017		vacuum	$^{13}C^{16}O$	0	1253,1254
6.5124170		vacuum	$^{12}C^{16}O$	0	1179,1187
6.5160626		vacuum	$^{12}C^{16}O$	0	1179,1187
6.5170156		vacuum	$^{13}C^{16}O$	0	1253,1254
6.5273627		vacuum	$^{12}C^{16}O$	0	1179,1187
6.5298788		vacuum	$^{12}C^{16}O$	0	1179,1187
6.5316249		vacuum	$^{13}C^{16}O$	0	1253,1254
6.5425180		vacuum	$^{12}C^{16}O$	0	1179,1187
6.5438988		vacuum	$^{12}C^{16}O$	0	1179,1187
6.5578840		vacuum	$^{12}C^{16}O$	0	1179,1187
6.5581240		vacuum	$^{12}C^{16}O$	0	1179,1187
6.5598797		vacuum	$^{13}C^{16}O$	0	1253,1254
6.5614352		vacuum	$^{13}C^{16}O$	0	1253,1254
6.5871970		vacuum	$^{12}C^{16}O$	0	1179,1187
6.6		vacuum	$^{12}CS_2$	0	1317
6.6020476		vacuum	$^{12}C^{16}O$	0	1179,1187
6.6028484		vacuum	$^{13}C^{16}O$	0	1253,1254
6.6077013		vacuum	$^{12}C^{16}O$	0	1179,1187
6.6171107		vacuum	$^{12}C^{16}O$	0	1179,1187
6.6175674		vacuum	$^{13}C^{16}O$	0	1253,1254
6.6216075		vacuum	$^{12}C^{16}O$	0	1179,1187
6.6357216		vacuum	$^{12}C^{16}O$	0	1179,1187
6.6478798		vacuum	$^{12}C^{16}O$	0	1179,1187
6.6500460		vacuum	$^{12}C^{16}O$	0	1179,1187
6.6645815		vacuum	$^{12}C^{16}O$	0	1179,1187
6.6793307		vacuum	$^{12}C^{16}O$	0	1179,1187
6.689		vacuum	$^{14}NH_3$	0	1203,1204

Table 4.1.1—*continued*
Neutral Atom, Ion, and Molecular Gas Lasers Arranged in Order of Increasing Wavelength

Wavelength (μm)	Uncertainty	Medium	Species	Charge	Reference
6.6942949		vacuum	$^{12}C^{16}O$	0	1179,1187
6.7020845		vacuum	$^{12}C^{16}O$	0	1179,1187
6.7094754		vacuum	$^{12}C^{16}O$	0	1179,1187
6.721966		vacuum	I	0	401,402
6.7248749		vacuum	$^{12}C^{16}O$	0	1179,1187
6.7302881		vacuum	$^{12}C^{16}O$	0	1179,1187
6.7404949		vacuum	$^{12}C^{16}O$	0	1179,1187
6.7447115		vacuum	$^{12}C^{16}O$	0	1179,1187
6.7463414		vacuum	Ar	0	601
6.7563372		vacuum	$^{12}C^{16}O$	0	1179,1187
6.7593509		vacuum	$^{12}C^{16}O$	0	1179,1187
6.7631543		vacuum	Te	0	155
6.7724038		vacuum	$^{12}C^{16}O$	0	1179,1187
6.7788016		vacuum	Ne	0	432
6.7892862		vacuum	$^{12}C^{16}O$	0	1179,1187
6.8045856		vacuum	$^{12}C^{16}O$	0	1179,1187
6.8129		air	Ar	0	1498
6.8175155		vacuum	O	0	215
6.8201087		vacuum	$^{12}C^{16}O$	0	1179,1187
6.8358573		vacuum	$^{12}C^{16}O$	0	1179,1187
6.8422596		vacuum	$^{12}C^{16}O$	0	1179,1187
6.848705		vacuum	Fe	0	155
6.8518334		vacuum	$^{12}C^{16}O$	0	1179,1187
6.8598868		vacuum	O	0	235,1698
6.8680385		vacuum	$^{12}C^{16}O$	0	1179,1187
6.8719695		vacuum	$^{12}C^{16}O$	0	1179,1187
6.874553		vacuum	O	0	215
6.8844754		vacuum	$^{12}C^{16}O$	0	1179,1187
6.8871614		vacuum	$^{12}C^{16}O$	0	1179,1187
6.8884755		vacuum	Ne	0	432
6.9		vacuum	$^{13}CS_2$	0	1325,1326
6.9025798		vacuum	$^{12}C^{16}O$	0	1179,1187
6.903502		vacuum	I	0	401,402
6.9182277		vacuum	$^{12}C^{16}O$	0	1179,1187
6.9341059		vacuum	$^{12}C^{16}O$	0	1179,1187

Table 4.1.1—*continued*
Neutral Atom, Ion, and Molecular Gas Lasers
Arranged in Order of Increasing Wavelength

Wavelength (μm)	Uncertainty	Medium	Species	Charge	Reference
6.9428408		vacuum	$^{12}C^{16}O$	0	1179,1187
6.9428548		vacuum	Ar	0	195,564
6.9448712		vacuum	Ar	0	195,564
6.9502170		vacuum	$^{12}C^{16}O$	0	1179,1187
6.9665623		vacuum	$^{12}C^{16}O$	0	1179,1187
6.9727640		vacuum	$^{12}C^{16}O$	0	1179,1187
6.9831451		vacuum	$^{12}C^{16}O$	0	1179,1187
6.9876797		vacuum	Ne	0	432
6.9880704		vacuum	$^{12}C^{16}O$	0	1179,1187
7.0036089		vacuum	$^{12}C^{16}O$	0	1179,1187
7.0193822		vacuum	$^{12}C^{16}O$	0	1179,1187
7.0353910		vacuum	$^{12}C^{16}O$	0	1179,1187
7.0516387		vacuum	$^{12}C^{16}O$	0	1179,1187
7.0580595		vacuum	Kr	0	195,564
7.0681263		vacuum	$^{12}C^{16}O$	0	1179,1187
7.0767557		vacuum	$^{12}C^{16}O$	0	1179,1187
7.0848565		vacuum	$^{12}C^{16}O$	0	1179,1187
7.0921770		vacuum	$^{12}C^{16}O$	0	1179,1187
7.093		vacuum	H_2O	0	1284,1329–31
7.098		vacuum	Ne	0	432
7.1018309		vacuum	$^{12}C^{16}O$	0	1179,1187
7.1078363		vacuum	$^{12}C^{16}O$	0	1179,1187
7.1190526		vacuum	$^{12}C^{16}O$	0	1179,1187
7.1237355		vacuum	$^{12}C^{16}O$	0	1179,1187
7.1398768		vacuum	$^{12}C^{16}O$	0	1179,1187
7.1562618		vacuum	$^{12}C^{16}O$	0	1179,1187
7.1728931		vacuum	$^{12}C^{16}O$	0	1179,1187
7.1764192		vacuum	Pb	0	175
7.1853791		vacuum	Cs	0	184–186,335
7.1897725		vacuum	$^{12}C^{16}O$	0	1179,1187
7.1996583		vacuum	$^{12}C^{16}O$	0	1179,1187
7.204		vacuum	H_2O	0	1284,1329–31
7.2069023		vacuum	$^{12}C^{16}O$	0	1179,1187
7.2154388		vacuum	$^{12}C^{16}O$	0	1179,1187
7.2168093		vacuum	Ar	0	559,564

Table 4.1.1—*continued*
Neutral Atom, Ion, and Molecular Gas Lasers Arranged in Order of Increasing Wavelength

Wavelength (μm)	Uncertainty	Medium	Species	Charge	Reference
7.2242852		vacuum	$^{12}C^{16}O$	0	1179,1187
7.2314653		vacuum	$^{12}C^{16}O$	0	1179,1187
7.240		vacuum	Xe	0	601
7.2419225		vacuum	$^{12}C^{16}O$	0	1179,1187
7.2477397		vacuum	$^{12}C^{16}O$	0	1179,1187
7.25		vacuum	HCN	0	1327
7.2642643		vacuum	$^{12}C^{16}O$	0	1179,1187
7.2810409		vacuum	$^{12}C^{16}O$	0	1179,1187
7.285		vacuum	H_2O	0	1284,1329–31
7.2932		air	Ar	0	428
7.297		vacuum	H_2O	0	1284,1329–31
7.2980715		vacuum	$^{12}C^{16}O$	0	1179,1187
7.3107084		vacuum	$^{12}C^{16}O$	0	1179,1187
7.3153591		vacuum	$^{12}C^{16}O$	0	1179,1187
7.3168036		vacuum	Xe	0	559,582,623
7.3228367		vacuum	Ne	0	195,564,570
7.3266112		vacuum	$^{12}C^{16}O$	0	1179,1187
7.3329051		vacuum	$^{12}C^{16}O$	0	1179,1187
7.3427655		vacuum	$^{12}C^{16}O$	0	1179,1187
7.3507122		vacuum	$^{12}C^{16}O$	0	1179,1187
7.3591743		vacuum	$^{12}C^{16}O$	0	1179,1187
7.3625232		vacuum	Kr	0	601
7.3687835		vacuum	$^{12}C^{16}O$	0	1179,1187
7.3758394		vacuum	$^{12}C^{16}O$	0	1179,1187
7.390		vacuum	H_2O	0	1284,1329–31
7.3927632		vacuum	$^{12}C^{16}O$	0	1179,1187
7.405131		vacuum	Ne	0	432
7.4099471		vacuum	$^{12}C^{16}O$	0	1179,1187
7.4222794		vacuum	Ne	0	195,564
7.4235357		vacuum	Ne	0	195,564
7.425		vacuum	H_2O	0	1284,1329–31
7.4273943		vacuum	$^{12}C^{16}O$	0	1179,1187
7.4313142		vacuum	Xe	0	601
7.435		air	F	0	155
7.4415636		vacuum	$^{12}C^{16}O$	0	1179,1187

Table 4.1.1—*continued*
Neutral Atom, Ion, and Molecular Gas Lasers Arranged in Order of Increasing Wavelength

Wavelength (μm)	Uncertainty	Medium	Species	Charge	Reference
7.4451067		vacuum	$^{12}C^{16}O$	0	1179,1187
7.453		vacuum	H_2O	0	1284,1329–31
7.4578473		vacuum	$^{12}C^{16}O$	0	1179,1187
7.45879		vacuum	H_2O	0	1284,1329–31
7.4699904		vacuum	Ne	0	195
7.4743918		vacuum	$^{12}C^{16}O$	0	1179,1187
7.4799887		vacuum	Ne	0	559,564,570
7.4911995		vacuum	$^{12}C^{16}O$	0	1179,1187
7.4995237		vacuum	Ne	0	185,195,559
7.5082719		vacuum	$^{12}C^{16}O$	0	1179,1187
7.5256113		vacuum	$^{12}C^{16}O$	0	1179,1187
7.5313000		vacuum	Ne	0	185,570
7.543		vacuum	H_2O	0	1284,1329–31
7.5432206		vacuum	$^{12}C^{16}O$	0	1179,1187
7.5709798		vacuum	Ne	0	432
7.5885028		vacuum	Ne	0	432
7.590		vacuum	H_2O	0	1284,1329–31
7.5936233		vacuum	$^{12}C^{16}O$	0	1179,1187
7.59659		vacuum	H_2O	0	1284,1329–31
7.6105742		vacuum	$^{12}C^{16}O$	0	1179,1187
7.6163805		vacuum	Ne	0	195,564,570
7.6277976		vacuum	$^{12}C^{16}O$	0	1179,1187
7.6452950		vacuum	$^{12}C^{16}O$	0	1179,1187
7.6458386		vacuum	Ne	0	195
7.6511521		vacuum	Ne	0	559,564,570
7.6630690		vacuum	$^{12}C^{16}O$	0	1179,1187
7.6811222		vacuum	$^{12}C^{16}O$	0	1198
7.6926284		vacuum	Ne	0	195
7.7016367		vacuum	Ne	0	559,564,570
7.70897		vacuum	H_2O	0	1284,1329–31
7.7342182		vacuum	$^{12}C^{16}O$	0	1198
7.740		vacuum	H_2O	0	1284,1329–31
7.7408259		vacuum	Ne	0	195
7.7515937		vacuum	$^{12}C^{16}O$	0	1198
7.7655499		vacuum	Ne	0	559,570

Table 4.1.1—*continued*
Neutral Atom, Ion, and Molecular Gas Lasers Arranged in Order of Increasing Wavelength

Wavelength (µm)	Uncertainty	Medium	Species	Charge	Reference
7.7665		air	Xe	0	1498
7.7692514		vacuum	$^{12}C^{16}O$	0	1198
7.7813		air	Xe	0	1498
7.7815561		vacuum	Ne	0	195,564
7.7871934		vacuum	$^{12}C^{16}O$	0	1198
7.8001903		vacuum	Ar	0	564
7.8026187		vacuum	Ar	0	564
7.8054215		vacuum	$^{12}C^{16}O$	0	1198
7.8065353		vacuum	Ar	0	195
7.8071691		vacuum	Te	0	155
7.809		vacuum	Ne	0	432
7.8239388		vacuum	$^{12}C^{16}O$	0	1198
7.8369230		vacuum	Ne	0	195,564,570
7.8716418		vacuum	Ne	0	432
7.8799387		vacuum	$^{12}C^{16}O$	0	1198
7.8953393		vacuum	K	0	131,142
7.8977594		vacuum	$^{12}C^{16}O$	0	1198
7.90202		vacuum	SiH_4	0	1186
7.9158714		vacuum	$^{12}C^{16}O$	0	1198
7.92198		vacuum	SiH_4	0	1186
7.9423392		vacuum	Pb	0	175
7.9429190		vacuum	Ne	0	432
7.94900		vacuum	SiH_4	0	1186
7.9529813		vacuum	$^{12}C^{16}O$	0	1198
7.95482		vacuum	SiH_4	0	1186
7.955		air	Ar	0	1498
7.96902		vacuum	SiH_4	0	1186
7.9846694		vacuum	Ne	0	432
7.99201		vacuum	SiH_4	0	1186
8.0088892		vacuum	Ne	0	195,564,570
8.0311235		vacuum	$^{12}C^{16}O$	0	1198
8.03406		vacuum	C_2H_2	0	1184–86,1317
8.03523		vacuum	C_2H_2	0	1184–86,1317
8.03561		vacuum	C_2H_2	0	1184–86,1317
8.03781		vacuum	C_2H_2	0	1184–86,1317

Table 4.1.1—*continued*
Neutral Atom, Ion, and Molecular Gas Lasers Arranged in Order of Increasing Wavelength

Wavelength (μm)	Uncertainty	Medium	Species	Charge	Reference
8.03800		vacuum	C_2H_2	0	1184–86,1317
8.04020		vacuum	C_2H_2	0	1184–86,1317
8.04091		vacuum	C_2H_2	0	1184–86,1317
8.04428		vacuum	C_2H_2	0	1184–86,1317
8.04453		vacuum	C_2H_2	0	1184–86,1317
8.0494092		vacuum	$^{12}C^{16}O$	0	1198
8.0621949		vacuum	Ne	0	195,564,570
8.0679977		vacuum	$^{12}C^{16}O$	0	1198
8.0868914		vacuum	$^{12}C^{16}O$	0	1198
8.1060932		vacuum	$^{12}C^{16}O$	0	1198
8.1151		air	Kr	0	1498
8.116		vacuum	Ne	0	432
8.1483		air	N_2	0	107,128
8.1736588		vacuum	Ne	0	432
8.1827		air	N_2	0	107,128
8.1893		vacuum	^{12}COS	0	1317–1319
8.1913		vacuum	^{12}COS	0	1317–1319
8.1934		vacuum	^{12}COS	0	1317–1319
8.1960		vacuum	^{12}COS	0	1317–1319
8.197		vacuum	C_2H_2	0	1184–86,1317
8.1981		vacuum	^{12}COS	0	1317–1319
8.2008		vacuum	^{12}COS	0	1317–1319
8.2028		vacuum	^{12}COS	0	1317–1319
8.2055		vacuum	^{12}COS	0	1317–1319
8.2069181		vacuum	$^{12}C^{16}O$	0	1198
8.2075		vacuum	^{12}COS	0	1317–1319
8.2102		vacuum	^{12}COS	0	1317–1319
8.2102		air	N_2	0	107,128
8.2122		vacuum	^{12}COS	0	1317–1319
8.2149		vacuum	^{12}COS	0	1317–1319
8.2169		vacuum	^{12}COS	0	1317–1319
8.2196		vacuum	^{12}COS	0	1317–1319
8.2217		vacuum	^{12}COS	0	1317–1319
8.2244		vacuum	^{12}COS	0	1317–1319
8.2260066		vacuum	$^{12}C^{16}O$	0	1198

Table 4.1.1—*continued*
Neutral Atom, Ion, and Molecular Gas Lasers Arranged in Order of Increasing Wavelength

Wavelength (μm)	Uncertainty	Medium	Species	Charge	Reference
8.2271		vacuum	^{12}COS	0	1317–1319
8.2291		vacuum	^{12}COS	0	1317–1319
8.2318		vacuum	^{12}COS	0	1317–1319
8.2338		vacuum	^{12}COS	0	1317–1319
8.2366		vacuum	^{12}COS	0	1317–1319
8.23886		vacuum	^{12}COS	0	1317–1319
8.24165		vacuum	^{12}COS	0	1317–1319
8.24389		vacuum	^{12}COS	0	1317–1319
8.2454111		vacuum	$^{12}C^{16}O$	0	1198
8.2467		vacuum	^{12}COS	0	1317–1319
8.2495		vacuum	^{12}COS	0	1317–1319
8.25178		vacuum	^{12}COS	0	1317–1319
8.25437		vacuum	^{12}COS	0	1317–1319
8.25709		vacuum	^{12}COS	0	1317–1319
8.25948		vacuum	^{12}COS	0	1317–1319
8.26228		vacuum	^{12}COS	0	1317–1319
8.26453		vacuum	^{12}COS	0	1317–1319
8.2651363		vacuum	$^{12}C^{16}O$	0	1198
8.26726		vacuum	^{12}COS	0	1317–1319
8.2699		vacuum	^{12}COS	0	1317–1319
8.2727		vacuum	^{12}COS	0	1317–1319
8.2754		vacuum	^{12}COS	0	1317–1319
8.2781		vacuum	^{12}COS	0	1317–1319
8.2809		vacuum	^{12}COS	0	1317–1319
8.2836		vacuum	^{12}COS	0	1317–1319
8.2857		vacuum	^{12}COS	0	1317–1319
8.299		vacuum	C_2H_2	0	1184–86,1317
8.3278		vacuum	^{12}COS	0	1317–1319
8.3306		vacuum	^{12}COS	0	1317–1319
8.3333		vacuum	^{12}COS	0	1317–1319
8.3361		vacuum	^{12}COS	0	1317–1319
8.3371447		vacuum	Ne	0	195,564
8.3389		vacuum	^{12}COS	0	1317–1319
8.3417		vacuum	^{12}COS	0	1317–1319
8.3452		vacuum	^{12}COS	0	1317–1319

Table 4.1.1—*continued*
Neutral Atom, Ion, and Molecular Gas Lasers Arranged in Order of Increasing Wavelength

Wavelength (μm)	Uncertainty	Medium	Species	Charge	Reference
8.3479		vacuum	^{12}COS	0	1317–1319
8.3496011		vacuum	Ne	0	195,564
8.3507		vacuum	^{12}COS	0	1317–1319
8.3535		vacuum	^{12}COS	0	1317–1319
8.3563		vacuum	^{12}COS	0	1317–1319
8.3598		vacuum	^{12}COS	0	1317–1319
8.36246		vacuum	^{12}COS	0	1317–1319
8.36540		vacuum	^{12}COS	0	1317–1319
8.36855		vacuum	^{12}COS	0	1317–1319
8.37156		vacuum	^{12}COS	0	1317–1319
8.37458		vacuum	^{12}COS	0	1317–1319
8.37788		vacuum	^{12}COS	0	1317–1319
8.38089		vacuum	^{12}COS	0	1317–1319
8.38392		vacuum	^{12}COS	0	1317–1319
8.38701		vacuum	^{12}COS	0	1317–1319
8.39004		vacuum	^{12}COS	0	1317–1319
8.39306		vacuum	^{12}COS	0	1317–1319
8.39616		vacuum	^{12}COS	0	1317–1319
8.39990		vacuum	^{12}COS	0	1317–1319
8.40237		vacuum	^{12}COS	0	1317–1319
8.4042		air	Xe	0	1498
8.40548		vacuum	^{12}COS	0	1317–1319
8.40852		vacuum	^{12}COS	0	1317–1319
8.41170		vacuum	^{12}COS	0	1317–1319
8.41468		vacuum	^{12}COS	0	1317–1319
8.41786		vacuum	^{12}COS	0	1317–1319
8.42134		vacuum	^{12}COS	0	1317–1319
8.42432		vacuum	^{12}COS	0	1317–1319
8.4274		vacuum	^{12}COS	0	1317–1319
8.4310		vacuum	^{12}COS	0	1317–1319
8.4345		vacuum	^{12}COS	0	1317–1319
8.4374		vacuum	^{12}COS	0	1317–1319
8.4410		vacuum	^{12}COS	0	1317–1319
8.4438		vacuum	^{12}COS	0	1317–1319
8.4474		vacuum	^{12}COS	0	1317–1319

Table 4.1.1—*continued*
Neutral Atom, Ion, and Molecular Gas Lasers Arranged in Order of Increasing Wavelength

Wavelength (μm)	Uncertainty	Medium	Species	Charge	Reference
8.4509		vacuum	^{12}COS	0	1317–1319
8.4538		vacuum	^{12}COS	0	1317–1319
8.4574		vacuum	^{12}COS	0	1317–1319
8.48		vacuum	HCN	0	1327
8.492785		vacuum	Fe	0	155
8.529294		vacuum	He	0	433
8.6		vacuum	^{13}COS	0	1317–1319
8.8414054		vacuum	Ne	0	195,564
8.8554154		vacuum	Ne	0	195,564
8.987673374		vacuum	$^{12}C^{18}O_2$	0	1640
8.9876733841		vacuum	$^{12}C^{18}O_2$	0	1640
8.994945420		vacuum	$^{12}C^{18}O_2$	0	1640
8.9949454256		vacuum	$^{12}C^{18}O_2$	0	1640
9.002380597		vacuum	$^{12}C^{18}O_2$	0	1640
9.0023805967		vacuum	$^{12}C^{18}O_2$	0	1640
9.0067086		vacuum	Xe	0	185,559,582
9.009980301		vacuum	$^{12}C^{18}O_2$	0	1640
9.0099803044		vacuum	$^{12}C^{18}O_2$	0	1640
9.017745954		vacuum	$^{12}C^{18}O_2$	0	1640
9.0177459554		vacuum	$^{12}C^{18}O_2$	0	1640
9.019572		vacuum	I	0	401,402
9.025678950		vacuum	$^{12}C^{18}O_2$	0	1640
9.0256789526		vacuum	$^{12}C^{18}O_2$	0	1640
9.033780694		vacuum	$^{12}C^{18}O_2$	0	1640
9.0337806978		vacuum	$^{12}C^{18}O_2$	0	1640
9.042052587		vacuum	$^{12}C^{18}O_2$	0	1640
9.0420525890		vacuum	$^{12}C^{18}O_2$	0	1640
9.050496018		vacuum	$^{12}C^{18}O_2$	0	1640
9.0504960196		vacuum	$^{12}C^{18}O_2$	0	1640
9.0591123781		vacuum	$^{12}C^{18}O_2$	0	1640
9.059112381		vacuum	$^{12}C^{18}O_2$	0	1640
9.067903046		vacuum	$^{12}C^{18}O_2$	0	1640
9.0679030476		vacuum	$^{12}C^{18}O_2$	0	1640
9.07505983		vacuum	$^{12}C^{18}O_2$	0	1640
9.0768694044		vacuum	$^{12}C^{18}O_2$	0	1640

Table 4.1.1—*continued*
Neutral Atom, Ion, and Molecular Gas Lasers Arranged in Order of Increasing Wavelength

Wavelength (μm)	Uncertainty	Medium	Species	Charge	Reference
9.076869410		vacuum	$^{12}C^{18}O_2$	0	1640
9.08316833		vacuum	$^{12}C^{18}O_2$	0	1640
9.0860128178		vacuum	$^{12}C^{18}O_2$	0	1640
9.086012820		vacuum	$^{12}C^{18}O_2$	0	1640
9.0894630		vacuum	Ne	0	195,564,570
9.0903830147		vacuum	$^{12}C^{17}O_2$	0	1640
9.09145210		vacuum	$^{12}C^{18}O_2$	0	1640
9.0934950		vacuum	$^{12}C^{16}O_2$	0	1640
9.0934950		vacuum	$^{12}C^{16}O_2$	0	1640
9.0943359524		vacuum	$^{12}C^{17}O_2$	0	1640
9.0953346501		vacuum	$^{12}C^{18}O_2$	0	1640
9.095334658		vacuum	$^{12}C^{18}O_2$	0	1640
9.0983367323		vacuum	$^{12}C^{17}O_2$	0	1640
9.0997598		vacuum	$^{12}C^{16}O_2$	0	1640
9.0997598		vacuum	$^{12}C^{16}O_2$	0	1640
9.09991271		vacuum	$^{12}C^{18}O_2$	0	1640
9.1023855930		vacuum	$^{12}C^{17}O_2$	0	1640
9.1048362549		vacuum	$^{12}C^{18}O_2$	0	1640
9.104836262		vacuum	$^{12}C^{18}O_2$	0	1640
9.10622908		vacuum	$^{12}C^{16}O_2$	0	1640
9.106232057		vacuum	$^{12}C^{16}O_2$	0	1640
9.1064827734		vacuum	$^{12}C^{17}O_2$	0	1640
9.10855153		vacuum	$^{12}C^{18}O_2$	0	1640
9.1106285129		vacuum	$^{12}C^{17}O_2$	0	1640
9.1129052508		vacuum	$^{12}C^{16}O_2$	0	1640
9.112905275		vacuum	$^{12}C^{16}O_2$	0	1640
9.1145189755		vacuum	$^{12}C^{18}O_2$	0	1640
9.114518985		vacuum	$^{12}C^{18}O_2$	0	1640
9.1148230497		vacuum	$^{12}C^{17}O_2$	0	1640
9.11737010		vacuum	$^{12}C^{18}O_2$	0	1640
9.1190666207		vacuum	$^{12}C^{17}O_2$	0	1640
9.1197909939		vacuum	$^{12}C^{16}O_2$	0	1640
9.119791005		vacuum	$^{12}C^{16}O_2$	0	1640
9.1233594639		vacuum	$^{12}C^{17}O_2$	0	1640
9.1243841479		vacuum	$^{12}C^{18}O_2$	0	1640

Table 4.1.1—*continued*
Neutral Atom, Ion, and Molecular Gas Lasers Arranged in Order of Increasing Wavelength

Wavelength (μm)	Uncertainty	Medium	Species	Charge	Reference
9.124384150		vacuum	$^{12}C^{18}O_2$	0	1640
9.12636993		vacuum	$^{12}C^{18}O_2$	0	1640
9.1268888458		vacuum	$^{12}C^{16}O_2$	0	1640
9.126888847		vacuum	$^{12}C^{16}O_2$	0	1640
9.1277018147		vacuum	$^{12}C^{17}O_2$	0	1640
9.1320939079		vacuum	$^{12}C^{17}O_2$	0	1640
9.134201369		vacuum	$^{12}C^{16}O_2$	0	1640
9.1342013707		vacuum	$^{12}C^{16}O_2$	0	1640
9.1344330951		vacuum	$^{12}C^{18}O_2$	0	1640
9.134433103		vacuum	$^{12}C^{18}O_2$	0	1640
9.1365359806		vacuum	$^{12}C^{17}O_2$	0	1640
9.1378432636		vacuum	$^{12}C^{16}O^{18}O$	0	1640
9.137843272		vacuum	$^{12}C^{16}O^{18}O$	0	1640
9.1410282644		vacuum	$^{12}C^{17}O_2$	0	1640
9.1417311231		vacuum	$^{12}C^{16}O_2$	0	1640
9.141731124		vacuum	$^{12}C^{16}O_2$	0	1640
9.1423917999		vacuum	$^{12}C^{16}O^{18}O$	0	1640
9.142391802		vacuum	$^{12}C^{16}O^{18}O$	0	1640
9.1446671323		vacuum	$^{12}C^{18}O_2$	0	1640
9.144667139		vacuum	$^{12}C^{18}O_2$	0	1640
9.14491873		vacuum	$^{12}C^{18}O_2$	0	1640
9.1455709948		vacuum	$^{12}C^{17}O_2$	0	1640
9.146991207		vacuum	$^{12}C^{16}O^{18}O$	0	1640
9.1469912065		vacuum	$^{12}C^{16}O^{18}O$	0	1640
9.149480647		vacuum	$^{12}C^{16}O_2$	0	1640
9.1494806499		vacuum	$^{12}C^{16}O_2$	0	1640
9.1501644027		vacuum	$^{12}C^{17}O_2$	0	1640
9.1516417199		vacuum	$^{12}C^{16}O^{18}O$	0	1640
9.151641723		vacuum	$^{12}C^{16}O^{18}O$	0	1640
9.15447066		vacuum	$^{12}C^{18}O_2$	0	1640
9.1548087203		vacuum	$^{12}C^{17}O_2$	0	1640
9.1550875622		vacuum	$^{12}C^{18}O_2$	0	1640
9.155087570		vacuum	$^{12}C^{18}O_2$	0	1640
9.156343568		vacuum	$^{12}C^{16}O^{18}O$	0	1640
9.1563435701		vacuum	$^{12}C^{16}O^{18}O$	0	1640

Table 4.1.1—*continued*
Neutral Atom, Ion, and Molecular Gas Lasers
Arranged in Order of Increasing Wavelength

Wavelength (μm)	Uncertainty	Medium	Species	Charge	Reference
9.157452485		vacuum	$^{12}C^{16}O_2$	0	1640
9.1574524861		vacuum	$^{12}C^{16}O_2$	0	1640
9.1595041789		vacuum	$^{12}C^{17}O_2$	0	1640
9.1610969896		vacuum	$^{12}C^{16}O^{18}O$	0	1640
9.161096994		vacuum	$^{12}C^{16}O^{18}O$	0	1640
9.16420944		vacuum	$^{12}C^{18}O_2$	0	1640
9.1642510074		vacuum	$^{12}C^{17}O_2$	0	1640
9.1656491546		vacuum	$^{12}C^{16}O_2$	0	1640
9.165649160		vacuum	$^{12}C^{16}O_2$	0	1640
9.1656956749		vacuum	$^{12}C^{18}O_2$	0	1640
9.165695685		vacuum	$^{12}C^{18}O_2$	0	1640
9.1659022105		vacuum	$^{12}C^{16}O^{18}O$	0	1640
9.165902211		vacuum	$^{12}C^{16}O^{18}O$	0	1640
9.1690494356		vacuum	$^{12}C^{17}O_2$	0	1640
9.170759460		vacuum	$^{12}C^{16}O^{18}O$	0	1640
9.1707594613		vacuum	$^{12}C^{16}O^{18}O$	0	1640
9.1738996911		vacuum	$^{12}C^{17}O_2$	0	1640
9.1740731659		vacuum	$^{12}C^{16}O_2$	0	1640
9.174073173		vacuum	$^{12}C^{16}O_2$	0	1640
9.17413644		vacuum	$^{12}C^{18}O_2$	0	1640
9.1756689718		vacuum	$^{12}C^{16}O^{18}O$	0	1640
9.175668974		vacuum	$^{12}C^{16}O^{18}O$	0	1640
9.1764927513		vacuum	$^{12}C^{18}O_2$	0	1640
9.176492758		vacuum	$^{12}C^{18}O_2$	0	1640
9.1788020012		vacuum	$^{12}C^{17}O_2$	0	1640
9.1791962		vacuum	K	0	131,142
9.1806309692		vacuum	$^{12}C^{16}O^{18}O$	0	1640
9.180630972		vacuum	$^{12}C^{16}O^{18}O$	0	1640
9.1827270164		vacuum	$^{12}C^{16}O_2$	0	1640
9.182727020		vacuum	$^{12}C^{16}O_2$	0	1640
9.1837565912		vacuum	$^{12}C^{17}O_2$	0	1640
9.18425316		vacuum	$^{12}C^{18}O_2$	0	1640
9.1856456819		vacuum	$^{12}C^{16}O^{18}O$	0	1640
9.185645690		vacuum	$^{12}C^{16}O^{18}O$	0	1640
9.1874800577		vacuum	$^{12}C^{18}O_2$	0	1640

Table 4.1.1—*continued*
Neutral Atom, Ion, and Molecular Gas Lasers Arranged in Order of Increasing Wavelength

Wavelength (μm)	Uncertainty	Medium	Species	Charge	Reference
9.187480064		vacuum	$^{12}C^{18}O_2$	0	1640
9.1887636870		vacuum	$^{12}C^{17}O_2$	0	1640
9.1907133355		vacuum	$^{12}C^{16}O^{18}O$	0	1640
9.190713345		vacuum	$^{12}C^{16}O^{18}O$	0	1640
9.1916131871		vacuum	$^{12}C^{16}O_2$	0	1640
9.191613190		vacuum	$^{12}C^{16}O_2$	0	1640
9.1938235127		vacuum	$^{12}C^{17}O_2$	0	1640
9.19456080		vacuum	$^{12}C^{18}O_2$	0	1640
9.1958341547		vacuum	$^{12}C^{16}O^{18}O$	0	1640
9.195834169		vacuum	$^{12}C^{16}O^{18}O$	0	1640
9.1986588474		vacuum	$^{12}C^{18}O_2$	0	1640
9.198658852		vacuum	$^{12}C^{18}O_2$	0	1640
9.1989362908		vacuum	$^{12}C^{17}O_2$	0	1640
9.2007341425		vacuum	$^{12}C^{16}O_2$	0	1640
9.200734148		vacuum	$^{12}C^{16}O_2$	0	1640
9.2010083628		vacuum	$^{12}C^{16}O^{18}O$	0	1640
9.201008374		vacuum	$^{12}C^{16}O^{18}O$	0	1640
9.2041022431		vacuum	$^{12}C^{17}O_2$	0	1640
9.20506088		vacuum	$^{12}C^{18}O_2$	0	1640
9.2062361840		vacuum	$^{12}C^{16}O^{18}O$	0	1640
9.206236198		vacuum	$^{12}C^{16}O^{18}O$	0	1640
9.2091716		vacuum	$^{12}C^{16}O_2$	0	1640
9.2093215901		vacuum	$^{12}C^{17}O_2$	0	1640
9.2100303627		vacuum	$^{12}C^{18}O_2$	0	1640
9.210030368		vacuum	$^{12}C^{18}O_2$	0	1640
9.2100923301		vacuum	$^{12}C^{16}O_2$	0	1640
9.210092333		vacuum	$^{12}C^{16}O_2$	0	1640
9.2115178379		vacuum	$^{12}C^{16}O^{18}O$	0	1640
9.211517856		vacuum	$^{12}C^{16}O^{18}O$	0	1640
9.2145945513		vacuum	$^{12}C^{17}O_2$	0	1640
9.21575458		vacuum	$^{12}C^{18}O_2$	0	1640
9.2168535462		vacuum	$^{12}C^{16}O^{18}O$	0	1640
9.216853566		vacuum	$^{12}C^{16}O^{18}O$	0	1640
9.2177733		vacuum	$^{12}C^{16}O_2$	0	1640
9.2196901801		vacuum	$^{12}C^{16}O_2$	0	1640

Table 4.1.1—*continued*
Neutral Atom, Ion, and Molecular Gas Lasers Arranged in Order of Increasing Wavelength

Wavelength (μm)	Uncertainty	Medium	Species	Charge	Reference
9.219690184		vacuum	$^{12}C^{16}O_2$	0	1640
9.2199213456		vacuum	$^{12}C^{17}O_2$	0	1640
9.2222435271		vacuum	$^{12}C^{16}O^{18}O$	0	1640
9.222243539		vacuum	$^{12}C^{16}O^{18}O$	0	1640
9.2253021895		vacuum	$^{12}C^{17}O_2$	0	1640
9.2266150		vacuum	$^{12}C^{16}O_2$	0	1640
9.22664333		vacuum	$^{12}C^{18}O_2$	0	1640
9.2276879987		vacuum	$^{12}C^{16}O^{18}O$	0	1640
9.227688015		vacuum	$^{12}C^{16}O^{18}O$	0	1640
9.2295301027		vacuum	$^{12}C^{16}O_2$	0	1640
9.229530105		vacuum	$^{12}C^{16}O_2$	0	1640
9.2307372990		vacuum	$^{12}C^{17}O_2$	0	1640
9.2331871778		vacuum	$^{12}C^{16}O^{18}O$	0	1640
9.233187192		vacuum	$^{12}C^{16}O^{18}O$	0	1640
9.2356997		vacuum	$^{12}C^{16}O_2$	0	1640
9.2362268897		vacuum	$^{12}C^{17}O_2$	0	1640
9.23772834		vacuum	$^{12}C^{18}O_2$	0	1640
9.2387412792		vacuum	$^{12}C^{16}O^{18}O$	0	1640
9.238741289		vacuum	$^{12}C^{16}O^{18}O$	0	1640
9.2393103478		vacuum	$^{12}C^{18}O_2$	0	1640
9.239310356		vacuum	$^{12}C^{18}O_2$	0	1640
9.2396144880		vacuum	$^{12}C^{16}O_2$	0	1640
9.239614496		vacuum	$^{12}C^{16}O_2$	0	1640
9.2417711746		vacuum	$^{12}C^{17}O_2$	0	1640
9.2443505168		vacuum	$^{12}C^{16}O^{18}O$	0	1640
9.244350527		vacuum	$^{12}C^{16}O^{18}O$	0	1640
9.2450296		vacuum	$^{12}C^{16}O_2$	0	1640
9.2473703656		vacuum	$^{12}C^{17}O_2$	0	1640
9.2499457048		vacuum	$^{12}C^{16}O_2$	0	1640
9.249945712		vacuum	$^{12}C^{16}O_2$	0	1640
9.2500151042		vacuum	$^{12}C^{16}O^{18}O$	0	1640
9.250015111		vacuum	$^{12}C^{16}O^{18}O$	0	1640
9.2513659810		vacuum	$^{12}C^{18}O_2$	0	1640
9.251365985		vacuum	$^{12}C^{18}O_2$	0	1640
9.2530246740		vacuum	$^{12}C^{17}O_2$	0	1640

Table 4.1.1—*continued*
Neutral Atom, Ion, and Molecular Gas Lasers Arranged in Order of Increasing Wavelength

Wavelength (μm)	Uncertainty	Medium	Species	Charge	Reference
9.2546080		vacuum	$^{12}C^{16}O_2$	0	1640
9.2557352512		vacuum	$^{12}C^{16}O^{18}O$	0	1640
9.255735258		vacuum	$^{12}C^{16}O^{18}O$	0	1640
9.2587343087		vacuum	$^{12}C^{17}O_2$	0	1640
9.2605261009		vacuum	$^{12}C^{16}O_2$	0	1640
9.260526108		vacuum	$^{12}C^{16}O_2$	0	1640
9.263619757		vacuum	$^{12}C^{18}O_2$	0	1640
9.2636197566		vacuum	$^{12}C^{18}O_2$	0	1640
9.2644362		vacuum	$^{12}C^{16}O_2$	0	1640
9.2644994798		vacuum	$^{12}C^{17}O_2$	0	1640
9.2703203918		vacuum	$^{12}C^{17}O_2$	0	1640
9.2713580020		vacuum	$^{12}C^{16}O_2$	0	1640
9.271358004		vacuum	$^{12}C^{16}O_2$	0	1640
9.2745176		vacuum	$^{12}C^{16}O_2$	0	1640
9.2760728413		vacuum	$^{12}C^{18}O_2$	0	1640
9.276072846		vacuum	$^{12}C^{18}O_2$	0	1640
9.2761972527		vacuum	$^{12}C^{17}O_2$	0	1640
9.2821302649		vacuum	$^{12}C^{17}O_2$	0	1640
9.2824437086		vacuum	$^{12}C^{16}O_2$	0	1640
9.282443709		vacuum	$^{12}C^{16}O_2$	0	1640
9.2848536		vacuum	$^{12}C^{16}O_2$	0	1640
9.2881196322		vacuum	$^{12}C^{17}O_2$	0	1640
9.2887263865		vacuum	$^{12}C^{18}O_2$	0	1640
9.288726393		vacuum	$^{12}C^{18}O_2$	0	1640
9.2937854986		vacuum	$^{12}C^{16}O_2$	0	1640
9.293785500		vacuum	$^{12}C^{16}O_2$	0	1640
9.2954478		vacuum	$^{12}C^{16}O_2$	0	1640
9.3			$^{14}NH_3$	0	1189,1210
9.3015815293		vacuum	$^{12}C^{18}O_2$	0	1640
9.301581533		vacuum	$^{12}C^{18}O_2$	0	1640
9.3053856241		vacuum	$^{12}C^{16}O_2$	0	1640
9.305385625		vacuum	$^{12}C^{16}O_2$	0	1640
9.3063026		vacuum	$^{12}C^{16}O_2$	0	1640
9.3146393934		vacuum	$^{12}C^{18}O_2$	0	1640
9.314639396		vacuum	$^{12}C^{18}O_2$	0	1640

Table 4.1.1—*continued*
Neutral Atom, Ion, and Molecular Gas Lasers Arranged in Order of Increasing Wavelength

Wavelength (μm)	Uncertainty	Medium	Species	Charge	Reference
9.3172463132		vacuum	$^{12}C^{16}O_2$	0	1640
9.317246315		vacuum	$^{12}C^{16}O_2$	0	1640
9.3174189		vacuum	$^{12}C^{16}O_2$	0	1640
9.32725182		vacuum	$^{12}C^{18}O_2$	0	1640
9.327901090		vacuum	$^{12}C^{18}O_2$	0	1640
9.3279010896		vacuum	$^{12}C^{18}O_2$	0	1640
9.3288003		vacuum	$^{12}C^{16}O_2$	0	1640
9.3293697650		vacuum	$^{12}C^{16}O_2$	0	1640
9.329369767		vacuum	$^{12}C^{16}O_2$	0	1640
9.34005619		vacuum	$^{12}C^{18}O_2$	0	1640
9.3404485		vacuum	$^{12}C^{16}O_2$	0	1640
9.341367710		vacuum	$^{12}C^{18}O_2$	0	1640
9.3413677122		vacuum	$^{12}C^{18}O_2$	0	1640
9.3417581550		vacuum	$^{12}C^{16}O_2$	0	1640
9.341758159		vacuum	$^{12}C^{16}O_2$	0	1640
9.346		vacuum	$^{14}NH_3$	0	1189,1210
9.3523664		vacuum	$^{12}C^{16}O_2$	0	1640
9.35306846		vacuum	$^{12}C^{18}O_2$	0	1640
9.354413628		vacuum	$^{12}C^{16}O_2$	0	1640
9.3544136288		vacuum	$^{12}C^{16}O_2$	0	1640
9.3550403444		vacuum	$^{12}C^{18}O_2$	0	1640
9.355040346		vacuum	$^{12}C^{18}O_2$	0	1640
9.3645559		vacuum	$^{12}C^{16}O_2$	0	1640
9.36628985		vacuum	$^{12}C^{18}O_2$	0	1640
9.3673383050		vacuum	$^{12}C^{16}O_2$	0	1640
9.367338308		vacuum	$^{12}C^{16}O_2$	0	1640
9.3689200543		vacuum	$^{12}C^{18}O_2$	0	1640
9.368920057		vacuum	$^{12}C^{18}O_2$	0	1640
9.3711993962		vacuum	$^{12}C^{17}O_2$	0	1640
9.3770190		vacuum	$^{12}C^{16}O_2$	0	1640
9.3779982255		vacuum	$^{12}C^{17}O_2$	0	1640
9.37972146		vacuum	$^{12}C^{18}O_2$	0	1640
9.3805342746		vacuum	$^{12}C^{16}O_2$	0	1640
9.380534281		vacuum	$^{12}C^{16}O_2$	0	1640
9.3830078982		vacuum	$^{12}C^{18}O_2$	0	1640

Table 4.1.1—*continued*
Neutral Atom, Ion, and Molecular Gas Lasers Arranged in Order of Increasing Wavelength

Wavelength (μm)	Uncertainty	Medium	Species	Charge	Reference
9.383007902		vacuum	$^{12}C^{18}O_2$	0	1640
9.3848562790		vacuum	$^{12}C^{17}O_2$	0	1640
9.3897578		vacuum	$^{12}C^{16}O_2$	0	1640
9.3917737361		vacuum	$^{12}C^{17}O_2$	0	1640
9.39336441		vacuum	$^{12}C^{18}O_2$	0	1640
9.39382		vacuum	H_2O	0	1284,1329–31
9.3940035981		vacuum	$^{12}C^{16}O_2$	0	1640
9.394003603		vacuum	$^{12}C^{16}O_2$	0	1640
9.3973049185		vacuum	$^{12}C^{18}O_2$	0	1640
9.397304920		vacuum	$^{12}C^{18}O_2$	0	1640
9.3987507727		vacuum	$^{12}C^{17}O_2$	0	1640
9.4027743		vacuum	$^{12}C^{16}O_2$	0	1640
9.4057875665		vacuum	$^{12}C^{17}O_2$	0	1640
9.40721976		vacuum	$^{12}C^{18}O_2$	0	1640
9.4118121458		vacuum	$^{12}C^{18}O_2$	0	1640
9.411812150		vacuum	$^{12}C^{18}O_2$	0	1640
9.4128842907		vacuum	$^{12}C^{17}O_2$	0	1640
9.4147245588		vacuum	$^{12}C^{16}O_2$	0	1640
9.414724562		vacuum	$^{12}C^{16}O_2$	0	1640
9.417384992		vacuum	$^{12}C^{16}O^{18}O$	0	1640
9.4173850048		vacuum	$^{12}C^{16}O^{18}O$	0	1640
9.4200411203		vacuum	$^{12}C^{17}O_2$	0	1640
9.42128859		vacuum	$^{12}C^{18}O_2$	0	1640
9.424619168		vacuum	$^{12}C^{16}O^{18}O$	0	1640
9.4246191839		vacuum	$^{12}C^{16}O^{18}O$	0	1640
9.4265305973		vacuum	$^{12}C^{18}O_2$	0	1640
9.426530601		vacuum	$^{12}C^{18}O_2$	0	1640
9.4272582269		vacuum	$^{12}C^{17}O_2$	0	1640
9.428886090		vacuum	$^{12}C^{16}O_2$	0	1640
9.4288860917		vacuum	$^{12}C^{16}O_2$	0	1640
9.431914062		vacuum	$^{12}C^{16}O^{18}O$	0	1640
9.4319140777		vacuum	$^{12}C^{16}O^{18}O$	0	1640
9.4345357789		vacuum	$^{12}C^{17}O_2$	0	1640
9.43557192		vacuum	$^{12}C^{18}O_2$	0	1640
9.439269841		vacuum	$^{12}C^{16}O^{18}O$	0	1640

Table 4.1.1—*continued*
Neutral Atom, Ion, and Molecular Gas Lasers Arranged in Order of Increasing Wavelength

Wavelength (μm)	Uncertainty	Medium	Species	Charge	Reference
9.4392698552		vacuum	$^{12}C^{16}O^{18}O$	0	1640
9.441461278		vacuum	$^{12}C^{18}O_2$	0	1640
9.4414612802		vacuum	$^{12}C^{18}O_2$	0	1640
9.4418739452		vacuum	$^{12}C^{17}O_2$	0	1640
9.4433279590		vacuum	$^{12}C^{16}O_2$	0	1640
9.443327961		vacuum	$^{12}C^{16}O_2$	0	1640
9.446686675		vacuum	$^{12}C^{16}O^{18}O$	0	1640
9.4466866846		vacuum	$^{12}C^{16}O^{18}O$	0	1640
9.4492728943		vacuum	$^{12}C^{17}O_2$	0	1640
9.45007080		vacuum	$^{12}C^{18}O_2$	0	1640
9.4505542		vacuum	$^{12}C^{16}O_2$	0	1640
9.454164717		vacuum	$^{12}C^{16}O^{18}O$	0	1640
9.4541647314		vacuum	$^{12}C^{16}O^{18}O$	0	1640
9.4566051900		vacuum	$^{12}C^{18}O_2$	0	1640
9.456605195		vacuum	$^{12}C^{18}O_2$	0	1640
9.4567327911		vacuum	$^{12}C^{17}O_2$	0	1640
9.4580520879		vacuum	$^{12}C^{16}O_2$	0	1640
9.458052092		vacuum	$^{12}C^{16}O_2$	0	1640
9.461704150		vacuum	$^{12}C^{16}O^{18}O$	0	1640
9.4617041611		vacuum	$^{12}C^{16}O^{18}O$	0	1640
9.4642537986		vacuum	$^{12}C^{17}O_2$	0	1640
9.46478631		vacuum	$^{12}C^{18}O_2$	0	1640
9.4648480		vacuum	$^{12}C^{16}O_2$	0	1640
9.4686955538		vacuum	$^{13}C^{18}O_2$	0	1640
9.469305123		vacuum	$^{12}C^{16}O^{18}O$	0	1640
9.4693051359		vacuum	$^{12}C^{16}O^{18}O$	0	1640
9.4718360798		vacuum	$^{12}C^{17}O_2$	0	1640
9.471963307		vacuum	$^{12}C^{18}O_2$	0	1640
9.4719633100		vacuum	$^{12}C^{18}O_2$	0	1640
9.4730603717		vacuum	$^{12}C^{16}O_2$	0	1640
9.473060373		vacuum	$^{12}C^{16}O_2$	0	1640
9.47472		vacuum	H_2O	0	1284,1329–31
9.4762231953		vacuum	$^{13}C^{18}O_2$	0	1640
9.476967805		vacuum	$^{12}C^{16}O^{18}O$	0	1640
9.4769678179		vacuum	$^{12}C^{16}O^{18}O$	0	1640

Table 4.1.1—*continued*
Neutral Atom, Ion, and Molecular Gas Lasers Arranged in Order of Increasing Wavelength

Wavelength (μm)	Uncertainty	Medium	Species	Charge	Reference
9.4794322		vacuum	$^{12}C^{16}O_2$	0	1640
9.4794797973		vacuum	$^{12}C^{17}O_2$	0	1640
9.47971944		vacuum	$^{12}C^{18}O_2$	0	1640
9.4839415976		vacuum	$^{13}C^{18}O_2$	0	1640
9.484692358		vacuum	$^{12}C^{16}O^{18}O$	0	1640
9.4846923680		vacuum	$^{12}C^{16}O^{18}O$	0	1640
9.4871851086		vacuum	$^{12}C^{17}O_2$	0	1640
9.487536611		vacuum	$^{12}C^{18}O_2$	0	1640
9.4875366157		vacuum	$^{12}C^{18}O_2$	0	1640
9.4883546772		vacuum	$^{12}C^{16}O_2$	0	1640
9.488354678		vacuum	$^{12}C^{16}O_2$	0	1640
9.4918527715		vacuum	$^{13}C^{18}O_2$	0	1640
9.492478928		vacuum	$^{12}C^{16}O^{18}O$	0	1640
9.4924789434		vacuum	$^{12}C^{16}O^{18}O$	0	1640
9.4943075		vacuum	$^{12}C^{16}O_2$	0	1640
9.49487115		vacuum	$^{12}C^{18}O_2$	0	1640
9.4949521722		vacuum	$^{12}C^{17}O_2$	0	1640
9.499958710		vacuum	$^{13}C^{18}O_2$	0	1640
9.4999587194		vacuum	$^{13}C^{18}O_2$	0	1640
9.500327691		vacuum	$^{12}C^{16}O^{18}O$	0	1640
9.5003277023		vacuum	$^{12}C^{16}O^{18}O$	0	1640
9.5027811452		vacuum	$^{12}C^{17}O_2$	0	1640
9.5033260720		vacuum	$^{12}C^{18}O_2$	0	1640
9.503326073		vacuum	$^{12}C^{18}O_2$	0	1640
9.5039368353		vacuum	$^{12}C^{16}O_2$	0	1640
9.503936837		vacuum	$^{12}C^{16}O_2$	0	1640
9.508238787		vacuum	$^{12}C^{16}O^{18}O$	0	1640
9.5082387994		vacuum	$^{12}C^{16}O^{18}O$	0	1640
9.508261425		vacuum	$^{13}C^{18}O_2$	0	1640
9.5082614354		vacuum	$^{13}C^{18}O_2$	0	1640
9.5094756		vacuum	$^{12}C^{16}O_2$	0	1640
9.51024239		vacuum	$^{12}C^{18}O_2$	0	1640
9.5106721823		vacuum	$^{12}C^{17}O_2$	0	1640
9.5162123445		vacuum	$^{12}C^{16}O^{18}O$	0	1640
9.516212378		vacuum	$^{12}C^{16}O^{18}O$	0	1640

Table 4.1.1—*continued*
Neutral Atom, Ion, and Molecular Gas Lasers Arranged in Order of Increasing Wavelength

Wavelength (μm)	Uncertainty	Medium	Species	Charge	Reference
9.516762896		vacuum	$^{13}C^{18}O_2$	0	1640
9.5167629029		vacuum	$^{13}C^{18}O_2$	0	1640
9.5186254374		vacuum	$^{12}C^{17}O_2$	0	1640
9.5193326343		vacuum	$^{12}C^{18}O_2$	0	1640
9.519332636		vacuum	$^{12}C^{18}O_2$	0	1640
9.519808647		vacuum	$^{12}C^{16}O_2$	0	1640
9.5198086483		vacuum	$^{12}C^{16}O_2$	0	1640
9.524248619		vacuum	$^{12}C^{16}O^{18}O$	0	1640
9.5242486252		vacuum	$^{12}C^{16}O^{18}O$	0	1640
9.5249398		vacuum	$^{12}C^{16}O_2$	0	1640
9.525465086		vacuum	$^{13}C^{18}O_2$	0	1640
9.5254650959		vacuum	$^{13}C^{18}O_2$	0	1640
9.52583417		vacuum	$^{12}C^{18}O_2$	0	1640
9.5266410620		vacuum	$^{12}C^{17}O_2$	0	1640
9.532347648		vacuum	$^{12}C^{16}O^{18}O$	0	1640
9.5323476576		vacuum	$^{12}C^{16}O^{18}O$	0	1640
9.534369974		vacuum	$^{13}C^{18}O_2$	0	1640
9.5343699763		vacuum	$^{13}C^{18}O_2$	0	1640
9.5347192086		vacuum	$^{12}C^{17}O_2$	0	1640
9.535557249		vacuum	$^{12}C^{18}O_2$	0	1640
9.5355572495		vacuum	$^{12}C^{18}O_2$	0	1640
9.535971884		vacuum	$^{12}C^{16}O_2$	0	1640
9.5359718858		vacuum	$^{12}C^{16}O_2$	0	1640
9.540509632		vacuum	$^{12}C^{16}O^{18}O$	0	1640
9.5405096371		vacuum	$^{12}C^{16}O^{18}O$	0	1640
9.5407011		vacuum	$^{12}C^{16}O_2$	0	1640
9.54164732		vacuum	$^{12}C^{18}O_2$	0	1640
9.5428600243		vacuum	$^{12}C^{17}O_2$	0	1640
9.543479489		vacuum	$^{13}C^{18}O_2$	0	1640
9.5434794930		vacuum	$^{13}C^{18}O_2$	0	1640
9.5487347124		vacuum	$^{12}C^{16}O^{18}O$	0	1640
9.548734713		vacuum	$^{12}C^{16}O^{18}O$	0	1640
9.5500053321		vacuum	$^{13}C^{16}O_2$	0	1640
9.5510636595		vacuum	$^{12}C^{17}O_2$	0	1640
9.5520008607		vacuum	$^{12}C^{18}O_2$	0	1640

Table 4.1.1—*continued*
Neutral Atom, Ion, and Molecular Gas Lasers Arranged in Order of Increasing Wavelength

Wavelength (μm)	Uncertainty	Medium	Species	Charge	Reference
9.552000862		vacuum	$^{12}C^{18}O_2$	0	1640
9.5524282869		vacuum	$^{12}C^{16}O_2$	0	1640
9.552428290		vacuum	$^{12}C^{16}O_2$	0	1640
9.552795581		vacuum	$^{13}C^{18}O_2$	0	1640
9.5527955805		vacuum	$^{13}C^{18}O_2$	0	1640
9.5567603		vacuum	$^{12}C^{16}O_2$	0	1640
9.557023028		vacuum	$^{12}C^{16}O^{18}O$	0	1640
9.5570230286		vacuum	$^{12}C^{16}O^{18}O$	0	1640
9.55768295		vacuum	$^{12}C^{18}O_2$	0	1640
9.5579534656		vacuum	$^{13}C^{16}O_2$	0	1640
9.5593302580		vacuum	$^{12}C^{17}O_2$	0	1640
9.5623201609		vacuum	$^{13}C^{18}O_2$	0	1640
9.562320165		vacuum	$^{13}C^{18}O_2$	0	1640
9.5653747346		vacuum	$^{12}C^{16}O^{18}O$	0	1640
9.565374739		vacuum	$^{12}C^{16}O^{18}O$	0	1640
9.5661640962		vacuum	$^{13}C^{16}O_2$	0	1640
9.56736		vacuum	H_2O	0	1284,1329–31
9.5676599674		vacuum	$^{12}C^{17}O_2$	0	1640
9.568664397		vacuum	$^{12}C^{18}O_2$	0	1640
9.5686643997		vacuum	$^{12}C^{18}O_2$	0	1640
9.569179556		vacuum	$^{12}C^{16}O_2$	0	1640
9.5691795595		vacuum	$^{12}C^{16}O_2$	0	1640
9.5720551385		vacuum	$^{13}C^{18}O_2$	0	1640
9.572055143		vacuum	$^{13}C^{18}O_2$	0	1640
9.5731220		vacuum	$^{12}C^{16}O_2$	0	1640
9.5737899740		vacuum	$^{12}C^{16}O^{18}O$	0	1640
9.573789978		vacuum	$^{12}C^{16}O^{18}O$	0	1640
9.57394187		vacuum	$^{12}C^{18}O_2$	0	1640
9.5746401430		vacuum	$^{13}C^{16}O_2$	0	1640
9.5760529309		vacuum	$^{12}C^{17}O_2$	0	1640
9.5820024033		vacuum	$^{13}C^{18}O_2$	0	1640
9.582002412		vacuum	$^{13}C^{18}O_2$	0	1640
9.5822688902		vacuum	$^{12}C^{16}O^{18}O$	0	1640
9.582268893		vacuum	$^{12}C^{16}O^{18}O$	0	1640
9.5833845196		vacuum	$^{13}C^{16}O_2$	0	1640

Table 4.1.1—*continued*
Neutral Atom, Ion, and Molecular Gas Lasers Arranged in Order of Increasing Wavelength

Wavelength (μm)	Uncertainty	Medium	Species	Charge	Reference
9.5845092900		vacuum	$^{12}C^{17}O_2$	0	1640
9.585548793		vacuum	$^{12}C^{18}O_2$	0	1640
9.5855487957		vacuum	$^{12}C^{18}O_2$	0	1640
9.586227377		vacuum	$^{12}C^{16}O_2$	0	1640
9.5862273766		vacuum	$^{12}C^{16}O_2$	0	1640
9.5897854		vacuum	$^{12}C^{16}O_2$	0	1640
9.5908116271		vacuum	$^{12}C^{16}O^{18}O$	0	1640
9.590811628		vacuum	$^{12}C^{16}O^{18}O$	0	1640
9.5921638277		vacuum	$^{13}C^{18}O_2$	0	1640
9.592163831		vacuum	$^{13}C^{18}O_2$	0	1640
9.5924001280		vacuum	$^{13}C^{16}O_2$	0	1640
9.5930291871		vacuum	$^{12}C^{17}O_2$	0	1640
9.599418318		vacuum	$^{12}C^{16}O^{18}O$	0	1640
9.5994183247		vacuum	$^{12}C^{16}O^{18}O$	0	1640
9.6			$^{14}NH_3$	0	1189,1210
9.6016127618		vacuum	$^{12}C^{17}O_2$	0	1640
9.6016898728		vacuum	$^{13}C^{16}O_2$	0	1640
9.601689891		vacuum	$^{13}C^{16}O_2$	0	1640
9.6025412654		vacuum	$^{13}C^{18}O_2$	0	1640
9.602541272		vacuum	$^{13}C^{18}O_2$	0	1640
9.6026549722		vacuum	$^{12}C^{18}O_2$	0	1640
9.602654976		vacuum	$^{12}C^{18}O_2$	0	1640
9.603573384		vacuum	$^{12}C^{16}O_2$	0	1640
9.6035733836		vacuum	$^{12}C^{16}O_2$	0	1640
9.6067533		vacuum	$^{12}C^{16}O_2$	0	1640
9.608089110		vacuum	$^{12}C^{16}O^{18}O$	0	1640
9.608089129		vacuum	$^{12}C^{16}O^{18}O$	0	1640
9.6102601521		vacuum	$^{12}C^{17}O_2$	0	1640
9.6112566121		vacuum	$^{13}C^{16}O_2$	0	1640
9.611256627		vacuum	$^{13}C^{16}O_2$	0	1640
9.6131365529		vacuum	$^{13}C^{18}O_2$	0	1640
9.613136556		vacuum	$^{13}C^{18}O_2$	0	1640
9.6189714956		vacuum	$^{12}C^{17}O_2$	0	1640
9.619983846		vacuum	$^{12}C^{18}O_2$	0	1640
9.6199838464		vacuum	$^{12}C^{18}O_2$	0	1640

Table 4.1.1—*continued*
Neutral Atom, Ion, and Molecular Gas Lasers Arranged in Order of Increasing Wavelength

Wavelength (μm)	Uncertainty	Medium	Species	Charge	Reference
9.6211032437		vacuum	$^{13}C^{16}O_2$	0	1640
9.621103251		vacuum	$^{13}C^{16}O_2$	0	1640
9.621219191		vacuum	$^{12}C^{16}O_2$	0	1640
9.6212191919		vacuum	$^{12}C^{16}O_2$	0	1640
9.6239515072		vacuum	$^{13}C^{18}O_2$	0	1640
9.623951515		vacuum	$^{13}C^{18}O_2$	0	1640
9.6240268		vacuum	$^{12}C^{16}O_2$	0	1640
9.6277469294		vacuum	$^{12}C^{17}O_2$	0	1640
9.6312326197		vacuum	$^{13}C^{16}O_2$	0	1640
9.631232623		vacuum	$^{13}C^{16}O_2$	0	1640
9.634987925		vacuum	$^{13}C^{18}O_2$	0	1640
9.6349879248		vacuum	$^{13}C^{18}O_2$	0	1640
9.6365865884		vacuum	$^{12}C^{17}O_2$	0	1640
9.637536335		vacuum	$^{12}C^{18}O_2$	0	1640
9.6375363360		vacuum	$^{12}C^{18}O_2$	0	1640
9.639		air	Kr	0	1498
9.6391663843		vacuum	$^{12}C^{16}O_2$	0	1640
9.639166385		vacuum	$^{12}C^{16}O_2$	0	1640
9.6416089		vacuum	$^{12}C^{16}O_2$	0	1640
9.641647583		vacuum	$^{13}C^{16}O_2$	0	1640
9.6416475855		vacuum	$^{13}C^{16}O_2$	0	1640
9.643		vacuum	$^{14}NH_3$	0	1774
9.6454906048		vacuum	$^{12}C^{17}O_2$	0	1640
9.646247580		vacuum	$^{13}C^{18}O_2$	0	1640
9.6462475823		vacuum	$^{13}C^{18}O_2$	0	1640
9.652350966		vacuum	$^{13}C^{16}O_2$	0	1640
9.6523509720		vacuum	$^{13}C^{16}O_2$	0	1640
9.6544591124		vacuum	$^{12}C^{17}O_2$	0	1640
9.6553133551		vacuum	$^{12}C^{18}O_2$	0	1640
9.655313356		vacuum	$^{12}C^{18}O_2$	0	1640
9.6574165094		vacuum	$^{12}C^{16}O_2$	0	1640
9.657416515		vacuum	$^{12}C^{16}O_2$	0	1640
9.657732230		vacuum	$^{13}C^{18}O_2$	0	1640
9.6577322334		vacuum	$^{13}C^{18}O_2$	0	1640
9.6594997		vacuum	$^{12}C^{16}O_2$	0	1640

Table 4.1.1—*continued*
Neutral Atom, Ion, and Molecular Gas Lasers
Arranged in Order of Increasing Wavelength

Wavelength (μm)	Uncertainty	Medium	Species	Charge	Reference
9.6609222307		vacuum	$^{13}C^{16}O^{18}O$	0	1640
9.663345593		vacuum	$^{13}C^{16}O_2$	0	1640
9.6633455966		vacuum	$^{13}C^{16}O_2$	0	1640
9.6634922422		vacuum	$^{12}C^{17}O_2$	0	1640
9.6665738784		vacuum	$^{13}C^{16}O^{18}O$	0	1640
9.669443611		vacuum	$^{13}C^{18}O_2$	0	1640
9.6694436138		vacuum	$^{13}C^{18}O_2$	0	1640
9.6722905548		vacuum	$^{13}C^{16}O^{18}O$	0	1640
9.6725901242		vacuum	$^{12}C^{17}O_2$	0	1640
9.674634252		vacuum	$^{13}C^{16}O_2$	0	1640
9.6746342589		vacuum	$^{13}C^{16}O_2$	0	1640
9.6759710879		vacuum	$^{12}C^{16}O_2$	0	1640
9.675971090		vacuum	$^{12}C^{16}O_2$	0	1640
9.6777025		vacuum	$^{12}C^{16}O_2$	0	1640
9.6780725408		vacuum	$^{13}C^{16}O^{18}O$	0	1640
9.681383425		vacuum	$^{13}C^{18}O_2$	0	1640
9.6813834329		vacuum	$^{13}C^{18}O_2$	0	1640
9.6817528890		vacuum	$^{12}C^{17}O_2$	0	1640
9.6839201195		vacuum	$^{13}C^{16}O^{18}O$	0	1640
9.686219735		vacuum	$^{13}C^{16}O_2$	0	1640
9.6862197410		vacuum	$^{13}C^{16}O_2$	0	1640
9.6898335717		vacuum	$^{13}C^{16}O^{18}O$	0	1640
9.6909806643		vacuum	$^{12}C^{17}O_2$	0	1640
9.693553374		vacuum	$^{13}C^{18}O_2$	0	1640
9.6935533804		vacuum	$^{13}C^{18}O_2$	0	1640
9.6948316089		vacuum	$^{12}C^{16}O_2$	0	1640
9.694831616		vacuum	$^{12}C^{16}O_2$	0	1640
9.6958131757		vacuum	$^{13}C^{16}O^{18}O$	0	1640
9.6962175		vacuum	$^{12}C^{16}O_2$	0	1640
9.698104805		vacuum	$^{13}C^{16}O_2$	0	1640
9.6981048064		vacuum	$^{13}C^{16}O_2$	0	1640
9.7			$^{14}NH_3$	0	1189,1210
9.7002735768		vacuum	$^{12}C^{17}O_2$	0	1640
9.7018592092		vacuum	$^{13}C^{16}O^{18}O$	0	1640
9.7031910		vacuum	Xe	0	559,582

Table 4.1.1—*continued*
Neutral Atom, Ion, and Molecular Gas Lasers Arranged in Order of Increasing Wavelength

Wavelength (μm)	Uncertainty	Medium	Species	Charge	Reference
9.705955110		vacuum	$^{13}C^{18}O_2$	0	1640
9.7059551215		vacuum	$^{13}C^{18}O_2$	0	1640
9.7079719484		vacuum	$^{13}C^{16}O^{18}O$	0	1640
9.7096317526		vacuum	$^{12}C^{17}O_2$	0	1640
9.710292190		vacuum	$^{13}C^{16}O_2$	0	1640
9.7102922000		vacuum	$^{13}C^{16}O_2$	0	1640
9.7139995302		vacuum	$^{12}C^{16}O_2$	0	1640
9.713999532		vacuum	$^{12}C^{16}O_2$	0	1640
9.7150460		vacuum	$^{12}C^{16}O_2$	0	1640
9.719055321		vacuum	$^{12}C^{17}O_2$	0	1640
9.722784635		vacuum	$^{13}C^{16}O_2$	0	1640
9.7227846476		vacuum	$^{13}C^{16}O_2$	0	1640
9.728544397		vacuum	$^{12}C^{17}O_2$	0	1640
9.7334762812		vacuum	$^{12}C^{16}O_2$	0	1640
9.733476288		vacuum	$^{12}C^{16}O_2$	0	1640
9.7341914		vacuum	$^{12}C^{16}O_2$	0	1640
9.735584845		vacuum	$^{13}C^{16}O_2$	0	1640
9.7355848547		vacuum	$^{13}C^{16}O_2$	0	1640
9.737		vacuum	$^{14}NH_3$	0	1640
9.747		vacuum	CH_3F	0	1218
9.748695495		vacuum	$^{13}C^{16}O_2$	0	1640
9.7486955049		vacuum	$^{13}C^{16}O_2$	0	1640
9.753263263		vacuum	$^{12}C^{16}O_2$	0	1640
9.7532632627		vacuum	$^{12}C^{16}O_2$	0	1640
9.7536530		vacuum	$^{12}C^{16}O_2$	0	1640
9.762119246		vacuum	$^{13}C^{16}O_2$	0	1640
9.7621192600		vacuum	$^{13}C^{16}O_2$	0	1640
9.764674832		vacuum	$^{13}C^{18}O_2$	0	1640
9.7646748371		vacuum	$^{13}C^{18}O_2$	0	1640
9.7733618462		vacuum	$^{12}C^{16}O_2$	0	1640
9.773361847		vacuum	$^{12}C^{16}O_2$	0	1640
9.775858749		vacuum	$^{13}C^{16}O_2$	0	1640
9.7758587613		vacuum	$^{13}C^{16}O_2$	0	1640
9.778380123		vacuum	$^{13}C^{18}O_2$	0	1640
9.7783801282		vacuum	$^{13}C^{18}O_2$	0	1640

Table 4.1.1—*continued*
Neutral Atom, Ion, and Molecular Gas Lasers Arranged in Order of Increasing Wavelength

Wavelength (μm)	Uncertainty	Medium	Species	Charge	Reference
9.789916613		vacuum	$^{13}C^{16}O_2$	0	1640
9.7899166255		vacuum	$^{13}C^{16}O_2$	0	1640
9.792327434		vacuum	$^{13}C^{18}O_2$	0	1640
9.7923274443		vacuum	$^{13}C^{18}O_2$	0	1640
9.7935337		vacuum	$^{12}C^{16}O_2$	0	1640
9.7937733740		vacuum	$^{12}C^{16}O_2$	0	1640
9.793773376		vacuum	$^{12}C^{16}O_2$	0	1640
9.806518258		vacuum	$^{13}C^{18}O_2$	0	1640
9.8065182635		vacuum	$^{13}C^{18}O_2$	0	1640
9.8139549		vacuum	$^{12}C^{16}O_2$	0	1640
9.8144991624		vacuum	$^{12}C^{16}O_2$	0	1640
9.814499164		vacuum	$^{12}C^{16}O_2$	0	1640
9.820954030		vacuum	$^{13}C^{18}O_2$	0	1640
9.8209540406		vacuum	$^{13}C^{18}O_2$	0	1640
9.835540498		vacuum	$^{12}C^{16}O_2$	0	1640
9.8355404997		vacuum	$^{12}C^{16}O_2$	0	1640
9.835636201		vacuum	$^{13}C^{18}O_2$	0	1640
9.8356362025		vacuum	$^{13}C^{18}O_2$	0	1640
9.850566153		vacuum	$^{13}C^{18}O_2$	0	1640
9.8505661530		vacuum	$^{13}C^{18}O_2$	0	1640
9.8568986484		vacuum	$^{12}C^{16}O_2$	0	1640
9.856898649		vacuum	$^{12}C^{16}O_2$	0	1640
9.8657452671		vacuum	$^{13}C^{18}O_2$	0	1640
9.865745271		vacuum	$^{13}C^{18}O_2$	0	1640
9.8730403614		vacuum	$^{13}C^{16}O_2$	0	1640
9.873040362		vacuum	$^{13}C^{16}O_2$	0	1640
9.8785748459		vacuum	$^{12}C^{16}O_2$	0	1640
9.878574860		vacuum	$^{12}C^{16}O_2$	0	1640
9.8811748957		vacuum	$^{13}C^{18}O_2$	0	1640
9.881174900		vacuum	$^{13}C^{18}O_2$	0	1640
9.8892297575		vacuum	$^{13}C^{16}O_2$	0	1640
9.889229768		vacuum	$^{13}C^{16}O_2$	0	1640
9.89073556		vacuum	$^{14}C^{18}O_2$	0	1640
9.89073556		vacuum	$^{14}C^{18}O_2$	0	1640
9.8968563624		vacuum	$^{13}C^{18}O_2$	0	1640

Table 4.1.1—*continued*
Neutral Atom, Ion, and Molecular Gas Lasers Arranged in Order of Increasing Wavelength

Wavelength (μm)	Uncertainty	Medium	Species	Charge	Reference
9.896856368		vacuum	$^{13}C^{18}O_2$	0	1640
9.898433863		vacuum	$^{14}C^{18}O_2$	0	1640
9.898433882		vacuum	$^{14}C^{18}O_2$	0	1640
9.9			$^{14}NH_3$	0	1189,1210
9.9005703050		vacuum	$^{12}C^{16}O_2$	0	1640
9.900570325		vacuum	$^{12}C^{16}O_2$	0	1640
9.9057563881		vacuum	$^{13}C^{16}O_2$	0	1640
9.905756396		vacuum	$^{13}C^{16}O_2$	0	1640
9.906357727		vacuum	$^{14}C^{18}O_2$	0	1640
9.906357736		vacuum	$^{14}C^{18}O_2$	0	1640
9.9127909665		vacuum	$^{13}C^{18}O_2$	0	1640
9.912790972		vacuum	$^{13}C^{18}O_2$	0	1640
9.914509593		vacuum	$^{14}C^{18}O_2$	0	1640
9.914509593		vacuum	$^{14}C^{18}O_2$	0	1640
9.917877180		vacuum	$^{14}C^{16}O_2$	0	1640
9.9178771801		vacuum	$^{14}C^{16}O_2$	0	1640
9.921		vacuum	$^{14}NH_3$	0	1189,1210
9.9226226651		vacuum	$^{13}C^{16}O_2$	0	1640
9.922622674		vacuum	$^{13}C^{16}O_2$	0	1640
9.9228862138		vacuum	$^{12}C^{16}O_2$	0	1640
9.922886254		vacuum	$^{12}C^{16}O_2$	0	1640
9.922891926		vacuum	$^{14}C^{18}O_2$	0	1640
9.922891936		vacuum	$^{14}C^{18}O_2$	0	1640
9.9236744988		vacuum	$^{13}C^{16}O^{18}O$	0	1640
9.9273316267		vacuum	$^{14}C^{16}O_2$	0	1640
9.927331635		vacuum	$^{14}C^{16}O_2$	0	1640
9.9289799801		vacuum	$^{13}C^{18}O_2$	0	1640
9.928979992		vacuum	$^{13}C^{18}O_2$	0	1640
9.9315071908		vacuum	$^{14}C^{18}O_2$	0	1640
9.931507202		vacuum	$^{14}C^{18}O_2$	0	1640
9.9319829837		vacuum	$^{13}C^{16}O^{18}O$	0	1640
9.9370891511		vacuum	$^{14}C^{16}O_2$	0	1640
9.937089163		vacuum	$^{14}C^{16}O_2$	0	1640
9.9398309766		vacuum	$^{13}C^{16}O_2$	0	1640
9.939830992		vacuum	$^{13}C^{16}O_2$	0	1640

Table 4.1.1—*continued*
Neutral Atom, Ion, and Molecular Gas Lasers Arranged in Order of Increasing Wavelength

Wavelength (μm)	Uncertainty	Medium	Species	Charge	Reference
9.9403578213		vacuum	$^{14}C^{18}O_2$	0	1640
9.9403661154		vacuum	$^{13}C^{16}O^{18}O$	0	1640
9.9454246503		vacuum	$^{13}C^{18}O_2$	0	1640
9.945424661		vacuum	$^{13}C^{18}O_2$	0	1640
9.9455237398		vacuum	$^{12}C^{16}O_2$	0	1640
9.94552379		vacuum	$^{12}C^{16}O_2$	0	1640
9.9471530250		vacuum	$^{14}C^{16}O_2$	0	1640
9.947153044		vacuum	$^{14}C^{16}O_2$	0	1640
9.9488241143		vacuum	$^{13}C^{16}O^{18}O$	0	1640
9.9494462448		vacuum	$^{14}C^{18}O_2$	0	1640
9.949446257		vacuum	$^{14}C^{18}O_2$	0	1640
9.9573571956		vacuum	$^{13}C^{16}O^{18}O$	0	1640
9.9573836833		vacuum	$^{13}C^{16}O_2$	0	1640
9.957383702		vacuum	$^{13}C^{16}O_2$	0	1640
9.9575265191		vacuum	$^{14}C^{16}O_2$	0	1640
9.957526539		vacuum	$^{14}C^{16}O_2$	0	1640
9.9587748736		vacuum	$^{14}C^{18}O_2$	0	1640
9.958774894		vacuum	$^{14}C^{18}O_2$	0	1640
9.9621262008		vacuum	$^{13}C^{18}O_2$	0	1640
9.962126218		vacuum	$^{13}C^{18}O_2$	0	1640
9.9659655768		vacuum	$^{13}C^{16}O^{18}O$	0	1640
9.9682128982		vacuum	$^{14}C^{16}O_2$	0	1640
9.968212914		vacuum	$^{14}C^{16}O_2$	0	1640
9.9683461051		vacuum	$^{14}C^{18}O_2$	0	1640
9.968346125		vacuum	$^{14}C^{18}O_2$	0	1640
9.9684846		vacuum	$^{12}C^{16}O_2$	0	1640
9.9684846		vacuum	$^{12}C^{16}O_2$	0	1640
9.9746494693		vacuum	$^{13}C^{16}O^{18}O$	0	1640
9.9752831248		vacuum	$^{13}C^{16}O_2$	0	1640
9.975283144		vacuum	$^{13}C^{16}O_2$	0	1640
9.9781623208		vacuum	$^{14}C^{18}O_2$	0	1640
9.978162343		vacuum	$^{14}C^{18}O_2$	0	1640
9.9790858269		vacuum	$^{13}C^{18}O_2$	0	1640
9.979085842		vacuum	$^{13}C^{18}O_2$	0	1640
9.9792154208		vacuum	$^{14}C^{16}O_2$	0	1640

Table 4.1.1—*continued*
Neutral Atom, Ion, and Molecular Gas Lasers Arranged in Order of Increasing Wavelength

Wavelength (μm)	Uncertainty	Medium	Species	Charge	Reference
9.979215440		vacuum	$^{14}C^{16}O_2$	0	1640
9.9834090856		vacuum	$^{13}C^{16}O^{18}O$	0	1640
9.98393525		vacuum	$^{13}C^{16}O_2$	0	1640
9.9882258833		vacuum	$^{14}C^{18}O_2$	0	1640
9.988225899		vacuum	$^{14}C^{18}O_2$	0	1640
9.9905373357		vacuum	$^{14}C^{16}O_2$	0	1640
9.990537353		vacuum	$^{14}C^{16}O_2$	0	1640
9.9917688		vacuum	$^{12}C^{16}O_2$	0	1640
9.9917688		vacuum	$^{12}C^{16}O_2$	0	1640
9.9922446352		vacuum	$^{13}C^{16}O^{18}O$	0	1640
9.9935316127		vacuum	$^{13}C^{16}O_2$	0	1640
9.993531627		vacuum	$^{13}C^{16}O_2$	0	1640
9.9963047030		vacuum	$^{13}C^{18}O_2$	0	1640
9.996304716		vacuum	$^{13}C^{18}O_2$	0	1640
9.9985391414		vacuum	$^{14}C^{18}O_2$	0	1640
9.998539153		vacuum	$^{14}C^{18}O_2$	0	1640
~10		vacuum	COF_2	0	1337
10.00115632		vacuum	$^{13}C^{16}O^{18}O$	0	1640
10.002016		vacuum	$^{13}C^{16}O_2$	0	1640
10.00218188		vacuum	$^{14}C^{16}O_2$	0	1640
10.0021819		vacuum	$^{14}C^{16}O_2$	0	1640
10.00910442		vacuum	$^{14}C^{18}O_2$	0	1640
10.0091044		vacuum	$^{14}C^{18}O_2$	0	1640
10.01014436		vacuum	$^{13}C^{16}O^{18}O$	0	1640
10.01213144		vacuum	$^{13}C^{16}O_2$	0	1640
10.0121314		vacuum	$^{13}C^{16}O_2$	0	1640
10.01378398		vacuum	$^{13}C^{18}O_2$	0	1640
10.0137840		vacuum	$^{13}C^{18}O_2$	0	1640
10.01415228		vacuum	$^{14}C^{16}O_2$	0	1640
10.0141523		vacuum	$^{14}C^{16}O_2$	0	1640
10.01538		vacuum	$^{12}C^{16}O_2$	0	1640
10.015380		vacuum	$^{12}C^{16}O_2$	0	1640
10.01920894		vacuum	$^{13}C^{16}O^{18}O$	0	1640
10.01992404		vacuum	$^{14}C^{18}O_2$	0	1640
10.0199240		vacuum	$^{14}C^{18}O_2$	0	1640

Table 4.1.1—*continued*
Neutral Atom, Ion, and Molecular Gas Lasers Arranged in Order of Increasing Wavelength

Wavelength (μm)	Uncertainty	Medium	Species	Charge	Reference
10.025912		vacuum	$^{12}C^{16}O_2$	0	1640
10.025912		vacuum	$^{12}C^{16}O_2$	0	1640
10.02645173		vacuum	$^{14}C^{16}O_2$	0	1640
10.0264517		vacuum	$^{14}C^{16}O_2$	0	1640
10.02835028		vacuum	$^{13}C^{16}O^{18}O$	0	1640
10.03100028		vacuum	$^{14}C^{18}O_2$	0	1640
10.0310003		vacuum	$^{14}C^{18}O_2$	0	1640
10.03108486		vacuum	$^{13}C^{16}O_2$	0	1640
10.0310849		vacuum	$^{13}C^{16}O_2$	0	1640
10.03152477		vacuum	$^{13}C^{18}O_2$	0	1640
10.0315248		vacuum	$^{13}C^{18}O_2$	0	1640
10.033468		vacuum	$^{12}C^{16}O_2$	0	1640
10.033468		vacuum	$^{12}C^{16}O_2$	0	1640
10.0375686		vacuum	$^{13}C^{16}O^{18}O$	0	1640
10.03908344		vacuum	$^{14}C^{16}O_2$	0	1640
10.0390835		vacuum	$^{14}C^{16}O_2$	0	1640
10.04039645		vacuum	$^{12}C^{16}O_2$	0	1640
10.0413232		vacuum	$^{12}C^{16}O_2$	0	1640
10.04132324		vacuum	$^{12}C^{16}O_2$	0	1640
10.04233542		vacuum	$^{14}C^{18}O_2$	0	1640
10.0423354		vacuum	$^{14}C^{18}O_2$	0	1640
10.0468640		vacuum	$^{13}C^{16}O^{18}O$	0	1640
10.0494775		vacuum	$^{12}C^{16}O_2$	0	1640
10.04952819		vacuum	$^{13}C^{18}O_2$	0	1640
10.0495282		vacuum	$^{13}C^{18}O_2$	0	1640
10.05039413		vacuum	$^{13}C^{16}O_2$	0	1640
10.0503941		vacuum	$^{13}C^{16}O_2$	0	1640
10.05205058		vacuum	$^{14}C^{16}O_2$	0	1640
10.0520506		vacuum	$^{14}C^{16}O_2$	0	1640
10.05393170		vacuum	$^{14}C^{18}O_2$	0	1640
10.0539317		vacuum	$^{14}C^{18}O_2$	0	1640
10.0562367		vacuum	$^{13}C^{16}O^{18}O$	0	1640
10.0579292		vacuum	$^{12}C^{16}O_2$	0	1640
10.05792925		vacuum	$^{12}C^{16}O_2$	0	1640
10.063422		vacuum	Ne	0	195,564

Table 4.1.1—*continued*
Neutral Atom, Ion, and Molecular Gas Lasers Arranged in Order of Increasing Wavelength

Wavelength (μm)	Uncertainty	Medium	Species	Charge	Reference
10.06535630		vacuum	$^{14}C^{16}O_2$	0	1640
10.0653563		vacuum	$^{14}C^{16}O_2$	0	1640
10.0656870		vacuum	$^{13}C^{16}O^{18}O$	0	1640
10.06579137		vacuum	$^{14}C^{18}O_2$	0	1640
10.0657914		vacuum	$^{14}C^{18}O_2$	0	1640
10.0666774		vacuum	$^{12}C^{16}O_2$	0	1640
10.06667739		vacuum	$^{12}C^{16}O_2$	0	1640
10.06779531		vacuum	$^{13}C^{18}O_2$	0	1640
10.0677953		vacuum	$^{13}C^{18}O_2$	0	1640
10.07006146		vacuum	$^{13}C^{16}O_2$	0	1640
10.0700615		vacuum	$^{13}C^{16}O_2$	0	1640
10.0752150		vacuum	$^{13}C^{16}O^{18}O$	0	1640
10.0757210		vacuum	$^{12}C^{16}O_2$	0	1640
10.07572104		vacuum	$^{12}C^{16}O_2$	0	1640
10.07791662		vacuum	$^{14}C^{18}O_2$	0	1640
10.0779166		vacuum	$^{14}C^{18}O_2$	0	1640
10.07900373		vacuum	$^{14}C^{16}O_2$	0	1640
10.0790037		vacuum	$^{14}C^{16}O_2$	0	1640
10.0848209		vacuum	$^{13}C^{16}O^{18}O$	0	1640
10.08505944		vacuum	$^{12}C^{16}O_2$	0	1640
10.0850599		vacuum	$^{12}C^{16}O_2$	0	1640
10.08632719		vacuum	$^{13}C^{18}O_2$	0	1640
10.0863272		vacuum	$^{13}C^{18}O_2$	0	1640
10.08761317		vacuum	$^{12}C^{16}O^{18}O$	0	1640
10.087613		vacuum	$^{12}C^{16}O^{18}O$	0	1640
10.0880271		vacuum	$^{12}C^{18}O_2$	0	1640
10.08802714		vacuum	$^{12}C^{18}O_2$	0	1640
10.09008904		vacuum	$^{13}C^{16}O_2$	0	1640
10.0900890		vacuum	$^{13}C^{16}O_2$	0	1640
10.09030964		vacuum	$^{14}C^{18}O_2$	0	1640
10.0903096		vacuum	$^{14}C^{18}O_2$	0	1640
10.092287		vacuum	$^{12}C^{16}O^{18}O$	0	1640
10.09228700		vacuum	$^{12}C^{16}O^{18}O$	0	1640
10.09299597		vacuum	$^{14}C^{16}O_2$	0	1640
10.0929960		vacuum	$^{14}C^{16}O_2$	0	1640

Table 4.1.1—*continued*
Neutral Atom, Ion, and Molecular Gas Lasers Arranged in Order of Increasing Wavelength

Wavelength (μm)	Uncertainty	Medium	Species	Charge	Reference
10.0945049		vacuum	$^{13}C^{16}O^{18}O$	0	1640
10.0946919		vacuum	$^{12}C^{16}O_2$	0	1640
10.09469196		vacuum	$^{12}C^{16}O_2$	0	1640
10.095			HF	0	1735
10.0960365		vacuum	$^{12}C^{18}O_2$	0	1640
10.09603650		vacuum	$^{12}C^{18}O_2$	0	1640
10.097035		vacuum	$^{12}C^{16}O^{18}O$	0	1640
10.09703546		vacuum	$^{12}C^{16}O^{18}O$	0	1640
10.101859		vacuum	$^{12}C^{16}O^{18}O$	0	1640
10.10185856		vacuum	$^{12}C^{16}O^{18}O$	0	1640
10.10297260		vacuum	$^{14}C^{18}O_2$	0	1640
10.1029726		vacuum	$^{14}C^{18}O_2$	0	1640
10.1029726		vacuum	$^{14}C^{18}O_2$	0	1640
10.1042671		vacuum	$^{13}C^{16}O^{18}O$	0	1640
10.1043452		vacuum	$^{12}C^{18}O_2$	0	1640
10.10434524		vacuum	$^{12}C^{18}O_2$	0	1640
10.1046181		vacuum	$^{12}C^{16}O_2$	0	1640
10.10461811		vacuum	$^{12}C^{16}O_2$	0	1640
10.10512486		vacuum	$^{13}C^{18}O_2$	0	1640
10.1051249		vacuum	$^{13}C^{18}O_2$	0	1640
10.10675630		vacuum	$^{12}C^{16}O^{18}O$	0	1640
10.1067563		vacuum	$^{12}C^{16}O^{18}O$	0	1640
10.10733613		vacuum	$^{14}C^{16}O_2$	0	1640
10.1073361		vacuum	$^{14}C^{16}O_2$	0	1640
10.11047906		vacuum	$^{13}C^{16}O_2$	0	1640
10.1104791		vacuum	$^{13}C^{16}O_2$	0	1640
10.11172872		vacuum	$^{12}C^{16}O^{18}O$	0	1640
10.1117287		vacuum	$^{12}C^{16}O^{18}O$	0	1640
10.1129535		vacuum	$^{12}C^{18}O_2$	0	1640
10.11295349		vacuum	$^{12}C^{18}O_2$	0	1640
10.11386934		vacuum	$^{12}C^{17}O_2$	0	1640
10.1148375		vacuum	$^{12}C^{16}O_2$	0	1640
10.11483754		vacuum	$^{12}C^{16}O_2$	0	1640
10.11590763		vacuum	$^{14}C^{18}O_2$	0	1640
10.1159076		vacuum	$^{14}C^{18}O_2$	0	1640

Table 4.1.1—*continued*
Neutral Atom, Ion, and Molecular Gas Lasers Arranged in Order of Increasing Wavelength

Wavelength (μm)	Uncertainty	Medium	Species	Charge	Reference
10.11677584		vacuum	$^{12}C^{16}O^{18}O$	0	1640
10.1167758		vacuum	$^{12}C^{16}O^{18}O$	0	1640
10.11882628		vacuum	$^{12}C^{17}O_2$	0	1640
10.1218615		vacuum	$^{12}C^{18}O_2$	0	1640
10.12186153		vacuum	$^{12}C^{18}O_2$	0	1640
10.12189771		vacuum	$^{12}C^{16}O^{18}O$	0	1640
10.1218977		vacuum	$^{12}C^{16}O^{18}O$	0	1640
10.12202725		vacuum	$^{14}C^{16}O_2$	0	1640
10.1220273		vacuum	$^{14}C^{16}O_2$	0	1640
10.12385803		vacuum	$^{12}C^{17}O_2$	0	1640
10.12418933		vacuum	$^{13}C^{18}O_2$	0	1640
10.1241893		vacuum	$^{13}C^{18}O_2$	0	1640
10.1253500		vacuum	$^{12}C^{16}O_2$	0	1640
10.12535001		vacuum	$^{12}C^{16}O_2$	0	1640
10.12615		vacuum	$^{12}C^{16}O_2$	0	1640
10.12709436		vacuum	$^{12}C^{16}O^{18}O$	0	1640
10.1270944		vacuum	$^{12}C^{16}O^{18}O$	0	1640
10.12896464		vacuum	$^{12}C^{17}O_2$	0	1640
10.12911687		vacuum	$^{14}C^{18}O_2$	0	1640
10.1291169		vacuum	$^{14}C^{18}O_2$	0	1640
10.1310697		vacuum	$^{12}C^{18}O_2$	0	1640
10.13106971		vacuum	$^{12}C^{18}O_2$	0	1640
10.13123365		vacuum	$^{13}C^{16}O_2$	0	1640
10.1312337		vacuum	$^{13}C^{16}O_2$	0	1640
10.13236586		vacuum	$^{12}C^{16}O^{18}O$	0	1640
10.1323659		vacuum	$^{12}C^{16}O^{18}O$	0	1640
10.13414614		vacuum	$^{12}C^{17}O_2$	0	1640
10.1361554		vacuum	$^{12}C^{16}O_2$	0	1640
10.13615541		vacuum	$^{12}C^{16}O_2$	0	1640
10.13624		vacuum	$^{12}C^{16}O_2$	0	1640
10.13771226		vacuum	$^{12}C^{16}O^{18}O$	0	1640
10.1377123		vacuum	$^{12}C^{16}O^{18}O$	0	1640
10.13940260		vacuum	$^{12}C^{17}O_2$	0	1640
10.1405785		vacuum	$^{12}C^{18}O_2$	0	1640
10.14057849		vacuum	$^{12}C^{18}O_2$	0	1640

Table 4.1.1—*continued*
Neutral Atom, Ion, and Molecular Gas Lasers Arranged in Order of Increasing Wavelength

Wavelength (μm)	Uncertainty	Medium	Species	Charge	Reference
10.14260239		vacuum	$^{14}C^{18}O_2$	0	1640
10.1426024		vacuum	$^{14}C^{18}O_2$	0	1640
10.14313362		vacuum	$^{12}C^{16}O^{18}O$	0	1640
10.1431336		vacuum	$^{12}C^{16}O^{18}O$	0	1640
10.14352161		vacuum	$^{13}C^{18}O_2$	0	1640
10.1435216		vacuum	$^{13}C^{18}O_2$	0	1640
10.14473407		vacuum	$^{12}C^{17}O_2$	0	1640
10.14662		vacuum	$^{12}C^{16}O_2$	0	1640
10.1472538		vacuum	$^{12}C^{16}O_2$	0	1640
10.14725379		vacuum	$^{12}C^{16}O_2$	0	1640
10.14863003		vacuum	$^{12}C^{16}O^{18}O$	0	1640
10.1486300		vacuum	$^{12}C^{16}O^{18}O$	0	1640
10.15014061		vacuum	$^{12}C^{17}O_2$	0	1640
10.1503884		vacuum	$^{12}C^{18}O_2$	0	1640
10.15038843		vacuum	$^{12}C^{18}O_2$	0	1640
10.15235497		vacuum	$^{13}C^{16}O_2$	0	1640
10.1523550		vacuum	$^{13}C^{16}O_2$	0	1640
10.15420157		vacuum	$^{12}C^{16}O^{18}O$	0	1640
10.1542016		vacuum	$^{12}C^{16}O^{18}O$	0	1640
10.15562232		vacuum	$^{12}C^{17}O_2$	0	1640
10.15636628		vacuum	$^{14}C^{18}O_2$	0	1640
10.1563663		vacuum	$^{14}C^{18}O_2$	0	1640
10.15730		vacuum	$^{12}C^{16}O_2$	0	1640
10.1586453		vacuum	$^{12}C^{16}O_2$	0	1640
10.15864527		vacuum	$^{12}C^{16}O_2$	0	1640
10.15984832		vacuum	$^{12}C^{16}O^{18}O$	0	1640
10.1598483		vacuum	$^{12}C^{16}O^{18}O$	0	1640
10.1605002		vacuum	$^{12}C^{18}O_2$	0	1640
10.16050021		vacuum	$^{12}C^{18}O_2$	0	1640
10.16117926		vacuum	$^{12}C^{17}O_2$	0	1640
10.16312266		vacuum	$^{13}C^{18}O_2$	0	1640
10.16557038		vacuum	$^{12}C^{16}O^{18}O$	0	1640
10.1655704		vacuum	$^{12}C^{16}O^{18}O$	0	1640
10.16681153		vacuum	$^{12}C^{17}O_2$	0	1640
10.16826		vacuum	$^{12}C^{16}O_2$	0	1640

Table 4.1.1—*continued*
Neutral Atom, Ion, and Molecular Gas Lasers Arranged in Order of Increasing Wavelength

Wavelength (μm)	Uncertainty	Medium	Species	Charge	Reference
10.17033015		vacuum	$^{12}C^{16}O_2$	0	1640
10.1703302		vacuum	$^{12}C^{16}O_2$	0	1640
10.1709146		vacuum	$^{12}C^{18}O_2$	0	1640
10.17091460		vacuum	$^{12}C^{18}O_2$	0	1640
10.17136785		vacuum	$^{12}C^{16}O^{18}O$	0	1640
10.1713679		vacuum	$^{12}C^{16}O^{18}O$	0	1640
10.17251922		vacuum	$^{12}C^{17}O_2$	0	1640
10.17384510		vacuum	$^{13}C^{16}O_2$	0	1640
10.1738452		vacuum	$^{13}C^{16}O_2$	0	1640
10.17724084		vacuum	$^{12}C^{16}O^{18}O$	0	1640
10.1772409		vacuum	$^{12}C^{16}O^{18}O$	0	1640
10.17830245		vacuum	$^{12}C^{17}O_2$	0	1640
10.17951		vacuum	$^{12}C^{16}O_2$	0	1640
10.18163248		vacuum	$^{12}C^{18}O_2$	0	1640
10.1816325		vacuum	$^{12}C^{18}O_2$	0	1640
10.1823088		vacuum	$^{12}C^{16}O_2$	0	1640
10.18230883		vacuum	$^{12}C^{16}O_2$	0	1640
10.18299344		vacuum	$^{13}C^{18}O_2$	0	1640
10.18318948		vacuum	$^{12}C^{16}O^{18}O$	0	1640
10.1831895		vacuum	$^{12}C^{16}O^{18}O$	0	1640
10.18416131		vacuum	$^{12}C^{17}O_2$	0	1640
10.18921389		vacuum	$^{12}C^{16}O^{18}O$	0	1640
10.1892139		vacuum	$^{12}C^{16}O^{18}O$	0	1640
10.19009593		vacuum	$^{12}C^{17}O_2$	0	1640
10.19105		vacuum	$^{12}C^{16}O_2$	0	1640
10.1926548		vacuum	$^{12}C^{18}O_2$	0	1640
10.19265483		vacuum	$^{12}C^{18}O_2$	0	1640
10.1945818		vacuum	$^{12}C^{16}O_2$	0	1640
10.19458184		vacuum	$^{12}C^{16}O_2$	0	1640
10.19531420		vacuum	$^{12}C^{16}O^{18}O$	0	1640
10.1953142		vacuum	$^{12}C^{16}O^{18}O$	0	1640
10.19570615		vacuum	$^{13}C^{16}O_2$	0	1640
10.19610644		vacuum	$^{12}C^{17}O_2$	0	1640
10.20149054		vacuum	$^{12}C^{16}O^{18}O$	0	1640
10.2014906		vacuum	$^{12}C^{16}O^{18}O$	0	1640

Table 4.1.1—*continued*
Neutral Atom, Ion, and Molecular Gas Lasers
Arranged in Order of Increasing Wavelength

Wavelength (μm)	Uncertainty	Medium	Species	Charge	Reference
10.202035		vacuum	$^{12}C^{18}O_2$	0	1640
10.20219297		vacuum	$^{12}C^{17}O_2$	0	1640
10.20288		vacuum	$^{12}C^{16}O_2$	0	1640
10.2039827		vacuum	$^{12}C^{18}O_2$	0	1640
10.20398273		vacuum	$^{12}C^{18}O_2$	0	1640
10.20676128		vacuum	$^{14}C^{18}O_2$	0	1640
10.2067613		vacuum	$^{14}C^{18}O_2$	0	1640
10.2071498		vacuum	$^{12}C^{16}O_2$	0	1640
10.20714981		vacuum	$^{12}C^{16}O_2$	0	1640
10.20774308		vacuum	$^{12}C^{16}O^{18}O$	0	1640
10.2077431		vacuum	$^{12}C^{16}O^{18}O$	0	1640
10.20835565		vacuum	$^{12}C^{17}O_2$	0	1640
10.213155		vacuum	$^{12}C^{18}O_2$	0	1640
10.21407196		vacuum	$^{12}C^{16}O^{18}O$	0	1640
10.2140720		vacuum	$^{12}C^{16}O^{18}O$	0	1640
10.21459464		vacuum	$^{12}C^{17}O_2$	0	1640
10.21501		vacuum	$^{12}C^{16}O_2$	0	1640
10.21561737		vacuum	$^{12}C^{18}O_2$	0	1640
10.2156174		vacuum	$^{12}C^{18}O_2$	0	1640
10.21794020		vacuum	$^{13}C^{16}O_2$	0	1640
10.2200135		vacuum	$^{12}C^{16}O_2$	0	1640
10.22001353		vacuum	$^{12}C^{16}O_2$	0	1640
10.22047734		vacuum	$^{12}C^{16}O^{18}O$	0	1640
10.2204774		vacuum	$^{12}C^{16}O^{18}O$	0	1640
10.22091010		vacuum	$^{12}C^{17}O_2$	0	1640
10.22180274		vacuum	$^{14}C^{18}O_2$	0	1640
10.2218028		vacuum	$^{14}C^{18}O_2$	0	1640
10.224579		vacuum	$^{12}C^{18}O_2$	0	1640
10.22695939		vacuum	$^{12}C^{16}O^{18}O$	0	1640
10.2269594		vacuum	$^{12}C^{16}O^{18}O$	0	1640
10.22730217		vacuum	$^{12}C^{17}O_2$	0	1640
10.22742		vacuum	$^{12}C^{16}O_2$	0	1640
10.22756007		vacuum	$^{12}C^{18}O_2$	0	1640
10.2275601		vacuum	$^{12}C^{18}O_2$	0	1640
10.23317390		vacuum	$^{12}C^{16}O_2$	0	1640

Table 4.1.1—*continued*
Neutral Atom, Ion, and Molecular Gas Lasers Arranged in Order of Increasing Wavelength

Wavelength (μm)	Uncertainty	Medium	Species	Charge	Reference
10.2331739		vacuum	$^{12}C^{16}O_2$	0	1640
10.23377105		vacuum	$^{12}C^{17}O_2$	0	1640
10.236308		vacuum	$^{12}C^{18}O_2$	0	1640
10.23713350		vacuum	$^{14}C^{18}O_2$	0	1640
10.2371335		vacuum	$^{14}C^{18}O_2$	0	1640
10.2398122		vacuum	$^{12}C^{18}O_2$	0	1640
10.23981221		vacuum	$^{12}C^{18}O_2$	0	1640
10.24013		vacuum	$^{12}C^{16}O_2$	0	1640
10.24031689		vacuum	$^{12}C^{17}O_2$	0	1640
10.24054929		vacuum	$^{13}C^{16}O_2$	0	1640
10.24370004		vacuum	$^{14}C^{16}O_2$	0	1640
10.2437001		vacuum	$^{14}C^{16}O_2$	0	1640
10.24663193		vacuum	$^{12}C^{16}O_2$	0	1640
10.2466319		vacuum	$^{12}C^{16}O_2$	0	1640
10.24693989		vacuum	$^{12}C^{17}O_2$	0	1640
10.248345		vacuum	$^{12}C^{18}O_2$	0	1640
10.2523753		vacuum	$^{12}C^{18}O_2$	0	1640
10.25237530		vacuum	$^{12}C^{18}O_2$	0	1640
10.25275548		vacuum	$^{14}C^{18}O_2$	0	1640
10.2527555		vacuum	$^{14}C^{18}O_2$	0	1640
10.25314		vacuum	$^{12}C^{16}O_2$	0	1640
10.25364024		vacuum	$^{12}C^{17}O_2$	0	1640
10.26038876		vacuum	$^{12}C^{16}O_2$	0	1640
10.2603888		vacuum	$^{12}C^{16}O_2$	0	1640
10.26041813		vacuum	$^{12}C^{17}O_2$	0	1640
10.260689		vacuum	$^{12}C^{18}O_2$	0	1640
10.26149432		vacuum	$^{14}C^{16}O_2$	0	1640
10.2614943		vacuum	$^{14}C^{16}O_2$	0	1640
10.26353546		vacuum	$^{13}C^{16}O_2$	0	1640
10.26525098		vacuum	$^{12}C^{18}O_2$	0	1640
10.2652510		vacuum	$^{12}C^{18}O_2$	0	1640
10.26643		vacuum	$^{12}C^{16}O_2$	0	1640
10.26727379		vacuum	$^{12}C^{17}O_2$	0	1640
10.26867054		vacuum	$^{14}C^{18}O_2$	0	1640
10.2686706		vacuum	$^{14}C^{18}O_2$	0	1640

Table 4.1.1—*continued*
Neutral Atom, Ion, and Molecular Gas Lasers Arranged in Order of Increasing Wavelength

Wavelength (μm)	Uncertainty	Medium	Species	Charge	Reference
10.273344		vacuum	$^{12}C^{18}O_2$	0	1640
10.27420741		vacuum	$^{12}C^{17}O_2$	0	1640
10.2744457		vacuum	$^{12}C^{16}O_2$	0	1640
10.27444568		vacuum	$^{12}C^{16}O_2$	0	1640
10.2784410		vacuum	$^{12}C^{18}O_2$	0	1640
10.27844096		vacuum	$^{12}C^{18}O_2$	0	1640
10.27966734		vacuum	$^{14}C^{16}O_2$	0	1640
10.2796674		vacuum	$^{14}C^{16}O_2$	0	1640
10.28002		vacuum	$^{12}C^{16}O_2$	0	1640
10.28121923		vacuum	$^{12}C^{17}O_2$	0	1640
10.28488054		vacuum	$^{14}C^{18}O_2$	0	1640
10.2848805		vacuum	$^{14}C^{18}O_2$	0	1640
10.28690074		vacuum	$^{13}C^{16}O_2$	0	1640
10.28880407		vacuum	$^{12}C^{16}O_2$	0	1640
10.2888041		vacuum	$^{12}C^{16}O_2$	0	1640
10.29016		vacuum	$^{14}NH_3$	0	1189,1210
10.29194709		vacuum	$^{12}C^{18}O_2$	0	1640
10.2919471		vacuum	$^{12}C^{18}O_2$	0	1640
10.29391		vacuum	$^{12}C^{16}O_2$	0	1640
10.29822186		vacuum	$^{14}C^{16}O_2$	0	1640
10.2982219		vacuum	$^{14}C^{16}O_2$	0	1640
10.3013873		vacuum	$^{14}C^{18}O_2$	0	1640
10.30138731		vacuum	$^{14}C^{18}O_2$	0	1640
10.30243		vacuum	$^{12}C^{16}O_2$	0	1640
10.30346545		vacuum	$^{12}C^{16}O_2$	0	1640
10.3034655		vacuum	$^{12}C^{16}O_2$	0	1640
10.3057713		vacuum	$^{12}C^{18}O_2$	0	1640
10.30577132		vacuum	$^{12}C^{18}O_2$	0	1640
10.30810		vacuum	$^{12}C^{16}O_2$	0	1640
10.3083028		vacuum	$^{14}N_2O$	0	1177–8,1331–5
10.3156527		vacuum	$^{14}N_2O$	0	1177–8,1331–5
10.31616		vacuum	$^{12}C^{16}O_2$	0	1640
10.31716065		vacuum	$^{14}C^{16}O_2$	0	1640
10.3171607		vacuum	$^{14}C^{16}O_2$	0	1640
10.31819264		vacuum	$^{14}C^{18}O_2$	0	1640

Table 4.1.1—*continued*
Neutral Atom, Ion, and Molecular Gas Lasers Arranged in Order of Increasing Wavelength

Wavelength (μm)	Uncertainty	Medium	Species	Charge	Reference
10.3181927		vacuum	$^{14}C^{18}O_2$	0	1640
10.31843147		vacuum	$^{12}C^{16}O_2$	0	1640
10.3184315		vacuum	$^{12}C^{16}O_2$	0	1640
10.32258		vacuum	$^{12}C^{16}O_2$	0	1640
10.3230518		vacuum	$^{14}N_2O$	0	1177–8,1331–5
10.33018		vacuum	$^{12}C^{16}O_2$	0	1640
10.3305000		vacuum	$^{14}N_2O$	0	1177–8,1331–5
10.33370389		vacuum	$^{12}C^{16}O_2$	0	1640
10.3337039		vacuum	$^{12}C^{16}O_2$	0	1640
10.3352983		vacuum	$^{14}C^{18}O_2$	0	1640
10.33529832		vacuum	$^{14}C^{18}O_2$	0	1640
10.3364864		vacuum	$^{14}C^{16}O_2$	0	1640
10.33648641		vacuum	$^{14}C^{16}O_2$	0	1640
10.33737		vacuum	$^{12}C^{16}O_2$	0	1640
10.3376		vacuum	$^{14}NH_3$	0	1630,1774–75
10.3379976		vacuum	$^{14}N_2O$	0	1177–8,1331–5
10.338			NH_3	0	1774
10.342			NH_3	0	1774
10.3423		vacuum	$^{14}NH_3$	0	1630,1774–75
10.34451		vacuum	$^{12}C^{16}O_2$	0	1640
10.3455446		vacuum	$^{14}N_2O$	0	1177–8,1331–5
10.3492846		vacuum	$^{12}C^{16}O_2$	0	1640
10.34928462		vacuum	$^{12}C^{16}O_2$	0	1640
10.35246		vacuum	$^{12}C^{16}O_2$	0	1640
10.35270609		vacuum	$^{14}C^{18}O_2$	0	1640
10.3527061		vacuum	$^{14}C^{18}O_2$	0	1640
10.3531411		vacuum	$^{14}N_2O$	0	1177–8,1331–5
10.35620187		vacuum	$^{14}C^{16}O_2$	0	1640
10.3562019		vacuum	$^{14}C^{16}O_2$	0	1640
10.3589		vacuum	$^{14}NH_3$	0	1630,1774–75
10.359			NH_3	0	1774
10.35912		vacuum	$^{12}C^{16}O_2$	0	1640
10.3607873		vacuum	$^{14}N_2O$	0	1177–8,1331–5
10.3651757		vacuum	$^{12}C^{16}O_2$	0	1640
10.36517570		vacuum	$^{12}C^{16}O_2$	0	1640

Table 4.1.1—*continued*
Neutral Atom, Ion, and Molecular Gas Lasers Arranged in Order of Increasing Wavelength

Wavelength (μm)	Uncertainty	Medium	Species	Charge	Reference
10.367			NH_3	0	1774
10.3670		vacuum	$^{14}NH_3$	0	1630,1774–75
10.36785		vacuum	$^{12}C^{16}O_2$	0	1640
10.3684833		vacuum	$^{14}N_2O$	0	1177–8,1331–5
10.37041767		vacuum	$^{14}C^{18}O_2$	0	1640
10.3704177		vacuum	$^{14}C^{18}O_2$	0	1640
10.37404		vacuum	$^{12}C^{16}O_2$	0	1640
10.3762292		vacuum	$^{14}N_2O$	0	1177–8,1331–5
10.37630968		vacuum	$^{14}C^{16}O_2$	0	1640
10.3763097		vacuum	$^{14}C^{16}O_2$	0	1640
10.38137928		vacuum	$^{12}C^{16}O_2$	0	1640
10.3813793		vacuum	$^{12}C^{16}O_2$	0	1640
10.38355		vacuum	$^{12}C^{16}O_2$	0	1640
10.3840251		vacuum	$^{14}N_2O$	0	1177–8,1331–5
10.38758858		vacuum	$^{12}C^{18}O_2$	0	1640
10.3875886		vacuum	$^{12}C^{18}O_2$	0	1640
10.38843476		vacuum	$^{14}C^{18}O_2$	0	1640
10.3884348		vacuum	$^{14}C^{18}O_2$	0	1640
10.38926		vacuum	$^{12}C^{16}O_2$	0	1640
10.3918711		vacuum	$^{14}N_2O$	0	1177–8,1331–5
10.3968125		vacuum	$^{14}C^{16}O_2$	0	1640
10.39681250		vacuum	$^{14}C^{16}O_2$	0	1640
10.39789766		vacuum	$^{12}C^{16}O_2$	0	1640
10.3978977		vacuum	$^{12}C^{16}O_2$	0	1640
10.39955		vacuum	$^{12}C^{16}O_2$	0	1640
10.3997674		vacuum	$^{14}N_2O$	0	1177–8,1331–5
10.40312		vacuum	O	0	235,1698
10.4035359		vacuum	$^{12}C^{18}O_2$	0	1640
10.40353594		vacuum	$^{12}C^{18}O_2$	0	1640
10.40477		vacuum	$^{12}C^{16}O_2$	0	1640
10.4067590		vacuum	$^{14}C^{18}O_2$	0	1640
10.40675902		vacuum	$^{14}C^{18}O_2$	0	1640
10.4077141		vacuum	$^{14}N_2O$	0	1177–8,1331–5
10.41105		vacuum	$^{14}N_2O$	0	1179
10.4157114		vacuum	$^{14}N_2O$	0	1177–8,1331–5

Table 4.1.1—*continued*
Neutral Atom, Ion, and Molecular Gas Lasers Arranged in Order of Increasing Wavelength

Wavelength (μm)	Uncertainty	Medium	Species	Charge	Reference
10.41587		vacuum	$^{12}C^{16}O_2$	0	1640
10.41771296		vacuum	$^{14}C^{16}O_2$	0	1640
10.4177130		vacuum	$^{14}C^{16}O_2$	0	1640
10.41887		vacuum	$^{14}N_2O$	0	1179
10.4198197		vacuum	$^{12}C^{18}O_2$	0	1640
10.41981973		vacuum	$^{12}C^{18}O_2$	0	1640
10.4232708		vacuum	$^{12}C^{16}O_2$	0	1640
10.42327084		vacuum	$^{12}C^{16}O_2$	0	1640
10.4237593		vacuum	$^{14}N_2O$	0	1177–8,1331–5
10.42539209		vacuum	$^{14}C^{18}O_2$	0	1640
10.4253921		vacuum	$^{14}C^{18}O_2$	0	1640
10.42674		vacuum	$^{14}N_2O$	0	1179
10.43178618		vacuum	$^{12}C^{16}O^{18}O$	0	1640
10.4317862		vacuum	$^{12}C^{16}O^{18}O$	0	1640
10.4318580		vacuum	$^{14}N_2O$	0	1177–8,1331–5
10.43466		vacuum	$^{14}N_2O$	0	1179
10.4364429		vacuum	$^{12}C^{18}O_2$	0	1640
10.43644294		vacuum	$^{12}C^{18}O_2$	0	1640
10.43901367		vacuum	$^{14}C^{16}O_2$	0	1640
10.4390137		vacuum	$^{14}C^{16}O_2$	0	1640
10.4400076		vacuum	$^{14}N_2O$	0	1177–8,1331–5
10.4405152		vacuum	$^{12}C^{16}O^{18}O$	0	1640
10.44051521		vacuum	$^{12}C^{16}O^{18}O$	0	1640
10.44058717		vacuum	$^{12}C^{16}O_2$	0	1640
10.4405872		vacuum	$^{12}C^{16}O_2$	0	1640
10.441778		vacuum	$^{12}C^{18}O_2$	0	1640
10.44263		vacuum	$^{14}N_2O$	0	1179
10.4443356		vacuum	$^{14}C^{18}O_2$	0	1640
10.44433560		vacuum	$^{14}C^{18}O_2$	0	1640
10.44700792		vacuum	$^{12}C^{17}O_2$	0	1640
10.4482084		vacuum	$^{14}N_2O$	0	1177–8,1331–5
10.4493292		vacuum	$^{12}C^{16}O^{18}O$	0	1640
10.44932925		vacuum	$^{12}C^{16}O^{18}O$	0	1640
10.45065		vacuum	$^{14}N_2O$	0	1179
10.451505		vacuum	Tl	0	175

Table 4.1.1—*continued*
Neutral Atom, Ion, and Molecular Gas Lasers Arranged in Order of Increasing Wavelength

Wavelength (μm)	Uncertainty	Medium	Species	Charge	Reference
10.4534086		vacuum	$^{12}C^{18}O_2$	0	1640
10.45340865		vacuum	$^{12}C^{18}O_2$	0	1640
10.45581237		vacuum	$^{12}C^{17}O_2$	0	1640
10.4564603		vacuum	$^{14}N_2O$	0	1177–8,1331–5
10.45803		vacuum	$^{12}C^{16}O_2$	0	1640
10.4582274		vacuum	$^{12}C^{16}O_2$	0	1640
10.45822742		vacuum	$^{12}C^{16}O_2$	0	1640
10.45822869		vacuum	$^{12}C^{16}O^{18}O$	0	1640
10.4582287		vacuum	$^{12}C^{16}O^{18}O$	0	1640
10.458482		vacuum	$^{12}C^{18}O_2$	0	1640
10.4587136		vacuum	$^{14}N_2O$	0	1179
10.4607172		vacuum	$^{14}C^{16}O_2$	0	1640
10.46071722		vacuum	$^{14}C^{16}O_2$	0	1640
10.4635911		vacuum	$^{14}C^{18}O_2$	0	1640
10.46359112		vacuum	$^{14}C^{18}O_2$	0	1640
10.46470237		vacuum	$^{12}C^{17}O_2$	0	1640
10.4647637		vacuum	$^{14}N_2O$	0	1177–8,1331–5
10.46684		vacuum	$^{14}N_2O$	0	1179
10.46721398		vacuum	$^{12}C^{16}O^{18}O$	0	1640
10.4672140		vacuum	$^{12}C^{16}O^{18}O$	0	1640
10.47072009		vacuum	$^{12}C^{18}O_2$	0	1640
10.4707201		vacuum	$^{12}C^{18}O_2$	0	1640
10.4731186		vacuum	$^{14}N_2O$	0	1177–8,1331–5
10.47367833		vacuum	$^{12}C^{17}O_2$	0	1640
10.47391309		vacuum	$^{13}C^{18}O_2$	0	1640
10.47501		vacuum	$^{14}N_2O$	0	1179
10.47545		vacuum	$^{12}C^{16}O_2$	0	1640
10.475528		vacuum	$^{12}C^{18}O_2$	0	1640
10.4761945		vacuum	$^{12}C^{16}O_2$	0	1640
10.47619452		vacuum	$^{12}C^{16}O_2$	0	1640
10.4762855		vacuum	$^{12}C^{16}O^{18}O$	0	1640
10.47628554		vacuum	$^{12}C^{16}O^{18}O$	0	1640
10.4815252		vacuum	$^{14}N_2O$	0	1177–8,1331–5
10.48274069		vacuum	$^{12}C^{17}O_2$	0	1640
10.48282618		vacuum	$^{14}C^{16}O_2$	0	1640

Table 4.1.1—*continued*
Neutral Atom, Ion, and Molecular Gas Lasers Arranged in Order of Increasing Wavelength

Wavelength (μm)	Uncertainty	Medium	Species	Charge	Reference
10.4828262		vacuum	$^{14}C^{16}O_2$	0	1640
10.4831602		vacuum	$^{14}C^{18}O_2$	0	1640
10.48316022		vacuum	$^{14}C^{18}O_2$	0	1640
10.48322654		vacuum	$^{13}C^{18}O_2$	0	1640
10.48324		vacuum	$^{14}N_2O$	0	1179
10.4854438		vacuum	$^{12}C^{16}O^{18}O$	0	1640
10.48544381		vacuum	$^{12}C^{16}O^{18}O$	0	1640
10.4883806		vacuum	$^{12}C^{18}O_2$	0	1640
10.48838063		vacuum	$^{12}C^{18}O_2$	0	1640
10.4899836		vacuum	$^{14}N_2O$	0	1177–8,1331–5
10.49151		vacuum	$^{14}N_2O$	0	1179
10.49188988		vacuum	$^{12}C^{17}O_2$	0	1640
10.49282497		vacuum	$^{13}C^{18}O_2$	0	1640
10.492919		vacuum	$^{12}C^{18}O_2$	0	1640
10.49319		vacuum	$^{12}C^{16}O_2$	0	1640
10.4944915		vacuum	$^{12}C^{16}O_2$	0	1640
10.49449153		vacuum	$^{12}C^{16}O_2$	0	1640
10.4946892		vacuum	$^{12}C^{16}O^{18}O$	0	1640
10.49468924		vacuum	$^{12}C^{16}O^{18}O$	0	1640
10.4984940		vacuum	$^{14}N_2O$	0	1177–8,1331–5
10.49983		vacuum	$^{14}N_2O$	0	1179
10.50112635		vacuum	$^{12}C^{17}O_2$	0	1640
10.50270827		vacuum	$^{13}C^{18}O_2$	0	1640
10.5027083		vacuum	$^{13}C^{18}O_2$	0	1640
10.5030444		vacuum	$^{14}C^{18}O_2$	0	1640
10.50304443		vacuum	$^{14}C^{18}O_2$	0	1640
10.5040223		vacuum	$^{12}C^{16}O^{18}O$	0	1640
10.50402230		vacuum	$^{12}C^{16}O^{18}O$	0	1640
10.5053431		vacuum	$^{14}C^{16}O_2$	0	1640
10.50534311		vacuum	$^{14}C^{16}O_2$	0	1640
10.50639376		vacuum	$^{12}C^{18}O_2$	0	1640
10.5063938		vacuum	$^{12}C^{18}O_2$	0	1640
10.5067		vacuum	$^{14}NH_3$	0	1630,1774–75
10.507			NH_3	0	1774
10.5070566		vacuum	$^{14}N_2O$	0	1177–8,1331–5

Table 4.1.1—*continued*
Neutral Atom, Ion, and Molecular Gas Lasers Arranged in Order of Increasing Wavelength

Wavelength (μm)	Uncertainty	Medium	Species	Charge	Reference
10.50821		vacuum	$^{14}N_2O$	0	1179
10.51045058		vacuum	$^{12}C^{17}O_2$	0	1640
10.510659		vacuum	$^{12}C^{18}O_2$	0	1640
10.51126		vacuum	$^{12}C^{16}O_2$	0	1640
10.5128764		vacuum	$^{13}C^{18}O_2$	0	1640
10.51287644		vacuum	$^{13}C^{18}O_2$	0	1640
10.51312168		vacuum	$^{12}C^{16}O_2$	0	1640
10.5131217		vacuum	$^{12}C^{16}O_2$	0	1640
10.51344346		vacuum	$^{12}C^{16}O^{18}O$	0	1640
10.5134435		vacuum	$^{12}C^{16}O^{18}O$	0	1640
10.5156714		vacuum	$^{14}N_2O$	0	1177–8,1331–5
10.51664		vacuum	$^{14}N_2O$	0	1179
10.51986303		vacuum	$^{12}C^{17}O_2$	0	1640
10.52295319		vacuum	$^{12}C^{16}O^{18}O$	0	1640
10.5229532		vacuum	$^{12}C^{16}O^{18}O$	0	1640
10.52324528		vacuum	$^{14}C^{18}O_2$	0	1640
10.5232453		vacuum	$^{14}C^{18}O_2$	0	1640
10.52332959		vacuum	$^{13}C^{18}O_2$	0	1640
10.5233296		vacuum	$^{13}C^{18}O_2$	0	1640
10.5243388		vacuum	$^{14}N_2O$	0	1177–8,1331–5
10.5247631		vacuum	$^{12}C^{18}O_2$	0	1640
10.52476313		vacuum	$^{12}C^{18}O_2$	0	1640
10.52512		vacuum	$^{14}N_2O$	0	1179
10.52827056		vacuum	$^{14}C^{16}O_2$	0	1640
10.5282706		vacuum	$^{14}C^{16}O_2$	0	1640
10.528750		vacuum	$^{12}C^{18}O_2$	0	1640
10.52936417		vacuum	$^{12}C^{17}O_2$	0	1640
10.52965		vacuum	$^{12}C^{16}O_2$	0	1640
10.53		vacuum	C_2H_4	0	1194
10.5320883		vacuum	$^{12}C^{16}O_2$	0	1640
10.53208833		vacuum	$^{12}C^{16}O_2$	0	1640
10.53255199		vacuum	$^{12}C^{16}O^{18}O$	0	1640
10.5325520		vacuum	$^{12}C^{16}O^{18}O$	0	1640
10.5330588		vacuum	$^{14}N_2O$	0	1177–8,1331–5
10.53366		vacuum	$^{14}N_2O$	0	1179

Table 4.1.1—*continued*
Neutral Atom, Ion, and Molecular Gas Lasers Arranged in Order of Increasing Wavelength

Wavelength (μm)	Uncertainty	Medium	Species	Charge	Reference
10.53406795		vacuum	$^{13}C^{18}O_2$	0	1640
10.5340680		vacuum	$^{13}C^{18}O_2$	0	1640
10.53895450		vacuum	$^{12}C^{17}O_2$	0	1640
10.5418316		vacuum	$^{14}N_2O$	0	1177–8,1331–5
10.54224		vacuum	$^{14}N_2O$	0	1179
10.54224035		vacuum	$^{12}C^{16}O^{18}O$	0	1640
10.5422404		vacuum	$^{12}C^{16}O^{18}O$	0	1640
10.5434925		vacuum	$^{12}C^{18}O_2$	0	1640
10.54349252		vacuum	$^{12}C^{18}O_2$	0	1640
10.54376425		vacuum	$^{14}C^{18}O_2$	0	1640
10.5437643		vacuum	$^{14}C^{18}O_2$	0	1640
10.5450918		vacuum	$^{13}C^{18}O_2$	0	1640
10.54509183		vacuum	$^{13}C^{18}O_2$	0	1640
10.5459		vacuum	$^{14}NH_3$	0	1630,1774–75
10.54594		vacuum	$^{14}NH_3$	0	1189,1210
10.546			NH_3	0	1774
10.547198		vacuum	$^{12}C^{18}O_2$	0	1640
10.54838		vacuum	$^{12}C^{16}O_2$	0	1640
10.54863453		vacuum	$^{12}C^{17}O_2$	0	1640
10.5506575		vacuum	$^{14}N_2O$	0	1177–8,1331–5
10.5508756		vacuum	$^{14}N_2O$	0	1179
10.5513950		vacuum	$^{12}C^{16}O_2$	0	1640
10.55139503		vacuum	$^{12}C^{16}O_2$	0	1640
10.5520188		vacuum	$^{12}C^{16}O^{18}O$	0	1640
10.55201880		vacuum	$^{12}C^{16}O^{18}O$	0	1640
10.55640169		vacuum	$^{13}C^{18}O_2$	0	1640
10.5564017		vacuum	$^{13}C^{18}O_2$	0	1640
10.55840475		vacuum	$^{12}C^{17}O_2$	0	1640
10.5595364		vacuum	$^{14}N_2O$	0	1177–8,1331–5
10.56188785		vacuum	$^{12}C^{16}O^{18}O$	0	1640
10.5618879		vacuum	$^{12}C^{16}O^{18}O$	0	1640
10.56258585		vacuum	$^{12}C^{18}O_2$	0	1640
10.5625859		vacuum	$^{12}C^{18}O_2$	0	1640
10.56460279		vacuum	$^{14}C^{18}O_2$	0	1640
10.5646028		vacuum	$^{14}C^{18}O_2$	0	1640

Table 4.1.1—*continued*
Neutral Atom, Ion, and Molecular Gas Lasers Arranged in Order of Increasing Wavelength

Wavelength (μm)	Uncertainty	Medium	Species	Charge	Reference
10.566004		vacuum	$^{12}C^{18}O_2$	0	1640
10.56744		vacuum	$^{12}C^{16}O_2$	0	1640
10.56799805		vacuum	$^{13}C^{18}O_2$	0	1640
10.5679981		vacuum	$^{13}C^{18}O_2$	0	1640
10.56826570		vacuum	$^{12}C^{17}O_2$	0	1640
10.56830		vacuum	$^{14}N_2O$	0	1179
10.5684688		vacuum	$^{14}N_2O$	0	1177–8,1331–5
10.5710454		vacuum	$^{12}C^{16}O_2$	0	1640
10.57104545		vacuum	$^{12}C^{16}O_2$	0	1640
10.5718480		vacuum	$^{12}C^{16}O^{18}O$	0	1640
10.57184804		vacuum	$^{12}C^{16}O^{18}O$	0	1640
10.5758593		vacuum	$^{13}C^{16}O^{18}O$	0	1640
10.57710		vacuum	$^{14}N_2O$	0	1179
10.5774546		vacuum	$^{14}N_2O$	0	1177–8,1331–5
10.57821790		vacuum	$^{12}C^{17}O_2$	0	1640
10.57988156		vacuum	$^{13}C^{18}O_2$	0	1640
10.5798816		vacuum	$^{13}C^{18}O_2$	0	1640
10.5818999		vacuum	$^{12}C^{16}O^{18}O$	0	1640
10.58189990		vacuum	$^{12}C^{16}O^{18}O$	0	1640
10.5820472		vacuum	$^{12}C^{18}O_2$	0	1640
10.58204723		vacuum	$^{12}C^{18}O_2$	0	1640
10.58205091		vacuum	$^{13}C^{16}O^{18}O$	0	1640
10.585175		vacuum	$^{12}C^{18}O_2$	0	1640
10.58576235		vacuum	$^{14}C^{18}O_2$	0	1640
10.5857624		vacuum	$^{14}C^{18}O_2$	0	1640
10.58595		vacuum	$^{14}N_2O$	0	1179
10.5864941		vacuum	$^{14}N_2O$	0	1177–8,1331–5
10.58684		vacuum	$^{12}C^{16}O_2$	0	1640
10.58826190		vacuum	$^{12}C^{17}O_2$	0	1640
10.58831203		vacuum	$^{13}C^{16}O^{18}O$	0	1640
10.5910255		vacuum	$^{12}C^{16}O_2$	0	1640
10.59104346		vacuum	$^{12}C^{16}O_2$	0	1640
10.5910435		vacuum	$^{12}C^{16}O_2$	0	1640
10.59204399		vacuum	$^{12}C^{16}O^{18}O$	0	1640
10.5920440		vacuum	$^{12}C^{16}O^{18}O$	0	1640

Table 4.1.1—*continued*
Neutral Atom, Ion, and Molecular Gas Lasers Arranged in Order of Increasing Wavelength

Wavelength (μm)	Uncertainty	Medium	Species	Charge	Reference
10.59205299		vacuum	$^{13}C^{18}O_2$	0	1640
10.5920530		vacuum	$^{13}C^{18}O_2$	0	1640
10.59464276		vacuum	$^{13}C^{16}O^{18}O$	0	1640
10.59485		vacuum	$^{14}N_2O$	0	1179
10.5955875		vacuum	$^{14}N_2O$	0	1177–8,1331–5
10.59839824		vacuum	$^{12}C^{17}O_2$	0	1640
10.6			$^{14}NH_3$	0	1189,1210
10.6006284		vacuum	$^{13}C^{16}O_2$	0	1640
10.60062840		vacuum	$^{13}C^{16}O_2$	0	1640
10.60104316		vacuum	$^{13}C^{16}O^{18}O$	0	1640
10.60188087		vacuum	$^{12}C^{18}O_2$	0	1640
10.6018809		vacuum	$^{12}C^{18}O_2$	0	1640
10.60228087		vacuum	$^{12}C^{16}O^{18}O$	0	1640
10.6022809		vacuum	$^{12}C^{16}O^{18}O$	0	1640
10.60380		vacuum	$^{14}N_2O$	0	1179
10.60451318		vacuum	$^{13}C^{18}O_2$	0	1640
10.6045132		vacuum	$^{13}C^{18}O_2$	0	1640
10.604713		vacuum	$^{12}C^{18}O_2$	0	1640
10.6047350		vacuum	$^{14}N_2O$	0	1177–8,1331–5
10.60658		vacuum	$^{12}C^{16}O_2$	0	1640
10.6072443		vacuum	$^{14}C^{18}O_2$	0	1640
10.60724433		vacuum	$^{14}C^{18}O_2$	0	1640
10.60751330		vacuum	$^{13}C^{16}O^{18}O$	0	1640
10.60862750		vacuum	$^{12}C^{17}O_2$	0	1640
10.61139307		vacuum	$^{12}C^{16}O_2$	0	1640
10.6113931		vacuum	$^{12}C^{16}O_2$	0	1640
10.6126111		vacuum	$^{12}C^{16}O^{18}O$	0	1640
10.61261113		vacuum	$^{12}C^{16}O^{18}O$	0	1640
10.61281		vacuum	$^{14}N_2O$	0	1179
10.6131040		vacuum	$^{13}C^{16}O_2$	0	1640
10.61310404		vacuum	$^{13}C^{16}O_2$	0	1640
10.6139368		vacuum	$^{14}N_2O$	0	1177–8,1331–5
10.61405328		vacuum	$^{13}C^{16}O^{18}O$	0	1640
10.6172631		vacuum	$^{13}C^{18}O_2$	0	1640
10.61726312		vacuum	$^{13}C^{18}O_2$	0	1640

Table 4.1.1—*continued*
Neutral Atom, Ion, and Molecular Gas Lasers Arranged in Order of Increasing Wavelength

Wavelength (μm)	Uncertainty	Medium	Species	Charge	Reference
10.61895025		vacuum	$^{12}C^{17}O_2$	0	1640
10.62066319		vacuum	$^{13}C^{16}O^{18}O$	0	1640
10.62187		vacuum	$^{14}N_2O$	0	1179
10.62209118		vacuum	$^{12}C^{18}O_2$	0	1640
10.6220912		vacuum	$^{12}C^{18}O_2$	0	1640
10.62303534		vacuum	$^{12}C^{16}O^{18}O$	0	1640
10.6230354		vacuum	$^{12}C^{16}O^{18}O$	0	1640
10.6231929		vacuum	$^{14}N_2O$	0	1177–8,1331–5
10.624624		vacuum	$^{12}C^{18}O_2$	0	1640
10.62584936		vacuum	$^{13}C^{16}O_2$	0	1640
10.6258494		vacuum	$^{13}C^{16}O_2$	0	1640
10.62666		vacuum	$^{12}C^{16}O_2$	0	1640
10.62734311		vacuum	$^{13}C^{16}O^{18}O$	0	1640
10.6290501		vacuum	$^{14}C^{18}O_2$	0	1640
10.62905014		vacuum	$^{14}C^{18}O_2$	0	1640
10.62936707		vacuum	$^{12}C^{17}O_2$	0	1640
10.63030388		vacuum	$^{13}C^{18}O_2$	0	1640
10.6303039		vacuum	$^{13}C^{18}O_2$	0	1640
10.63209848		vacuum	$^{12}C^{16}O_2$	0	1640
10.6320985		vacuum	$^{12}C^{16}O_2$	0	1640
10.6325038		vacuum	$^{14}N_2O$	0	1177–8,1331–5
10.6335541		vacuum	$^{12}C^{16}O^{18}O$	0	1640
10.63355411		vacuum	$^{12}C^{16}O^{18}O$	0	1640
10.63409317		vacuum	$^{13}C^{16}O^{18}O$	0	1640
10.6388646		vacuum	$^{13}C^{16}O_2$	0	1640
10.63886462		vacuum	$^{13}C^{16}O_2$	0	1640
10.63987856		vacuum	$^{12}C^{17}O_2$	0	1640
10.64091348		vacuum	$^{13}C^{16}O^{18}O$	0	1640
10.6418694		vacuum	$^{14}N_2O$	0	1177–8,1331–5
10.6426827		vacuum	$^{12}C^{18}O_2$	0	1640
10.64268271		vacuum	$^{12}C^{18}O_2$	0	1640
10.64363665		vacuum	$^{13}C^{18}O_2$	0	1640
10.6436367		vacuum	$^{13}C^{18}O_2$	0	1640
10.64416804		vacuum	$^{12}C^{16}O^{18}O$	0	1640
10.6441681		vacuum	$^{12}C^{16}O^{18}O$	0	1640

Table 4.1.1—*continued*
Neutral Atom, Ion, and Molecular Gas Lasers Arranged in Order of Increasing Wavelength

Wavelength (μm)	Uncertainty	Medium	Species	Charge	Reference
10.64710		vacuum	$^{12}C^{16}O_2$	0	1640
10.64780415		vacuum	$^{13}C^{16}O^{18}O$	0	1640
10.65048533		vacuum	$^{12}C^{17}O_2$	0	1640
10.6511811		vacuum	$^{14}C^{18}O_2$	0	1640
10.65118113		vacuum	$^{14}C^{18}O_2$	0	1640
10.65215015		vacuum	$^{13}C^{16}O_2$	0	1640
10.6521502		vacuum	$^{13}C^{16}O_2$	0	1640
10.65316406		vacuum	$^{12}C^{16}O_2$	0	1640
10.6531641		vacuum	$^{12}C^{16}O_2$	0	1640
10.653167		vacuum	$^{13}C^{16}O_2$	0	1640
10.65476531		vacuum	$^{13}C^{16}O^{18}O$	0	1640
10.65487776		vacuum	$^{12}C^{16}O^{18}O$	0	1640
10.6548778		vacuum	$^{12}C^{16}O^{18}O$	0	1640
10.6572627		vacuum	$^{13}C^{18}O_2$	0	1640
10.65726273		vacuum	$^{13}C^{18}O_2$	0	1640
10.6607661		vacuum	$^{14}N_2O$	0	1177–8,1331–5
10.66118800		vacuum	$^{12}C^{17}O_2$	0	1640
10.66179709		vacuum	$^{13}C^{16}O^{18}O$	0	1640
10.66366019		vacuum	$^{12}C^{18}O_2$	0	1640
10.6636602		vacuum	$^{12}C^{18}O_2$	0	1640
10.6656839		vacuum	$^{12}C^{16}O^{18}O$	0	1640
10.66568390		vacuum	$^{12}C^{16}O^{18}O$	0	1640
10.6657064		vacuum	$^{13}C^{16}O_2$	0	1640
10.66570643		vacuum	$^{13}C^{16}O_2$	0	1640
10.666054		vacuum	$^{13}C^{16}O_2$	0	1640
10.66789		vacuum	$^{12}C^{16}O_2$	0	1640
10.66889965		vacuum	$^{13}C^{16}O^{18}O$	0	1640
10.6702974		vacuum	$^{14}N_2O$	0	1177–8,1331–5
10.6711835		vacuum	$^{13}C^{18}O_2$	0	1640
10.67118352		vacuum	$^{13}C^{18}O_2$	0	1640
10.67198720		vacuum	$^{12}C^{17}O_2$	0	1640
10.67459436		vacuum	$^{12}C^{16}O_2$	0	1640
10.6745944		vacuum	$^{12}C^{16}O_2$	0	1640
10.67607312		vacuum	$^{13}C^{16}O^{18}O$	0	1640
10.6765871		vacuum	$^{12}C^{16}O^{18}O$	0	1640

Table 4.1.1—*continued*
Neutral Atom, Ion, and Molecular Gas Lasers Arranged in Order of Increasing Wavelength

Wavelength (μm)	Uncertainty	Medium	Species	Charge	Reference
10.67658710		vacuum	$^{12}C^{16}O^{18}O$	0	1640
10.679209		vacuum	$^{13}C^{16}O_2$	0	1640
10.6795340		vacuum	$^{13}C^{16}O_2$	0	1640
10.67953401		vacuum	$^{13}C^{16}O_2$	0	1640
10.6798844		vacuum	$^{14}N_2O$	0	1177–8,1331–5
10.68288358		vacuum	$^{12}C^{17}O_2$	0	1640
10.68331766		vacuum	$^{13}C^{16}O^{18}O$	0	1640
10.6850285		vacuum	$^{12}C^{18}O_2$	0	1640
10.68502852		vacuum	$^{12}C^{18}O_2$	0	1640
10.6854005		vacuum	$^{13}C^{18}O_2$	0	1640
10.68540053		vacuum	$^{13}C^{18}O_2$	0	1640
10.68565		vacuum	$^{12}C^{16}O_2$	0	1640
10.6875880		vacuum	$^{12}C^{16}O^{18}O$	0	1640
10.68758802		vacuum	$^{12}C^{16}O^{18}O$	0	1640
10.68904		vacuum	$^{12}C^{16}O_2$	0	1640
10.6895273		vacuum	$^{14}N_2O$	0	1177–8,1331–5
10.69063344		vacuum	$^{13}C^{16}O^{18}O$	0	1640
10.692633		vacuum	$^{13}C^{16}O_2$	0	1640
10.69363357		vacuum	$^{13}C^{16}O_2$	0	1640
10.6936336		vacuum	$^{13}C^{16}O_2$	0	1640
10.69387779		vacuum	$^{12}C^{17}O_2$	0	1640
10.6963941		vacuum	$^{12}C^{16}O_2$	0	1640
10.69639414		vacuum	$^{12}C^{16}O_2$	0	1640
10.69802061		vacuum	$^{13}C^{16}O^{18}O$	0	1640
10.6986873		vacuum	$^{12}C^{16}O^{18}O$	0	1640
10.69868734		vacuum	$^{12}C^{16}O^{18}O$	0	1640
10.6992262		vacuum	$^{14}N_2O$	0	1177–8,1331–5
10.6999154		vacuum	$^{13}C^{18}O_2$	0	1640
10.69991540		vacuum	$^{13}C^{18}O_2$	0	1640
10.7			$^{14}NH_3$	0	1189,1210
10.70497050		vacuum	$^{12}C^{17}O_2$	0	1640
10.70547937		vacuum	$^{13}C^{16}O^{18}O$	0	1640
10.706327		vacuum	$^{13}C^{16}O_2$	0	1640
10.70652		vacuum	$^{12}C^{16}O_2$	0	1640
10.70679276		vacuum	$^{12}C^{18}O_2$	0	1640

Table 4.1.1—*continued*
Neutral Atom, Ion, and Molecular Gas Lasers Arranged in Order of Increasing Wavelength

Wavelength (μm)	Uncertainty	Medium	Species	Charge	Reference
10.7067928		vacuum	$^{12}C^{18}O_2$	0	1640
10.70800589		vacuum	$^{13}C^{16}O_2$	0	1640
10.7080059		vacuum	$^{13}C^{16}O_2$	0	1640
10.7089814		vacuum	$^{14}N_2O$	0	1177–8,1331–5
10.7098857		vacuum	$^{12}C^{16}O^{18}O$	0	1640
10.70988572		vacuum	$^{12}C^{16}O^{18}O$	0	1640
10.71055		vacuum	$^{12}C^{16}O_2$	0	1640
10.71300988		vacuum	$^{13}C^{16}O^{18}O$	0	1640
10.71472987		vacuum	$^{13}C^{18}O_2$	0	1640
10.7147299		vacuum	$^{13}C^{18}O_2$	0	1640
10.718			NH_3	0	1774
10.7182		vacuum	$^{14}NH_3$	0	1630,1774–75
10.7185683		vacuum	$^{12}C^{16}O_2$	0	1640
10.71856831		vacuum	$^{12}C^{16}O_2$	0	1640
10.7187931		vacuum	$^{14}N_2O$	0	1177–8,1331–5
10.720291		vacuum	$^{13}C^{16}O_2$	0	1640
10.72118388		vacuum	$^{12}C^{16}O^{18}O$	0	1640
10.721184		vacuum	$^{12}C^{16}O^{18}O$	0	1640
10.7226518		vacuum	$^{13}C^{16}O_2$	0	1640
10.72265182		vacuum	$^{13}C^{16}O_2$	0	1640
10.72775		vacuum	$^{12}C^{16}O_2$	0	1640
10.7286616		vacuum	$^{14}N_2O$	0	1177–8,1331–5
10.72895816		vacuum	$^{12}C^{18}O_2$	0	1640
10.7289582		vacuum	$^{12}C^{18}O_2$	0	1640
10.732			NH_3	0	1774
10.7322		vacuum	$^{14}NH_3$	0	1190
10.73243		vacuum	$^{12}C^{16}O_2$	0	1640
10.732582		vacuum	$^{12}C^{16}O^{18}O$	0	1640
10.73258251		vacuum	$^{12}C^{16}O^{18}O$	0	1640
10.734525		vacuum	$^{13}C^{16}O_2$	0	1640
10.73488		vacuum	$^{14}N_2O$	0	1179
10.737			NH_3	0	1774
10.73729		vacuum	$^{14}NH_3$	0	1630,1774–75
10.73757236		vacuum	$^{13}C^{16}O_2$	0	1640
10.7375724		vacuum	$^{13}C^{16}O_2$	0	1640

Table 4.1.1—*continued*
Neutral Atom, Ion, and Molecular Gas Lasers Arranged in Order of Increasing Wavelength

Wavelength (μm)	Uncertainty	Medium	Species	Charge	Reference
10.7385869		vacuum	$^{14}N_2O$	0	1177–8,1331–5
10.7411220		vacuum	$^{12}C^{16}O_2$	0	1640
10.74112203		vacuum	$^{12}C^{16}O_2$	0	1640
10.74394		vacuum	$^{14}NH_3$	0	1190
10.744			NH_3	0	1775
10.74466		vacuum	$^{14}N_2O$	0	1179
10.7485694		vacuum	$^{14}N_2O$	0	1177–8,1331–5
10.749031		vacuum	$^{13}C^{16}O_2$	0	1640
10.74934		vacuum	$^{12}C^{16}O_2$	0	1640
10.75153017		vacuum	$^{12}C^{18}O_2$	0	1640
10.7515302		vacuum	$^{12}C^{18}O_2$	0	1640
10.75276857		vacuum	$^{13}C^{16}O_2$	0	1640
10.7527686		vacuum	$^{13}C^{16}O_2$	0	1640
10.75387		vacuum	$^{14}NH_3$	0	1190
10.754			NH_3	0	1774
10.75449		vacuum	$^{14}N_2O$	0	1179
10.75468		vacuum	$^{12}C^{16}O_2$	0	1640
10.7586093		vacuum	$^{14}N_2O$	0	1177–8,1331–5
10.762			NH_3	0	1774
10.7624		vacuum	$^{14}NH_3$	0	1630,1774–75
10.763809		vacuum	$^{13}C^{16}O_2$	0	1640
10.764060534		vacuum	$^{12}C^{16}O_2$	0	1640
10.7640606		vacuum	$^{12}C^{16}O_2$	0	1640
10.7643831		vacuum	$^{14}N_2O$	0	1179
10.767			NH_3	0	1774
10.76710		vacuum	$^{14}NH_3$	0	1190
10.7682416		vacuum	$^{13}C^{16}O_2$	0	1640
10.76824164		vacuum	$^{13}C^{16}O_2$	0	1640
10.7687068		vacuum	$^{14}N_2O$	0	1177–8,1331–5
10.77130		vacuum	$^{12}C^{16}O_2$	0	1640
10.77433		vacuum	$^{14}N_2O$	0	1177–8,1331–5
10.7745144		vacuum	$^{12}C^{18}O_2$	0	1640
10.77451441		vacuum	$^{12}C^{18}O_2$	0	1640
10.77731		vacuum	$^{12}C^{16}O_2$	0	1640
10.778861		vacuum	$^{13}C^{16}O_2$	0	1640

Table 4.1.1—*continued*
Neutral Atom, Ion, and Molecular Gas Lasers Arranged in Order of Increasing Wavelength

Wavelength (μm)	Uncertainty	Medium	Species	Charge	Reference
10.7788621		vacuum	$^{14}N_2O$	0	1177–8,1331–5
10.7837		vacuum	$^{14}NH_3$	0	1630,1774–75
10.78399284		vacuum	$^{13}C^{16}O_2$	0	1640
10.7839929		vacuum	$^{13}C^{16}O_2$	0	1640
10.784			NH_3	0	1775
10.78434		vacuum	$^{14}N_2O$	0	1179
10.7873896		vacuum	$^{12}C^{16}O_2$	0	1640
10.78738965		vacuum	$^{12}C^{16}O_2$	0	1640
10.789			NH_3	0	1775
10.789		vacuum	$^{15}NH_3$	0	1635,1775
10.7890756		vacuum	$^{14}N_2O$	0	1177–8,1331–5
10.7890778		vacuum	$^{12}C^{16}O_2$	0	1640
10.79362		vacuum	$^{12}C^{16}O_2$	0	1640
10.794187		vacuum	$^{13}C^{16}O_2$	0	1640
10.79440		vacuum	$^{14}N_2O$	0	1179
10.7979167		vacuum	$^{12}C^{18}O_2$	0	1640
10.79791670		vacuum	$^{12}C^{18}O_2$	0	1640
10.7993473		vacuum	$^{14}N_2O$	0	1177–8,1331–5
10.80002356		vacuum	$^{13}C^{16}O_2$	0	1640
10.8000236		vacuum	$^{13}C^{16}O_2$	0	1640
10.80032		vacuum	$^{12}C^{16}O_2$	0	1640
10.80452		vacuum	$^{14}N_2O$	0	1179
10.8096777		vacuum	$^{14}N_2O$	0	1177–8,1331–5
10.809789		vacuum	$^{13}C^{16}O_2$	0	1640
10.81111488		vacuum	$^{12}C^{16}O_2$	0	1640
10.8111149		vacuum	$^{12}C^{16}O_2$	0	1640
10.81470		vacuum	$^{14}N_2O$	0	1179
10.81632		vacuum	$^{12}C^{16}O_2$	0	1640
10.81633528		vacuum	$^{13}C^{16}O_2$	0	1640
10.8163353		vacuum	$^{13}C^{16}O_2$	0	1640
10.81846594		vacuum	$^{13}C^{18}O_2$	0	1640
10.820067		vacuum	$^{14}N_2O$	0	1177–8,1331–5
10.82372		vacuum	$^{12}C^{16}O_2$	0	1640
10.82494		vacuum	$^{14}N_2O$	0	1179
10.825668		vacuum	$^{13}C^{16}O_2$	0	1640

Table 4.1.1—*continued*
Neutral Atom, Ion, and Molecular Gas Lasers Arranged in Order of Increasing Wavelength

Wavelength (μm)	Uncertainty	Medium	Species	Charge	Reference
10.830515		vacuum	$^{14}N_2O$	0	1177–8,1331–5
10.83292957		vacuum	$^{13}C^{16}O_2$	0	1640
10.8329296		vacuum	$^{13}C^{16}O_2$	0	1640
10.8352265		vacuum	$^{14}N_2O$	0	1179
10.8352423		vacuum	$^{12}C^{16}O_2$	0	1640
10.83524231		vacuum	$^{12}C^{16}O_2$	0	1640
10.8355947		vacuum	$^{13}C^{18}O_2$	0	1640
10.83559471		vacuum	$^{13}C^{18}O_2$	0	1640
10.83941		vacuum	$^{12}C^{16}O_2$	0	1640
10.841023		vacuum	$^{14}N_2O$	0	1177–8,1331–5
10.841826		vacuum	$^{13}C^{16}O_2$	0	1640
10.84559		vacuum	$^{14}N_2O$	0	1179
10.84751		vacuum	$^{12}C^{16}O_2$	0	1640
10.8498081		vacuum	$^{13}C^{16}O_2$	0	1640
10.84980812		vacuum	$^{13}C^{16}O_2$	0	1640
10.851590		vacuum	$^{14}N_2O$	0	1177–8,1331–5
10.8530426		vacuum	$^{13}C^{18}O_2$	0	1640
10.85304265		vacuum	$^{13}C^{18}O_2$	0	1640
10.8548		vacuum	$^{14}NH_3$	0	1630,1774–75
10.855			NH_3	0	1774
10.85600		vacuum	$^{14}N_2O$	0	1179
10.858263		vacuum	$^{13}C^{16}O_2$	0	1640
10.85977817		vacuum	$^{12}C^{16}O_2$	0	1640
10.8597782		vacuum	$^{12}C^{16}O_2$	0	1640
10.862216		vacuum	$^{14}N_2O$	0	1177–8,1331–5
10.86647		vacuum	$^{14}N_2O$	0	1179
10.8669727		vacuum	$^{13}C^{16}O_2$	0	1640
10.86697272		vacuum	$^{13}C^{16}O_2$	0	1640
10.8708127		vacuum	$^{13}C^{18}O_2$	0	1640
10.87081266		vacuum	$^{13}C^{18}O_2$	0	1640
10.87171		vacuum	$^{12}C^{16}O_2$	0	1640
10.872903		vacuum	$^{14}N_2O$	0	1177–8,1331–5
10.874982		vacuum	$^{13}C^{16}O_2$	0	1640
10.87700		vacuum	$^{14}N_2O$	0	1179
10.883651		vacuum	$^{14}N_2O$	0	1177–8,1331–5

Table 4.1.1—*continued*
Neutral Atom, Ion, and Molecular Gas Lasers Arranged in Order of Increasing Wavelength

Wavelength (μm)	Uncertainty	Medium	Species	Charge	Reference
10.88442525		vacuum	$^{13}C^{16}O_2$	0	1640
10.8844252		vacuum	$^{13}C^{16}O_2$	0	1640
10.8847289		vacuum	$^{12}C^{16}O_2$	0	1640
10.88472894		vacuum	$^{12}C^{16}O_2$	0	1640
10.88759		vacuum	$^{14}N_2O$	0	1179
10.88890777		vacuum	$^{13}C^{18}O_2$	0	1640
10.8889078		vacuum	$^{13}C^{18}O_2$	0	1640
10.8901847		vacuum	$^{12}C^{16}O_2$	0	1640
10.891985		vacuum	$^{13}C^{16}O_2$	0	1640
10.89446		vacuum	$^{14}N_2O$	0	1177–8,1331–5
10.89631		vacuum	$^{12}C^{16}O_2$	0	1640
10.89824		vacuum	$^{14}N_2O$	0	1179
10.900267		vacuum	$^{14}C^{18}O_2$	0	1640
10.900268		vacuum	$^{14}C^{18}O_2$	0	1640
10.9009647		vacuum	$^{12}C^{16}O_2$	0	1640
10.90216771		vacuum	$^{13}C^{16}O_2$	0	1640
10.9021677		vacuum	$^{13}C^{16}O_2$	0	1640
10.90533		vacuum	$^{14}N_2O$	0	1177–8,1331–5
10.90733115		vacuum	$^{13}C^{18}O_2$	0	1640
10.9073312		vacuum	$^{13}C^{18}O_2$	0	1640
10.90895		vacuum	$^{14}N_2O$	0	1179
10.909272		vacuum	$^{13}C^{16}O_2$	0	1640
10.9101013		vacuum	$^{12}C^{16}O_2$	0	1640
10.91010135		vacuum	$^{12}C^{16}O_2$	0	1640
10.911233		vacuum	$^{14}C^{18}O_2$	0	1640
10.9112334		vacuum	$^{14}C^{18}O_2$	0	1640
10.91626		vacuum	$^{14}N_2O$	0	1177–8,1331–5
10.91972		vacuum	$^{14}N_2O$	0	1179
10.92133		vacuum	$^{12}C^{16}O_2$	0	1640
10.9214691		vacuum	$^{12}C^{16}O_2$	0	1640
10.9224632		vacuum	$^{14}C^{18}O_2$	0	1640
10.922463		vacuum	$^{14}C^{18}O_2$	0	1640
10.92527598		vacuum	$^{13}C^{16}O^{18}O$	0	1640
10.92608610		vacuum	$^{13}C^{18}O_2$	0	1640
10.9260861		vacuum	$^{13}C^{18}O_2$	0	1640

Table 4.1.1—*continued*
Neutral Atom, Ion, and Molecular Gas Lasers Arranged in Order of Increasing Wavelength

Wavelength (μm)	Uncertainty	Medium	Species	Charge	Reference
10.92725		vacuum	$^{14}N_2O$	0	1177–8,1331–5
10.93055		vacuum	$^{14}N_2O$	0	1179
10.9307080		vacuum	$^{12}C^{16}O_2$	0	1640
10.9337		air	Kr	0	1498
10.9339572		vacuum	$^{14}C^{18}O_2$	0	1640
10.933957		vacuum	$^{14}C^{18}O_2$	0	1640
10.93476009		vacuum	$^{13}C^{16}O^{18}O$	0	1640
10.9359024		vacuum	$^{12}C^{16}O_2$	0	1640
10.93590238		vacuum	$^{12}C^{16}O_2$	0	1640
10.93830		vacuum	$^{14}N_2O$	0	1177–8,1331–5
10.94144		vacuum	$^{14}N_2O$	0	1179
10.9423513		vacuum	$^{12}C^{16}O_2$	0	1640
10.94432366		vacuum	$^{13}C^{16}O^{18}O$	0	1640
10.94517607		vacuum	$^{13}C^{18}O_2$	0	1640
10.9451761		vacuum	$^{13}C^{18}O_2$	0	1640
10.9457153		vacuum	$^{14}C^{18}O_2$	0	1640
10.945715		vacuum	$^{14}C^{18}O_2$	0	1640
10.94676		vacuum	$^{12}C^{16}O_2$	0	1640
10.94942		vacuum	$^{14}N_2O$	0	1177–8,1331–5
10.9514868		vacuum	$^{12}C^{16}O_2$	0	1640
10.95240		vacuum	$^{14}N_2O$	0	1179
10.95396711		vacuum	$^{13}C^{16}O^{18}O$	0	1640
10.9577377		vacuum	$^{14}C^{18}O_2$	0	1640
10.957738		vacuum	$^{14}C^{18}O_2$	0	1640
10.96059		vacuum	$^{14}N_2O$	0	1177–8,1331–5
10.9621393		vacuum	$^{12}C^{16}O_2$	0	1640
10.96213927		vacuum	$^{12}C^{16}O_2$	0	1640
10.96341		vacuum	$^{14}N_2O$	0	1179
10.96369084		vacuum	$^{13}C^{16}O^{18}O$	0	1640
10.96460463		vacuum	$^{13}C^{18}O_2$	0	1640
10.9646046		vacuum	$^{13}C^{18}O_2$	0	1640
10.970024		vacuum	$^{14}C^{18}O_2$	0	1640
10.9700243		vacuum	$^{14}C^{18}O_2$	0	1640
10.97183		vacuum	$^{14}N_2O$	0	1177–8,1331–5
10.9726149		vacuum	$^{12}C^{16}O_2$	0	1640

Table 4.1.1—*continued*
Neutral Atom, Ion, and Molecular Gas Lasers Arranged in Order of Increasing Wavelength

Wavelength (μm)	Uncertainty	Medium	Species	Charge	Reference
10.97349525		vacuum	$^{13}C^{16}O^{18}O$	0	1640
10.97449		vacuum	$^{14}N_2O$	0	1179
10.98		vacuum	C_2H_4	0	1194
10.981641		vacuum	Ne	0	195,564,570
10.982576		vacuum	$^{14}C^{18}O_2$	0	1640
10.9825756		vacuum	$^{14}C^{18}O_2$	0	1640
10.98314		vacuum	$^{14}N_2O$	0	1177–8,1331–5
10.98338078		vacuum	$^{13}C^{16}O^{18}O$	0	1640
10.98437552		vacuum	$^{13}C^{18}O_2$	0	1640
10.9843755		vacuum	$^{13}C^{18}O_2$	0	1640
10.9852663		vacuum	$^{12}C^{16}O_2$	0	1640
10.98563		vacuum	$^{14}N_2O$	0	1179
10.98565546		vacuum	$^{13}C^{16}O_2$	0	1640
10.9856555		vacuum	$^{13}C^{16}O_2$	0	1640
10.9888196		vacuum	$^{12}C^{16}O_2$	0	1640
10.98881958		vacuum	$^{12}C^{16}O_2$	0	1640
10.99334786		vacuum	$^{13}C^{16}O^{18}O$	0	1640
10.994		vacuum	$^{12}C^{16}O_2$	0	1640
10.99450		vacuum	$^{14}N_2O$	0	1177–8,1331–5
10.9953918		vacuum	$^{14}C^{18}O_2$	0	1640
10.995392		vacuum	$^{14}C^{18}O_2$	0	1640
11.00339692		vacuum	$^{13}C^{16}O^{18}O$	0	1640
11.00449265		vacuum	$^{13}C^{18}O_2$	0	1640
11.0044927		vacuum	$^{13}C^{18}O_2$	0	1640
11.00503296		vacuum	$^{13}C^{16}O_2$	0	1640
11.0050330		vacuum	$^{13}C^{16}O_2$	0	1640
11.00593		vacuum	$^{14}N_2O$	0	1177–8,1331–5
11.0073078		vacuum	$^{12}C^{16}O_2$	0	1640
11.0084736		vacuum	$^{14}C^{18}O_2$	0	1640
11.008474		vacuum	$^{14}C^{18}O_2$	0	1640
11.009542		vacuum	$^{13}C^{16}O_2$	0	1640
11.01080		vacuum	$^{14}NH_3$	0	1190
11.011			NH_3	0	1775
11.0111		vacuum	$^{14}NH_3$	0	1181
11.01352843		vacuum	$^{13}C^{16}O^{18}O$	0	1640

Table 4.1.1—*continued*
Neutral Atom, Ion, and Molecular Gas Lasers Arranged in Order of Increasing Wavelength

Wavelength (μm)	Uncertainty	Medium	Species	Charge	Reference
11.0159341		vacuum	$^{12}C^{16}O_2$	0	1640
11.0159511		vacuum	$^{12}C^{16}O_2$	0	1640
11.01595112		vacuum	$^{12}C^{16}O_2$	0	1640
11.01743		vacuum	$^{14}N_2O$	0	1177–8,1331–5
11.021816		vacuum	$^{14}C^{18}O_2$	0	1640
11.021822		vacuum	$^{14}C^{18}O_2$	0	1640
11.02374283		vacuum	$^{13}C^{16}O^{18}O$	0	1640
11.02471598		vacuum	$^{13}C^{16}O_2$	0	1640
11.0247160		vacuum	$^{13}C^{16}O_2$	0	1640
11.02496004		vacuum	$^{13}C^{18}O_2$	0	1640
11.0249601		vacuum	$^{13}C^{18}O_2$	0	1640
11.028734		vacuum	$^{13}C^{16}O_2$	0	1640
11.02898		vacuum	$^{14}N_2O$	0	1177–8,1331–5
11.0297448		vacuum	$^{12}C^{16}O_2$	0	1640
11.03404059		vacuum	$^{13}C^{16}O^{18}O$	0	1640
11.035436		vacuum	$^{14}C^{18}O_2$	0	1640
11.0354362		vacuum	$^{14}C^{18}O_2$	0	1640
11.039		vacuum	$^{12}C^{16}O_2$	0	1640
11.04061		vacuum	$^{14}N_2O$	0	1177–8,1331–5
11.0415		air	Ar	0	1498
11.043542		vacuum	$^{12}C^{16}O_2$	0	1640
11.04354201		vacuum	$^{12}C^{16}O_2$	0	1640
11.04442221		vacuum	$^{13}C^{16}O^{18}O$	0	1640
11.04470736		vacuum	$^{13}C^{16}O_2$	0	1640
11.0447074		vacuum	$^{13}C^{16}O_2$	0	1640
11.0457819		vacuum	$^{13}C^{18}O_2$	0	1640
11.04578190		vacuum	$^{13}C^{18}O_2$	0	1640
11.048228		vacuum	$^{13}C^{16}O_2$	0	1640
11.0493184		vacuum	$^{14}C^{18}O_2$	0	1640
11.049319		vacuum	$^{14}C^{18}O_2$	0	1640
11.053		vacuum	$^{12}C^{16}O_2$	0	1640
11.05488816		vacuum	$^{13}C^{16}O^{18}O$	0	1640
11.061		vacuum	$^{12}C^{16}O_2$	0	1640
11.063469		vacuum	$^{14}C^{18}O_2$	0	1640
11.0634691		vacuum	$^{14}C^{18}O_2$	0	1640

Table 4.1.1—*continued*
Neutral Atom, Ion, and Molecular Gas Lasers
Arranged in Order of Increasing Wavelength

Wavelength (μm)	Uncertainty	Medium	Species	Charge	Reference
11.06501005		vacuum	$^{13}C^{16}O_2$	0	1640
11.0650101		vacuum	$^{13}C^{16}O_2$	0	1640
11.06543894		vacuum	$^{13}C^{16}O^{18}O$	0	1640
11.06696259		vacuum	$^{13}C^{18}O_2$	0	1640
11.0669626		vacuum	$^{13}C^{18}O_2$	0	1640
11.068028		vacuum	$^{13}C^{16}O_2$	0	1640
11.07160069		vacuum	$^{12}C^{16}O_2$	0	1640
11.071601		vacuum	$^{12}C^{16}O_2$	0	1640
11.076		vacuum	$^{12}C^{16}O_2$	0	1640
11.07607506		vacuum	$^{13}C^{16}O^{18}O$	0	1640
11.0778893		vacuum	$^{14}C^{18}O_2$	0	1640
11.077890		vacuum	$^{14}C^{18}O_2$	0	1640
11.0836301		vacuum	$^{12}C^{16}O_2$	0	1640
11.0856271		vacuum	$^{13}C^{16}O_2$	0	1640
11.08562711		vacuum	$^{13}C^{16}O_2$	0	1640
11.08679705		vacuum	$^{13}C^{16}O^{18}O$	0	1640
11.088135		vacuum	$^{13}C^{16}O_2$	0	1640
11.08850666		vacuum	$^{13}C^{18}O_2$	0	1640
11.0885067		vacuum	$^{13}C^{18}O_2$	0	1640
11.0925799		vacuum	$^{14}C^{18}O_2$	0	1640
11.092580		vacuum	$^{14}C^{18}O_2$	0	1640
11.09760542		vacuum	$^{13}C^{16}O^{18}O$	0	1640
11.100		vacuum	$^{12}C^{16}O_2$	0	1640
11.100135		vacuum	$^{12}C^{16}O_2$	0	1640
11.100135		vacuum	$^{12}C^{16}O_2$	0	1640
11.10656174		vacuum	$^{13}C^{16}O_2$	0	1640
11.1065618		vacuum	$^{13}C^{16}O_2$	0	1640
11.106994		vacuum	$^{14}C^{16}O_2$	0	1640
11.106994		vacuum	$^{14}C^{16}O_2$	0	1640
11.107		vacuum	$^{12}C^{16}O_2$	0	1640
11.1075423		vacuum	$^{14}C^{18}O_2$	0	1640
11.107543		vacuum	$^{14}C^{18}O_2$	0	1640
11.10850073		vacuum	$^{13}C^{16}O^{18}O$	0	1640
11.108552		vacuum	$^{13}C^{16}O_2$	0	1640
11.11041879		vacuum	$^{13}C^{18}O_2$	0	1640

Table 4.1.1—*continued*
Neutral Atom, Ion, and Molecular Gas Lasers Arranged in Order of Increasing Wavelength

Wavelength (μm)	Uncertainty	Medium	Species	Charge	Reference
11.1104188		vacuum	$^{13}C^{18}O_2$	0	1640
11.11948352		vacuum	$^{13}C^{16}O^{18}O$	0	1640
11.119855		vacuum	$^{14}C^{16}O_2$	0	1640
11.120968		vacuum	$^{14}C^{16}O_2$	0	1640
11.1227776		vacuum	$^{14}C^{18}O_2$	0	1640
11.122778		vacuum	$^{14}C^{18}O_2$	0	1640
11.124		vacuum	$^{12}C^{16}O_2$	0	1640
11.12781727		vacuum	$^{13}C^{16}O_2$	0	1640
11.1278173		vacuum	$^{13}C^{16}O_2$	0	1640
11.129155		vacuum	$^{12}C^{16}O_2$	0	1640
11.129155		vacuum	$^{12}C^{16}O_2$	0	1640
11.129283		vacuum	$^{13}C^{16}O_2$	0	1640
11.13055436		vacuum	$^{13}C^{16}O^{18}O$	0	1640
11.131		vacuum	$^{12}C^{16}O_2$	0	1640
11.13270389		vacuum	$^{13}C^{18}O_2$	0	1640
11.1327039		vacuum	$^{13}C^{18}O_2$	0	1640
11.1351978		vacuum	$^{14}C^{16}O_2$	0	1640
11.135198		vacuum	$^{14}C^{16}O_2$	0	1640
11.1382873		vacuum	$^{14}C^{18}O_2$	0	1640
11.138288		vacuum	$^{14}C^{18}O_2$	0	1640
11.1417138		vacuum	$^{13}C^{16}O^{18}O$	0	1640
11.148		vacuum	$^{12}C^{16}O_2$	0	1640
11.14939715		vacuum	$^{13}C^{16}O_2$	0	1640
11.1493972		vacuum	$^{13}C^{16}O_2$	0	1640
11.1496835		vacuum	$^{14}C^{16}O_2$	0	1640
11.149684		vacuum	$^{14}C^{16}O_2$	0	1640
11.150331		vacuum	$^{13}C^{16}O_2$	0	1640
11.15296216		vacuum	$^{13}C^{16}O^{18}O$	0	1640
11.1540728		vacuum	$^{14}C^{18}O_2$	0	1640
11.154073		vacuum	$^{14}C^{18}O_2$	0	1640
11.1553670		vacuum	$^{13}C^{18}O_2$	0	1640
11.15536702		vacuum	$^{13}C^{18}O_2$	0	1640
11.15867		vacuum	$^{12}C^{16}O_2$	0	1640
11.158670		vacuum	$^{12}C^{16}O_2$	0	1640
11.1643009		vacuum	$^{13}C^{16}O^{18}O$	0	1640

Table 4.1.1—*continued*
Neutral Atom, Ion, and Molecular Gas Lasers Arranged in Order of Increasing Wavelength

Wavelength (μm)	Uncertainty	Medium	Species	Charge	Reference
11.164426		vacuum	$^{14}C^{16}O_2$	0	1640
11.1644260		vacuum	$^{14}C^{16}O_2$	0	1640
11.17130499		vacuum	$^{13}C^{16}O_2$	0	1640
11.1713050		vacuum	$^{13}C^{16}O_2$	0	1640
11.171700		vacuum	$^{13}C^{16}O_2$	0	1640
11.1757298		vacuum	$^{13}C^{16}O^{18}O$	0	1640
11.1784134		vacuum	$^{13}C^{18}O_2$	0	1640
11.17841341		vacuum	$^{13}C^{18}O_2$	0	1640
11.1794258		vacuum	$^{14}C^{16}O_2$	0	1640
11.179426		vacuum	$^{14}C^{16}O_2$	0	1640
11.18868		vacuum	$^{12}C^{16}O_2$	0	1640
11.188680		vacuum	$^{12}C^{16}O_2$	0	1640
11.193392		vacuum	$^{13}C^{16}O_2$	0	1640
11.1935445		vacuum	$^{13}C^{16}O_2$	0	1640
11.19354451		vacuum	$^{13}C^{16}O_2$	0	1640
11.194684		vacuum	$^{14}C^{16}O_2$	0	1640
11.1946840		vacuum	$^{14}C^{16}O_2$	0	1640
11.20184854		vacuum	$^{13}C^{18}O_2$	0	1640
11.2018486		vacuum	$^{13}C^{18}O_2$	0	1640
11.20879		vacuum	$^{14}NH_3$	0	1190,1191
11.209			NH_3	0	1775
11.210201		vacuum	$^{14}C^{16}O_2$	0	1640
11.2102013		vacuum	$^{14}C^{16}O_2$	0	1640
11.212			NH_3	0	1774
11.2123		vacuum	$^{14}NH_3$	0	1630,1774–75
11.215411		vacuum	$^{13}C^{16}O_2$	0	1640
11.2161196		vacuum	$^{13}C^{16}O_2$	0	1640
11.21611960		vacuum	$^{13}C^{16}O_2$	0	1640
11.2256780		vacuum	$^{13}C^{18}O_2$	0	1640
11.22567803		vacuum	$^{13}C^{18}O_2$	0	1640
11.2259788		vacuum	$^{14}C^{16}O_2$	0	1640
11.225979		vacuum	$^{14}C^{16}O_2$	0	1640
11.237762		vacuum	$^{13}C^{16}O_2$	0	1640
11.23903428		vacuum	$^{13}C^{16}O_2$	0	1640
11.2390343		vacuum	$^{13}C^{16}O_2$	0	1640

Table 4.1.1—*continued*
Neutral Atom, Ion, and Molecular Gas Lasers
Arranged in Order of Increasing Wavelength

Wavelength (μm)	Uncertainty	Medium	Species	Charge	Reference
11.2420175		vacuum	$^{14}C^{16}O_2$	0	1640
11.242018		vacuum	$^{14}C^{16}O_2$	0	1640
11.2499077		vacuum	$^{13}C^{18}O_2$	0	1640
11.24990774		vacuum	$^{13}C^{18}O_2$	0	1640
11.257			NH_3	0	1775
11.257		vacuum	$^{15}NH_3$	0	1635,1775
11.2583176		vacuum	$^{14}C^{16}O_2$	0	1640
11.258318		vacuum	$^{14}C^{16}O_2$	0	1640
11.260			NH_3	0	1774
11.2603		vacuum	$^{14}NH_3$	0	1630,1774–75
11.260448		vacuum	$^{13}C^{16}O_2$	0	1640
11.261		vacuum	$^{14}NH_3$	0	1191,1192
11.26229275		vacuum	$^{13}C^{16}O_2$	0	1640
11.2622928		vacuum	$^{13}C^{16}O_2$	0	1640
11.2628		vacuum	$^{14}NH_3$	0	1630,1774–75
11.263			NH_3	0	1774
11.27488285		vacuum	$^{14}C^{16}O_2$	0	1640
11.274883		vacuum	$^{14}C^{16}O_2$	0	1640
11.283472		vacuum	$^{13}C^{16}O_2$	0	1640
11.28589935		vacuum	$^{13}C^{16}O_2$	0	1640
11.2858994		vacuum	$^{13}C^{16}O_2$	0	1640
11.291712		vacuum	$^{14}C^{16}O_2$	0	1640
11.29171205		vacuum	$^{14}C^{16}O_2$	0	1640
11.298683		vacuum	Xe	0	559,564
11.299632		vacuum	$^{14}C^{18}O_2$	0	1640
11.299632		vacuum	$^{14}C^{18}O_2$	0	1640
11.306840		vacuum	$^{13}C^{16}O_2$	0	1640
11.308807		vacuum	$^{14}C^{16}O_2$	0	1640
11.30880736		vacuum	$^{14}C^{16}O_2$	0	1640
11.30985858		vacuum	$^{13}C^{16}O_2$	0	1640
11.3098586		vacuum	$^{13}C^{16}O_2$	0	1640
11.318130		vacuum	$^{14}C^{18}O_2$	0	1640
11.318130		vacuum	$^{14}C^{18}O_2$	0	1640
11.326170		vacuum	$^{14}C^{16}O_2$	0	1640
11.32617017		vacuum	$^{14}C^{16}O_2$	0	1640

Table 4.1.1—*continued*
Neutral Atom, Ion, and Molecular Gas Lasers Arranged in Order of Increasing Wavelength

Wavelength (μm)	Uncertainty	Medium	Species	Charge	Reference
11.330556		vacuum	$^{13}C^{16}O_2$	0	1640
11.3341751		vacuum	$^{13}C^{16}O_2$	0	1640
11.33417514		vacuum	$^{13}C^{16}O_2$	0	1640
11.3369254		vacuum	$^{14}C^{18}O_2$	0	1640
11.336926		vacuum	$^{14}C^{18}O_2$	0	1640
11.34380199		vacuum	$^{14}C^{16}O_2$	0	1640
11.343802		vacuum	$^{14}C^{16}O_2$	0	1640
11.354624		vacuum	$^{13}C^{16}O_2$	0	1640
11.356021		vacuum	$^{14}C^{18}O_2$	0	1640
11.3560210		vacuum	$^{14}C^{18}O_2$	0	1640
11.35885386		vacuum	$^{13}C^{16}O_2$	0	1640
11.3588539		vacuum	$^{13}C^{16}O_2$	0	1640
11.361704		vacuum	$^{14}C^{16}O_2$	0	1640
11.36170437		vacuum	$^{14}C^{16}O_2$	0	1640
11.3754195		vacuum	$^{14}C^{18}O_2$	0	1640
11.375420		vacuum	$^{14}C^{18}O_2$	0	1640
11.379049		vacuum	$^{13}C^{16}O_2$	0	1640
11.37987894		vacuum	$^{14}C^{16}O_2$	0	1640
11.379879		vacuum	$^{14}C^{16}O_2$	0	1640
11.38389979		vacuum	$^{13}C^{16}O_2$	0	1640
11.3838998		vacuum	$^{13}C^{16}O_2$	0	1640
11.395124		vacuum	$^{14}C^{18}O_2$	0	1640
11.3951240		vacuum	$^{14}C^{18}O_2$	0	1640
11.398327		vacuum	$^{14}C^{16}O_2$	0	1640
11.39832741		vacuum	$^{14}C^{16}O_2$	0	1640
11.403836		vacuum	$^{13}C^{16}O_2$	0	1640
11.40931817		vacuum	$^{13}C^{16}O_2$	0	1640
11.4093182		vacuum	$^{13}C^{16}O_2$	0	1640
11.4151374		vacuum	$^{14}C^{18}O_2$	0	1640
11.415138		vacuum	$^{14}C^{18}O_2$	0	1640
11.4161330		vacuum	$^{13}C^{16}O_2$	0	1625,1640
11.41705156		vacuum	$^{14}C^{16}O_2$	0	1640
11.417052		vacuum	$^{14}C^{16}O_2$	0	1640
11.4273891		vacuum	$^{13}C^{16}O_2$	0	1625,1640
11.428989		vacuum	$^{13}C^{16}O_2$	0	1640

Table 4.1.1—*continued*
Neutral Atom, Ion, and Molecular Gas Lasers Arranged in Order of Increasing Wavelength

Wavelength (μm)	Uncertainty	Medium	Species	Charge	Reference
11.43511443		vacuum	$^{13}C^{16}O_2$	0	1640
11.435115		vacuum	$^{13}C^{16}O_2$	0	1640
11.435463		vacuum	$^{14}C^{18}O_2$	0	1640
11.4354630		vacuum	$^{14}C^{18}O_2$	0	1640
11.436053		vacuum	$^{14}C^{16}O_2$	0	1640
11.43605325		vacuum	$^{14}C^{16}O_2$	0	1640
11.4376785		vacuum	$^{13}C^{16}O_2$	0	1625,1640
11.4492864		vacuum	$^{13}C^{16}O_2$	0	1625,1640
11.454515		vacuum	$^{13}C^{16}O_2$	0	1640
11.455334		vacuum	$^{14}C^{16}O_2$	0	1640
11.45533442		vacuum	$^{14}C^{16}O_2$	0	1640
11.456104		vacuum	$^{14}C^{18}O_2$	0	1640
11.4561042		vacuum	$^{14}C^{18}O_2$	0	1640
11.4595534		vacuum	$^{13}C^{16}O_2$	0	1640
11.460			NH_3	0	1774
11.46044		vacuum	$^{14}NH_3$	0	1189–95
11.461294		vacuum	$^{13}C^{16}O_2$	0	1640
11.46129419		vacuum	$^{13}C^{16}O_2$	0	1640
11.471			NH_3	0	1774
11.47135		vacuum	$^{14}NH_3$	0	1191
11.4715433		vacuum	$^{13}C^{16}O_2$	0	1625,1640
11.474897		vacuum	$^{14}C^{16}O_2$	0	1640
11.47489707		vacuum	$^{14}C^{16}O_2$	0	1640
11.476		vacuum	$^{12}CS_2$	0	1321–1323
11.477064		vacuum	$^{14}C^{18}O_2$	0	1640
11.4770642		vacuum	$^{14}C^{18}O_2$	0	1640
11.480419		vacuum	$^{13}C^{16}O_2$	0	1640
11.4817608		vacuum	$^{13}C^{16}O_2$	0	1625,1640
11.4823		vacuum	$^{12}CS_2$	0	1321–1323
11.487863		vacuum	$^{13}C^{16}O_2$	0	1640
11.48786331		vacuum	$^{13}C^{16}O_2$	0	1640
11.4893		vacuum	$^{12}CS_2$	0	1321–1323
11.4941634		vacuum	$^{13}C^{16}O_2$	0	1625,1640
11.494743		vacuum	$^{14}C^{16}O_2$	0	1640
11.49474329		vacuum	$^{14}C^{16}O_2$	0	1640

Table 4.1.1—*continued*
Neutral Atom, Ion, and Molecular Gas Lasers Arranged in Order of Increasing Wavelength

Wavelength (μm)	Uncertainty	Medium	Species	Charge	Reference
11.4962		vacuum	$^{12}CS_2$	0	1321–1323
11.4983468		vacuum	$^{14}C^{18}O_2$	0	1640
11.498347		vacuum	$^{14}C^{18}O_2$	0	1640
11.5031		vacuum	$^{12}CS_2$	0	1321–1323
11.5043041		vacuum	$^{13}C^{16}O_2$	0	1625,1640
11.506706		vacuum	$^{13}C^{16}O_2$	0	1640
11.5099		vacuum	$^{12}CS_2$	0	1321–1323
11.5166		vacuum	$^{12}CS_2$	0	1321–1323
11.5171508		vacuum	$^{13}C^{16}O_2$	0	1625,1640
11.5199556		vacuum	$^{14}C^{18}O_2$	0	1640
11.519956		vacuum	$^{14}C^{18}O_2$	0	1640
11.52074		vacuum	$^{14}NH_3$	0	1189–1195
11.521			NH_3	0	1774,1775
11.5212		vacuum	$^{14}NH_3$	0	1189–5,1243–4
11.5237		vacuum	$^{12}CS_2$	0	1321–1323
11.52446		vacuum	$^{14}NH_3$	0	1243,1244
11.5271866		vacuum	$^{13}C^{16}O_2$	0	1625,1640
11.5307		vacuum	$^{12}CS_2$	0	1321–1323
11.5376		vacuum	$^{12}CS_2$	0	1321–1323
11.5405096		vacuum	$^{13}C^{16}O_2$	0	1625,1640
11.541894		vacuum	$^{14}C^{18}O_2$	0	1640
11.5418945		vacuum	$^{14}C^{18}O_2$	0	1640
11.5446		vacuum	$^{12}CS_2$	0	1321–1323
11.5504119		vacuum	$^{13}C^{16}O_2$	0	1625,1640
11.553		vacuum	$^{12}CS_2$	0	1321–1323
11.55466		vacuum	$^{14}NH_3$	0	1203
11.560		vacuum	$^{12}CS_2$	0	1321–1323
11.5641676		vacuum	$^{14}C^{18}O_2$	0	1640
11.564168		vacuum	$^{14}C^{18}O_2$	0	1640
11.5642439		vacuum	$^{13}C^{16}O_2$	0	1625,1640
11.568		vacuum	$^{12}CS_2$	0	1321–1323
11.5739835		vacuum	$^{13}C^{16}O_2$	0	1625,1640
11.582		vacuum	$^{12}CS_2$	0	1321–1323
11.5821		air	Xe	0	1498
11.586			NH_3	0	1775

Table 4.1.1—*continued*
Neutral Atom, Ion, and Molecular Gas Lasers Arranged in Order of Increasing Wavelength

Wavelength (μm)	Uncertainty	Medium	Species	Charge	Reference
11.586		vacuum	$^{15}NH_3$	0	1635,1775
11.586779		vacuum	$^{14}C^{18}O_2$	0	1640
11.5867790		vacuum	$^{14}C^{18}O_2$	0	1640
11.5883582		vacuum	$^{13}C^{16}O_2$	0	1625,1640
11.5979053		vacuum	$^{13}C^{16}O_2$	0	1625,1640
11.609067		vacuum	$^{14}C^{16}O_2$	0	1640
11.60906705		vacuum	$^{14}C^{16}O_2$	0	1640
11.609733		vacuum	$^{14}C^{18}O_2$	0	1640
11.6097330		vacuum	$^{14}C^{18}O_2$	0	1640
11.6128570		vacuum	$^{13}C^{16}O_2$	0	1625,1640
11.6221811		vacuum	$^{13}C^{16}O_2$	0	1625,1640
11.63081284		vacuum	$^{14}C^{16}O_2$	0	1640
11.630813		vacuum	$^{14}C^{16}O_2$	0	1640
11.633034		vacuum	$^{14}C^{18}O_2$	0	1640
11.6330341		vacuum	$^{14}C^{18}O_2$	0	1640
11.6377451		vacuum	$^{13}C^{16}O_2$	0	1625,1640
11.6468149		vacuum	$^{13}C^{16}O_2$	0	1640
11.652860		vacuum	$^{14}C^{16}O_2$	0	1640
11.65286043		vacuum	$^{14}C^{16}O_2$	0	1640
11.656687		vacuum	$^{14}C^{18}O_2$	0	1640
11.6566870		vacuum	$^{14}C^{18}O_2$	0	1640
11.6630274		vacuum	$^{13}C^{16}O_2$	0	1625,1640
11.6718107		vacuum	$^{13}C^{16}O_2$	0	1625,1640
11.67521261		vacuum	$^{14}C^{16}O_2$	0	1640
11.675213		vacuum	$^{14}C^{16}O_2$	0	1640
11.6806966		vacuum	$^{14}C^{18}O_2$	0	1640
11.680697		vacuum	$^{14}C^{18}O_2$	0	1640
11.6887088		vacuum	$^{13}C^{16}O_2$	0	1625,1640
11.6971729		vacuum	$^{13}C^{16}O_2$	0	1625,1640
11.697872		vacuum	$^{14}C^{16}O_2$	0	1640
11.69787227		vacuum	$^{14}C^{16}O_2$	0	1640
11.7050679		vacuum	$^{14}C^{18}O_2$	0	1640
11.705068		vacuum	$^{14}C^{18}O_2$	0	1640
11.712			NH_3	0	1775
11.71208		vacuum	$^{14}NH_3$	0	1630,1774–75

Table 4.1.1—*continued*
Neutral Atom, Ion, and Molecular Gas Lasers Arranged in Order of Increasing Wavelength

Wavelength (μm)	Uncertainty	Medium	Species	Charge	Reference
11.7147947		vacuum	$^{13}C^{16}O_2$	0	1625,1640
11.71582		vacuum	$^{14}NH_3$	0	1630,1774–75
11.716			NH_3	0	1774
11.720842		vacuum	$^{14}C^{16}O_2$	0	1640
11.72084238		vacuum	$^{14}C^{16}O_2$	0	1640
11.7229056		vacuum	$^{13}C^{16}O_2$	0	1625,1640
11.727			NH_3	0	1774
11.72712		vacuum	$^{14}NH_3$	0	1190,1191
11.729806		vacuum	$^{14}C^{18}O_2$	0	1640
11.7298063		vacuum	$^{14}C^{18}O_2$	0	1640
11.7412905		vacuum	$^{13}C^{16}O_2$	0	1625,1640
11.744126		vacuum	$^{14}C^{16}O_2$	0	1640
11.74412603		vacuum	$^{14}C^{16}O_2$	0	1640
11.746			NH_3	0	1775
11.74637		vacuum	$^{14}NH_3$	0	1190,1191
11.7490134		vacuum	$^{13}C^{16}O_2$	0	1640
11.763			NH_3	0	1775
11.763		vacuum	$^{15}NH_3$	0	1635,1775
11.767726		vacuum	$^{14}C^{16}O_2$	0	1640
11.76772638		vacuum	$^{14}C^{16}O_2$	0	1640
11.7682017		vacuum	$^{13}C^{16}O_2$	0	1625,1640
11.7755010		vacuum	$^{13}C^{16}O_2$	0	1625,1640
11.79164670		vacuum	$^{14}C^{16}O_2$	0	1640
11.791647		vacuum	$^{14}C^{16}O_2$	0	1640
11.794			NH_3	0	1774
11.7942		vacuum	$^{14}NH_3$	0	1630,1774–75
11.796		vacuum	$^{14}NH_3$	0	1191
11.798			NH_3	0	1775
11.798		vacuum	$^{15}NH_3$	0	1635,1775
11.7983		vacuum	$^{14}NH_3$	0	1630,1774–75
11.80167		vacuum	$^{14}NH_3$	0	1191,1195
11.815890		vacuum	$^{14}C^{16}O_2$	0	1640
11.81589039		vacuum	$^{14}C^{16}O_2$	0	1640
11.83		vacuum	H_2O	0	1328
11.84046091		vacuum	$^{14}C^{16}O_2$	0	1640

Table 4.1.1—*continued*
Neutral Atom, Ion, and Molecular Gas Lasers
Arranged in Order of Increasing Wavelength

Wavelength (μm)	Uncertainty	Medium	Species	Charge	Reference
11.840461		vacuum	$^{14}C^{16}O_2$	0	1640
11.859		vacuum	$^{15}NH_3$	0	1635,1775
11.8605		vacuum	Ne	0	564
11.863			HF	0	1735
11.86536187		vacuum	$^{14}C^{16}O_2$	0	1640
11.865362		vacuum	$^{14}C^{16}O_2$	0	1640
11.866			NH_3	0	1775
11.866		vacuum	$^{15}NH_3$	0	1635,1775
11.890597		vacuum	$^{14}C^{16}O_2$	0	1640
11.8905970		vacuum	$^{14}C^{16}O_2$	0	1640
11.9020		vacuum	Ne	0	195
11.916170		vacuum	$^{14}C^{16}O_2$	0	1640
11.9161700		vacuum	$^{14}C^{16}O_2$	0	1640
11.942085		vacuum	$^{14}C^{16}O_2$	0	1640
11.9420850		vacuum	$^{14}C^{16}O_2$	0	1640
11.959		vacuum	$^{13}CS_2$	0	1323
11.96		vacuum	H_2O	0	1328
11.963		vacuum	$^{13}CS_2$	0	1323
11.968346		vacuum	$^{14}C^{16}O_2$	0	1640
11.9683460		vacuum	$^{14}C^{16}O_2$	0	1640
11.97859		vacuum	$^{14}NH_3$	0	1189–1192
11.979			NH_3	0	1774
11.983		vacuum	$^{13}CS_2$	0	1323
11.990			NH_3	0	1774
11.99025		vacuum	$^{14}NH_3$	0	1189–1192
11.994957		vacuum	$^{14}C^{16}O_2$	0	1640
11.9949572		vacuum	$^{14}C^{16}O_2$	0	1640
12.010			NH_3	0	1775
12.01008		vacuum	$^{14}NH_3$	0	1189–1196
12.0219229		vacuum	$^{14}C^{16}O_2$	0	1640
12.021923		vacuum	$^{14}C^{16}O_2$	0	1640
12.03872		vacuum	$^{14}NH_3$	0	1189–1196
12.039			NH_3	0	1774
12.0492476		vacuum	$^{14}C^{16}O_2$	0	1640
12.049248		vacuum	$^{14}C^{16}O_2$	0	1640

Table 4.1.1—*continued*
Neutral Atom, Ion, and Molecular Gas Lasers Arranged in Order of Increasing Wavelength

Wavelength (μm)	Uncertainty	Medium	Species	Charge	Reference
12.063			NH_3	0	1775
12.063		vacuum	$^{15}NH_3$	0	1635,1775
12.076936		vacuum	$^{14}C^{16}O_2$	0	1640
12.0769360		vacuum	$^{14}C^{16}O_2$	0	1640
12.079			NH_3	0	1775
12.07912		vacuum	$^{14}NH_3$	0	1630,1774–75
12.0797		vacuum	$^{14}NH_3$	0	1189–1199
12.080			NH_3	0	1774
12.0997		vacuum	$^{14}NH_3$	0	1630,1774–75
12.100			NH_3	0	1774
12.1049927		vacuum	$^{14}C^{16}O_2$	0	1640
12.104993		vacuum	$^{14}C^{16}O_2$	0	1640
12.11418		vacuum	$^{14}NH_3$	0	1203,1204
12.1334228		vacuum	$^{14}C^{16}O_2$	0	1640
12.133423		vacuum	$^{14}C^{16}O_2$	0	1640
12.140	±0.0012	vacuum	Ar	0	1452
12.1405		vacuum	Ar	0	195,559,564
12.1464		vacuum	Ar	0	195,559,564
12.147	±0.0012	vacuum	Ar	0	1452
12.148			NH_3	0	1775
12.148		vacuum	$^{15}NH_3$	0	1211
12.15575		vacuum	$^{14}NH_3$	0	1203
12.18444		vacuum	$^{14}NH_3$	0	1203
12.214		vacuum	$^{13}CS_2$	0	1323
12.237		vacuum	$^{13}CS_2$	0	1323
12.245			NH_3	0	1775
12.24521		vacuum	$^{14}NH_3$	0	1189–1199
12.247		vacuum	$^{13}CS_2$	0	1323
12.249			NH_3	0	1774
12.24911		vacuum	$^{14}NH_3$	0	1189–1192
12.261			NH_3	0	1774
12.26105		vacuum	$^{14}NH_3$	0	1189–1192
12.262			HF	0	1618
12.266	±0.0037	vacuum	Xe	0	1452
12.2663		vacuum	Xe	0	559,582

Table 4.1.1—*continued*
Neutral Atom, Ion, and Molecular Gas Lasers Arranged in Order of Increasing Wavelength

Wavelength (μm)	Uncertainty	Medium	Species	Charge	Reference
12.281			NH_3	0	1775
12.28136		vacuum	$^{14}NH_3$	0	1189–1192
12.299			NH_3	0	1775
12.299		vacuum	$^{15}NH_3$	0	1635,1775
12.31072		vacuum	$^{14}NH_3$	0	1189–1192
12.311			NH_3	0	1774
12.336			NH_3	0	1775
12.336		vacuum	$^{15}NH_3$	0	1635,1775
12.350			NH_3	0	1774
12.35002		vacuum	$^{14}NH_3$	0	1189–1192
12.37821		vacuum	$^{14}NH_3$	0	1191
12.384			NH_3	0	1775
12.38425		vacuum	$^{14}NH_3$	0	1191
12.39560		vacuum	$^{14}NH_3$	0	1191
12.4027		vacuum	$^{14}NH_3$	0	1630,1774–75
12.403			NH_3	0	1774
12.447		vacuum	$^{15}NH_3$	0	1635,1775
12.52781		vacuum	$^{14}NH_3$	0	1189–2,1243–4
12.528			NH_3	0	1774
12.53999		vacuum	$^{14}NH_3$	0	1189–2,1243–4
12.540			NH_3	0	1774
12.56068		vacuum	$^{14}NH_3$	0	1189–2,1243–4
12.561			NH_3	0	1775
12.5688		vacuum	K	0	131,142
12.59		vacuum	$^{14}NH_3$	0	1631–1634
12.59059		vacuum	$^{14}NH_3$	0	1189–2,1243–4
12.591			NH_3	0	1774
12.616			NH_3	0	1775
12.616		vacuum	$^{15}NH_3$	0	1635,1775
12.63063		vacuum	$^{14}NH_3$	0	1189–2,1243–4
12.631			NH_3	0	1774
12.678			HF	0	1618
12.682			NH_3	0	1775
12.68213		vacuum	$^{14}NH_3$	0	1189–92
12.697			NH_3	0	1774

Table 4.1.1—*continued*
Neutral Atom, Ion, and Molecular Gas Lasers Arranged in Order of Increasing Wavelength

Wavelength (μm)	Uncertainty	Medium	Species	Charge	Reference
12.69716		vacuum	$^{14}NH_3$	0	1191
12.701			HF	0	1618
12.7196		vacuum	$^{14}NH_3$	0	1630,1774–75
12.720			NH_3	0	1774
12.739			NH_3	0	1775
12.739		vacuum	$^{15}NH_3$	0	1635,1775
12.80959		vacuum	$^{14}NH_3$	0	1205
12.811			NH_3	0	1775
12.81145		vacuum	$^{14}NH_3$	0	1189–04,1244
12.81532		vacuum	$^{14}NH_3$	0	1189–92
12.82765		vacuum	$^{14}NH_3$	0	1190–91,1244
12.835	±0.0039	vacuum	Ne	0	583,586,1452
12.8353		vacuum	Ne	0	195,564
12.84863		vacuum	$^{14}NH_3$	0	1189–1192
12.849			NH_3	0	1775
12.85		vacuum	HCN	0	1328
12.867		vacuum	$^{15}NH_3$	0	1635,1775
12.87890		vacuum	$^{14}NH_3$	0	1189–1192
12.905			NH_3	0	1775
12.905		vacuum	$^{15}NH_3$	0	1635,1775
12.917	±0.0039		vacuum	Xe	1452
12.9173		vacuum	Xe	0	559,582
12.91946		vacuum	$^{14}NH_3$	0	1189–1192
12.967			NH_3	0	1775
12.97163		vacuum	$^{14}NH_3$	0	1189–1192
12.976			NH_3	0	1775
12.977			NH_3	0	1775
12.977		vacuum	$^{15}NH_3$	0	1635,1775
12.98352		vacuum	$^{14}NH_3$	0	1205
13.024			NH_3	0	1774
13.0241		vacuum	$^{14}NH_3$	0	1630,1774–75
13.030			NH_3	0	1775
13.030		vacuum	$^{15}NH_3$	0	1635,1775
13.03715		vacuum	$^{14}NH_3$	0	1189–1192
13.0505		vacuum	$^{14}NH_3$	0	1630,1774–75

Table 4.1.1—*continued*
Neutral Atom, Ion, and Molecular Gas Lasers Arranged in Order of Increasing Wavelength

Wavelength (μm)	Uncertainty	Medium	Species	Charge	Reference
13.051			NH_3	0	1774
13.11233		vacuum	$^{14}NH_3$	0	1189,1210
13.12477		vacuum	$^{14}NH_3$	0	1189,1210
13.144		vacuum	$^{12}C^{16}O_2$	0	1640
13.14593		vacuum	$^{14}NH_3$	0	1630,1774–75
13.146			NH_3	0	1775
13.154		vacuum	$^{12}C^{16}O_2$	0	1640
13.160		vacuum	$^{12}C^{16}O_2$	0	1640
13.17643		vacuum	$^{14}NH_3$	0	1189,1210
13.188			HF	0	1618
13.1887		vacuum	Cd	0	155
13.201			HF	0	1618
13.204		vacuum	$^{15}NH_3$	0	1635,1775
13.21725		vacuum	$^{14}NH_3$	0	1189,1210
13.221			HF	0	1618
13.23390		vacuum	$^{14}NH_3$	0	1196
13.26978		vacuum	$^{14}NH_3$	0	1189,1210
13.270			NH_3	0	1775
13.323		vacuum	$^{15}NH_3$	0	1635,1775
13.33580		vacuum	$^{14}NH_3$	0	1210
13.411		vacuum	$^{14}NH_3$	0	1210
13.415			NH_3	0	1774
13.4153		vacuum	$^{14}NH_3$	0	1630,1774–75
13.453			NH_3	0	1774
13.4534		vacuum	$^{14}NH_3$	0	1630,1774–75
13.473		vacuum	$^{15}NH_3$	0	1635,1775
13.475		air	Ar	0	1498
13.54		vacuum	CF_3I	0	1181
13.541		vacuum	$^{12}C^{16}O_2$	0	1640
13.57		vacuum	CF_3I	0	1181
13.57749		vacuum	$^{14}NH_3$	0	1210
13.578			NH_3	0	1774
13.63		vacuum	CF_3I	0	1181
13.65533		vacuum	$^{14}NH_3$	0	1196
13.72555		vacuum	$^{14}NH_3$	0	1204,1205

Table 4.1.1—*continued*
Neutral Atom, Ion, and Molecular Gas Lasers Arranged in Order of Increasing Wavelength

Wavelength (μm)	Uncertainty	Medium	Species	Charge	Reference
13.728			HF	0	1618
13.7583		vacuum	Ne	0	195,559,564
13.7584		vacuum	Ne	0	195,559,564
13.759	±0.0041	vacuum	Ne	0	583,586,1452
13.784			HF	0	1618
13.826			NH_3	0	1774
13.82608		vacuum	$^{14}NH_3$	0	1210
13.87		vacuum	$^{12}C^{16}O_2$	0	1640
13.872			HCl	0	1755
13.910		vacuum	$^{15}NH_3$	0	1636
14.099			HCl	0	1755
14.1		vacuum	$^{12}C^{16}O_2$	0	1640
14.1681		vacuum	$^{2}C^{16}O_2$	0	1640
14.2001		vacuum	$^{12}C^{16}O_2$	0	1640
14.2325		vacuum	$^{12}C^{16}O_2$	0	1640
14.288			HF	0	1618
14.3		vacuum	$^{15}NH_3$	0	1211
14.441			HF	0	1618
14.5820		vacuum	Cd	0	155
14.78		vacuum	$^{14}NH_3$	0	1206
14.78			NH_3	0	1811
14.8		vacuum	$^{15}NH_3$	0	1211
14.930		vacuum	Ne	0	564
15.022	±0.0015	vacuum	Ar	0	1452
15.037	±0.0015	vacuum	Ar	0	1452
15.0370		air	Ar	0	559,564
15.039	±0.0015	vacuum	Ar	0	1452
15.04		vacuum	$^{14}NH_3$	0	1206
15.04			NH_3	0	1811
15.0421		vacuum	Ar	0	559,564
15.061			HF	0	1618
15.08		vacuum	$^{14}NH_3$	0	1206
15.08			NH_3	0	1811
15.19		vacuum	NSF	0	1239
15.2		vacuum	$^{15}NH_3$	0	1211

Table 4.1.1—*continued*
Neutral Atom, Ion, and Molecular Gas Lasers Arranged in Order of Increasing Wavelength

Wavelength (μm)	Uncertainty	Medium	Species	Charge	Reference
15.26		vacuum	NSF	0	1239
15.29		vacuum	$^{13}CF_4$	0	1217,1337
15.306		vacuum	$^{12}CF_4$	0	1181,1212
15.32		vacuum	$^{13}CF_4$	0	1217,1337
15.34		vacuum	NSF	0	1239
15.39		vacuum	$^{14}CF_4$	0	1337,1488
15.40		vacuum	$^{12}CF_4$	0	1181,1212
15.41		vacuum	$^{14}NH_3$	0	1206
15.41			NH_3	0	1811
15.42		vacuum	NSF	0	1239
15.42		vacuum	$^{13}CF_4$	0	1217,1337
15.44		vacuum	$^{13}CF_4$	0	1217,1337
15.45		vacuum	$^{13}CF_4$	0	1213
15.46		vacuum	$^{12}CF_4$	0	1181
15.46		vacuum	$^{14}CF_4$	0	1337,1488
15.47		vacuum	$^{13}CF_4$	0	1217,1337
15.47		vacuum	NSF	0	1239
15.47		vacuum	$^{14}NH_3$	0	1206
15.47			NH_3	0	1811
15.48		vacuum	$^{12}CF_4$	0	1212–13,1488
15.49		vacuum	$^{12}CF_4$	0	1488
15.51		vacuum	$^{14}CF_4$	0	1337,1488
15.53		vacuum	$^{13}CF_4$	0	1217,1337
15.53		vacuum	CH_3CCH	0	1637
15.54		vacuum	$^{14}CF_4$	0	1337,1488
15.54		vacuum	CH_3CCH	0	1637
15.54		vacuum	$^{14}CF_4$	0	1337,1488
15.54		vacuum	$^{13}CF_4$	0	1213,1217
15.547		vacuum	$^{12}CF_4$	0	1181,1212–13
15.57		vacuum	$^{12}CF_4$	0	1181,1488
15.58		vacuum	$^{14}CF_4$	0	1337,1488
15.58		vacuum	$^{12}CF_4$	0	1213
15.59		vacuum	$^{13}CF_4$	0	1217,1337
15.5936		vacuum	$^{12}C^{16}O_2$	0	1640
15.60		vacuum	$^{14}CF_4$	0	1337,1488

Table 4.1.1—*continued*
Neutral Atom, Ion, and Molecular Gas Lasers Arranged in Order of Increasing Wavelength

Wavelength (μm)	Uncertainty	Medium	Species	Charge	Reference
15.60		vacuum	$^{12}CF_4$	0	1213
15.607		vacuum	$^{12}CF_4$	0	1181,1212–13
15.61		vacuum	NSF	0	1239
15.61		vacuum	$^{14}CF_4$	0	1337,1488
15.62		vacuum	$^{13}CF_4$	0	1217,1337
15.62		vacuum	$^{14}CF_4$	0	1337,1488
15.6311		vacuum	$^{12}C^{16}O_2$	0	1640
15.64		vacuum	$^{14}CF_4$	0	1337,1488
15.65		vacuum	$^{14}CF_4$	0	1337,1488
15.6688		vacuum	$^{12}C^{16}O_2$	0	1640
15.68		vacuum	$^{14}NH_3$	0	1206
15.7		vacuum	$^{15}NH_3$	0	1211
15.70		vacuum	$^{14}NH_3$	0	1631–1634
15.7067		vacuum	$^{12}C^{16}O_2$	0	1640
15.71		vacuum	$^{12}CF_4$	0	1181,1488
15.71		vacuum	$^{14}CF_4$	0	1337,1488
15.711		vacuum	CH_3CCH	0	1246
15.716		vacuum	CH_3CCH	0	1246
15.72		vacuum	CH_3CCH	0	1637
15.72		vacuum	CH_3CCH	0	1246
15.721		vacuum	CH_3CCH	0	1246
15.74		vacuum	$^{13}CF_4$	0	1213
15.74		vacuum	$^{12}CF_4$	0	1488
15.7449		vacuum	$^{12}C^{16}O_2$	0	1640
15.75		vacuum	$^{13}CF_4$	0	1213
15.78		vacuum	$^{13}CF_4$	0	1217,1337
15.78148		vacuum	$^{14}NH_3$	0	1196
15.7833		vacuum	$^{12}C^{16}O_2$	0	1640
15.79		vacuum	$^{14}CF_4$	0	1337,1488
15.80		vacuum	CH_3CCH	0	1637
15.80		vacuum	$^{13}CF_4$	0	1488
15.81600		vacuum	$^{14}NH_3$	0	1205
15.82		vacuum	$^{13}CF_4$	0	1213
15.82		vacuum	$^{14}CF_4$	0	1337,1488
15.8220		vacuum	$^{12}C^{16}O_2$	0	1640

Table 4.1.1—*continued*
Neutral Atom, Ion, and Molecular Gas Lasers Arranged in Order of Increasing Wavelength

Wavelength (μm)	Uncertainty	Medium	Species	Charge	Reference
15.83		vacuum	$^{12}CF_4$	0	1212
15.83		vacuum	CH_3CCH	0	1637
15.84		vacuum	$^{12}CF_4$	0	1488
15.838		vacuum	NSF	0	1239
15.84		vacuum	$^{12}CF_4$	0	1212
15.844		vacuum	$^{12}CF_4$	0	1212,1213
15.845		vacuum	$^{12}CF_4$	0	1181
15.847		vacuum	$^{12}CF_4$	0	1181,1212
15.85		vacuum	$^{12}CF_4$	0	1212
15.85726		vacuum	$^{14}NH_3$	0	1196
15.8609		vacuum	$^{12}C^{16}O_2$	0	1640
15.87758		vacuum	$^{14}NH_3$	0	1196
15.89		vacuum	$^{13}CF_4$	0	1217,1337
15.89		vacuum	NSF	0	1239
15.9000		vacuum	$^{12}C^{16}O_2$	0	1640
15.905		vacuum	SF_6	0	1221
15.91		vacuum	$^{12}CF_4$	0	1213,1488
15.91292		vacuum	$^{14}NH_3$	0	1205
15.913		vacuum	$^{14}NH_3$	0	1631–1634
15.92		vacuum	$^{12}CF_4$	0	1488
15.93		vacuum	$^{14}CF_4$	0	1337,1488
15.94		vacuum	$^{14}CF_4$	0	1337,1488
15.9394		vacuum	$^{12}C^{16}O_2$	0	1640
15.94		vacuum	$^{12}CF_4$	0	1488
15.94		vacuum	$^{14}CF_4$	0	1337,1488
15.94637		vacuum	$^{14}NH_3$	0	1203,1204
15.95		vacuum	$^{14}CF_4$	0	1337,1488
15.9680		vacuum	K	0	131,142
15.98		vacuum	NSF	0	1239
15.9790		vacuum	$^{12}C^{16}O_2$	0	1640
15.99		vacuum	$^{14}CF_4$	0	1337,1488
16.0		vacuum	$^{15}NH_3$	0	1211
16.00		vacuum	$^{13}CF_4$	0	1217
16.00		vacuum	$^{12}CF_4$	0	1488
16.0			CF_4	0	1768

Table 4.1.1—*continued*
Neutral Atom, Ion, and Molecular Gas Lasers Arranged in Order of Increasing Wavelength

Wavelength (μm)	Uncertainty	Medium	Species	Charge	Reference
16.02		vacuum	$^{12}CF_4$	0	1213
16.022			HF	0	1618
16.03		vacuum	NSF	0	1239
16.04		vacuum	$^{14}CF_4$	0	1337,1488
16.04		vacuum	$^{14}NH_3$	0	1631–1634
16.06		vacuum	$^{12}CF_4$	0	1488
16.06		vacuum	CH_3CCH	0	1637
16.07		vacuum	$^{12}CF_4$	0	1488
16.08		vacuum	$^{14}CF_4$	0	1337,1488
16.09		vacuum	$^{12}CF_4$	0	1488
16.09		vacuum	$^{13}CF_4$	0	1488
16.10		vacuum	CH_3CCH	0	1637
16.10		vacuum	$^{13}CF_4$	0	1488
16.11		vacuum	$^{13}CF_4$	0	1488
16.11		vacuum	$^{12}CF_4$	0	1212,1488
16.121		vacuum	CH_3CCH	0	1246
16.14		vacuum	$^{14}CF_4$	0	1337,1488
16.15		vacuum	$^{13}CF_4$	0	1213
16.15		vacuum	$^{14}CF_4$	0	1337,1488
16.15		vacuum	NSF	0	1239
16.16		vacuum	$^{13}CF_4$	0	1217
16.178		vacuum	$^{12}CF_4$	0	1181,1212–13
16.1819		vacuum	$^{12}C^{16}O_2$	0	1640
16.1827		vacuum	$^{12}C^{16}O_2$	0	1640
16.1836		vacuum	$^{12}C^{16}O_2$	0	1640
16.1847		vacuum	$^{12}C^{16}O_2$	0	1640
16.1860		vacuum	$^{12}C^{16}O_2$	0	1640
16.1874		vacuum	$^{12}C^{16}O_2$	0	1640
16.1891		vacuum	$^{12}C^{16}O_2$	0	1640
16.19		vacuum	NSF	0	1239
16.1908		vacuum	$^{12}C^{16}O_2$	0	1640
16.1928		vacuum	$^{12}C^{16}O_2$	0	1640
16.1949		vacuum	$^{12}C^{16}O_2$	0	1640
16.1972		vacuum	$^{12}C^{16}O_2$	0	1640
16.20		vacuum	$^{13}CF_4$	0	1217

Table 4.1.1—*continued*
Neutral Atom, Ion, and Molecular Gas Lasers Arranged in Order of Increasing Wavelength

Wavelength (μm)	Uncertainty	Medium	Species	Charge	Reference
16.21		vacuum	$^{12}CF_4$	0	1212
16.21		vacuum	$^{14}CF_4$	0	1337,1488
16.213			HCl	0	1755
16.22		vacuum	$^{14}CF_4$	0	1337,1488
16.23		vacuum	$^{13}CF_4$	0	1213,1337
16.24		vacuum	$^{12}CF_4$	0	1212
16.25		vacuum	$^{12}CF_4$	0	1488
16.25		vacuum	$^{13}CF_4$	0	1488
16.25		vacuum	$^{14}CF_4$	0	1337,1488
16.259		vacuum	$^{12}CF_4$	0	1181,1212–13
16.26		vacuum	$^{13}CF_4$	0	1337
16.27		vacuum	$^{12}CF_4$	0	1212
16.27		vacuum	$^{13}CF_4$	0	1488
16.28		vacuum	$^{14}CF_4$	0	1337,1488
16.29		vacuum	$^{12}CF_4$	0	1212
16.29		vacuum	$^{13}CF_4$	0	1217
16.30		vacuum	$^{14}CF_4$	0	1337,1488
16.30		vacuum	$^{13}CF_4$	0	1213
16.32		vacuum	$^{13}CF_4$	0	1337
16.32		vacuum	$FClO_3$	0	1245
16.33		vacuum	$^{14}CF_4$	0	1337,1488
16.33		vacuum	$FClO_3$	0	1245
16.340		vacuum	$^{12}CF_4$	0	1181,1212–13
16.35		vacuum	$^{14}CF_4$	0	1337,1488
16.35		vacuum	$FClO_3$	0	1245
16.36		vacuum	$^{13}CF_4$	0	1337
16.3670		vacuum	$^{12}C^{16}O_2$	0	1640
16.39		vacuum	CH_3CCH	0	1246
16.4	±0.1	vacuum	NOCl	0	1180,1181
16.40		vacuum	$^{12}CF_4$	0	1212
16.4091		vacuum	$^{12}C^{16}O_2$	0	1640
16.42		vacuum	CH_3CCH	0	1246
16.44		vacuum	$^{14}CF_4$	0	1337,1488
16.444			HF	0	1735
16.45		vacuum	$FClO_3$	0	1245

Table 4.1.1—*continued*
Neutral Atom, Ion, and Molecular Gas Lasers Arranged in Order of Increasing Wavelength

Wavelength (μm)	Uncertainty	Medium	Species	Charge	Reference
16.4515		vacuum	$^{12}C^{16}O_2$	0	1640
16.48		vacuum	$^{13}CF_4$	0	1337
16.49		vacuum	$FClO_3$	0	1245
16.49		vacuum	$FClO_3$	0	1245
16.4941		vacuum	$^{12}C^{16}O_2$	0	1640
16.50		vacuum	$FClO_3$	0	1245
16.52		vacuum	$FClO_3$	0	1245
16.52		vacuum	NOCl	0	1180,1181
16.52		vacuum	$FClO_3$	0	1245
16.5370		vacuum	$^{12}C^{16}O_2$	0	1640
16.56		vacuum	$FClO_3$	0	1245
16.56		vacuum	$FClO_3$	0	1245
16.57		vacuum	NOCl	0	1180,1181
16.5801		vacuum	$^{12}C^{16}O_2$	0	1640
16.585		vacuum	$^{12}C^{16}O_2$	0	1640
16.596		vacuum	$^{12}C^{16}O^{18}O$	0	1640
16.597		vacuum	$^{12}C^{16}O_2$	0	1640
16.609			HCl	0	1755
16.61		vacuum	$FClO_3$	0	1245
16.6235		vacuum	$^{12}C^{16}O_2$	0	1640
16.638	±0.005	vacuum	Ne	0	583,586,1452
16.638076		vacuum	Ne	0	195,564
16.644			HCl	0	1735
16.655			HF	0	1735
16.66		vacuum	$FClO_3$	0	1245
16.6672		vacuum	$^{12}C^{16}O_2$	0	1640
16.667472		vacuum	Ne	0	564
16.668	±0.005	vacuum	Ne	0	583,586,1452
16.69		vacuum	NOCl	0	1180,1181
16.7	±0.1	vacuum	NOCl	0	1180,1181
16.7111		vacuum	$^{12}C^{16}O_2$	0	1640
16.73		vacuum	$FClO_3$	0	1245
16.75		vacuum	NOCl	0	1180,1181
16.75		vacuum	$FClO_3$	0	1245
16.7553		vacuum	$^{12}C^{16}O_2$	0	1640

Table 4.1.1—*continued*
Neutral Atom, Ion, and Molecular Gas Lasers Arranged in Order of Increasing Wavelength

Wavelength (μm)	Uncertainty	Medium	Species	Charge	Reference
16.76		vacuum	$FClO_3$	0	1245
16.76		vacuum	$^{12}C^{16}O^{18}O$	0	1640
16.76		vacuum	$FClO_3$	0	1245
16.765			HCl	0	1735
16.77		vacuum	$FClO_3$	0	1245
16.79		vacuum	$FClO_3$	0	1245
16.7998		vacuum	$^{12}C^{16}O_2$	0	1640
16.82		vacuum	$FClO_3$	0	1245
16.86		vacuum	NOCl	0	1180,1181
16.893	±0.005	vacuum	Ne	0	583,586,1452
16.8932		vacuum	Ne	0	559,564
16.9	±0.1	vacuum	NOCl	0	1180,1181
16.927		vacuum	$^{12}C^{16}O^{18}O$	0	1640
16.93		vacuum	$FClO_3$	0	1245
16.931			H_2O	0	1713
16.936		vacuum	$^{14}NH_3$	0	1631–1634
16.946194		vacuum	Ne	0	195,559,564
16.947	±0.005	vacuum	Ne	0	583,586,1452
16.95		vacuum	$^{14}NH_3$	0	1631–1634
16.970		vacuum	$^{12}C^{16}O^{18}O$	0	1640
16.975			HF	0	1735
16.99		vacuum	NOCl	0	1180,1181
17.000		vacuum	$FClO_3$	0	1245
17.023		vacuum	$^{12}C^{16}O_2$	0	1640
17.029		vacuum	$^{12}C^{16}O_2$	0	1640
17.034			HCl	0	1755
17.036		vacuum	$^{12}C^{16}O_2$	0	1640
17.048		vacuum	$^{12}C^{16}O_2$	0	1640
17.0709		air	Kr	0	1498
17.125			HCl	0	1735
17.15		vacuum	$FClO_3$	0	1245
17.1572		vacuum	Ne	0	559,564
17.158	±0.0051	vacuum	Ne	0	583,586,1452
17.188155		vacuum	Ne	0	564
17.189	±0.0052	vacuum	Ne	0	583,586,1452

Table 4.1.1—*continued*
Neutral Atom, Ion, and Molecular Gas Lasers Arranged in Order of Increasing Wavelength

Wavelength (μm)	Uncertainty	Medium	Species	Charge	Reference
17.19		vacuum	$FClO_3$	0	1245
17.22		vacuum	$FClO_3$	0	1245
17.2328		air	Kr	0	1498
17.26		vacuum	$FClO_3$	0	1245
17.28		vacuum	$FClO_3$	0	1245
17.32		vacuum	$FClO_3$	0	1245
17.325			HF	0	1735
17.36		vacuum	$FClO_3$	0	1245
17.370		vacuum	$^{12}C^{16}O_2$	0	1640
17.376		vacuum	$^{12}C^{16}O_2$	0	1640
17.390		vacuum	$^{12}C^{16}O_2$	0	1640
17.44		vacuum	$FClO_3$	0	1245
17.45		vacuum	C_2D_2	0	1242,1801–02
17.46		vacuum	$FClO_3$	0	1245
17.463		vacuum	$^{12}C^{18}O_2$	0	1640
17.492			HCl	0	1755
17.498		vacuum	C_2D_2	0	1242,1801–02
17.56		vacuum	C_2D_2	0	1242,1801–02
17.575			HCl	0	1735
17.58		vacuum	$FClO_3$	0	1245
17.61		vacuum	C_2D_2	0	1242,1801–02
17.610		vacuum	C_2D_2	0	1242,1801–02
17.645			HF	0	1735
17.665		vacuum	C_2D_2	0	1242,1801–02
17.71		vacuum	$FClO_3$	0	1245
17.71		vacuum	$^{12}C_2HD$	0	1627
17.722		vacuum	C_2D_2	0	1242,1801–02
17.730		vacuum	$^{12}C^{18}O_2$	0	1640
17.775		vacuum	$^{12}C^{18}O_2$	0	1640
17.778		vacuum	C_2D_2	0	1242,1801–02
17.8		vacuum	$^{15}NH_3$	0	1211
17.8035		vacuum	Ne	0	559,564
17.804	±0.0053	vacuum	Ne	0	583,586,1452
17.821		vacuum	$^{12}C^{18}O_2$	0	1640
17.835		vacuum	C_2D_2	0	1242,1801–02

Table 4.1.1—*continued*
Neutral Atom, Ion, and Molecular Gas Lasers Arranged in Order of Increasing Wavelength

Wavelength (μm)	Uncertainty	Medium	Species	Charge	Reference
17.839876		vacuum	Ne	0	559,564
17.841	±0.0054	vacuum	Ne	0	583,586,1452
17.888	±0.0054	vacuum	Ne	0	583,586,1452
17.888255		vacuum	Ne	0	195,559,564
17.893		vacuum	$^{12}C_2D_2$	0	1628
17.915		vacuum	$^{12}C^{18}O_2$	0	1640
17.962		vacuum	$^{12}C^{18}O_2$	0	1640
17.987			HCl	0	1755
17.997			HCl	0	1735
18.010		vacuum	$^{12}C^{18}O_2$	0	1640
18.035			HCl	0	1735
18.046		vacuum	$^{14}NH_3$	0	1631–1634
18.053		vacuum	$^{12}C^{18}O_2$	0	1640
18.085			HF	0	1735
18.09		vacuum	$^{12}C_2HD$	0	1627
18.203		vacuum	$^{14}NH_3$	0	1631–1634
18.21			NH_3	0	1811
18.21		vacuum	$^{14}NH_3$	0	1631–1634
18.3		vacuum	BCl_3	0	1183,1431
18.30	±0.090	vacuum	$^{10}BCl_3$	0	1183,1431
18.395067		vacuum	Ne	0	195,559,564
18.396	±0.0055	vacuum	Ne	0	583,586,1452
18.42		vacuum	$^{12}C^{16}O_2$	0	1642
18.45		vacuum	$^{12}C_2HD$	0	1627
18.505324		vacuum	Xe	0	559,564
18.506	±0.0056	vacuum	Xe	0	1452
18.522			HCl	0	1755
18.555			HCl	0	1735
18.593			HCl	0	1735
18.64		vacuum	$^{12}C_2HD$	0	1627
18.67		vacuum	C_2D_2	0	1242,1801–02
18.79		vacuum	C_2D_2	0	1242,1801–02
18.79		vacuum	C_2D_2	0	1242,1801–02
18.798		vacuum	$^{14}NH_3$	0	1631–1634
18.8		vacuum	BCl_3	0	1183,1431

Table 4.1.1—*continued*
Neutral Atom, Ion, and Molecular Gas Lasers Arranged in Order of Increasing Wavelength

Wavelength (μm)	Uncertainty	Medium	Species	Charge	Reference
18.80	±0.092	vacuum	$^{11}BCl_3$	0	1183,1431
18.801			HF	0	1618
18.84		vacuum	$^{12}C_2D_2$	0	1628
18.84		vacuum	C_2D_2	0	1242,1801–02
18.85		vacuum	C_2D_2	0	1242,1801–02
18.92674		vacuum	$^{14}NH_3$	0	1203,1204
18.960		vacuum	C_2D_2	0	1242,1801–02
18.97		vacuum	C_2D_2	0	1242,1801–02
18.983		vacuum	^{12}COS	0	1320
19.03		vacuum	C_2D_2	0	1242,1801–02
19.03		vacuum	C_2D_2	0	1242,1801–02
19.03		vacuum	C_2D_2	0	1242,1801–02
19.057		vacuum	^{12}COS	0	1320
19.081		vacuum	C_2D_2	0	1242,1801–02
19.1		vacuum	BCl_3	0	1183,1431
19.10±0.094		vacuum	$^{10}BCl_3$	0	1183,1431
19.113			HF	0	1618
19.122			HCl	0	1735
19.13		vacuum	$^{12}C_2HD$	0	1627
19.13		vacuum	C_2D_2	0	1242,1801–02
19.145			HCl	0	1735
19.18		vacuum	$^{12}C_2HD$	0	1627
19.183			HCl	0	1735
19.20		vacuum	C_2D_2	0	1242,1801–02
19.27		vacuum	C_2D_2	0	1242,1801–02
19.29		vacuum	$^{14}NH_3$	0	1631–1634
19.33		vacuum	$^{12}C_2HD$	0	1627
19.37		vacuum	$^{12}C_2HD$	0	1627
19.399			HBr	0	1735
19.4		vacuum	BCl_3	0	1183,1431
19.40		vacuum	$^{12}C_2HD$	0	1627
19.40		vacuum	$^{12}C_2HD$	0	1627
19.40	±0.095	vacuum	$^{10}BCl_3$	0	1183,1431
19.401		vacuum	$^{14}NH_3$	0	1631–1634
19.48		vacuum	$^{12}C_2HD$	0	1627

Table 4.1.1—*continued*
Neutral Atom, Ion, and Molecular Gas Lasers Arranged in Order of Increasing Wavelength

Wavelength (μm)	Uncertainty	Medium	Species	Charge	Reference
19.511		vacuum	$^{12}C_2D_2$	0	1627
19.52			CH_3OH	0	1797
19.55019		vacuum	$^{14}NH_3$	0	1203,1204
19.634		vacuum	$^{12}C_2D_2$	0	1628
19.67		vacuum	C_2D_2	0	1242,1801–02
19.700			HCl	0	1755
19.758		vacuum	$^{12}C_2D_2$	0	1628
19.783			HCl	0	1735
19.821			HCl	0	1735
19.884		vacuum	$^{12}C_2D_2$	0	1628
19.947		vacuum	C_2D_2	0	1242,1801–02
19.988			HBr	0	1735
20.134			HF	0	1618
20.2	±0.099	vacuum	BCl_3	0	1183,1431
20.346			HCl	0	1755
20.351			HF	0	1618
20.360			HBr	0	1735
20.38798		vacuum	$^{14}NH_3$	0	1205
20.411			HCl	0	1755
20.480	±0.0051	vacuum	Ne	0	583,586,1452
20.6	±0.10	vacuum	BCl_3	0	1183,1431
20.896			HBr	0	1735
20.939			HF	0	1618
20.949			HBr	0	1735
20.999			HCl	0	1755
21.047			HCl	0	1755
21.05409		vacuum	$^{14}NH_3$	0	1205
21.156			HCl	0	1755
21.333			NH_3	0	1631–1634
21.46			NH_3	0	1811
21.471		vacuum	$^{14}NH_3$	0	1206–1208
21.501			HBr	0	1735
21.546			HBr	0	1735
21.699			HF	0	1618
21.752	±0.0054	vacuum	Ne	0	583,586,1452

Table 4.1.1—*continued*
Neutral Atom, Ion, and Molecular Gas Lasers Arranged in Order of Increasing Wavelength

Wavelength (μm)	Uncertainty	Medium	Species	Charge	Reference
21.789			HF	0	1618
21.813			HCl	0	1755
21.971			HCl	0	1755
22.136			HBr	0	1735
22.226			HBr	0	1735
22.4	±0.11	vacuum	BCl_3	0	1183,1431
22.54			NH_3	0	1811
22.542		vacuum	$^{14}NH_3$	0	1206–1208
22.563		vacuum	$^{14}NH_3$	0	1206–1208
22.651			HCl	0	1755
22.71			NH_3	0	1206,1811
22.836	±0.0057	vacuum	Ne	0	583,586,1452
22.855			HBr	0	1735
22.864			HCl	0	1755
23.0	±0.11	vacuum	BCl_3	0	1183,1431
23.359			H_2O	0	1713
23.436			HBr	0	1735
23.571			HCl	0	1755
23.675		vacuum	$^{14}NH_3$	0	1206–1208
23.68			NH_3	0	1759
23.849			HCl	0	1755
23.86			NH_3	0	1206,1759
24.318			HCl	0	1755
24.583			HCl	0	1755
24.618			HCl	0	1755
24.918		vacuum	$^{14}NH_3$	0	1206–1208
24.92			NH_3	0	1759
24.937			HCl	0	1755
25.12			NH_3	0	1206,1759
25.423	±0.0051	vacuum	Ne	0	583,586,1452
25.704			HCl	0	1755
26.146			HCl	0	1755
26.247			HCl	0	1735
26.27			NH_3	0	1759
26.282		vacuum	$^{14}NH_3$	0	1206–1208

Table 4.1.1—*continued*
Neutral Atom, Ion, and Molecular Gas Lasers Arranged in Order of Increasing Wavelength

Wavelength (μm)	Uncertainty	Medium	Species	Charge	Reference
26.666			H_2O	0	1713
26.933	±0.0027	vacuum	Ar	0	1452
26.936	±0.0027	vacuum	Ar	0	1452
26.944			Ar	0	1452
27.508			HCl	0	1735
27.9707534	±0.00000025	vacuum	H_2O	0	1369,1407
28.053	±0.0056	vacuum	Ne	0	583,586,1452
28.054			H_2O	0	1713
28.273			H_2O	0	1713
28.356			H_2O	0	1713
29.786			HBr	0	1735
30.445			HBr	0	1735
30.69			NH_3	0	1206,1759
30.948			HBr	0	1735
31.368			HBr	0	1735
31.47			NH_3	0	759,1206
31.553	±0.0047	vacuum	Ne	0	583,586,1452
31.849			HBr	0	1735
31.92			NH_3	0	1759
31.928	±0.0048	vacuum	Ne	0	583,586,1452
31.951		vacuum	$^{14}NH_3$	0	1206–1208
32.016	±0.0048	vacuum	Ne	0	583,586,1452
32.13			NH_3	0	1206,1759
32.469			HBr	0	1735
32.516	±0.0049	vacuum	Ne	0	583,586,1452
32.799			HBr	0	1735
32.830	±0.0066	vacuum	Ne	0	583,586,1452
32.929			H_2O	0	1713
33.029	±0.0033	vacuum	H_2O	0	1368
33.033			H_2O	0	1713
33.409			HBr	0	1735
33.47			H_2S	0	1806
33.64			H_2S	0	1806
33.896			D_2O	0	1713
34.552	±0.0052	vacuum	Ne	0	583,586,1452

Table 4.1.1—*continued*
Neutral Atom, Ion, and Molecular Gas Lasers Arranged in Order of Increasing Wavelength

Wavelength (μm)	Uncertainty	Medium	Species	Charge	Reference
34.679	±0.0052	vacuum	Ne	0	583,586,1452
35.000			H_2O	0	1713
35.090			D_2O	0	1713
35.602	±0.0053	vacuum	Ne	0	583,586,1452
35.841			H_2O	0	1713
36.319			D_2O	0	1713
36.524			D_2O	0	1713
36.619			H_2O	0	1713
37.231	±0.0056	vacuum	Ne	0	583,586,1452
37.791			D_2O	0	1713
37.859			H_2O	0	1713
38.094			H_2O	0	1713
39.698			H_2O	0	1713
40.526			HBr	0	1735
40.629			H_2O	0	1713
40.994			D_2O	0	1713
41.741	±0.0042	vacuum	Ne	0	583,586,1452
45.523			H_2O	0	1713
47.244	±0.0047	vacuum	H_2O	0	1368
47.251			H_2O	0	1713
47.46315	±0.000095	vacuum	H_2O	0	1368,1383
47.687	±0.0048	vacuum	H_2O	0	1368
47.693			H_2O	0	1713
48.677			H_2O	0	1713
49.62			H_2S	0	1746
50.705	±0.0051	vacuum	Ne	0	583,586,1452
52.030	±0.0083	vacuum	NH_3	0	1461–4,1393
52.40			H_2S	0	1746
52.425	±0.0052	vacuum	Ne	0	583,586,1452
52.800	±0.0084	vacuum	NH_3	0	1393,1461–4
53.486	±0.0053	vacuum	Ne	0	583,586,1452
53.906			H_2O	0	1713
54.019	±0.0054	vacuum	Ne	0	583,586,1452
54.117	±0.0054	vacuum	Ne	0	583,586,1452
54.460	±0.0087	vacuum	NH_3	0	1393,1461–4

Table 4.1.1—*continued*
Neutral Atom, Ion, and Molecular Gas Lasers Arranged in Order of Increasing Wavelength

Wavelength (μm)	Uncertainty	Medium	Species	Charge	Reference
55.077			H_2O	0	1713
55.088	±0.0055	vacuum	H_2O	0	1368
55.537	±0.0056	vacuum	Ne	0	583,586,1452
56.84			H_2S	0	1746
56.845			D_2O	0	1713
57.040	±0.0097	vacuum	NH_3	0	1393,1461–4
57.355	±0.0057	vacuum	Ne	0	583,586,1452
57.660			H_2O	0	1713
60.29			H_2S	0	1746
61.50			H_2S	0	1746
67.177			H_2O	0	1713
68.329	±0.0068	vacuum	Ne	0	583,586,1452
71.899			HCN	0	1445
71.965			D_2O	0	1713
72.108	±0.0072	vacuum	Ne	0	583,586,1452
72.429			D_2O	0	1713
72.747780	±0.000022	vacuum	D_2O	0	1368,1455
72.748			D_2O	0	1455
73.101			HCN	0	1445
73.337			D_2O	0	1713
73.401	±0.0073	vacuum	H_2O	0	1368
73.402			H_2O	0	1713
73.52			H_2S	0	1746
74.545			D_2O	0	1713
75.578			Xe	0	1765
76.093			HCN	0	1445
76.305			D_2O	0	1713
77.001			HCN	0	1445
78.443327	±0.000024	vacuum	H_2O	0	1368,1455
79.091010	±0.000024	vacuum	H_2O	0	1368,1455
80.50			H_2S	0	1746
81.554			HCN	0	1445
83.43			H_2S	0	1746
84.111			D_2O	0	1713
85.047	±0.0085	vacuum	Ne	0	583,586,1452

Table 4.1.1—*continued*
Neutral Atom, Ion, and Molecular Gas Lasers Arranged in Order of Increasing Wavelength

Wavelength (μm)	Uncertainty	Medium	Species	Charge	Reference
86.962	±0.0087	vacuum	Ne	0	583,586,1452
87.47			H_2S	0	1746
88.471	±0.0088	vacuum	Ne	0	583,586,1452
89.775			H_2O	0	1713
89.859	±0.0090	vacuum	Ne	0	583,586,1452
92.00			H_2S	0	1746
93.0	±0.22	vacuum	Ne	0	583,586,1452
95.763			He	0	1756
95.788	±0.0018	vacuum	He	0	434,435
96.38			H_2S	0	1746
96.401			HCN	0	1445
98.693			HCN	0	1445
101.257			HCN	0	1445
101.9			H_2CO	0	1804
103.3			H_2S	0	1746
106.07	±0.053	vacuum	Ne	0	583,586,1452
107.720			D_2O	0	1455
108.8			H_2S	0	1746
110.240			HCN	0	1445
112.066			HCN	0	1445
113.311			HCN	0	1445
115.420			H_2O	0	1713
116.132			HCN	0	1445
116.8			H_2S	0	1746
118.5910			H_2O	0	1713
119.6			H_2CO	0	1734
120.080			H_2O	0	1713
122.8			H_2CO	0	1734
123			OCS	0	1753
124.4	±0.30	vacuum	Ne	0	583,586,1452
125.9			H_2CO	0	1734
126.1	±0.30	vacuum	Ne	0	583,586,1452
126.164			HCN	0	1447
126.2			H_2S	0	1746
128.629			HCN	0	1445

Table 4.1.1—*continued*
Neutral Atom, Ion, and Molecular Gas Lasers
Arranged in Order of Increasing Wavelength

Wavelength (μm)	Uncertainty	Medium	Species	Charge	Reference
129.1			H_2S	0	1746
130.8			H_2S	0	1746
130.839			HCN	0	1447
132			OCS	0	1753
132.8	±0.32	vacuum	Ne	0	583,586,1452
134.933			HCN	0	1447
135.5			H_2S	0	1746
138.768			HCN	0	1445
140.6			H_2S	0	1746
140.89	±0.042	vacuum	SO_2	0	1395,1468
155.1			H_2CO	0	1734
157.6			H_2CO	0	1734
159.5			H_2CO	0	1734
162.4			H_2S	0	1746
163.8			H_2CO	0	1734
165.150			HCN	0	1445
170.2			H_2CO	0	1734
171.67	±0.017	vacuum	D_2O	0	1368,1393
184.4			H_2CO	0	1734
189.9490	±0.00038	vacuum	DCN	0	1423,1447
192.9			H_2S	0	1746
194.7027	±0.00039	vacuum	DCN	0	1423,1447
194.7644	±0.00039	vacuum	DCN	0	1423,1447
201.059			HCN	0	1445
204.3872	±0.00041	vacuum	DCN	0	1423,1447
211.00	±0.017	vacuum	HCN	0	1445–1449
216.12			He	0	1756
216.3	±0.43	vacuum	He	0	435
220.2279	±0.00022	vacuum	H_2O	0	1368,1457
222.949			HCN	0	1445
225.3			H_2S	0	1746
284.000			HCN	0	1422
309.7140	±0.00031	vacuum	HCN	0	1422,1447–9
310.8870	±0.00031	vacuum	HCN	0	1421,1447–9
335.1831	±0.00034	vacuum	HCN	0	1422,1447–9

Table 4.1.1—*continued*
Neutral Atom, Ion, and Molecular Gas Lasers Arranged in Order of Increasing Wavelength

Wavelength (μm)	Uncertainty	Medium	Species	Charge	Reference
336.5578	±0.00034	vacuum	HCN	0	1421,1447–9
372.528		vacuum	HCN	0	1422
791.06			H_2O	0	1746

Section 4.2
OPTICALLY PUMPED FAR INFRARED AND MILLIMETER WAVE GAS LASERS

Introduction to the Table

Optically pumped far infrared and millimeter wave lasers are listed in order of increasing wavelength in Table 4.2.1. Most of the lasers operate in a continuous wave mode. The uncertainty in the wavelength determination, if stated in the reference, is noted in the second column. Accurate measurements, typically 1 part in 10^5 or better, refer to vacuum since they are calculated from frequency measurements. Interferometric wavelength measurements may refer to vacuum, the laser medium, or air but are of low accuracy, ranging from a few percent to (rarely) 1 part in 10^4. Thus within the measurement uncertainties almost all measurements may be considered to refer to vacuum. The lasing molecule is given in the third column. References with titles or descriptions of the contents are given in Section 4.3. The references generally include the original report of lasing and other reports relevant to the identification of the transition and laser operation.

Pumping transitions used for lasing are given in the book by Douglas or the tabulations of Inguscio et al., Knight (1982), and Tobin (see Further Reading below).

Further Reading

Chang, T.-Y., Vibrational Transition Lasers, in *Handbook of Laser Science and Technology, Vol. II: Gas Lasers*, CRC Press, Boca Raton, FL (1982), p. 313 and *Suppl. 1: Lasers*, CRC Press, Inc., Boca Raton, FL (1991), p. 387.

Cheo, P. K., Ed., *Handbook of Molecular Lasers*, Marcel Dekker Inc., New York (1987).

Douglas, N. G., *Millimetre and Submillimetre Wavelength Lasers: A Handbook of CW Measurements*, Springer-Verlag, Berlin Heidelberg (1989).

Evans, J. D., Ed., *Selected Papers on CO_2 Lasers*, SPIE Milestone Series, Vol. MS22, SPIE Optical Engineering Press, Bellingham, WA (1990).

Inguscio, M., Moruzzi, G., Evenson, K. M. and Jennings, D. A., A review of frequency measurements of optically pumped lasers from 0.1 to 8 THz, *J. Appl. Phys.* 60, R161 (1986).

Jacobsson, S., Optically pumped far infrared lasers, *Infrared Phys.* 29, 853 (1989).

Knight, D. J. E., Far-Infrared CW Gas Lasers in *Handbook of Laser Science and Technology, Vol. II: Gas Lasers,* CRC Press, Boca Raton, FL (1982), p. 411 and *Suppl. 1: Lasers,* CRC Press, Boca Raton, FL (1991), p. 415.

Moruzzi, G., Winnewisser, B. P., Winnewisser, M., Mukkopadhyay, I. and Strumia, F., Microwave, infrared and laser transitions of methanol: atlas of assigned lines from 0 to 1258 cm^{-1}, CRC Press, Boca Raton, FL (1995).

Tobin, M. S., A review of optically pumped NMMW lasers, *Proc. IEEE* 73, 61 (1985).

Witteman, W. J., *The CO_2 Laser*, Springer Verlag, Berlin (1987).

See, also, *International Journal of Infrared and Millimeter Waves* (proceedings of the Infrared and Millimeter Wave Conference series are published in this journal), *Infrared Physics and Technology, Journal of Molecular Spectroscopy, Journal of Applied Physics, IEEE Journal of Quantum Electronics, and Russian Quantum Electronics* for reports of new infrared lasers.

Table 4.2.1
Optically Pumped Far Infrared and Millimeter Wave Lasers Arranged in Order of Increasing Wavelength

Wavelength (μm)	Uncertainty	Molecule	Reference
20.008		$^{14}NH_3$	1632
20.010		C_2D_2	827
20.073		C_2D_2	827
20.13		C_2D_2	827
20.15		$^{12}C_2HD$	826
20.202		C_2D_2	827
20.267		C_2D_2	1801
20.332		C_2D_2	827
20.358		$^{14}NH_3$	1632
20.38		$^{12}C_2HD$	826
20.38798		$^{14}NH_3$	1205
20.44		C_2D_2	1802
20.47		$^{12}C_2HD$	826
20.48		$^{14}NH_3$	1632
20.604		$^{14}NH_3$	1632
20.622		$^{14}NH_3$	1632
21.05409		$^{14}NH_3$	1205
21.333		$^{14}NH_3$	1632
21.38		$^{12}C_2HD$	826
21.471		$^{14}NH_3$	1208,1209
22.542		$^{14}NH_3$	1208,1209
22.563		$^{14}NH_3$	1208,1209
23.675		$^{14}NH_3$	1208,1209
24.4		$^{13}CH_3OH$	1706
24.78		SiF_4	1219
24.918		$^{14}NH_3$	1208,1209
25.270		CH_3OH	179
25.31		SiF_4	1219
25.36		SiF_4	1219
25.40		SiF_4	1219
25.67		SiF_4	1219
25.68		SiF_4	1219
25.77		SiF_4	1219
25.79		SiF_4	1219

Table 4.2.1—*continued*
Optically Pumped Far Infrared and Millimeter Wave Lasers Arranged in Order of Increasing Wavelength

Wavelength (μm)	Uncertainty	Molecule	Reference
26.01		SiF_4	1219
26.14		SiF_4	1219
26.282		$^{14}NH_3$	1208,1209
27.7		CD_3OH	1720
27.9707534	±0.00000025	H_2O	1369
30.7		CD_3OH	1720
31.1		CD_3OH	1720
31.951		$^{14}NH_3$	1208,1209
33.029	±0.0033	H_2O	1368
33.500		CH_3OH	179
33.6		CH2CHF	1730
34.2		CD_3OH	1720
34.60		$CH_3{}^{18}OH$	1741
34.79	±0.010	$^{13}CH_3OH$	1415,1417
34.8	±0.17	CD_3OH	1391
35.0	±0.70	CD_3OD	1425
35.00		$CH_3{}^{18}OH$	1741
35.7		CD_3OH	1720
35.860		CH_3OH	179
35.968		CH_3OH	179
36.666		CH_3OH	179
37.1		CD_3OH	1720
37.6	±0.18	CD_3OH	1391
37.85421	±0.000019	CH_3OH	1481
39.785		CH_3OH	823
39.92423	±0.0000205	CH_3OH	1481
40.00		$CH_3{}^{18}OH$	1741
40.1	±0.20	CD_3OH	1391
41.0	±0.82	CD_3OD	1425,1432
41.034		CH_3OH	179
41.06		CH_3OH	829
41.171		CH_3OH	179
41.25	±0.021	CD_3OH	1405
41.33		CH_3OH	830
41.35487	±0.000021	CD_3OH	1481

Table 4.2.1—*continued*
Optically Pumped Far Infrared and Millimeter Wave Lasers Arranged in Order of Increasing Wavelength

Wavelength (μm)	Uncertainty	Molecule	Reference
41.355		CD_3OH	826
41.40	±0.021	CD_3OH	1391,1405
41.46		CD_3OH	1718
41.500		CH_3OH	823
41.8		CHD_2OH	663
41.871		CH_3OH	179
41.90	±0.013	$^{13}CH_3OH$	1415,1417
42.15908	±0.000021	CH_3OH	1481
42.31	±0.042	CH_3OH	1414,1420
42.400		CH_3OH	179
42.5		CD_3OH	1720
42.5		CD_3OH	1720
42.50	±0.094	CH_2DOH	1467
42.6		CD_3OH	1720
42.92		CD_3OH	1718
42.953		CH_3OH	179
43.1		CD_3OH	1720
43.1	±0.21	CH_3OH	1414,1427
43.47	±0.013	CH_3OH	1414,1427
43.697		CD_3OH	826
43.69729	±0.000022	CD_3OH	1481
43.70		CD_3OH	1718
43.70		$CH_3{}^{18}OH$	1741
43.784		CH_3OH	823
44.0	±0.48	CH_2DOH	1467
44.24		CH_3OH	179
44.3		CD_3OH	1720
44.307		CH_3OH	823
44.55		CD_3OH	1718
44.70		CD_3OH	1718
44.70		CD_3OH	1718
45.00		CD_3OH	1718
45.1		CHD_2OH	663
45.66		CD_3OH	1718
46.164		CH_3OH	179

Table 4.2.1—*continued*
Optically Pumped Far Infrared and Millimeter Wave Lasers Arranged in Order of Increasing Wavelength

Wavelength (μm)	Uncertainty	Molecule	Reference
46.6		CH_3OD	113
46.7	±0.47	CH_3OD	1367,1370
47.1		CD_3OH	1720
47.244	±0.0047	H_2O	1368
47.46315	±0.000095	H_2O	1368,1383
47.687	±0.0048	H_2O	1368
47.910	±0.0067	NH_3	1393,1461–1464
48.363		CH_3OH	179
48.40		$CH_3{}^{18}OH$	1741
48.6		CD_3OH	1720
48.630		CH_3OH	179
48.7		CD_3OH	1720
48.760		CH_3OH	823
49.07		CD_3OH	1718
49.50		$CH_3{}^{18}OH$	1741
49.78		CD_3OH	1718
49.8	±0.24	CD_3OH	1410
50.1		CD_3OH	1720
50.224		CH_3OH	179
50.30		CD_3OH	1718
51.207		CH_3OH	823
51.240		CH_3OH	823
51.85		CH_3OH	829
52.		NH_3	816
52.030	±0.0083	NH_3	1393,1461,1464
52.1		$^{13}CD_3OH$	1703
52.2		$^{13}CD_3OH$	1703
52.27		CH_3OH	830
52.4		CD_3OD	1714
52.48		CH_3OH	829
52.70		$CH_3{}^{18}OH$	1741
52.800	±0.0084	NH_3	1393,1461,1464
53.10		CD_3OH	1718
53.5	±0.26	CH_3OH	1414,1427
53.6		CD_3OD	1716

Table 4.2.1—*continued*
Optically Pumped Far Infrared and Millimeter Wave Lasers Arranged in Order of Increasing Wavelength

Wavelength (μm)	Uncertainty	Molecule	Reference
53.60		$CH_3{}^{18}OH$	1741
53.82	±0.027	CD_3OH	1405,1410
53.831		CH_3OH	823
53.988		CH_3OH	823
54.1		CD_3OH	1720
54.460	±0.0087	NH_3	1393,1461–1464
54.7		CD_3OH	1719
55.088	±0.0055	H_2O	1368
55.370041	±0.000028	CH_3OH	1481
55.56	±0.028	CD_3OH	1405
55.59		CH_3OD	121
55.6		CHD_2OH	663
55.8		$^{13}CD_3OH$	1703
56.23		CH_3OH	829
56.5		CD_3OH	1720
56.7		CD_3OH	1720
56.730		CH_3OH	179
56.87		CD_3OH	1718
57.0	±0.57	CH_3OD	1433
57.040	±0.0097	NH_3	1393,1461–1464
57.24		CH_3OD	121
57.3		$^{13}CD_3OD$	453
57.355	±0.0057	Ne	583,586,1452
57.4		NH_3	975
57.9		CHD_2OH	663
58.		NH_3	816
58.010	±0.0099	NH_3	1393,1461–1464
58.5		CD_3OD	1715
59.368	±0.010	NH_3	1393,1461–1464
59.60		CD_3OH	1718
60.00		$^{13}CH_3OH$	1707
60.10	±0.030	CD_3OH	1405,1410
60.173273	0.000030	CH_3OH	1484
60.35		CH_3OH	829
60.58		CH_3OH	829

Table 4.2.1—*continued*
Optically Pumped Far Infrared and Millimeter Wave Lasers Arranged in Order of Increasing Wavelength

Wavelength (μm)	Uncertainty	Molecule	Reference
60.6		CD_3OD	1716
61.40		CD_3OH	1718
61.613300	±0.000031	CH_3OH	1484
61.7		CD_3OH	1720
62.171		CH_3OH	823
62.965968	±0.000031	CH_3OH	1484
63.006		CH_3OH	823
63.096391	±0.000032	$^{13}CH_3OH$	1415
63.369541	±0.000032	CH_3OH	1481
63.681		CH_3OH	823
63.88		CH_3OH	830
63.948		CH_3OH	823
64.397		CH_3OH	823
64.499	±0.012	NH_3	1393,1461–1464
64.9		H_2O	975
65.4		$^{13}CD_3OH$	1703
65.509	±0.013	NH_3	1393,1461–1464
65.544		CH_3OH	823
65.55		$CH_3{}^{18}OH$	1741
65.87		CD_3OH	1718
66.21		CH_3OH	830
66.249		CH_3OH	179
66.40		CD_3OH	1718
66.78		CD_3OD	1717
66.80		CD_3OH	1718
67.224		CH_3OH	823
67.240	±0.013	NH_3	1393,1461–1464
67.3		$^{15}NH_3$	905
67.430		CH_3OH	179
67.479		CD_3OH	826
67.479411	±0.000034	CD_3OH	1481
67.495361	±0.000034	CH_3OH	1414,1427,1481
67.751	±0.014	NH_3	1393,1461–1464
67.8		$^{13}CD_3OH$	1703
68.		CH_3NH_2	1752

Table 4.2.1—*continued*
Optically Pumped Far Infrared and Millimeter Wave Lasers Arranged in Order of Increasing Wavelength

Wavelength (μm)	Uncertainty	Molecule	Reference
68.45		CD_3OH	1718
68.698		CH_3OH	179
68.70		CD_3OH	1718
68.93		CD_3OD	1717
69.18		CD_3OH	1718
69.5	±0.70	CH_3OD	1440
69.679560	±0.000035	CH_3OH	1484
69.90		$CH_3{}^{18}OH$	1741
69.95		CH_3OH	830
70.00		$^{13}CH_3OH$	1707
70.3	±0.28	CH_3OD	1440
70.511628	±0.000014	CH_3OH	1371,1484
70.989		CD_3OH	826
71.0	±0.35	CD_3OH	1391
71.50		CD_3OH	1718
71.70		$^{13}CH_3OH$	1707
71.70		CD_3OH	1718
72.00		$^{13}CH_3OH$	1707
72.747780	±0.000022	D_2O	1368,1393,1455
72.9		$^{13}CD_3OH$	1703
73.20		CH_3OH	829
73.306420	±0.000037	CH_3OH	1484
73.401	±0.0073	H_2O	1368
73.467		$^{13}CD_3OH$	1703
73.467		$^{13}CD_3OH$	1703
73.8		CD_3OD	1714
74.1		CHD_2OH	663
74.38		CH_3OD	121
74.384		CH_3OH	179
74.8		CHD_2OH	663
75.06		CH_3OH	829
75.275		$^{13}CD_3OD$	1702
75.5		$^{13}CD_3OD$	1702
75.821		CH_3OH	179
75.932		CH_3OH	179

Table 4.2.1—*continued*
Optically Pumped Far Infrared and Millimeter Wave Lasers Arranged in Order of Increasing Wavelength

Wavelength (μm)	Uncertainty	Molecule	Reference
76.00		CD_3OH	1718
76.1	±0.37	CD_3OH	1391
76.3		CD_3OD	1716
76.4		$^{15}NH_3$	905
76.90		CD_3OH	1718
76.93		CD_3OH	1718
77.07	±1.5	NH_2D	1437
77.405648	±0.000039	CH_3OH	1484
77.487		CH_3OH	823
77.489387	±0.000039	$^{13}CH_3OH$	1415
77.65		$CH_3{}^{18}OH$	1741
77.84		CH_3OH	830
77.904888	±0.000039	CH_3OH	1484
77.93		CH_3OD	121
78.0	±0.78	CD_3OD	1425
78.272	±0.018	NH_3	1393,1461–1464
78.39		CH_3OH	830
78.443327	±0.000024	H_2O	1368,1455
78.60		CD_3OH	1718
78.78		CD_3OH	1718
79.05		CD_3OD	1717
79.091010	±0.000024	H_2O	1368,1455
79.618	±0.019	NH_3	1393,1461–1464
79.98		CH_3OH	829
80.	±2.4	CH_3OH	1420
80.0	±0.16	CH_3OD	1397,1436
80.1		CD_3OD	1715
80.3		$^{13}CH_3OH$	1706
80.44		CD_3OH	1718
80.5		CD_3OD	1714
80.5122.304		CD_3OD	1714
80.6	±0.39	CH_3OH	1473,1475
80.8		CHD_2OH	663
80.843		CH_3OH	179
81.01		$^{13}CH_3OH$	1707

Table 4.2.1—*continued*
Optically Pumped Far Infrared and Millimeter Wave Lasers Arranged in Order of Increasing Wavelength

Wavelength (μm)	Uncertainty	Molecule	Reference
81.480	±0.020	NH_3	1393,1461–1464
81.500	±0.049	NH_3	1376,1393,1464
81.557101	±0.000041	CD_3OH	1391,1481
81.8		$^{13}CD_3OD$	453
81.9	±0.40	CH_3OD	1451
81.903		CH_3OH	179
82.1		$^{13}CD_3OD$	1702
82.2		CD_3OD	1714
82.4		$^{13}CD_3OD$	1702
82.6		CD_3OD	1715
82.7		CD_3OH	1719
83.6		CD_3OD	1715
83.7		CHD_2OH	663
83.70		CD_3OH	1718
83.77		PH_3	815
83.9		CD_3OH	1719
83.9		CHD_2OH	663
84.005		CH_3OH	179
84.278893	±0.000025	D_2O	1368,1393,1455
84.4		$^{13}CD_3OD$	1702
84.406		$^{13}CD_3OH$	1703
84.5		CD_3OH	1719
84.7		$^{13}CD_3OD$	453
84.908		CH_3OH	823
84.913		CH_3OH	179
85.317287	±0.000043	$^{13}CH_3OH$	1415
85.5		CD_3OD	1715
85.600931	±0.000043	CH_3OH	1484
85.79	±0.026	$^{13}CH_3OH$	1415,1417
85.83		CH_3OH	829
85.90	±1.7	NH_2D	1437
86.111788	±0.000043	$^{13}CH_3OH$	1415
86.239385	±0.000043	CH_3OH	1481
86.30		CD_3OH	1718
86.4	±0.42	CD_3OH	1391

Table 4.2.1—*continued*
Optically Pumped Far Infrared and Millimeter Wave Lasers Arranged in Order of Increasing Wavelength

Wavelength (μm)	Uncertainty	Molecule	Reference
86.5		CD_3OD	1714
86.741		CD_3OH	826
86.90	±1.7	ND_3	1437
87.		CH_3NH_2	1752
87.09	±0.022	NH_3	1393,1461–1464
87.10	±0.096	CH_2DOH	1467
87.20		$CH_3{}^{18}OH$	1741
87.3		CD_3OD	1714
87.4	±0.43	NH_3	1393,1461–1464
87.65		$CH_3{}^{18}OH$	1741
87.8		CD_3OH	1719
87.9		CD_3OH	1719
87.90	±0.026	$^{13}CH_3OH$	1415,1417
87.90	±0.097	CH_2DOH	1467
88.		NH_3	816
88.059	±0.018	NH_3	1393,1461–1464
88.819		CH_3OH	179
89.		NH_3	816
89.00		$^{13}CH_3OH$	1707
89.31		CH_3OH	830
89.6	±0.44	CH_3OD	1451
89.98		CH_3OH	830
90.		NH_3	816
90.16		CD_3OH	1718
90.303		CH_3OH	823
90.4		$^{15}NH_3$	905
90.40	±0.099	CH_2DOH	1467
90.934	±0.025	NH_3	1393,1461–1464
90.97		$CH_3{}^{18}OH$	1741
91.08		CH_3OD	121
92.		CH_3NH_2	1752
92.543913	±0.000046	CH_3OH	1484
92.60		$CH_3{}^{18}OH$	1741
92.664287	±0.000046	CH_3OH	1481
92.876	±0.026	NH_3	1393,1461–1464

Table 4.2.1—*continued*
Optically Pumped Far Infrared and Millimeter Wave Lasers Arranged in Order of Increasing Wavelength

Wavelength (μm)	Uncertainty	Molecule	Reference
92.9		$^{13}CD_3OD$	453
93.0		CHD_2OH	663
93.40		$CH_3{}^{18}OH$	1741
93.6		$^{13}CD_3OD$	1702
93.6		NH_3	827
93.88		CD_3OH	1718
93.96		CH_3OH	829,830
94.3		CD_3OD	1716
94.447	±0.026	NH_3	1393,1461–1464
94.52	±0.46	D_2O	1368,1393
94.90		CD_3OH	1718
95.10		CH_3OH	830
95.24		$^{13}CH_3OH$	1707
95.25		CH_3OH	829
95.551		CH_2F_2	1738
95.551057	±0.000048	CH_2F_2	1392,1454,1466
96.522395	±0.000015	CH_3OH	1455
96.674	±0.028	NH_3	1393,1461–1464
96.812		CH_3OH	823
97.29		CH_3OH	830
97.518534	±0.000049	CH_3OH	1481
97.800		CH_3OH	179
98.	±2.9	CH_3OH	1471
98.0		$^{13}CH_3OH$	1706
98.5		$^{13}CD_3OH$	1703
98.65		$CH_3{}^{18}OH$	1741
98.862		CH_3OH	823
99.14		$CH_3{}^{18}OH$	1741
99.861		CH_3OH	834
100.		CH_3NH_2	1751
100.	±2.0	CH_3NH_2	1437
100.00	±0.10	CH_2DOH	1467
100.010		CH_3OH	179
100.166		CH_3OH	834
100.80647	±0.000050	CH_3OH	1484

Table 4.2.1—*continued*
Optically Pumped Far Infrared and Millimeter Wave Lasers Arranged in Order of Increasing Wavelength

Wavelength (μm)	Uncertainty	Molecule	Reference
101.3		$^{13}CH_3OH$	1706
101.6	±0.50	CH_3OD	1436,1440,1451
101.9		$^{13}CD_3OD$	453
102.		CH_3NH_2	1752
102.02349	±0.000051	CH_2DOH	1467
102.2		CD_3OD	1715
102.3		CH_3OH	829
102.6	±0.50	CD_3OH	1391
102.60		CH_3OH	830
103.0		CD_3OH	1719
103.0		CHD_2OH	663
103.00		$^{13}CH_3OH$	1707
103.12463	±0.000018	CH_3OD	1397,1483
103.48079	±0.000052	$^{13}CH_3OH$	1415
103.58629	±0.000052	$^{13}CH_3OH$	1415
104.00		PH_3	815
104.3		CD_3OD	1714
104.6		CHD_2OH	663
104.60		$CH_3{}^{18}OH$	1741
105.		CH_3NH_2	1752
105.0		CHD_2OH	663
105.07		$^{13}CH_3OH$	1707
105.14719	±0.000053	$^{13}CH_3OH$	1415
105.35	±0.034	NH_3	1393,1461–1464
105.35	±0.034	NH_3	1393,1461–1464
105.4		CD_3OD	1715
105.518		CH_2F_2	1738
105.51827	±0.000053	CH_2F_2	1392,1454,1466
106.	±2.1	CH_3OD	1436
106.400		$^{13}CH_2F_2$	1704
107.20		CD_3OH	1718
107.4		CHD_2OH	663
107.538		CD_3OD	1714
107.72019	±0.000043	D_2O	1368,1393,1455
107.8		$^{13}CH_3OH$	1706

Table 4.2.1—*continued*
Optically Pumped Far Infrared and Millimeter Wave Lasers Arranged in Order of Increasing Wavelength

Wavelength (μm)	Uncertainty	Molecule	Reference
107.8	±2.2	NH_2D	1437
108.6		CH_3OH	829
108.668		CD_3OH	826
108.66842	±0.000054	CD_3OH	1481
108.7		CD_3OD	1714
108.81775	±0.000054	CH_2DOH	1467,1478
108.94124	±0.000054	CH_2DOH	1467
109.		CH_3NH_2	1752
109.1		CD_3OH	1719
109.29579	±0.000055	CH_2F_2	1392,1454,1466
109.296		CH_2F_2	1738
109.3		CHD_2OH	663
109.30		$CH_3{}^{18}OH$	1741
109.4		CH_3OH	829
109.926		$^{13}CD_3OD$	1702
109.938		$^{13}CD_3OD$	1702
110.	±1.1	CH_3OD	1433,1436
110.0		$^{13}CD_3OH$	1703
110.43238	±0.000055	$^{13}CH_3OH$	1415
110.7	±0.54	CH_3OD	1436,1451
110.716		CH_3OH	834
110.9		$^{13}CH_3OH$	1706
111.3		CD_3OD	1715
111.4		CHD_2OH	663
111.40		CD_3OH	1718
111.6		CHD_2OH	663
111.60		$CH_3{}^{18}OH$	1741
111.9		CHD_2OH	663
111.9	±0.55	$^{15}NH_3$	1390
112.000		$^{13}CH_2F_2$	1704
112.10		CD_3OH	1718
112.22	±0.038	NH_3	1461,1479
112.3		NH_3	827
112.3	±0.55	CD_3OH	1391
112.53224	±0.000056	CH_2DOH	1467

Table 4.2.1—*continued*
Optically Pumped Far Infrared and Millimeter Wave Lasers Arranged in Order of Increasing Wavelength

Wavelength (μm)	Uncertainty	Molecule	Reference
112.6	±0.55	D_2O	1368,1393
112.946		CH_3OH	834
113.1	±2.3	NH_2D	1437
113.4		$^{13}CH_3OH$	1706
113.450		CH_3OH	179
113.5		$^{15}NH_3$	905
113.5		CH_3OD	113
113.60		$^{13}CH_3OH$	1707
113.73188	±0.000057	CH_3OH	1484
113.8	±0.56	CH_3OD	1451
114.20		$CH_3{}^{18}OH$	1741
114.29	±0.040	NH_3	1393,1461–1464
114.40		CD_3OH	1718
115.0		N_2D_4	735
115.00		$^{13}CH_3OH$	1707
115.2		$^{13}CD_3OD$	453
115.32	±0.012	H_2O	1368
115.70		$CH_3{}^{18}OH$	1741
115.80		$CH_3{}^{18}OH$	1741
115.82318	±0.000058	$^{13}CH_3OH$	1415
115.935		CH_2F_2	1739
116.		CH_3NH_2	1751
116.		CH_3SH	485
116.	±2.3	CH_3NH_2	1437
116.	±2.3	CH_3SH	1435
116.00		CH_3OH	830
116.27	±0.041	NH_3	1393,1461–1464
116.5		CD_3OH	1719
117.		CH_3SH	485
117.	±2.3	CH_3SH	1435
117.000		CH_3OH	179
117.08507	±0.000059	CH_2DOH	1467
117.22707	±0.000022	CH_3OD	1397,1483
117.3		CHD_2OH	663
117.4		CH_3OH	829

Table 4.2.1—*continued*
Optically Pumped Far Infrared and Millimeter Wave Lasers Arranged in Order of Increasing Wavelength

Wavelength (μm)	Uncertainty	Molecule	Reference
117.62		CD_3OH	1718
117.630		CH_3OH	179
117.727		CH_2F_2	1738
117.72748	±0.000059	CH_2F_2	1392,1454,1466
117.92		$^{13}CH_3OH$	1707
117.95948	±0.000059	CH_3OH	1481
118.01308	±0.000059	$^{13}CH_3OH$	1415
118.553		$^{13}CD_3OD$	1702
118.5910	±0.00012	H_2O	1368,1457
118.8		CD_3OH	1719
118.83409	±0.000024	CH_3OH	1427,1455
119.		CH_3NH_2	1751
119.	±1.2	CD_3OD	1425
119.	±2.4	CH_3NH_2	1437
119.02	±0.042	NH_3	1393,1461–1464
119.057		CD_3OD	1714
119.1		$^{13}CD_3OH$	1703
119.4		$^{13}CD_3OH$	1703
119.7		CH_3OH	829
119.800		CH_3OH	179
119.84		$CH_3{}^{18}OH$	1741
119.9		CD_3OH	1718
120.		CH_3NH_2	1751
120.		CHFO	820
120.	±2.4	CH_3NH_2	1437
120.3		CD_3OH	1719
120.45		CD_3OH	1718
120.469		CD_2F_2	1710
120.661		CD_3OH	826
120.9		CHD_2OH	663
120.902		CH_3OH	834
121.	±2.8	CH_3OH	1427–1429,1471
121.0		CH2CHF	1730
121.20	±0.036	$^{13}CH_3OH$	1415,1417
121.270		CH_3OH	179

Table 4.2.1—*continued*
Optically Pumped Far Infrared and Millimeter Wave Lasers Arranged in Order of Increasing Wavelength

Wavelength (μm)	Uncertainty	Molecule	Reference
122.00		$^{13}CH_3OH$	1707
122.154		CD_3OH	826
122.46551	±0.000061	CH_2F_2	1392,1454,1466
122.466		CH_2F_2	1738
122.466		CH_2F_2	1738
123.26	±0.037	$^{13}CH_3OH$	1415,1417
123.55		CD3OH	1718
123.640		CH_3OH	179
123.7		$^{13}CD_3OD$	453
123.8		CHD_2OH	663
123.85		$CH_3{}^{18}OH$	1741
123.9		CH_3OH	829
123.9		CHD_2OH	663
123.9	±2.5	NH_2D	1437
123.90		$CH_3{}^{18}OH$	1741
123.90		$CH_3{}^{18}OH$	1741
124.		CH_3SH	485
124.		DFCO	684
124.	±2.5	CH_3SH	1435
124.253		$^{13}CD_3OD$	1702
124.3		$^{13}CD_3OH$	1703
124.4		CHD_2OH	663
124.43170	±0.000062	CH_2DOH	1467,1478
124.7		CH_3OH	829
124.798		CD_3OD	1714
125.4		CHD_2OH	663
126.	±2.5	CH_3NH_2	1437
126.1		$^{13}CD_3OH$	1703
126.2		$^{13}CD_3OD$	1702
126.2		$^{13}CD_3OD$	1702
126.545		CH_2F_2	1739
126.6		CH_3OH	829
127.		CH_3SH	485
127.	±2.5	CH_3SH	1435
127.021		$^{13}CD_3OH$	1703

Table 4.2.1—*continued*
Optically Pumped Far Infrared and Millimeter Wave Lasers Arranged in Order of Increasing Wavelength

Wavelength (μm)	Uncertainty	Molecule	Reference
127.021		$^{13}CD_3OH$	1703
127.3		CD_3OH	1719
127.30		CH_2F_2	808
127.4		CHD_2OH	663
127.656		$^{13}CD_3OH$	1703
127.656		$^{13}CD_3OH$	1703
127.77		$CH_3{}^{18}OH$	1741
128.		CH_3NH_2	1751
128.		CH_3SH	485
128.		CHFO	821
128.	±2.6	CH_3NH_2	1437
128.	±2.6	CH_3SH	1435
128.0	±0.26	CH_3OD	1397
128.034		CD_3OH	826
128.1		$^{13}CD_3OD$	1702
128.1		SO_2	812
128.5		CHD_2OH	663
128.629	±0.0063	HCN	1445
128.7	±0.63	CD_3OH	1391
129.10		CH_2F_2	808
129.2		$^{13}CD_3OD$	1702
129.5497	±0.00013	CH_3OH	1414,1427–1429
129.6		CD_3OD	1715
130.		CH_3NH_2	1752
131.2		CH_3OD	113
131.560		CH_3OH	823
131.56276	±0.000066	CD_3OH	1481
131.563		CD_3OH	826
131.6		CH_3OH	829
131.69		$CH_3{}^{18}OH$	1741
132.2		CHD_2OH	663
132.7		$^{13}CD_3OD$	453
132.9		CH_3OH	829
133.1196	±0.00013	CH_3OH	1427–1429,1481
133.7		$^{13}CH_3OH$	1706

Table 4.2.1—*continued*
Optically Pumped Far Infrared and Millimeter Wave Lasers Arranged in Order of Increasing Wavelength

Wavelength (μm)	Uncertainty	Molecule	Reference
133.7		CD_3OH	1719
133.9	±0.27	HCOOH	1366,1384
133.99765	±0.000067	CH_2F_2	1392,1454,1466
133.998		CH_2F_2	1738
134.0		N_2D_4	735
134.0	±0.54	CH_3OD	1425,1436,1440
134.2		SO_2	810
134.60		$CH_3{}^{18}OH$	1741
134.900		CH_2F_2	1739
135.	±1.4	CH_3OD	1367
135.0		CHD_2OH	663
135.000		CH_3OH	179
135.17175	±0.000068	CH_2DOH	1467
135.17256	±0.000068	CH_2DOH	1467
135.269		CH_2F_2	1738
135.26932	±0.000068	CH_2F_2	1392,1454,1466
135.4		CD_3OH	1719
135.5		$^{15}NH_3$	905
135.523		$^{13}CH_2F_2$	1704
135.83350	±0.000068	CH_2DOH	1467
135.9		CD_3OD	1716
135.94		PH_3	815
136.	±1.4	CH_3OD	1425,1436
136.120		CH_3OH	179
136.2		CHD_2OH	663
136.5		CD_3OH	1719
136.627		CD_3OH	826
136.62721	±0.000068	CD_3OH	1481
137.		CH_3NH_2	1752
137.	±2.7	CH_3OD	1436
137.0		CHD_2OH	663
138.281		$^{13}CH_2F_2$	1704
138.4		CD_3OH	1718
139.0		$^{13}CD_3OD$	453
139.266		CD_2F_2	1710

Table 4.2.1—*continued*
Optically Pumped Far Infrared and Millimeter Wave Lasers Arranged in Order of Increasing Wavelength

Wavelength (μm)	Uncertainty	Molecule	Reference
140.0		CD_3OH	1719
140.30	±0.098	CH_2DOH	1467
140.4		CH_3OD	113
140.405		$^{13}CH_2F_2$	1704
140.89	±0.042	SO_2	1395,1426,1468
140.9		$^{13}CH_3OH$	1706
140.95		CD_3OH	1718
141.	±2.8	CH_3OD	1436
141.3		CD_3OD	1714
141.7		CD_3OD	1715
142.		CH_3NH_2	1751
142.		CH_3NH_2	1752
142.	±2.8	CH_3NH_2	1437
142.1		SO_2	812
142.43		$CH_3{}^{18}OH$	1741
142.6		CH2CHF	1732
142.80		$CH_3{}^{18}OH$	1741
142.9		CHD_2OH	663
143.186		CH_2F_2	1739
143.3		H_2O	975
143.64		$CH_3{}^{18}OH$	1741
143.8		CD_3OD	1717
143.8		CD_3OD	1717
143.80		CD3OH	1718
144.		DFCO	684
144.11787	±0.000072	CD_3OH	1391,1481
144.118		CD_3OH	826
144.18		$CH_3{}^{18}OH$	1741
144.40		CD_3OH	1718
144.8		CHD_2OH	663
145.		CH_3NH_2	1751
145.	±2.9	CH_3NH_2	1437
145.0		CH_3OH	829
145.081		CH_2F_2	1739
145.252		CH_3OH	834

Table 4.2.1—*continued*
Optically Pumped Far Infrared and Millimeter Wave Lasers Arranged in Order of Increasing Wavelength

Wavelength (μm)	Uncertainty	Molecule	Reference
145.3		CHD_2OH	663
145.5	±0.71	CH_3OH	1427,1473
145.563		$^{13}CD_3OH$	1703
145.6	±0.71	CH_3OD	1451
145.66171	±0.000022	CH_3OD	1370,1433
145.7		CD_3OH	1719
146.		CH_3NH_2	1751
146.	±2.9	CH_3NH_2	1437
146.09738	±0.000073	$^{13}CH_3OH$	1415
146.2		SO_2	812
146.326		$^{13}CD_3OH$	1703
146.6		$^{15}NH_3$	905
147.		CH_3NH_2	1751
147.		CH_3SH	485
147.	±2.9	CH_3NH_2	1437
147.	±2.9	CH_3SH	1435
147.15	±0.065	NH_3	1393,1461–1464
147.28		CD_3OH	1718
147.349		CD_3OH	826
147.65		CD_3OH	1718
147.84469	±0.000030	CH_3NH_2	1437,1483
147.97	±0.044	$^{13}CH_3OH$	1415,1417
148.		C_3H_2O	681
148.0		CD_3OD	829,1717
148.2		CH2CHF	1732
148.3		$^{13}CD_3OH$	1703
148.59041	±0.000074	$^{13}CH_3OH$	1415,1417,1481
148.617		$^{13}CD_3OD$	1702
148.94		CD_3OH	1718
149.0		$SiHF_3$	814
149.00		$CH_3{}^{18}OH$	1741
149.2		$^{13}CD_3OD$	453
149.27228	±0.000075	$^{13}CH_3OH$	1415
149.38792	±0.000075	CH_2DOH	1467
149.61284	±0.000075	CH_2DOH	1467

Table 4.2.1—*continued*
Optically Pumped Far Infrared and Millimeter Wave Lasers Arranged in Order of Increasing Wavelength

Wavelength (μm)	Uncertainty	Molecule	Reference
149.7		SO_2	812
149.8		CH_3OH	829
149.800		CH_3OH	179
150.		CH_3NH_2	1752
150.	±1.5	CD_3OD	1425
150.2		$^{13}CD_3OH$	1703
150.3		$^{13}CD_3OD$	1702
150.438		CD_2F_2	1710
150.5		CD_3OD	1715
150.57167	±0.000075	CH_2DOH	1467
150.8		CD_3OH	1719
150.81629	±0.000075	CH_2DOH	1467,1478
151.		CH_3OH	203
151.0		$^{13}CD_3OD$	1702,1703
151.25369	±0.000076	CH_3OH	1481
151.3		CD_3OD	1715
151.49	±0.068	NH_3	1393,1461–1464
151.65		$CH_3{}^{18}OH$	1741
151.8		$^{13}CD_3OD$	1702
151.8		$^{15}NH_3$	905
151.80		CD_3OH	1718
152.	±3.0	CH_3OH	1427,1436,1473
152.0		CH_3OH	820
152.07569	±0.000076	$^{13}CH_3OH$	1415,1417,1481
152.3		CD_3OD	1715
152.6		CHD_2OH	663
152.670		CH_3OH	823
152.7	±0.11	CH_2DOH	1467
152.9	±0.31	$^{15}NH_3$	1390,1476,1477
153.195		CH_2F_2	1739
153.54		$CH_3{}^{18}OH$	1741
153.694		$^{13}CD_3OH$	1703
153.70		CD_3OH	1718
154.160		CH_2F_2	1739
154.2		CD_3OD	1717

Table 4.2.1—*continued*
Optically Pumped Far Infrared and Millimeter Wave Lasers Arranged in Order of Increasing Wavelength

Wavelength (μm)	Uncertainty	Molecule	Reference
155.0		$^{13}CH_3OH$	1706
155.28	±0.071	NH_3	1393,1461–1464
155.6	±0.45	CD_3F	1472,1485
156.		C_3H_2O	681
156.0		$^{13}CD_3OH$	1703
156.5		$^{13}CD_3OD$	453
156.510		CH_3OH	823
157.		CH_2CHCl	1830
157.92848	±0.000079	$^{13}CH_3OH$	1415
158.0	±0.77	CD_3OH	1410
158.513		CH_2F_2	1738
158.51348	±0.000079	CH_2F_2	1392,1454,1466
158.9		CD_3OH	1719
158.960		CH_2F_2	1738
158.96020	±0.000080	CH_2F_2	1392,1454,1466
159.2	±0.78	CH_3OH	1475
159.21794	±0.000080	CH_2DOH	1467
159.400		CD_3OH	826
159.5		N_2D_4	735
159.5		SO_2	811
159.67569	±0.000080	CH_3OH	1481
161.		CH_3SH	485
161.	±3.2	CH_3SH	1435
161.10		CD_3OH	1718
161.3		CD_3OD	1715
161.530		CH_3OH	179
161.8		CHFCHF	665
162.218		CH_3OH	834
162.670		CH_3OH	179
162.70	±0.098	CH_2DOH	1467
162.85		CD_3OH	1718
163.		CHD_2F	662
163.03353	±0.000046	CH_3OH	1455
163.120		CH_2F_2	1739
163.574		CH_3OH	834

Table 4.2.1—*continued*
Optically Pumped Far Infrared and Millimeter Wave Lasers Arranged in Order of Increasing Wavelength

Wavelength (μm)	Uncertainty	Molecule	Reference
164.		DFCO	684
164.0	±0.80	CH_3OH	1427
164.2		CH_3OD	113
164.4		CHD_2OH	663
164.40		CH_3OH	830
164.5076	±0.00016	CH_3OH	1418,1429,1481
164.56421	±0.000082	CH_3OH	1484
164.60038	±0.000082	CH_3OH	1375,1427,1481
164.656		$^{13}CH_2F_2$	1704
164.69747	±0.000082	CH_3OH	1484
164.74645	±0.000082	CH_2DOH	1467
164.7832	±0.00016	CH_3OH	1418,1429,1481
164.815		$^{13}CH_2F_2$	1704
165.		CH_3NH_2	1751
165.	±1.7	CD_3OD	1425
165.	±3.3	CH_3NH_2	1437
165.0	±0.81	CHD_2OH	1478
165.1		CHD_2OH	663
165.10		$CH_3{}^{18}OH$	1741
165.2		SO_2	812
165.3		CD_3OD	1715
165.604		CD_3OD	1714
166.		CH_3NH_2	1752
166.	±3.3	CH_3NH_2	1437
166.28		$^{13}CH_3OH$	1707
166.631		CH_2F_2	1738
166.63105	±0.000083	CH_2F_2	1392,1454,1466
166.67665	±0.000083	CH_2F_2	1392,1454,1466
166.677		CH_2F_2	1738
166.76		CD_3OH	1718
166.8		SO_2	810
166.879		CD_2F_2	1710
167.1		CH_3OH	829
167.35235	±0.000084	CH_2DOH	1467
167.54117	±0.000084	CH_2DOH	1467,1478

Table 4.2.1—*continued*
Optically Pumped Far Infrared and Millimeter Wave Lasers Arranged in Order of Increasing Wavelength

Wavelength (μm)	Uncertainty	Molecule	Reference
167.58700	±0.000084	CH_3OH	1484
167.810		CH_3OH	179
168.0	±0.82	CHD_2OH	1478
168.083		CD_3OH	826
168.1	±0.34	CH_3OD	1397
168.84	±0.051	$^{13}CH_3OH$	1415,1417
169.		CH_3NH_2	1751
169.	±3.4	CH_3NH_2	1437
169.	±3.4	CH_3OD	1436
169.0		SiH_2F_2	814
169.6		SO_2	812
170.10		$CH_3{}^{18}OH$	1741
170.18		$CH_3{}^{18}OH$	1741
170.57637	±0.000048	CH_3OH	1397,1427,1455,1469
170.9		$^{13}CD_3OD$	453
171.1		CHD_2OH	663
171.3	±0.84	CH_3OH	1473,1475
171.4		SO_2	812
171.67	±0.017	D_2O	1368,1393
171.75758	±0.000086	$^{13}CH_3OH$	1415
171.8	±0.10	CH_2DOH	1467
172.0		CD_3OD	1715
172.000		CH_3OH	179
172.020		CH_3OH	823
172.1		CHD_2OH	663
172.4		CHD_2OH	663
172.6		CH_3OD	121
172.620272.30		CD_3OH	1718
172.7		CH_3OD	121
172.8		C_2H_3Cl	1730
172.8	±0.50	CD_3F	1472,1485
172.84620	±0.000086	CH_2DOH	1467,1478
173.5		CH_3OD	113
173.637		$^{13}CD_3OD$	1702
173.637		$^{13}CD_3OD$	1702

Table 4.2.1—*continued*
Optically Pumped Far Infrared and Millimeter Wave Lasers Arranged in Order of Increasing Wavelength

Wavelength (μm)	Uncertainty	Molecule	Reference
174.00		CD_3OH	1718
174.3		CD_3OD	1717
174.7		SO_2	810
175.1		$^{13}CD_3OD$	1702
175.5		SiH_2F_2	814
176.		C_2H_4O	1748
176.		CH_2ClF	1737
176.		ClO_2	1823
176.	±3.5	CH_3OH	1427,1436
176.000		CH_3OH	179
176.45		$CH_3{}^{18}OH$	1741
176.5		CH_3OH	829
176.80		CD_3OH	1718
177.0		CH_3OH	829
177.00		CD_3OH	1718
177.4		CD_3OH	1719
177.6		$^{13}CD_3OH$	1703
178.		CH_3NH_2	1751
178.	±3.6	CH_3NH_2	1437
178.	±3.6	CH_3OH	1427,1436
178.0		CD_2Cl_2	1709
178.410		CH_3OH	823
178.6		CD_3OD	1715
179.		CH_3NH_2	1751
179.	±3.6	CH_3NH_2	1437
179.0	±0.36	CH_3OD	1397,1437
179.0	±0.88	CD_3OH	1410
179.0	±0.88	CHD_2OH	1478
179.72791	±0.000090	CH_3OH	1484
179.8		CHD_2OH	663
179.80		$CH_3{}^{18}OH$	1741
180.	±3.6	CH_3NH_2	1437,1459
180.0		SO_2	812
180.4	±0.36	CH_3OH	1397,1427
180.600		$^{13}CH_2F_2$	1704

Table 4.2.1—*continued*
Optically Pumped Far Infrared and Millimeter Wave Lasers Arranged in Order of Increasing Wavelength

Wavelength (μm)	Uncertainty	Molecule	Reference
180.600		CH_3OH	179
180.74051	±0.000090	CD_3OH	1391,1481
180.741		CD_3OH	826
180.75		CD_3OH	1718
181.10		$CH_3{}^{18}OH$	1741
181.2		$^{13}CH_3OH$	1706
181.20		$CH_3{}^{18}OH$	1741
181.580		CH_3OH	823
181.60		$CH_3{}^{18}OH$	1741
181.711		CD_3OH	826
181.92643	±0.000055	N_2H_4	1483
182.	±3.6	CH_3OD	1436
182.0		SO_2	811
182.1	±0.89	CH_3OD	1451
182.10	±0.091	CH_2DOH	1467
182.19		$CH_3{}^{18}OH$	1741
182.2		CH_3OD	113
182.381		$^{13}CH_2F_22$	1704
182.566		CD_3OH	826
182.56629	±0.000091	CD_3OH	1481
183.289		$1^{13}CH_2F_2$	1704
183.36		$CH_3{}^{18}OH$	1741
183.4		$^{13}CD_3F$	1699
183.62132	±0.000092	CH_2DOH	1467
183.680		CH_3OH	823
184.	±3.7	CD_3OD	1425,1432
184.0	±0.90	CD_3OH	1402
184.1		SO_2	810
184.2		CD_3OD	1715
184.30590	±0.000092	CH_2F_2	1392,1454,1466
184.306		CH_2F_2	1738
184.4		CD_2Cl_2	1708
184.5		SiH_2F_2	814
184.766		CD_3OD	1714
184.80		$CH_3{}^{18}OH$	1741

Table 4.2.1—*continued*
Optically Pumped Far Infrared and Millimeter Wave Lasers Arranged in Order of Increasing Wavelength

Wavelength (μm)	Uncertainty	Molecule	Reference
185.		CH_3NH_2	1751
185.		CH_3SH	485
185.	±1.9	CHF_2	1362,1389
185.	±3.7	CH_3NH_2	1437
185.	±3.7	CH_3SH	1435
185.0		CD_3OH	1719
185.1		SO_2	810
185.50040	±0.000093	CH_3OH	1375,1427,1481
185.9	±0.37	CH_3OH	1397,1427
186.		CH_2CHCl	1830
186.	±3.7	CH_3OD	1436
186.04219	±0.000093	CH_3OH	1397,1427,1481
186.043		$^{13}CH_2F_2$	1704
186.4		CH_3OD	113
187.		CH_3NH_2	1752
187.0		SiH_3F	813
187.05		CD_3OH	1718
187.20		CH_3OH	830
187.5		CHD_2OH	663
187.819		CD_2F_2	1710
188.41111	±0.000094	CH_2DOH	1467
188.42390	±0.000094	CD_3OH	1481
188.424		CD_3OH	826
188.9		CD_3OH	1719
188.96		$^{13}CH_3OH$	1707
189.190		CH_3OH	179
189.2		CD_3OD	1716
189.30	±0.095	CH_2DOH	1467
189.730		CD_3OH	826
189.832		CD_2F_2	1710
189.9		CD_3OD	1717
189.9490	±0.00038	DCN	1423,1447
190.	±1.9	CHFF	1362,1389
190.0090	±0.00038	DCN	1423,1447
190.270		CH_3OH	823

Table 4.2.1—*continued*
Optically Pumped Far Infrared and Millimeter Wave Lasers Arranged in Order of Increasing Wavelength

Wavelength (μm)	Uncertainty	Molecule	Reference
190.3		$^{13}CH_3OH$	1706
190.5		SiH_2F_2	814
190.72590	±0.000095	CH_3OH	1375,1427,1481
191.04		$CH_3{}^{18}OH$	1741
191.356		CD_3OH	826
191.5		CD_2Cl_2	1709
191.5	±0.38	CH_3OH	1397,1427
191.61960	±0.000096	CH_3OH	1484
191.683		CH_3OH	179
191.848		CH_2F_2	1738
191.84803	±0.000096	CH_2F_2	1392,1454,1466
191.9	±0.94	CD_3OH	1391
192.0		SiH_2F_2	814
192.0		SO_2	810
192.000		CD_2F_2	1710
192.5		CD_3OD	1714
192.72	±0.058	SO_2	1395,1426,1468
192.78	±0.048	CH_3F	1374,1377,1393
192.790		CD_2F_2	1710
192.9072	±0.00035	N_2H_4	1398,1483
193.0		SiH_2F_2	814
193.1		SO_2	812
193.14158	±0.000097	CH_3OH	1375,1427,1481
193.173		CH_2F_2	1739
193.25		$CH_3{}^{18}OH$	1741
193.3		$^{13}CD_3OD$	453
193.4		CHD_2F	662
193.497		$1{}^{13}CH_2F_2$	1704
193.5		CH_3OH	829
193.55		$CH_3{}^{18}OH$	1741
193.904		CH_2F_2	1738
193.90445	±0.00010	CH_2F_2	1392,1454,1466
194.00		PH_3	815
194.06320	±0.000097	CH_3OH	1484
194.2		CH2CHF	1732

Table 4.2.1—*continued*
Optically Pumped Far Infrared and Millimeter Wave Lasers Arranged in Order of Increasing Wavelength

Wavelength (μm)	Uncertainty	Molecule	Reference
194.260		CH_3OH	823
194.300		CH_3OH	179
194.44761	±0.00010	CH_2F_2	1392,1454,1466
194.448		CH_2F_2	1738
194.5		SO_2	810
194.7027	±0.00039	DCN	1423,1447
194.7644	±0.00039	DCN	1423,1447
195.0	±0.96	HDCO	1380,1382
195.158		$^{13}CH_2F_2$	1704
195.3		CH_2Cl_2	1736
195.49558	±0.0000980	CH_2DOH	1467
195.5		SiH_2F_2	814
196.		CHFO	820,821
196.		ClO_2	1823
196.0	±0.96	HDCO	1380,1382
196.10	±0.098	CH_2DOH	1467
196.2		$^{13}CD_3OH$	1703
196.3		CHFCHF	665
196.5	±0.39	HCOOH	1366,1384
196.564		CH_3OH	834
196.6		CD_3OH	1719
196.8		CHD_2OH	663
196.95		CD_3OH	1718
197.046		$^{13}CD_3OH$	1703
197.388		$^{13}CH_2F_2$	1704
198.		DFCO	684
198.	±2.0	CHFF	1362,1389
198.0	±0.40	CH_3NH_2	1396,1437
198.66433	±0.000099	CH_3OH	1375,1427,1481
198.682		CD_3OH	826
198.79		$^{13}CH_3OH$	1707
199.		CH_3NH_2	1752
199.50		CD_3OH	1718
199.81		CD_3OH	1718
199.90		$CH_3{}^{18}OH$	1741

Table 4.2.1—*continued*
Optically Pumped Far Infrared and Millimeter Wave Lasers Arranged in Order of Increasing Wavelength

Wavelength (μm)	Uncertainty	Molecule	Reference
200.0	±0.48	CD_3F	1472,1485
200.0	±0.48	CH_2DOH	1467
200.210		CH_3OH	179
200.295		$^{13}CH_2F_2$	1704
200.87		CD_3OH	1718
200.9		CH_3OH	830
201.	±0.98	CD_3OH	1410
201.5	±0.48	CD_3F	1472,1485
201.9		CH2CHF	1732
202.0		CD_3OD	1715
202.40	±0.051	CH_3OH	1375,1427
202.4649	±0.00010	CH_2F_2	1392,1454,1466
202.465		CH_2F_2	1738
202.6		$^{13}CD_3OD$	453
202.6		CHD_2OH	663
203.		CH2CHF	1733
203.		CH_2CHF	1733
203.		CH_3NH_2	1752
203.1		CHD_2OH	663
203.300		CD_2F_2	1710
203.5		CD_3OH	1719
203.6358	±0.00010	$^{13}CH_3OH$	1415
203.80		$CH_3{}^{18}OH$	1741
203.96		$^{13}CH_3OH$	1707
204.		ClO_2	1823
204.0		CHD_2F	662
204.3872	±0.00041	DCN	1423,1447
204.7		CHD_2OH	663
204.8		CD_3OD	1717
205.		CH_3SH	485
205.	±1.0	CH_3OH	1427
205.	±4.1	CH_3SH	1435
205.00		$^{13}CH_3OH$	1707
205.1		CH_3OH	829
205.3		SO_2	812

Table 4.2.1—*continued*
Optically Pumped Far Infrared and Millimeter Wave Lasers Arranged in Order of Increasing Wavelength

Wavelength (μm)	Uncertainty	Molecule	Reference
205.8		CD_3OH	1719
205.981		CH_2F_2	1739
206.0	±0.49	CD_3F	1472,1485
206.043		$^{13}CH_2F_2$	1704
206.60		$CH_3{}^{18}OH$	1741
206.6874	±0.00010	CH_2DOH	1467,1478
206.90	±0.062	CH_3OH	1414,1427,1481
207.		ClO_2	1827
207.835		CD_2F_2	1710
208.	±2.1	CH_3OH	1427
208.0		SO_2	811
208.3		CD_3OD	1715
208.3		CH_2Cl_2	1736
208.4121	±0.00010	$^{13}CH_3OH$	1415
208.8		SO_2	810
208.950		CH_3OH	823
209.0		$^{13}CD_3OH$	1703
209.1		$^{13}CD_3F$	1699
209.223		$^{13}CD_3OD$	1702
209.3		CH_3OCH_3	1754
209.6		CH_3OD	121
209.9302	±0.00010	CH_3OH	1484
210.00		CH_2F_2	808
210.5		CD_3OD	1714,1715
211.0		CH_3OH	829
211.00	±0.017	HCN	1445
211.2629	±0.00011	CH_3OH	1427,1471,1481
211.3148	±0.00011	CH_3OH	1484
212.	±2.1	CH_3OD	1436,1440
212.2		CH_3OH	829
212.5	±0.11	CH_2DOH	1467
212.9		CH_3OD	121
212.9		CHD_2OH	663
213.3		CHFCHF	665
213.351		$^{13}CH_2F_2$	1704

Table 4.2.1—*continued*
Optically Pumped Far Infrared and Millimeter Wave Lasers Arranged in Order of Increasing Wavelength

Wavelength (μm)	Uncertainty	Molecule	Reference
213.4625	±0.00011	CH_3OH	1397,1427,1481
214.20		$CH_3{}^{18}OH$	1741
214.3		$^{13}CH_3OH$	1706
214.35	±0.064	CH_3OH	1414,1427,1481
214.579		CH_2F_2	1738
214.5791	±0.00011	CH_2F_2	1386,1454,1466
214.597		$^{13}CH_2F_2$	1704
214.714		CD_2F_2	1710
215.		ClO_2	1827
215.01	±0.14	NH_3	1393,1461–1464
215.081		CD_3OH	826
215.0812	±0.00011	CD_3OH	1481
215.25		CD_3OH	1718
215.3		SO_2	810
215.37246	±0.000032	CH_3OD	1370,1433
215.6		CD_3OH	1719
215.80		$CH_3{}^{18}OH$	1741
216		ClO_2	1823
216.100		CH_3OH	823
216.356		$^{13}CD_3OD$	1702
216.5		$^{13}CH_3OH$	1706
216.5		CH_3OH	829
216.8	±0.11	CH_2DOH	1467
216.9		CD_3OD	1715
217.0		N_2D_4	735
217.20		CD_3OH	1718
217.9		CH_3OD	121
217.9		CHD_2OH	663
218.		CH_2ClF	1737
218.0	±0.11	CH_2DOH	1467
218.0	±0.44	CH_3NH_2	1396,1437
218.0	±0.44	DCOOD	1384,1403
218.0	±1.1	$^{15}NH_3$	1390
218.2		SO_2	810
218.22	±0.065	CH_3OH	1427

Table 4.2.1—*continued*
Optically Pumped Far Infrared and Millimeter Wave Lasers Arranged in Order of Increasing Wavelength

Wavelength (μm)	Uncertainty	Molecule	Reference
218.267		CD_2F_2	1710
218.28	±0.14	NH_3	1393,1461–1464
218.28	±0.14	NH_3	1393,1461–1464
218.500		CH_3OH	179
218.6		NH_3	827
218.70		$CH_3{}^{18}OH$	1741
219.0		CD_3OH	1719
219.0960	±0.00011	CH_2DOH	1467
219.3		CHFCHF	665
219.4		CD_2Cl_2	1708
219.5		CHFCHF	665
219.600		CD_2F_2	1710
219.70		CD_3OH	1718
219.80		$CH_3{}^{18}OH$	1741
219.90		$CH_3{}^{18}OH$	1741
220.		CH_3NH_2	1751
220.		CHFO	821
220.	±1.1	CD_3OH	1391
220.	±4.4	CH_3NH_2	1437
220.0		$^{15}NH_3$	905
220.1		CH_3OD	121
220.2279	±0.00022	H_2O	1368,1457
220.27		$CH_3{}^{18}OH$	1741
220.3		CH_3OCH_3	1754
220.7		CHFCHF	665
221.		CH_2ClF	1737
221.		CH_3NH_2	1751
221.	±4.4	CH_3NH_2	1437
221.0		$^{13}CD_3OH$	1703
221.0		SiH_3F	813
221.2		CHD_2OH	663
221.2		SO_2	810
221.5		CH_3OD	113
221.86		$CH_3{}^{18}OH$	1741
221.88		CD_3OH	1718

Table 4.2.1—*continued*
Optically Pumped Far Infrared and Millimeter Wave Lasers Arranged in Order of Increasing Wavelength

Wavelength (μm)	Uncertainty	Molecule	Reference
221.9		CD_3OH	1719
222.	±1.1	CD_3OH	1402
222.0		CD_3OD	1715
222.1		$^{13}CD_3OD$	453
222.217		CD_3OH	826
222.3		CH2CHF	1732
222.50		$CH_3{}^{18}OH$	1741
222.70		CD_3OH	1718
222.8		$^{13}CH_3OH$	1706
223.	±1.1	CD_3OH	1410
223.50	±0.056	CH_3OH	1375,1427
223.57		CH_2F_2	808
223.7		CH_3OD	113
223.800		CH_3OH	823
223.840		CH_3OH	179
223.91	±0.15	NH_3	1393,1461–1464
224.		CH_3SH	485
224.	±1.1	CD_3Cl	1388,1401
224.	±4.5	CH_3OD	1436
224.	±4.5	CH_3SH	1435
224.2256	±0.00011	CH_2DOH	1467
224.7		CH_3OH	829
225.	±2.3	CH_3OD	1436,1440
225.0		CD_3OH	1719
225.07	±0.15	NH_3	1393,1461–1464
225.2		CH_3OD	121
225.5159	±0.00011	CH_3OH	1397,1427,1481
225.8		CD_3OH	1719
226.		CH_3NH_2	1752
226.2974	±0.00011	CH_2DOH	1467
226.3		CH_3OD	113
226.8		CHD_2OH	663
226.9		CD_3OH	1719
227.0		$^{13}CD_3OD$	1702
227.00		$CH_3{}^{18}OH$	1741

Table 4.2.1—*continued*
Optically Pumped Far Infrared and Millimeter Wave Lasers Arranged in Order of Increasing Wavelength

Wavelength (μm)	Uncertainty	Molecule	Reference
227.15	±0.10	CH_3Cl	1378,1385
227.5		CHD_2OH	663
227.657		CH_2F_2	1738
227.6570	±0.00011	CH_2F_2	1392,1454,1466
227.661		CD_3OD	1714
228.	±2.3	CHFF	1362,1389
228.1		CHFCHF	665
228.3		CD_3OH	1718
228.7		CHD_2OH	663
229.	±1.1	CH_3OD	1367
229.	±4.6	CD_3OD	1425,1432
229.067		CD_2F_2	1710
229.10		CD_3OH	1718
229.3		CH2CHF	1730
229.40		$CH_3{}^{18}OH$	1741
230.1059	±0.00012	CH_2F_2	1392,1454,1466
230.106		CH_2F_2	1738
230.70		$CH_3{}^{18}OH$	1741
231.0		CH_2Cl_2	1736
231.0		CHD_2F	662
231.0	±0.51	CHFF	1389
231.1		CD_3OH	1719
231.1		CHFCHF	665
232.	±1.1	CD_3OH	1402
232.0		CD_2Cl_2	1709
232.0		CHD_2F	662
232.08		CH_3OH	179
232.1		CD_3OH	1719
232.4		CD_3OD	1714
232.65		$CH_3{}^{18}OH$	1741
232.7	±0.47	CH_3OH	1397,1427
232.7884	±0.00012	CH_3OH	1484
232.8		CHFCHF	665
232.9		SO_2	810
232.93906	±0.000091	CH_3OH	1397,1427,1455

Table 4.2.1—*continued*
Optically Pumped Far Infrared and Millimeter Wave Lasers Arranged in Order of Increasing Wavelength

Wavelength (μm)	Uncertainty	Molecule	Reference
233.		CH_3OH	203
233.		ClO_2	1827
233.	±1.1	D_2CO	1380,1382
233.4		CH_3OH	829
233.685		CD_2F_2	1710
233.9157	±0.00019	N_2H_4	1398,1483
234.		CH_3SH	485
234.	±1.1	CH_3OD	1451
234.	±4.7	CH_3SH	1435
234.0	±0.47	N_2H_4	1398
234.610		CH_3OH	823
234.8		CD_3OH	1719
234.800		$^{13}CH_2F_2$	1704
235.2		CD_3OD	1717
235.3		$^{13}CD_3OD$	453
235.40		CD_3OH	1718
235.5		CH_2Cl_2	1736
235.654		CH_2F_2	1738
235.6541	±0.00012	CH_2F_2	1394,1454,1466
235.7		CD_3OD	1715
235.80		CD_3OH	1718
235.9		CH_3OD	113
236.	±1.2	CD_3OH	1410
236.0		SiH_3F	813
236.1		CD_3OD	1715
236.108		CD_2F_2	1710
236.25	±0.097	CH_3Cl	1378,1385
236.5303	±0.00012	$^{13}CH_3OH$	1415,1417,1481
236.5915	±0.00012	CH_2F_2	1392,1454,1466
236.592		CH_2F_2	1738
236.599		CH_2F_2	1739
236.6008	±0.00012	CH_2F_2	1392,1454,1466
236.601		CH_2F_2	1738
237		CH_2F_2	1739
237.1		CD_3OH	1719

Table 4.2.1—*continued*
Optically Pumped Far Infrared and Millimeter Wave Lasers Arranged in Order of Increasing Wavelength

Wavelength (μm)	Uncertainty	Molecule	Reference
237.5		CD_2Cl_2	1709
237.5230	±0.00012	$^{13}CH_3OH$	1415
237.60	±0.048	CH_3OH	1375,1427
238.	±1.2	CD_3OH	1391
238.	±1.2	CH_3OD	1451
238.	±1.2	CHD_2OH	1478
238.	±2.4	CH_3OD	1436,1440
238.5227	±0.00012	$^{13}CH_3OH$	1415
238.65		CD_3OH	1718
240.		CH_3OH	203
240.0	±0.48	HCOOD	1384,1403
240.1		$^{13}CH_3OH$	1706
240.290		CH_3OH	179
240.4		CH_3OH	829
240.98	±0.096	CH_3Cl	1378,1385
241.	±1.2	CH_3OD	1436,1451
241.1		CD_2Cl_2	1708
241.2	±0.48	DCOOD	1384,1403
241.5		CHFCHF	665
241.50		$CH_3{}^{18}OH$	1741
241.6		$^{13}CD_3OD$	1702
241.75		$CH_3{}^{18}OH$	1741
242.310		CH_3OH	823
242.47		$CH_3{}^{18}OH$	1741
242.4727	±0.00012	CH_3OH	1427,1473,1481
242.5	±0.49	CH_3OH	1397,1427
242.6	±0.49	CHFF	1389
242.79	±0.024	CH_3OH	1427,1474
242.847		CH_3OH	834
242.9		CD_3OD	1715
243.0		$^{13}CD_3OD$	1702
243.356		CH_2F_2	1739
243.5		CH_3OD	113
244.		CH_2ClF	1737
244.		CH_3OH	203

Table 4.2.1—*continued*
Optically Pumped Far Infrared and Millimeter Wave Lasers Arranged in Order of Increasing Wavelength

Wavelength (μm)	Uncertainty	Molecule	Reference
244.	±4.9	D_2CO	1382,1438,1439
244.0		N_2D_4	735
244.1		CH2CHF	1732
244.89.		CH_3NH_2	1752
245.		CH_3NH_2	1752
245.	±1.2	CD_3Cl	1388,1401
245.	±1.2	D_2CO	1380,1382
245.	±4.9	D_2CO	1382,1438,1439
245.0	±0.098	CH_3Br	1378
245.652		$^{13}CH_2F_2$	1704
246.		CH_2ClF	1737
246.		CH_3NH_2	1751
246.	±1.2	CD_3Cl	1388,1401
246.	±4.9	CH_3NH_2	1437
246.1		CHD_2OH	663
246.33		CH_2F_2	808
246.5	±0.49	N_2H_4	1398
246.8		CHD_2OH	663
247.		ClO_2	1827
247.0		$^{13}CD_3OD$	1702
247.3	±0.49	CD_3F	1472,1485
247.4		CH_3OH	829
247.5		CD_3OD	1715
247.5	±0.50	CD_3F	1472,1485
247.679		CH_2F_2	1739
248.1220	±0.00012	CH_2DOH	1467
248.606		$^{13}CH_2F_2$	1704
248.620		CH_3OH	823
249.	±1.2	CD_2Cl_2	1478
249.	±1.2	CD_3Cl	1388,1401
249.0		N_2D_4	735
249.1		$^{13}CH_3OH$	1706
249.392		CD_2F_2	1710
249.6		CHD_2OH	663
249.7		CH_3OD	113

Table 4.2.1—*continued*
Optically Pumped Far Infrared and Millimeter Wave Lasers Arranged in Order of Increasing Wavelength

Wavelength (μm)	Uncertainty	Molecule	Reference
249.7		CHD_2OH	663
249.7204	±0.00012	CH_2DOH	1467,1478
249.800		CD_2F_2	1710
249.9		$^{13}CD_3F$	1699
250.		C_2H_3N	1750
250.06	±0.19	NH_3	1393,1461–1464
250.1		CD_3OH	1719
250.5	±0.50	N_2H_4	1398
250.7813	±0.00010	CH_3OH	1455
250.97		CH_2F_2	808
251.		CH_3NH_2	1751
251.	±2.8	CH_3OH	1427,1471,1472
251.	±5.0	CH_3NH_2	1437
251.1398	±0.00010	CH_3OH	1455
251.18.		CH_3NH_2	1752
251.3	±0.50	CH_3NH_2	1396,1437
251.40		CD_3OH	1718
251.4324	±0.00013	CH_3OH	1484
251.90		$CH_3{}^{18}OH$	1741
251.91	±0.050	CH_3F	1374,1377,1393
251.912		CH_3OH	834
252.0		N_2D_4	735
252.3		CD_3OH	1719
252.336		CH_2F_2	1739
253.1		CD_3OH	1719
253.50		$^{13}CH_3OH$	1707
253.5530	±0.00013	CH_3OH	1375,1427,1481
253.60	±0.051	CH_3OH	1375,1427
253.7196	±0.00013	CD_3OH	1391,1481
253.720		CD_3OH	826
253.8		CD_3OH	1719
253.8		CD_3OH	1719
254.		CH_3NH_2	1751
254.	±1.2	CD_2Cl_2	1478
254.	±5.1	CH_3NH_2	1437

Table 4.2.1—*continued*
Optically Pumped Far Infrared and Millimeter Wave Lasers Arranged in Order of Increasing Wavelength

Wavelength (μm)	Uncertainty	Molecule	Reference
254.230		CH_3OH	823
254.3		CHD_2OH	663
254.5	±0.51	HCOOH	1384,1400,1403
254.7		CH_2Cl_2	1736
254.7		CH_3OH	830
254.802		$^{13}CH_2F_2$	1704
255.		ClO_2	1827
255.	±5.1	CD_3OD	1432,1425
255.0		CD_3OD	1715
255.000		$H^{13}COOH$	683
255.2		CD_3OH	1719
255.3		CD_3OD	1714
255.3		CHD_2OH	663
256.	±5.1	D_2CO	1382,1438,1439
256.027		CH_2F_2	1738
256.0270	±0.00013	CH_2F_2	1392,1454,1466
256.4		CD_3OD	1715
257.13	±0.20	NH_3	1393,1461–1464
257.4		CH_2CF_2	1725
258.		CHFO	821
258.	±2.6	CH_2Cl_2	1430
258.0		SO_2	811
258.30		CD_3OH	1718
258.425		$H^{13}COOH$	683
258.4356	±0.00013	CD_3OH	1402,1481
258.436		CD_3OH	826
259.		CH_3OH	203
259.		NH_3	816
259.9		CHD_2OH	663
260.		CHFO	821
260.	±1.3	CHD_2OH	1478
260.	±1.3	$H^{13}COOH$	1384
260.0		CHD_2F	662
260.042		$^{13}CH_2F_22$	1704
260.1	±0.47	CHFF	1362,1389

Table 4.2.1—*continued*
Optically Pumped Far Infrared and Millimeter Wave Lasers Arranged in Order of Increasing Wavelength

Wavelength (μm)	Uncertainty	Molecule	Reference
261.	±5.2	CH_3OH	1427,1436
261.03	±0.099	CH_3Cl	1378,1385
261.2		$^{13}CD_3OD$	453
261.200		CH_3OH	823
261.5		SiH_2F_2	814
261.729		CH_2F_2	1738
261.7292	±0.00013	CH_2F_2	1392,1454,1466
262.		CH_3OH	203
262.		CH_3SH	485
262.	±2.6	CHFF	1362,1389
262.	±5.2	CH_3SH	1435
262.0	±0.52	N_2H_4	1398
262.1		CD_3OD	1715
262.248		CH_2F_2	1739
262.4		CH_3OH	829
262.40		$CH_3{}^{18}OH$	1741
263.0		SiH_2F_2	814
263.2		CD_3OD	1716
263.40	±0.053	NH_3	1376,1393,1464
263.44	±0.21	NH_3	1393,1461–1464
263.5		CH2CHF	1732
263.70	±0.053	CH_3OH	1375,1427
264.		ClO_2	1827
264.		ClO_2	1827
264.1	±0.10	$CH_3{}^{79}Br$	1378,1450
264.5		SiH_3F	813
264.5359	±0.00013	CH_3OH	1375,1427,1481
264.7	±0.10	CH3CH2F	1459,1474
264.70		CD_3OH	1718
264.759		CD_3OH	826
264.8014	±0.00024	N_2H_4	1398,1483
264.9		CH_2NOH	1740
265.	±1.3	CD_3OH	1402
265.	±5.0	CD_3F	1472,1485
265.0		CD_3OH	1719

Table 4.2.1—*continued*
Optically Pumped Far Infrared and Millimeter Wave Lasers Arranged in Order of Increasing Wavelength

Wavelength (μm)	Uncertainty	Molecule	Reference
265.0	±0.53	N_2H_4	1398
265.1	±0.53	DCOOH	1403
265.3		CD_3OH	1719
265.6		CD_2Cl_2	1708
266.	±1.3	CD_3OH	1402
266.0		CD_3OH	1719
266.0		CDF_3	1672
266.1	±0.53	DCOOD	1384,1403
266.5		CD_2Cl_2	1709
266.7352	±0.00013	CH_2DOH	1467
266.866		$^{13}CH_2F_22$	1704
266.9		CDF_3	1672
267.	±1.3	CD_3OH	1402
267.20		CD_3OH	1718
267.4432	±0.00013	CH_3OH	1427,1481
267.823		CD_2F_2	1710
267.9		CD_2Cl_2	1708
268.	±1.3	CD_3OH	1402
268.062		CH_2F_2	1739
268.30		$CH_3{}^{18}OH$	1741
268.5722	±0.00013	$^{13}CH_3OH$	1415
268.6		$^{13}CH_3OH$	1706
268.6		CD_3OH	1719
268.82	±0.22	NH_3	1393,1461–1464
269.9		$^{13}CH_3OH$	1706
270.		CH_3NH_2	1751
270.	±1.3	CH_3OH	1427
270.	±5.4	CH_3NH_2	1437
270.0		CHD_2OH	663
270.005		CH_2F_2	1738
270.0055	±0.00014	CH_2F_2	1392,1454,1466
270.6	±0.54	CH_2CHCN	1396
270.700		CH_3OH	179
270.733		CD_3OD	1714
271.		CH_3NH_2	1751

Table 4.2.1—*continued*
Optically Pumped Far Infrared and Millimeter Wave Lasers Arranged in Order of Increasing Wavelength

Wavelength (μm)	Uncertainty	Molecule	Reference
271.	±5.4	CH_3NH_2	1437
271.3	±0.10	CH_3Cl	1378,1385
271.5		CH_3OH	829
271.5	±0.54	N_2H_4	1398
272.	±1.3	CD_3I	1401
272.0	±0.54	DCOOH	1403
272.1		CHFCHF	665
272.2516	±0.00014	CH_2DOH	1467,1478
272.3389	±0.00014	CH_2F_2	1392,1454,1466
272.339		CH_2F_2	1738
272.5		CD_3OD	1715
272.9		CH_3OH	829
272.958		$^{13}CD_3OD$	1702
273.0		CH_3OH	830
273.0037	±0.00014	CH_2DOH	1467
273.36	±0.22	NH_3	1393,1461–1464
273.4		NH_3	827
273.400		CH_2F_2	1739
273.764		$^{13}CH_2F_22$	1704
274.	±2.7	CH_3OH	1427,1471
274.245		CH_3OH	834
274.776		CD_2F_2	1710
275.		CH_3OH	203
275.0		N_2D_4	735
275.00	±0.096	CH_3Cl	1378,1385
275.09	±0.096	CH_3Cl	1378,1385
275.1		CH_3OD	121
275.5		CH2CHF	1732
275.61		$^{13}CH_3OH$	1707
276.1	±0.55	DCOOD	1384,1403
276.6		CD_3OH	1719
276.7157	±0.00014	CD_3OH	1402,1481
276.716		CD_3OH	826
276.79	±0.23	NH_3	1393,1461–1464
276.9		CD_3OH	1719

Table 4.2.1—*continued*
Optically Pumped Far Infrared and Millimeter Wave Lasers Arranged in Order of Increasing Wavelength

Wavelength (μm)	Uncertainty	Molecule	Reference
277.		C_2H_3N	1750
277.		NH_2OH	682
277.	±1.4	CD_3OH	1402
277.00		$CH_3{}^{18}OH$	1741
277.2		CD_2Cl_2	1708
278.	±1.4	CD_3OH	1402
278.0		N_2D_4	735
278.3		CH_2NOH	1740
278.4		CHD_2OH	663
278.5	±0.56	HCOOH	1384,1400,1403
278.8048	±0.00014	CH_3OH	1375,1427,1481
279.		$(H_2CO)_3$	809
279.	±1.4	D_2CO	1380,1382
279.0		CHD_2OH	663
279.014		$^{13}CH_2F_2$	1704
279.32	±0.23	NH_3	1393,1461–1464
279.4		CH_3OD	113
279.4	±0.56	CH_3OD	1397,1436
279.8		CD_2Cl_2	1708
279.8		CH_3OH	830
279.8	±0.098	CH_3Br	1378
279.9		SO_2	810
280.		CH_3NC	1823
280.		CHFO	821
280.	±1.4	CH_3OD	1451
280.2183	±0.00014	$^{13}CH_3OH$	1415,1417,1481
280.2397	±0.00014	$^{13}CH_3OH$	1415,1417,1481
280.5		SiH_3F	813
280.512		CD_2F_2	1710
280.8		$^{13}CD_3F$	1699
280.8		CHD_2OH	663
280.9341	±0.00014	CH_3OH	1484
281.		CH_3NH_2	1752
281.0		$^{13}CH_3OH$	1706
281.053		CH_2F_2	1739

Table 4.2.1—*continued*
Optically Pumped Far Infrared and Millimeter Wave Lasers Arranged in Order of Increasing Wavelength

Wavelength (μm)	Uncertainty	Molecule	Reference
281.18	±0.098	CH_3CN	1365,1378
281.6		CH_2CF_2	1725
281.6		CH2CHF	1732
281.7	±0.099	CH_3Cl	1378,1385
281.98	±0.099	CH_3CN	1365,1378
282.		CHFO	821
282.0		CH_3OD	121
282.1		SO_2	812
282.80		CD_3OH	1718
282.900		CH_2F_2	1739
282.96		$^{13}CH_3OH$	1707
283.		CH_3NH_2	1751
283.	±5.7	CH_3NH_2	1437
283.0	±0.57	C_2H_3Br	1404
283.1	±0.57	DCOOD	1384,1403
283.2		CH_3OD	113
283.75		CD_3OH	1718
283.783		CH_2F_2	1739
284.		C_2H_3N	807
284.		CH_2ClF	1737
284.15		$CH_3{}^{18}OH$	1741
284.3		CD_3OH	1719
284.330		CH_3OH	179
284.354		CH_2F_2	1739
284.40		CD_3OH1718	
284.50		$CH_3{}^{18}OH$	1741
284.6		CHFCHF	665
284.90		$CH_3{}^{18}OH$	1741
285.		CH_3OH	203
285.		ClO_2	1827
285.0		N_2D_4	735
285.1		CHD_2F	662
285.25		$CH_3{}^{18}OH$	1741
285.3		C_2H_5OH	1746
285.5		N_2D_4	735

Table 4.2.1—*continued*
Optically Pumped Far Infrared and Millimeter Wave Lasers Arranged in Order of Increasing Wavelength

Wavelength (μm)	Uncertainty	Molecule	Reference
286.	±2.9	CH_3OH	1427,1471
286.0		N_2D_4	735
286.155		CH_3OH	834
286.197		CD_3OH	826
286.1974	±0.00014	CD_3OH	1402,1481
286.2		CD_3OH	1719
286.3		CDF_3	1672
286.398		CD_2F_2	1710
286.5	±0.46	CHFF	1389
286.724		CD_3OH	826
286.7242	±0.00014	CD_3OH	1402,1481
286.8		CDF_3	1672
286.8	±0.10	CH_3Cl	1378,1385
286.9	±0.10	CH_3CN	1365,1378
287.3076	±0.00014	CD_3OH	1402,1481
287.308		CD_3OH	826
287.667		CH_2F_2	1738
287.6672	±0.00014	CH_2F_2	1392,1454,1466
287.908		CH_2F_2	1739
287.95		CD_3OH	1718
288.		CH_3NC	1823
288.	±1.4	CD_3Cl	1388,1401
288.	±5.8	CH_3NH_2	1437,1459
288.3		CHD_2OH	663
288.5	±0.58	CF_2CH_2	1387,1388,1396
288.51	±0.25	NH_3	1393,1461–1464
289.1		$^{15}NH_3$	905
289.139		CH_2F_2	1739
289.35	±0.25	NH_3	1393,1461–1464
289.4		CD_2Cl_2	1708
289.4999	±0.00014	CH_2F_2	1392,1454,1466
289.5		CD_3OD	1715
289.500		CH_2F_2	1738
289.6		CD_3OD	121,1715
289.8		CHFCHF	665

Table 4.2.1—*continued*
Optically Pumped Far Infrared and Millimeter Wave Lasers Arranged in Order of Increasing Wavelength

Wavelength (μm)	Uncertainty	Molecule	Reference
289.8		CH_2CF_2	1725
290.		CD_3Br	1712
290.		CH2CHF	1733
290.		CH_3OH	203
290.		NH_2OH	682
290.	±1.4	CD_3OH	1391
290.	±5.8	CD_3Br	1438,1439
290.0		N_2D_4	735
290.2		CHD_2OH	663
290.2	±1.4	NH_3	1393,1461–1464
290.44	±0.25	NH_3	1393,1461–1464
290.62	±0.087	CH_3OH	1427
290.812		CH_2F_2	1739
290.95	±0.25	NH_3	1461,1479
290.95	±0.25	NH_3	1393,1461–1464
291.0		$^{13}CD_3OH$	1703
291.3		CH_2CF_2	1725
291.3		CH_2NOH	1740
291.3		CHD_2OH	663
291.61	±0.087	$^{13}CH_3OH$	1415,1417
291.9	±0.58	HCOOD	1384,1403
292.		CH_2ClF	1737
292.		NH_2OH	682
292.1415	±0.00015	CH_3OH	1375,1427,1481
292.50	±0.050	CH_3OH	1375,1427
292.7		CHD_2F	662
293.0		N_2D_4	735
293.13		$^{13}CH_3I$	1705
293.4		CH2CHF	1730
293.6480	±0.00059	CD_3Cl	1388,1401
293.8		C_2H_3Cl	1728
293.8		CH_2CF_2	1725
293.8		CH_2CHCl	1728
293.8217	±0.00015	CH_3OH	1484
293.901		CH_2F_2	1738

Table 4.2.1—*continued*
Optically Pumped Far Infrared and Millimeter Wave Lasers Arranged in Order of Increasing Wavelength

Wavelength (μm)	Uncertainty	Molecule	Reference
293.9015	±0.00015	CH_2F_2	1392,1454,1466
294.	±5.9	D_2CO	1382,1438,1439
294.04		$^{13}CH_3OH$	1707
294.3		CH_3OD	121
294.3	±0.10	CH_3Br	1378
294.30		$CH_3{}^{18}OH$	1741
294.6		CD_3OD	1715
294.6		CH_2Cl_2	1736
294.7		CH_3OD	113
294.81098	±0.000029	CH_3OD	1370,1433
295.3967	±0.00015	CH_2DOH	1467,1478
295.6394	±0.00015	CH_2DOH	1467,1478
296.		CH_2ClF	1737
296.0		N_2D_4	735
296.480		CH_3OH	179
297.		CD_3Br	1712
297.	±1.5	CD_3OH	1402
297.	±1.5	CD_3OH	1402
297.	±5.9	CD_3Br	1438,1439
297.0		CD_3OD	1716
297.09		COF_2	828
297.1		CD_3OH	1719
297.7		CH_3OH	829
298.		CH2CHF	1733
298.		CH_3SH	485
298.	±6.0	CH_3SH	1435
298.0		CD_3I	1701
298.0	±0.60	DCOOD	1384,1403
298.211		CH_2F_2	1738
298.2910	±0.00015	CH_2F_2	1392,1454,1466
298.470		CH_2F_2	1739
298.5		CH_2Cl_2	1736
298.6		$^{13}CD_3OD$	453
298.736		CD_3OD	1714
298.9		SO_2	810

Table 4.2.1—*continued*
Optically Pumped Far Infrared and Millimeter Wave Lasers Arranged in Order of Increasing Wavelength

Wavelength (µm)	Uncertainty	Molecule	Reference
299.	±1.5	CD_3OH	1402
299.	±6.0	CD_3OD	1425,1432
299.5		$^{13}CD_3F$	1699
299.9		CH_2CF_2	1725
300.		ClO_2	1827
300.0	±0.48	CH_2DOH	1467
300.1		CD_3OD	1716
300.233		$^{13}CH_2F_22$	1704
300.246		$^{13}CH_2F_22$	1704
300.476		CH_2F_2	1739
300.5		$^{13}CD_3OD$	453
300.60		$CH_3{}^{18}OH$	1741
301.	±1.5	CD_3I	1401
301.0		CHD_2F	662
301.0		N_2D_4	735
301.0		$SiHF_3$	814
301.1		CD_2Cl_2	1708
301.2		CH_2NOH	1740
301.2754	±0.00045	N_2H_4	1398,1483
301.3	±0.27	NH_3	1393,1461–1464
301.37		COF_2	828
301.654		$^{13}CH_2F_22$	1704
301.9943	±0.00015	CH_3OH	1414,1427,1481
302.2781	±0.00030	HCOOH	1381,1384,1439
303.		CH_3OH	203
303.800		CD_2F_2	1710
304.0832	±0.00030	DCOOD	1384,1403
304.1	±0.61	HCOOD	1384,1403
304.3		CH_3OCH_3	1754
304.35		COF_2	828
305.24		COF_2	828
305.6		CHD_2OH	663
305.72611	±0.000031	CH_3OD	1370,1433
306.		CHFO	821
306.3	±0.28	NH_3	1393,1461–1464

Table 4.2.1—*continued*
Optically Pumped Far Infrared and Millimeter Wave Lasers Arranged in Order of Increasing Wavelength

Wavelength (μm)	Uncertainty	Molecule	Reference
306.5		$^{13}CH_3OH$	1706
306.5		CH_3OD	113
306.7		CH_2CF_2	1725
306.993		$^{13}CH_2F_2$	1704
307.07		$^{13}CH_3OH$	1707
307.20		$CH_3{}^{18}OH$	1741
307.5		CH_3OD	113
307.5	±0.49	CHFF	1362,1389
307.65	±0.098	CH_3Cl	1378,1385
307.78	±0.092	$^{13}CH_3OH$	1415
308.		CH_2ClF	1737
308.		FCN	1823
308.0		CD_3OD	1716
308.0405	±0.00015	CH_2DOH	1467,1478
308.2957	±0.00015	CH_2DOH	1467
308.5		CD_3OH	1719
309.	±1.5	CD_3OH	1402
309.193		CH_2F_2	1739
309.5		CD_2Cl_2	1709
309.5		CH2CHF	1730
309.5	±0.29	NH_3	1393,1461–1464
309.5	±0.62	HCOOH	1366,1384
309.7		CD_2Cl_2	1708
310.		CH_3OH	203
310.	±1.5	CD_3OH	1402
310.	±3.1	CHFF	1362,1389
310.0	±0.62	DCOOD	1384,1403
310.000		$H^{13}COOH$	683
310.10		CD_3OH	1718
310.35		CD_3OH	1718
310.7		CD_3OH	1719
310.8	±0.50	CHFF	1389
310.80		CD_3OH	1718
311.		CH_3NO_2	1753
311.0		CD_3OD	1715

Table 4.2.1—*continued*
Optically Pumped Far Infrared and Millimeter Wave Lasers Arranged in Order of Increasing Wavelength

Wavelength (μm)	Uncertainty	Molecule	Reference
311.0		N_2D_4	735
311.07	±0.10	CH_3Br	1378
311.0747	±0.00031	N_2H_4	1398,1483
311.1		$^{13}CH_3OH$	1706
311.10	±0.10	CH_3Br	1378
311.2	±0.62	CH_3OH	1397,1427
311.20	±0.10	$CH_3{}^{81}Br$	1378,1450
311.21	±0.10	CH_3Br	1378
311.213		$^{13}CH_2F_2$	1704
311.554	±0.0015	HCOOH	1381,1384,1439
311.9		C_2H_5OH	1746
312.		CH_3OH	203
312.	±6.2	CD_3OD	1425,1432
312.0	±0.62	DCOOH	1403
312.1		CH_3OH	829
312.1		SO_2	811
312.276		$^{13}CH_2F_2$	1704
312.50		CD_3OH	1718
312.7		CD_3OD	1715
312.9		CD_3OH	1719
312.91		COF_2	828
313.	±1.5	$H^{13}COOH$	1384
313.5		CH_3OH	830
313.797		$H^{13}COOH$	683
313.88		CH_3OH	179
314.1		$^{13}CD_3OD$	453
314.646		CD_2F_2	1710
314.841		CD_3OD	1714
314.8469	±0.00047	CH_3NH_2	1437,1459,1483
316.		CH_3SH	485
316.	±6.3	CH_3SH	1435
316.329		$^{13}CH_2F_2$	1704
316.6		CDF_3	1672
316.7		CH_2CF_2	1725
317.0		CHD_2OH	663

Table 4.2.1—*continued*
Optically Pumped Far Infrared and Millimeter Wave Lasers Arranged in Order of Increasing Wavelength

Wavelength (μm)	Uncertainty	Molecule	Reference
317.052		CD_2F_2	1710
317.5		SiH_2F_2	814
318.		CH_3NO_2	1753
318.	±1.6	CD_3Cl	1388,1401
318.080		$^{13}CH_2F_2$	1704
318.600		CD_2F_2	1710
319.		CH_3OH	203
319.		CH_3SH	485
319.	±6.4	CH_3SH	1435
319.0		CH_3CHF_2	1747
319.4		CD_3OD	1715
319.7		$^{13}CH_3OH$	1706
319.9	±0.64	HCOOH	1366,1384,1403
320.	±6.4	D_2CO	1382,1438,1439
320.O	±0.64	CH_3OD	1397
320.4		$^{13}CH_3OH$	1706
320.597		CD_2F_2	1710
320.7		CH_3OD	121
321.	±1.6	CD_3OH	1402
321.0		CH2CHF	1732
321.140		$^{13}CD_3OD$	1702
322.10		CD_3OH	1718
322.35		CD_3OH	1718
322.4		$^{13}CD_3OD$	453
322.4522	±0.00016	CH_2DOH	1467,1478
322.5		$^{13}CD_3OD$	453
322.5		$SiHF_3$	814
322.8		CH2CHF	1732
323.1	±0.65	DCOOD	1384,1403
323.179		CD_2F_2	1710
323.3	±0.45	CD_3F	1472,1485
323.5		CD_2Cl_2	1709
323.50		CD_3OH	1718
324.		CH_2ClF	1737
324.		CH_3SH	485

Table 4.2.1—*continued*
Optically Pumped Far Infrared and Millimeter Wave Lasers Arranged in Order of Increasing Wavelength

Wavelength (μm)	Uncertainty	Molecule	Reference
324.	±6.5	CH_3SH	1435
324.	±6.5	D_2CO	1382,1438,1439
324.1	±0.65	HCOOD	1384,1403
324.140		$^{13}CD_3OD$	1702
325.17	±0.098	$^{13}CH_3OH$	1415,1417
325.2	±0.65	DCOOD	1384,1403
325.3		CH_2CF_2	1725
325.9		$^{13}CD_3F$	1699
325.9	±0.65	HCOOD	1384,1403
326.423		CH_2F_2	1738
326.4230	±0.00016	CH_2F_2	1392,1454,1466
326.5		CD_2Cl_2	1708
326.6		CHFCHF	665
327.0	±0.65	N_2H_4	1398
327.50		$CH_3{}^{18}OH$	1741
327.6		C_2H_5Br	1744
327.770		CH_3OH	179
327.8		CD_3OD	1714
328.		C_2H_4O	1748
328.4570	±0.00033	DCOOH	1403
328.9		$^{13}CH_3OH$	1706
329.		CH_3OH	203
329.2		CD_3OD	1715
329.5		CD_3OH	1719
329.5		CH_2CF_2	1725
329.8		CH2CHF	1730
329.9		CD_3OD	1715
330.	±1.6	CH_3OD	1367
330.0		SiH_2F_2	814
330.0		SiH_3F	813
330.0		$SiHF_3$	814
330.019		CDF3	1721
330.1		CH2CHF	1732
330.991		CD_2F_2	1710
331.5	±0.66	N_2H_4	1398

Table 4.2.1—*continued*
Optically Pumped Far Infrared and Millimeter Wave Lasers Arranged in Order of Increasing Wavelength

Wavelength (μm)	Uncertainty	Molecule	Reference
331.6694	±0.00036	N_2H_4	1398,1483
331.7		CD_3OD	1716
331.79		CD_3I	1713
331.79		CD_3I	1713
332.		CH_3OH	203
332.6		CH_3OD	121
332.6034	±0.00017	$^{13}CH_3OH$	1415,1417,1481
332.86	±0.10	CH_3Br	1378
333.15	±0.10	CH_3Br	1378
333.261		$^{13}CD_3OH$	1703
333.90		CD_3OH	1718
333.926		$^{13}CH_2F_2$	1704
334.0		$SiHF_3$	814
334.0	±0.10	CH_3Cl	1378,1385
334.6		$^{13}CH_3OH$	1706
335.		CH2CHF	1733
335.1		$^{13}CD_3OD$	453
335.6		$^{13}CD_3OD$	453
335.7087	±0.00034	DCOOD	1384,1403
335.85		COF_2	828
336.		C_3H_2O	681
336.		CH2CHF	1733
336.	±1.6	CD_3OH	1402
336.0	±0.67	N_2H_4	1398
336.2461	±0.00017	CH_2DOH	1467
336.3	±0.67	HCOOH	1384,1400,1403
336.5		$^{13}CD_3F$	1699
336.5		$^{13}CD_3OH$	1703
336.5		CD_3OH	1719
336.6	±0.44	CD_3F	1472,1485
336.7	±0.10	CH3CH2F	1459,1474
336.8		CD_3OH	1719
337.		ClO_2	1827
337.		NH_3	816
337.040		CH_3OH	179

Table 4.2.1—*continued*

Optically Pumped Far Infrared and Millimeter Wave Lasers Arranged in Order of Increasing Wavelength

Wavelength (μm)	Uncertainty	Molecule	Reference
337.3		CD_3OH	1719
337.5		CD_3OH	1719
338.9		CH_3OCH_3	1754
338.9638	±0.00017	$^{13}CH_3OH$	1415
339.	±6.8	CD_3OD	1425,1432
339.0		CHFCHF	665
339.1		CH_2CF_2	1725
339.3		CH_2CF_2	1725
339.8		CH_2CHCN	1731
339.9		$^{13}CH_3OH$	1706
339.9	±0.68	HCOOD	1384,1403
340.		CH_3NO_2	1753
340.		ClO_2	1827
340.0		SiH_3F	813
340.00		$^{13}CH_3OH$	1707
340.300		$CHCl_2F$	487
340.3566	±0.00017	CH_2DOH	1467
340.6		CD_2Cl_2	1708
340.627		$^{13}CD_3OH$	1703
340.7		CD_3OD	1715
340.7		CD_3OD	1715
341.		CD_3Br	1712
341.		CH_3SH	485
341.	±6.8	CD_3Br	1438,1439
341.	±6.8	CH_3SH	1435
341.	±6.8	D_2CO	1382,1438,1439
341.8	±0.68	DCOOH	1403
342.	±1.7	CD_2Cl_2	1478
342.127		CD_2F_2	1710
342.7		CD_3OD	1715
342.8		CD_3OD	1715
342.80		$CH_3{}^{18}OH$	1741
343.		C_2H_4O	1748
343.0		SiH_2F_2	814
343.26		C_2H_3CN	1731

Table 4.2.1—*continued*
Optically Pumped Far Infrared and Millimeter Wave Lasers Arranged in Order of Increasing Wavelength

Wavelength (μm)	Uncertainty	Molecule	Reference
343.26		CH_2CHCN	1731
343.3		$^{13}CD_3OD$	453
343.5		SiH_3F	813
344.		CH_2ClF	1737
344.		CH_3NO_2	1753
344.521		$^{13}CH_2F_2$	1704
344.778		CD_3OD	1714
344.8		CH2CHF	1730
344.9		CHD_2OH	663
345.0		$SiHF_3$	814
345.5		CH2CHF	1732
345.50		COF_2	828
345.8		CDF_3	1672
346.		CH_3OH	203
346.	±1.7	CD_3OH	1410
346.	±1.7	CHD_2OH	1478
346.	±6.9	D_2CO	1382,1438,1439
346.3	±0.10	CH_3CN	1365,1378
346.4875	±0.00017	CH_3OH	1397,1427,1481
346.67		$^{13}CH_3I$	1675
346.67		$^{13}CH_3I$	1705
346.67		CD_3I	1713
347.0		$^{13}CD_3OD$	453
347.0	±0.69	HCOOD	1384,1403
347.640		CH_3OH	179
348.1		CD_3OD	1715
348.3		CH_3OH	829
348.6		CD_3OD	1717
349.		CH_2ClF	1737
349.		CH_3NH_2	1751
349.	±7.0	CH_3NH_2	1437
349.0	±0.45	CD_3F	1472,1485
349.1		SO_2	812
349.3	±0.10	CH_3Cl	1378,1385,1430
349.5		CH_2CF_2	1725

Table 4.2.1—*continued*
Optically Pumped Far Infrared and Millimeter Wave Lasers Arranged in Order of Increasing Wavelength

Wavelength (μm)	Uncertainty	Molecule	Reference
350.	±1.7	CD_3OH	1402
350.0		CH_2CF_2	1726
350.0		CHFCHF	665
350.2	±0.70	DCOOD	1384,1403
350.5		CD_3OH	1719
351.		CH_3NH_2	1751
351.		CH_3NO_2	1753
351.		CH_3SH	485
351.	±1.7	CD_3OH	1402
351.	±7.0	CH_3NH_2	1437
351.	±7.0	CH_3SH	1435
351.0		CH_2CF_2	1725
351.0		CH2CHF	1730
351.0		CH_3OH	830
351.0	±0.70	HCOOD	1384,1403
351.1		CH_3OH	829
351.2		CD_3OH	1719
351.3		CH_3OD	113
351.4		CD_3OH	1719
351.8		$^{13}CD_3OD$	453
351.9	±0.70	DCOOD	1384,1403
351.9	±0.70	HCOOD	1384,1403
352.	±1.7	CD_3OH	1402
352.3		CD_3OH	1719
352.5		SiH_2F_2	814
352.5	±0.71	CH_3OD	1397
352.503		CD_3OH	826
352.75	±0.099	CH_3Br	1378
352.8		CH2CHF	1730
352.902		CD_2F_2	1710
353.	±1.7	CD_3OH	1402
353.0		CH2CHF	1730
353.1		$^{13}CD_3OD$	1702
353.1	±0.71	HCOOD	1384,1403
353.8		CD_3OD	1715

Table 4.2.1—*continued*
Optically Pumped Far Infrared and Millimeter Wave Lasers Arranged in Order of Increasing Wavelength

Wavelength (μm)	Uncertainty	Molecule	Reference
354.		DFCO	684
354.	±3.5	CD_3OD	1425
354.176		CD_3OD	1714
354.5		CH2CHF	1730
354.5		N_2D_4	735
354.63		COF_2	828
355.		CH2CHF	1733
355.	±1.7	CHD_2OH	1478
355.0		SiH_2F_2	814
355.126		CH_2F_2	1738
355.1261	±0.00018	CH_2F_2	1392,1454,1466
355.2	±0.71	HCOOD	1384,1403
355.5		$SiHF_3$	814
355.55		$^{13}CH_3I$	1705
355.9		CD_2Cl_2	1708
356.0		CH2CHF	1730
356.0	±0.71	C_2H_3Br	1404
356.0	±0.71	HCOOD	1384,1403
356.4		CD_3OD	1715
356.5		CD_3OD	1715
356.5		CH_3OH	830
356.6		CH2CHF	1730
357.867		$^{13}CH_2F_2$	1704
357.901		CH_2F_2	1739
358.		DFCO	684
358.0		CH_2CF_2	1725
358.1		CD_2Cl_2	1708
358.111		COF_2	998
358.2	±0.72	HCOOD	1384,1403
358.4		$^{13}CD_3OD$	1702
358.9	±0.11	$13CH_3OH$	1415,1417
359.20		$CH_3{}^{18}OH$	1741
359.362		$^{13}CH_2F_2$	1704
359.9	±0.72	HCOOH	1366,1384,1403
360.0		CD_2Cl_2	1708

Table 4.2.1—*continued*
Optically Pumped Far Infrared and Millimeter Wave Lasers Arranged in Order of Increasing Wavelength

Wavelength (μm)	Uncertainty	Molecule	Reference
360.00		CD_3I	1713
360.053		CH_2F_2	1739
360.2		CD_2Cl_2	1708
360.5		CHFCHF	665
360.504		$^{13}CH_2F_2$	1704
360.9		CH2CHF	1730
361.2		CD_2Cl_2	1708
361.2	±0.72	HCOOD	1384,1403
361.231		CDF3	1721
361.5		$SiHF_3$	814
361.8		CH2CHF	1730
362.1	±0.72	DCOOH	1403
362.2		CH2CHF	1732
362.423		CDF3	1721
362.65		$CH_3{}^{18}OH$	1741
362.8		CD_3OH	1719
362.8		CH2CHF	1732
362.8		CH_2CHF	1732
363.		$C_2H_3O_2D$	1749
363.		CH_3OH	203
363.	±1.8	CHD_2OH	1478
363.86		$CH_3{}^{18}OH$	1741
363.9		CH_2CF_2	1725
364.30		$CH_3{}^{18}OH$	1741
364.50		$CH_3{}^{18}OH$	1741
364.6		CH_2CF_2	1726
365.2	±0.73	DCOOH	1403
365.725		$CHCl_2F$	487
365.830		CH_3OH	823
365.866		CD_2F_2	1710
366.		$(H_2CO)_3$	809
366.420		CH_3OH	823
366.625		CD_3Br	1711
366.9	±0.73	DCOOD	1384,1403
366.92		$^{13}CH_3I$	1675

Table 4.2.1—*continued*
Optically Pumped Far Infrared and Millimeter Wave Lasers Arranged in Order of Increasing Wavelength

Wavelength (μm)	Uncertainty	Molecule	Reference
366.92		$^{13}CH_3I$	1705
367.	±7.3	CD_3Br	1438,1439
367.399		CD_2F_2	1710
367.6		CH_2CF_2	1725
368.		CH_3OH	203
368.4	±0.48	CD_3F	1472,1485
368.862	±0.0023	N_2H_4	1398,1483
368.9		CH_3OH	830
369.0		SiH_3F	813
369.1		$C_2H_3F_3$	1743
369.1137	±0.00023	CH_3OH	1375,1427,1455
369.4		CD_2Cl_2	1708
369.550124.93		CD_3OH	1718
369.62		COF_2	828
369.70		CD_3OH	1718
369.9678	±0.00037	HCOOD	1384,1403
370.		CH_3SH	485
370.	±1.8	CD_3OH	1402
370.	±7.4	CH_3SH	1435
370.0		CD_2Cl_2	1709
370.0	±0.74	C_2H_3Br	1404
370.483		CD_3OH	826
370.8		CH_3CHF_2	1747
371.3		CH_3CHF_2	1747
372.		CH2CHF	1733
372.		CH_3OH	203
372.0		CD_3OD	1715
372.0	±0.74	HCOOD	1384,1403
372.36		CD_3OH	1718
372.4		CH_3OD	113
372.5	±0.75	N_2H_4	1398
372.68	±0.045	CH_3F	1374,1377,1393
372.80		$^{13}CH_3I$	1705
372.87	±0.045	CH_3CN	1365,1459,1469
373.0		CD_2Cl_2	1709

Table 4.2.1—*continued*
Optically Pumped Far Infrared and Millimeter Wave Lasers Arranged in Order of Increasing Wavelength

Wavelength (μm)	Uncertainty	Molecule	Reference
373.0	±0.75	N_2H_4	1398
373.4		$^{15}NH_3$	905
373.4		CD_3OD	1715
373.4		NH_3	827
373.8	±0.75	HCOOD	1384,1403
374.0861	±0.00019	CH_2DOH	1467,1478
375.		NH_3	816
375.	±1.8	$^{15}NH_3$	1390,1476
375.1		CH_2CF_2	1725
375.3		CD_3OD	1715
375.4		CH_2CF_2	1725
375.407		CHD_2F	662
375.5		SiH_2F_2	814
375.5449	±0.00045	CF_2CH_2	1483
375.980		$CHCl_2F$	487
376.		CH_3NO_2	1753
376.6		CH_2CF_2	1725
376.7		CHFCHF	665
376.8		$^{13}CD_3F$	1699
376.9		CD_2Cl_2	1708
377.		CH_3NH_2	1752
377.45	±0.094	CH_3I	1378,1399,1409
377.5		CH_2CF_2	1725
377.718		$^{13}CH_2F_2$	1704
378.		CH_3NO_2	1753
378.2		CH_3OCH_3	1754
378.4	±0.76	OCS	1436
378.5		CH_2CF_2	1725
378.57	±0.095	CH_3Cl	1378,1385
378.880		CD_2F_2	1710
379.		CH_3SH	485
379.	±7.6	CH_3SH	1435
379.242		COF_2	668
379.59		COF_2	828
380.		ClO_2	1827

Table 4.2.1—*continued*
Optically Pumped Far Infrared and Millimeter Wave Lasers Arranged in Order of Increasing Wavelength

Wavelength (μm)	Uncertainty	Molecule	Reference
380.02	±0.095	CH_3Br	1378
380.5654	±0.00038	DCOOD	1384,1403
380.8		CD_3OH	1719
381.		$(H_2CO)_3$	809
381.615		$H^{13}COOH$	683
381.820		CH_3OH	179
381.9956	±0.00019	CH_2F_2	1392,1454,1466
381.996		CH_2F_2	1738
382.0		CD_2Cl_2	1709
382.357		$H^{13}COOH$	683
382.639		CH_2F_2	1738
382.6392	±0.00019	CH_2F_2	1392,1454,1466
382.88		$CH_3{}^{18}OH$	1741
383.0		CH_2CF_2	1726
383.2		$C_2H_3F_3$	1743
383.2845	±0.00077	CD_3Cl	1388,1401
384.		CH_3SH	485
384.		DFCO	684
384.	±1.9	$(H_2CO)_3$	1380
384.	±7.7	CH_3SH	1435
384.319		CHD_2F	662
384.7	±0.46	CD_3F	1472,1485
384.869		$(H_2CO)_3$	809
384.916		COF_2	998
385.		C_2H_4O	1748
385.	±1.9	CD_3OH	1402
385.4		CD_2Cl_2	1708
385.4		CH_2CF_2	1726
385.4		CHD_2OH	663
385.7		CD_3OH	1719
385.8		CH_2CHCN	1731
385.80		C_2H_3CN	1731
385.9092	±0.00046	H_2CCl	1375,1483
386.	±1.9	CD_3OH	1402
386.037		CD_3OH	826

Table 4.2.1—*continued*
Optically Pumped Far Infrared and Millimeter Wave Lasers Arranged in Order of Increasing Wavelength

Wavelength (μm)	Uncertainty	Molecule	Reference
386.3392	±0.00019	CH_3OH	1484
386.4	±0.097	CH_3CN	1365,1378
386.5		CHFCHF	665
386.5		N_2D_4	735
386.6		CD_3OH	1719
386.9		CD_3OH	1719
387.		CH_3NH_2	1752
387.2		$^{13}CD_3OH$	1703
387.31	±0.097	CH_3CN	1365,1378,1459
387.5		CD_3OD	1716
387.5591	±0.00019	CH_2DOH	1467
387.8		CH_3CHF_2	1747
387.8	±0.78	HCOOD	1384,1403
388.0		CD_3OH	1719
388.273		CDF3	1721
388.4	±0.097	CH_3CN	1365,1378
388.652		CDF3	1721
388.9		$C_2H_3F_3$	1743
389.		$(H_2CO)_3$	809
389.0		N_2D_4	735
389.6		$^{13}CD_3OH$	1703
389.9		CD_2Cl_2	1708
389.9070	±0.00039	DCOOD	1384,1403
390.		CH_3OH	203
390.	±1.9	CD_3I	1401
390.1	±0.78	CH_3OH	1473,1475
390.2		CH_3OH	830
390.4		C_2H_3Cl	1728
390.4		CH_2CHCl	1728
390.5	±0.098	CH_3I	1378,1409
390.780		COF_2	668
391.461		$^{13}CH_2F_2$	1704
391.6886	±0.00039	HCOOD	1384,1403
392.0		CD_2Cl_2	1708
392.0		CH_3OD	113

Table 4.2.1—*continued*
Optically Pumped Far Infrared and Millimeter Wave Lasers Arranged in Order of Increasing Wavelength

Wavelength (μm)	Uncertainty	Molecule	Reference
392.0687	±0.00026	CH_3OH	1427,1455
392.48	±0.098	CH_3I	1378,1409
392.9	±0.79	HCOOD	1384,1403
393.		$(H_2CO)_3$	809
393.000		CD_2F_2	1710
393.3		$C_2H_3F_3$	1743
393.33		COF_2	828
393.485		$H^{13}COOH$	683
393.6311	±0.00016	HCOOH	1379,1384,1434
394.2	±0.79	HCOOH	1384,1400,1439
394.7009	±0.00020	CH_2F_2	1392,1454,1466
394.701		CH_2F_2	1738
395.0	±0.79	HCOOD	1384,1403
395.0	±0.79	HCOOD	1384,1403
395.00		$^{13}CH_3I$	1705
395.1488	±0.00040	DCOOD	1384,1403
395.7124	±0.00040	HCOOD	1384,1403
396.	±4.0	C_2H_5OH	1430
396.0	±0.79	C_2H_3Br	1404
396.0	±0.79	DCOOD	1384,1403
396.00	±0.099	CH_2DOH	1467,1478
396.4		CD_3OH	1719
397.1	±0.79	DCOOD	1384,1403
397.7		CH_3CHF_2	1747
398.		CH_3NO_2	1753
398.	±2.0	CD_3OH	1402
398.1	±0.80	HCOOD	1384,1403
398.96		C_2H_3CN	1731
398.96		CH_2CHCN	1731
399.288		$^{13}CH_2F_2$	1704
399.42		C_2H_3CN	1731
399.42		CH_2CHCN	1731
399.8		$^{13}CD_3OH$	1703
399.80		C_2H_3CN	1731
400.1		$^{13}CH_3OH$	1706

Table 4.2.1—*continued*
Optically Pumped Far Infrared and Millimeter Wave Lasers Arranged in Order of Increasing Wavelength

Wavelength (μm)	Uncertainty	Molecule	Reference
400.2		CD_2Cl_2	1708
401.2		CH_2CHCN	1731
401.25		C_2H_3CN	1731
401.3		CH_2CF_2	1725
401.444		CH_2F_2	1739
402.		C_2H_3N	807
402.3		CH_2CF_2	1726
402.915		COF_2	668
403.		CH_3SH	485
403.	±8.1	CH_3SH	1435
403.710		CH_2F_2	1739
403.777		$^{13}CH_2F_2$	1704
404.		$(H_2CO)_3$	809
404.		CH_3NC	1823
404.	±0.97	CH3CH2F	1459,1474
404.0	±0.81	HCOOH	1366,1384,1403
404.3		CD_3OD	1715
404.6		CHD_2OH	663
404.7	±0.45	NH_3	1393,1461–1464
405.0	±0.81	HCOOH	1384,1400,1403
405.5044	±0.00053	CH3CH2F	1459,1483
405.5848	±0.00039	HCOOH	1403,1483
405.95		C_2H_3CN	1731
405.95		CH_2CHCN	1731
406.	±8.1	CD_3OD	1425,1432
406.36		C_2H_3CN	1731
406.36		CH_2CHCN	1731
406.878		CHD_2F	662
407.	±2.0	CD_3OH	1402
407.1		$^{13}CD_3OD$	1702
407.2937	±0.00053	CF_2CH_2	1420,1465,1483
407.5		$CH_3{}^{18}OH$	1741
407.6		CH2CHF	1730
407.7	±0.10	CH_3Br	1378
407.9		CD_3OH	1719

Table 4.2.1—*continued*
Optically Pumped Far Infrared and Millimeter Wave Lasers Arranged in Order of Increasing Wavelength

Wavelength (μm)	Uncertainty	Molecule	Reference
408.8		CD_3OD	1715
409.		ClO_2	1827
409.	±2.0	CD_3OH	1402
409.10		CD_3OH	1718
409.3		CH_2CF_2	1725
409.6		CD_2Cl_2	1708
409.8		CD_2Cl_2	1708
410.	±2.0	CD_3OH	1402
410.1		$C_2H_3F_3$	1743
410.2		CD_3OD	1716
410.712		CD_3OD	1714
411.0	±0.82	C_2H_3Br	1404
411.1		CHFCHF	665
411.2		$C_2H_3F_3$	1743
411.2	±0.82	HCOOD	1384,1403
411.6		CD_3OD	1715
412.	±2.0	CD_3OH	1402
412.0		$SiHF_3$	814
412.2		CH2CHF	1732
414.		CH_3NO_2	1753
414.	±4.1	CD_3OD	1425,1432
414.1	±0.83	DCOOD	1384,1403
414.98	±0.10	$CH_3{}^{79}Br$	1378,1450
415.		C_2H_4O	1748
415.2	±0.83	DCOOD	1384,1403
415.363		$^{13}CH_2F_2$	1704
416.0	±0.83	C_2H_3Br	1404
416.5223	±0.00042	CH_3OH	1427,1473,1481
416.7		CH_3OD	113
416.71		CH_3OH	179
417.	±2.0	CH_3OD	1367
417.0	±0.83	HCOOD	1384,1403
417.00		CD_3OH	1718
417.244		CD_2F_2	1710
417:3		$^{13}CD_3OD$	1702

Table 4.2.1—*continued*
Optically Pumped Far Infrared and Millimeter Wave Lasers Arranged in Order of Increasing Wavelength

Wavelength (μm)	Uncertainty	Molecule	Reference
417.4		CD_3OD	1716
418.		CH_3OH	203
418.		ClO_2	1827
418.0827	±0.00021	CH_3OH	1375,1427,1481
418.1		CD_3OH	1719
418.1		CD_3OH	1719
418.1	±0.84	HCOOH	1384,1403,1439
418.2		CD_3OD	1716
418.270		CH_2F_2	1738
418.2703	±0.00021	CH_2F_2	1392,1454,1466
418.3	±0.10	CH_3Br	1378
418.6129	±0.00042	HCOOH	1381,1384,1439
418.7118	±0.00021	CD_3OH	1402,1481
418.712		CD_3OH	826
418.79		CH_3OH	179
419.		CH_3OH	203
419.0	±0.84	C_2H_3Br	1404
419.3		CH_3OH	829
419.839		$(H_2CO)_3$	809
420.		CH2CHF	1733
420.2		CH_3OD	113
420.3		CD_3OH	1719
420.311		CDF3	1721
420.3911	±0.00042	HCOOH	1384,1403,1439
420.980		CDF3	1721
421.	±2.1	CD_3OH	1402
421.053		$^{13}CH_2F_2$	1704
421.3		CH_3CHF_2	1747
421.8		CH2CHF	1730
422.	±2.1	CD_3OH	1402
422.0		$C_2H_3F_3$	1743
422.1512	±0.00021	CH_2DOH	1467
422.5		CHFCHF	665
422.8	±0.10	CH_3Br	1378
423.0		CH2CHF	1732

Table 4.2.1—*continued*
Optically Pumped Far Infrared and Millimeter Wave Lasers Arranged in Order of Increasing Wavelength

Wavelength (μm)	Uncertainty	Molecule	Reference
423.354		CH_2CHCl	1711
424.		CH_2CHCl	1829
424.		CH_3NO_2	1753
424.	±8.5	H_2CCl	1459
424.0		CHFCHF	665
424.0	±0.85	C_2H_3Br	1404
424.13		COF_2	828
424.55		CD_3I	1713
425.2	±0.85	DCOOD	1384,1403
425.65		C_2H_3CN	1731
425.65		CH_2CHCN	1731
425.8		$^{13}CH_3OH$	1706
425.87		C_2H_3CN	1731
425.87		CH_2CHCN	1731
426.		CH_3NO_2	1753
426.	±2.1	CHD_2OH	1478
426.8		CH_2CF_2	1725
427.0	±0.85	C_2H_3Br	1404
427.04	±0.098	CH_3CN	1365,1378
427.1		CHD_2OH	663
427.2		$^{13}CD_3OD$	453
427.2	±0.13	CH_2DOH	1467
427.3		$^{13}CD_3OD$	453
428.		CD_3Br	1712
428.	±8.6	CD_3Br	1438,1439
428.87	±0.099	CH_3CCH	1377,1378
429.6898	±0.00043	HCOOD	1384,1403
429.9		CH2CHF	1730
430.		CD_3Br	1712
430.		CH2CHF	1733
430.		CH_3OH	203
430.	±8.6	CD_3Br	1438,1439
430.1		CH_2CF_2	1725
430.4380	±0.00043	HCOOD	1384,1403
430.55	±0.052	CH_3CN	1364,1365,1377

Table 4.2.1—*continued*
Optically Pumped Far Infrared and Millimeter Wave Lasers Arranged in Order of Increasing Wavelength

Wavelength (μm)	Uncertainty	Molecule	Reference
430.91		COF_2	828
430.927		CD_3OH	826
431.14		C_2H_3CN	1731
431.14		CH_2CHCN	1731
431.4		CD_3OH	1719
431.736		CD_3Br	1711
432.		CHFO	821
432.1094	±0.00043	HCOOH	1381,1403,1439
432.6313	±0.00013	HCOOH	1384,1434,1439
432.6665	±0.00043	HCOOH	1366,1384,1403
433.		$C_2H_3O_2D$	1749
433.		CH2CHF	1729
433.	±2.1	$(H_2CO)_3$	1380
433.1036	±0.00087	CD_3I	1401
433.2	±0.87	DCOOH	1403
433.2	±0.87	HCOOD	1384,1403
433.2353	±0.00043	DCOOH	1403
433.5		CH2CHF	1730
433.6		CD_3OH	1719
433.8		CHFCHF	665
433.9		CH_3CHF_2	1747
434.0		N_2D_4	735
434.95		$CH_3{}^{18}OH$	1741
434.951		CH_2F_2	1738
434.9514	±0.00022	CH_2F_2	1392,1454,1466
435.	±2.1	CD_3OH	1410
435.10		CD_3OH	1718
435.30		CD_3OH	1718
435.427		CHD_2F	662
435.5		$SiHF_3$	814
435.7718	±0.00031	N_2H_4	1398,1483
435.9		C_2H_3Cl	1730
435.9		CH_3CHF_2	1747
437.4		CHD_2OH	663
437.4510	±0.00044	HCOOH	1366,1384,1403

Table 4.2.1—*continued*
Optically Pumped Far Infrared and Millimeter Wave Lasers Arranged in Order of Increasing Wavelength

Wavelength (μm)	Uncertainty	Molecule	Reference
437.6		CHFCHF	665
438.		CH_2CHCl	1728
438.0		C_2H_3Cl	1728
438.022		$^{13}CH_2F_22$	1704
438.10		$CH_3{}^{18}OH$	1741
438.4		CD_2Cl_2	1708
438.5069	±0.00044	C_2H_3Br	1404
438.8		CD_3OD	1715
438.87		CD_3OH	1718
439.0		$SiHF_3$	814
439.063		CH_2F_2	1739
440.		CD_3Br	1712
440.	±8.8	CD_3Br	1438,1439
440.01		C_2H_3CN	1731
440.01		CH_2CHCN	1731
440.2		CD_2Cl_2	1708
440.884		CD_2F_2	1710
441.1	±0.10	CH_3CN	1365,1378
441.3		CH_2CF_2	1726
441.3		CH_3OCH_3	1754
441.674		CD_3Br	1711
441.7		CH2CHF	1732
442.1		CHFCHF	665
442.1678	±0.00066	H_2CCl	1483
442.8	±0.89	DCOOD	1384,1403
443.0		SiH_2F_2	814
443.2645	±0.00089	CD_3Cl	1388,1401
443.5	±0.89	C_2H_3Br	1404
443.6		CH_3CHF_2	1747
443.8		CH_3OH	830
444.3862	±0.00089	CD_3I	1401
444.745		COF_2	668
444.8	±0.89	HCOOH	1366,1384,1403
445.		CH_2CHCl	1829
445.	±8.9	H_2CCl	1459

Table 4.2.1—*continued*
Optically Pumped Far Infrared and Millimeter Wave Lasers Arranged in Order of Increasing Wavelength

Wavelength (μm)	Uncertainty	Molecule	Reference
445.0		CH_2CF_2	1726
445.0	±0.89	C_2H_3Br	1404
445.2		CH2CHF	1730
445.4		CH_3OD	113
445.663		CDF3	1721
445.8996	±0.00045	HCOOH	1366,1384,1403
446.5054	±0.00031	HCOOH	1400,1483
446.7		CH2CHF	1730,1732
446.8	±0.89	HCOOD	1384,1403
446.8730	±0.00045	HCOOH	1366,1384,1403
447.		C_2H_5Cl	1745
447.080		CH_3OH	179
447.1421	±0.00089	CH_3I	1378,1401,1409
447.6		CD_2Cl_2	1708
448.3		CH_3OH	829
448.455		CH_3OH	834
448.5335	±0.00045	$H^{13}COOH$	1384
448.534		$H^{13}COOH$	683
448.7		CD_3I	1701
449.0		C_2H_5OH	1746
449.2		CH_2CF_2	1726
449.3		CD_3OD	1716
449.3		CH_3CHF_2	1747
449.5		CH_3CHF_2	1747
449.7997	±0.00090	CD_3Cl	1388,1401
450.		CH_3NO_2	1753
450.		DFCO	684
450.1	±0.90	HCOOD	1384,1403
450.7		CD_3OD	1716
450.9799	±0.00045	HCOOD	1384,1403
451.		$C_2H_3O_2D$	1749
451.4754	±0.00023	CH_2DOH	1467
452.	±9.0	CH3CH2F	1483
452.2	±0.90	DCOOD	1384,1403
452.38		CD_3I	1713

Table 4.2.1—*continued*
Optically Pumped Far Infrared and Millimeter Wave Lasers Arranged in Order of Increasing Wavelength

Wavelength (μm)	Uncertainty	Molecule	Reference
452.4		$^{13}CH_3OH$	1706
452.40	±0.10	CH_2DOH	1467
452.425		$^{13}CH_2F_2$	1704
452.5		CHD_2OH	663
452.9		CD_3OH	1719
453.1		CD_3OD	1715
453.3974	±0.00068	CH_3CN	1378,1459,1483
453.57		C_2H_3CN	1731
453.57		CH_2CHCN	1731
453.6		C_2H_5Br	1744
453.6		CD_3OD	1715
453.8		C_2H_3Br	1727
454.		C_2H_3N	807
454.		CH_3NO_2	1753
454.0		N_2D_4	735
454.3		CH2CHF	1732
454.5		C_2H_3Cl	1728
454.5		CH_2CF_2	1725
454.6		$C_2H_3F_3$	1743
454.8		$C_2H_3F_3$	1743
455.	±2.2	CD_3OH	1402
455.5		$SiHF_3$	814
455.6		CD_3OH	1719
456.		CH2CHF	1729
456.		CH_2CHF	1729
456.		CH_3SH	485
456.	±9.1	CH_3SH	1435
456.1		CD_3OD	1715
456.200		CD_2F_2	1710
457.2	±0.10	CH_3I	1378,1409
457.3		CH_2CF_2	1725
457.3410	±0.00046	DCOOD	1384,1403
457.5		CD_3OD	1715
457.9		C_2H_3Br	1730
458.		CH2CHF	1729

Table 4.2.1—*continued*
Optically Pumped Far Infrared and Millimeter Wave Lasers Arranged in Order of Increasing Wavelength

Wavelength (μm)	Uncertainty	Molecule	Reference
458.0		CH2CHF	1732
458.0	±0.92	CF_2CH_2	1387,1388,1465
458.5229	±0.00069	HCOOH	1381,1384,1439
459.		CH_2CHCN	1731
459.00		C_2H_3CN	1731
459.2	±0.10	CH_3I	1378,1409
459.31		C_2H_3CN	1731
459.31		CH_2CHCN	1731
459.4		C_2H_3Cl	1728
459.4		CDF_3	1668
459.4		CDF_3	1722
459.4		CH_2CHCl	1728
459.428		$(H_2CO)_3$	809
459.6		CDF_3	1668
459.6		CDF_3	1722
459.8		CH2CHF	1732
459.886		ClO_2	1827
460.	±14.	CH_3CHF_2	1420
460.	±2.3	$(H_2CO)_3$	1380
460.440		CH_3OH	179
460.5619	±0.00092	CD_3I	1401
461.		CH_3OH	203
461.0		CH2CHF	1732
461.0718	±0.00069	N_2H_4	1398,1483
461.2	±0.10	CH_3Cl	1378,1385
461.2610	±0.00046	HCOOD	1384,1403
461.3847	±0.00023	$^{13}CH_3OH$	1415
462.8		$^{13}CD_3OH$	1703
463.0		$C_2H_3F_3$	1743
464.3	±0.93	CF_2CH_2	1387,1388,1465
464.40		C_2H_3CN	1731
464.40		CH_2CHCN	1731
464.412		CH_2F_2	1738
464.412		CH_2F_2	1738
464.4123	±0.00023	CH_2F_2	1392,1454,1466

Table 4.2.1—*continued*
Optically Pumped Far Infrared and Millimeter Wave Lasers Arranged in Order of Increasing Wavelength

Wavelength (μm)	Uncertainty	Molecule	Reference
464.627		$H^{13}COOH$	683
464.7		$^{13}CD_3OD$	1702
464.7567	±0.00093	CD_3Cl	1388,1401
464.8		CH_3CHF_2	1747
465.		$C_2H_3O_2D$	1749
465.		NH_3	816
465.0		$SiHF_3$	814
465.50		CD_3I	1713
465.50		CD_3I	1713
465.50		$CH_3{}^{18}OH$	1741
465.500		CD_2F_2	1710
465.70		$CH_3{}^{18}OH$	1741
466.00		CD_3I	1713
466.00		CD_3I	1713
466.25	±0.093	CH_3CN	1365,1378
466.5461	±0.00047	DCOOH	1403
466.643		CD_3Br	1711
467.		$(H_2CO)_3$	809
467.2		CH2CHF	1730
467.515		$CHCl_2F$	487
467.850		CH_3OH	179
468.2359	±0.00023	CH_2DOH	1467,1478
468.965		$^{13}CD_3OH$	1703
469.	±2.3	CD_2Cl_2	1478
469.0233	±0.00037	CH_3OH	1427,1455,1473
469.2	±0.94	DCOOD	1384,1403
470.		CH_2CHCN	1731
470.		CH_3NO_2	1753
470.		CH_3OH	203
470.00		C_2H_3CN	1731
470.065		$^{13}CD_3F$	1700
470.386		$CHCl_2F$	487
471.0		SiH_2F_2	814
471.2		$C_2H_3F_3$	1743
471.5		CD_2Cl_2	1708

Table 4.2.1—*continued*
Optically Pumped Far Infrared and Millimeter Wave Lasers Arranged in Order of Increasing Wavelength

Wavelength (μm)	Uncertainty	Molecule	Reference
472.		CH_3NO_2	1753
472.	±2.3	CD_3OH	1402
472.1	±0.94	HCOOD	1384,1403
472.4		CD_3OH	1719
472.9	±0.95	HCOOD	1384,1403
473.68		CD_3I	1713
474.6		C_2H_3Cl	1728
474.6		CH_2CHCl	1728
475.1		$^{13}CD_3OD$	453
475.1		CH_2CF_2	1725,1726
475.3		C_2H_3Br	1730
476.0		CH2CHF	1730
476.00		CD_3I	1713
476.25		CD_3OH	1718
476.3		CH_2CF_2	1726
476.8		C_2H_3Cl	1728
476.8		CH_2CHCl	1728
477.		CH2CHF	1733
477.1		$C_2H_3F_3$	1743
477.3		CD_3OH	1719
477.3		CH_2CF_2	1725
477.4	±0.95	HCOOD	1384,1403
477.9	±0.096	CH_3I	1378,1409
477.963		$H^{13}COOH$	683
478.072		COF_2	668
478.9	±0.96	DCOOD	1384,1403
479.123		$^{13}CH_2F_2$	1704
479.150		CH_3OH	179
479.9040	±0.00048	DCOOH	1403
480.	±2.4	CD_3OH	1402
480.	±2.4	$H^{13}COOH$	1384
480.0	±0.096	CH_3CN	1365,1378
480.000		$H^{13}COOH$	683
480.3101	±0.00096	CD_3Cl	1388,1401
481.		C_2H_3N	807

Table 4.2.1—*continued*
Optically Pumped Far Infrared and Millimeter Wave Lasers Arranged in Order of Increasing Wavelength

Wavelength (μm)	Uncertainty	Molecule	Reference
482.12		$CH_3{}^{18}OH$	1741
482.2		CH_3CHF_2	1747
482.5		CD_3OD	1715
482.7		CD_3OH	1719
482.9		CHD_2OH	663
482.9615	±0.00048	C_2H_3Br	1404
483.	±2.4	CD_3OH	1402
483.	±2.4	CHD_2OH	1478
483.0		CD_3OD	1716
483.16		CD_3I	1713
483.5		CD_2Cl_2	1708
483.5		CD_3OD	1716
483.5	±0.97	N_2H_4	1398
483.8		CH2CHF	1732
484.3		CD_2Cl_2	1708
484.4		CHD_2OH	663
485.27		COF_2	828
485.4		$C_2H_3F_3$	1743
485.6		$C_2H_3F_3$	1743
485.8		$C_2H_3F_3$	1743
486.	±9.7	CH3CH2F	1459
486.1		$C_2H_3F_3$	1743
486.1		CH_2CF_2	1725
486.1	±0.97	CH_3OH	1427,1473,1475
486.5		CD_3OD	1714
487.		CH_2CHCl	1829
487.		CH_3NO_2	1753
487.	±9.7	H_2CCl	1459
487.0		$SiHF_3$	814
487.2260	±0.00097	CD_3I	1401
487.4		CD_3OD	1715
487.5		CH2CHF	1730
487.5		CH_2CHF	1730
487.6		CH_3OD	113
487.7		CH2CHF	1732

Table 4.2.1—*continued*
Optically Pumped Far Infrared and Millimeter Wave Lasers Arranged in Order of Increasing Wavelength

Wavelength (μm)	Uncertainty	Molecule	Reference
487.8		C_2H_3Cl	1728
487.8		CH_2CHCl	1728
487.8		CH2CHF	1732
488.0		$SiHF_3$	814
488.11		COF_2	828
488.276		CD_2F_2	1710
488.528		CDF3	1721
489.		CH_2CHCN	1829
489.		CH_3NO_2	1753
489.	±9.8	CH_2CHCN	1459
489.238		CD_2F_2	1710
489.3		CH2CHF	1730
490.		CH2CHF	1733
490.0		CH2CHF	1732
490.0829	±0.00049	C_2H_3Br	1404
490.3909	±0.00098	CD_3I	1401
490.7		CH_2CF_2	1726
491.2		CD_3OD	1715
491.376		$(H_2CO)_3$	809
491.8		$C_2H_3D_2OH$	1742
491.8906	±0.00049	DCOOD	1384,1403
492.040		$CHCl_2F$	487
493.0		CD_2Cl_2	1709
493.1562	±0.00049	HCOOD	1384,1403
493.5		CD_3OD	1716
493.5		CH2CHF	1730
493.541		CH_3OH	834
494.0		SiH_2F_2	814
494.1		CD_3OD	1715
494.6461	±0.00074	CH_3CN	1378,1459,1483
495.	±2.4	CD_3OH	1402
495.	±2.9	CH_3OH	1471,1484
495.	±5.0	CD_3OD	1425
495.963		$CHCl_2F$	487
496.072	±0.0024	CH_3F	1374,1393,1434

Table 4.2.1—*continued*
Optically Pumped Far Infrared and Millimeter Wave Lasers Arranged in Order of Increasing Wavelength

Wavelength (μm)	Uncertainty	Molecule	Reference
496.1009	±0.00040	CH_3F	1393,1434
496.3		$^{13}CH_3OH$	1706
496.4		$^{13}CH_3OH$	1706
496.5		CH_3OCH_3	1754
496.660		$^{13}CH_2F_2$	1704
497.		$(H_2CO)_3$	809
497.0		CD_2Cl_2	1709
497.2		CD_3OD	1715
497.3		CD_2Cl_2	1708
497.4		CH_3OCH_3	1754
497.5		C_2H_3Br	1727
497.5		C_2H_3Br	1727
497.677		CD_2F_2	1710
498.	±2.4	CD_3OH	1391
498.0	±1.0	CH_3OD	1397
498.0	±1.0	HCOOD	1384,1403
498.5		$SiHF_3$	814
498.7		CH_2CF_2	1726
500.	±10.	CH_2CHCN	1459
500.0		$(H_2CO)_3$	809
500.577		CD_2F_2	1710
500.577		CD_2F_2	1710
501.164		$(H_2CO)_3$	809
501.6		$C_2H_3F_3$	1743
501.9		CHD_2OH	663
502.2623	±0.00080	CH3CH2F	1459,1483
503.		CH_2CHCN	1829
503.0567	±0.00025	CH_2F_2	1392,1454,1466
504.5		C_2H_3Br	1730
504.752		CDF_3	1721
505.0		CH_3CHF_2	1747
505.0		SO_2	810
505.829		COF_2	668
506.	±1.0	C_2H_3Br	1404
506.3		C_2H_3F	1732

Table 4.2.1—*continued*
Optically Pumped Far Infrared and Millimeter Wave Lasers Arranged in Order of Increasing Wavelength

Wavelength (μm)	Uncertainty	Molecule	Reference
507.480		CH_3OH	179
507.584		CH_2CHCl	1711
507.5840	±0.00081	H_2CCl	1375,1483
507.591		CH_2CHCl	1711
508.	±1.0	DCOOD	1384,1403
508.	±2.5	CD_3OH	1402
508.33		CH_2CHCN	1731
508.4	±0.10	CH_3I	1378,1409
508.5	±0.10	CH_3Br	1378
508.7911	±0.00051	DCOOD	1384,1403
509.		C_2H_4O	1748
509.16		CH_2CHCN	1731
509.3717	±0.00025	CH_2DOH	1467
509.44		COF_2	828
509.5		CD_3OH	1719
509.7		C_2H_3Br	1727
509.859		ClO_2	1827
509.890		$(H_2CO)_3$	809
510.2	±0.10	$CH_3C\acute{N}$	1365,1377,1378
510.4		$C_2H_3F_3$	1743
510.7		$C_2H_3F_3$	1743
511.4451	±0.00026	CH_2F_2	1392,1454,1466
511.9		CH_3OCH_3	1754
511.9	±0.10	CH_3Cl	1378,1385
512.	±2.5	$(H_2CO)_3$	1380
512.8		CHD_2OH	663
513.0022	±0.00077	HCOOH	1379,1384,1439
513.0157	±0.00051	HCOOH	1403,1439
513.4		CH_3CHF_2	1747
513.7572	±0.00051	HCOOD	1384,1403
514.		CH_3NO_2	1753
514.		DFCO	684
514.9507	±0.00051	DCOOD	1384,1403
515.1695	±0.00052	HCOOH	1384,1403,1439
515.8		CH_3OD	113

Table 4.2.1—*continued*
Optically Pumped Far Infrared and Millimeter Wave Lasers Arranged in Order of Increasing Wavelength

Wavelength (μm)	Uncertainty	Molecule	Reference
516.		C_3H_2O	681
516.		CH_3OH	203
516.0		SiH_3F	813
516.382		COF_2	668
516.5		CD_3OH	1719
516.77	±0.098	CH_3CCH	1377,1378
517.	±2.5	CD_3OH	1402
517.33	±0.098	CH_3I	1378,1409
517.5		C_2H_3Br	1727
517.8		CHD_2OH	663
518.	±2.5	CHD_2OH	1478
518.4		C_2H_3F	1732
518.8		$C_2H_3F_3$	1743
518.9		CH_3CHF_2	1747
519.		CH_2CHCl	1829
519.075	±0.0013	CH3CH2F	1459,1483
519.2		$C_2H_3F_3$	1743
519.303	±0.0010	CD_3Cl	1388,1401
519.6		C_2H_3F	1735
519.6		CH_2CHF	1730
520.	±10.	CF_2CH_2	1387,1459,1465
520.	±10.	H_2CCl	1459
520.	±2.5	CD_2Cl_2	1478
520.3		CD_3OH	1719
521.11		$^{13}CH_3I$	1675
521.237		CDF_3	1721
521.4		CH_3CHF_2	1747
521.6		$^{13}CD_3OD$	453
522.7		$^{13}CD_3OD$	453
523.0914	±0.00026	CH_2DOH	1467
523.120		CH_3OH	179
523.406	±0.0010	CD_3I	1401
524.		CH_3NO_2	1753
524.6		CD_3OH	1719
524.8		CH_2CHCl	1730

Table 4.2.1—*continued*
Optically Pumped Far Infrared and Millimeter Wave Lasers Arranged in Order of Increasing Wavelength

Wavelength (μm)	Uncertainty	Molecule	Reference
524.9		NH_3	827
525.		$C_2H_3O_2D$	1749
525.		ClO_2	1827
525.3		SO_2	810
525.32	±0.10	CH_3I	1378,1409
525.56		CH_2CHCN	1731
526.3		CH_3OCH_3	1754
526.4856	±0.00053	DCOOD	1384,1403
527.2146	±0.00053	DCOOD	1384,1403
527.8730	±0.00090	N_2H_4	1483
527.9		C_2H_5Br	1744
528.4965	±0.00053	C_2H_3Br	1404
529.3		C_2H_5OH	1746
529.3	±0.10	CH_3I	1378,1409
530.		CH_3NO_2	1753
530.	±11.	CD_3Br	1438,1439
530.	±11.	H_2CCl	1459
530.	±16.	CH_3CHF_2	1420
530.533		CH_2CHCl	1711
530.7		CH_3OCH_3	1754
530.854		$CHCl_2F$	487
531.	±1.1	HCOOD	1384,1403
531.1	±0.10	CH_3Br	1378
531.1	±0.10	CH_3CCH	1377,1378
533.0		N_2D_4	735
533.33		CD_3I	1713
533.655	±0.0048	N_2H_4	1398,1483
533.6783	±0.00053	HCOOH	1384,1403,1439
533.7006	±0.00053	HCOOH	1384,1403,1439
534.2		CH_3CHF_2	1747
535.	±1.1	HCOOH	1384,1400,1439
536.096		$H^{13}COOH$	683
537.06		CH_2CHCN	1731
537.410		$^{13}CD_3F$	1700
537.65		CH_2CHCN	1731

Table 4.2.1—*continued*
Optically Pumped Far Infrared and Millimeter Wave Lasers Arranged in Order of Increasing Wavelength

Wavelength (μm)	Uncertainty	Molecule	Reference
538.		CH_2CHCl	1829
538.415		COF_2	668
539.10		COF_2	828
540.	±11.	H_2CCl	1459
540.	±2.6	CD_3I	1401
540.736		CDF_3	1721
540.9864	±0.00027	CH_2F_2	1392,1454,1466
542.99	±0.092	CH_3I	1378,1409
543.2		CHFCHF	665
544.1		C_2H_3Br	1730
545.		CH_2CHCN	1731
545.		NH_2OH	682
545.21	±0.093	$CH_3{}^{81}Br$	1378,1450
545.39	±0.093	CH_3Br	1378
545.5		CH_2CHCl	1728
545.56		CD_3I	1713
545.88		CD_3I	1713
546.8		CHFCHF	665
546.80		$CH_3{}^{18}OH$	1741
547.529		$CHCl_2F$	487
548.843		$H^{13}COOH$	683
549.258		$CHCl_2F$	487
549.5		CHFCHF	665
549.686		CH_2CHCN	1711
550.		CH_3NO_2	1753
550.	±1.1	CH_2CHCN	1396
550.	±11.	CD_3Br	1438,1439
550.0		CH_3OD	113
550.1		CD_3OD	1715
550.2		CD_3OH	1719
551.	±2.7	CD_3OH	1402
551.2		CH_2CF_2	1726
552.		CH_3NO_2	1753
552.0		C_2H_5OH	1746
552.0		N_2D_4	735

Table 4.2.1—*continued*
Optically Pumped Far Infrared and Millimeter Wave Lasers Arranged in Order of Increasing Wavelength

Wavelength (μm)	Uncertainty	Molecule	Reference
552.4		CD_3OD	1715
552.6		CH_3OH	830
552.94		COF_2	828
553.	±2.7	CD_3OH	1402
553.0		CD_3OH	1719
553.6962	±0.00055	C_2H_3Br	1404
554.	±2.7	CD_3OH	1402
554.0		CD_3OH	1719
554.365	±0.0021	CF_2CH_2	1483
554.56		CH_2CHCN	1731
554.7		$^{13}CD_3I$	1701
555.1		C_2H_3Br	1727
555.75		$CH_3{}^{18}OH$	1741
556.47		CH_2CHCN	1731
556.8		CH_2CHCl	1728
556.8		CH_2CHCl	1728
556.876	±0.0011	CD_3I	1401
557.0		CHFCHF	665
558.5		CD_3OD	1715
558.577		$(H_2CO)_3$	809
558.8		$C_2H_3F_3$	1743
558.82		$^{13}CH_3I$	1675
560.	±11.	CD_3Br	1438,1439
560.0		CH_3OD	113
560.703		CDF_3	1721
560.803		CDF_3	1721
561.028		$CHCl_2F$	487
561.2939	±0.00056	DCOOD	1384,1403
561.4	±0.095	CH_3CN	1365,1377,1378
563.		C_2H_3F	1733
563.		CH_2CHF	1733
563.44		CH_2CHCN	1731
564.		CH_3NO_2	1753
564.		CH_3OH	203
564.7		CH_3OCH_3	1754

Table 4.2.1—*continued*
Optically Pumped Far Infrared and Millimeter Wave Lasers Arranged in Order of Increasing Wavelength

Wavelength (μm)	Uncertainty	Molecule	Reference
564.7	±0.096	$CH_3{}^{81}Br$	1378,1450
564.70		CH_2CHCN	1731
565.		C_2H_3F	1734
565.		CH_2CHF	1729
566.1		C_2H_5OH	1746
566.4	±0.096	CH_3CCH	1377,1378
566.750		CH_3OH	179
567.1065	±0.00057	HCOOD	1384,1403
567.5316	±0.00028	CH_2F_2	1392,1454,1466
567.8		CD_3OD	1715
567.8683	±0.00057	DCOOD	1384,1403
567.946	±0.0010	H_2CCl	1459,1483
568.8	±0.097	CH_3Cl	1378,1385
569.		DFCO	684
569.4		CH_3CHF_2	1747
569.477	±0.0011	CD_3I	1401
569.7		CH_3OH	830
570.	±17.	CF_2CH_2	1388,1420,1465
570.3		SO_2	810
570.5687	±0.00055	CH_3OH	1375,1427,1455
572.330		$H^{13}COOH$	683
572.51		COF_2	828
572.692		CH_2CHCN	1711
573.		CH_2CHF	1729
573.75		$^{13}CH_3I$	1675
574.		CH_2CHCl	1829
574.	±1.1	CH_2CHCN	1396
574.027		CH_2CHCN	1711
574.38		CH_2CHCN	1731
574.6		$^{13}CD_3I$	1701
575.3		C_2H_5OH	1746
576.17	±0.098	CH_3I	1378,1409
577.8		CD_3OD	1715
578.		CH_2CHCN	1829
578.90	±0.098	CH_3I	1378,1409

Table 4.2.1—*continued*
Optically Pumped Far Infrared and Millimeter Wave Lasers Arranged in Order of Increasing Wavelength

Wavelength (μm)	Uncertainty	Molecule	Reference
579.		C_2H_3F	1734
579.		CH_2CHF	1729
579.761		CH_2CHCl	1711
580.	±12.	CH_2CHCN	1459
580.3872	±0.00058	HCOOH	1381,1384,1439
580.6		$C_2H_3F_3$	1743
580.8		$C_2H_3F_3$	1743
580.8		CH_2CHCl	1728
580.8010	±0.00058	HCOOH	1384,1403,1439
580.869		$CHCl_2F$	487
581.3		CH_2CHCl	1728
581.6		$C_2H_3D_2OH$	1742
581.984		CDF_3	1721
582.1		CDF_3	1668
582.5		CH_2CHF	1732
582.5		CH_3CHF_2	1747
582.5536	±0.00058	HCOOD	1384,1403
583.	±2.9	CD_3OH	1402
583.1		C_2H_3F	1735
583.1		CH_2CHF	1730
583.3		CD_3OH	1719
583.3		CH_3CHF_2	1747
583.7		CHFCHF	665
583.77	±0.099	CH_3CCH	1377,1378
583.87	±0.099	CH_3I	1378,1409
583.872		CH_2CHCN	1711
584.	±1.2	CH_2CHCN	1396
584.0		CH_2CHCl	1730
585.5		CD_3OD	1715
585.72	±0.10	$CH_3{}^{81}Br$	1378,1450
585.8		C_2H_3F	1732
586.		C_2H_3F	1734
586.		CH_2CHF	1729
586.382		CH_2CHCN	1711
586.72		CH_2CHCN	1731

Table 4.2.1—*continued*
Optically Pumped Far Infrared and Millimeter Wave Lasers Arranged in Order of Increasing Wavelength

Wavelength (μm)	Uncertainty	Molecule	Reference
586.8		CH_2CHCN	1731
587.	±1.2	CH_2CHCN	1396
587.5		N_2D_4	735
588.0276	±0.00029	CH_2F_2	1392,1454,1466
588.44		CH_2CHCN	1731
590.		C_2HF	681
590.		HCCF	1823
590.	±1.2	HCOOD	1384,1403
590.0		C_2H_3Br	1727
590.369		CH_2CHCl	1711
591.1		$^{13}CD_3OD$	453
591.6157	±0.00059	DCOOD	1384,1403
593.		$(H_2CO)_3$	809
593.	±1.2	DCOOD	1384,1403
593.1		CD_3OH	1719
593.279		CD_2F_2	1710
593.506	±0.0012	CH3CH2F	1459,1483
593.9		CH_2CHCl	1730
594.		CH_3NO_2	1753
594.	±1.2	HCOOD	1384,1403
594.7286	±0.00059	C_2H_3Br	1404
597.00		CH_2CHCN	1731
597.00		CH_2CHCN	1731
597.33		CH_2CHCN	1731
598.		CH_3NO_2	1753
598.3		CH_2CHCl	1730
598.3		CHD_2OH	663
598.4		CD_3OD	1715
598.4		CD_3OD	1715
598.6		CD_3OH	1719
599.	±2.9	CD_3OH	1402
599.0		CH_2CHCN	1731
599.550	±0.0012	CD_3I	1401
600.		CH_3NH_2	1752
601.67		COF_2	828

Table 4.2.1—*continued*
Optically Pumped Far Infrared and Millimeter Wave Lasers Arranged in Order of Increasing Wavelength

Wavelength (μm)	Uncertainty	Molecule	Reference
601.897	±0.0012	H_2CCl	1459,1483
602.4870	±0.00030	CH_3OH	1397,1427,1481
602.50		CD_3I	1713
605.4		CH_3CHF_2	1747
605.6		CDF_3	1668
605.7		C_2H_3Br	1730
606.		CH_2CHCl	1728
606.6		C_2H_3F	1735
606.6		CH_2CHF	1730
606.7		C_2H_3Br	1727
606.7		CH_2CHCl	1728
606.7		CHD_2OH	663
606.8		$C_2H_3F_3$	1743
607.3		CHD_2OH	663
607.714		$(H_2CO)_3$	809
608.		DFCO	684
610.3		CD_3OH	1719
613.0		SiH_2F_2	814
613.5		CH_3OD	113
614.110	±0.0012	CD_3I	1401
614.2851	±0.00031	CH_3OH	1427,1481
614.3	±0.74	CF_2Cl_2	1442
616.3351	±0.00031	CH_2DOH	1467,1478
617.0		C_2H_3F	1732
617.7		C_2H_3Br	1727
617.7		CH_2CF_2	1726
618.4462	±0.00062	C_2H_3Br	1404
619.	±3.0	$(H_2CO)_3$	1380
620.		CH_2CHCl	1728
620.		CH_3NO_2	1753
620.	±12.	CH_2CHCN	1459
620.3		C_2H_5OH	1746
620.34		CH_2CHCN	1731
620.40	±0.099	CH3CH2F	1459,1474
621.		CH_2CHCl	1728

Table 4.2.1—*continued*
Optically Pumped Far Infrared and Millimeter Wave Lasers Arranged in Order of Increasing Wavelength

Wavelength (μm)	Uncertainty	Molecule	Reference
622.		CH_3OH	203
622.0		SiH_3F	813
622.3		CH_2CHCl	1728
623.		C_3H_3F	682
623.		CH_2CHCN	1829
624.0958	±0.00062	C_2H_3Br	1404
624.4301	±0.00031	CH_3OH	1427,1481
625.7		C_2H_3Br	1727
628.		$C_2H_3O_2D$	1749
628.00		CD_3I	1713
629.00		CD_3I	1713
629.3		$C_2H_3F_3$	1743
629.8442	±0.00031	$13CH_3OH$	1415,1417,1481
630.	±13.	CH_2CHCN	1459
630.1661	±0.00063	HCOOD	1384,1403
630.7		C_2H_3Br	1730
631.		CH_2CHCN	1829
631.		CH_3NO_2	1753
631.	±3.1	CD_2Cl_2	1478
631.9	±0.10	CH_3Br	1378
632.0	±0.10	CH_3Br	1378
632.00		CD_3I	1713
632.9		CH_3CHF_2	1747
633.4		$C_2H_3F_3$	1743
634.		CH_3NO_2	1753
634.0		$C_2H_3F_3$	1743
634.471		CH_2CHCl	1711
634.471	±0.0013	H_2CCl	1375,1483
634.7		$C_2H_3F_3$	1743
635.3548	±0.00064	C_2H_3Br	1404
636.3		CH_2CHCl	1728
637.1		CH_2CHCl	1730
637.5		CH_3CHF_2	1747
638.		CH_2CHCl	1829
638.4	±0.57	CF_2Cl_2	1442

Table 4.2.1—*continued*
Optically Pumped Far Infrared and Millimeter Wave Lasers Arranged in Order of Increasing Wavelength

Wavelength (μm)	Uncertainty	Molecule	Reference
639.1282	±0.00064	DCOOH	1403
639.7	±0.10	CH_3I	1378,1409
640.	±13.	H_2CCl	1459
640.	±3.1	CD_3I	1401
640.35		COF_2	828
640.7		C_2H_3Br	1727
641.0		N_2D_4	735
641.43		CH_2CHCN	1731
642.28		CH_2CHCN	1731
642.5999	±0.00032	CH_2F_2	1392,1454,1466
643.516		CD_2F_2	1710
643.86		CH_2CHCN	1731
644.	±3.2	CD_3I	1401
644.5		CH_3CHF_2	1747
644.64		CH_2CHCN	1731
644.64		CH_2CHCN	1731
645.		CH_2CHCN	1731
645.	±1.3	DCOOD	1384,1403
645.289		CH_2CHCl	1711
645.5		CH_3OCH_3	1754
646.		CH_3NO_2	1753
646.	±1.3	C_2H_3Br	1404
646.	±3.2	CD_3OH	1402
646.477		CD_3OH	826
646.5		CH_3CHF_2	1747
647.3485	±0.00065	DCOOH	1403
647.89	±0.049	CH_3CCH	1377
648.	±3.2	CD_3OH	1402
648.6		C_2H_3Br	1730
649.4255	±0.00065	C_2H_3Br	1404
650.70		COF_2	828
651.79		CH_2CHCN	1731
651.9		CD_3OD	1715
652.68	±0.098	CH_3CN	1365,1378,1459
654.		CHFO	821

Table 4.2.1—*continued*
Optically Pumped Far Infrared and Millimeter Wave Lasers Arranged in Order of Increasing Wavelength

Wavelength (μm)	Uncertainty	Molecule	Reference
654.920		CH_3OH	179
655.0		C_2H_3F	1735
655.0		CH_2CHF	1730
655.4		CH_2CHCl	1728
655.9		CH_2CHCN	1731
656.		$(H_2CO)_3$	809
656.		CH_3NO_2	1753
657.	±1.3	HCOOD	1384,1403
657.2391	±0.00033	CH_2F_2	1392,1454,1466
657.59		CH_2CHCN	1731
657.938		CDF_3	1721
657.989		CDF_3	1721
658.152		CDF_3	1721
658.26		CH_2CHCN	1731
658.5		N_2D_4	735
658.53	±0.099	$CH_3{}^{81}Br$	1378,1450
658.57		CD_3I	1713
658.57		CD_3I	1713
659.		NH_2OH	682
659.69		CH_2CHCN	1731
660.	±1.3	HCOOD	1384,1403
660.	±13.	CH3CH2F	1483
660.2		C_2H_3F	1735
660.2		CH_2CHF	1730
660.34		CH_2CHCN	1731
660.582	±0.0013	CD_3I	1401
660.70	±0.099	CH_3Br	1378
661.		$(H_2CO)_3$	809
661.153		$CHCl_2F$	487
662.0		CH_2CF_2	1726
662.816	±0.0015	CF_2CH_2	1483
663.0		CD_3OD	1715
663.08		$^{13}CH_3I$	1675
663.670		CH_3OH	179
664.		DFCO	684

Table 4.2.1—*continued*
Optically Pumped Far Infrared and Millimeter Wave Lasers Arranged in Order of Increasing Wavelength

Wavelength (μm)	Uncertainty	Molecule	Reference
665.70		COF_2	828
666.	±1.3	DCOOD	1384,1403
666.604		CH_2CHCl	1711
667.232	±0.0013	CD_3I	1401
668.	±1.3	HCOOD	1384,1403
668.1		CH_2CHCl	1728
669.531	±0.0010	HCOOH	1379,1403,1439
670.094	±0.0013	CD_3I	1401
670.114	±0.0013	CD_3I	1401
670.4		CD_3OD	1715
670.4		CD_3OD	1715
670.7		$^{13}CD_3OD$	453
670.79		CH_2CHCN	1731
671.0		C_2H_3F	1732
671.0	±0.10	CH_3I	1378,1409
671.15		CH_2CHCN	1731
671.43		CH_2CHCN	1731
673.		CH_3NO_2	1753
674.8		CH_3OH	829
675.		$C_2H_3O_2D$	1749
675.		CH_3NO_2	1753
675.3	±0.10	CH_3CCH	1377,1378
676.		$C_2H_3O_2D$	1749
676.7		$C_2H_3F_3$	1743
678.57		CD_3I	1713
679.766		$(H_2CO)_3$	809
680.	±3.3	$(H_2CO)_3$	1380
680.	±3.3	CD_3OH	1402
680.0		CH_3CHF_2	1747
680.00		CD_3I	1713
680.5414	±0.00068	C_2H_3Br	1404
681.4		C_2H_3Br	1727
681.5		CH_2CHCl	1728
682.6	±0.10	CH_2DOH	1467
683.6		C_2H_3Br	1727

Table 4.2.1—*continued*
Optically Pumped Far Infrared and Millimeter Wave Lasers Arranged in Order of Increasing Wavelength

Wavelength (μm)	Uncertainty	Molecule	Reference
683.738		CH_2CHCl	1711
684.3		CD_3OD	1715
684.5		CD_3OH	1719
684.6		C_2H_3Br	1730
684.7		CD_3OD	1715
684.7	±0.48	CF_2Cl_2	1442
684.7	±0.75	CF_2Cl_2	1442
685.	±3.4	CD_3OH	1402
685.19		CH_2CHCN	1731
687.2		CHFCHF	665
687.837		CDF_3	1721
688.3		CH_2CF_2	1726
689.0		SiH_3F	813
689.9981	±0.00069	HCOOD	1384,1403
690.	±3.4	$13CD_3I$	1388
690.0		CD_3OH	1719
690.4		CH_3CHF_2	1747
691.119	±0.0014	CD_3I	1401
692.	±1.4	HCOOD	1384,1403
693.0		C_2H_3Br	1727
693.1396	±0.00069	C_2H_3Br	1404
694.		NH_3	816
694.	±2.8	CH_3OH	1427,1471
694.1893	±0.00035	CH_3OH	1427,1481
694.428		$(H_2CO)_3$	809
695.		$(H_2CO)_3$	809
695.		NH_3	816
695.	±3.4	CD_3OH	1402
695.202		CH_2CHCl	1711
695.3499	±0.00035	CH_3OH	1427,1481
695.6720	±0.00070	HCOOD	1384,1403
696.	±3.4	$(H_2CO)_3$	1380
697.		CH_3NO_2	1753
697.		CH_3OH	203
697.4552	±0.00070	DCOOH	1403

Table 4.2.1—*continued*
Optically Pumped Far Infrared and Millimeter Wave Lasers Arranged in Order of Increasing Wavelength

Wavelength (μm)	Uncertainty	Molecule	Reference
698.		C_2H_5Cl	1745
698.555	±0.0014	CD_3Cl	1388,1401
698.6		CH_3CHF_2	1747
699.		CH_2CHCl	1829
699.0		CD_3OH	1719
699.0		N_2D_4	735
699.4226	±0.00084	CH_3OH	1375,1427,1455
700.	±14.	H_2CCl	1459
700.4		CH_2CHCl	1728
701.		$C_2H_3O_2D$	1749
701.1		C_2H_3Br	1727
701.5		CD_3OH	1719
702.	±3.4	CD_3OH	1402
703.	±3.4	CD_3OH	1402
704.53	±0.099	CH_3CN	1364,1365,1378
704.925.		CH_2CHCl	1711
705.	±1.4	HCOOH	1384,1403,1439
705.0		CHFCHF	665
705.3		C_2H_3Br	1727
706.6		CD_3OH	1719
707.1		C_2H_3Br	1730
707.2210	±0.00071	C_2H_3Br	1404
707.8		C_2H_5Br	1744
709.2		$C_2H_3F_3$	1743
709.5		$C_2H_3F_3$	1743
709.8		$C_2H_3F_3$	1743
710.	±1.4	DCOOH	1403
710.	±14.	H_2CCl	1459
710.0		$(H_2CO)_3$	809
711.	±3.5	CD_3OH	1402
711.752		$(H_2CO)_3$	809
712.	±1.4	C_2H_3Br	1404
712.	±3.5	$(H_2CO)_3$	1380
712.76		CH_2CHCN	1731
713.1056	±0.00071	DCOOH	1403

Table 4.2.1—*continued*
Optically Pumped Far Infrared and Millimeter Wave Lasers Arranged in Order of Increasing Wavelength

Wavelength (μm)	Uncertainty	Molecule	Reference
713.72	±0.050	CH_3CN	1364,1365,1377
715.3		CH_3OD	113
715.4		CH_2CF_2	1726
715.4	±0.10	$CH_3{}^{79}Br$	1378,1450
717.		CH_3NO_2	1753
718.0		CH_3CHF_2	1747
719.3	±0.10	CH_3I	1378,1409
720.	±14.	CH_2CHCN	1459
721.	±1.4	N_2H_4	1398
722.	±3.5	CD_3OH	1402
722.		CH_2CHCN	1829
724.0		N_2D_4	735
724.1399	±0.00072	C_2H_3Br	1404
724.9203	±0.00036	CH_2F_2	1392,1454,1466
726.9203	±0.00073	DCOOD	1384,1403
727.		$(H_2CO)_3$	809
727.570		CH_2CHCN	1711
727.9491	±0.00073	HCOOD	1384,1403
730.323	±0.0015	CD_3I	1401
733.		$(H_2CO)_3$	809
733.	±1.5	HCOOD	1384,1403
733.5740	±0.00073	D_2CO	1380,1382
734.1616	±0.00088	N_2H_4	1483
734.262	±0.0015	CD_3I	1401
734.6		CH_3OD	113
734.8		C_2H_3Br	1727
735.		CH_3NO_2	1753
735.130	±0.0015	CD_3Cl	1388,1401
737.	±1.5	DCOOD	1384,1403
738.		CH_2CHCN	1829
740.	±15.	CH_2CHCN	1459
741.1149	±0.00074	C_2H_3Br	1404
741.62	±0.089	CH_3CN	1364,1365,1378
742.572	±0.0015	HCOOH	1379,1384,1439

Table 4.2.1—*continued*
Optically Pumped Far Infrared and Millimeter Wave Lasers Arranged in Order of Increasing Wavelength

Wavelength (μm)	Uncertainty	Molecule	Reference
744.050	±0.0015	HCOOH	1379,1384,1439
745.	±3.7	CD_3I	1401
745.	±3.7	CD_3OH	1402
745.5		$^{13}CD_3I$	1701
746.4		CH_3CHF_2	1747
749.29	±0.090	$CH_3{}^{79}Br$	1378,1450
749.36	±0.090	$CH_3{}^{79}Br$	1378,1450
749.372		$(H_2CO)_3$	809
750.		$(H_2CO)_3$	1380
750.		DFCO	684
750.	±15.	CH_3CN	1364,1365,1483
750.	±3.7	$(H_2CO)_3$	1380
750.38		CH_2CHCN	1731
750.606		$(H_2CO)_3$	809
751.4	±0.60	CF_2Cl_2	1442
751.831		CH_2CHCN	1711
752.6808	±0.00075	D_2CO	1380,1382
752.7485	±0.00075	DCOOH	1403
754.		CH_2CHCN	1731
756.		$C_2H_3O_2D$	1749
758.2		CH_2CHCl	1728
758.46		CH_2CHCN	1731
759.6		CH_3OD	113
760.	±15	HCOOH	1366,1439,1459
760.	±3.7	CD_3OH	1402
761.2		CH_3CHF_2	1747
761.67		CH_2CHCN	1731
761.7617	±0.00076	DCOOD	1384,1403
762.50	±0.092	CH_2DOH	1467
764.	±3.7	CF_2CH_2	1367,1388,1465
764.2		C_2H_3F	1732
765.2	±0.54	CF_2Cl_2	1442
765.42		COF_2	828
766.6		$C_2H_3F_3$	1743
767.8		$C_2H_3F_3$	1743

Table 4.2.1—*continued*
Optically Pumped Far Infrared and Millimeter Wave Lasers Arranged in Order of Increasing Wavelength

Wavelength (μm)	Uncertainty	Molecule	Reference
768.8		CH_2CHCl	1730
769.053		CH_2CHCN	1711
769.1		C_2H_3Br	1730
769.8		C_2H_5Br	1744
770.	±15.	CF_2CH_2	1387,1388,1465
770.00		CH_2CHCN	1731
770.4		CH_2CHCl	1728
771.038		$(H_2CO)_3$	809
771.2		CH_2CHCl	1728
771.3		CH_2CHCl	1728
774.	±3.8	CD_3OH	1402
775.		CH_2CHCN	1829
775.		ClO_2	1827
777.92		CH_2CHCN	1731
778.		CH_3NO_2	1753
779.8744	±0.00078	DCOOD	1384,1403
780.		CH_3NO_2	1753
780.	±16.	CH_2CHCN	1459
780.1330	±0.00078	C_2H_3Br	1404
780.83		CH_2CHCN	1731
781.		CH_3OH	203
781.000		CH_3OH	179
782.		$(H_2CO)_3$	809
782.7		$C_2H_3F_3$	1743
783.2		$C_2H_3F_3$	1743
783.3		$^{13}CD_3OD$	453
783.7		C_2H_3Br	1730
784.2681	±0.00078	C_2H_3Br	1404
784.4		CH_3OD	113
784.5		$C_2H_3F_3$	1743
784.6		C_2H_3Br	1730
786.1617	±0.00079	HCOOH	1381,1384,1439
786.9419	±0.00079	HCOOH	1384,1403,1439
787.50		CH_2CHCN	1731
788.		DFCO	684

Table 4.2.1—*continued*
Optically Pumped Far Infrared and Millimeter Wave Lasers Arranged in Order of Increasing Wavelength

Wavelength (μm)	Uncertainty	Molecule	Reference
788.33		CH_2CHCN	1731
788.482	±0.0016	CD_3I	1401
788.9192	±0.00079	$H^{13}COOH$	1384
789.		$(H_2CO)_3$	809
789.4203	±0.00079	DCOOD	1384,1403
789.8396	±0.00079	HCOOH	1384,1403,1439
790.	±16.	CH_2CHCN	1459
790.8		CH_3CHF_2	1747
792.	±3.9	CD_3Cl	1388,1401
793.		CH_2CHCN	1829
793.2		CH_2CF_2	1726
795.	±1.6	DCOOD	1384,1403
795.	±1.6	N_2H_4	1398
796.3		C_2H_3F	1735
796.3		CH_2CHF	1730
796.5		C_2H_3F	1732
796.7		C_2H_3Br	1727
797.5		CH_3OD	113
799.17		COF_2	828
802.	±1.6	N_2H_4	1398
802.5		C_2H_3Br	1730
805.8		$C_2H_3F_3$	1743
806.	±3.9	$^{13}CD_3I$	1388
809.		CH_3NO_2	1753
812.	±1.6	DCOOD	1384,1403
812.6		CD_3OH	1719
813.		$(H_2CO)_3$	809
813.654		$(H_2CO)_3$	809
815.	±4.0	$(H_2CO)_3$	1380
815.123		CH_2CHCl	1711
816.195		CH_2CHCN	1711
817.50		COF_2	828
819.	±1.6	HCOOD	1384,1403
820.0		CH_3OD	113
820.00		$^{13}CH_3I$	1675

Table 4.2.1—*continued*
Optically Pumped Far Infrared and Millimeter Wave Lasers Arranged in Order of Increasing Wavelength

Wavelength (μm)	Uncertainty	Molecule	Reference
822.3		CH_3CHF_2	1747
823.		C_2H_3N	1750
823.5		CF_3Br	1456
824.	±1.4	CF_3Br	1444,1456
825.0		CH_3OD	113
826.	±1.7	HCOOD	1384,1403
826.9443	±0.00083	C_2H_3Br	1404
828.		CH_2CHCl	1829
828.		CH_2CHCN	1829
829.	±4.1	CD_2Cl_2	1478
829.54		CH_2CHCN	1731
830.	±17.	CH_2CHCN	1459
830.	±17.	H_2CCl	1459
830.45		CH_2CHCN	1731
831.13	±0.10	$CH_3{}^{81}Br$	1378,1450
832.7		C_2H_3Br	1727
832.757		$CHCl_2F$	487
832.77		CH_2CHCN	1731
833.2		CH_2CHCl	1730
833.3		$C_2H_3F_3$	1743
835.	±1.7	DCOOD	1384,1403
836.8		C_2H_3Br	1730
837.27		COF_2	828
837.73		CH_2CHCN	1731
838.3		C_2H_5Br	1744
838.369		$H^{13}COOH$	683
839.40		COF_2	828
841.		CH_3NO_2	1753
843.2369	±0.00084	DCOOD	1384,1403
845.		CH_3NO_2	1753
848.00		$^{13}CH_3I$	1675
851.0		$C_2H_3F_3$	1743
851.9	±0.10	CH3CH2F	1459,1474
852.5		$C_2H_3F_3$	1743
853.3		$C_2H_3F_3$	1743

Table 4.2.1—*continued*
Optically Pumped Far Infrared and Millimeter Wave Lasers Arranged in Order of Increasing Wavelength

Wavelength (μm)	Uncertainty	Molecule	Reference
853.4380	±0.00085	C_2H_3Br	1404
854.4	±0.10	CH_3CN	1364,1365,1378
858.254		CD_3OH	826
858.7	±0.52	CF_2Cl_2	1442
859.5		C_2H_3Br	1727
862.0		CD_3OH	1719
863.1		CH_2CHCl	1728
865.5		CH_3CHF_2	1747
866.4		CH_2CF_2	1726
869.		CH_3NO_2	1753
869.	±4.3	CD_3OD	1388,1425
870.80	±0.096	CH_3Cl	1378,1385
871.36		CH_2CHCN	1731
871.5850	±0.00044	CD_3OH	1402,1481
872.27		CH_2CHCN	1731
875.		C_2H_3F	1734
875.		CH_2CHF	1729
875.0		CH_2CHCN	1731
876.8		CH_2CHCl	1730
877.2		CH_3CHF_2	1747
877.3		C_2H_3Br	1730
877.5481	±0.00088	DCOOD	1384,1403
878.1		$C_2H_3F_3$	1743
878.5		$C_2H_3F_3$	1743
880.	±26.	CF_2CH_2	1388,1420,1465
880.41		CH_2CHCN	1731
883.		CF_3Br	1456
883.	±2.4	CF_3Br	1444,1456
883.598	±0.0018	CD_3Cl	1388,1401
889.466		$(H_2CO)_3$	809
889.716		$(H_2CO)_3$	809
890.	±1.8	CF_2CH_2	1388,1396,1465
890.	±4.4	$(H_2CO)_3$	1380
890.2		CH_2CF_2	1828
891.	±4.4	$(H_2CO)_3$	1380

Table 4.2.1—*continued*
Optically Pumped Far Infrared and Millimeter Wave Lasers Arranged in Order of Increasing Wavelength

Wavelength (μm)	Uncertainty	Molecule	Reference
891.087		$H^{13}COOH$	683
892.0		CH_3CHF_2	1747
894.0		C_2H_3F	1732
895.	±4.4	CD_3I	1401
899.		CH_2CHCN	1731
900.	±9.0	C_2H_5Cl	1430
900.1338	±0.00090	C_2H_3Br	1404
901.191		CH_2CHCN	1711
901.3		$^{13}CD_3I$	1701
902.5		CH_2CHCl	1728
903.00		CH_2CHCN	1731
905.428		$CHCl_2F$	487
906.		DFCO	684
906.5		C_2H_3Br	1730
910.		CH_2CHCN	1829
910.	±18.	CH_2CHCN	1459
912.5		C_2H_3Br	1730
914.721		ClO_2	1827
914.735		ClO_2	1827
914.755		ClO_2	1827
914.780		ClO_2	1827
918.610	±0.0018	CD_3I	1401
919.9355	±0.00092	HCOOD	1384,1403
921.5		CH_2CHCl	1730
925.5	±0.10	$CH_3{}^{79}Br$	1378,1450
926.2087	±0.00093	HCOOD	1384,1403
926.66		CD_3I	1713
927.		CH_2CHCN	1731
927.9814	±0.00093	DCOOD	1384,1403
929.		CH_2CHCN	1731
929.8		$^{13}CD_3I$	1701
930.	±19	HCOOH	1381,1483
934.0		CH_2CHCl	1730
934.2		CH_3OCH_3	1754
934.2230	±0.00093	C_2H_3Br	1404

Table 4.2.1—*continued*
Optically Pumped Far Infrared and Millimeter Wave Lasers Arranged in Order of Increasing Wavelength

Wavelength (μm)	Uncertainty	Molecule	Reference
935.		CH_2CHCl	1829
935.0095	±0.00094	DCOOD	1384,1403
936.1590	±0.00094	C_2H_3Br	1404
936.6023	±0.00094	DCOOD	1384,1403
937.9		CH_3CHF_2	1747
938.		C_2H_3N	1750
939.0		C_2H_3F	1735
939.0		CH_2CHF	1730
939.5		CH_3CHF_2	1747
939.50		CH_2CHCN	1731
939.50		CH_2CHCN	1731
940.		CH_2CHCN	1829
940.	±19.	CH_2CHCN	1459
940.	±19.	H_2CCl	1459
944.0	±0.10	CH_3Cl	1378,1385,1430
945.0		CH_2CHCl	1730
948.250		$(H_2CO)_3$	809
948.9247	±0.00095	$(H_2CO)_3$	1380
948.925		$(H_2CO)_3$	1380
949.0		C_2H_3F	1735
949.0		CH_2CHF	1730
949.685		ClO_2	1827
952.		$(H_2CO)_3$	809
953.880	±0.0019	CD_3I	1401
957.		CH_2CHCN	1731
958.25	±0.096	CH_3Cl	1378,1385
959.0		CH_2CHCN	1731
961.		CH_2CHF	1729
961.0		C_2H_3F	1735
963.4873	±0.00096	C_2H_3Br	1404
964.	±4.7	CH_3I	1401,1409
967.5		C_2H_3Br	1727
967.8		C_2H_3Br	1730
967.9		$C_2H_3F_3$	1743
968.	±4.7	CD_3OH	1402

Table 4.2.1—*continued*
Optically Pumped Far Infrared and Millimeter Wave Lasers Arranged in Order of Increasing Wavelength

Wavelength (μm)	Uncertainty	Molecule	Reference
968.9		$C_2H_3F_3$	1743
969.0		$C_2H_3F_3$	1743
971.		C_2H_3F	1734
971.		CH_2CHF	1729
971.8064	±0.00097	DCOOH	1403
972.		NH_3	816
972.0		CD_3OD	1715
972.0		CD_3OD	1715
973.		CH_3NO_2	1753
973.		CH_3OH	203
973.0		$C_2H_3F_3$	1743
973.0		CH_3OH	830
976.8		CH_2CHCl	1728
980.	±9.8	CF_2Cl_2	1442
981.1		C_2H_3F	1735
981.1		CH_2CHF	1730
981.709	±0.0020	CD_3I	1401
985.8588	±0.00099	C_2H_3Br	1404
986.3125	±0.00099	HCOOD	1384,1403
986.349		CH_2CHCN	1711
988.		FCN	682
988.1		CD_3OH	1719
988.259		CH_2CHCl	1711
988.695		CH_2CHCl	1711
989.1904	±0.00099	C_2H_3Br	1404
990.	±2.0	CF_2CH_2	1388,1396,1465
990.5		CH_2CHCN	1731
990.51	±0.099	CH_3Br	1378
990.6303	±0.00099	C_2H_3Br	1404
992.	±4.9	CH_3F	1374,1377,1393
995.		CH_2CHCl	1829
998.5140	±0.0010	DCOOD	1384,1403
1000.	±20.	H_2CCl	1459
1001.		CH_3NO_2	1753
1005.		CH_2CHCl	1728

Table 4.2.1—*continued*
Optically Pumped Far Infrared and Millimeter Wave Lasers Arranged in Order of Increasing Wavelength

Wavelength (μm)	Uncertainty	Molecule	Reference
1005.		DFCO	684
1005.230	±0.0016	CH3CH2F	1459,1483
1005.348	±0.0020	CD_3I	1401
1005.8		CH_3CHF_2	1747
1006.		C_3H_3F	682
1007.	±2.0	N_2H_4	1398
1008.558		CDF_3	1721
1009.409	±0.0010	DCOOD	1384,1403
1014.		SiH_3F	813
1014.9	±0.10	CH_3CN	1365,1377,1378
1016.009		CH_2CHCN	1711
1016.3	±0.10	CH_3CN	1365,1377,1378
1016.7		C_2H_3Br	1727
1017.8		CH_2CHCN	1731
1018.4		C_2H_3Br	1730
1020.	±31.	CF_2CH_2	1388,1420,1465
1025.	±1.0	CF_2Cl_2	1442
1026.680		CH_2CHCl	1711
1026.709		CH_2CHCl	1711
1028.		C_2HF	681
1028.		HCCF	1823
1028.3		C_2H_3Br	1730
1030.378	±0.0010	$H^{13}COOH$	1384
1035.6		$C_2H_3F_3$	1743
1039.855		CH_2CHCl	1711
1040.	±11.	CF_3Br	1443,1444,1456
1040.	±21.	H_2CCl	1459
1041.		CH_2CHCl	1829
1042.2		CH_3CHF_2	1747
1043.		CF_3Br	1444
1044.8		CH_3CHF_2	1747
1047.00		CD_3I	1713
1047.2		CH_2CHCl	1730
1047.579	±0.0010	DCOOH	1403
1053.0		SiH_2F_2	814

Table 4.2.1—*continued*
Optically Pumped Far Infrared and Millimeter Wave Lasers Arranged in Order of Increasing Wavelength

Wavelength (μm)	Uncertainty	Molecule	Reference
1053.477		$(H_2CO)_3$	809
1055.		$(H_2CO)_3$	809
1056.		SiH_3F	813
1058.		SiH_3F	813
1059.0896.5		C_2H_5Br	1744
1062.0		CH_2CF_2	1828
1063.29	±0.095	CH_3I	1378,1409
1065.0		CH_3CHF_2	1747
1070.		CH_3NO_2	1753
1070.	±21.	CH3CH2F	1459
1070.231	±0.0011	DCOOD	1384,1403
1071.3		CH_2CHCl	1730
1080.537		CDF_3	1721
1083.		CF_3Br	1444
1083.	±4.3	CF_3Br	1443,1444,1456
1086.89	±0.097	CH_3CN	1365,1377,1378
1092.8		CD_3OH	1719
1093.1		CH_2CHCl	1730
1094.		CH_2CF_2	1726
1097.11	±0.098	CH_3CCH	1377,1378
1099.544	±0.0022	CD_3I	1401
1100.		CH_2CF_2	1726
1100.	±5.4	CD_3OH	1402
1116.483		$H^{13}COOH$	683
1127.752		CH_2CHCN	1711
1133.8		CH_3CHF_2	1747
1134.0		CH_3OD	113
1134.113		ClO_2	1827
1135.070		COF_2	668
1137.50		CD_3I	1713
1146.	±5.6	CD_3OH	1402
1151.		CF_3Br	1444
1151.	±4.6	CF_3Br	1443,1444,1456
1155.5		CD_3OH	1718
1156.		CH_2CHCN	1829

Table 4.2.1—*continued*
Optically Pumped Far Infrared and Millimeter Wave Lasers Arranged in Order of Increasing Wavelength

Wavelength (μm)	Uncertainty	Molecule	Reference
1157.318	±0.0012	HCOOD	1384,1403
1158.	±2.3	DCOOD	1384,1403
1160.	±23.	CH_2CHCN	1459
1161.676	±0.0012	HCOOD	1384,1403
1162.2		CD_3OD	1715
1162.2		CD_3OD	1715
1164.	±2.3	CF_2Cl_2	1442
1164.8		CH_2CHCl	1728
1164.8	±0.10	CH_3CN	1365,1377,1378
1170.7		C_2H_3F	1735
1170.7		CH_2CHF	1730
1173.7		CH_2CHCN	1731
1173.9		C_2H_3Br	1727
1174.87	±0.049	CH_3CCH	1377
1180.	±24.	CH_2CHCN	1459
1182.2		$^{13}CD_3I$	1701
1184.		CH_2CHCN	1829
1184.38		COF_2	828
1185.079		$(H_2CO)_3$	809
1194.0		$^{13}CD_3OD$	453
1194.3		CH_2CHCl	1730
1197.1		CH_2CHCN	1731
1198.6		C_2H_3Br	1730
1201.4		CH_3CHF_2	1747
1202.20		CH_2CHCN	1731
1204.3		C_2H_3Br	1730
1205.	±2.4	CF_2Cl_2	1442
1210.7		CH_3CHF_2	1747
1213.362	±0.0012	HCOOH	1384,1403,1439
1218.6		CH_2CHCN	1731
1218.6		CH_2CHCN	1731
1221.79	±0.048	$^{13}CH_3F$	1377,1461
1223.858	±0.0024	CH_3OH	1427,1470,1483
1234.3		C_2H_3F	1735
1234.3		CH_2CHF	1730

Table 4.2.1—*continued*
Optically Pumped Far Infrared and Millimeter Wave Lasers Arranged in Order of Increasing Wavelength

Wavelength (μm)	Uncertainty	Molecule	Reference
1237.1		CH_3CHF_2	1747
1237.966	±0.0012	DCOOH	1403
1239.480	±0.0025	CD_3Cl	1388,1401
1245.71		$^{13}CH_3I$	1675
1247.594	±0.0012	C_2H_3Br	1404
1253.738	±0.0025	CH_3I	1378,1401,1409
1257.1		CH_2CHCN	1731
1260.561		CDF_3	1721
1264.3		C_2H_3F	1735
1264.3		$C_2H_3F_3$	1743
1264.3		CH_2CHF	1730
1267.1		CH_2CHCN	1731
1278.6		C_2H_3Br	1730
1281.649	±0.0013	DCOOD	1384,1403
1286.		SiH_3F	813
1290.	±6.3	CD_3OH	1402
1292.1		CH_3CHF_2	1747
1292.2		CH_2CHCl	1730
1292.743		$(H_2CO)_3$	809
1306.		C_2H_5Cl	1745
1310.4	±0.10	$CH_3{}^{79}Br$	1378,1450
1310.748		ClO_2	1827
1315.		CH_2CHCN	1731
1324.3		$C_2H_3F_3$	1743
1325.0		C_2H_3Br	1730
1325.0		CH_3OD	113
1350.	±14.	C_2H_5Cl	1430
1351.8	±0.11	CH_3CN	1365,1377,1378
1372.		CH_2CF_2	1726
1374.2		CH_2CHCN	1731
1377.		CDF_3	1668
1383.882	±0.0014	C_2H_3Br	1404
1388.30		CH_2CHCN	1731
1394.063	±0.0014	C_2H_3Br	1404
1400.	±14.	C_2H_5Cl	1430

Table 4.2.1—*continued*
Optically Pumped Far Infrared and Millimeter Wave Lasers Arranged in Order of Increasing Wavelength

Wavelength (μm)	Uncertainty	Molecule	Reference
1405.0		CH_3CHF_2	1747
1406.		CH_2CHCl	1728
1427.50		CH_2CHCN	1731
1432.5		CH_2CHCN	1731
1434.		CH_2CF_2	1726
1440.	±29.	CH3CH2F	1483
1448.0958	±0.00072	CH_2F_2	1392,1454,1466
1450.	±19	CD_3F	1472,1485
1480.0		CD_3I	1713
1490.	±10	CD_3F	1472,1485
1491.846		$H^{13}COOH$	683
1492.5		C_2H_3Br	1727
1503.4		CH_3CHF_2	1747
1504.0		C_2H_3Br	1730
1521.376	±0.0037	CH3CH2F	1459,1483
1523.		C_2H_3F	1732
1523.		CH_2CHF	1732
1526.		CF_3Br	1444
1530.	±11.	CF_3Br	1443,1444,1456
1541.750	±0.0015	HCOOD	1384,1403
1543.00		CD_3I	1713
1547.		C_3H_3F	682
1549.505	±0.0031	CD_3I	1401
1550.	±4.5	CF_3Br	1443,1444,1456
1555.		CH_2CHCN	1731
1556.		CF_3Br	1444
1570.2		SO_2	810
1572.64	±0.093	$CH_3{}^{79}Br$	1378,1450
1577.		CH_2CHCN	1731
1579.		CH_2CHCN	1731
1579.903		$(H_2CO)_3$	809
1581.705		$(H_2CO)_3$	809
1600.		CH_2CF_2	1726
1612.		C_2H_3F	1734
1612.		CH_2CHF	1729

Table 4.2.1—*continued*
Optically Pumped Far Infrared and Millimeter Wave Lasers Arranged in Order of Increasing Wavelength

Wavelength (μm)	Uncertainty	Molecule	Reference
1613.0		$C_2H_3F_3$	1743
1614.888	±0.0016	C_2H_3Br	1404
1624.00		$^{13}CH_3I$	1675
1650.312		COF_2	668
1669.		C_2H_5Cl	1745
1671.0		CH_3OD	113
1676.		CD_3OH	1719
1676.		CD_3OH	1719
1678.		CH_2CHF	1732
1687.		CF_3Br	1444
1692.	±4.9	CF_3Br	1443,1444,1456
1714.13		CD_2F_2	1710
1714.130		CD_2F_2	1710
1720.	±17.	C_2H_5Cl	1430
1730.833	±0.0017	HCOOD	1384,1403
1733.		C_2H_3F	1734
1733.		CH_2CHF	1729
1814.37	±0.049	CH_3CN	1364,1365,1377
1827.424		ClO_2	1827
1880.0		CH_2CHCl	1728
1886.87	±0.092	CH_3Cl	1378,1385
1890.	±9.3	CF_3Br	1443,1444,1456
1891.062		COF_2	668
1895.		CF_3Br	1444
1899.889	±0.0019	C_2H_3Br	1404
1900.		COF_2	1659
1930.		CD_3OH	1719
1930.		CD_3OH	1719
1965.34	±0.096	$CH_3{}^{79}Br$	1378,1450
1990.757	±0.0040	CD_3Cl	1388,1401
2031.281		$(H_2CO)_3$	809
2042.5		CH_3CHF_2	1747
2085.0		CD_3I	1713
2140.		C_2H_3N	1750
2140.		CF_3Br	1444

Table 4.2.1—*continued*
Optically Pumped Far Infrared and Millimeter Wave Lasers Arranged in Order of Increasing Wavelength

Wavelength (μm)	Uncertainty	Molecule	Reference
2140.	±19.	CF_3Br	1443,1444,1456
2206.		CH_2CHCl	1728
2216.		DFCO	684
2347.5		C_2H_3Br	1727
2356.4		C_2H_3Br	1727
2388.8		CH_3CHF_2	1747
2453.		C_2H_3F	1732
2453.		CH_2CHF	1732
2650.	±13.	$CH_3{}^{81}Br$	1450
2906.		N_2D_4	735
2923.		CD_3OD	1715
2923.0		CD_3OD	1715
3030.		CD_3OH	1719
3898.3		CHFCHF	665

Section 4.3

REFERENCES

1. Hughes, W. M., Shannon, J., Hunter, R., *Appl. Phys. Lett.*, 24, 488 (1974), 126.1 nm argon laser.
2. Golde, M. F., Thrush, B. A., *Chem. Phys. Lett.*, 29, 486 (1974), Vacuum UV emission from reactions of metastable inert-gas atoms: chemiluminescence of ArO and ArCl.
3. Champagne, L. F., *Appl. Phys. Lett.*, 33, 523 (1978), Efficient operation of the electron-beam-pumped XeCl laser.
4. Burnham, R., *Opt. Commun.*, 24, 161 (1978), Improved performance of the discharge-pumped XeCl laser.
5. Bichkov, Y. I., Gorbatenko, A. I., Mesyats, G. A., Tarasenko, V. F., *Opt. Commun.*, 30, 224 (1979), Effective XeCl-laser performance conditions with combined pumping.
6. Sur, A., Hui, A. K., Tellinghuisen, J., *J. Mol. Spectrosc.*, 74, 465 (1979), Noble gas halides.
7. Burnham, R., Powell, F. X., Djeu, N., *Appl. Phys. Lett.*, 29, 30 (1976), Efficient electric discharge lasers in XeF and KrF.
8. Ault, E. R., Bradford, R. S., Jr., Bhaumik, M. L., *Appl. Phys. Lett.*, 27, 413 (1975), High-power xenon fluoride laser.
9. Brau, C. A., Ewing, J. J., *Appl. Phys. Lett.*, 27, 435 (1975), 354-nm laser action on XeF.
10. Bischel, W. K., Nakano, H. H., Eckstrom, D. J., Hill, R. M., Huestis, D., *Appl. Phys. Lett.*, 34, 565 (1979), A new blue-green excimer laser in XeF.
11. Ernst, W. E., Tittel, F. K., *Appl. Phys. Lett.*, 35, 36 (1979), A new electron-beam pumped XeF laser at 486 nm.
12. Tellinghuisen, J., Tellinghuisen, P. C., Tisone, G. C., Hoffman, J. M., *J. Chem. Phys.*, 68, 5177 (1978), Spectroscopic studies of diatomic noble gas halides. III. Analysis of XeF 3500 Å, band system.
13. Velazco, J. E., Kolts, J. H., Setser, D. W., *J. Chem. Phys.*, 65, 3468 (1976), Quenching rate constants for metastable argon, krypton, and xenon atoms, by fluorine containing molecules and branching ratios for XeF* and KrF*.
14. Basov, N. G., Babeiko, Y. A., Zuev, V. S., Mesyats, G. A., Orlov, V. K., *Sov. J. Quantum Electron.*, 6, 505 (1976), Laser emission from the XeO molecule under optical pumping conditions.
15. Basov, N. G., Danilychev, V. A., Popov, Y. M., *Sov. J. Quantum Electron.*, 1, 18 (1971), Stimulated emission in the vacuum-ultraviolet region.
16. Turner, C. E., Jr., *Appl. Phys. Lett.*, 31, 659 (1977), Near-atmospheric-pressure xenon excimer laser.
17. Hughes, W. M., Shannon, J., Kolb, A., Ault, E., Bhaumik, M., *Appl. Phys. Lett.*, 23, 385 (1973), High-power ultraviolet laser radiation from molecular xenon.
18. Hoff, P. W., Swingle, J. C., Rhodes, C. K., *Opt. Commun.*, 8, 128 (1973), Demonstration of temporal coherence, spatial coherence, and threshold effects in the molecular xenon laser.
19. Bradley, D. J., Hull, D. R., Hutchinson, M. H. R., McGeoch, M. W., *Opt. Commun.*, 14, 1 (1975), Co-axially pumped, narrow band, continuously tunable, high power VUV xenon laser.
20. Koehler, H. A., Ferderber, L. J., Redhead, D. L., Ebert, P. J., *Appl. Phys. Lett.*, 21, 198 (1972), Stimulated VUV emission in high-pressure xenon excited by relativistic electron beams.
21. Tittel, F. K., Wilson, W. L., Stickel, R. E., Marowsky, G., Ernst, W. E., *Appl. Phys. Lett.*, 36, 405 (1980), A triatomic Xe_2Cl excimer laser in the visible.

22. Tang, K. Y., Lorents, D. C., Huestis, D. L., *Appl. Phys. Lett.*, 36, 347 (1980), Gain measurements on the triatomic excimer Xe_2Cl.
23. Waynant, R. W., *Appl. Phys. Lett.*, 30, 234 (1977), A discharge-pumped ArCl superfluorescent laser at 175.0 nm.
24. Jacobs, R. R., Krupke, W. F., *Appl. Phys. Lett.*, 32, 31 (1978), Optical gain at 1.06 μm in the neodymium chloride-aluminum chloride vapor complex.
25. Jacobs, R. R., Krupke, W. F., *Appl. Phys. Lett.*, 34, 497 (1979), Excited state kinetics for $Nd(thd)_3$ and $Tb(thd)_3$ chelate vapors and prospects as fusion laser media.
26. Jacobs, R. R., Krupke, W. F., *Appl. Phys. Lett.*, 35, 126 (1979), Kinetics and fusion laser potential for the terbium aluminum chloride vapor complex.
27. Jacobs, R. R., Weber, M. J., Pearson, R. K., *Chem. Phys. Lett.*, 34, 80 (1975), Nonradiative intramolecular deactivation of Tb^{+3} fluorescence in a vapor phase, terbium (III) complex.
28. Ginter, M. L., Battino, R., *J. Chem. Phys.*, 52, 4469 (1970), Potential-energy curves for the He_2 molecule.
29. Ninomiya, H. and Hirata, K., *J. Appl. Phys.* 68, 5378 (1990), Visible laser action in N_2 laser pumped Ti vapor.
30. Hirata, K., Yoshino, S. and Ninomiya, H., *J. Appl. Phys.* 68, 1460 (1990), Characteristics of an optically pumped titanium vapor laser.
31. Hirata, K., Yoshino, S. and Ninomiya, H., *J. Appl. Phys.* 67, 45 (1990), Optically pumped titanium laser at 551.4 nm.
32. Johnson, R. O., Perram, G. P. and Roh, W. B., *Appl. Phys. B* 65, 5 (1997), Dynamics of a $Br(4^2P_{1/2} \rightarrow 4^2P_{3/2})$ pulsed laser and a Br $(^2P_{1/2})$–NO($\nu=2 \rightarrow \nu=1$) transfer laser driven by photolysis of iodine monobromine.
33. Golde, M. F., *J. Mol. Spectrosc.*, 58, 261 (1975), Interpretation of the oscillatory spectra of the inert-gas halides.
34. Lorents, D. C., Huestis, D. L., McCusker, R. V., Nakano, H. H., Hill, R., *J. Chem. Phys.*, 68, 4657 (1978), Optical emissions of triatomic rare gas halides.
35. Hoffman, J. M., Hays, A. K., Tisone, G. C., *Appl. Phys. Lett.*, 28, 538 (1976), High-power noble-gas-halide lasers.
36. Sze, R. C., Scott, P. B., *Appl. Phys. Lett.*, 33, 419 (1978), Intense lasing in discharge-excited noble-gas monochlorides.
37. Burnham, R., Djeu, N., *Appl. Phys. Lett.*, 29, 707 (1976), Ultraviolet-preionized discharge-pumped lasers in XeF, KrF, and ArF.
38. Rokni, M., Jacob, J. H., Mangano, J. H., *Phys. Rev.* A, 16, 2216 (1977), Dominant formation and quenching processes in e-beam pumped ArF* and KrF* lasers.
39. Powell, H. T., Murray, J. R., Rhodes, C. K., *Appl. Phys. Lett.*, 25, 730 (1974), Laser oscillation on the greenbands of XeO and KrO.
40. Lorents, D. C., *Physica*, 82C, 19 (1976), The physics of electron-beam excited rare-gases at high densities.
41. Murray, J. R., Rhodes, C. K., *J. Appl. Phys.*, 47, 5041 (1976), The possibility of high-energy-storage lasers using the auroral and transauroral transitions of column-VI elements.
42. Rockwood, S. D., LAUR 73-1031(1973), Mechanisms for Achieving Lasing on the 5577 Â Line of Atomic Oxygen, Los Alamos Scientific Laboratory.
43. Murray, J. R., Powell, H. T., Schlitt, L. G., Toska, J., UCRL-50021-76(1976), Laser Program Annual Report, Lawrence Livermore Laboratory.
44. Julienne, P. S., Krauss, M., Stevens, W., *Chem. Phys. Lett.*, 38, 374 (1976), Collision-induced O^1D_2 - $^1S_0$ emission near 5577 Â in argon.
45. Krauss, M., Mies, F. H., *Excimer Lasers* (1979), Springer-Verlag, New York, Electronic structure and radiative transitions of excimer systems.

46. Hughes, W. M., Olson, N. T., Hunter, R., *Appl. Phys. Lett.*, 28, 81 (1976), Experiments on the 558 nm argon oxide laser system.
47. Cunningham, D. L., Clark, K. C., *J. Chem. Phys.* 61, 1118 (1974), Rates of collision-induced emission from metastable O^1S atoms.
48. Wellegehausen, B., Friede, D., Steger, G., *Opt. Commun.*, 26, 391 (1978), Optically pumped continuous Bi_2 and Te_2 lasers.
49. West, W. P., Broida, H. P., *Chem. Phys. Lett.*, 56, 283 (1978), Optically pumped vapor phase Bi_2 laser.
50. Murray, J. R., Swingle, J. C., Turner, C. E., Jr., *Appl. Phys. Lett.*, 28, 530 (1976), Laser oscillation of the 292 nm band system of Br_2.
51. Koehler, H. A., Ferderber, L. J., Redhead, D. L., Ebert, P. J., *Phys. Rev.* A, 9, 768 (1974), Vacuum-ultraviolet emission from high-pressure xenon and argon excited by high-current relativistic electron-beams.
52. Ewing, J. J., Jacob, J. H., Mangano, J. A., Brown, H. A., *Appl. Phys. Lett.*, 28, 656 (1976), Discharge pumping of the Br_2^* laser.
53. Hunter, R. O., unpublished (1975), ARPA Review Meeting, Stanford Research Institute.
54. Veukateswarlu, P., Verma, R. D., *Proc. Indian Acad. Sci.*, 46, 251 (1957), Emission spectrum of bromine excited in the presence of argon - Part I.
55. Hays, A. K., *Laser Focus*, 14, 28 (1978), Cl_2 laser emitting 96 mJ at 258 nm seen promising for iodine-laser, pump.
56. Diegelmann, M., Hohla, K., Kompa, K. L., *Opt. Commun.*, 29, 334 (1979), Interhalogen UV laser on the 285 nm line band of ClF*.
57. Baboshii, V. N., Dobychin, S. L., Zuev, V. S., Mikheev, L. D., Pavlov, A., *Sov. J. Quantum Electron.*, 7, 1183 (1977), Laser utilizing an electronic transition in CN radicals pumped by radiation from an open, high current discharge.
58. West, G. A., Berry, M. J., *J. Chem. Phys.*, 61, 4700 (1974), CN photodissociation and predissociation chemical lasers: molecular electronic and vibrational laser emissions.
59. Mathias, L. E. S., Crocker, A., *Phys. Lett.*, 7, 194 (1963), Visible laser oscillations from carbon monoxide.
60. Hodgson, R. T., *J. Chem. Phys.*, 55, 5378 (1971), Vacuum-ultraviolet lasing observed in CO: 1800-2000 Â.
61. Waller, R. A., Collins, C. B., Cunningham, A. J., *Appl. Phys. Lett.*, 27, 323 (1975), Stimulated emission from CO^+ pumped by charge transfer from He_2^+ in the afterglow of an e-beam discharge.
62. Wilkinson, P. G., Byram, E. T., *Appl. Opt.*, 4, 581 (1965), Rare-gas light sources for the vacuum ultraviolet.
63. Rice, J. K., Hays, A. K., Woodworth, J. R., *Appl. Phys. Lett.*, 31, 31 (1977), Vacuum-UV emissions from mixtures of F_2 and the noble gases - a molecular F_2 laser at 1575 Â.
64. Pummer, H., Hohla, H., Diegelmann, M., Reilly, J. P., *Opt. Commun.*, 28, 104 (1979), Discharge pumped F_2 laser at 1580 Â.
65. Woodworth, J. K., Rice, J. K., *J. Chem. Phys.*, 69, 2500 (1978), An efficient high-power F_2 laser near 157 nm.
66. Dreyfus, R. W., *Phys. Rev.* A, 9, 2635 (1974), Molecular hydrogen laser: 1098-1613 Â.
67. Knyazev, I. N., Letokhov, V. S., Movshev, V. G., *IEEE J. Quantum Electron.*, QE-11, 805 (1975), Efficient and practical hydrogen vacuum ultraviolet laser.
68. Waynant, R. W., Ali, A. W., Julienne, P. S., *J. Appl. Phys.*, 42, 3406 (1971), Experimental observations and calculated bond strengths for the D_2 Lyman band laser.
69. Waynant, R. W., Shipman, J. D., Jr., Elton, R. C., Ali, A. W., *Appl. Phys. Lett.*, 17, 383 (1970), VUV laser emission from molecular hydrogen.

70. Hodgson, R. T., *Phys. Rev. Lett.*, 25, 494 (1970), VUV laser action observed in the Lyman bands of molecular hydrogen.
71. Bockasten, K., Lundholm, T., Andrede, D., *J. Opt. Soc. Am.*, 56, 1260 (1966), Laser lines in atomic and molecular hydrogen.
72. Bazhulin, P. A., Knyazev, I. N., Petrash, G. G., *Sov. Phys. JETP*, 20, 1068 (1965), Pulsed laser action in molecular hydrogen.
73. Hoff, P. W., Swingle, J. C., Rhodes, C. K., *Appl. Phys. Lett.*, 23, 245 (1973), Observations of stimulated emission from high-pressure krypton and argon xenon mixtures.
74. Bazhulin, P. A., Knyazev, I. N., Petrash, G. G., *Sov. Phys. JETP*, 22, 11 (1966), Stimulated emission from hydrogen and deuterium in the infrared.
75. Whitney, W. T., *Appl. Phys. Lett.*, 32, 239 (1978), Sustained discharge excitation of HgCl and HgBr $1B^2\Sigma^+{}_{1/2} \rightarrow X^2\Sigma^+{}_{1/2}$ lasers.
76. Parks, J. H., *Appl. Phys. Lett.*, 31, 297 (1977), Laser action on the $B^2\Sigma^+{}_{1/2} \rightarrow X^2\Sigma^+{}_{1/2}$ band of HgBr at 5018 Â.
77. Schimitschek, E. J., Celto, J. E., Trias, J. A., *Appl. Phys. Lett.*, 31, 608 (1977), Mercuric bromide photodissociation laser.
78. Burnham, R., *Appl. Phys. Lett.*, 33, 156 (1978), Discharge pumped mercuric halide dissociation lasers.
79. Maya, J., *J. Chem. Phys.*, 67, 4976 (1977), Ultraviolet absorption cross-sections of HgI_2, $HgBr_2$, and tin (II) halide vapors.
80. Parks, J. H., *Appl. Phys. Lett.*, 31, 192 (1977), Laser action on the $B^2\Sigma^+{}_{1/2}$ $X^2\Sigma^+{}_{1/2}$ band of HgCl at 5576 Â.
81. Tang, K. Y., Hunter, R. O., Oldenettel, J., Howton, C., Huestis, D., Eckstrom, D., Perry, B. and McCusker, M., *Appl. Phys. Lett.*, 32, 226 (1978), Electron-beam controlled HgCl* laser.
82. Eden, J. G., *Appl. Phys. Lett.*, 33, 495 (1978), VUV-pumped HgCl laser.
83. Eden, J. G., *Appl. Phys. Lett.*, 31, 448 (1977), Green HgCl ($B^2\Sigma^+ \rightarrow X^2\Sigma^+$) laser.
84. Werner, C. W., George, E. V., Hoff, P. W., Rhodes, C. K., *Appl. Phys. Lett.*, 25, 235 (1974), Dynamic model of high-pressure rare-gas excimer lasers.
85. Bradford, R. S., Jr., Ault, E. R., Bhaumik, M. L., *Appl. Phys. Lett.*, 27, 546 (1975), High-power I_2 laser in the 342 nm band system.
86. Hays, A. K., Hoffman, J. M., Tisone, G. C., *Chem. Phys. Lett.*, 39, 353 (1976), Molecular iodine laser.
87. Mikheev, L. D., Shirokikh, A. P., Startsev, A. V., Zuev, V. S., *Opt. Commun.*, 26, 237 (1978), Optically pumped molecular iodine laser on the 342 nm band.
88. Byer, R. L., Herbst, R. L., Kildal, H., *Appl. Phys. Lett.*, 20, 463 (1972), Optically pumped molecular iodine vapor-phase laser.
89. Koffend, J. B., Field, R. W., *J. Appl. Phys.*, 48, 4468 (1977), CW optically pumped molecular iodine laser.
90. Wellegehausen, B., Stephan, K. H., Friede, D., Welling, H., *Opt. Commun.*, 23, 157 (1977), Optically pumped continuous I_2 molecular laser.
91. Hartmann, B., Kleman, B., Steinvall, O., *Opt. Commun.*, 21, 33 (1977), Quasi-tunable I_2-laser for absorption measurements in the near infrared.
92. Hanko, L., Benard, D. J., Davis, S. J., *Opt. Commun.*, 30, 63 (1979), Observation of super-fluorescent emission of the B-X system in I_2.
93. Murray, J. R., Powell, H. T., *Appl. Phys. Lett.*, 29, 252 (1976), KrCl laser oscillation at 222 nm.
94. Eden, J. G., Searles, S. K., *Appl. Phys. Lett.*, 29, 350 (1976), Observation of stimulated emission in KrCl.
95. Gedanken, A., Jortner, J., Raz, B., Szoke, A., *J. Chem. Phys.*, 57, 3456 (1972), Electronic energy transfer phenomena in rare gases.

96. Hay, P. J., Dunning, T. H., Jr., *J. Chem. Phys.*, 66, 1306 (1977), The electronic states of KrF.
97. Tisone, G. C., Hays, A. K., Hoffman, J. M., *Opt. Commun.*, 15, 188 (1975), 100 mW, 248.8 nm KrF laser excited by an electron beam.
98. Ewing, J. J., Brau, C. A., *Appl. Phys. Lett.*, 27, 350 (1975), Laser action on the $^2\Sigma^+{}_{1/2} \rightarrow {}^2\Sigma^+{}_{1/2}$ bands of KrF and XeCl.
99. Mangano, J. A., Jacob, J. H., *Appl. Phys. Lett.*, 27, 495 (1975), Electron-beam-controlled discharge pumping of the KrF laser.
100. Jacob, J. H., Hsia, J. C., Mangano, J. A., Rokni, M., *J. Appl. Phys.*, 50, 5130 (1979), Pulse shape and laser-energy extraction from e-beam-pumped KrF.
101. Fahlen, T. S., *J. Appl. Phys.*, 49, 455 (1978), High-pulse-rate 10-W KrF laser.
102. Hawkins, R. T., Egger, H., Bokor, J., Rhodes, C. K., *Appl. Phys. Lett.*, 36, 391 (1980), A tunable, ultrahigh spectral brightness KrF* excimer laser source.
103. Smith, D., Dean, A. G., Plumb, I. C., *J. Phys. B*, 5, 2134 (1972), Three-body conversion reactions in pure rare gases.
104. Petit, A., Launay, F., Rostas, J., *Appl. Opt.*, 17, 3081 (1978), Spectroscopic analysis of the transverse excited $C^3\Pi_u \rightarrow B, {}^3\Pi_g$ (0,0) UV laser band of N_2 at room temperature.
105. Massone, C. A., Garavaglia, M., Gallardo, M., Calatroni, J. A. E., Tagliaferri, A. A., *Appl. Opt.*, 11, 1317 (1972), Investigation of a pulsed molecular nitrogen laser at low temperature.
106. Verkovtseva, E. T., Ovechkin, A. E., Fogel, Y. M., *Chem. Phys. Lett.*, 30, 120 (1975), The vacuum-uv spectra of supersonic jets of Ar-Kr-Xe mixtures excited by an electron beam.
107. McFarlane, R. A., *Phys. Rev.*, 140, 1070 (1965), Observation of $a^1\Pi$ - $^1\Sigma^-$ transition in the nitrogen molecule.
108. McFarlane, R. A., *IEEE J. Quantum Electron.*, QE-2, 229 (1966), Precision spectroscopy of new infrared emission systems of molecular nitrogen.
109. Ault, E. R., Bhaumik, M. L., Olson, N. T., *IEEE J. Quantum Electron.*, QE-10, 624 (1974), High-power Ar-N_2 transfer laser at 3577 Â.
110. Black, G., Sharpless, R. L., Slanger, T. G., Lorents, D. C., *J. Chem. Phys.*, 62, 4266 (1975), Quantum yields for the production of O(^{1}S), N(^{2}D), and $N_2(A^2\Sigma^+{}_u)$ from VUV, photolysis of N_2O.
111. Searles, S. K., Hart, G. A., *Appl. Phys. Lett.*, 25, 79 (1974), Laser emission at 3577 and 3805 Â in electron-beam-pumped Ar-N_2 mixtures.
112. Ischenko, V. N., Lisitsyn, V. N., Razhev, A. M. et al., *Opt. Commun.*, 13, 231 (1975), The $N^+{}_2$ laser.
113. Ninomiya, H., Takashima, N. and Hirata, K., *J. Appl. Phys.* 69, 67 (1991), Measurement of a population inversion on the 551.4 nm transition in an optically pumped Ti vapor laser.
114. Rothe, D. E., Tan, K. O., *Appl. Phys. Lett.*, 30, 152 (1977), High-power $N^+{}_2$ laser pumped by charge transfer in a high-pressure pulsed glow discharge.
115. Collins, C. B., Cunningham, A. J., Stockton, M., *Appl. Phys. Lett.*, 25, 344 (1974), A nitrogen ion laser pumped by charge transfer.
116. McDaniel, E. W., Flannery, M. R., Ellis, H. W., Eisele, F. L., Pope, W., Tech. Rep. H-78-1(1978), U.S. Army Missile Research and Development Command, Compilation of data relevant to rare gas-rare gas and rare gas-monohalide, excimer lasers.
117. Wellegehauser, B., Shahdin, S., Friede, D., Welling, H., *IEEE J. Quantum Electron.*, QE-13, 65D (1977), Continuous laser oscillation in alkali dimers.
118. Henesian, M. A., Herbst, R. L., Byer, R. L., *J. Appl. Phys.*, 47, 1515 (1976), Optically pumped superfluorescent Na_2 molecular laser.

119. Itoh, H., Uchiki, H., Matsuoka, M., *Opt. Commun.*, 18, 271 (1976), Stimulated emission from molecular sodium.
120. Leone, S. R., Kosnik, K. G., *Appl. Phys. Lett.*, 30, 346 (1977), A tunable visible and ultraviolet laser on S_2.
121. Gerasimov, V. A. and Yunzhakov, B. P., *Sov. J. Quantum. Electron.* 19, 1532 (1989), Investigation of a thulium vapor laser.
122. Zuev, V. S., Mikheev, L. D., Yalovi, V. I., *Sov. J. Quantum Electron.*, 5, 442 (1975), Photochemical laser utilizing the $^1\Sigma^+{}_g - {}^3\Sigma^-{}_g$ vibronic transition in S_2.
123. Searles, S. K., Hart, G. A., *Appl. Phys. Lett.*, 27, 243 (1975), Stimulated emission at 281.8 nm from XeBr.
124. Sze, R. C., Scott, P. B., *Appl. Phys. Lett.*, 32, 479 (1978), High-energy lasing of XeBr in an electric discharge.
125. Tellinghuisen, J., Hays, A. K., Hoffman, J. M., Tisone, G. C., *J. Chem. Phys.*, 65, 4473 (1976), Spectroscopic studies of diatomic noble gas halides. II. Analysis of bound-free emission, from XeBr, XeI, and KrF.
126. Velazco, J. E., Setser, D. W., *J. Chem. Phys.*, 62, 1990 (1975), Bound-free emission spectra of diatomic xenon halides.
127. Basov, N. G., Vasil'ev, L. A., Danilychev, V. A., Dolgov-Saval'ev, G. G., *Sov. J. Quantum Electron.*, 5, 869 (1975), High-pressure N_2^+ laser emitting violet radiation.
128. McFarlane, R. A., unpublished work.
129. Tittel, F. K., unpublished (private communication).
130. Dezenberg, G. J., Willett, C. S., *IEEE J. Quantum Electron.*, QE-7, 491-493 (1971), New unidentified high-gain oscillation at 486.1 and 434.0 nm in the presence of neon.
131. Grishkowsky, D. R., Sorokin, P. P., Lankard, J. R., *Opt. Commun.*, 18, 205-206 (1977), An atomic 16 micron laser.
132. Isaev, A. A., Kazaryan, M. A., Petrash, G. G., *Kratk. Soobshch. Fiz.*, 3, 3 (1972), unspecified.
133. Fahlen, T. S., *IEEE J. Quantum Electron.*, QE-12, 200-201 (1976), Self-heated, multiple-metal-vapor laser.
134. Markova, S. V., Cherezov, V. M., *Sov. J. Quantum Electron.*, 7, 339-342 (1977), Investigation of pulse stimulated emission from gold vapor.
135. Markova, S. V., Petrash, G. G., Cherezov, V. M., *Sov. J. Quantum Electron.*, 8, 904-906 (1978), Ultraviolet-emitting gold vapor laser.
136. Cahuzac, P., *IEEE J. Quantum Electron.*, QE-8, 500 (1972), New infrared laser lines in Mg vapor.
137. Trainor, D. W., Mani, S. A., *Appl. Phys. Lett.*, 33, 648-650 (1978), Atomic calcium laser: pumped via collision-induced absorption.
138. Baron, K. U., Stadler, B., unpublished (June, 1976), Hollow cathode-excited laser transitions in calcium, strontium and barium.
139. Deech, J. S., Sanders, J. H., *IEEE J. Quantum Electron.*, QE-4, 474 (1968), New self-terminating laser transitions in calcium and stronium.
140. Klimkin, V. M., Kolbycheva, P. D., *Sov. J. Quantum Electron.*, 7, 1037-1039 (1977), Tunable single-frequency calcium-hydrogen laser emitting at 5.54 μm.
141. Klimkin, V. M., Monastyrev, S. S., Prokop'ev, V. E., *JETP Lett.*, 20, 110-111 (1974), Selective relaxation of long-lived states of metal atoms in a gas discharge plasma. Stationary generation on $^1P^0{}_1 - {}^1D_2$ transitions.
142. Grishkowsky, D. R., Lankard, J. R., Sorokin, P. P., *IEEE J. Quantum Electron.*, QE-13, 392-396 (1977), An atomic Rydberg state 16-μ laser.
143. Platanov, A. V., Soldatov, A. N., Filonov, A. G., *Sov. J. Quantum Electron.*, 8, 120-121 (1978), Pulsed strontium vapor laser.

144. Bokhan, P. A., Burlakov, V. D., *Sov. J. Quantum Electron.*, 9, 374-376 (1979), Mechanism of laser action due to $4d^3D_{1/2} \rightarrow 5p^3P^0_2$ transitions in a strontium atom.

145. Baron, K. U., Stadler, B., *IEEE J. Quantum Electron.*, QE-11, 852-853 (1975), New visible laser transitions in Ba I and Ba II.

146. Cahuzac, P., *Phys. Lett.*, 32a, 150-151 (1970), New infrared laser lines in barium vapor.

147. Isaev, A. A., Kazaryan, M. A., Markova, S. V., Petrash, G. G., *Sov. J. Quantum Electron.*, 5, 285-287 (1975), Investigation of pulse infrared stimulated emission from barium vapor.

148. Bricks, B. G., Karras, T. W., Anderson, R. S., *J. Appl. Phys.*, 49, 38-40 (1978), An investigation of a discharge-heated barium laser.

149. Isaev, A. A., Kazaryan, M. A., Petrash, G. G., *Sov. J. Quantum Electron.*, 3, 358-359 (1974), Emission of laser pulses due to transitions from a resonance to a metastable level in barium vapor.

150. Cross, L. A., Gokay, M. C., *IEEE J. Quantum Electron.*, QE-14, 648 (1978), A pulse repetition frequency scaling law for the high repetition rate neutral barium laser.

151. Bokhan, P. A., Solomonov, V. I., *Sov. Tech. Phys. Lett.*, 4, 486-487 (1978), Barium vapor laser with a high average output power.

152. Dubrovin, A. N., Tibilov, A. S., Shevtsov, M. K., *Opt. Spectrosc.*, 32, 685 (1972), Lasing on Cd, Zn and Mg lines and possible applications.

153. Tibilov, A. S., *Opt. Spectrosc.*, 19, 463-464 (1965), Generation of radiation in He-Cd and Ne-Cd mixtures.

154. Silfvast, W. T., Szeto, L. H., Wood, O. R., II, *Opt. Lett.* 4, 271-273 (1979), Recombination lasers in Nd and CO_2 laser-produced cadmium plasmas.

155. Chou, M. S., Cool, T. A., *J. Appl. Phys.*, 48, 1551-1555 (1977), Laser operation by dissociation of metal complexes. II. New transitions in Cd, Fe, Ni, Se, Sn, Te, V and Zn.

156. Komine, H., Byer, R. L., *J. Appl. Phys.*, 48, 2505-2508 (1977), Optically pumped atomic mercury photodissociation laser.

157. Djeu, N., Burnham, R., *Appl. Phys. Lett.*, 25, 350-351 (1974), Optically pumped CW Hg laser at 546.1 nm.

158. Artusy, M., Holmes, N., Siegman, A. E., *Appl. Phys. Lett.*, 28, 1331-1334 (1976), D.C.-Excited and sealed-off operation of the optically pumped 546.1 nm Hg laser.

159. Holmes, N. C., Siegman, A. E., *J. Appl. Phys.*, 49, 3155-3170 (1978), The optically pumped mercury vapor laser.

160. Bloom, A. L., Bell, W. E., Lopez, F. O., *Phys. Rev.*, 135, A578-A579 (1964), Laser spectroscopy of a pulsed mercury-helium discharge.

161. Bockasten, K., Garavaglia, M., Lengyel, B. A., Lundholm, T., *J. Opt. Soc. Am.*, 55, 1051-1053 (1965), Laser lines in Hg I.

162. Heard, H. G., Peterson, J., *Proc. IEEE,* 52, 414 (1964), Laser action in mercury rare gas mixtures.

163. Heard, H. G., Peterson, J., *Proc. IEEE,* 52, 1049-1050 (1964), Mercury-rare gas visible-UV laser.

164. Rigden, J. D., White, A. D., *Nature (London)*, 198, 774 (1963), Optical laser action in iodine and mercury discharges.

165. Paananen, R. A., Tang, C. L., Horrigan, F. A., Statz, H., *J. Appl. Phys.*, 34, 3148-3149 (1963), Optical laser action in He-Hg rf discharges.

166. Armand, M., Martinot-Lagarde, P., *C. R. Acad. Sci. Ser.* B, 258, 867-868 (1964), Effect laser sur la vapeur de mercure dans un melange He-Hg.

167. Doyle, W. M., *J. Appl. Phys.*, 35, 1348-1349 (1964), Use of time resolution in identifying laser transitions in mercury rare gas discharge.

168. Chebotayev, V. P., *Opt. Spectrosc.*, 25, 267-268 (1968), Isotopic structure of the 1.5295 millimicron laser line of mercury.
169. Convert, G., Armand, M., Martinot-Lagarde, P., *C. R. Acad. Sci. Ser.* B, 257, 3259-3260 (1964), Effect laser dans des melanges mercure-gas rares.
170. Beterov, I. M., Klement'ev, V. M., Chebotaev, V. P., *Radio Eng. Electron. Phys.* (USSR), 14, 1790-1792 (1969), A mercury laser secondary frequency standard in the microwave region.
171. Bikmukhametov, K. A., Klement'ev, V. M., Chebotaev, V. P., *Sov. J. Quantum Electron.*, 2, 254-256 (1972), Investigation of the stability of the oscillation frequency of a mercury, laser emitting at $\lambda = 1.53\ \mu$.
172. Bikmukhametov, K. A., Klement'ev, V. M., Chebotaev, V. P., *Opt. Spectrosc.*, 34, 616-617 (1973), Collision broadening of the 1.53 μm line of mercury in an Hg-He, Hg-Ne mixture.
173. Klement'ev, V. M., Solov'ev, M. V., *J. Appl. Spectrosc.*, 18, 29-32 (1973), Mercury-vapor laser.
174. Bikmukhametov, K. A., Klement'ev, V. M., Chebotaev, V. P., *Sov. J. Quantum Electron.*, 5, 278-281 (1975), Experimental investigation of the dependences of the collision broadening and shift of the emission line of a mercury laser on He and Ne pressure.
175. Chou, M. S., Cool, T. A., *J. Appl. Phys.*, 47, 1055-1061 (1976), Laser operation by dissociation of metal complexes: new transitions in As, Bi, Ga, Hg, In, Pb, Sb, and Tl.
176. Stricker, J., Bauer, S. H., *IEEE J. Quantum Electron.*, QE-11, 701-702 (1975), An atomic boron laser-pumping by incomplete autoionization or ion-electron.
177. Hemmati, H., Collins, G. J., *Appl. Phys. Lett.*, 34, 844-845 (1979), Atomic gallium photodissociation laser.
178. Burnham, R., *Appl. Phys. Lett.*, 30, 132-133 (1977), Atomic indium photodissociation laser at 451 nm.
179. Gerasimov, V. A. and Yunzhakov, B. P., *Sov. J. Quantum. Electron.* 19, 1323 (1989), Stimulated emission from thulium bromide vapor.
180. Ehrlich, D. J., Maya, J., Osgood, R. M., Jr., *Appl. Phys. Lett.*, 33, 931-933 (1978), Efficient thallium photodissociation laser.
181. Isaev, A. A., Ischenko, P. I., Petrash, G. G., *JETP Lett.*, 6, 118-121 (1967), Super-radiance at transitions terminating at metastable levels of helium and thallium.
182. Isaev, A. A., Petrash, G. G., *JETP Lett.*, 7, 156-158 (1968), Pulsed superradiance at the green line of thallium in TlI-vapor.
183. Sorokin, P. P., Lankard, J. R., *J. Chem. Phys.*, 51, 2929-2931 (1969), Infrared lasers resulting from photodissociation of Cs_2 and Rb_2,.
184. Jacobs, S., Rabinowitz, P., Gould, G., *Phys. Rev. Lett.*, 7, 415-417 (1961), Coherent light amplification in optically pumped Cs vapor.
185. Bennett, W. R., Jr., *Appl. Opt., Suppl. Chem. Lasers*, 3-33 (1965), Inversion mechanisms in gas lasers.
186. Rabinowitz, P., Jacobs, S., *Quantum Electronics III* (1964), 489-498, The optically pumped cesium laser, Columbia University Press, New York .
187. Isaev, A. A., Kazaryan, M. A., Petrash, G. G., *Opt. Spectrosc.*, 31, 180-183 (1971), Mechanism of pulsed lasing of the green thallium line in a thallium iodide vapor discharge.
188. Korolev, F. A., Odintsov, A. I., Turkin, N. G., Yakunin, V. P., *Sov. J. Quantum Electron.*, 5, 237-239 (1975), Spectral structure of pulse superluminescence lines of gases.
189. Chilukuri, S., *Appl. Phys. Lett.*, 34, 284-286 (1979), Selective optical excitation and inversions via the excimer channel: superradiance at the thallium green line.
190. Tunitskii, L. N., Cherkasov, E. M., *Sov. Phys. Tech. Phys.*, 13, 1696-1697 (1969), New oscillation lines in the spectra of NI and Cl.

191. Atkinson, J. B., Sanders, J. H., *J. Phys. B*, 1, 1171-1179 (1968), Laser action in C and N following dissociative excitation transfer.
192. Cooper, G. W., Verdeyen, J. T., *J. Appl. Phys.*, 48, 1170-1175 (1977), Recombination pumped atomic nitrogen and carbon afterglow lasers.
193. Voitovich, A. P., Dubovik, M. V., *J. Appl. Spectrosc.*, 27, 1399-1403 (1978), Time and power characteristics of pulsed atomic gas lasers in a magnetic field.
194. Boot, H. A. H., Clunie, D. M., *Nature (London)*, 197, 173-174 (1963), Pulsed gaseous maser.
195. Patel, C. K. N., McFarlane, R. A., Faust, W. L., *Quantum Electronics III* (1964), 561-572, Further infrared spectroscopy using stimulated emission techniques, Columbia University Press, New York .
196. Patel, C. K. N., McFarlane, R. A., Faust, W. L., *Phys. Rev.*, 133, A1244-A1248 (1964), Optical maser action in C, N, O, S and Br on dissociation of diatomic and polyatomic molecules.
197. English, J. R., II, Gardner, H. C., Merritt, J. A., *IEEE J. Quantum Electron.*, QE-8, 843-844 (1972), Pulsed stimulated emission from N, C, Cl and F atoms.
198. DePoorter, G. C., Balog, G., *IEEE J. Quantum Electron.*, QE-8, 917-918 (1972), New infrared laser line in OCS and new method for C atom lasing.
199. Shimazu, M., Suzaki, Y., *Jpn. J. Appl. Phys.*, 4, 819 (1965), Laser oscillations in silicon tetrachloride vapor.
200. Cooper, H. G., Cheo, P. K., *IEEE J. Quantum Electron.*, QE-2, 785 (1966), Laser transitions in BII, BrII, and Sn.
201. Carr, W. C., Grow, R. W., *Proc. IEEE,* 55, 1198 (1967), A new laser line in tin using stannic chloride vapor.
202. Zhukov, V. V., Latush, E. L., Mikhalevskii, V. S., Sem, M. F., *Sov. J. Quantum Electron.*, 5, 468-469 (1975), New laser transitions in the spectrum of tin and population-inversion mechanism.
203. Xing, D., Ueda, K. and Takuma, H., *Appl. Phys. Lett.* 60, 2961 (1992), K_2 yellow-band and Rb_2 orange-band excimer emissions by electron-beam excitation.
204. Isaev, A. A., Petrash, G. G., *JETP Lett.*, 10, 119-121 (1969), New generation and superradiance lines of lead vapor.
205. Fowles, G.R., Silfvast, W. T., *Appl. Phys. Lett.*, 6, 236-237 (1965), High gain laser transition in lead vapor.
206. Silfvast, W. T., Deech, J. S., *Appl. Phys. Lett.*, 11, 97-99 (1967), Six db/cm single pass gain at 7229 A in lead vapor.
207. Anderson, R. S., Bricks, B. G., Karras, T. W., Springer, L. W., *IEEE J. Quantum Electron.*, QE-12, 313-315 (1976), Discharge-heated lead vapor laser.
208. Kirilov, A. E., Kukharev, V. N., Soldatov, A. N., Tarasenko, V. F., *Sov. Phys. J.*, 20, 1381-1384 (1977), Lead vapor lasers.
209. Feldman, D. W., Liu, C. S., Pack, J. L., Weaver, L. A., *J. Appl. Phys.*, 49, 3679-3683 (1978), Long-lived lead-vapor lasers.
210. Kirilov, A. E., Kukharev, V. N., Soldatov, V. N., *Sov. J. Quantum Electron.*, 9, 285-287 (1979), Investigation of a pulsed λ = 722.9 nm Pb laser with a double-section gas-discharge tube.
211. Piltch, M., Gould, G., *Rev. Sci. Instrum.*, 37, 925-927 (1966), High temperature alumina discharge tube for pulsed metal vapor lasers.
212. Heard, H. G., Peterson, J., *Proc. IEEE,* 52, 1258 (1964), Visible laser transitions in ionized oxygen, nitrogen and carbon monoxide.
213. Hitt, J. S., Haswell, W. T., II, *IEEE J. Quantum Electron.*, QE-2, xlii (1966), Stimulated emission in the theta pinch discharge.
214. Chou, M. S., Zawadzkas, G. A., *Opt. Commun.*, 26, 92 (1978), Observation of new atomic nitrogen laser transition at 9046 A.
215. McFarlane, R. A., *Physics of Quantum Electronics* (1966), 655-663. Stimulated emission spectroscopy of some diatomic molecules, McGraw-Hill, New York.

216. DeYoung, R. J., Wells, W. E., Miley, G. H., Verdeyen, J. T., *Appl. Phys. Lett.*, 28, 519-521 (1976), Direct nuclear pumping of a Ne-N_2 laser.
217. Janney, G. M., *IEEE J. Quantum Electron.*, QE-3, 133 (1967), New infrared laser oscillations in atomic nitrogen.
218. Janney, G. M., *IEEE J. Quantum Electron.*, QE-3, 339 (1) (1967), Correction to near infrared laser oscillations in atomic nitrogen.
219. Sutton, D. G., *IEEE J. Quantum Electron.*, QE-12, 315-316 (1976), New laser oscillation in the N atom quartet manifold.
220. Cheo, P. K., Cooper, H. G., *Appl. Phys. Lett.*, 7, 202-204 (1965), UV and visible laser oscillations in fluorine, phosphorus and chlorine.
221. Fowles, G.R., Zuryk, J. A., Jensen, R. C., *IEEE J. Quantum Electron.*, QE-10, 394-395 (1974), Infrared laser lines in neutral atomic phosphorus.
222. Fowles, G.R., Zuryk, J. A., Jensen, R. C., *IEEE J. Quantum Electron.*, QE-10, 849 (1974), Infrared laser lines in arsenic vapor.
223. Markova, S. V., Petrash, G. G., Cherezov, V. M., *Sov. J. Quantum Electron.*, 7, 657 (1977), Pulse stimulated emission of the 472.2 nm line of the bismuth atom.
224. Powell, H. T., Murray, J. R., Rhodes, C. K., *Appl. Phys. Lett.*, 25, 730-732 (1974), Laser oscillation on the green bands of XeO and KrO.
225. Hughes, W. M., Olson, N. T., Hunter, R., *Appl. Phys. Lett.*, 28, 81-83 (1976), Experiments on 558-nm argon oxide laser systems.
226. Bennett, W. R., Jr., Faust, W. L., McFarlane, R. A., Patel, C. K. N., *Phys. Rev. Lett.*, 8, 470-473 (1962), Dissociative excitation transfer and optical maser oscillation in Ne-O_2 and, Ar-O_2 rf discharges.
227. Tunitskii, L. N., Cherkasov, E. M., *Sov. Phys. Tech. Phys.*, 12, 1500-1501 (1968), Method for varying the frequency of a gas laser.
228. Tunitskii, L. N., Cherkasov, E. M., *J. Opt. Soc. Am.*, 56, 1783-1784 (1966), Interpretation of oscillation lines in Ar-Br_2 laser.
229. Rautian, S. G., Rubin, P. L., *Opt. Spectrosc.*, 18, 180-181 (1965), On some features of gas lasers containing mixtures of oxygen and rare gases.
230. Tunitskii, L. N., Cherkasov, E. M., *Opt. Spectrosc.*, 27, 344-346 (1969), Pulsed mode generation in an argon-oxygen laser.
231. Feld, M. S., Feldman, B. J., Javan, A., *Bull. Am. Phys. Soc.*, 12, 15 (1967), Frequency shifts of the fine structure oscillations of the 8446-Â atomic oxygen laser.
232. Tunitskii, L. N., Cherkasov, E. M., *Opt. Spectrosc.*, 23, 154-157 (1967), The mechanism of laser action in oxygen-inert gas mixtures.
233. Tunitskii, L. N., Cherkasov, E. M., *Sov. Phys. Tech. Phys.*, 13, 993-994 (1969), Pure oxygen laser.
234. Kolpakova, I. V., Redko, T. P., *Opt. Spectrosc.*, 23, 351-352 (1967), Some remarks on the operation of the neon-oxygen gas laser.
235. Feld, M. S., Feldman, B. J., Javan, A., Domash, L. H., *Phys. Rev.*, A7, 257-262 (1973), Selective reabsorption leading to multiple oscillations in the 8446Â atomic oxygen laser.
236. Sutton, D. G., Galvan, L., Suchard, S. N., *IEEE J. Quantum Electron.*, QE-11, 92 (1975), New laser oscillation in the oxygen atom.
237. Powell, F. X., Djeu, N. I., *IEEE J. Quantum Electron.*, QE-7, 176-177 (1971), CW atomic oxygen laser at 4.56 μ.
238. Powell, H. T., Prosnitz, D., Schleicher, B. R., *Appl. Phys. Lett.*, 34, 571-573 (1979), Sulfur $^1S_0$-$^1D_2$ laser by OCS photodissociation.
239. Martinelli, R. U., Gerritsen, H. J., *J. Appl. Phys.*, 37, 444-445 (1966), Laser action in sulphur using hydrogen sulphide.
240. Ultee, C. J., *J. Appl. Phys.*, 44, 1406 (1973), Infrared laser emission from discharges through gaseous sulfur compounds.
241. Hocker, L. O., *J. Appl. Phys.*, 48, 3127-3128 (1977), New infrared laser transitions in neutral sulfur.

242. Hubner, G., Wittig, C., *J. Opt. Soc. Am.*, 61, 415-416 (1971), Some new infrared laser transitions in atomic oxygen and sulfur.
243. Cooper, H. G., Cheo, P. K., *Physics of Quantum Electronics* (1966), 690-697, McGraw-Hill, New York, Ion laser oscillations in sulphur.
244. Davis, C. C., King, T. A., *Advances in Quantum Electronics*, Vol. 3 (1977), 169-454, Academic Press, London, Gaseous ion lasers.
245. Powell, H. T., Ewing, J. J., *Appl. Phys. Lett.*, 33, 165-167 (1978), Photodissociation lasers using forbidden transitions of selenium atoms.
246. Cahuzac, P., *Phys. Lett.*, 27A, 473-474 (1968), Emission laser infrarouges dans les vapeurs de thulium et d'ytterbium.
247. Cahuzac, P., *Phys. Lett.*, 31A, 541-542 (1970), Infrared laser emission from rare-earth vapors.
248. Chapovsky, P. L., Kochubei, S. A., Lisitsyn, V. N., Razhev, A. M., *Appl. Phys.*, 14, 231-233 (1977), Excimer ArF/XeF lasers providing high-power stimulated radiation in Ar/Xe and F lines.
249. Lisitsyn, V. N., Razhev, A. M., *Sov. Tech. Phys. Lett.*, 3, 350-351 (1977), High-power, high-pressure laser based on red fluorine lines.
250. Loree, T. R., Sze, R. C., *Opt. Commun.*, 21, 255-257 (1977), The atomic fluorine laser: spectral pressure dependence.
251. Bigio, I. J., Begley, R. F., *Appl. Phys. Lett.*, 28, 263-264 (1976), High power visible laser action in neutral atomic fluorine.
252. Hocker, L. O., Phi, T. B., *Appl. Phys. Lett.*, 29, 493-494 (1976), Pressure dependence of the atomic fluorine transition intensities.
253. Kovacs, M. A., Ultee, C. J., *Appl. Phys. Lett.*, 17, 39-40 (1970), Visible laser action in fluorine I.
254. Jeffers, W. Q., Wiswall, C. E., *Appl. Phys. Lett.*, 17, 444-447 (1970), Laser action in atomic fluorine based on collisional dissociation of HF.
255. Florin, A. E., Jensen, R. J., *IEEE J. Quantum Electron.*, QE-7, 472 (1971), Pulsed laser oscillation at 0.7311 μ from F atoms.
256. Sumida, S., Obara, M., Fujioka, T., *J. Appl. Phys.*, 50, 3884-3887 (1979), Novel neutral atomic fluorine laser lines in a high-pressure mixture of F_2 and He.
257. Lawler, J. E., Parker, J. W., Anderson, L. W., Fitzsimmons, W. A., *IEEE J. Quantum Electron.*, QE-15, 609-613 (1979), Experimental investigation of the atomic fluorine laser.
258. Paananen, R. A., Horrigan, F. A., *Proc. IEEE,* 52, 1261-1262 (1964), Near infra-red lasering in $NeCl_2$ and $He\text{-}Cl_2$.
259. Shimazu, M., Suzaki, Y., *Jpn. J. Appl. Phys.*, 4, 381-382 (1965), Laser oscillation in the mixtures of freon and rare gases.
260. Jarrett, S. M., Nunez, J., Gould, G., *Appl. Phys. Lett.*, 8, 150-151 (1966), Laser oscillation in atomic Cl in HCl and HI gas discharges.
261. Trusty, G. L., Yin, P. K., Koozekanani, S. K., *IEEE J. Quantum Electron.*, QE-3, 368 (1967), Observed laser lines in freon-helium mixtures.
262. Paananen, R. A., Tang, C. L., Horrigan, F. A., *Appl. Phys. Lett.*, 3, 154-155 (1963), Laser action in Cl_2 and $He\text{-}Cl_2$.
263. Bockasten, K., *Appl. Phys. Lett.*, 4, 118-119 (1964), On the classification of laser lines in chlorine and iodine.
264. Dauger, A. B., Stafsudd, O. M., *IEEE J. Quantum Electron.*, QE-6, 572-573 (1970), Observation of CW laser action in chlorine, argon and helium gas mixtures.
265. Jarrett, S. M., Nunez, J., Gould, G., *Appl. Phys. Lett.*, 7, 294-296 (1965), Infrared laser oscillation in HBr and HI gas discharges.
266. Jensen, R. C., Fowles, G.R., *Proc. IEEE,* 52, 1350 (1964), New laser transitions in iodine-inert-gas mixtures.
267. Kasper, J. V. V., Pimentel, G. C., *Appl. Phys. Lett.*, 5, 231-233 (1964), Atomic iodine photodissociation laser.

268. Kasper, J. V. V., Parker, J. H., Pimentel, G. C., *J. Chem. Phys.*, 43, 1827-1828 (1965), Iodine-atom laser emission in alkyl iodide photolysis.
269. Pollack, M. A., *Appl. Phys. Lett.*, 8, 36-38 (1966), Pressure dependence of the iodine photodissociation laser peak output.
270. Andreeva, T. L., Dudkin, V. A., Malyshev, V. I., Mikhailov, G. V., Sorok, *Sov. Phys. JETP*, 22, 969-970 (1966), Gas laser excited in the process of photodissociation.
271. DeMaria, A. J., Ultee, C. J., *Appl. Phys. Lett.*, 9, 67-69 (1966), High-energy atomic iodine photodissociation laser.
272. Gregg, D. W., Kidder, R. E., Dobler, C. V., *Appl. Phys. Lett.*, 13, 297-298 (1968), Zeeman splitting used to increase energy from a Q-switched laser.
273. Ferrar, C. M., *Appl. Phys. Lett.*, 12, 381-383 (1968), Q-switching and mode locking of a CF_3I photolysis laser.
274. Zalesskii, V. Yu, Venediktov, A. A., *Sov. Phys. JETP*, 28, 1104-1107 (1969), Mechanism of generation termination at the $5^2P_{1/2}$–$5^2P_{3/2}$ transition.
275. O'Brien, D. E., Bowen, J. R., *J. Appl. Phys.*, 40, 4767-4769 (1969), Kinetic model for the iodine photodissociation laser.
276. Andreeva, T. L., Malyshev, V. I., Maslov, A. I., Sobel'man, I. I., Sorokin, V. N., *JETP Lett.*, 10, 271-274 (1969), Possibility of obtaining excited iodine atoms as a result of chemical reactions.
277. Zalesskii, V. Yu., Moskalev, E. I., *Sov. Phys. JETP*, 30, 1019-1023 (1970), Optical probing of a photodissociation laser.
278. Belousova, I. M., Danilov, O. B., Sinitsina, I. A., Spiridonov, V. V., *Sov. Phys. JETP*, 31, 791-793 (1970), Investigation of the optical inhomogeneities of the active medium of a CF_3I photodissociation laser.
279. Velikanov, S. D., Kormer, S. B., Nikolaev, V. D., Sinitsyn, M. V., Solov'ev, Yu A. and Urlin. V. D., *Sov. Phys. Dokl.*, 15, 478-480 (1970), Lower limit of the luminescence spectral linewidth of the $5^2P_{1/2} - 5^2P_{3/2}$ transition in atomic iodine in a photodissociation laser.
280. Gensel, P., Hohla, K., Kompa, K. L., *Appl. Phys. Lett.*, 18, 48-50 (1971), Energy storage of CF_3I photodissociation laser.
281. Belousova, I. M., Danilov, O. B., Kladovikova, N. S., Yachnev, I. L., *Sov. Phys. Tech. Phys.*, 15, 1212-1213 (1971), Quenching of excited atoms in a photodissociation laser.
282. O'Brien, D. E., Bowen, J. R., *J. Appl. Phys.*, 42, 1010-1015 (1971), Parametric studies of the iodine photodissociation laser.
283. DeWolf Lanzerotti, M. Y., *IEEE J. Quantum Electron.*, QE-7, 207-208 (1971), Iodine-atom laser emission in 2-2-2 trifluoroethyliodide.
284. Zalesskii, V. Yu., Krupenikova, T. I., *Opt. Spectrosc.*, 30, 439-443 (1971), Deactivation of metastable iodine atoms by collision with perfluoroalkyl iodide molecules.
285. Andreeva, T. L., Kuznetsova, S. V., Maslov, A. I., Sobel'man, I. I., Sorokin, V. N., *JETP Lett.*, 13, 449-452 (1971), Investigation of reactions of excited iodine atoms with the aid of a photodissociation laser.
286. Hohla, K., IPP Report IV/3(Dec, 1971), Max-Planck-Institut fur Plasma Physik, Photochemical iodine laser: kinetic foundations for giant pulse operation.
287. Belousova, I. M., Kiselev, V. M., Kurzenkov, V. N., *Opt. Spectrosc.*, 33, 112-114 (1972), Induced emission spectrum of atomic iodine due to the hyperfine structure of the transition $^2P_{1/2} - {}^2P_{3/2}$ (7603 cm^{-1}).
288. Belousova, I. M., Kiselev, V. M., Kurzenkov, V. N., *Opt. Spectrosc.*, 33, 115-116 (1972), Line width for induced emission due to the $^2P_{1/2} - {}^2P_{3/2}$ transition of atomic iodine.
289. Zalesskii, V. Yu., *Sov. Phys. JETP*, 34, 474-480 (1972), Kinetics of a CF_3I photodissociation laser.

290. Hwang, W. C., Kasper, J. V. V., *Chem. Phys. Lett.*, 13, 511-514 (1972), Zeeman effects in the hyperfine structure of atomic iodine photodissociation laser emission.

291. Hohla, K., Kompa, K. L., *Chem. Phys. Lett.*, 14, 445-448 (1972), Energy transfer in a photochemical iodine laser.

292. Hohla, K., Kompa, K. L., *Z. Naturforsch.*, 27a, 938-947 (1972), Kinetische prozesse in einem photochemischen jodlaser.

293. Filyukov, A. A., Karpov, Ya., *Sov. Phys. JETP*, 35, 63-65 (1972), A criterion for probable laser quenching.

294. Gavrilina, L. K., Karpov, V. Ya., Leonov, Yu. S., Sautkin, V. A., Filyukov, A. A., *Sov. Phys. JETP*, 35, 258-259 (1972), Selective pumping effect of a photodissociative laser.

295. Hohla, K., Kompa, K. L., *Appl. Phys. Lett.*, 22, 77-78 (1973), Gigawatt photochemical laser.

296. Aldridge, F. T., *Appl. Phys. Lett.*, 22, 180-182 (1973), High-pressure iodine photodissociation laser.

297. Alekseev, V. A., Andreeva, T. L., Volkov, V. N., Yukov, E. A., *Sov. Phys. JETP*, 36, 238-242 (1973), Kinetics of the generation spectrum of a photodissociation laser.

298. Yukov, E. A., *Sov. J. Quantum Electron.*, 3, 117-120 (1973), Elementary processes in the active medium of an iodine photodissociation laser.

299. Gusinow, M. A., Rice, J. K., Padrick, T. D., *Chem. Phys. Lett.*, 21, 197-199 (1973), The apparent late-time gain in a photodissociation iodine laser.

300. Hohla, K., *Laser Interaction and Related Plasma Phenomena*, Vol. 3A(1974), Plenum Press, New York, The iodine laser, a high power gas laser.

301. Birich, G. N., Drozd, G. I., Sorokin, V. N., Struk, I. I., *JETP Lett.*, 19, 27-29 (1974), Photodissociation iodine laser using compounds containing group-V atoms.

302. Belousova, I. M., Gorshkov, N. G., Danilov, O. B., Zalesskii, V. Yu., Yachnev, I. L., *Sov. Phys. JETP*, 38, 254-257 (1974), Accumulation of iodine molecules in flash photolysis of CF_3I and n-C_3F_7I vapor.

303. Belousova, I. M., Bobrov, B. D., Kiselev, V. M., Kurzenkov, V. N., Krepostnov, P. I., *Sov. Phys. JETP*, 38, 258-263 (1974), Photodissociative I^{127} laser in a magnetic field.

304. Basov, N. G., Golubev, L. E., Zuev, V. S., Katulin, V. A., Netemin, V. N., *Sov. J. Quantum Electron.*, 3, 524 (1974), Iodine laser emitting short pulses of 50 J energy and 5 nsec duration.

305. Golubev, L. E., Zuev, V. S., Katulin, V. A., Nosach, V. Yu.., Nosach, O., *Sov. J. Quantum Electron.*, 3, 464-467 (1974), Investigation of optical inhomogeneities which appear in an active medium of a photodissociation laser during coherent emission.

306. Kuznetsova, S. V., Maslov, A. I., *Sov. J. Quantum Electron.*, 3, 468-471 (1974), Investigation of the reactions of atomic iodine in a photodissociation laser using n-C_3F_7I and i-C_3F_7I molecules.

307. Palmer, R. E., Gusinow, M. A., *J. Appl. Phys.*, 45, 2174-2178 (1974), Late-time gain of the CF_3I iodine photodissociation laser.

308. Belousova, I. M., Bobrov, B. D., Kiselev, V. M., Kurzenkov, V. N., Krepostnov, P. I., *Opt. Spectrosc.*, 37, 20-24 (1974), I^{127} atom in a magnetic field.

309. Palmer, R. E., Gusinow, M. A., *IEEE J. Quantum Electron.*, QE-10, 615-616 (1974), Gain versus time in the CF_3I iodine photodissociation laser.

310. Silfvast, W. T., Szeto, L. H., Wood, O. R., II, *Appl. Phys. Lett.*, 25, 593-595 (1974), C_3F_7I photodissociation laser initiated by a CO_2-laser-produced plasma.

311. Belousova, I. M., Bobrov, B. D., Kiselev, V. M., Kurzenkov, V. N., *Sov. J. Quantum Electron.*, 4, 767-769 (1974), Characteristics of the stimulated emission from iodine atoms in pulsed magnetic fields.

312. Hohla, K., Fuss, W., Volk, R., Witte, K. J., *Opt. Commun.*, 13, 114-116 (1975), Iodine laser oscillator in gain switch mode for ns pulses.
313. Hohla, K., Brederlow, G., Fuss, W., Kompa, K. L., Raeder, J., Volk, R., *J. Appl. Phys.*, 46, 808-809 (1975), 60J 1-nsec iodine laser.
314. Zalesskii, V. Yu., *Sov. J. Quantum Electron.*, 4, 1009-1014 (1975), Analytic estimate of the maximum duration of stimulated emission from a CF_3I photodissociation laser.
315. Butcher, R. J., Donovan, R. J., Fotakis, C., Fernie, D., Rae, A. G. A., *Chem. Phys. Lett.*, 30, 398-402 (1975), Photodissociation laser isotope effects.
316. Baker, H. J., King, T. A., *J. Phys. D*, 8, L31-L33 (1975), Mode-beating in gain-switch iodine photodissociation laser.
317. Baker, H. J., King, T. A., *J. Phys. D*, 8, 609-619 (1975), Iodine photodissociation laser oscillator characteristics.
318. Aldridge, F. T., *IEEE J. Quantum Electron.*, QE-11, 215-217 (1975), Stimulated emission cross section and inversion lifetime in a three-atmophere iodine photodissociation laser.
319. Antonov, A. V., Basov, N. G., Zuev, V. S., Katulin, V. A., Korol'kov, K., *Sov. J. Quantum Electron.*, 5, 123 (1975), Amplifier with a stored energy over 700 J designed for a short-pulse iodine laser.
320. Ishii, S., Ahlborn, B., Curzon, F. L., *Appl. Phys. Lett.*, 27, 118-119 (1975), Gain switching and Q spoiling of iodine laser with a shock wave.
321. Borovich, B. L., Zuev, V. S., Katulin, V. A., Nosach, V. Yu., Nosach, O., *Sov. J. Quantum Electron.*, 5, 695-702 (1975), Characteristics of iodine laser short-pulse amplifier.
322. Zalesskii, V. Yu., Polikarpov, S. S., *Sov. J. Quantum Electron.*, 5, 826-831 (1975), Investigation of the conditions governing the stimulated emission threshold of a CF_3I (iodine) laser.
323. Pirkle, R. J., Davis, C. C., McFarlane, R. A., *J. Appl. Phys.*, 46, 4083-4085 (1975), Self-mode-locking of an iodine photodissociation laser.
324. Pleasance, L. D., Weaver, L. A., *Appl. Phys. Lett.*, 27, 407-409 (1975), Laser emission at 1.32 μm from atomic iodine produced by electrical dissociation of CF_3I.
325. Beverly, R. E., II, *Opt. Commun.*, 15, 204-208 (1975), Pressure-broadened iodine-laser-amplifier kinetics and a comparison of diluent effectiveness.
326. Gusinow, M. A., *Opt. Commun.*, 15, 190-192 (1975), The enhancement of the near UV flashlamp spectra with special emphasis on the iodine photodissociation laser.
327. Skribanowitz, N., Kopainsky, B., *Appl. Phys. Lett.*, 27, 490-492 (1975), Pulse shortening and pulse deformation in a high-power iodine laser amplifier.
328. Pirkle, R. J., Jr., Davis, C. C., McFarlane, R. A., *Chem. Phys. Lett.*, 36, 805-807 (1975), Comparative performance of CF_3I, CD_3I and CH_3I in an atomic iodine photodissociation laser.
329. Basov, N. G., Zuev, V. S., *Nuovo Cimento*, 31, 129-151 (1976), Short-pulsed iodine laser.
330. Brederlow, G., Witte, K. J., Fill, E., Hohla, K., Volk, R., *IEEE J. Quantum Electron.*, QE-12, 152-155 (1976), The Asterix III pulsed high-power iodine laser.
331. Jones, E. D., Palmer, M. A., Franklin, F. R., *Opt. Quantum Electron.*, 8, 231-235 (1976), Subnanosecond high-pressure iodine photodissociation laser oscillator.
332. Katulin, V. A., Nosach, V. Yu., Petrov, A. L., *Sov. J. Quantum Electron.*, 6, 205-208 (1976), Iodine laser with active Q switching.
333. Ishii, S., Ahlborn, B., *J. Appl. Phys.*, 47, 1076-1078 (1976), Elimination of compression waves induced by pump light in iodine lasers.
334. Swingle, J. C., Turner, C. E., Jr., Murray, J. R., George, E. V., Krupke, W. F., *Appl. Phys. Lett.*, 28, 387-388 (1976), Photolytic pumping of the iodine laser by $XeBr^*$.

335. Rabinowitz, P., Jacobs, S., Gould, G., *Appl. Opt.*, 1, 511-516 (1962), Continuously optically pumped Cs laser.
336. Walter, W. T., Solimene, N., Piltch, M., Gould, G., *IEEE J. Quantum Electron.*, QE-2, 474-479 (1966), Efficient pulsed gas discharge lasers.
337. Walter, W. T., Solimene, N., Piltch, M., Gould, G., *Bull. Am. Phys. Soc.*, 11, 113 (1966), Pulsed-laser action in atomic copper vapor.
338. Walter, W. T., *Bull. Am. Phys. Soc.*, 12, 90 (1967), 40-kW pulsed copper laser.
339. Bockasten, K., Lundholm, T., Andrade, O., *J. Opt. Soc. Am.*, 56, 1260-1261 (1966), Laser lines in atomic and molecular hydrogen.
340. Leonard, D. A., *IEEE J. Quantum Electron.*, QE-3, 380-381 (1967), A theoretical description of the 5106-A pulsed copper vapor laser.
341. Walter, W. T., *IEEE J. Quantum Electron.*, QE-4, 355-356 (1968), Metal vapor lasers.
342. Asmus, J. E., Moncur, N., *Appl. Phys. Lett.*, 13, 384-385 (1968), Pulse broadening in a MHD copper vapor laser.
343. Isaev, A. A., Kazaryan, M. A., Petrash, G. G., *JETP Lett.*, 16, 27-29 (1972), Effective pulsed copper-vapor laser with high average generation power.
344. Russell, G. R., Nerheim, M. M., Pivirotto, T. J., *Appl. Phys. Lett.*, 21, 656-657 (1972), Supersonic electrical-discharge copper vapor laser.
345. Liu, C. S., Sucov, E. W., Weaver, L. A., *Appl. Phys. Lett.*, 23, 92 (1973), Copper superradiant emission from pulsed discharges in copper iodide vapor.
346. Ferrar, C. M., *IEEE J. Quantum Electron.*, QE-9, 856-857 (1973), Copper-vapor laser with closed-cycle transverse vapor flow.
347. Chen, C. J., Nerheim, N. M., Russell, G. R., *Appl. Phys. Lett.*, 23, 514-515 (1973), Double-discharge copper vapor laser with copper chloride as a lasant.
348. Isaev, A. A., Kazaryan, M. A., Petrash, G. G., *Opt. Spectrosc.*, 35, 307-308 (1973), Copper vapor pulsed laser with a repetition frequency of 10 kHz.
349. Weaver, L. A., Liu, C. S., Sucov, E. W., *IEEE J. Quantum Electron.*, QE-10, 140-147 (1974), Superradiant emission at 5106, 5700 and 5782 Å in pulsed copper iodide discharges.
350. Ehrlich, D. J., Osgood, R. M., *Appl. Phys. Lett.*, 34, 655-658 (1979), Alkali-metal resonance-line lasers based on photodissociation.
351. Fahlen, T. S., *J. Appl. Phys.*, 45, 4132-4133 (1974), Hollow-cathode copper-vapor laser.
352. Gal'pern, M. G., Gorbachev, V. A., Katulin, V. A. et al., *Sov. J. Quantum Electron.*, 5, 1384-1385 (1975), Bleachable filter for the iodine laser emitting at $\lambda = 1.35\ \mu m$.
353. Brederlow, G., Fill, E., Fuss, W., Hohla, K., Volk, R., Witte, K. J., *Sov. J. Quantum Electron.*, 6, 491-495 (1976), High-power iodine laser development at the Institut fur Plasmaplysik, Garching.
354. Davis, C. C., Pirkle, R. J., McFarlane, R. A., Wolga, G. J., *IEEE J. Quantum Electron.*, QE-12, 334-352 (1976), Output mode spectra, comparative parametric operation, quenching, photolytic reversibility, and short-pulse generation in atomic iodine photodissociation laser.
355. Arkhipova, E. V., Borovich, B. L., Zapol'skii, A. K., *Sov. J. Quantum Electron.*, 6, 686-696 (1976), Accumulation of excited iodine atoms in iodine photodissociation laser. Analysis of kinetic equations.
356. Liberman, I., Babcock, R. V., Liu, C. S., George, T. V., Weaver, L. A., *Appl. Phys. Lett.*, 25, 334-335 (1974), High-repetition-rate copper iodide laser.
357. Nosach, O. Yu., Orlov, E. P., *Sov. J. Quantum Electron.*, 6, 770-777 (1976), Some features of formation of the angular spectrum of the stimulated radiation emitted from iodine laser.

358. Andreeva, T. L., Birich, G. N., Sorokin, V. N., Struk, I. I., *Sov. J. Quantum Electron.*, 6, 781-789 (1976), Investigations of photodissociation iodine lasers utilizing molecules with bonds between iodine atoms and group V elements. I. Experimental investigation.
359. Zalesskii, V. Yu., Kokushkin, A. M., *Sov. J. Quantum Electron.*, 6, 813-817 (1976), Tracing chemical changes in the active media of iodine photodissociation laser.
360. Katulin, V. A., Nosach, V. Yu., Petrov, A. L., *Sov. J. Quantum Electron.*, 6, 998-999 (1976), Nanosecond iodine laser with an output energy of 200J.
361. Fuss, W., Hohla, K., *Z. Naturforsch.* Teil A, 31, 569-577 (1976), Pressure broadening of the 1.3 μm iodine laser line.
362. Fuss, W., Hohla, K., *Opt. Commun.*, 18, 427-430 (1976), A closed cycle iodine laser.
363. Fill, E., Hohla, K., *Opt. Commun.*, 18, 431-436 (1976), A saturable absorber for the iodine laser.
364. Kamrukov, A. S., Kashnikov, G. N., Kozlov, N. P., Malashchenko, V. A. et al., *Sov. J. Quantum Electron.*, 6, 1101-1104 (1976), Investigation of an iodine laser excited optically by high-current plasmadynamic discharges.
365. Mukhtar, E. S., Baker, H. J., King, T. A., *Opt. Commun.*, 19, 193-196 (1976), Selection of oscillation frequency in the 1.315 μm iodine laser.
366. Alekhin, B. V., Lazhintsev, B. V., Norarevyan, V. A., Petrov, N. N. et al., *Sov. J. Quantum Electron.*, 6, 1290-1292 (1976), Short-pulse photodissociation laser with magnetic-field modulation of gain.
367. Bokhan, P. A., Solomonov, V. I., *Sov. J. Quantum Electron.*, 3, 481-483 (1974), Mechanism of laser action in copper vapor.
368. Baker, H. J., King, T. A., *J. Phys. D*, 9, 2433-2445 (1976), Line broadening and saturation parameters for short-pulse high-pressure iodine photodissociation laser systems.
369. Padrick, T. D., Palmer, R. E., *J. Appl. Phys.*, 47, 5109-5110 (1976), Use of titanium doped quartz to eliminate carbon deposits in an atomic iodine photodissociation laser.
370. Fill, E., Hohla, K., Schappert, G. T., Volk, R., *Appl. Phys. Lett.*, 29, 805-807 (1976), 100-ps pulse generation and amplification in the iodine laser.
371. Olsen, J. N., *J. Appl. Phys.*, 47, 5360-5364 (1976), Pulse shaping in the iodine laser.
372. Belousova, I. M., Bobrov, B. D., Grenishin, A. S., Kiselev, V. M., *Sov. J. Quantum Electron.*, 7, 249-255 (1977), Magnetic-field control of the duration of pulses emitted from an iodine, photodissociation laser.
373. Andreeva, T. L., Birich, G. N., Sobel'man, I. I., Sorokin, V. N., et al., *Sov. J. Quantum Electron.*, 7, 1230-1234 (1977), Continuously pumped continuous-flow iodine laser.
374. Mukhtar, E. S., Baker, H. J., King, T. A., *Opt. Commun.*, 24, 167-169 (1978), Pressure-induced frequency shifts in the atomic iodine laser.
375. Kiselev, V. M., Bobrov, B. D., Grenishin, A. S., Kotlikova, T. N., *Sov. J. Quantum Electron.*, 8, 181-184 (1978), Faraday rotation in the active medium of an iodine photodissociation laser oscillator or amplifier.
376. Babkin, V. I., Kuznetsova, S. V., Maslov, A. I., *Sov. J. Quantum Electron.*, 8, 285-289 (1978), Simple method for determination of stimulated emission cross section of $^2P_{1/2}(F = 3) \rightarrow {}^2P_{3/2}$ (F' = 4) transition in atomic iodine.
377. Akitt, D. P. and and Wittig, C. F., *J. Appl. Phys.* 40, 902 (1969), Laser emission in ammonia.
378. Ferrar, C. M., *IEEE J. Quantum Electron.*, QE-10, 655-656 (1974), Buffer gas effects in a rapidly pulsed copper vapor laser.
379. Saito, H., Uchiyama, T., Fujioka, T., *IEEE J. Quantum Electron.*, QE-14, 302-309 (1978), Pulse propagation in the amplifier of a high-power iodine laser.

380. Gaidash, V. A., Mochalov, M. R., Shemyakin, V. I., Shurygin, V. K., *Sov. J. Quantum Electron.*, 8, 530-531 (1978), Modulation of the transmission of an atomic iodine switch.

381. McDermott, W. E., Pchelkin, N. R., Benard, D. G., Bousek, R. R., *Appl. Phys. Lett.*, 32, 469-470 (1978), An electronic transition chemical laser.

382. Antonov, A. S., Belousova, I. M., Gerasimov, V. A. et al., *Sov. Tech. Phys. Lett.*, 4, 459 (1978), Flashlamp-excited photodissociation 1000 J laser with 1.4 percent efficiency.

383. Silfvast, W. T., Szeto, L. H., Wood, O. R., II, *Appl. Phys. Lett.*, 36, 615 (1980), Simple metal vapor recombination lasers using segmented plasma excitation.

384. Sutton, D. G., Galvan, L., Suchard, S. N., *IEEE J. Quantum Electron.*, QE-11, 312 (1975), Two-electron laser transition in Sn(I)?.

385. Katulin, V. A., Nosach, V. Yu., Petrov, A. L., *Sov. J. Quantum Electron.*, 8, 380-382 (1978), Q-Switched iodine laser emitting at two frequencies.

386. Burnham, R., unpublished (May, 1978), Optically pumped bismuth lasers at 472 and 475 nm.

387. Murray, J. R., Powell, H. T., Rhodes, C. K., unpublished (June1974), Inversion of the auroral green transition of atomic oxygen by argon excimer transfer to nitrous oxide.

388. Powell, H. T., Murray, J. R., Rhodes, C. K., unpublished (May, 1975), Collision-induced auroral line lasers.

389. Bennett, W. R., Jr., *Appl. Optics Supplement on Optical Masers* (1962), 24-61, Gaseous optical masers.

390. Powell, H. T., Ewing, J. J., unpublished (May, 1978), Forbidden transition selenium atom photodissociation lasers.

391. Pollack, M. A., unpublished (private communication to C. S. Willett).

392. Vinokurov, G. N., Zalesskii, V. Yu., *Sov. J. Quantum Electron.*, 8, 1191-1197 (1978), Chemical kinetics and gas dynamics of a Q-switch iodine laser with an optically thick active medium.

393. Benard, D. J., McDermott, W. E., Pchelkin, N. R., Bousek, R. R., *Appl. Phys. Lett.*, 34, 40-41 (1979), Efficient operation of a 100-W transverse-flow oxygen-iodine chemical laser.

394. Richardson, R. J., Wiswall, C. E., *Appl. Phys. Lett.*, 35, 138-139 (1979), Chemically pumped iodine laser.

395. Witte, K. J., Burkhard, P., Luthi, H. R., *Opt. Commun.*, 28, 202-206 (1979), Low pressure mercury lamp pumped atomic iodine laser of high efficiency.

396. Riley, M. E., Padrick, T. D., Palmer, R. E., *IEEE J. Quantum Electron.*, QE-15, 178-189 (1979), Multilevel paraxial Maxwell-Bloch equation description of short pulse amplification in the atomic iodine laser.

397. Zuev, V. S., Netemin, V. N., Nosach, O. Yu., *Sov. J. Quantum Electron.*, 9, 522-524 (1979), Wavefront instability of iodine laser radiation and dynamics of optical inhomogeneity evolution in the active medium.

398. Anderson, R. S., Springer, L., Bricks, B. G., Karras, T. W., *IEEE J. Quantum Electron.*, QE-11, 172-174 (1975), A discharge heated copper vapor laser.

399. Palmer, R. E., Padrick, T. D., Palmer, M. A., *Opt. Quantum Electron.*, 11, 61-70 (1979), Diffraction-limited atomic iodine photodissociation laser.

400. Djeu, N., Powell, F. X., *IEEE J. Quantum Electron.*, QE-7, 537-538 (1971), More infrared laser transitions in atomic iodine.

401. Kim, H. H., Marantz, H., *Appl. Opt.*, 9, 359-368 (1970), A study of the neutral atomic iodine laser.

402. Kim, H., Paananen, R., Hanst, P., *IEEE J. Quantum Electron.*, QE-4, 385-386 (1968), Iodine infrared laser.

403. Brandelik, J. E., Smith, G. A., *IEEE J. Quantum Electron.*, QE-16, 7-10 (1980), Br, C, Cl, S and Si laser action using a pulsed microwave discharge.

404. Klimkin, V. M., *Sov. J. Quantum Electron.*, 5, 326-329 (1975), Investigation of an ytterbium vapor laser.
405. Prelas, M. A., Akerman, M. A., Boody, F. P., Miley, G. H., *Appl. Phys. Lett.*, 31, 428-430 (1977), A direct nuclear pumped 1.45 μ atomic carbon laser in mixtures of He-CO and He-CO_2.
406. Piltch, M., Walter, W. T., Solimene, N., Gould, G., Bennett, W. R., Jr., *Appl. Phys. Lett.*, 7, 309-310 (1965), Pulsed laser transitions in manganese vapor.
407. Silfvast, W. T., Fowles, G. R., *J. Opt. Soc. Am.*, 56, 832-833 (1966), Laser action on several hyperfine transitions in MnI.
408. Chen, C. J., Russell, G. R., *Appl. Phys. Lett.*, 26, 504-505 (1975), High-efficiency multiply pulsed copper vapor laser utilizing copper chloride as a lasant.
409. Chen, C. J., *J. Appl. Phys.*, 44, 4246-4247 (1973), Manganese laser.
410. Chen, C. J., *Appl. Phys. Lett.*, 24, 499-500 (1974), Manganese laser using manganese chloride as lasant.
411. Isakov, V. K., Kapugin, M. M., Potapov, S. E., *Sov. Tech. Phys. Lett.*, 2, 292-293 (1976), $MnCl_2$-vapor laser (energy characteristics).
412. Isakov, V. K., Kalugin, M. M., Potapov, S. E., *Sov. Tech. Phys. Lett.*, 4, 333-334 (1978), Output spectrum of a manganese-chloride laser.
413. Bokhan, P. A., Burlakov, V. D., Gerasimov, V. A., Solomonov, V. I., *Sov. J. Quantum Electron.*, 6, 672-675 (1976), Stimulated emission mechanism and energy characteristics of manganese vapor laser.
414. Isaev, A. A., Kazaryan, M. A., Petrash, G. G., Cherezov, V. M., *Sov. J. Quantum Electron.*, 6, 978-980 (1976), An investigation of pulse manganese vapor laser.
415. Trainor, D. W., Mani, S. A., *J. Chem. Phys.*, 68, 5481-5485 (1978), Pumping iron: a KrF laser pumped atomic iron laser.
416. Linevsky, M. J., Karras, T. W., *Appl. Phys. Lett.*, 33, 720-721 (1978), An iron-vapor laser.
417. Solanki, R., Collins, G. J., Fairbank, W. M., Jr., *IEEE J. Quantum Electron.*, QE-15, 525 (1979), IR laser transitions in a nickel hollow cathode discharge.
418. Bokhan, P. A., Klimkin, V. M., Prokop'ev, V. E., Solomonov, V. I., *Sov. J. Quantum Electron.*, 7, 81-82 (1977), Investigation of a laser utilizing self-terminating transitions in europium atoms and ions.
419. Bokhan, P. A., Nikolaev, V. N., Solomonov, V. I., *Sov. J. Quantum Electron.*, 5, 96-98 (1975), Sealed Copper vapor laser.
420. Cahuzac, P., Sontag, H., Toschek, P. E., *Opt. Commun.*, 31, 37-41 (1979), Visible superfluorescence from atomic europium.
421. Breichignac, C., Cahuzac, P., (unpublished).
422. Pixton, R. M., Fowles, G. R., *Phys. Lett.*, 29A, 654-655 (1969), Visible laser oscillations in helium at 7065 Â.
423. Schuebel, W. K., *Appl. Phys. Lett.*, 30, 516-519 (1977), Laser action in AlII and HeI in a slot cathode discharge.
424. Abrams, R. L., Wolga, G. J., *IEEE J. Quantum Electron.*, QE-5, 368 (1967), Near infrared laser transitions in pure helium.
425. Abrams, R. L., Wolga, G. J., *Phys. Rev. Lett.*, 19, 1411-1414 (1967), Direct demonstration of the validity of the Wigner spin rule for helium-helium collisions.
426. Cagnard, R., der Agobian, R., Otto, J. L., Echard, R., *C. R. Acad. Sci. Ser.* B, 257, 1044-1047 (1963), L'emission stimulee de quelque transitions infrarouges de l'helium et du neon.
427. der Agobian, R., Otto, J. L., Cagnard, R., Echard, R., *J. Phys.*, 25, 887-897 (1964), Emission stimulee de nouvelles transitions infrarouges dans les gaz rares.
428. Wood, O. R., Burkhardt, E. G., Pollack, M. A., Bridges, T. J., *Appl. Phys. Lett.*, 18, 261-264 (1971), High pressure laser action in 13 gases with transverse excitation.

429. Patel, C. K. N., Bennett, W. R., Jr., Faust, W. L., McFarlane, R. A., *Phys. Rev. Lett.*, 9, 102-104 (1962), Infrared spectroscopy using stimulated emission techniques.
430. Smilanski, I., Levin, L. A., Erez, G., *IEEE J. Quantum Electron.*, QE-11, 919-920 (1975), A copper laser using CuI vapor.
431. Bennett, W. R., Jr., Kindlmann, P. J., *Bull. Am. Phys. Soc.*, 8, 87 (1963), Collision cross sections and optical maser considerations for helium.
432. Brochard, J., Liberman, S., *C. R. Acad. Sci. Ser.* B, 260, 6827-6829 (1965), Emission stimulee de nouvelles transitions infrarouge de l'helium et du neon.
433. Brochard, J., Lespritt, J. F., Liberman, L., *C. R. Acad. Sci. Ser.* B, 600-602 (1970), Measurement of the isotopic separation of two infrared laser lines of HeI.
434. Mathias, L. E. S., Crocker, A., Wills, M. S., *IEEE J. Quantum Electron.*, QE-3, 170 (1967), Pulsed laser emission from helium at 95 μm.
435. Levine, J. S., Javan, A., *Appl. Phys. Lett.*, 14, No. 11, 348-350 (1969), Far infrared continuous wave oscillation in pure helium.
436. Turner, R., Murphy, R. A., *Infrared Phys.*, 16, 197-200 (1976), The far infrared helium laser.
437. Giuliano, C. R., Hess, L. D., *J. Appl. Phys.*, 40, 2428-2430 (1969), Reversible photodissociative laser system.
438. Campbell, J. D., Kasper, J. V. V., *Chem. Phys. Lett.*, 10, 436-437 (1971), Hyperfine structure in the CF_3Br photodissociation laser.
439. Spencer, D. J., Wittig, C., *Opt. Lett.*, 4, 1-3 (1979), Atomic bromine electronic-transition chemical laser.
440. Hocker, L. O., *J. Opt. Soc. Am.*, 68, 262-265 (1978), High resolution study of the helium-fluorine laser.
441. Shukhtin, M., Fedotov, G. A., Mishakov, V. G., *Opt. Spectrosc.*, 39, 681 (1976), Lasing with CuI lines using copper bromide vapor.
442. Bell, W. E., Bloom, A. L., Goldsborough, J. P., *IEEE J. Quantum Electron.*, QE-2, 154 (1966), New laser transitions in antimony and tellurium.
443. Webb, C. E., *IEEE J. Quantum Electron.*, QE-4, 426-427 (1968), New pulsed laser transitions in TeII.
444. Clunie, D. M., Thorn, R. S. A., Trezise, K. E., *Phys. Lett.*, 14, 28-29 (1965), Asymmetric visible super-radiant emission from a pulsed neon discharge.
445. Leonard, D. A., Neal, R. A., Gerry, E. T., *Appl. Phys. Lett.*, 7, 175 (1965), Observation of a super-radiant self-terminating green laser transition in neon.
446. Leonard, D. A., *IEEE J. Quantum Electron.*, QE-3, 133-135 (1967), The 5401 Â pulsed neon laser.
447. Isaev, A. A., Kazaryan, M. A., Petrash, G. G., *Sov. J. Quantum Electron.*, 2, 49-51 (1972), Shape and duration of superradiance pulses corresponding to neon lines.
448. Magda, I. I., Tkach, Yu. V., Lemberg, E. A., Skachek, G. V., Gadetskii, N., *Sov. J. Quantum Electron.*, 3, 260-261 (1973), High power nitrogen and neon pulsed gas lasers.
449. Perry, D. L., *IEEE J. Quantum Electron.*, QE-7, 102 (1971), CW laser oscillation at 5433 Â in neon.
450. Bridges, W. B., Chester, A. N., *Appl. Opt.*, 4, 573-580 (1965), Visible and uv laser oscillation at 118 wavelengths in ionized neon, argon, krypton, xenon, oxygen and other gases.
451. White, A. D., Rigden, J. D., *Appl. Phys. Lett.*, 2, 211-212 (1963), The effect of super-radiance at 3.39 μ on the visible transitions in the He-Ne maser.
452. Sovero, E., Chen, C. J., Culick, F. E. C., *J. Appl. Phys.*, 47, 4538-4542 (1976), Electron temperature measurements in a copper chloride laser utilizing a microwave radiometer.
453. Viscovini, R. C., Scalabrin, A. and Pereira, D., *IEEE J. Quantum Electron.* 33, 916 (1997), $^{13}CD_3OD$ optically pumped by a waveguide CO_2 laser: new FIR laser lines.

454. Ericsson, K. G., Lidholt, L. R., *IEEE J. Quantum Electron.*, QE-3, 94 (1967), Super-radiant transitions in argon, krypton and xenon.
455. Rosenberger, D., *Phys. Lett.*, 14, 32 (1965), Superstrahlung in gepulsten argon-, krypton- und xenon-entladungen.
456. Bloom, A. L., *Appl. Phys. Lett.*, 2, 101-102 (1963), Observation of new visible gas laser transition by removal of dominance.
457. Ericsson, G., Lidholt, R., *Ark. Fys.*, 37, 557-568 (1967), Generation of short light pulses by superradiance in gases.
458. Rosenberger, D., *Phys. Lett.*, 13, 228-229 (1964), Laser-ubergange and superstrahlung bei 6143 A in einer gepulsten Neon-entladungen.
459. Abrosimov, G. V., *Opt. Spectrosc.*, 31, 54-56 (1971), Spatial and temporal coherence of the radiation of pulsed neon and thallium gas lasers.
460. Isaev, A. A., Kazaryan, M. A., Petrash, G. G., *Sov. J. Quantum Electron.*, 2, 49-52 (1972), Shape and duration of superradiance pulses corresponding to neon lines.
461. Odintsov, A. I., Turkin, N. G., Yakunin, V. P., *Opt. Spectrosc.*, 38, 244-245 (1975), Spatial coherence and angular divergence of pulsed superradiance of neon.
462. Isaev, A. A., Petrash, G. G., *Sov. Phys. JETP*, 29, 607-614 (1969), Mechanism of pulsed superradiance from 2p–1s transitions in neon.
463. White, J. C., *Appl. Phys. Lett.*, 33, 325-327 (1978), Inversion of the Na resonance line by selective photodissociation of NaI.
464. Shukhtin, A. M., Fedotov, G. A., Mishakov, V. G., *Opt. Spectrosc.*, 40, 237-238 (1976), Stimulated emission on copper lines during pulsed production of vapor without the use of a heating element.
465. White, A. D., Rigden, J. D., *Proc. IEEE,* 50, 1796 (1962), Continuous gas maser operation in the visible.
466. Rigden, J. D., White, A. D., *Proc. IEEE,* 51, 943 (1963), The interaction of visible and infrared maser transitions in the helium-neon system.
467. White, A. D., Gordon, E. I., *Appl. Phys. Lett.*, 3, 197-198 (1963), Excitation mechanism and current dependence of population inversion in He-Ne lasers.
468. Gordon, E. I., White, A. D., *Appl. Phys. Lett.*, 3, 199-201 (1963), Similarity laws for the effect of pressure and discharge diameter on gain of He-Ne lasers.
469. Labuda, E. F., Gordon, E. I., *J. Appl. Phys.*, 35, 1647-1648 (1964), Microwave determination of average electron energy and density in He-Ne discharges.
470. Young, R. T., Willett, C. S., Maupin, R. T., *J. Appl. Phys.*, 41, 2936-2941 (1970), The effect of helium on population inversion in the He-Ne laser.
471. Young, R. T., Jr., Willett, C. S., Maupin, R. T., *Bull. Am. Phys. Soc.*, 13, 206 (1968), An experimental determination of the relative contributions of resonance and electron impact collision to the excitation of Ne atom in He-Ne laser.
472. Jones, C. R., Robertson, W. W., *Bull. Am. Phys. Soc.*, 13, 198 (1968), Temperature dependence of reaction of the helium metastable atom.
473. Korolev, F. A., Odintsov, A. I., Mitsai, V. N., *Opt. Spectrosc.*, 19, 36-39 (1965), A study of certain characteristics of a He-Ne laser.
474. Suzuki, N., *Jpn. J. Appl. Phys.*, 4, 285-291 (1965), Vacuum uv measurements of helium-neon laser discharge.
475. Abrosimov, G. V., Vasil'tsov, V. V., Voloshin, V. N., Korneev, A. V. et al., *Sov. Tech. Phys. Lett.*, 2, 162-163 (1976), Pulsed laser action on self-limiting transitions of copper atoms in copper halide vapor.
476. Field, R. L., Jr., *Rev. Sci. Instrum.*, 38, 1720-1722 (1967), Operating characteristics of dc-excited He-Ne gas lasers.
477. Suzuki, N., *Jpn. J. Appl. Phys.*, 4, 642-647 (1965), Spectroscopy of He-Ne laser discharges.
478. Gonchukov, G. A., Ermakov, G A., Mikhenko, G. A., Protsenko, E. D., *Opt. Spectrosc.*, 20, 601-602 (1966), Temperature effects in the He-Ne laser.
479. Alekseeva, A. N., Gordeev, D. V., *Opt. Spectrosc.*, 23, 520-524 (1967), The effect of longitudinal magnetic field on the output of a helium-neon laser.

480. Smith, P. W., *IEEE J. Quantum Electron.*, QE-2, 62-68 (1966), The output power of a 6328-Â He-Ne gas laser.

481. Smith, P. W., *IEEE J. Quantum Electron.*, QE-2, 77-79 (1966), On the optimum geometry of a 6328 Â laser oscillator.

482. Herziger, G., Holzapfel, W., Seelig, W., *Z. Phys.*, 189, 385-400 (1966), Verstarkung einer He-Ne gasentladung fur die laser wellenlange, λ=6328 Â.

483. Bell, W. E., Bloom, A. L., *Appl. Opt.*, 3, 413-415 (1964), Zeeman effect at 3.39 microns in a helium-neon laser.

484. Belusova, I. M., Danilov, O. B., Elkina, I. A., Kiselev, K. M., *Opt. Spectrosc.*, 16, 44-47 (1969), Investigation of the causes of gas temperature effects on the generation of power of a He-Ne laser at 6328 Â.

485. Landsberg, B. M., *IEEE Quantum Electron.* QE-16, 684 (1980), Optically pumped CW submillimeter emission lines from methyl mercaptan CH_3SH.

486. Akirtava, O. S., Dzhikiya, V. L., Oleinik, Yu. M., *Sov. J. Quantum Electron.*, 5, 1001-1002 (1976), Laser utilizing CuI transitions in copper halide vapors.

487. Vasconcellos, E. C. C., Wyss, J. C., Petersen, F. R. and Evenson, K. M., *Int. J. Infrared Millimeter Waves,* 4, 401 (1983), Frequency measurements of far infrared cw lasing lines in optically pumped $CHCl_2F$.

488. White, A. D., *Proc. IEEE,* 52, 721 (1964), Anomalous behaviour of the 6402.84 Â gas laser.

489. Bloom, A. L., Hardwick, D. L., *Phys. Lett.*, 20, 373-375 (1966), Operation of He-Ne lasers in the forbidden resonator region.

490. Zitter, R. N., *Bull. Am. Phys. Soc.*, 9, 500 (1964), 2s–2p and 3s–3p neon transitions in a very long laser.

491. Patel, C. K. N., *Lasers*, Vol. 1(1968), 39-50, Marcel Dekker, New York, Gas lasers.

492. Henningsen, J. O., *Int. J. Infrared Millimeter Waves* 7, 1605 (1986), Methanol laser lines from torsionally excited CO stretch states and from OH-bend, CH_3-rock, and CH_3-deformation states.

493. Uchida, T., *Appl. Opt.*, 4, 129-131 (1965), Frequency spectra of the He-Ne optical masers with external concave mirrors.

494. Massey, G. A., Oshman, M. K., Targ, R., *Appl. Phys. Lett.*, 6, 10-11 (1965), Generation of single-frequency light using the FM laser.

495. Steier, W. H., *Proc. IEEE,* 54, 1604-1606 (1966), Coupling of high peak power pulses from He-Ne lasers.

496. Kuznetsov, A. A., Mash, D. I., Skuratova, N. V., *Radio Eng. Electron. Phys.* (USSR), 12, 140-143 (1967), Effect of an axial magnetic field on the output power of a neon-helium laser.

497. Subotinov, N. V., Kalchev, S. D., Telbizov, P. K., *Sov. J. Quantum Electron.*, 5, 1003-1004 (1976), Copper vapor laser operating at a high pulse repetition frequency.

498. Lee, P. H., Skolnick, M. L., *Appl. Phys. Lett.*, 10, 303-305 (1967), Saturated neon absorption inside a 6328 Â laser.

499. Hochuli, U., Haldemann, P., Hardwick, D., *IEEE J. Quantum Electron.*, QE-3, 612-614 (1967), Cold cathodes for He-Ne gas lasers.

500. Carlson, F. P., *IEEE J. Quantum Electron.*, QE-4, 98-99 (1968), On the optimal use of dc and RF excitation in He-Ne lasers.

501. Suzuki, T., *Jpn. J. Appl. Phys.*, 9, 309-310 (1970), Discharge current noise in He-Ne laser and its suppression.

502. Ehlers, K. W., Brown, I. G., *Rev. Sci. Instrum.*, 41, 1505-1506 (1970), Regeneration of helium neon lasers.

503. Leontov, V. G., Ostapchenko, E. P., Sedov, G. S., *Opt. Spectrosc.*, 31, 418-419 (1971), Optimum lasing conditions of a He-Ne laser operating in the TEM_{00} axial mode.

504. Sakurai, T., *Jpn. J. Appl. Phys.*, 11, 1832-1836 (1972), Discharge current dependence of saturation parameter of a He-Ne gas laser.
505. Wang, S. C., Siegman, A. E., *Appl. Phys.*, 2, 143-150 (1973), Hollow-cathode transverse discharge He-Ne and He-Cd^+ lasers.
506. Zborovskii, V. A., Molchanov, M. I., Turkin, A. A., Yaroshenko, N. G., *Opt. Spectrosc.*, 34, 704-705 (1973), Measurement of the natural line width of a travelling-wave He-Ne laser in the 0.63 μm region.
507. Dote, T., Yamaguchi, N., Nakamura, T., *Phys. Lett.*, 45A, 29-30 (1973), Effects of RF electric field on He-Ne laser output.
508. Aleksandrov, I. S., Babeiko, Yu. A., Babaev, A. A., Buzhinskii, O. I. et al., *Sov. J. Quantum Electron.*, 5, 1132-1133 (1976), Stimulated emission from a transverse discharge in copper vapor.
509. Belousova, I. M., Znamemskii, V. B., *J. Appl. Spectrosc.*, 25, 1109-1114 (1976), Properties of the lasing mechanism of a helium-neon mixture excited by a discharge in a hollow cathode.
510. Young, R. T., Jr., *J. Appl. Phys.*, 36, 2324-2325 (1965), Calculation of average electron energies in He-Ne discharges.
511. Crisp, M. D., *Opt. Commun.*, 19, 316-319 (1976), A magnetically polarized He-Ne laser.
512. Ihjima, T., Kuroda, K., Ogura, I., *J. Appl. Phys.*, 48, 437-439 (1977), Radial profiles of upper- and lower-laser-level emission in an oscillating He-Ne laser.
513. Muller, Ya. N., Geller, V. M., Lisitsyna, L. I., Grif, G. I., *Sov. J. Quantum Electron.*, 7, 1013-1015 (1977), Microwave-pumped helium-neon laser.
514. Leontov, V. G., Ostapchenko, E. P., *Opt. Spectrosc.*, 43, 321-325 (1977), Effect of excitation conditions on the radial distribution of population inversion in the active element of a He-Ne laser.
515. Honda, T., Endo, M., *IEEE J. Quantum Electron.*, QE-14, 213-214 (1978), International intercomparison of laser power at 633 nm..
516. Otieno, A. V., *Opt. Commun.*, 26, 207-210 (1978), Homogeneous saturation of the 6328 Â neon laser transition due to collisions in the weak collision model.
517. Ferguson, J. B., Morris, R. H., *Appl. Opt.*, 17, 2924-2929 (1978), Single-mode collapse in 6328 Â He-Ne lasers.
518. Chance, D. A., Chastang, J. -C. A., Crawford, V. S., Horstmann, R. E., L, *IBM J. Res. Develop.*, 23, 108-118 (1979), HeNe parallel plate laser development.
519. Alaev, M. A., Baranov, A. I., Vereshchagin, N. M., Gnedin, I. N. et al., *Sov. J. Quantum Electron.*, 6, 610-611 (1976), Copper vapor laser with a pulse repetition frequency of 100 kHz.
520. Chance, D. A., Brusic, V., Crawford, V. S., Macinnes, R. D., *IBM J. Res. Develop.*, 23, 119-127 (1979), Cathodes for HeNe lasers.
521. Ahearn, W. E., Horstmann, R. E., *IBM J. Res. Develop.*, 23, 128-131 (1979), Nondestructive analysis for HeNe lasers.
522. Schuocker, D., Reif, W., Lagger, H., *IEEE J. Quantum Electron.*, QE-15, 232-239 (1979), Theoretical description of discharge plasma and calculation of maximum output intensity of He-Ne waveguide lasers as a function of discharge tube.
523. Tobias, I., Strouse, W. M., *Appl. Phys. Lett.*, 10, 342-344 (1967), The anomalous appearance of laser oscillation at 6401 Â.
524. Schlie, L. A., Verdeyn, J. T., *IEEE J. Quantum Electron.*, QE-5, 21-29 (1969), Radial profile of Ne $1s_5$ atoms in a He-Ne discharge and their lens effect on lasing at 6401 Â.
525. Sanders, J. H., Thomson, J. E., *J. Phys. B*, 6, 2177-2183 (1973), New high-gain laser transitions in neon.
526. Zitter, R. N., *J. Appl. Phys.*, 35, 3070-3071 (1964), 2s–2p and 3p–2s transitions of neon in a laser ten meters long.
527. Schearer, L. D., *Phys. Lett.*, 27a, 544-545 (1968), Polarization transfer between oriented metastable helium atoms and neon atoms.

528. Chebotayev, V. P., Vasilenko, L. S., *Sov. Phys. JETP*, 21, 515-516 (1965), Investigation of a neon-hydrogen laser at large discharge current.
529. Chebotayev, V. P., *Radio Eng. Electron. Phys.* (USSR), 10, 316-318 (1965), Effect of hydrogen and oxygen on the operation of a neon maser.
530. Anderson, R. S., Bricks, B. G., Karras, T. W., *Appl. Phys. Lett.*, 29, 187-189 (1976), Copper oxide as the metal source in a discharge heated copper vapor laser.
531. McClure, R. M., Pizzo, R., Schiff, M., Zarowin, C. B., *Proc. IEEE,* 52, 851 (1964), Laser oscillation at 1.06 microns in He-Ne.
532. Shtyrkov, E. I., Subbes, E. V., *Opt. Spectrosc.*, 21, 143-144 (1966), Characteristics of pulsed laser action in helium-neon and helium-argon mixtures.
533. Itzkan, I., Pincus, G., *Appl. Opt.*, 5, 349 (1966), 1.0621-μ He-Ne laser.
534. McFarlane, R. A., Patel, C. K. N., Bennett, W. R., Jr., Faust, W. L., *Proc. IEEE,* 50, 2111-2112 (1962), New helium-neon optical maser transitions.
535. Chebotayev, V. P., Pokasov, V. V., *Radio Eng. Electron. Phys.* (USSR), 10, 817-819 (1965), Operation of a laser on a mixture of He-Ne with discharge in a hollow cathode.
536. Petrash, G. G., Knyazev, I. N., *Sov. Phys. JETP*, 18, 571-575 (1964), Study of pulsed laser generation in neon and mixtures of neon and helium.
537. der Agobian, R., Otto, J. L., Cagnard, R., Echard, R., *C. R. Acad. Sci. Ser.* B, 259, 323-326 (1964), Cascades de transitions stimulees dans le neon pur.
538. Andrade, O., Gallardo, M., Bockasten, K., *Appl. Phys. Lett.*, 11, 99-100 (1967), High-gain laser lines in noble gases.
539. Javan, A., Bennett, W. R., Jr., Herriott, D. R., *Phys. Rev. Lett.*, 6, 106-110 (1961), Population inversion and continuous optical maser oscillation in a gas discharge containing a He-Ne mixture.
540. Herriott, D. R., *J. Opt. Soc. Am.*, 52, 31-37 (1962), Optical properties of a continuous helium-neon optical maser.
541. Isaev, A. A., Kazaryan, M. A., Lemmerman, G. Yu., Petrash, G. G. et al., *Sov. J. Quantum Electron.*, 6, 976-977 (1976), Pulse stimulated emission due to transitions in copper atoms excited by discharges in cuprous bromide and chloride vapors.
542. der Agobian, R., Cagnard, R., Echard, R., Otto, J. L., *C. R. Acad. Sci. Ser.* B, 258, 3661-3663 (1964), Nouvelle cascade de transitions stimulee du neon.
543. Bennett, W. R., Jr., Knutson, J. W., Jr., *Proc. IEEE,* 52, 861-862 (1964), Simultaneous laser oscillation on the neon doublet at 1.1523 μ.
544. Patel, C. K. N., *J. Appl. Phys.*, 33, 3194-3195 (1962), Optical power output in He-Ne and pure Ne maser.
545. Chebotayev, V. P., *Radio Eng. Electron. Phys.* (USSR), 10, 314-316 (1965), Operating condition of an optical maser containing a helium-neon mixture.
546. Boot, H. A. H., Clunie, D. M., Thorn, R. S. A., *Nature (London)*, 198, 773-774 (1963), Pulsed laser operation in a high-pressure helium-neon mixture.
547. Smith, J., *J. Appl. Phys.*, 35, 723-724 (1964), Optical maser action in the negative glow region of a cold cathode glow discharge.
548. Cool, T. A., *Appl. Phys. Lett.*, 9, 418-420 (1966), A fluid mixing laser.
549. Mustafin, K. S., Seleznev, V. A., Shtyrkov, E. I., *Opt. Spectrosc.*, 21, 429-430 (1966), Stimulated emission in the negative glow region of a glow discharge.
550. Kuznetsov, A. A., Mash, D. I., Milinkis, B. M., Chirina, L. P., *Radio Eng. Electron. Phys.* (USSR), 9, 1576 (1964), Operating conditions of an optical quantum generator (laser) in helium-neon and xenon-helium gas mixtures.
551. Lisitsyn, V. N., Fedchenko, A. I., Chebotayev, V. O., *Opt. Spectrosc.*, 27, 157-161 (1969), Generation due to upper neon transitions in a He-Ne discharge optically pumped by a helium lamp.
552. Babeiko, Yu. A., Vasil'ev, L. A., Orlov, V. K., Sokolov, A. V. et al., *Sov. J. Quantum Electron.*, 6, 1258–1259 (1976), Stimulated emission from copper vapor in a radial transverse discharge.

553. der Agobian, R., Otto, J. L., Echard, R., Cagnard, R., *C. R. Acad. Sci. Ser.* B, 257, 3844–3847 (1963), Emission stimulee de nouvelles transitions infrarouges du neon.
554. Blau, E. J., Hochheimer, B. F., Massey, J. T., Schulz, A. G., *J. Appl. Phys.*, 34, 703 (1963), Identification of lasing energy levels by spectroscopic techniques.
555. Smith, D. S., Riccius, H. D., *IEEE J. Quantum Electron.*, QE-13, 366 (1977), Observation of new helium-neon laser transitions near 1.49 μm.
556. Gires, F., Mayer, H., Pailette, M., *C. R. Acad. Sci. Ser.* B, 256, 3438–3439 (1963), Sur quelques transitions presentant l'effet laser dans le melange helium-neon.
557. McFarlane, R. A., Faust, W. L., Patel, C. K. N., *Proc. IEEE,* 51, 468 (1963), Oscillation of f–d transitions in neon in a gas optical maser.
558. Lisitsyn, V. N., Chebotayev, V. P., *Opt. Spectrosc.*, 20, 603–604 (1966), The generation of laser action in the 4f–3d transitions of neon by optical pumping of a He-Ne discharge with helium lamp.
559. McFarlane, R. A., Faust, W. L., Patel, C. K. N., Garrett, C. G. B., *Quantum Electronics III*(1964), 573–586, Gas maser operation at wavelengths out to 28 micron, Columbia University Press, New York.
560. Rosenberger, D., *Phys. Lett.*, 9, 29–31 (1964), Oscillation of three 3p–2s transitions in a He-Ne mixtures.
561. Bennett, W. R., Jr., Pawilkowski, A. T., *Bull. Am. Phys. Soc.*, 9, 500 (1964), Additional cascade laser transitions in He-Ne mixtures.
562. Smiley, V. N., *Appl. Phys. Lett.*, 4, 123–124 (1964), New He-Ne and Ne laser lines in the infra-red.
563. Fedorov, A. I., Sergeenko, V. P., Tarasenko, V. F., Sedoi, V. S., *Sov. Phys. J.*, 20, 251–253 (1976), Copper vapor laser with pulse production of the vapor.
564. Faust, W. L., McFarlane, R. A., Patel, C. K. N., Garrett, C. G. B., *Phys. Rev.*, 133, A1476–A1478 (1964), Noble gas optical maser lines at wavelengths between 2 and 35 μ.
565. Otto, J. L., Cagnard, R., Echard, R., der Agobian, R., *C. R. Acad. Sci. Ser.* B, 258, 2779–2780 (1964), Emission stimulee de nouvelles transition infrarouges dan les gas rares.
566. Grudzinski, R., Pailette, M. R., Becrelle, J., *C. R. Acad. Sci. Ser.* B, 258, 1452–1454 (1964), Etude des transitions laser complees dans un melange helium-neon.
567. Bergman, K., Demtroder, W., *Phys. Lett.*, 29a, 94–95 (1969), A new cascade laser transition in He-Ne mixture.
568. Gerritsen, H. J., Goedertier, P. V., *Appl. Phys. Lett.*, 4, 20–21 (1964), A gaseous (He-Ne) cascade laser.
569. Smiley, V. N., *Quantum Electronics III*(1964), 587–591, A long gas phase optical maser cell, Columbia University Press, New York.
570. McMullin, P. G., *Appl. Opt.*, 3, 641–642 (1964), Precise wavelength measurement of infrared optical maser lines.
571. Brunet, H., Laures, P., *Phys. Lett.*, 12, 106–107 (1964), New infrared gas laser transitions by removal of dominance.
572. Bloom, A. L., Bell, W. E., Rempel, R. C., *Appl. Opt.*, 2, 317–318 (1963), Laser action at 3.39 μ in a helium-neon mixtures.
573. Tibilov, A. S., Shukhtin, A. M., *Opt. Spectrosc.*, 21, 69-70 (1966), Laser action with sodium lines.
574. Smilanski, I., Kerman, A., Levin, L. A., Erez, G., *IEEE J. Quantum Electron.*, QE-13, 24-36 (1977), A hollow-cathode copper halide laser.
575. Herceg, J. E., Miley, G. H., *J. Appl. Phys.*, 39, 2147-2149 (1968), A laser utilizing a low-voltage arc discharge in helium-neon.
576. Konovalov, I. P., Popov, A. I., Protsenko, E. D., *Opt. Spectrosc.*, 33, 6-10 (1972), Measurement of spectral characteristics of the $5s'[1/2]^0_1 \rightarrow 4p[3/2]_2$ Ne (3.39 μm) transition.

577. Mazanko, I. P., Ogurok, N. D. D., Sviridov, M. V., *Opt. Spectrosc.*, 35, 327-328 (1973), Measurement of the saturation parameter of a neon-helium mixture at the 3.39-μm wavelength.
578. Watanabe, S., Chihara, M., Ogura, I., *Jpn. J. Appl. Phys.*, 13, 164-169 (1974), Decay rate measurements of upper laser levels in He-Ne and He-Se lasers.
579. Balakin, V. A., Konovalov, I. P., Ocheretyanyi, A. I., Popov, A. I. et al., *Sov. J. Quantum Electron.*, 5, 230-231 (1975), Switching of the emission wavelength of a helium-neon laser in the 3.39 μ region.
580. Balakin, V. A., Konovalov, I. P., Protsenko, E. D., *Sov. J. Quantum Electron.*, 5, 581-583 (1975), Measurement of the spectral characteristics of the 3.3912 μ ($3s_2$-$3p_2$) Ne line.
581. Popov, A. I., Protsenko, E. D., *Sov. J. Quantum Electron.*, 5, 1153-1154 (1976), Laser gain due to $5s'[1/2]^0{}_1 - 4p'[3/2]_2$ transition in neon at $\lambda = 3.39$ μ.
582. Faust, W. L., McFarlane, R. A., Patel, C. K. N., Garrett, C. G. B., *Appl. Phys. Lett.*, 4, 85-88 (1962), Gas maser spectroscopy in the infrared.
583. Patel, C. K. N., McFarlane, R. A., Garrett, C. G. B., *Appl. Phys. Lett.*, 4, 18-19 (1964), Laser action up to 57.355 μ in gaseous discharges (Ne, He-Ne).
584. Liu, G. S., Feldman, D. W., Pack, J. C., Weaver, L. A., *J. Appl. Phys.*, 48, 194-195 (1977), Axial cataphoresis effects in continuously pulsed copper halide lasers.
585. Patel, C. K. N., McFarlane, R. A., Garrett, C. G. B., *Bull. Am. Phys. Soc.*, 9, 65 (1964), Optical-maser action up to 57.355 μm in neon.
586. Patel, C. K. N., Faust, W. L., McFarlane, R. A., Garrett, C. G. B., *Proc. IEEE*, 52, No. 6, 713 (1964), CW optical-maser action up to 133 μ (0.133 mm) in neon discharges.
587. McFarlane, R. A., Faust, W. L., Patel, C. K. N., Garrett, C. G. B., *Proc. IEEE*, 52, 318 (1964), Neon gas maser lines at 68.329 μ and 85.047 μ.
588. Chapovsky, P. L., Lisitsyn, V. N., Sorokin, A. R., *Opt. Commun.*, 26, 33-36 (1976), High-pressure gas lasers on ArI, XeI and KrI transitions.
589. Kochubei, S. A., Lisitsyn, V. N., Sorokin, A. R., Chapovskii, P. L., *Sov. J. Quantum Electron.*, 7, 1142-1144 (1978), High-pressure tunable atomic gas lasers.
590. Bockasten, K., Lundholm, T., Andrade, O., *Phys. Lett.*, 22, 145-146 (1966), New near infrared laser lines in argon 1.
591. Brisbane, A. D., *Nature (London)*, 214, 75 (1967), High gain pulsed laser.
592. Bockasten, K., Andrade, O., *Nature (London)*, 215, 382 (1967), Identification of high gain laser lines in argon.
593. Sutton, D. G., Galvan, L., Valenzuela, P. R., Suchard, S. N., *IEEE J. Quantum Electron.*, QE-11, 54-57 (1975), Atomic laser action in rare gas-SF_6 mixture.
594. Horrigan, F. A., Koozekanani, S. H., Paananen, R. A., *Appl. Phys. Lett.*, 6, 41-43 (1965), Infrared laser action and lifetimes in argon II.
595. Nerheim, N. M., *J. Appl. Phys.*, 48, 1186-1190 (1977), A parametric study of the copper chloride laser.
596. Dauger, A. B., Stafsudd, O. M., *Appl. Opt.*, 10, 2690-2697 (1971), Characteristics of the continuous wave neutral argon laser.
597. Willett, C. S., *Appl. Opt.*, 11, 1429-1431 (1972), Comments on characteristics of the continuous wave neutral argon laser.
598. Dauger, A. B., Stafsudd, O. M., *IEEE J. Quantum Electron.*, QE-8, 912-913 (1972), Line competition in the neutral argon laser.
599. Jalufka, N. W., DeYoung, R. J., Hohl, F., Williams, M. D., *Appl. Phys. Lett.*, 29, 188-190 (1976), Nuclear-pumped ^{3}He-Ar laser excited by the ^{3}He(n,p)^{3}H reaction.
600. Wilson, J. W., DeYoung, R. J., Harries, W. L., *J. Appl. Phys.*, 50, 1226-1234 (1979), Nuclear-pumped ^{3}He-Ar laser modelling.
601. Liberman, S., *C. R. Acad. Sci. Ser.* B, 261, 2601-2604 (1965), Emission stimulee de nouvelles transitions infrarouge de l'argon, du krypton et du xenon.

602. Linford, G. J., *IEEE J. Quantum Electron.*, QE-9, 611-612 (1973), New pulsed and CW laser lines in the heavy noble gases.
603. Brochard, J., Cahuzac, P., Vetter, R., *C. R. Acad. Sci. Ser.* B, 265, 467-470 (1967), Mesure des ecouts isotopiques de six raies laser infrarouges dans l'argon.
604. Linford, G. J., *IEEE J. Quantum Electron.*, QE-8, 477-482 (1972), High-gain neutral laser lines in pulsed noble-gas discharges.
605. Linford, G. J., *IEEE J. Quantum Electron.*, QE-9, 610-611 (1973), New pulsed laser lines in krypton.
606. Anderson, R. S., Bricks, B. G., Karras, T. W., *IEEE J. Quantum Electron.*, QE-13, 115-117 (1977), Steady multiply pulsed discharge-heated copper-vapor laser with copper halide lasant.
607. Walter, W. T., Jarrett, J. M., *Appl. Opt.*, 3, 789-790 (1964), Strong 3.27 μ oscillation in xenon.
608. DeYoung, R. J., Jalufka, N. W., Hohl, F., *Appl. Phys. Lett.*, 30, 19-21 (1977), Nuclear-pumped lasing of ^{3}He-Xe and ^{3}He-Kr.
609. Sinclair, D. C., *J. Opt. Soc. Am.*, 55, 571 (1965), Near-infrared oscillations in pulsed noble-gas-ion lasers.
610. Courville, G. E., Walsh, P. J., Wasko, J. H., *J. Appl. Phys.*, 35, 2547-2548 (1964), Laser action in Xe in two distinct current regions of ac and dc discharges.
611. Clark, P. O., *Phys. Lett.*, 17, 190-192 (1965), Pulsed operation of the neutral xenon laser.
612. Newman, L. A., DeTemple, T. A., *Appl. Phys. Lett.*, 27, 678-680 (1975), High-pressure infrared Ar-Xe laser system: ionizer-sustainer mode of excitation.
613. Lawton, S. A., Richards, J. B., Newman, L. A., Specht, L., DeTemple, T., *J. Appl. Phys.*, 50, 3888-3898 (1979), The high-pressure neutral infrared xenon laser.
614. Patel, C. K. N., Faust, W. L., McFarlane, R. A., *Appl. Phys. Lett.*, 1, 84-85 (1962), High gain gaseous (Xe-He) optical maser.
615. Tang, C. L., *Proc. IEEE,* 219-220 (1963), Relative probabilities for the xenon laser transitions.
616. Fork, R. L., Patel, C. K. N., *Appl. Phys. Lett.*, 2, 180-181 (1963), Broadband magnetic field tuning of optical masers.
617. Vetter, A. A., Nerheim, N. M., *Appl. Phys. Lett.*, 30, 405-407 (1977), Addition of HCl to the double-pulse copper chloride laser.
618. Aisenberg, S., *Appl. Phys. Lett.*, 2, 187-189 (1963), The effect of helium on electron temperature and electron density in rare gas lasers.
619. Patel, C. K. N., *Phys. Rev.*, 131, 1582-1584 (1963), Determination of atomic temperature and Doppler broadening in a gaseous discharge with population inversion.
620. Patel, C. K. N., McFarlane, R. A., Faust, W. L., *Quantum Electronics III*(1964), Columbia University Press, New York, p. 507-514, High gain medium for gaseous optical masers.
621. Faust, W. L., McFarlane, R. A., *J. Appl. Phys.*, 35, 2010-2015 (1964), Line strengths for noble-gas maser transitions; calculations of gain/inversion at various wavelengths.
622. Smiley, V. N., Lewis, A. L., Forbes, D. K., *J. Opt. Soc. Am.*, 55, 1552-1553 (1965), Gain and bandwidth narrowing in a regenerative He-Xe laser amplifier.
623. Kuznetsov, A. A., Mash, D. I., *Radio Eng. Electron. Phys. (USSR)*, 10, 319-320 (1965), Operating conditions of an optical maser with a helium-xenon mixture in the middle infrared region of the spectrum.
624. Moskalenko, V. F., Ostapchenko, E. P., Pugnin, V. I., *Opt. Spectrosc.*, 23, 94-95 (1967), Mechanism of xenon-level population inversion in the positive column of a helium-xenon mixture.
625. Schwarz, S. E., DeTemple, T. A., Targ, R., *Appl. Phys. Lett.*, 17, 305-306 (1970), High-pressure pulsed xenon laser.

626. Shafer, J. H., *Phys. Rev.*, A3, 752-757 (1971), Optical heterodyne measurement of xenon isotope shifts.
627. Targ, R., Sasnett, M. W., *IEEE J. Quantum Electron.*, QE-8, 166-169 (1972), High-repetition-rate xenon laser with transverse excitation.
628. Abrosimov, G. V., Vasil'tsov, V. V., *Sov. J. Quantum Electron.*, 7, 512-513 (1977), Stimulated emission due to transitions in copper atoms formed in transverse discharge in copper halide vapors.
629. Dandawate, V. D., Thomas, G. C., Zembrod, A., *IEEE J. Quantum Electron.*, QE-8, 918-919 (1972), Time behavior of a TEA xenon laser.
630. Fahlen, T. S., Targ, R., *IEEE J. Quantum Electron.*, QE-9, 609 (1973), High-average-power xenon laser.
631. Mansfield, C. R., Bird, P. F., Davis, J. F., Wimett, T. F., Helmick, H., *Appl. Phys. Lett.*, 30, 640-641 (1977), Direct nuclear pumping of a ^{3}He-Xe laser.
632. Liberman, L., *C. R. Acad. Sci. Ser.* B, 266, 236-239 (1968), Sur la structure hyperfine de quelques raies laser infrarouges de xenon 129.
633. Armstrong, D. R., *IEEE J. Quantum Electron.*, QE-4, 968-969 (1968), A method for the control of gas pressure in the xenon laser.
634. Culshaw, W., Kannelaud, J., *Phys. Rev.*, 156, 308-319 (1967), Mode interaction in a Zeeman laser.
635. Paananen, R. A., Bobroff, D. L., *Appl. Phys. Lett.*, 2, 99-100 (1963), Very high gain gaseous (Xe-He) optical maser at 3.5 μ.
636. Bridges, W. B., *Appl. Phys. Lett.*, 3, 45-47 (1963), High optical gain at 3.5 μ in pure xenon.
637. Markin, E. P., Nikitin, V. V., *Opt. Spectrosc.*, 17, 519 (1964), The 3.5 μ Xe-Ne laser.
638. Clark, P. O., *IEEE J. Quantum Electron.*, QE-1, 109-113 (1965), Investigation of the operating characteristics of the 3.5 μ xenon laser.
639. Gabay, S., Smilanski, I., Levin, L. A., Erez, G., *IEEE J. Quantum Electron.*, QE-13, 364-366 (1977), Comparison of CuCl, CuBr, and CuI as lasants for copper-vapor lasers.
640. Kluver, J. W., *J. Appl. Phys.*, 37, 2987-2999 (1966), Laser amplifier noise at 3.5 microns in helium-xenon.
641. Freiberg, R. J., Weaver, L. A., *J. Appl. Phys.*, 38, 250-262 (1967), Effects of lasering upon the electron gas and excited-state populations in xenon discharges.
642. Aleksandrov, E. B., Kulyasov, V. N., *Sov. Phys. JETP*, 28, 396-400 (1969), Determination of the elementary-emitter spectrum latent in an inhomogeneously broadened spectral line.
643. Fork, R. L., Dienes, A., Kluver, J. W., *IEEE J. Quantum Electron.*, QE-5, 607-616 (1969), Effects of combined RF and optical fields on a laser medium.
644. Wang, S. C., Byer, R. L., Siegman, A. E., *Appl. Phys. Lett.*, 17, 120-122 (1970), Observation of an enhanced Lamb dip with a pure Xe gain cell inside a 3.51 μ He-Xe laser.
645. Kasuya, T., *Appl. Phys.*, 2, 339-343 (1973), Broad band frequency tuning of a He-Xe laser with a superconducting solenoid.
646. Linford, G. J., Peressini, E. R., Sooy, W. R., Spaeth, M. L., *Appl. Opt.*, 13, 379-390 (1974), Very long lasers.
647. Aleksandrov, E. B., Kulyasov, V. N., Kharnang, K., *Opt. Spectrosc.*, 38, 439-440 (1975), Tunable single-frequency xenon laser operating at the two infrared transitions, $\lambda = 5.57$ μm and 3.51 μm.
648. Wolff, P. A., Abraham, N. B., Smith, S. R., *IEEE J. Quantum Electron.*, QE-13, 400-403 (1977), Measurement of radial variation of 3.51 μm gain in xenon discharge tubes.
649. Vetter, R., *Phys. Lett.*, 42A, 231-232 (1972), Ecarts isotopiques dans la transition laser a $\lambda = 3.99$ μm du xenon, Accroissement de l'effet de volume pour les couples.

650. Fahlen, T. S., *IEEE J. Quantum Electron.*, QE-13, 546-547 (1977), High pulse rate, mode-locked copper vapor laser.
651. Olson, R. A., Bletzinger, P., Garscadden, A., *IEEE J. Quantum Electron.*, QE-12, 316-317 (1976), New pulsed Xe-neutral laser line.
652. Petrov, Yu. N., Prokhorov, A. M., *Sov. Phys. Tech. Phys.*, 1, 24-25 (1965), 75-micron laser.
653. Cheo, P. K., Cooper, H. G., *J. Appl. Phys.*, 36, 1862-1865 (1965), Ultraviolet ion laser transitions between 2300 Å and 4000 Å.
654. Dreyfus, R. W., Hodgson, R. T., *IEEE J. Quantum Electron.*, QE-8, 537-538 (1972), Electron-beam gas laser excitation.
655. Dana, L., Laures, P., Rocherolles, R., *C. R.*, 260, 481-484 (1965), Raies laser ultraviolettes dans le neon, l'argon et le xenon.
656. Cottrell, T. H. E., Sinclair, D. C., Forsyth, J. M., *IEEE J. Quantum Electron.*, QE-2, 703 (1966), New laser wavelengths in krypton.
657. Akitt, D. P., Wittig, C. F., *J. Appl. Phys.*, 40, 902-903 (1969), Laser emission in ammonia.
658. Turner, R., Murphy, R. A., *Infrared Phys.* 16, 197(1976)The far infrared helium laser.
659. Willett, C. S., *Handbook of Lasers*, R. Pressley, Ed., CRC Press, Boca Raton, FL (1971), Neutral gas lasers.
660. Willett, C. S., *Progress in Quantum Electronics*, Vol. 1, Pergamon Press, New York (1971), Laser lines in atomic species.
661. Isaev, A. A., Lemmerman, G. Yu., *Sov. J. Quantum Electron.*, 7, 799-801 (1977), Investigation of a copper vapor pulsed laser at elevated powers.
662. Tobin, M. S., *IEEE Quantum Electron.* QE-20, 825 (1984); p.985., SMMW laser emission and frequency measurements in doubly deuterated methyl fluoride (CHD_2F)O.
663. Facin, J. A., Pereira, D., Vasconcellos, E. C. C., Scalabrin, A. and Ferrari, C. A., *Appl. Phys. B*, 48, 245 (1989), New FIR laser lines from CHD_2OH optically pumped by a cw CO_2 laser.
664. McFarlane, R. A., *Appl. Phys. Lett.*, 5, 91-93 (1964), Laser oscillation on visible and ultraviolet transitions of singly and multiply ionized oxygen, carbon and nitrogen.
665. Bennett, A. S. and Herman, H., *IEEE Quantum Electron.*, QE-18, 323 (1982), Optically pumped far-infrared emission in the cis 1,2-difluoroethene laser.
666. Beaulieu, A. J., *Appl. Phys. Lett.*, 16, 504-505 (1970), Transversely excited atmospheric pressure CO_2 laser.
667. Dumanchin, R., Michon, M., Farcy, J. C., Boudinet, G., Rocca-Serra, J., *IEEE J. Quantum Electron.*, QE-8, 163-165 (1972), Extension of TEA CO_2 laser capabilities.
668. Petrov, Yu. N. and Prokhorov, A. M., *JETP Lett.*, 1, 24 (1965), 75- micron laser.
669. Batenin, V. M., Burmakin, V. A., Vokhmin, P. A., Evtyunin, A. I., Klimov, *Sov. J. Quantum Electron.*, 7, 891-893 (1977), Time dependence of the electron density in a copper vapor laser.
670. Lamberton, H. M., Pearson, P. R., *Electron. Lett.*, 7, 141-142 (1971), Improved excitation techniques for atmospheric pressure CO_2 lasers.
671. Pearson, P. R., Lamberton, H. M., *IEEE J. Quantum Electron.*, QE-8, 145-149 (1972), Atmospheric pressure CO_2 lasers giving high output energy per unit volume.
672. Richardson, M. C., Alcock, A. J., Leopold, K., Burtyn, P., *IEEE J. Quantum Electron.*, QE-9, 236-243 (1973), A 300-J multigigawatt CO_2 laser.
673. Reid, R. D., McNeil, J. R., Collins, G. J., *Appl. Phys. Lett.*, 29, 666-668 (1976), New ion laser transitions in He-Au mixtures.

674. Borodin, V. S., Kagan, Yu. M., *Sov. Phys. Tech. Phys.*, 11, 131-134 (1966), Investigations of hollow-cathode discharges. I. Comparison of the electrical characteristics of a hollow cathode and a positive column.

675. Hodgson, R. T., Dreyfus, R. W., *Phys. Rev. Lett.*, 28, 536-539 (1972), Vacuum-uv laser action observed in H_2 Werner Bands: 1161-1240 Â.

676. Silfvast, W. T., *Appl. Phys. Lett.*, 15, 23-25 (1969), Efficient CW laser oscillation at 4416 Â in Cd(II).

677. Goldsborough, J. P., *Appl. Phys. Lett.*, 15, 159-161 (1969), Stable long life CW excitation of helium-cadmium lasers by dc cataphoresis.

678. Goldsborough, J. P., *IEEE J. Quantum Electron.*, QE-5, 133 (1969), Continuous laser oscillation at 3250 Â in cadmium ion.

679. Pogorelyi, P. A., Tibilov, A. S., *Opt. Spectrosc.*, 25, 301-305 (1968), On the mechanism of laser action in Na-H_2 mixtures.

680. Bokhan, P. A., Solomonov, V. I., Shcheglov, V. B., *Sov. J. Quantum Electron.*, 7, 1032-1033 (1977), Investigation of the energy characteristics of a copper vapor laser with a longitudinal discharge.

681. Davies, P. B. and Jones, H., *Appl. Phys.*, 22, 53 (1980), New cw far infrared molecular lasers from ClO_2, HCCF, FCN, CH_3NC, CH_3F and propynal.

682. Telle, J., *IEEE Quantum Electron.* QE-19, 1469 (1983), Continuous wave 16 μm CF_4 laser.

683. Dangoisse, E. J. and Glorieux, P., *J. Mol. Spec.* 92, 283 (1982), Optically pumped continuous wave submillimeter emission from $H^{13}COOH$: measurements and assignments.

684. Jones, H., Davies, P. B. and Lewis-Bevan, W., *Appl. Phys.* 30, 1 (1983), New FIR laser lines from optically pumped DCOF.

685. Andriakin, V. M., Velikhov, E. P., Golubev, S. A., Krasil'nikov, S. S., *JETP Lett.*, 8, 214 (1968), Increase of CO_2 laser power under the influence of a beam of fast protons.

686. Gancey, T., Verdeyen, J. T., Miley, G. H., *Appl. Phys. Lett.*, 18, 568-569 (1971), Enhancement of CO_2 laser power and efficiency by neutron irradiation.

687. Rhoads, H. S., Schneider, R. T., *Trans. Am. Nucl. Soc.*, 14, 429 (1971), Nuclear enhancement of CO_2 laser output.

688. DeYoung, R. J., Wells, W. E., Miley, G. H., *Trans. Am. Nucl. Soc.*, 19, 66 (1974), Enhancement of He-Ne lasers by nuclear radiation.

689. McArthur, D. A., Tollefsrud, P. B., *Appl. Phys. Lett.*, 26, 187-190 (1975), Observation of laser action in CO gas excited only by fission fragments.

690. Helmick, H. K., Fuller, J. L., Schneider, R. T., *Appl. Phys. Lett.*, 26, 327-328 (1975), Direct nuclear pumping of a helium-xenon laser.

691. Warner, B. E., Gerstenberger, D. C., Reid, R. D., McNeil, J. R. et al. *IEEE J. Quantum Electron.*, QE-14, 568-570 (1978), 1 W operation of singly ionized silver and copper lasers.

692. Dorgela, H. B., Alting, H., Boers, J., *Physica*, 2, 959-967 (1935), Electron temperature in the positive column in mixtures of neon and argon or mercury.

693. Chen, C. H., Haberland, H., Lee, Y. T., *J. Chem. Phys.*, 61, 3095-3103 (1974), Interaction potential and reaction dynamics of He(2^1S, 2^3S) + Ne, Ar by the crossed molecular beam method.

694. Leasure, E. L., Mueller, C. R., *J. Appl. Phys.*, 47, 1062-1064 (1976), Crossed-molecular beams investigation of the excitation of ground-state neon atoms by 4.6-eV helium metastables.

695. Brion, C. E., Olsen, L. A. R., *J. Phys. B*, 3, 1020-1033 (1970), Threshold electron impact excitation of the rare gases.

696. Holstein, T., *Phys. Rev.*, 72, 1212-1233 (1947), Imprisonment of resonance radiation in gases.

697. Holstein, T., *Phys. Rev.*, 83, 1159-1168 (1951), Imprisonment of resonance radiation in gases. II.
698. Donohue, T., Wiesenfeld, J. R., *Chem. Phys. Lett.*, 33, 176-180 (1975), Relative yields of electronically excited iodine atoms, I $5^2P_{1/2}$, in the photolysis of alkyl iodides.
699. Donohue, T., Wiesenfeld, J. R., *J. Chem. Phys.*, 63, 3130-3135 (1975), Photodissociation of alkyl iodides.
700. Ershov, L. S., Zalesskii, V. Yu., Sokolov, V. N., *Sov. J. Quantum Electron.*, 8, 494-501 (1978), Laser photolysis of perfluoroalkyl iodides.
701. Derwent, R. G., Thrush, B. A., *Faraday Disc. Chem. Soc.*, 53, 16-167 (1972), Excitation of iodine by singlet molecular oxygen.
702. Pirkle, R. J., Wiesenfeld, J. R., Davis, C. C., Wolga, G. J., et al., *IEEE J. Quantum Electron.*, QE-11, 834-838 (1975), Production of electronically excited iodine atoms, I ($^2P_{1/2}$) following injection of HI into a flow of discharged oxygen.
703. Beaty, E. C., Patterson, P. L., *Phys. Rev.*, A137, 346-357 (1965), Mobilities and reaction rates of ions in helium.
704. Deese, J. E., Hassan, H. A., *AIAA J.*, 14, 1589-1597 (1976), Analysis of nuclear induced plasmas.
705. DeYoung, R. J., Winters, P. A., *J. Appl. Phys.*, 48, 3600-3602 (1977), Power deposition in He from the volumetric $^3He(n,p)^3H$ reaction.
706. Wilson, J. W., DeYoung, R. J., *J. Appl. Phys.*, 49, 980-988 (1978), Power density in direct nuclear-pumped ^3He lasers.
707. Carter, B. D., Rowe, M. J., Schneider, R. T., *Appl. Phys. Lett.*, 36, 115-117 (1980), Nuclear-pumped CW lasing of the ^3He-Ne system.
708. Falcone, R. W., Zdasiuk, G. A., *Opt. Lett.*, 5, 155-157 (1980), Pair-absorption-pumped barium laser.
709. Ramsay, J. V., Tanaka, K., *Jpn. J. Appl. Phys.*, 5, 918-923 (1966), Construction of single-mode dc operated He/Ne lasers.
710. Sakurai, T., *Jpn. J. Appl. Phys.*, 11, 1826-1831 (1972), Dependence of He-Ne laser output power on discharge current, gas pressure and tube radius.
711. Brunet, H., *Appl. Opt.*, 4, 1354 (1965), Laser gain measurements in a xenon-krypton discharge.
712. Mash, D. I., Papulovskii, V. F., Chirina, L. P., *Opt. Spectrosc.*, 17, 431-432 (1964), On the operation of a xenon-krypton laser.
713. Allen, L., Jones, D. G. C., Schofield, D. G., *J. Opt. Soc. Am.*, 59, 842-847 (1969), Radiative lifetimes and collisional cross sections for XeI and II.
714. Allen, L., Peters, G. I., *Phys. Lett.*, 31A, 95-96 (1970), Superradiance, coherence brightening and amplified spontaneous emission.
715. Shukhtin, A. M., Mishakov, V. G., Fedotov, G. A., *Sov. Tech. Phys. Lett.*, 3, 304-305 (1977), Production of Cu vapor from Cu_2O dust in a pulsed discharge.
716. Peters, G. I., Allen, L., *J. Phys. A*, 4, 238-243 (1971), Amplified spontaneous emission. I. The threshold condition.
717. Allen, L., Peters, G. I., *J. Phys. A*, 4, 377-381 (1971), Amplified spontaneous emission. II. The connection with laser theory.
718. Allen, L., Peters, G. I., *J. Phys. A*, 4, 564-573 (1971), Amplified spontaneous emission. III. Intensity and saturation.
719. Peters, G. I., Allen, L., *J. Phys. A*, 5, 546-554 (1972), Amplified spontaneous emission. IV. Beam divergence and spatial coherence.
720. Shelton, R. A. J., *Trans. Faraday. Soc.*, 57, 2113-2118 (1961), Vapour pressures of the solid copper (I) halides.
721. Tobin, R. C., *Opt. Commun.*, 32, 325-330 (1980), Rapid differential decay of metastable populations in a copper halide laser.
722. Rothe, D. E., Tan, K. O., *Appl. Phys. Lett.*, 30, 152-154 (1977), High power N_2^+ laser pumped by change transfer in a high-pressure pulsed glow discharge.

723. Olson, R. A., Grosjean, D., Sarka, B., Jr., Garscadden, A. et al., *Rev. Sci. Instrum.*, 47, 677-683 (1976), High-repetition-rate closed-cycle rare gas electrical discharge laser.

724. Hasle, E. K., *Opt. Commun.*, 31, 206-210 (1979), Polarization properties of He-Ne lasers.

725. Nerheim, N. M., *J. Appl. Phys.*, 48, 3244-3250 (1977), Measurement of copper ground-state and metastable level population densities in copper-chloride laser.

726. Majer, J. R., Simons, J. P., *Adv. Photochem.*, 2, 137-182 (1964), Photochemical process in halogenated compounds.

727. Riley, S. J., Wilson, K. R., *Faraday Disc. Chem. Soc.*, 53, 132-146 (1972), Excited fragments from excited molecules: energy partitioning in the photodissociation of alkyl iodides.

728. Harris, G. M., Willard, J. E., *J. Am. Chem. Soc.*, 76, 4678-4687 (1954), Photochemical reactions in the system methyl iodide-iodine-methane: the reaction, $C^{14}H_3 + CH_4 - C^{14}H_4 + CH_3$.

729. Jaseja, T. S., Javan, A., Townes, C. H., *Phys. Rev. Lett.*, 10, 165-167 (1963), Frequency stability of He-Ne masers and measurements of length.

730. Jaseja, T. S., Javan, A., Murray, J., Townes, C. H., *Phys. Rev.* A, 133, 1221-1225 (1964), Test of special relativity or of the isotropy of space by use of infrared masers.

731. McFarlane, R. A., Bennett, W. R., Jr., Lamb, W. E., Jr., *Appl. Phys. Lett.*, 2, 189-190 (1963), Single mode tuning dip in the power output of an He-Ne optical maser.

732. Tobias, I., Skolnick, M., Wallace, R. A., Polanyi, T. G., *Appl. Phys. Lett.*, 6, 198-201 (1965), Deviation of a frequency-sensitive signal from a gas laser in an axial magnetic field.

733. Andrews, A. J., Webb, C. E., Tobin, R. C., Denning, R. G., *Opt. Commun.*, 22, 272-274 (1977), A copper vapor laser operating at room temperature.

734. Wallard, A. J., *J. Phys. E*, 6, 793-807 (1973), The frequency stabilization of gas lasers.

735. Shevirev, A. S., Dyubko, S. F., Efimenko, M. N. and Fesenko, L. D., *Zh. Prikl. Specktosk.* 42, 480 (1985).

736. White, A. D., *IEEE J. Quantum Electron.*, QE-1, 349-357 (1965), Frequency stabilization of gas lasers.

737. Hochuli, U. E., Haldemann, P., Li, H. A., *Rev. Sci. Instrum.*, 45, 1378-1381 (1974), Factors influencing the relative frequency stability of He-Ne laser structures.

738. Hanes, G. R., Dahlstrom, C. E., *Appl. Phys. Lett.*, 14, 362-364 (1969), Iodine hyperfine structure observed in saturated absorption at 633 nm.

739. VanOorschot, B. D. J., VanderHoeven, C. J., *J. Phys. E*, 12, 51-55 (1979), A recently developed iodine-stabilized laser.

740. Schweitzer, W. G., Jr., Kessler, E. G, Jr., Deslattes, R. D., Layer, H. P., *Appl. Opt.*, 12, 2927-2938 (1973), Description, performance and wavelengths of iodine stabilized lasers.

741. Cerez, P., Brillet, A., Hartmann, F., *IEEE Trans. Inst. Meth.*, IM-23, 526-528 (1974), Metrological properties of the R(127) line of iodine studied by laser saturated absorption.

742. Fedorov, A. J., Sergeenko, V. P., Tarasenko, V. F., *Sov. J. Quantum Electron.*, 7, 1166-1167 (1977), Apparatus for investigating stimulated emission from explosively formed metal vapors.

743. Helmcke, J., Bayer-Helms, F., *IEEE Trans. Inst. Meth.*, IM-23, 529-531 (1974), He-Ne laser stabilized by saturated absorption in I_2.

744. Wallard, A. J., *IEEE Trans. Inst. Meth.*, IM-23, 52-535 (1974), The reproducibility of 633 nm lasers stabilized by $^{127}I_2$.

745. Cole, J. B., Bruce, C. F., *Appl. Opt.*, 14, 1303-1310 (1975), Iodine stabilized laser with three internal mirrors.
746. Cerez, P., Brillet, A., Hajdukovic, S., Man, N., *Opt. Commun.*, 21, 332-336 (1977), Iodine stabilized He-Ne laser with a hot wall iodine cell.
747. Melnikov, N. A., Privalov, V. E., Fofanov, Ya. A., *Opt. Spectrosc.*, 42, 425-428 (1977), Experimental investigation of He-Ne lasers stabilized by saturation absorption in iodine.
748. Layer, H. P., *Proc. Soc. Photo Opt. Instrum. Eng.*, 129, 9-11 (1977), The iodine stabilized laser as a realization of the length unit.
749. Chartier, J. M., Helmcke, J., Wallard, A. J., *IEEE Trans. Inst. Meth.*, IM-25, 450-453 (1976), International intercomparison of the wavelength of iodine-stabilized lasers.
750. Bagaev, S. N., Baklanov, E. V., Chebotaev, V. O., *JETP Lett.*, 16, 243-246 (1972), Anomalous decrease of the shift of the center of the Lamb dip in low-pressure, molecular gases.
751. Koshelyaevskii, N. B., Tatarenkov, V. M., Titov, A. N., *Sov. J. Quantum Electron.*, 6, 222-226 (1976), Power shift of the frequency of an He-Ne-CH_4 laser.
752. Liu, C. S., Feldman, D. W., Pack, J. L., Weaver, L. A., *IEEE J. Quantum Electron.*, QE-13, 744-751 (1977), Kinetic Processes in Continuously Pulsed Copper Halide Lasers.
753. Jolliffe, B. W., Kramer, G., Chartier, J. M., *IEEE Trans. Inst. Meth.*, IM-25, 447-450 (1976), Methane-stabilized He-Ne laser intercomparisons 1976.
754. Alekseev, V. A., Malyugin, A. V., *Sov. J. Quantum Electron.*, 7, 1075-1081 (1977), Influence of the hyperfine structure on the frequency reproducibility of an He-Ne laser with a methane absorption cell.
755. Nakazawa, M., Musha, T., Tako, T., *J. Appl. Phys.*, 50, 2544-2547 (1979), Frequency-stabilized 3.39 μm He-Ne laser with no frequency modulation.
756. Evenson, K. M., Day, G. W., Wells, J. S., Mullen, L. O., *Appl. Phys. Lett.*, 20, 133-134 (1972), Extension of absolute frequency measurements to the CW He-Ne laser at 88 THz (3.39 μ).
757. Evenson, K. M., Wells, J. S., Petersen, F. R., Danielson, B. L., Day, G., *Appl. Phys. Lett.*, 22, 192-195 (1973), Absolute frequencies of molecular transitions used in laser stabilization: the 3.39 μm transition in CH_4 and the 9.33- and 10.18-μm.
758. Barger, R. L., Hall, J. L., *Appl. Phys. Lett.*, 22, 196-199 (1973), Wavelength of the 3.39 μm laser-saturated absorption line of methane.
759. Jennings, D. A., Petersen, F. R., Evenson, K. M., *Opt. Lett.*, 4, 129-130 (1979), Frequency measurements of the 260-THz (1.15 μm) He-Ne laser.
760. Baird, K. M., Evenson, K. M., Hanes, G. R., Jennings, D. A., Peterson, F., *Opt. Lett.*, 4, 263-264 (1979), Extension of absolute-frequency measurements to the visible: frequencies of ten, hyperfine components of iodine.
761. Jennings, D. A., Petersen, F. R., Evenson, K. M., *Appl. Phys. Lett.*, 510-511 (1975), Extension of absolute frequency measurements to 148THz: frequencies of the 2.0- and, 3.5-μ Xe laser.
762. Gokay, M. C., Jenkins, R. S., Cross, L. A., *J. Appl. Phys.*, 48, 4395-4396 (1977), Output characteristics of the CuCl double-pulse laser at small pumping pulse delays.
763. Baird, K. M., Smith, D. S., Whitford, B. G., *Opt. Commun.*, 31, 367-368 (1979), Confirmation of the currently accepted value 299792458 metres per second for the speed of light.
764. Evenson, K. M., Wells, J. S., Petersen, F. R., Danielson, B. L., Day, G., *Phys. Rev. Lett.*, 29, 1346-1349 (1972), Speed of light from direct frequency and wavelength measurements of the methane-stabilized laser.
765. Knight, T. G., Rowley, W. R. C., Shotton, K. C., Woods, P. T., *Nature (London)*, 251, 46 (1974), Measurement of the speed of light.

766. Blaney, T. G., Bradley, C. C., Edwards, G. J., Knight, D. J. E., Woods, P. T., *Nature (London)*, 244, 504 (1973), Absolute frequency measurement of the R(12) transition of CO_2 at 9.3 μm.

767. Jolliffe, B. W., Rowley, W. R. C., Shotton, K. C., Wallard, A. J., Woods, P. T., *Nature (London)*, 251, 46-47 (1974), Accurate wavelength measurement on up-converted CO_2 laser radiation.

768. Woods, P. T., Shotton, K. C., Rowley, W. R. C., *Appl. Opt.*, 17, 1048-1054 (1978), Frequency determination of visible laser light by interferometric comparison.

769. Markova, S. V., Petrash, G. G., Cherezov, V. M., *Sov. J. Quantum Electron.*, 9, 707-711 (1979), Investigation of the stimulated emission mechanism in a pulsed bismuth vapor laser.

770. Fill, E. E., Thieme, W. H., Volk, R., *J. Phys. D.*, 12, L41-L45 (1979), A tunable iodine laser.

771. Silfvast, W. T., Szeto, L. H., Wood, O. R., II, *Appl. Phys. Lett.*, 31, 334-337 (1977), Recombination lasers in expanding CO_2 laser-produced plasmas of argon, krypton, and xenon.

772. Tibilov, A. S., Shukhtin, A. M., *Opt. Spectrosc.*, 25, 221-224 (1968), Investigation of generation of radiation in the Na-H_2 mixture.

773. Vetter, A. A., *IEEE J. Quantum Electron.*, QE-13, 889-891 (1977), Quantitative effect of initial current rise on pumping the double-pulsed copper chloride.

774. Silfvast, W. T., Szeto, L. H., Wood, O. R., II, *Appl. Phys. Lett.*, 34, 213-215 (1979), Ultra-high-gain laser-produced plasma laser in xenon using periodic pumping.

775. Silfvast, W. T., Szeto, L. H., Wood, O. R., II, *Appl. Phys. Lett.*, 36, 500-502 (1980), Power output enhancement of a laser-produced Cd plasma recombination laser by plasma confinement.

776. Kneipp, H., Rentsch, M., *Sov. J. Quantum Electron.*, 7, 1454-1455 (1977), Discharge-heated copper vapor laser.

777. Vetter, A. A., Nerheim, N. M., *IEEE J. Quantum Electron.*, QE-14, 73-74 (1978), Effect of dissociation pulse circuit inductance on the CuCl laser.

778. Cross, L. A., Jenkins, R. S., Gokay, M. C., *J. Appl. Phys.*, 49, 453-454 (1978), The effects of a weak axial magnetic field on the total energy output of the CuCl, double-pulse laser.

779. Nerheim, N. M., Vetter, A. A., Russell, G. R., *J. Appl. Phys.*, 49, 12-15 (1978), Scaling a double-pulsed copper chloride laser to 10 mJ.

780. Gordon, E. B., Egorov, V. G., Pavcenko, V. S., *Sov. J. Quantum Electron.*, 8, 266-268 (1978), Excitation of metal vapor lasers by pulse trains.

781. Bokhan, P. A., Shcheglov, V. B., *Sov. J. Quantum Electron.*, 8, 219-222 (1978), Investigation of a transversely excited pulsed copper vapor laser.

782. Zemskov, K. I., Kazaryan, M. A., Mokerov, V. G., Petrash, G. G. et al., *Sov. J. Quantum Electron.*, 8, 245-247 (1978), Coherent properties of a copper vapor laser and dynamic holograms in vanadium dioxide films.

783. Burmakin, V. A., Evtyunin, A.N., Lesnoi, M. A., Bylkin, V.I., *Sov. J. Quantum Electron.*, 8, 574-576 (1978), Long-life sealed copper vaper laser.

784. Gridnev, A. G., Gorblinova, T. M., Elaev, V. F., Evtushenko, G. S. et al., *Sov. J. Quantum Electron.*, 8, 656-658 (1978), Spectroscopic investigation of a gas discharge plasma of a Cu + Ne laser.

785. Mishakov, V. G., Tibilov, A. S., Shukhtin, A. M., *Opt. Spectrosc.*, 31, 176-177 (1971), Laser action in Na-H_2 and K-H_2 mixtures with pulsed injection of metal vapor into a gas-discharge plasma.

786. Tenenbaum, J., Smilanski, I., Gabay, S., Erez, G., Levin, L. A., *J. Appl. Phys.*, 49, 2662-2665 (1978), Time dependence of copper-atom concentration in ground and metastable states in a pulsed CuCl laser.

787. Piper, J. A., *IEEE J. Quantum Electron.*, QE-14, 405-407 (1978), A transversely excited copper halide laser with large active volume.
788. Chen, C. J., Bhanji, A. M., Russell, G. R., *Appl. Phys. Lett.*, 33, 146-148 (1978), Long duration high-efficiency operation of a continuously pulsed copper laser utilizing copper bromide as a lasant.
789. Gokay, M. C., Soltanoalkotabi, M., Cross, L. A., *J. Appl. Phys.*, 49, 4357-4358 (1978), Copper acetylacetonate as a source in the 5106 - A neutral copper laser.
790. Nerheim, N. M., Bhanji, A. M., Russell, G. R., *IEEE J. Quantum Electron.*, QE-14, 686-693 (1978), A continuously pulsed copper halide laser with a cable-capacitor Blumlein discharge circuit.
791. Babeiko, Yu. A., Vasil'ev, L. A., Sokolov, A. V., Sviridov, A. V., et al., *Sov. J. Quantum Electron.*, 8, 1153-1154 (1978), Coaxial copper-vapor laser with a buffer gas at above atmospheric pressure.
792. Kushner, M. J., Culick, F. E. C., *Appl. Phys. Lett.*, 33, 728-731 (1978), Extrema of electron density and output pulse energy in a CuCl/Ne discharge and a Cu/CuCl double-pulsed laser.
793. Bokhan, P. A., Gerasimov, V. A., Solomonov, V. I., Shcheglov, V. B., *Sov. J. Quantum Electron.*, 8, 1220-1227 (1978), Stimulated emission mechanism of a copper vapor laser.
794. Kazaryan, M. A., Trofimov, A. N., *Sov. J. Quantum Electron.*, 8, 1390-1391 (1978), Gas-discharge tubes for metal halide vapor lasers.
795. Bokhan, P. A., Gerasimov, V. A., *Sov. J. Quantum Electron.*, 9, 273-275 (1979), Optimization of the excitation conditions in a copper vapor laser.
796. Sorokin, P. P., Lankard, J. R., *J. Chem. Phys.*, 54, 2184-2190 (1971), Infrared lasers resulting from giant-pulse laser excitation of alkali metal molecules.
797. Smilanski, I., Erez, G., Kerman, A., Levin, L. A., *Opt. Commun.*, 30, 70-74 (1979), High-power, high-pressure, discharge-heated copper vapor laser.
798. Kushner, M. J., Culick, F. E. C., *IEEE J. Quantum Electron.*, QE-15, 835-837 (1979), A continuous discharge improves the performance of the Cu/CuCl double pulse laser.
799. Tennenbaum, J., Smilanski, I., Gabay, S., Levin, L. A., Erez, G., *J. Appl. Phys.*, 50, 57-61 (1979), Laser power variation and time dependence of populations in a burst-mode CuBr laser.
800. Miller, J. L., Kan, T., *J. Appl. Phys.*, 50, 3849-3851 (1979), Metastable decay rates in a Cu-metal-vapor laser.
801. Kan, T., Ball, D., Schmitt, E., Hill, J., *Appl. Phys. Lett.*, 35, 676-677 (1979), Annular discharge copper vapor laser.
802. Gokay, M. C., Cross, L. A., *IEEE J. Quantum Electron.*, QE-15, 65-66 (1979), Comparison of copper acetylacetonate, copper (II) acetate, and copper chloride as lasants for copper vapor lasers.
803. Kazaryan, M. A., Trofimov, A. N., *Sov. J. Quantum Electron.*, 9, 148-152 (1979), Kinetics of metal salt vapor lasers.
804. Babeiko, Yu. A., Vasil'ev, L. A., Sviridov, A. V., Sokolov, A. V. et al., *Sov. J. Quantum Electron.*, 9, 651-653 (1979), Efficiency of a copper vapor laser.
805. Hargrove, R. S., Grove, R., Kan, T., *IEEE J. Quantum Electron.*, QE-15, 1228-1233 (1979), Copper vapor laser unstable resonator oscillator and oscillator-amplifier characteristics.
806. Solanki, R., Latush, E. L., Fairbank, W. M., Jr., Collins, G. J., *Appl. Phys. Lett.*, 34, 568-570 (1979), New infrared laser transitions in copper and silver hollow cathode discharges.
807. Landsberg, B. M., Shafki, M. S., Butcher, R. J., *Int. J. Infrared Millimeter Waves* 2, 49 (1981).
808. Ioli, N., Moretti, A., Moruzzi, G., Strumia, F., D'Amato, F., New CH_2F_2 FIR laser lines pumped by a tunable WG CO_2 laser, *Int. J. Infrared Millimeter Waves* 6, 1017 (1985).

809. Dangoisse, E. J., Wasscat, J., Colmont, J. M., Assignment of laser lines in an optically pumped submillimeter and near millimeter laser $(H_2CO)_3$, *Int. J. Infrared Millimeter Waves* 2, 1177 (1981).
810. Bugaev, V. A., Shliteris, E. P., *Sov. J. Quantum Electron.* 14, 1331 (1984), Submillimeter lasing transitions in isotopic modifications of SO_2.
811. Bugaev, V. A., Shliteris, E. P., *Sov. J. Quantum Electron.* 11, 742 (1981), Sulfur dioxide laser pumped by CO_2 laser radiation and identification of the transitions.
812. Sattler, J. P., Lafferty, W. J., *Reviews of Infrared and Millimeter Waves 2*, Button, K. J., Inguscio, M.,, Strumia, F., Eds., Plenum Press, New York.
813. Davies, P. B., Stern, D. P., *Int. J. Infrared Millimeter Waves* 3, 909 (1982), New cw FIR laser lines from optically pumped silyl fluoride (SiH_3F).
814. Davies, P. B., Ferguson, A. H., Stern, D. P., *Infrared Phys.* 25, 87 (1985), New optically pumped lasers containing silicon.
815. Shafik, S., Crocker, D., Landsberg, B. M., Butcher, R. J., *IEEE Quantum Electron.* QE-17, 115 (1981), Phosphine far-infrared CW laser transitions: optical pumping at more than 11 MHz from resonance.
816. Gerasimov, W. G., Dyubkko, S. F., Efimenko, L. D., Fesenko, L. D., Jarcev, W. I., *Ukr. Fiz. Zh.* 28, 1323 (1983).
817. Siemsen, K. J., Reid, J., Danagher, D. J., *Appl. Opt.*, 25, 86 (1986), Improved cw lasers in the 11-13 μm wavelength region produced by optically pumping NH_3.
818. Kroeker, D. F., Reid, J., *Appl. Opt.*, 25, 2929 (1986), Line-tunble cw *ortho-* and *para*-NH_3 lasers operating at wavelengths of 11 and 14 μm.
819. Jones, H., Taubmann, G., Takami, M., *IEEE Quantum Electron.*, QE-18, 1997 (1982), The optically pumped hydrazin FIR laser: assignments and new laser lines.
820. Landsberg, B. M., *IEEE Quantum Electron.,* QE-16,704 (1980), New optically pumped CW submillimeter emission lines from OCS, CH_3OH, and CH_3OD.
821. Davies, P. B., Jones, H., *IEEE Quantum Electron.,* QE-17, 13 (1981), A powerful new optically pumped FIR laser—formylfluoride.
822. Dumanchin, R., Rocca-Serra, J., unpublished (Sept. 1970), High power density pulsed molecular laser.
823. Tang, F., Olafsson, A., Hennignsen, J. O., *Appl. Phys. B,* 47, 47 (1988), A study of the methanol laser with a 500 MHz tunable CO_2 pump laser.
824. Allario, F., Schneider, R. T., *Research on Uranium Plasmas*, NASA SP-236, No. 236(1971), Enhancement of laser output by nuclear reactions, National Aeronautics and Space Administration.
825. Horrigan, F. A., unpublished (1966), Raytheon Company, Estimated lifetimes in Neon I and Xenon I.
826. Saykally, R. J., Evenson, K., Jennings, D. A., Zink, L. R., Scalabrin, A., New FIR laser line and frequency measurement for optically pumped CD_3OH, *Int. J. Infrared Millimeter Waves* 8, 653 (1987).
827. Baldacci, S. A., Ghersetti, S. H., Jurlock, S. C., Rao, K. N., *J. Mol. Spectrosc.* 42, 327 (1972), Spectrum of dideuteroacetylene near 18.6 microns.
828. Gastaud, C., Redon, M., Fourrier, M., *Appl. Phys. B,* 47, 303 (1988).
829. Petersen, J. C., Duxbury G., *Appl. Phys. B* 27, 19 (1982), Observation and assignment of submillimetre laser lines from CH_3OH pumped by isotopic CO_2 lasers.
830. Petersen, J. C., Duxbury G., *Appl. Phys. B* 34, 17 (1984).
831. Coleman, C. D., Bozman, W. R., Meggers, W. F., Monograph, Vol. 1 & 2, No. 3 (1960), *Table of Wave Numbers*, U. S. National Bureau of Standards.
832. Jones, D. R., Little, C. E., *IEEE J. Quantum Electron.,* 28, 590 (1992), A lead bromide laser operating at 722.9 and 406.2 nm.
833. Liou, H.-T., Yang, H., Dan, P., *Appl. Phys. B,* 54, 221 (1992), Laser induced lasing in CS_2 vapor.

834. Vasconcellos, E. C. C., Wyss, J., Evenson, K. M., Frequency measurements of far infrared $^{13}CH_3OH$ laser lines, *Int. J. Infrared Millimeter Waves* 8, 647 (1987).

835. Rutt, H. N., Green, J. M., *Opt. Commun.* 19, 320 (1978), Optically pumped laser action in dideuteroacetylene.

836. Deka, B. K., Dyer, P. E., Winfield, R. J., *Opt. Lett.* 5, 194 (1980), New 17-21 μm laser lines in C_2D_2 using a continuously tunable CO_2 laser pump.

837. Hall, J. C., *Atomic Physics*, 615(1973), Plenum Press, New York, Saturated absorption spectroscopy with applications to the 3.39 μm methane transition.

838. Jones, D. R., Little, C. E., *Opt. Commun.* 91, 223 (1992), A 472.2 nm bismuth halide laser.

839. Thom, K., Schneider, R. T., *AIAA* J., 10, 400-406 (1972), Nuclear pumped gas lasers.

840. Birnbaum, G., *Proc. IEEE,* 55, 1015-1026 (1967), Frequency stabilization of gas lasers.

841. Silfvast, W. T., Macklin, J. J., Wood II, O. R., *Opt. Lett.,* 8, 551 (1983), High-gain inner-shell photoionization laser in Cd vapor pumped by soft-X-ray radiation from a laser-produced plasma source.

842. Silfvast, W. T., Wood II, O. R., *J. Opt. Soc. Am. B,* 4, 609 (1987), Photoionization lasers pumped by broadband soft-X-ray flux from laser-produced plasmas.

843. Hube, M., Kumkar, M., Dieckmann, M., Beigang, R., Willegehausen, B., *Optics Commun.,* 66, 107 (1988), Potassium photoionization laser produced by innershell ionization of excited potassium.

844. Lundberg, H., Macklin, J. J., Silfvast, W. T., Wood II, O. R., *Appl. Phys. Lett.* 45, 335 (1984), High-gain soft-x-ray-pumped photoionization laser in zinc vapor.

845. Hooker, S. M., Haxell, A. M., Webb, C. E., *Appl. Phys. B* 54, 119 (1992), Observation of new laser transitions and saturation effects in optically pumped NO.

846. Berkeliev, B. M., Dolgikh, V. A., Rudoi, I. G., Soroka, A. M., *Sov. J. Quantum Electron.* 21, 250 (1991), Efficient stimulated emission of ultraviolet and infrared radiation from a mixture of nitrogen and light rare gases; simultaneous lasing at 1790 and 358 nm.

847. Willett, C. S., Gleason, T. J., Kruger, J. S., unpublished (1970).

848. Hochuli, U. E., Haldemann, P., private communication.

849. Laures, P., *Onde Electr.*, No. 469, (1966), Stabilisation de la frequence des lasers a gaz.

850. Walker, D. J., Barty, C. P. J., Yin, G. Y., Young, J. F., Harris, S. E., *Opt. Lett.* 12, 894 (1987), Observation of super Coster-Kronig-pumped gain in Zn III.

851. Silfvast, W. T., Wood II, O. R., Al-Salameh, D. Y., In *Short Wavelength Coherent Radiation: Generation and Applications* AIP Conf. Proc. 147, Attwood, D. T., Bokor, J., Eds., AIP, New York (1986), p. 134., Direct photoionization pumping of VUV, UV and visible inversions in helium, cadmium and argon via two-electron shakeup and of sodium via the output from the LLNL soft-X-ray laser.

852. Silfvast, W. T., Wood II, O. R., Lundberg, H., Macklin, J. J., *Opt. Lett.* 10, 122 (1985), Stimulated emission in the ultraviolet by optical pumping from photoionization-produced inner-shell states in Cd^+.

853. Hube, M., Brinkmann, R., Willing, H., Beigang, R., Willegehausen, B., *Appl. Phys. B* 45, 197(1988), Potassium photoionization laser produced by laser induced plasma radiation from a, multi foci device.

854. Klimovski, I. I., Vokhmin, P. A., *Proc. 13th Int. Conf. Phenomena in Ionized Gases* (Sept. 1977), Connection of the copper vapor laser emission pulse characteristics with plasma parameters, East German Physical Society .

855. Rodin, A. V., Zemtsov, Yu. L., *Proc. 13th Int. Conf. Phenomena in Ionized Gases* (Sept. 1977), Electron energy distribution function for copper vapor, East German Physical Society.

856. Elaev, V. F., Kirilov, A. E., Polunin, Yu. P., Soldatov, A. N., Fedorov, A. I., *Proc. 13th Int. Conf. Phenomena in Ionized Gases* (Sept. 1977), Experimental investigation of the pulse discharge in Cu + Ne mixture in high repetition rate regime, East German Physical Society.

857. Bloom, A. L., unpublished, private communication.

858. Russell, G. R., *Research on Uranium Plasmas*, NASA SP-236, No. 236 (1971), National Aeronautics and Space Administration, Feasibility of a nuclear laser excited by fission fragments produced in pulsed nuclear reactor.

859. DeShong, J. A., ANL Report, No. 7310(1966), Argonne National Lab, Nuclear Pumped Carbon Dioxide Gas Lasers - Model I Experiment.

860. Boguslovskii, A. A., Guryev, T. T., Didrikell, L. N., Novikova, V. A. K., *Electronnaya Tekhn.*, 8 (1967), unspecified.

861. der Agobian, R., Otto, J. L., Cagnard, R., Barthelemy, J., Echard, R., *C.R. Acad. Sci. Paris*, 260, 6327-6329 (1965), Emission stimulee en regime permanent dans le spectre visible du krypton ionise.

862. Banse, K., Herziger, G., Schafer, G., Seelig, W., *Phys. Lett.*, 27A, 682-683 (1968), Continuous UV-laser power in the watt range.

863. Fendley, J. R. Jr.., O'Grady, J. J., Tech. Rep. ECOM-0246-F (1970), RCA, Development, construction, and demonstration of a 100W cw argon-ion laser.

864. Ferrario, A., *Opt. Commun.*, 7, 375-378 (1973), Excitation mechanism in Hg^+ ion laser.

865. Fowles, G.R., Jensen, R. C., *Proc. IEEE,* 52, 851-852 (1964), Visible laser transitions in the spectrum of singly ionized iodine.

866. Fowles, G.R., Jensen, R. C., *Appl. Opt.*, 3, 1191-1192 (1964), Visible laser transitions in ionized iodine.

867. Fowles, G.R., Jensen, R. C., *Phys. Rev. Lett.*, 14, 347-348 (1965), Laser oscillation on a single hyperfine transition in iodine.

868. Fowles, G.R., Silfvast, W. T., *IEEE J. Quantum Electron.*, QE-1, 131 (1965), Laser action in the ionic spectra of zinc and cadmium.

869. Fowles, G.R., Silfvast, W. T., Jensen, R. C., *IEEE J. Quantum Electron.*, QE-1, 183-184 (1965), Laser action in ionized sulfur and phosphorus.

870. Fowles, G.R., Hopkins, B. D., *IEEE J. Quantum Electron.*, QE-3, 419 (1967), CW laser oscillation at 4416 Â in cadmium.

871. Fukuda, S., Miya, M., *Jpn. J. Appl. Phys.*, 13, 667-674 (1974), A metal-ceramic He-Cd II laser with sectioned hollow cathodes and output power characteristics of simultaneous oscillations.

872. Gadetskii, N. P., Tkach, Yu V., Slezov, V. V., Bessarab, Ya. et al., *JETP Lett.*, 14, 101-105 (1971), New mechanism of coherent-radiation generation in the visible region of the spectrum in ionized oxygen and nitrogen.

873. Gadetskii, N. P., Tkach, Yu V., Bessarab, Ya. Ya., Sidel'nikova, A. V., *Sov. J. Quantum Electron.*, 3, 168-169 (1973), Stimulated emission of visible light due to transitions in singly ionized chlorine and iodine atoms.

874. Gallardo, M., Garavaglia, M., Tagliaferri, A. A., Gallego Lluesma, E., *IEEE J. Quantum Electron.*, QE-6, 745-747 (1970), About unidentified ionized Xe laser lines.

875. Gallardo, M., Massone, C. A., Tagliaferri, A. A., Garavaglia, M., *Phys. Scripta*, 19, 538-544 (1979), $5s^25p^3(^4S)nl$ Levels of Xe III.

876. Gerritsen, H. J., Goedertier, P. V., *J. Appl. Phys.*, 35, 3060-3061 (1964), Blue gas laser using Hg^{2+}.

877. Birnbaum, M., Stocker, T. L., unpublished (1965), private communication.

878. Gerstenberger, D. C., Reid, R. D., Collins, G. J., *Appl. Phys. Lett.*, 30, 466-468 (1977), Hollow-cathode aluminum ion laser.

879. Gill, P., Webb, C. E., *J. Phys. D.*, 11, 245-254 (1978), Radial profiles of excited ions and electron density in the hollow cathode He/Zn laser.

880. Gilles, M., *Ann. Phys.*, 15, 267-410 (1931), Recherches sur la Structure des Spectres du Soufre.
881. Cem Gokay, M., Soltanolkotabi, M., Cross, L. A., *IEEE J. Quantum Electron.*, QE-14, 1004-1007 (1978), Single- and double-pulse experiments on the Sr^+ cyclic ion laser.
882. Goldsborough, J. P., Hodges, E. B., Bell, W. E., *Appl. Phys. Lett.*, 8, 137-139 (1966), RF induction excitation of CW visible laser transitions in ionized gases.
883. Lacy, R. A., Byer, R. L., Silfvast, W. T., Wood II, O. R., Svanberg, S., In *Short Wavelength Coherent Radiation: Generation and Applications*, AIP Conf. Proc. 147, Attwood, D. T., Bokor, J., Eds., AIP, New York (1986), p. 96., Optical gain at 185 nm in a laser ablated, inner-shell ionization-pumped indium plasma.
884. Goldsborough, J. P., Bloom, A. L., *IEEE J. Quantum Electron.*, QE-3, 96 (1) (1967), New CW ion laser oscillation in microwave-excited xenon.
885. Goldsborough, J. P., Hodges, E. B., *IEEE J. Quantum Electron.*, QE-5, 361-367 (1969), Stable long-life operation of helium-cadmium lasers at 4416 Å and 3250 Å.
886. Goldsborough, J. P., Bloom, A. L., *IEEE J. Quantum Electron.*, QE-5, 459-460 (1969), Near-infrared operating characteristics of the mercury ion laser.
887. Goldsmith, S., Kaufman, A. S., *Proc. Phys. Soc.*, 81, 544-552 (1963), The spectra of Ne IV, Ne V, and Ne VI: a further analysis.
888. Gordon, E. I., Labuda, E. F., Bridges, W. B., *Appl. Phys. Lett.*, 4, 178-180 (1964), Continuous visible laser action in singly ionized argon, krypton, and xenon.
889. Gordon, E. I., Labuda, E. F., Miller, R. C., Webb, C. E., *Proc. Phys. of QE Conf.*, 664-673(1966), Excitation mechanisms of the argon-ion laser, McGraw-Hill, New York.
890. Gorog, I., Spong, F. W., *Appl. Phys. Lett.*, 9, 61-63 (1966), High pressure, high magnetic field effects in continuous argon ion lasers.
891. Goto, T., Kano, H., Yoshino, N., Mizeraczyk, J. K., Hattori, S., *J. Phys. B*, 10, 292-295 (1977), Construction of a practical sealed-off He-I^+ laser device.
892. Kapteyn, H. C., Falcone, R. W., *Phys. Rev. A* 37, 2033 (1988), Auger-pumped short-wavelength lasers in xenon and krypton.
893. Green, J. M., Collins, G. J., Webb, C. E., *J. Phys. B*, 6, 1545-1555 (1973), Collisional excitation and destruction of excited Zn II levels in a helium afterglow.
894. Green, J. M., Webb, C. E., *J. Phys. B*, 7, 1698-1711 (1974), The production of excited metal ions in thermal energy charge transfer and Penning reactions.
895. Green, W. R., Falcone, R. W., *Opt. Lett.*, 2, 115-116 (1978), Inversion of the resonance line of Sr^+ produced by optically pumped Sr atoms.
896. Sher, M. H., Macklin, J. J., Young, J. F. and Harris, S. E., *Optics Lett.* 12, 89 (1987), Saturation of the Xe III 109-nm laser using traveling-wave laser-produced-plasma, excitation.
897. Grozeva, M. G., Sabotinov, N. V., Vuchkov, N. K., *Opt. Commun.*, 29, 339-340 (1979), CW laser generation on Tl II in a hollow-cathode Ne-Tl discharge.
898. Grozeva, M. G., Sabotinov, N. V., Telbizov, P. K., Vuchkov, N. K., *Opt. Commun.*, 31, 211-213 (1979), CW laser oscillation on transitions of Tl in a hollow-cathode Ne-Tl halide discharge.
899. Gunderson, M., Harper, C. D., *IEEE J. Quantum Electron.*, QE-9, 1160 (1973), A high-power pulsed xenon ion laser.
900. Hagen, L., Martin, W. C., Special Publication 363(1972), *Bibliography on Atomic Energy Levels and Spectra*, July 1968 through June 1971, U. S. National Bureau of Standards.
901. Hagen, L., Special Publication 363, Supplement 1(1977), *Bibliography on Atomic Energy Levels and Spectra*, July 1971 through June 1975, U. S. National Bureau of Standards.
902. Hallin, R., *Ark. Fys.*, 32, 201-210 (1966), The spectrum of N IV.

903. Barty, C. P. J., King, D. A., Yin, G. Y., Hahn, K. H., Field, J. E., Young, J. F., Harris, S. E., *Phys. Rev. Lett.* 61, 2201 (1988), 12.8-eV laser in neutral cesium.
904. Handrup, M. B., Mack, J. E., *Physica*, 30, 1245-1275 (1964), On the spectrum of ionized tellurium, Te II.
905. Tachikawa, M., Evenson, K. M., *Opt. Lett.* 21, 1247 (1996), Sequential optical pumping of a far-infrared ammonia laser.
906. Harper, C. D., Gunderson, M., *Rev. Sci. Instrum.*, 45, 400-402 (1974), Construction of a high power xenon ion laser.
907. Hashino, Y., Katsuyama, Y., Fukuda, K., *Jpn. J. Appl. Phys.*, 11, 907 (1972), Laser oscillation of multiply ionized Ne, Ar, and N ions in a Z-pinch discharge.
908. Hashino, Y., Katsuyama, Y., Fukuda, K., *Jpn. J. Appl. Phys.*, 12, 470 (1973), Laser oscillation of O V in Z-pinch discharge.
909. Hattori, S., Kano, H., Tokutome, K., Collins, G. J., Goto, T., *IEEE J. Quantum Electron.*, QE-10, 530-531 (1974), CW iodine-ion laser in a positive-column discharge.
910. Hattori, S., Kano, H., Goto, T., *IEEE J. Quantum Electron.*, QE-10, 739-740 (1974), A continuous positive-column He-I^+ laser using a sealed-off tube.
911. Heard, H. G., Peterson, J., *Proc. IEEE,* 52, 1050 (1964), Orange through blue-green transitions in a pulsed-CW xenon gas laser.
912. Hernqvist, K. G., Pultorak, D. C., *Rev. Sci. Instrum.*, 41, 696-697 (1970), Simplified construction and processing of a helium-cadmium laser.
913. Bell, W. E., unpublished (1964), private communication.
914. Hernqvist, K. G., *Appl. Phys. Lett.*, 16, 464-466 (1970), Stabilization of He-Cd laser.
915. Hernqvist, K. G., Pultorak, D. C., *Rev. Sci. Instrum.*, 43, 290-292 (1972), Study of He-Se laser performance.
916. Hernqvist, K. G., *IEEE J. Quantum Electron.*, QE-8, 740-743 (1972), He-Cd lasers using recirculation geometry.
917. Hernqvist, K. G., *RCA Rev.*, 34, 401-407 (1973), Vented-bore He-Cd lasers.
918. Hernqvist, K. G., *IEEE J. Quantum Electron.*, QE-13, 929 (1977), Continuous laser oscillation at 2703 A in copper ion.
919. Herziger, G., Seelig, W., *Z. Phys.*, 219, 5-31 (1969), Ionenlaser hober leistung.
920. Hodges, D. T., *Appl. Phys. Lett.*, 18, 454-456 (1971), CW laser oscillation in singly ionized magnesium.
921. Hodges, D. T., Tang, C. L., *IEEE J. Quantum Electron.*, QE-6, 757-758 (1970), New CW ion laser transitions in argon, krypton, and xenon.
922. Hoffmann, V., Toschek, P., *IEEE J. Quantum Electron.*, QE-6, 757 (1970), New laser emission from ionized xenon.
923. Hoffmann, V., Toschek, P. E., *J. Opt. Soc. Am.*, 66, 152-154 (1976), On the ionic assignment of xenon laser lines.
924. Bell, W. E., *Appl. Phys. Lett.*, 4, 34-35 (1964), Visible laser transitions in Hg^+.
925. Humphreys, C. J., Meggers, W. F., de Bruin, T. L., *J. Res. Natl. Bur. Stand.*, 23, 683-699 (1939), Zeeman effect in the second and third spectra of xenon.
926. Humphreys, C. J., unpublished, Line list for ionized krypton.
927. McFarlane, R. A., unpublished (private communication).
928. Iijima, T., Sugawara, Y., *J. Appl. Phys.*, 45, 5091-5092 (1974), New CW laser oscillation in He-Zn hollow cathode laser.
929. Illingworth, R., *Appl. Phys.*, 3, 924-930 (1970), Laser action and plasma properties of an argon Z-pinch discharge.
930. Isaev, A. A., Petrash, G. G., *J. Appl. Spectrosc. (USSR)*, 12, 835-837 (1970), New superradiance on the violet line of the mercury ion.
931. Ivanov, I. G., Sem, M. F., *J. Appl. Spectrosc. (USSR)*, 19, 1092-1093 (1973), New lasing lines in thallium.

932. Ivanov, I. G., Il'yushko, V. G., Sem, M. F., *Sov. J. Quantum Electron.*, 4, 589-593 (1974), Dependences of the gain of cataphoretic lasers on helium pressure and discharge-tube diameter.
933. Bell, W. E., *Appl. Phys. Lett.*, 7, 190-191 (1965), Ring discharge excitation of gas ion lasers.
934. Jain, K., *Opt. Commun.*, 28, 207-208 (1979), Cw laser oscillation at 8096 Å in Cu II in a hollow cathode discharge.
935. Jain, K., *Appl. Phys. Lett.*, 34, 398-399 (1979), New ion laser transitions in copper, silver, and gold.
936. Jain, K., *Appl. Phys. Lett.*, 34, 845-846 (1979), A nickel-ion laser.
937. Jain, K., *Appl. Phys. Lett.*, 36, 10-11 (1980), A milliwatt-level cw laser source at 224 nm.
938. Jain, K., *IEEE J. Quantum Electron.*, QE-16, 387-388 (1980), New UV and IR transitions in gold, copper, and cadmium hollow cathode lasers.
939. Jain, K., *Appl. Phys. Lett.*, 37, 362-364 (1980), Laser action in chromium vapor.
940. Janossy, M., Csillag, L., Rozsa, K., Salamon, T., *Phys. Lett.*, 46A, 379-380 (1974), CW laser oscillation in a hollow cathode He-Kr discharge.
941. Janossy, M., Rozsa, K., Csillag, L., Bergou, J., *Phys. Lett.*, 68A, 317-318 (1978), New cw laser lines in a noble gas mixture high voltage hollow cathode discharge.
942. Jarrett, S. M., Barker, G. C., *IEEE J. Quantum Electron.*, QE-5, 166 (1969), High-power output at 5353 Å from a pulsed xenon ion laser.
943. Bell, W. E., Bloom, A. L., Goldsborough, J. P., *IEEE J. Quantum Electron.*, QE-1, 400 (1965), Visible laser transitions in ionized selenium, arsenic, and bromine.
944. Jensen, R. C., Bennett, W. R., Jr., *IEEE J. Quantum Electron.*, QE-4, 356 (1968), Role of charge exchange in the zinc ion laser.
945. Jensen, R. C., Collins, G. J., Bennett, W. R., Jr., *Phys. Rev. Lett.*, 23, 363-367 (1969), Charge-exchange excitation and cw oscillation in the zinc-ion laser.
946. Jensen, R. C., Collins, G. J., Bennett, W. R., Jr., *Appl. Phys. Lett.*, 18, 50-51 (1971), Low-noise CW hollow-cathode zinc-ion laser.
947. Johnson, A. M., Webb, C. E., *IEEE J. Quantum Electron.*, QE-3, 369 (1967), New CW laser wavelength in Kr II.
948. Johnson, W. L., McNeil, J. R., Collins, G. J., Persson, K. B., *Appl. Phys. Lett.*, 29, 101-102 (1976), CW laser action in the blue-green spectral region from Ag II.
949. Johnston, T. F., Jr., Kolb, W. P., *IEEE J. Quantum Electron.*, QE-12, 482-493 (1976), The self-heated 442-nm He-Cd laser: optimizing the power output, and the origin of beam noise.
950. Kano, H., Goto, T., Hattori, S., *IEEE J. Quantum Electron.*, QE-9, 776-778 (1973), Electron temperature and density in the He-CdI_2 positive column used for an I^+ laser.
951. Kano, H., Goto, T., Hattori, S., *J. Phys. Soc. Jpn.*, 38, 596 (1975), CW laser oscillation of visible and near-infrared Hg(II) lines in a He-Hg positive, column discharge.
952. Kano, H., Shay, T., Collins, G. J., *Appl. Phys. Lett.*, 27, 610-612 (1975), A second look at the excitation mechanism of the He-Hg^+ laser.
953. Karabut, E. K., Mikhalevskii, V. S., Papakin, V. F., Sem, M. F., *Sov. Phys. Tech. Phys.*, 14, 1447-1448 (1970), Continuous generation of coherent radiation in a discharge in Zn and Cd vapors obtained by cathode sputtering.
954. Karabut, E. K., Kravchenko, V. F., Papakin, V. F., *J. Appl. Spectrosc. (USSR),* 19, 938-939 (1973), Excitation of the Ag II Lines by a pulsed discharge in a mixture of silver vapor and helium.
955. Keeffe, W. M., Graham, W. J., *Appl. Phys. Lett.*, 7, 263-264 (1965), Laser oscillation in the visible spectrum of singly ionized pure bromine vapor.
956. Keeffe, W. M., Graham, W. J., *Phys. Lett.*, 20, 643 (1966), Observation of new Br II laser transitions.

957. Keiden, V. F., Mikhalevskii, V. S., *J. Appl. Spectrosc. (USSR)*, 9, 1154 (1968), Pulsed generation in bismuth vapor.

958. Keiden, V. F., Mikhalevskii, V. S., Sem, M. F., *J. Appl. Spectrosc. (USSR)*, 15, 1089-1090 (1971), Generation from ionic transitions of selenium.

959. Kiess, C. C., de Bruin, T. L., *J. Res. Natl. Bur. Stand.*, 23, 443-470 (1939), Second spectrum of chlorine and its structure.

960. Kitaeva, V. F., Odintsov, A. N., Sobolev, N. N., *Sov. Phys. Usp.*, 12, 699-730 (1970), Continuously operating argon ion lasers.

961. Klein, M. B., Ph.D. dissertation (1969), University of California, Radiation Trapping Processes in the Pulsed Ion Laser.

962. Klein, M. B., *Appl. Phys. Lett.*, 17, 29-32 (1970), Time-resolved temperature measurements in the pulsed argon ion laser.

963. Klein, M. B., Silfvast, W. T., *Appl. Phys. Lett.*, 18, 482-485 (1971), New CW laser transitions in Se II.

964. Akirtava, O. S., Bogus, A. M., Dzhilkiya, V. L., Oleinik, Yu. M., *Sov. J. Quantum Electron.*, 3, 519-520 (1974), Quasicontinuous emission from ion lasers in electrodeless high-frequency discharges.

965. Bennett, W. R., Jr., Knutson, J. W., Jr., Mercer, G. N., Detch, J. L., *Appl. Phys. Lett.*, 4, 180-182 (1964), Super-radiance, excitation mechanisms, and quasi-cw oscillation in the visible Ar^+ laser.

966. Kobayashi, S., Izawa, T., Kawamura, K., Kamiyama, M., *IEEE J. Quantum Electron.*, QE-2, 699-700 (1966), Characteristics of a pulsed Ar II ion laser using the external spark gap.

967. Kobayashi, S., Kurihara, K., Kamiyama, M., *Oyo Butsuri*, 38, 766-768 (1969), New laser oscillation in ionized xenon at 5592 Â.

968. Koval'chuk, V. M., Petrash, G. G., *JETP Lett.*, 4, 144-146 (1966), New generation lines of a pulsed iodine-vapor laser.

969. Kruithof, A. A., Penning, F. M., *Physica*, 4, 430-449 (1937), Determination of the Townsend ionization coefficient alpha for mixtures of neon and argon.

970. Kulagin, S. G., Likhachev, V. M., Markuzon, E. V., Rabinovich, M. S. et al., *JETP Lett.*, 3, 6-8 (1966), States with population inversion in a self-compressed discharge.

971. Labuda, E. F., Gordon, E. I., Miller, R. C., *IEEE J. Quantum Electron.*, QE-1, 273-279 (1965), Continuous-duty argon ion lasers.

972. Labuda, E. F., Webb, C. E., Miller, R. C., Gordon, E. I., *Bull. Am. Phys. Soc.*, 11, 497 (1966), A study of capillary discharges in noble gases at high current densities.

973. Labuda, E. F., Johnson, A. M., *IEEE J. Quantum Electron.*, QE-2, 700-701 (1966), Threshhold properties of continuous duty rare gas ion laser transitions.

974. Latimer, I. D., *Appl. Phys. Lett.*, 13, 333-335 (1968), High power quasi-CW ultra-violet ion laser.

975. Schneider, M., Evenson, K. M., Johns, J. W. C., *Opt. Lett.* 21, 1038 (1996), Far-infrared continuous-wave laser emission from H_2O and from NH_3 optically pumped by a CO laser.

976. Latush, E. L., Sem, M. F., *Sov. J. Quantum Electron.*, 3, 216-219 (1973), Stimulated emission due to transitions in alkaline-earth metal ions.

977. Latush, E. L., Mikhalevskii, V. S., Sem, M. F., Tolmachev, G. N. et al., *JETP Lett.*, 24, 69-71 (1976), Metal-ion transition lasers with transverse HF excitation.

978. Laures, P., Dana, L., Frapard, C., *C.R. Acad. Sci. Paris*, 258, 6363-6365 (1964), Nouvelles transitions laser dans le domaine 0.43-0.52 μ obtenues a partir du spectre du krypton ionise.

979. Laures, P., Dana, L., Frapard, C., *C.R. Acad. Sci. Paris*, 259, 745-747 (1964), Nouvelles Raies Laser Visibles dans le Xenon Ionise.

980. Levinson, G. R., Papulovskiy, V. F., Tychinskiy, V. P., *Radio Eng. Electron. Phys.* (USSR), 13, 578-582 (1968), The mechanism of inversion of the populations of the various levels in multivalent argon ions.
981. Li, H., Andrew, K. L., *J. Opt. Soc. Am.*, 61, 96-109 (1971), First spark spectrum of arsenic.
982. Littlewood, I. M., Piper, J. A., Webb, C. E., *Opt. Commun.*, 16, 45-49 (1976), Excitation mechanisms in CW He-Hg lasers.
983. Lluesma, E. G., Tagliaferri, A. A., Massone, C. A., Garavaglia, A. et al., *J. Opt. Soc. Am.*, 63, 362-364 (1973), Ionic assignment of unidentified xenon laser lines.
984. Luthi, H. R., Seelig, W., Steinger, J., Lobsiger, W., *IEEE J. Quantum Electron.*, QE-13, 404-405 (1977), Continuous 40-W UV laser.
985. Luthi, H. R., Seelig, W., Steinger, J., *Appl. Phys. Lett.*, 31, 670-672 (1977), Power enhancement of continuous ultraviolet lasers.
986. Birnbaum, M., Tucker, A. W., Gelbwachs, J. A., Fincher, C. L., *IEEE J. Quantum Electron.*, QE-7, 208 (1971), New O II 6640 Å laser line.
987. Luthi, H. R., Steinger, J., *Opt. Commun.*, 27, 435-438 (1978), Continuous operation of a high power neon ion laser.
988. Manley, J. H., Duffendack, O. S., *Phys. Rev.*, 47, 56-61 (1935), Collisions of the second kind between magnesium and neon.
989. Marling, J. B., *IEEE J. Quantum Electron.*, QE-11, 822-834 (1975), Ultraviolet ion laser performance and spectroscopy. I. New strong noble-gas transitions below 2500 Â.
990. Marling, J. B., Lang, D. B., *Appl. Phys. Lett.*, 31, 181-184 (1977), Vacuum ultraviolet lasing from highly ionized noble gases.
991. Marling, J. B., *IEEE J. Quantum Electron.*, QE-14, 4-6 (1978), Ultraviolet ion laser performance and spectroscopy for sulfur, fluorine, chlorine, and bromine.
992. Martin, W. C., *J. Opt. Soc. Am.*, 49, 1071-1085 (1959), Atomic energy levels and spectra of neutral and singly ionized phosphorus (P I and P II).
993. Martin, W. C., Corliss, C. H., *J. Res. Natl. Bur. Stand.*, 64A, 443-477 (1960), The spectrum of singly ionized atomic iodine (I.II).
994. Martin, W. C., Kaufman, V., *J. Res. Natl. Bur. Stand.*, 74A, 11-22 (1970), New vacuum ultraviolet wavelengths and revised energy levels in the second spectrum, of zinc (Zn II).
995. Davies, P. B., Jones, H., *Appl. Phys.* 22, 53 (1980).
996. Gerritsen, H. J., unpublished (1965), private communication.
997. McFarlane, R. A., *Appl. Opt.*, 3, 1196 (1964), Optical maser oscillation on iso-electronic transitions in Ar III and Cl II.
998. Davis, I. H., Pharaoh, K. I., Knight, D. J. E., *Appl. Phys.* B 35, 127 (1987).
999. McNeil, J. R., Collins, G. J., Persson, K. B., Franzen, D. L., *Appl. Phys. Lett.*, 28, 207-209 (1976), Ultraviolet laser action from Cu II in the 2500 Â region.
1000. McNeil, J. R., Collins, G. J., *IEEE J. Quantum Electron.*, QE-12, 371-372 (1976), Additional ultraviolet laser transitions in Cu II.
1001. McNeil, J. R., Johnson, W. L., Collins, G. J., Persson, K. B., *Appl. Phys. Lett.*, 29, 172-174 (1976), Ultraviolet laser action in He-Ag and Ne-Ag mixtures.
1002. Meggers, W. F., *J. Res. Natl. Bur. Stand.*, 71A, 396-544 (1967), The second spectrum of ytterbium (Yb II).
1003. Petersen, A. B., *Handbook of Laser Science and Technology,* Vol. II, Gas Lasers, CRC Press, Boca Raton, FL (1991), p. 335, Ionized gas lasers.
1004. Minnhagen, L., Stigmark, L., *Ark. Fys.*, 13, 27-36 (1957), The excitation of ionic spectra by 100 kw high frequency pulses.
1005. Minnhagen, L., *Ark. Fys.*, 25, 203-284 (1963), The spectrum of singly ionized argon, Ar II.
1006. Ninomiya, H., Abe, M., Takashima, N., *Appl. Phys. Lett.* 58, 18191 (1991), Laser action of optically pumped atomic vanadium vapor.

1007. Minnhagen, L., Strihed, H., Petersson, B., *Ark. Fys.*, 39, 471-493 (1969), Revised and extended analysis of singly ionized krypton, Kr II.

1008. Dyubko, S. F., Svich, V. A., Fesenko, L. D., *Sov. Tech. Phys. Lett*, 1, 192 (1975).

1009. Moore, C. E., Merrill, P. W., *NSRDS-NBS*23(1968), Partial Grotrian Diagrams of AstroPhysical Interest, U. S. National Bureau of Standards.

1010. Moskalenko, V. F., Ostapchenko, E. P., Perchurina, S. V., Stepanov, V. A, *Opt. Spectrosc.*, 30, 201-202 (1971), Radiation of a pulsed ion laser.

1011. Myers, R. A., Wieder, H., Pole, R. V., *IEEE J. Quantum Electron.*, QE-2, 270-275 (1966), 9A5 - Wide field active imaging.

1012. Neusel, R. H., *IEEE J. Quantum Electron.*, QE-2, 70 (1966), A new xenon laser oscillation at 5401 Â.

1013. Neusel, R. H., *IEEE J. Quantum Electron.*, QE-2, 106 (1966), A new krypton laser oscillation at 5016.4 Â.

1014. Neusel, R. H., *IEEE J. Quantum Electron.*, QE-2, 334 (1) (1966), New laser oscillations in krypton and xenon.

1015. Neusel, R. H., *IEEE J. Quantum Electron.*, QE-2, 758 (1966), New laser oscillations in xenon and krypton.

1016. Neusel, R. H., *IEEE J. Quantum Electron.*, QE-3, 207-208 (1967), New laser oscillations in Ar, Kr, Xe, and N.

1017. Olme, A., *Phys. Scripta*, 1, 256-260 (1970), The spectrum of singly ionized boron, B II.

1018. Paananen, R., *Appl. Phys. Lett.*, 9, 34-35 (1966), Continuously operated ultraviolet laser.

1019. Pacheva, Y., Stefanova, M., Pramatarov, P., *Opt. Commun.*, 27, 121-122 (1978), Cw laser oscillations on the Kr II 4694 Â and Kr II 4318 Â lines in a hollow-cathode, He-Kr discharge.

1020. Palenius, H. P., *Appl. Phys. Lett.*, 8, 82 (1966), The identification of some Si and Cl laser lines observed by Cheo and Cooper.

1021. Palenius, H. P., *Ark. Fys.*, 39, 15-64 (1969), Spectrum and term system of singly ionized fluorine, F II.

1022. Palenius, H. P., *Phys. Scripta*, 1, 113-135 (1970), Spectrum and term system of doubly ionized fluorine, F III.

1023. Papakin, V. F., Sem, M. F., *Sov. Phys. J.*, 13, 230-231 (1970), Use of isotopes in cadmium and zinc vapor lasers.

1024. Papayoanou, A., Buser, R. G., Gumeiner, I. M., *IEEE J. Quantum Electron.*, QE-9, 580-585 (1973), Parameters in a dynamically compressed xenon plasma laser.

1025. Papayoanou, A., Gumeiner, I., *Appl. Phys. Lett.*, 16, 5-8 (1970), High power xenon laser action in high current pinched discharges.

1026. Pappalardo, R., *J. Appl. Phys.*, 45, 3547-3553 (1974), Observation of afterglow character and high gain in the laser lines of O(III).

1027. Bloom, A. L., Goldsborough, J. P., *IEEE J. Quantum Electron.*, QE-6, 164 (1970), New CW laser transitions in cadmium and zinc ion.

1028. Pappalardo, R., *IEEE J. Quantum Electron.*, QE-10, 897-898 (1974), Some observations on multiply ionized xenon laser lines.

1029. Penning, F. M., *Physica*, 1, 1028-1044 (1934), The starting potential of the glow discharge in neon-argon mixtures between large parallel plates.

1030. Persson, W., *Phys. Scripta*, 3, 133-155 (1971), The spectrum of singly ionized neon, Ne II.

1031. Petersen, A. B., Birnbaum, M., *IEEE J. Quantum Electron.*, QE-10, 468 (1974), The singly ionized carbon laser at 6783, 6578, and 5145 Â.

1032. Piper, J. A., Collins, G. J., Webb, C. E., *Appl. Phys. Lett.*, 21, 203-205 (1972), CW laser oscillation in singly ionized iodine.

1033. Piper, J. A., Webb, C. E., *J. Phys. B*, 6, L116-L120 (1973), Continuous-wave laser oscillation in singly ionized arsenic.

1034. Piper, J. A., Webb, C. E., *J. Phys. D.*, 6, 400-407 (1973), A hollow cathode device for CW helium-metal vapor laser systems.

1035. Piper, J. A., Webb, C. E., *Opt. Commun.*, 13, 122-125 (1975), Power limitations of the CW He-Hg laser.

1036. Piper, J. A., Gill, P., *J. Phys. D.*, 8, 127-134 (1975), Output characteristics of the He-Zn laser.

1037. Piper, J. A., Webb, C. E., *IEEE J. Quantum Electron.*, QE-12, 21-25 (1976), High-current characteristics of the continuous-wave hollow-cathode He-I_2 laser.

1038. Bockasten, K., *Ark. Fys.*, 9, 457-481 (1955), A study of C III by means of a sliding vacuum spark.

1039. Piper, J. A., Brandt, M., *J. Appl. Phys.*, 48, 4486-4494 (1977), Cw laser oscillation on transitions of Cd^+ and Zn^+ in He-Cd-halide and He-Zn-halide discharges.

1040. Piper, J. A., Neely, D. F., *Appl. Phys. Lett.*, 33, 621-623 (1978), Cw laser oscillation on transitions of Cu^+ in He-Cu-halide gas discharges.

1041. Piper, J. A., *Opt. Commun.*, 31, 374-376 (1979), CW laser oscillation on transitions of As^+ in He-AsH_3 gas discharges.

1042. Pugnin, V. I., Rudelev, S. A., Stepanov, A. F., *J. Appl. Spectrosc. (USSR)*, 18, 667-668 (1973), Laser generation with ionic transitions of iodine.

1043. Rao, Y. B., *Indian J. Phys.*, 32, 497-515 (1958), Structure of the spectrum of singly ionized bromine.

1044. Reid, R. D., Johnson, W. L., McNeil, J. R., Collins, G. J., *IEEE J. Quantum Electron.*, QE-12, 778-779 (1976), New infrared laser transitions in Ag II.

1045. Reid, R. D., Gerstenberger, D. C., McNeil, J. R., Collins, G. J., *J. Appl. Phys.*, 48, 3994 (1977), Investigations of unidentified laser transitions in Ag II.

1046. Risberg, P., *Ark. Fys.*, 9, 483-494 (1955), The spectrum of singly ionized magnesium, Mg II.

1047. Riseberg, L. A., Schearer, L. B., *IEEE J. Quantum Electron.*, QE-7, 40-41 (1971), On the excitation mechanism of the He-Zn laser.

1048. Bockasten, K., *Ark. Fys.*, 10, 567-582 (1956), A study of C IV: term values, series formulae, and stark effect.

1049. Rozsa, K., Janossy, M., Bergou, J., Csillag, L., *Opt. Commun.*, 23, 15-18 (1977), Noble gas mixture CW hollow cathode laser with internal anode system.

1050. Rozsa, K., Janossy, M., Csillag, L., Bergou, J., *Phys. Lett.*, 63A, 231-232 (1977), CW aluminum ion laser in a high voltage hollow cathode discharge.

1051. Rozsa, K., Janossy, M., Csillag, L., Bergou, J., *Opt. Commun.*, 23, 162-164 (1977), CW Cu II laser in a hollow anode-cathode discharge.

1052. Rudko, R. I., Tang, C. L., *Appl. Phys. Lett.*, 9, 41-44 (1966), Effects of cascade in the excitation of the Ar II laser.

1053. Rudko, R. I., Tang, C. L., *J. Appl. Phys.*, 38, 4731-4739 (1967), Spectroscopic studies of the Ar^+ laser.

1054. Russell, H. N., Albertson, W., Davis, D. N., *Phys. Rev.*, 60, 641-656 (1941), The spark spectrum of europium, Eu II.

1055. Schearer, L. D., Padovani, F. A., *J. Chem. Phys.*, 52, 1618-1619 (1970), De-excitation cross section of metastable helium by Penning collisions with cadmium atoms.

1056. Schearer, L. D., *IEEE J. Quantum Electron.*, QE-11, 935-937 (1975), A high-power pulsed xenon ion laser as a pump source for a tunable dye laser.

1057. Schuebel, W. K., *Appl. Phys. Lett.*, 16, 470-472 (1970), New cw Cd-vapor laser transitions in a hollow-cathode structure.

1058. Schuebel, W. K., *IEEE J. Quantum Electron.*, QE-6, 574-575 (1970), Transverse-discharge slotted hollow-cathode laser.

1059. Schuebel, W. K., *IEEE J. Quantum Electron.*, QE-6, 654-655 (1970), New CW laser transitions in singly-ionized cadmium and zinc.

1060. Schuebel, W. K., *IEEE J. Quantum Electron.*, QE-7, 39-40 (1971), Continuous visible and near-infrared laser action in Hg II.

1061. Shay, T., Kano, H., Hattori, S., Collins, G. J., *J. Appl. Phys.*, 48, 4449-4453 (1977), Time-resolved double-probe study in a He-Hg afterglow plasma.

1062. Shenstone, A. G., *Proc. Roy. Soc.*, A261, 153-174 (1961), The second spectrum of silicon.

1063. Shenstone, A. G., *Proc. Roy. Soc.*, A276, 293-307 (1963), The second spectrum of germanium.

1064. Silfvast, W. T., Fowles, G.R., Hopkins, B. D., *Appl. Phys. Lett.*, 8, 318-319 (1966), Laser action in singly ionized Ge, Sn, Pb, In, Cd, and Zn.

1065. Silfvast, W. T., *Appl. Phys. Lett.*, 13, 169-171 (1968), Efficient cw laser oscillation at 4416 Å in Cd(II).

1066. Silfvast, W. T., Szeto, L. H., *Appl. Opt.*, 9, 1484-1485 (1970), A simple high temperature system for CW metal vapor lasers.

1067. Aleinikov, V. S., *Opt. Spectra*, 28, 15-17 (1970), Use of an electron gun to determine the nature of collisions of the second kind in a mercury-helium mixture.

1068. Boersch, H., Boscher, J., Hoder, D., Schafer, G., *Phys. Lett.*, 31A, 188-189 (1970), Saturation of laser power of CW ion laser with large bored tubes and high power CW UV.

1069. Silfvast, W. T., Klein, M. B., *Appl. Phys. Lett.*, 17, 400-403 (1970), CW laser action on 24 visible wavelengths in Se II.

1070. Silfvast, W. T., Szeto, L. H., *Appl. Phys. Lett.*, 19, 445-447 (1971), Simplified low-noise He-Cd laser with segmented bore.

1071. Silfvast, W. T., *Phys. Rev. Lett.*, 27, 1489-1492 (1971), Penning ionization in a He-Cd DC discharge.

1072. Silfvast, W. T., unpublished (1971), private communication.

1073. Silfvast, W. T., Klein, M. B., *Appl. Phys. Lett.*, 20, 501-504 (1972), CW laser action on 31 transitions in tellurium vapor.

1074. Simmons, W. W., Witte, R. S., *IEEE J. Quantum Electron.*, QE-6, 466-469 (1970), High power pulsed xenon ion lasers.

1075. Sinclair, D. C., *J. Opt. Soc. Am.*, 55, 571-572 (1965), Near-infrared oscillation in pulsed noble-gas-ion lasers.

1076. Sinclair, D. C., *J. Opt. Soc. Am.*, 56, 1727-1731 (1966), Polarization characteristics of an ionized-gas laser in a magnetic field.

1077. Solanki, R., Latush, E. L., Gerstenberger, D. C., Fairbank, W. M., Jr., *Appl. Phys. Lett.*, 35, 317-319 (1979), Hollow-cathode excitation of ion laser transitions in noble-gas mixtures.

1078. Bokhan, P. A., Klimkin, V. M., Prokop'ev, V. E., *JETP Lett.*, 18, 44-45 (1973), Gas laser using ionized europium.

1079. Sosnowski, T. P., *J. Appl. Phys.*, 40, 5138-5144 (1969), Cataphoresis in the helium-cadmium laser discharge tube.

1080. Sosnowski, T. P., Klein, M. B., *IEEE J. Quantum Electron.*, QE-7, 425-426 (1971), Helium cleanup in the helium-cadmium laser discharge.

1081. Sattler, J. P., Worchesky, T. L., Tobin, M. S., Ritter, K. J., Daley, T. W., Submillimeter emission assignments for 1,1-difluoroethylene, *Int. J. Infrared Millimeter Waves* 1, 127 (1980).

1082. Sugawara, Y., Tokiwa, Y., *Technology Reports of Seikei University*, 9 (1970), Seikei University, CW hollow cathode laser oscillation in Zn^+ and Cd^+.

1083. Sugawara, Y., Tokiwa, Y., *Jpn. J. Appl. Phys.*, 9, 588-589 (1970), CW laser oscillations in Zn II and Cd II in hollow cathode discharges.

1084. Sugawara, Y., Tokiwa, Y., Iijima, T., *Int. Quantum Electronics Conf.*, 320-321(1970), Excitation mechanisms of CW laser oscillations in Zn II and Cd II in hollow cathode.

1085. Sugawara, Y., Tokiwa, Y., Iijima, T., *Jpn. J. Appl. Phys.*, 9, 1537 (1970), New CW laser oscillations in Cd-He and Zn-He hollow cathode lasers.

1086. Szeto, L. H., Silfvast, W. T., Wood, O. R., II, *IEEE J. Quantum Electron.*, QE-15, 1332-1334 (1979), High-gain laser in Pb^+ populated by direct electron excitation from the neutral ground states.

1087. Bokhan, P. A., Klimkin, V. M., Prokop'ev, V. E., *Sov. J. Quantum Electron.*, 4, 752-754 (1974), Collision gas-discharge laser utilizing europium vapor. I. Observation of self-terminating oscillations and transition from cyclic to quasicontinuous.

1088. Tio, T. K., Luo, H. H., Lin, S.-C., *Appl. Phys. Lett.*, 29, 795-797 (1976), High cw power ultraviolet generation from wall-confined noble gas ion lasers.

1089. Tolkachev, V. A., *J. Appl. Spectrosc. (USSR),* 8, 449-451 (1968), Super-radiant transitions in Ar and Kr.

1090. Toresson, Y. G., *Ark. Fys.*, 17, 179-192 (1959), Spectrum and term system of trebly ionized silicon, Si IV.

1091. Toresson, Y. G., *Ark. Fys.*, 18, 389-416 (1960), Spectrum and term system of doubly ionized silicon, Si III.

1092. Tucker, A. W., Birnbaum, M., *IEEE J. Quantum Electron.*, QE-10, 99-100 (1974), Pulsed-ion laser performance in nitrogen, oxygen, krypton, xenon, and argon.

1093. Turner-Smith, A. R., Green, J. M., Webb, C. E., *J. Phys. B*, 6, 114-130 (1973), Charge transfer into excited states in thermal energy collisions.

1094. Vuchkov, N. K., Sabotinov, N. V., *Opt. Commun.*, 25, 199-200 (1978), Pulse generation of Cl II in vapours of CuCl and $FeCl_2$.

1095. Wang, C. P., Lin, S.-C., *J. Appl. Phys.*, 43, 5068-5073 (1972), Experimental study of argon ion laser discharge at high current.

1096. Bridges, W. B., *Appl. Phys. Lett.*, 4, 128-130 (1964), Laser oscillation in singly ionized argon in the visible spectrum.

1097. Wang, C. P., Lin, S.-C., *J. Appl. Phys.*, 44, 4681-4682 (1973), Performance of a large-bore high-power argon ion laser.

1098. Watanabe, S., Chihara, M., Ogura, I., *Jpn. J. Appl. Phys.*, 11, 600 (1972), New continuous oscillation at 5700 Å in He-Te laser.

1099. Waynant, R. W., *Appl. Phys. Lett.*, 22, 419-420 (1973), Vacuum ultraviolet laser emission from C IV.

1100. Webb, C. E., Turner-Smith, A. R., Green, J. M., *J. Phys. B*, 3, L134-L138 (1970), Optical excitation in charge transfer and Penning ionization.

1101. Wheeler, J. P., *IEEE J. Quantum Electron.*, QE-7, 429 (1971), New xenon laser line observed.

1102. Wieder, H., Myers, R. A., Fisher, C. L., Powell, C. G., Colombo, J., *Rev. Sci. Instrum.*, 38, 1538-1546 (1967), Fabrication of wide bore hollow cathode Hg^+ lasers.

1103. Willett, C. S., Heavens, O. S., *Opt. Acta,* 13, 271-274 (1966), Laser transition at 651.6 nm in ionized iodine.

1104. Willett, C. S., *IEEE J. Quantum Electron.*, QE-3, 33 (1967), New laser oscillations in singly-ionized iodine.

1105. Willett, C. S., Heavens, O. S., *Opt. Acta*, 14, 195-197 (1967), Laser oscillation on hyperfine transitions in ionized iodine.

1106. Willett, C. S., *IEEE J. Quantum Electron.*, QE-6, 469-471 (1970), Note on near-infrared operating characteristics of the mercury ion laser.

1107. Zarowin, C. B., *Appl. Phys. Lett.*, 9, 241-242 (1966), New visible CW laser lines in singly ionized chlorine.

1108. Zhukov, V. V., Il'yushko, V. G., Latush, E. L., Sem, M. F., *Sov. J. Quantum Electron.*, 5, 757-760 (1975), Pulse stimulated emission from beryllium vapor.

1109. Zhukov, V. V., Kucherov, V. S., Latush, E. L., Sem, M. F., *Sov. J. Quantum Electron.*, 7, 708-714 (1977), Recombination lasers utilizing vapors of chemical elements II. Laser action due to transitions in metal ions.

1110. Toyoda, K., Kobiyama, M., Namba, S., *Jpn. J. Appl. Phys.*, 15, 2033-2034 (1976), Laser oscillation at 5378 Â by laser-produced cadmium plasma.

1111. Bennett, J., Bennett, W. R., Jr., *IEEE J. Quantum Electron.*, QE-15, 842-843 (1979), CW oscillation on a new argon ion laser line at 5062 Â and relation to laser Raman spectroscopy.

1112. Kato, I., Nakaya, M., Satake, T., Shimizu, T., *Jpn. J. Appl. Phys.*, 14, 2001-2004 (1975), Output power characteristics of microwave-pulse-excited He-Kr^+ ion laser.

1113. Radford, H. E., *IEEE J. Quantum Electron.,* 11, 213 (1992), New cw lines from a submillimeter waveguide laser.

1114. Anon., unpublished, LGI-37 High Power Pulsed Optical Quantum Oscillator.

1115. Bridges, W. B., unpublished (1964), private communication.

1116. Bridges, W. B., *Proc. IEEE,* 52, 843-844 (1964), Laser action in singly ionized krypton and xenon.

1117. Bridges, W. B., Chester, A. N., *IEEE J. Quantum Electron.*, QE-1, 66-84 (1965), Spectroscopy of ion lasers.

1118. Bridges, W. B., Halsted, A. S., *IEEE J. Quantum Electron.*, QE-2, 84 (1966), New CW laser transitions in argon, krypton, and xenon.

1119. Bridges, W. B., Clark, P. O., Halsted, A. S., Tech. Rep. AFAL-TR-66-369, DDC No. AD-807363(1967), Hughes Research Laboratories, High power gas laser research.

1120. Alferov, G. N., Donin, V. I., Yurshin, B. Ya., *JETP Lett.*, 18, 369-370 (1973), CW argon laser with 0.5 kW output power.

1121. Bridges, W. B., Halsted, A. S., Tech. Rep. AFAL-TR-67-89; DDC No. AD-814897(1967), Hughes Research Laboratories, Gaseous Ion Laser Research.

1122. Bridges, W. B., Mercer, G. N., *IEEE J. Quantum Electron.*, QE-5, 476-477 (1969), CW operation of high ionization states in a xenon laser.

1123. Bridges, W. B., Mercer, G. N., Tech. Rep. ECOM-0229-F, DDC No.AD-861927(1969), Hughes Research Laboratories, Ultraviolet Ion Laser.

1124. Bridges, W. B., Chester, A. N., unpublished data (1969 and 1970), private communication.

1125. Bridges, W. B., Chester, A. N., *Handbook of Lasers* R. Pressley, Ed., CRC Press, Boca Raton, FL (1971), 242-297. Ionized gas lasers.

1126. Bridges, W. B., Chester, A. N., Halsted, A. S., Parker, J. V., *Proc. IEEE,* 59, 724-737 (1971), Ion laser plasmas.

1127. Bridges, W. B., Chester, A. N., *IEEE J. Quantum Electron.*, QE-7, 471-472 (1971), Comments on the identification of some xenon ion laser lines.

1128. Bridges, W. B., unpublished (1975), private communication.

1129. Bridges, W. B., *Methods of Experimental Physics*, Vol. 15A (1979), Academic Press, London, Atomic and ionic gas lasers.

1130. Bromander, J., *Ark. Fys.*, 40, 257-274 (1969), The spectrum of triply-ionized oxygen, O IV.

1131. Burkhard, P., Luthi, H. R., Seelig, W., *Opt. Commun.*, 18, 485-487 (1976), Quasi-CW laser action from Hg-III lines.

1132. Byer, R. L., Bell, W. E., Hodges, E., Bloom, A. L., *J. Opt. Soc. Am.*, 55, 1598-1602 (1965), Laser emission in ionized mercury: isotope shift, linewidth, and precise wavelength.

1133. Cahuzac, Ph., *J. Phys.*, 32, 499-505 (1971), Raies laser infrarouges dans les vapeurs de terres rares et D'Alcalino-Terreux.

1134. Carr, W. C., Grow, R. W., *Proc. IEEE,* 55, 726 (1967), Silicon and chlorine laser oscillations in $SiCl_4$.

1135. Chester, A. N., *Phys. Rev.*, 169, 172-184 (1968), Gas pumping in discharge tubes.

1136. Chester, A. N., *Phys. Rev.*, 169, 184-193 (1968), Experimental measurements of gas pumping in an argon discharge.

1137. Clunie, D. M., unpublished (1966), private communication.

1138. Collins, G. J., Ph.D. thesis(1970), Cw Oscillation and Charge Exchange Excitation in the Zinc-Ion Laser, Yale University.
1139. Collins, G. J., Jensen, R. C., Bennett, W. R., Jr., *Appl. Phys. Lett.*, 18, 282-284 (1971), Excitation mechanisms in the zinc ion laser.
1140. Collins, G. J., Jensen, R. C., Bennett, W. R., Jr., *Appl. Phys. Lett.*, 19, 125-130 (1971), Charge-exchange excitation in the He-Cd laser.
1141. Collins, G. J., Kuno, H., Hattori, S., Tokutome, K., Ishikawa, M., Kamid, *IEEE J. Quantum Electron.*, QE-8, 679-680 (1972), Cw laser oscillation at 6127 Å in singly ionized iodine.
1142. Convert, G., Armand, M., Martinot-Lagrade, P., *C.R. Acad. Sci. Paris*, 258, 3259-3260 (1964), Effet laser dans des melanges mercure-gaz rares.
1143. Wernsman, B., Prabhuram, T., Lewis, K., Gonzalez, F., Villagran, M., Rocca, J. J., *IEEE J. Quantum Electron.* 24, 1554 (1988), CW silver ion laser with electron beam excitation.
1144. Landberg, B. M., Shafik, M. S., Butcher, J. R., CW optically pumped far-infrared mission from acetaldehyde, vinyl chloride, and methyl isocyanide, *IEEE J. Quantum Electron.* 17, 828 (1992).
1145. Crooker, A. M., Dick, K. A., *Can. J. Phys.*, 46, 1241-1251 (1968), Extensions to the spark spectra of Zinc I, Zinc II and Zinc IV.
1146. Csillag, L., Janossy, M., Kantor, K., Rozsa, K., Salamon, T., *Appl. Phys.*, 3, 64-68 (1970), Investigation on a continuous wave 4416 Å cadmium ion laser.
1147. Csillag, L., Janossy, M., Salamon, T., *Phys. Lett.*, 31A, 532-533 (1970), Time delay of laser oscillation of the green transition of a pulsed He-Cd laser.
1148. Csillag, L., Itagi, V. V., Janossy, M., Rozsa, K., *Phys. Lett.*, 34A, 110-111 (1971), Laser oscillation at 4416 Â in a Ne-Cd discharge.
1149. Csillag, L., Janossy, M., Rozsa, K., Salamon, T., *Phys. Lett.*, 50A, 13-14 (1974), Near infrared Cw laser oscillation in Cu II.
1150. Dahlquist, J. A., *Appl. Phys. Lett.*, 6, 193-194 (1965), New line in a pulsed xenon laser.
1151. Dana, L., Laures, P., *Proc. IEEE,* 53, 78-79 (1965), Stimulated emission in krypton and xenon ions by collisions with metastable atoms.
1152. Kielkopf, J., *J. Opt. Soc. Am. B* 8, 212 (1991), Lasing in the aluminum and indium resonance lines following photoionization, and recombination in the presence of H_2.
1153. Dana, L., Laures, P., Rocherolles, R., *C. R. Acad. Sci. Paris*, 260, 481-484 (1965), Raies laser ultraviolettes dans le neon, l'argon, et le xenon.
1154. Davis, C. C., King, T. A., *Phys. Lett.*, 36A, 169-170 (1971), Time-resolved gain measurements and excitation mechanisms of the pulsed argon ion laser.
1155. Davis, C. C., King, T. A., *IEEE J. Quantum Electron.*, QE-8, 755-757 (1972), Laser action on unclassified xenon transitions in a highly ionized plasma.
1156. Davis, C. C., King, T. A., *Advances in Quantum Electronics*, Vol. 3(1975), 169, Academic Press, London, Gaseous ion lasers.
1157. de Bruin, T. L., Humphreys, C. J., Meggers, W. F., *J. Res. Natl. Bur. Stand.*, 11, 409-440 (1933), The second spectrum of krypton.
1158. Demtroder, W., *Phys. Lett.*, 22, 436-438 (1966), Excitation mechanisms of pulsed argon ion lasers at 4880 Â.
1159. Donin, V. I., *Sov. Phys. JETP*, 35, 858-864 (1972), Output power saturation with a discharge current in powerful continuous argon lasers.
1160. Donin, V. I., Shipilov, A. F., Grigor'ev, V. A., *Sov. J. Quantum Electron.*, 9, 210-212 (1979), High-power cw ion lasers with an improved service life.
1161. Auschwitz, B., Eichler, H. J., Wittwer, W., *Appl. Phys. Lett.*, 36, 804-805 (1980), Extension of the operating period of an UV Cu II-laser by a mixture of argon.

1162. Duffendack, O. S., Thomson, K., *Phys. Rev.*, 43, 106-111 (1933), Some factors affecting action cross section for collisions of the second kind between atoms and ions.

1163. Duffendack, O. S., Gran, W. H., *Phys. Rev.*, 51, 804-809 (1937), Regularity along a series in the variation of the action cross section with energy discrepancy in impacts of the second kind.

1164. Dunn, M. H., Ross, J. N., *Progr. Quantum Electron.*, 4, 233-269 (1976), The argon ion laser.

1165. Dyson, D. J., *Nature (London)*, 207, 361-363 (1965), Mechanism of population inversion at 6149 Â in the mercury ion laser.

1166. Edlen, B., Risberg, P., *Ark. Fys.*, 10, 553-566 (1956), The spectrum of singly-ionized calcium, Ca II.

1167. Engelhard, E., Spieweck, F., *Z. Naturforsch.*, 25a, 156 (1969), Ein ionen-laser fur metrologische anwendungen mitteilung aus der physika, lisch-technischen bundesanstalt braunschweig.

1168. Fendley, J. R. Jr.., Gorog, I., Hernqvist, K. G., Sun, C., *IEEE J. Quantum Electron.*, QE-6, 8 (1970), Characteristics of a sealed-off He^3-Cd^{114} laser.

1169. Fendley, J. R. Jr.., *IEEE J. Quantum Electron.*, QE-4, 627-631 (1968), Continuous UV lasers.

1170. Eriksson, K. B. S., *Ark. Fys.*, 13, 303-329 (1958), The spectrum of the singly-ionized nitrogen atom.

1171. Tell, B., Martin, R. J., McNair, D., *IEEE J. Quantum Electron.*, QE-3, 96 (2) (1967), CW laser oscillation in ionized xenon at 9697 Å.

1172. Bridges, W. B., Freiberg, R. J., Halsted, A. S., *IEEE J. Quantum Electron.*, QE-3, 339 (2) (1967), New continuous UV ion laser transitions in neon, argon, and krypton.

1173. Allen, R. B., Starnes, R. B., Dougal, A. A., *IEEE J. Quantum Electron.*, QE-2, 334 (2) (1966), A new pulsed ion laser transition in nitrogen at 3995 Å.

1174. McNeil, J. R., Collins, G. J., Persson, K. B., Franzen, D. L., *Appl. Phys. Lett.*, 27, 595-598 (1975), CW laser oscillation in Cu II.

1175. Coleman, C. D., Bozman, W. R., Meggers, W. F., Monograph, No. 3(1960), *Table of Wavenumbers II*, U. S. National Bureau of Standards.

1176. Gregg, D. W., Thomas, S. J., *J. Appl. Phys.*, 39, 4399 (1968), Analysis of the CS_2-O_2 chemical laser showing new lines and selective excitation.

1177. Patel, C. K. N., *Appl. Phys. Lett.*, 6, 12 (1965), CW laser action in N_2O (N_2-N_2O system).

1178. Mathias, L. E. S., Crocker, A., Wills, M. S., *Phys. Lett.*, 13, 303 (1964), Laser oscillations from nitrous oxide at wavelengths around 10.9 micrometers.

1179. Siemsen, K., Reid, J., *Opt. Commun.*, 20, 284(1977), New N_2O laser band in the 10-μm wavelength region.

1180. Tiee, J. J., Wittig, C., *Appl. Phys. Lett.*, 30, 420 (1977), CF_4 and NOCl molecular lasers operating in the 16-μm region.

1181. Tiee, J. J., Wittig, C., *J. Appl. Phys.*, 49, 61 (1978), Optically pumped molecular lasers in the 11- to 17-μm region.

1182. Karlov, N. V., Konev, Yu. B., Petrov, Yu. N., Prokhorov, A. M. et al., *JETP Lett.*, 8, 12 (1968), Laser based on boron trichloride.

1183. Akitt, D. P., Yardley, J. T., *IEEE J. Quantum Electron.*, QE-6, 113 (1970), Far-infrared laser emission in gas discharges containing boron trihalides.

1184. Shelton, C. F., Byrne, F. T., *Appl. Phys. Lett.*, 17, 436 (1970), Laser emission near 8 μ from a H_2-C_2H_2-He mixture.

1185. Kildal, H., Deutsch, T. F., *Appl. Phys. Lett.*, 27, 500 (1975), Optically pumped infrared V-V transfer lasers.

1186. Nelson, L. Y., Fisher, C. H., Hoverson, S. J., Byron, S. R., O'Neill, F., *Appl. Phys. Lett.*, 30, 192 (1977), Electron-beam-controlled discharge excitation of a CO-C_2H_2 energy transfer laser.
1187. Puerta, J., Herrmann, W., Bourauel, G., Urban, W., *Appl. Phys.*, 19, 439 (1979), Extended spectral distribution of lasing transitions in a liquid-nitrogen cooled CO-laser.
1188. Rutt, H. N., Green, J. M., *Opt. Commun.*, 26, 422 (1978), Optically pumped laser action in dideuteroacetylene.
1189. Fry, S. M., *Opt. Commun.*, 19, 320 (1976), Optically pumped multiline NH_3 laser.
1190. Yamabayashi, N., Yoshida, T., Myazaki, K., Fujisawa, K., *Opt. Commun.*, 30, 245 (1979), Infrared multiline NH_3 laser and its application for pumping an InSb laser.
1191. Tashiro, H., Suzuki, K., Toyoda, K., Namba, S., *Appl. Phys.*, 21, 237 (1980), Wide-range line-tunable oscillation of an optically pumped NH_3 laser.
1192. Grasiuk, A. Z., *Appl. Phys.*, 21, 173 (1980), High-power tunable ir Raman and optically pumped molecular lasers for spectroscopy.
1193. Danielewicz, E. J., Malk, E. G., Coleman, P. D., *Appl. Phys. Lett.*, 29, 557 (1976), High-power vibration-rotation emission from $^{14}NH_3$ optically pumped off reasonance.
1194. Chang, T. Y., McGee, J. D., *Appl. Phys. Lett.*, 29, 725 (1976), Off-resonant infrared laser action in NH_3 and C_2H_4 without population inversion.
1195. Mochizuki, T., Yamanaka, M., Morikawa, M., Yamanaka, C., *Jpn. J. Appl. Phys.*, 17, 1295 (1978), unspecified.
1196. Bobrovskii, A. N., Vedenov, A. A., Kozhevnikov, A. V., Sobolenkó, D. N., *JETP Lett.*, 29, 537 (1979), NH_3 laser pumped by two CO_2 lasers.
1197. Baranov, V. Yu., Kazakov, S. A., Pis'mennyi, V. D., Starodubtsev, A. I., *Appl. Phys.*, 17, 317 (1978), Multiwatt optically pumped ammonia laser operation in the 12- to 13-µm region.
1198. Roh, W. B., Rao, K. N., *J. Mol. Spectrosc.*, 49, 317 (1974), CO laser spectra.
1199. Yoshida, T., Yamabayashi, N., Miyazaki, K., Fujisawa, K., *Opt. Commun.*, 26, 410 (1978), Infrared and far-infrared laser emissions from a TE CO_2 laser pumped NH_3 gas.
1200. Chang, T. Y., McGee, J. D., *Appl. Phys. Lett.*, 28, 526 (1976), Laser action at 12.812 µm in optically pumped NH_3.
1201. Shaw, E. D., Patel, C. K. N., *Opt. Commun.*, 27, 419 (1978), Improved pumping geometry for high-power NH_3 lasers.
1202. Gupta, P. K., Kar, A. K., Taghizadeh, M. R., Harrison, R. G., *Appl. Phys. Lett.*, 39, 32 (1981), 12.8-µ NH_3 laser emission with 40-60% power conversion and up to 28% energy conversion efficiency.
1203. Lee, W., Kim, D., Malk, E., Leap, J., *IEEE J. Quantum Electron.*, QE-15, 838 (1979), Hot-band lasing in NH_3.
1204. Jacobs, R. R., Prosnitz, D., Bischel, W. K., Rhodes, C. K., *Appl. Phys. Lett.*, 29, 710 (1976), Laser generation from 6 to 35 µm following two-photon excitation of ammonia.
1205. Eggleston, J., Dallarosa, J., Bischel, W. K., Bokor, J., Rhodes, C. K., *J. Appl. Phys.*, 50, 3867 (1979), Generation of 16-µm radiation in $^{14}NH_3$ by two-quantum excitation of the, $2\nu^{-2}(7.5)$ state.
1206. Akitt, D. P., Wittig, C. F., *J. Appl. Phys.*, 40, 902 (1969), Laser emission in ammonia.
1207. Mathias, L. E. S., Crocker, A., Wills, M. S., *Phys. Lett.*, 14, 33 (1965), Laser oscillations at wavelengths between 21 and 32 µ from a pulsed discharge through ammonia.

1208. Lide, D. R., Jr., *Phys. Lett.* A, 24, 599 (1967), Interpretation of the far-infrared laser oscillation in ammonia.

1209. Patel, C. K. N., Kerl, R. J., *Appl. Phys. Lett.*, 5, 81 (1964), Laser oscillation on $X'\Sigma^+$ vibrational rotational transitions of CO.

1210. Urban, S., Spirko, V., Papousek, D., McDowell, R S., Nereson, N. G. et al., *J. Mol. Spectrosc.* 79, 455 (1980), Coriolis and 1-type interactions in the ν_2, $2\nu_2$, and ν_4 states of $^{14}NH_3$.

1211. Jones, C. R., Buchwald, M. I., Gundersen, M., Bushnell, A. H., *Opt. Commun.*, 24, 27 (1978), Ammonia laser optically pumped with an HF laser.

1212. McDowell, R S., Patterson, C. W., Jones, C. R., Buchwald, M. I., Telle, J. M., *Opt. Lett.*, 4, 274 (1979), Spectroscopy of the CF_4 laser.

1213. Averim, V. G., Alimpiev, S. S., Baronov, G. S., Karlov, N. V. et al., *Sov. Tech. Phys. Lett.*, 4, 527 (1978), Spectroscopic characteristics of an optically pumped carbon tetrafluoride laser.

1214. Alimpiev, S. S., Baronov, G. S., Karlov, N. V., Karchevskii, A. I. et al., *Sov. Tech. Phys. Lett.*, 4, 69 (1978), Tuning and stabilization of an optically pumped carbon tetrafluoride laser.

1215. Lomaev, M. I., Nagornyi, D. Yu., Tarasenko, V. F., Fedenev, A. V., Kirillin, G. V., *Sov. J. Quantum Electron.* 19, 1321 (1989), Lasing due to atomic transitions of rare gases in mixtures with NF_3.

1216. Jones, C. R., Telle, J. M., Buchwald, M. I., *J. Opt. Soc. Am.*, 68, 671 (1978), Optically pumped isotopic CF_4 lasers.

1217. Knyazev, I. N., Letokhov, V. S., Lobko, V. V., *Opt. Commun.*, 29, 73 (1979), Weakly forbidden vibration-rotation transitions ΔR not= 0 in CF_4 laser.

1218. Prosnitz, D., Jacobs, R. R., Bischel, W. K., Rhodes, C. K., *Appl. Phys. Lett.*, 32, 221 (1978), Stimulated emission at 9.75 μm following two-photon excitation of methyl fluoride.

1219. Green, J. M., Rutt, H. N., *Proc. 2nd Int. Conf. on Infrared Physics*, 205 (Mar, 1979), Optically pumped laser action in silicon tetrafluoride.

1220. Patel, C. K. N., *Phys. Rev.*, 141, 71 (1966), Vibrational-rotational laser action in carbon monoxide.

1221. Barch, W. E., Fetterman, H. R., Schlossberg, H. R., *Opt. Commun.*, 15, 358 (1975), Optically pumped 15.90 μm SF_6 laser.

1222. Dunham, J. L., *Phys. Rev.*, 41, 721 (1932), The energy levels of a rotating vibrator.

1223. Osgood, R. M., Jr., Eppers, W. C., Jr., *Appl. Phys. Lett.*, 13, 409 (1968), High-power CO-N_2-He laser.

1224. Treanor, C. E., Rich, J. W., Rehm, R. G., *J. Chem. Phys.*, 48, 1798 (1968), unspecified.

1225. Jeffers, W. Q., Wiswall, C. E., *J. Appl. Phys.*, 42, 5059 (1971), Analysis of pulsed CO lasers.

1226. Rich, J. W., Thompson, H. M., *Appl. Phys. Lett.*, 19, 3 (1971), Infrared sidelight studies in the high-power carbon monoxide laser.

1227. Amat, G., *J. Chim. Phys.*, 64, 91 (1967), Discussion following a paper by C. K. N. Patel.

1228. Tyte, D. C., *Advances in Quantum Electronics*, Vol. 1(1970), Academic Press, London, Carbon dioxide lasers.

1229. Cheo, P. K., *Lasers,* Vol. 3(1971), Marcel Dekker, New York, CO_2 lasers.

1230. Patel, C. K. N., *Appl. Phys. Lett.*, 7, 246 (1965), CW laser on vibrational-rotational transitions of CO.

1231. Nigham, W. L., *Phys. Rev.* A, 2, 1989 (1970), Electron energy distribution and collision rates in electrically excited N_2, CO, and CO_2.

1232. Burkhardt, E. G., Bridges, T. J., Smith, P. W., *Opt. Commun.*, 6, 193 (1972), BeO capillary CO_2 waveguide laser.

1233. Wood, O. R., II, *Proc. IEEE,* 62, 355 (1974), High-pressure pulsed molecular lasers.
1234. Taylor, R. S., Alcock, A. J., Sarjeant, W. J., Leopold, K. E., *IEEE J. Quantum Electron.*, QE-15, 1131 (1979), Electrical and gain characteristics of a multiatmosphere uv-preionized CO_2 laser.
1235. Chang, T. Y., Wood, O. R., II, *IEEE J. Quantum Electron.*, QE-13, 907 (1977), Optically pumped continuously tunable high-pressure molecular lasers.
1236. Javan, A., *Phys. Rev.*, 107, 1579 (1957), Theory of a three-level maser.
1237. Siemsen, K. J., *Opt. Lett.*, 6, 114 (1981), Sequence bands of the isotope $^{12}C^{18}O_2$ laser.
1238. Siemsen, K. J., *Opt. Commun.*, 34, 447 (1980), The sequence bands of the carbon-13 isotope CO_2 laser.
1239. Fischer, T. A., Tiee, J. J., Wittig, C., *Appl. Phys. Lett.*, 37, 592 (1980), Optically pumped NSF molecular laser.
1240. Deka, B. K., Dyer, P. E., Winfield, R. J., *Opt. Lett.*, 5, 194 (1980), New 17-21-μm laser lines in C_2D_2 using a continuously tunable CO_2 laser pump.
1241. Bhaumik, M. L., *Appl. Phys. Lett.*, 17, 188 (1970), High efficiency CO laser at room temperature.
1242. Baldacci, A., Ghersetti, S., Hurlock, S. C., Rao, K. N., *J. Mol. Spectrosc.*, 42, 327 (1972), Spectrum of dideuteroacetylene near 18.6 microns.
1243. Znotins, T. A., Reid, J., Garside, B. K., Ballik, E. A., *Opt. Lett.*, 5, 528 (1980), 12-μm NH_3 laser pumped by a sequence CO_2 laser.
1244. Deka, B. K., Dyer, P. E., Winfield, R. J., *Opt. Commun.*, 33, 206 (1980), Optically pumped NH_3 laser using a continuously tunable CO_2 laser.
1245. Rutt, H. N., *Opt. Commun.*, 34, 434 (1980), Optically pumped laser action in perchloryl fluoride.
1246. Fischer, T. A., Wittig, C., *Appl. Phys. Lett.*, 39, 6 (1981), 16-μm laser oscillation in propyne.
1247. Schmid, W. E., *High-Power Lasers and Applications* (1979), 148, Springer-Verlag, New York, A simple high energy TEA CO laser.
1248. Osgood, R. M., Jr., Nichols, E. R., Eppers, W. C., Jr., Petty, R. D., *Appl. Phys. Lett.*, 15, 69 (1969), Q switching of the carbon monoxide laser.
1249. Bhaumik, M. L., Lacina, W. B., Mann, M. M., *IEEE J. Quantum Electron.*, QE-6, 576 (1970), Enhancement of CO laser efficiency by addition of Xenon.
1250. Qi, N., Krishnan, M., *Phys. Rev. Lett.* 59, 2051 (1987), Photopumping of a C III ultraviolet laser by Mn VI line radiation.
1251. Daiber, J. W., Thompson, H. M., *IEEE J. Quantum Electron.*, QE-13, 10 (1977), Performance of a large, cw, preexcited CO supersonic laser.
1252. Boness, M. J. W., Center, R. E., *J. Appl. Phys.*, 48, 2705 (1977), High-pressure pulsed electrical CO laser.
1253. Johns, J. W. C., McKellar, A. R. W., Weitz, D., *J. Mol. Spectrosc.*, 51, 539 (1974), Wavelength measurements of $^{13}C^{16}O$ laser transitions.
1254. Ross, A. H. M., Eng, R. S., Kildal, H., *Opt. Commun.*, 12, 433 (1974), Heterodyne measurements of $^{12}C^{18}O$, $^{13}C^{16}O$, and $^{13}C^{18}O$ laser frequencies;, mass dependence of Dunham coefficients.
1255. Deutsch, T. F., *IEEE J. Quantum Electron.*, QE-3, 419 (2) (1967), New infrared laser transitions in HCl, HBr, DCl, and DBr.
1256. Wood, O. R., Chang, T. Y., *Appl. Phys. Lett.*, 20, 77 (1972), Transverse-discharge hydrogen halide lasers.
1257. Rutt, H. N., *J. Phys. D.*, 12, 345 (1979), A high-energy hydrogen bromide laser.
1258. Keller, F. L., Nielsen, A. H., *J. Chem. Phys.*, 22, 294 (1954), The infrared spectrum and molecular constants of DBr.
1259. Rank, D. H., Rao, B. S., Wiggins, T. A., *J. Mol. Spectrosc.*, 17, 122 (1965), Molecular constants of HCl^{35}.

1260. Webb, D. U., Rao, K. N., *Appl. Opt.*, 5, 1461 (1966), A heated absorption cell for studying infrared absorption bands.

1261. Diemer, U., Demtröder, W., *Chem. Phys. Lett.* 175, 135 (1991), Infrared atomic Cs laser based on optical pumping of Cs_2 molecules.

1262. Kasper, J. V. V., Pimentel, G. C., *Phys. Rev. Lett.*, 14, 352 (1965), HCl chemical laser.

1263. Corneil, P. H., Pimentel, G. C., *J. Chem. Phys.*, 49, 1379 (1968), Hydrogen-chloride explosion laser II DCl.

1264. Henry, A., Bourcin, F., Arditi, I., Charneau, R., Menard, J., *C. R. Acad. Sci. Ser. B*, 267, 616 (1968), Effect laser par reaction chimique de l'hydrogene sur du chlore ou du chlorure de nitrosyle.

1265. Pickworth, J., Thompson, H. W., *Proc. R. Soc. London, Ser. A*, 218, 37 (1953), The fundamental vibration-rotation band of deuterium chloride.

1266. Webb, D. U., Rao, K. N., *J. Mol. Spectrosc.*, 28, 121 (1968), Vibration rotation bands of heated hydrogen halides.

1267. Skribanowitz, N., Herman, I. P., Osgood, R. M., Jr., Feld, M. S., Javan, A., *Appl. Phys. Lett.*, 20, 428 (1972), Anisotropic ultrahigh gain emission observed in rotational transitions in optically pumped HF gas.

1268. Skribanowitz, N., Herman, I. P., Feld, M. S., *Appl. Phys. Lett.*, 21, 466 (1972), Laser oscillation and anisotropic gain in the $1 \rightarrow 0$ vibrational band of optically pumped HF gas.

1269. Ultee, C. J., *Rev. Sci. Instrum.*, 42, 1174 (1971), Compact pulsed HF lasers.

1270. Suchard, S. N., Gross, R. W. F., Whittier, J. S., *Appl. Phys. Lett.*, 19, 411 (1971), Time-resolved spectroscopy of a flash-imitated H_2-F_2 laser.

1271. Kwok, M. A., Giedt, R. R., Gross, R. W. F., *Appl. Phys. Lett.*, 16, 386 (1970), Comparison of HF and DF continuous chemical lasers. II. Spectroscopy.

1272 Goldsmith, J. E. M., *J. Opt. Soc. Am. B* 6, 1979 (1989), Two photon excited stimulated emission from atomic hydrogen in flames.

1273. Kwok, M. A., Giedt, R. R., Varwig, R. L., *AIAA J.*, 14, 1318 (1976), Medium diagnostics for a 10-kW cw HF chemical laser.

1274. Deutsch, T. F., *Appl. Phys. Lett.*, 10, 234 (1967), Molecular laser action in hydrogen and deuterium halides.

1275. Gerber, R. A., Patterson, E. L., *J. Appl. Phys.*, 47, 3524 (1976), Studies of a high-energy HF laser using an electron-beam-excited mixture of high-pressure F_2 and H_2.

1276. Eng, R. S., Spears, D. L., *Appl. Phys. Lett.*, 27, 650 (1975), Frequency stabilization and absolute frequency measurements of a cw HF/DF laser.

1277. Pummer, H., Proch, D., Schmailzl, U., Kompa, K. L., *Opt. Commun.*, 19, 273 (1976), The generation of partial and total vibrational inversion in colliding molecular systems initiated by ir-laser absorption.

1278. Gensel, P., Kompa, K. L., Wanner, J., *Chem. Phys. Lett.*, 5, 179 (1970), IF_5-H_2 hydrogen fluoride chemical laser involving a chain reaction.

1279. Dolgov-Savel'ev, G. G., Zharov, V. F., Neganov, Yu. S., Chumak, G. M., *Sov. Phys. JETP*, 34, 34 (1972), Vibrational-rotational transitions in an $H_2 + F_2$ chemical laser.

1280. Mayer, S. W., Taylor, D., Kwok, M. A., *Appl. Phys. Lett.*, 23, 434 (1973), HF chemical lasing at higher vibrational levels.

1281. Suchard, S. N., Pimentel, G. C., *Appl. Phys. Lett.*, 18, 530 (1971), Deuterium fluoride vibrational overtone chemical laser.

1282. Ultee, C. J., *IEEE J. Quantum Electron.*, QE-8, 820 (1972), Compact pulsed deuterium fluoride laser.

1283. Spencer, D. J., Mirels, H., Jacobs, T. A., *Appl. Phys. Lett.*, 16, 384 (1970), Comparison of HF and DF continuous chemical lasers. I. Power.

1284. Petersen, A. B., Braverman, L. W., Wittig, C., *J. Appl. Phys.*, 48, 230 (1977), H_2O, NO, and N_2O infrared lasers pumped directly and indirectly by electronic-vibrational energy transfer.
1285. Deutsch, T. F., *Appl. Phys. Lett.*, 9, 295 (1966), NO molecular laser.
1286. Pollack, M. A., *Appl. Phys. Lett.*, 9, 94 (1966), Molecular laser action in nitric oxide by photodissociation of NOCl.
1287. Giuliano, C. R., Hess, L. D., *J. Appl. Phys.*, 38, 4451 (1967), Chemical reversibility and solar excitation rates of the nitrosyl chloride photodissociative laser.
1288. Callear, A. B., Van den Bergh, H. E., *Chem. Phys. Lett.*, 8, 17 (1971), An hydroxyl radical infrared laser.
1289. Wanchop, T. S., Schiff, H. I., Welge, K. H., *Rev. Sci. Instrum.*, 45, 653 (1974), Pulsed-discharge infrared OH laser.
1290. Rice, W. W., Jensen, R. J., *Appl. Phys. Lett.*, 22, 67 (1973), Aluminum fluoride exploding-wire laser.
1291. Jensen, R. J., *Handbook of Chemical Lasers* (1976), 703, John Wiley & Sons, New York, Metal-atom oxidation laser.
1292. Rice, W. W., Beattie, W. H., Oldenborg, R. C., Johnson, S. E., Scott, P., *Appl. Phys. Lett.*, 28, 444 (1976), Boron fluoride and aluminum fluoride infrared lasers from quasicontinuous supersonic mixing flames.
1293. Bergman, R. C., Rich, J. W., *Appl. Phys. Lett.*, 31, 597 (1977), Overtone bands lasing at 2.7-3μm in electrically excited CO.
1294. Znotis, T. A., Reid, J., Garside, B. K., Ballik, E. A., *Opt. Lett.*, 4, 253 (1979), 4.3-μm TE CO_2 lasers.
1295. Guelachvili, G., *J. Mol. Spectrosc.*, 79, 72 (1980), High-resolution Fourier spectra of carbon dioxide and three of its isotopic species near 4.3 μm.
1296. Petersen, A. B., Wittig, C., *J. Appl. Phys.*, 48, 3665 (1977), Line-tunable CO_2 laser operating in the region 2280-2360 cm^{-1} pumped by energy transfer from Br ($4^2P_{1/2}$).
1297. Rao, D. R., Hocker, L. O., Javan, A., Knable, K., *J. Mol. Spectrosc.*, 25, 410 (1968), Spectroscopic studies of 4.3 μ transient laser oscillation in CO_2.
1298. Freed, C., Bradley, L. C., O'Donnell, R. G., *IEEE J. Quantum Electron.*, QE-16, 1195 (1980), Absolute frequencies of lasing transitions in seven CO_2 isotopic species.
1299. Ernst, G. J., Witteman, W. J., *IEEE J. Quantum Electron.*, QE-7, 484 (1971), Transition selection with adjustable outcoupling for a laser device applied to CO_2.
1300. Patel, C. K. N., *Phys. Rev.*, 136, 1187 (1964), Continuous-wave laser action on vibrational-rotational transitions of CO_2.
1301. Brown, C. O., Davis, J. W., *Appl. Phys. Lett.*, 21, 480 (1972), Closed-cycle performance of a high-power electric discharge laser.
1302. Gerry, E. T., *Laser Focus*, 27, (1970), The gas dynamic laser.
1303. Richardson, M. C., Alcock, A. J., Leopold, K., Burtyn, P., *IEEE J. Quantum Electron.*, QE-9, 236 (1973), A 300-J multigigawatt CO_2 laser.
1304. Kildal, H., Eng, R. S., Ross, A. H. M., *J. Mol. Spectrosc.*, 53, 479 (1974), Heterodyne measurements of $^{12}C^{16}O$ laser frequencies and improved Dunham coefficients.
1305. Schappert, G. T., Singer, S., Ladish, J., Montgomery, M. D., *J. Opt. Soc. Am.*, 68, 668 (1978), Comparison of theory and experiment on the performance of the Los Alamos, Scientific Laboratory eight-beam 10 kJ CO_2 laser.
1306. Siemsen, K. J., Whitford, B. G., *Opt. Commun.*, 22, 11 (1977), Heterodyne frequency measurements of CO_2 laser sequence-band transitions.
1307. Whitford, B. G., Siemsen, K. J., Reid, J., *Opt. Commun.*, 22, 261 (1977), Heterodyne frequency measurements of CO_2 laser hot-band transitions.

1308. Hartmann, B., Kleman, B., *Can. J. Phys.*, 44, 1609 (1966), Laser lines from CO_2 in the 11 to 18 micron region.

1309. Kasner, W. H., Pleasance, L. D., *Appl. Phys. Lett.*, 31, 82 (1977), Laser emission from the 13.9-μm $10^0 \rightarrow 01^10$ CO_2 transition in pulsed electrical discharges.

1310. Paso, R., Kauppinen, J., Anttila, R., *J. Mol. Spectrosc.*, 79, 236 (1980), Infrared spectrum of CO_2 in the region of the bending fundamental ν_2.

1311. Osgood, R. M., Jr., *Appl. Phys. Lett.*, 32, 564 (1978), 1-mJ line-tunable optically pumped 16-μm laser.

1312. Buchwald, M. I., Jones, C. R., Fetterman, H. R., Schlossberg, H. R., *Appl. Phys. Lett.*, 29, 300 (1976), Direct optically pumped multi-wavelength CO_2 laser.

1313. Freed, C., Ross, A. H. M., O'Donnell, R. G., *J. Mol. Spectrosc.*, 49, 439 (1974), Determination of laser line frequencies and vibrational-rotational constants of the $^{12}C^{18}O_2$, $^{13}C^{16}O_2$, and $^{13}C^{18}O_2$.

1314. Siddoway, J. C., *J. Appl. Phys.*, 39, 4854 (1968), Calculated and observed laser transitions using $C^{14}C^{16}O_2$.

1315. Sadie, F. G., Buger, P. A., Malan, O. G., *J. Appl. Phys.*, 43, 2906 (1972), Continuous-wave overtone bands in a CS_2-O_2 chemical laser.

1316. DePoorter, G. L., Balog, G., *IEEE J. Quantum Electron.*, QE-8, 917 (1972), New infrared laser lines in OCS and new method for C atom lasing.

1317. Kildal, H., Deutsch, T. F., *Tunable Lasers and Applications* (1976), 367, Springer-Verlag, New York, Optically pumped gas lasers.

1318. Maki, A. G., Plyler, E. K., Tidwell, E. D., *J. Res. Natl. Bur. Stand.* Sect. A, 66, 163 (1962), Vibration-rotation bands of carbonyl sulfide.

1319. Deutsch, T. F., *Appl. Phys. Lett.*, 8, 334 (1966), OCS molecular laser.

1320. Schlossberg, H. R., Fetterman, H. R., *Appl. Phys. Lett.*, 26, 316 (1975), Optically pumped vibrational transition laser in OCS.

1321. Deutsch, T. F., Kildal, H., *Chem. Phys. Lett.*, 40, 484 (1976), A reexamination of the CS_2 laser.

1322. Patel, C. K. N., *Appl. Phys. Lett.*, 7, 273 (1965), CW laser oscillation in an N_2-CS_2 system.

1323. Nelson, L. Y., Fisher, C. H., Byron, S. R., *Appl. Phys. Lett.*, 25, 517 (1974), unspecified.

1324. Basov, N. G., Kazakevich, V. S., Kovsh, I. B., *Sov. J. Quantum Electron.*, 10, 1131 (1980), Electron-beam-controlled laser utilizing the first overtones of the vibrational-rotational transitions in the CO molecule, I and II.

1325. Bushnell, A. H., Jones, C. R., Buchwald, M. I., Gundersen, M., *IEEE J. Quantum Electron.*, QE-15, 208 (1979), New HF laser pumped molecular lasers in the middle infrared.

1326. Jones, C. R., *Laser Focus*, 68, (1978), Optically pumped mid-ir lasers.

1327. Petersen, A. B., Wittig, C., Leone, S. R., *Appl. Phys. Lett.*, 27, 305 (1975), Infrared molecular lasers pumped by electronic-vibrational energy transfer from $Br(4^2P_{1/2})$: CO_2, N_2O, HCN, and C_2H_2.

1328. Turner, R., Poehler, T. O., *Phys. Lett.* A, 27, 479 (1968), Emission from HCN and H_2O lasers in the 4- to 13-μm region.

1329. Benedict, W. S., Pollack, M. A., Tomlinson, W. J., *IEEE J. Quantum Electron.*, QE-5, 108 (1969), The water-vapor laser.

1330. Hartmann, B., Kleman, B., Spangstedt, G., *IEEE J. Quantum Electron.*, QE-4, 296 (1968), Water vapor laser lines in the 7-μm region.

1331. Wood, O. R., Burkhardt, E. G., Pollack, M. A., Bridges, T. J., *Appl. Phys. Lett.*, 18, 112 (1971), High-pressure laser action in 13 gases with transverse excitation.

1332. Djeu, N., Wolga, G. J., *IEEE J. Quantum Electron.*, QE-5, 50 (1969), Observation of new laser transitions in N_2O.

1333. Howe, J. A., *Phys. Lett.*, 17, 252 (1965), R-branch laser action in N_2O.

1334. Moeller, G., Rigden, J. D., *Appl. Phys. Lett.*, 8, 69 (1966), Observation of laser action in the R-branch of CO_2 and N_2O vibrational spectra.
1335. Whitford, B. G., Siemsen, K. J., Riccius, H. D., Hanes, G. R., *Opt. Commun.*, 14, 70 (1975), Absolute frequency measurements of N_2O laser transitions.
1336. Auyeung, R. C. Y., Cooper, D. G., Kim, S., Feldman, B. J., *Opt. Comm.* 79, 207 (1990), Stimulated emission in atomic hydrogen at 656 nm.
1337. Jones, C. R., *Laser Focus*, 70, (1978), Optically pumped mid-ir lasers.
1338. Cord, M. S., Peterson, J. D., Lojko, M. S., Haas, R. H., Monograph 70, Vol. 4(1968), U. S. National Bureau of Standards, Microwave spectral tables.
1339. DeTemple, T. A., *Infrared and Millimeter Waves, Sources of Radiation*, Vol. 1(1979), Academic Press, London, Pulsed optically pumped FIR lasers.
1340. Henningsen, J. O., Jensen, H. G., *IEEE J. Quantum Electron.*, QE-11, 248 (1975), The optically pumped FIR laser.
1341. DeTemple, T. A., Danielewicz, E. J., *IEEE J. Quantum Electron.*, QE-12, 40 (1976), CW CH_3F waveguide laser at 496 μm: theory and experiment.
1342. Temkin, R. J., *IEEE J. Quantum Electron.*, QE-13, 450 (1977), Theory of optically pumped submillimeter lasers.
1343. Kim, K. J., Coleman, P. D., *IEEE J. Quantum Electron.*, QE-16, 1341 (1980), Calculated-experimental evaluation of the gain/absorption spectra of several optically pumped NH_3 systems.
1344. Curtis, J., Ph.D. Thesis(1974), Ohio State University, Vibration-Rotation Bands of NH_3 in the Region 670-1860 cm^{-1}.
1345. Henningsen, J. O., *Infrared and Millimeter Waves, Sources of Radiation* (1981), Academic Press, London, Spectroscopy of molecules by far IR laser emission.
1346. Woods, D. R., Ph.D. Thesis(1970), The High Resolution Infrared Spectra of Normal and Deuterated Methanol Between 400 and 1300 cm^{-1}, University of Michigan.
1347. Berdnikov, A. A., Derzhiev, V. I., Murav'ev, I. I., Yakovlenko, S. I., Yancharina, A. M., *Sov. J. Quantum Electron.* 17, 1400 (1987), Penning plasma laser utilizing new transitions in the helium atom resulting in the emission of visible light.
1348. Zolotarev, V. A., Kryukov, P. G., Podmar'kov, Yu. P., Frolov, M. P., Shcheglov, V. A., *Sov. J. Quantum Electron.* 18, 643 (1988), Optically pumped pulsed IF (B→X) laser utilizing a CF_3I-NF_2-He mixture.
1349. Hocker, L. O., Javan, A., Rao, D. R., Frenkel, L., Sullivan, T., *Appl. Phys. Lett.*, 10, 147 (1967), Absolute frequency measurement and spectroscopy of gas laser transitions in the far IR.
1350. Frenkel, L., *Appl. Phys. Lett.*, 11, 344 (1967), Absolute frequency measurement of the 118.6 μm water vapor laser transition.
1351. Evenson, K. M., *Appl. Phys. Lett.*, 20, 133 (1972), Extension of absolute frequency measurements to the CW He-Ne laser at 88THz (3.39 μm).
1352. Hodges, D. T., Hartwick, T. S., *Appl. Phys. Lett.*, 23, 252 (1973), Waveguide lasers for the far infrared pumped by a CO_2 laser.
1353. Danielewicz, E. J., Coleman, P. D., *Appl. Opt.*, 15, 761 (1976), Hybrid metal mesh-dielectric mirrors for optically pumped far IR lasers.
1354. Fetterman, H. R., Tannenwald, P. E., Parker, C. D., Melngailas, J. et al., *Appl. Phys. Lett.*, 34, 123 (1979), Real-time spectral analysis of far IR laser pulses using a SAW dispersive delay line.
1355. Mathis, L. E. S., Parker, J. T., *Appl. Phys. Lett.*, 3, 16 (1963), Stimulated emission in band spectrum of nitrogen.
1356. Lide, D. R., *Appl. Phys. Lett.*, 11, 62 (1967), On the explanation of the so called CN laser.
1357. Chang, T. Y., Budgee, T., *Opt. Commun.*, 1, 423 (1970), Laser action at 452, 496, and 541 μm in optically pumped CH_3F.

1358. Hodges, D. T., *Infrared Phys.*, 18, 375 (1978), A review of advances in optically pumped far IR lasers.
1359. Tucker, J. R., *IEEE Conference Digest 17, Int. Conf. on Submillimeter Waves and Their Applications*, Theory of a FIR gas laser.
1360. Chang, T. Y., *IEEE Trans. Microwave Theory Tech.*, MTT-22, 983 (1974), Optically pumped submillimeter-wave sources.
1361. Coleman, P. D., *J. Opt. Soc. Am.*, 67, 894 (1977), Present and future problems concerning lasers in the far IR spectra range.
1362. Amos, K. B., Davis, J. A., *IEEE J. Quantum Electron.*, QE-16, No. 5, 574-575 (1980), Additional CW far-infrared laser lines from CO_2 laser pumped cis 12 $C_2H_2F_2$.
1363. Arimondo, E., Inguscio, M., *J. Mol. Spectrosc.*, 75, 81-86 (1979), The rotation-vibration constants of the $^{12}CH_3F$ ν_3 band.
1364. Arimondo, E., Inguscio, A., *Digest for the 4th Ann. Conf. on Infrared and Millimeter Waves, IEEE,* Assignments of laser lines in optically pumped CH_3CN, IEEE, Piscataway, NJ.
1365. Arimondo, E., Inguscio, A., *Int. J. Infrared Millimeter Waves*, 1, No. 3, 437-458 (1980), Assignments of laser lines in optically pumped CH_3CN.
1366. Baskakov, O. I., Dyubko, S. F., Moskienko, M. V., Fesenko, L. D., *Sov. J. Quantum Electron.*, 7, No. 4, 445-449 (1977), Identification of active transitions in a formic acid vapor laser.
1367. Bean, B. L., Perkowitz, S., *Opt. Lett.*, 1, No. 6, 202-204 (1977), Complete frequency coverage for submillimetre laser spectroscopy with optically pumped CH_3OH, CH_3OD, CD_3OD and CH_2CF_2.
1368. Benedict, W. S., Pollack, M. A., Tomlinson, W. J., III, *IEEE J. Quantum Electron.*, QE-5, No. 2, 108-124 (1969), The water-vapor laser.
1369. Blaney, T. G., Bradley, C. C., Edwards, G. J., Knight, D. J. E., *Phys. Lett.* A, 43, No. 5, 471-472 (1973), Absolute frequency measurement of a Lamb-dip stabilised water vapour laser oscillating at 10.7 THz (28 μm).
1370. Blaney, T. G., Knight, D. J. E., Murray-Lloyd, E., *Opt. Commun.*, 25, No. 2, 176-178 (1978), Frequency measurements of some optically-pumped laser lines in CH_3OD.
1371. Blaney, T. G., Cross, N. R., Knight, D. J. E., Edwards, G. J., Pearce, P., *J. Phys. D.*, 13, No. 8, 1365-1370 (1980), Frequency measurement at 4.25 THz (70.5 μm) using a Josephson harmonic mixer and phase-lock techniques.
1372. Belland, P., Veron, D., Whitbourn, B., *Appl. Opt.*, 15, No. 12, 3047-3053 (1976), Scaling laws for cw 337-μm HCN waveguide lasers.
1373. Belland, P., Veron, D., *IEEE J. Quantum Electron.*, QE-16, No. 8, 885-889 (1980), Amplifying medium characteristics in optimised 190 μm/195 μm DCN waveguide lasers.
1374. Chang, T. Y., Bridges, T. J., *Opt. Commun.*, 1, No. 9, 423-426 (1970), Laser action at 452, 496 and 541 μm in optically pumped CH_3F.
1375. Chang, T. Y., Bridges, T. J., Burkhardt, E. J., *Appl. Phys. Lett.*, 17, No. 6, 249-251 (1970), cw submillimeter laser action in optically pumped methyl fluoride, methyl alcohol and vinyl chloride gases.
1376. Chang, T. Y., Bridges, T. J., Burkhardt, E. J., *Appl. Phys. Lett.*, 17, No. 9, 357-358 (1970), cw laser action at 81.5 and 263.4 μm in optically pumped ammonia gas.
1377. Chang, T. Y., McGee, J. D., *Appl. Phys. Lett.*, 19, No. 4, 103-105 (1971), Millimeter and submillimeter wave laser action in symmetric top molecules optically pumped via parallel absorption bands.
1378. Chang, T. Y., McGee, J. D., *IEEE J. Quantum Electron.*, QE-12, No. 1, 62-65 (1976), Millimeter and submillimeter wave laser action in symmetric top molecules optically pumped via perpendicular absorption bands.

1379. Dangoisse, D., Deldalle, A., Splingard, J-P., Bellet, J., *C. R. Acad. Sci. Ser.* B, 283, 115-118 (1976), Mesure precise des emissions continues du laser submillimetrique a acide formique.

1380. Dangoisse, D., Deldalle, A., Splingard, J-P., Bellet, J., *IEEE J. Quantum Electron.*, QE-13, No. 9, 730-731 (1977), CW optically pumped laser action in D_2CO, HDCO and $(H_2CO)_3$.

1381. Dangoisse, D., Willemot, E., Deldalle, A., Bellet, J., *Opt. Commun.*, 28, No. 1, 111-116 (1979), Assignment of the HCOOH cw-submillimeter laser.

1382. Dangoisse, D., Duterage, B., Glorieux, P., *IEEE J. Quantum Electron.*, QE-16, No. 3, 296-300 (1980), Assignment of laser lines in optically pumped submillimetre lasers: HDCO, D_2CO.

1383. Daneu, V., Hocker, L. O., Javan, A., Rao, D. R., Szoke, A., Zernike, F., *Phys. Lett.* A, 29, No. 6, 319-320 (1969), Accurate laser wavelength measurements in the infrared and far infrared using a Michelson interferometer.

1384. Deldalle, A., Dangoisse, D., Splingard, J-P., Bellet, J., *Opt. Commun.*, 22, No. 3, 333-336 (1977), Accurate measurements of cw optically pumped FIR laser lines of formic acid molecule and its isotopic species $H^{13}COOH$, HCOOD and DCOOD.

1385. Deroche, J-C., *J. Mol. Spectrosc.*, 69, 19-24 (1978), Assignment of submillimeter laser lines in methyl chloride.

1386. Danielewicz, E. J., Galantowicz, T. A., Foote, F. B., Reel, R. D., Hodges, D. T., *Opt. Lett.*, 4, No. 9, 280-282 (1979), High performance at new FIR wavelengths from optically pumped CH_2F_2.

1387. Duxbury, G., Gamble, T. J., Herman, H., *IEEE Trans. Microwave Theory Tech.*, MTT-22, No. 12, 1108-1109 (1974), Assignments of optically pumped laser lines of 11 difluoroethylene.

1388. Duxbury, G., Herman, H., *J. Phys. B*, 11, No. 5, 935-949 (1978), Optically pumped millimetre lasers.

1389. Danielewicz, E. J., Reel, R. D., Hodges, D. T., *IEEE J. Quantum Electron.*, QE-16, No. 4, 402-405 (1980), New far-infrared CW optically pumped cis- $C_2H_2F_2$ laser.

1390. Danielewicz, E. J., Weiss, C. O., *IEEE J. Quantum Electron.*, QE-14, No. 4, 222-223 (1978), Far infrared emission from $^{15}NH_3$ optically pumped by a CW sequence band CO_2 laser.

1391. Danielewicz, E. J., Weiss, C. O., *IEEE J. Quantum Electron.*, QE-14, No. 7, 458-459 (1978), New CW far-infrared laser lines from CO_2 laser-pumped CD_3OH.

1392. Danielewicz, E. J., Weiss, C. O., *IEEE J. Quantum Electron.*, QE-14, No. 10, 705-707 (1978), New efficient CW far-infrared optically pumped CH_2F_2 laser.

1393. Danielewicz, E. J., Weiss, C. O., *Opt. Commun.*, 27, No. 1, 98-100 (1978), New cw far-infrared D_2O, $^{12}CH_3F$ and $^{14}NH_3$ laser lines.

1394. Danielewicz, E. J., *Digest for the 4th Ann. Conf. on Infrared and Millimeter Waves,* IEEE, Molecular parameters determining the performance of CW optically pumped, FIR lasers, IEEE, Piscataway, NJ.

1395. Dyubko, S. F., Svich, V. A., Valitov, R. A., *JETP Lett.*, 7, No. 11, 320 (1968), SO_2 submillimetre laser generating at wavelengths 0.141 and 0.193 mm.

1396. Dyubko, S. F., Svich, V. A., Fesenko, L. D., *JETP Lett.*, 16, No. 11, 418-419 (1972), Submillimeter-band gas laser pumped by a CO_2 laser.

1397. Dyubko, S. F., Svich, V. A., Fesenko, L. D., *Sov. Phys. Tech. Phys.*, 18, No. 8, 1121 (1974), Submillimeter CH_3OH and CH_3OD lasers with optical pumping.

1398. Dyubko, S. F., Svich, V. A., Fesenko, L. D., *Zh. Prikl. Spectrosk. (USSR)*, 20, No. 4, 718-719 (1974), Stimulated emission of submillimeter waves by hydrazine excited by a CO_2 laser.

1399. Dyubko, S. F., Svich, V. A., Fesenko, L. D., *Opt. Spectrosc.*, 37, No. 1, 118 (1974), Submillimeter laser emission of CH_3I molecules excited by CO_2.

1400. Dyubko, S. F., Svich, V. A., Fesenko, L. D., *Sov. J. Quantum Electron.*, 3, No. 5, 446 (1974), Submillimeter laser using formic acid vapor pumped with carbon dioxide laser radiation.

1401. Dyubko, S. F., Fesenko, L. D., Baskakov, O. I., Svich, V. A., *Zh. Prikl. Spectrosk. (USSR)*, 23, No. 2, 317-320 (1975), Use of CD_3I, CH_3I and CD_3Cl molecules as active substances for submillimeter lasers with optical pumping.

1402. Dyubko, S. F., Svich, V. A., Fesenko, L. D., *Izv. Vuz. Radiofiz. (USSR)*, 18, No. 10, 1434-1437 (1975), An experimental study of the radiation spectrum of submillimeter laser on CD_3OH molecules.

1403. Dyubko, S. F., Svich, V. A., Fesenko, L. D., *Sov. Phys. Tech. Phys.*, 20, No. 11, 1536-1538 (1975), Submillimeter HCOOH, DCOOH, HCOOD, and DCOOD laser.

1404. Dyubko, S. F., Efimenko, M. N., Svich, V. A., Fesenko, L. D., *Sov. J. Quantum Electron.*, 6, No. 5, 600-601 (1976), Stimulated emission of radiation from optically pumped vinyl bromide molecules.

1405. Edwards, G. J., unpublished (1978), private communication, National Physical Laboratory.

1406. Epton, P. J., Wilson, W. L., Jr., Tittel, F. K., Rabson, R. A., *Appl. Opt.*, 18, No. 11, 1704-1705 (1979), Frequency measurement of the formic acid laser 311-μm line.

1407. Evenson, K. M., Wells, J. S., Matarrese, L. M., Elwell, J. B., *Appl. Phys. Lett.*, 16, No. 4, 159-162 (1970), Absolute frequency measurements of the 28- and 78-μm CW water vapor laser lines.

1408. Fetterman, H. R., Schlossberg, H. R., Parker, C. D., *Appl. Phys. Lett.*, 23, No. 12, 684-686 (1973), cw submillimeter laser generation in optically pumped Stark-tuned NH_3.

1409. Graner, G., *Opt. Commun.*, 14, No. 1, 67-69 (1975), Assignment of submillimeter laser lines in CH_3I.

1410. Grinda, M., Weiss, C. O., *Opt. Commun.*, 26, No. 1, 91 (1978), New far infrared laser lines from CD_3OH.

1411. Hard, T. M., *Appl. Phys. Lett.*, 14, No. 4, 130 (1969), Sulfur dioxide submillimeter laser.

1412. Heppner, J., Weiss, C. O., Plainchamp, P., *Opt. Commun.*, 23, No. 3, 381-384 (1977), Far infrared gain measurements in optically pumped CH_3OH.

1413. Henningsen, J. O., *IEEE J. Quantum Electron.*, QE-13, No. 6, 435-441 (1977), Assignment of laser lines in optically pumped CH_3OH.

1414. Henningsen, J. O., *IEEE J. Quantum Electron.*, QE-14, No. 12, 958-962 (1978), New FIR laser lines from optically pumped CH_3OH measurements and assignments.

1415. Henningsen, J. O., Petersen, J. C., *Infrared Phys.*, 18, No. 5, 6, 475-479 (1978), Observation and assignment of far-infrared laser lines from optically pumped $^{13}CH_3OH$.

1416. Henningsen, J. O., *Digest for the 4th Ann. Conf. on Infrared and Millimeter Waves*, Spectroscopy of the lasing CH_3OH molecule, IEEE, Piscataway, NJ.

1417. Henningsen, J. O., Petersen, J. C., Petersen, F. R., Jennings, D. A., Evenson, K. M., *J. Mol. Spectrosc.*, 77, No. 2, 298-309 (1979), High resolution spectroscopy of vibrationally excited $^{13}CH_3OH$ by frequency measurement of FIR laser emission.

1418. Henningsen, J. O., *J. Mol. Spectrosc.*, 83, No. 1, 70-93 (1980), Stark effect in the CO_2 laser pumped CH_3OH far infrared laser as a technique for high resolution infrared spectroscopy.

1419. Henningsen, J. O., *J. Mol. Spectrosc.*, 85, No. 2, 282-300 (1981), Improved molecular constants and empirical corrections for the torsional ground state of the C-O stretch fundamental of CH_3OH.

1420. Hodges, D. T., Reel, R. D., Barker, D. H., *IEEE J. Quantum Electron.*, QE-9, No. 12, 1159-1160 (1973), Low-threshold CW submillimeter- and millimeter-wave laser action in CO_2-laser-pumped $C_2H_4F_2$, $C_2H_2F_2$, and CH_3OH.

1421. Hocker, L. O., Javan, A., Ramachandra-Rao, D., Frenkel, L., Sullivan, T., *Appl. Phys. Lett.*, 10, No. 5, 147-149 (1967), Absolute frequency measurement and spectroscopy of gas laser transitions in the far infrared.

1422. Hocker, L. O., Javan, A., *Phys. Lett.* A, 25, No. 7, 489-490 (1967), Absolute frequency measurements on new cw HCN submillimeter laser lines.

1423. Hocker, L. O., Javan, A., *Appl. Phys. Lett.*, 12, No. 4, 124-125 (1968), Absolute frequency measurements of new cw DCN submillimeter laser lines.

1424. Hocker, L. O., Ramachandra-Rao, D., Javan, A., *Phys. Lett.* A, 24, No. 12, 690-691 (1967), Absolute frequency measurement of the 190 μm and 194 μm gas laser transitions.

1425. Herman, H., Prewer, B. E., *Appl. Phys.*, 19, 241-242 (1979), New FIR laser lines from optically pumped methanol analogues.

1426. Hubner, G., Hassler, J. C., Coleman, P. D., Steenbeckeliers, G., *Appl. Phys. Lett.*, 18, No. 11, 511-513 (1971), Assignments of the far-infrared SO_2 laser lines.

1427. Inguscio, M., Moretti, A., Strumia, F., *Opt. Commun.*, 32, No. 1, 87-90 (1980), New laser lines from optically-pumped CH_3OH measurements and assignments.

1428. Inguscio, M., Moretti, A., Strumia, F., *Digest for the 4th Ann. Conf. on Infrared and Millimeter Waves*, IEEE, IR-FIR Stark spectroscopy of CH_3OH around the 9-P(34) CO_2 laser line, Piscataway, NJ.

1429. Bionducci, G., Inguscio, M., Moretti, A., Strumia, F., *Infrared Phys.*, 19, 297-308 (1979), Design of FIR molecular lasers frequency tunable by Stark effect: electric breakdown of CH_3OH, CH_3F, CH_3I, and CH_3CN.

1430. Jennings, D. A., Evenson, K. M., Jimenez, J. J., *IEEE J. Quantum Electron.*, QE-11, No. 8, 637 (1975), New CO_2 pumped CW far infrared laser lines.

1431. Karlov, N. V., Petrov, Yu. B., Prokhorov, A. M., Stel'makh, O. M., *JETP Lett.*, 8, 12-14 (1968), Laser based on boron trichloride.

1432. Kon, S., Hagiwara, E., Yano, T., Hirose, H., *Jpn. J. Appl. Phys.*, 14, No. 5, 731-732 (1975), Far infrared laser action in optically pumped CD_3OD.

1433. Kon, S., Yano, T., Hagiwara, E., Hirose, H., *Jpn. J. Appl. Phys.*, 14, No. 11, 1861-1862 (1975), Far infrared laser action in optically pumped CH_3OD.

1434. Kramer, G., Weiss, C. O., *Appl. Phys.*, 10, 187-188 (1976), Frequencies of some optically pumped submillimetre laser lines.

1435. Landsberg, B. M., *IEEE J. Quantum Electron.*, QE-16, No. 6, 684-685 (1980), Optically pumped CW submillimetre emission lines from methyl mercaptan CH_3SH.

1436. Landsberg, B. M., *IEEE J. Quantum Electron.*, QE-16, No. 7, 704-706 (1980), New optically pumped CW submillimetre emission lines from OCS, CH_3OH and CH_3OD.

1437. Landsberg, B. M., *Appl. Phys.*, 23, 127-130 (1980), New CW FIR laser lines from optically pumped ammonia analogues.

1438. Landsberg, B. M., unpublished (June 1980), See reference 1439.

1439. Landsberg, B. M., *Appl. Phys.*, 23, 345-348 (1980), New CW optically pumped FIR emissions in HCOOH, D_2CO and CD_3Br.

1440. Lund, M. W., Davis, J. A., *IEEE J. Quantum Electron.*, QE-15, No. 7, 537-538 (1979), New CW far-infrared laser lines from CO_2 laser-pumped CH_3OD.

1441. Lide, D. R., Maki, A. G., *Appl. Phys. Lett.*, 11, No. 2, 62-64 (1967), On the explanation of the so-called CN laser.

1442. Lourtioz, J-M., Pontnau, J., Morillon-Chapey, M., Deroche, J-C., *Int. J. Infrared Millimeter Waves*, 2, No. 1, 49-63 (1981), Submillimetre Laser Action of CW Optically Pumped CF_2Cl_2 (fluorocarbon 12).

1443. Lourtioz, J-M., unpublished (July 1980), private communication.

1444. Lourtioz, J-M., Pontnau, J., Meyer, C., *Int. J. Infrared Millimeter Waves*, 2, No. 3, 525-532 (1981), Optically pumped CW CF_3Br FIR laser. New emission lines and tentative assignments.

1445. Mathias, L. E. S., Crocker, A., Wills, M. S., *IEEE J. Quantum Electron.*, QE-4, No. 4, 205-208 (1968), Spectroscopic measurements of the laser emission from discharges in compounds of hydrogen, carbon and nitrogen.

1446. Muller, W. M., Flesher, G. T., *Appl. Phys. Lett.*, 10, No. 3, 93-94 (1967), Continuous-wave submillimeter oscillation in discharges containing C, N and H or D.

1447. Maki, A. G., *Appl. Phys. Lett.*, 12, No. 4, 122-124 (1968), Assignment of some DCN and HCN laser lines.

1448. Maki, A. G., *J. Appl. Phys.*, 49, No. 1, 7-11 (1978), Further assignments for the far-infrared laser transitions of HCN and $HC^{15}N$.

1449. Maki, A. G., Olson, W. B., Sams, R. L., *J. Mol. Spectrosc.*, 36, No. 3, 433-447 (1970), HCN rotational-vibrational energy levels and intensity anomalies determined from infrared measurements.

1450. Moskienko, M. V., Dyubko, S. F., *Izv. Vuz. Radiofiz. (USSR)*, 21, No. 7, 951-960 (1978), Identification of generation lines of submillimeter methyl bromide and acetonitrile molecule laser.

1451. Ni, Y. C., Heppner, J., *Opt. Commun.*, 32, No. 3, 459-460 (1980), New cw laser lines from CO_2 laser pumped CH_3OD.

1452. Patel, C. K. N., *Lasers: A Series of Advances*, Vol. 2 (1968), Marcel Dekker, New York.

1453. Petersen, F. R., U. S. National Bureau of Standards, unpublished (Nov, 1978), private communications.

1454. Petersen, F. R., Scalabrin, A., Evenson, K. M., *Int. J. Infrared Millimeter Waves*, 1, No. 1, 111-115 (1980), Frequencies of cw FIR laser lines from optically pumped CH_2F_2.

1455. Petersen, F. R., Evenson, K. M., Jennings, D. A., Wells, J. S., Goto, K., *IEEE J. Quantum Electron.*, QE-11, No. 10, 838-843 (1975), Far infrared frequency synthesis with stabilized CO_2 lasers: accurate measurements of the water vapor and methyl alcohol laser frequencies.

1456. Pontnau, J., Lourtioz, J-M., Meyer, C., *IEEE J. Quantum Electron.*, QE-15, No. 10, 1088-1090 (1979), Submillimetre laser action of CW optically pumped CF_3Br.

1457. Pollack, M. A., Frenkel, L., Sullivan, T., *Phys. Lett.* A, 26, No. 8, 381-382 (1968), Absolute frequency measurement of the 220 μm water vapor laser transition.

1458. Pollack, M. A., Bridges, T. J., Tomlinson, W. J., *Appl. Phys. Lett.*, 10, No. 9, 253-256 (1967), Competitive and cascade coupling between transitions in the cw water vapor laser.

1459. Radford, H. E., *IEEE J. Quantum Electron.*, QE-11, No. 5, 213-214 (1975), New CW lines from a submillimeter waveguide laser.

1460. Reid, J., Oka, T., *Phys. Rev. Lett.*, 38, No. 2, 67-70 (1977), Direct observation of velocity-tuned multiphonon processes in the laser cavity.

1461. Redon, M., Gastaud, C., Fourrier, M., *IEEE J. Quantum Electron.*, QE-15, No. 6, 412-414 (1979), New CW far-infrared lasing in $^{14}NH_3$ using Stark tuning.

1462. Redon, M., Gastaud, C., Fourrier, M., *Opt. Commun.*, 30, No. 1, 95-98 (1979), Far-infrared emission in NH_3 using "forbidden" transitions pumped by a CO_2 laser.

1463. Redon, M., Gastaud, C., Fourrier, M., *Infrared Phys.*, 20, 93-98 (1980), Far-infrared emissions in ammonia by infrared pumping using a N_2O laser.

1464. Redon, M., Gastaud, C., Fourrier, M., *Int. J. Infrared Millimeter Waves,* 1, No. 1, 95-100 (1980), New CW FIR laser lines obtained in ammonia pumped by a CO_2 laser using the Stark tuning method.

1465. Sattler, J. P., Worchesky, T. L., Tobin, M. S., Ritter, K. J., Daley, T., *Int. J. Infrared Millimeter Waves*, 1, No. 1, 127-138 (1980), Submillimeter-wave emission assignments for 11 difluoroethylene.

1466. Scalabrin, A., Evenson, K. M., *Opt. Lett.*, 4, No. 9, 277-279 (1979), Additional cw FIR laser lines from optically pumped CH_2F_2.

1467. Scalabrin, A., Petersen, F. R., Evenson, K. M., Jennings, D. A., *Int. J. Infrared Millimeter Waves*, 1, No. 1, 117-126 (1980), Optically pumped cw CH_2DOH FIR laser: new lines and frequency measurements.

1468. Steenbeckeliers, G., Bellet, J., *J. Appl. Phys.*, 46, No. 6, 2620-2626 (1975), New interpretation of the far-infrared SO_2 laser spectrum.

1469. Tanaka, A., Tanimoto, A., Murata, N., Yamanaka, M., Yoshinaga, H., *Opt. Commun.*, 22, No. 1, 1721 (1977), CW efficient optically pumped far-infrared waveguide NH_3 lasers.

1470. Tanaka, A., Tanimoto, A., Murata, N., Yamanaka, M., Yoshinaga, H., *Jpn. J. Appl. Phys.*, 13, No. 9, 1491-1492 (1974), Optically pumped far infrared and millimeter wave waveguide lasers.

1471. Tanaka, A., Yamanaka, M., Yoshinaga, H., *IEEE J. Quantum Electron.*, QE-11, No. 10, 853-854 (1975), New far-infrared laser lines from CO_2 laser pumped CH_3OH gas by using a copper waveguide cavity.

1472. Tobin, M. S., Sattler, J. P., Wood, G. L., *Digest for the 4th Ann. Conf. on Infrared and Millimeter Waves*, IEEE, CD_3F optically-pumped near millimeter laser, IEEE, Piscataway, NJ.

1473. Worchesky, T. L., *Opt. Lett.*, 3, No. 12, 232-234 (1978), Assignments of methyl alcohol submillimeter laser transitions.

1474. Wagner, R. J., Zelano, A. J., Ngai, L. H., *Opt. Commun.*, 8, No. 1, 46-47 (1973), New submillimeter laser lines in optically pumped gas molecules.

1475. Weiss, C. O., Grinda, M., Siemsen, K., *IEEE J. Quantum Electron.*, QE-13, No. 11, 892 (1977), FIR laser lines of CH_3OH pumped by CO_2 laser sequence lines.

1476. Wood, R. A., Davis, B. W., Vass, A., Pidgeon, C. R., *Opt. Lett.*, 5, No. 4, 153-154 (1980), Application of an isotopically enriched $^{13}C^{16}O_2$ laser to an optically-pumped FIR laser.

1477. Wood, R. A., Vass, A., Pidgeon, C. R., Colles, M. J., Norris, B., *Opt. Commun.*, 33, No. 1, 89-90 (1980), High power FIR lasing in $^{15}NH_3$ optically pumped with an isotopically enriched, CO_2 laser.

1478. Ziegler, G., Durr, U., *IEEE J. Quantum Electron.*, QE-14, No. 10, 708 (1978), Submillimeter laser action of CW optically pumped CD_2Cl_2, CH_2DOH and CHD_2OH.

1479. Redon, M., Universite Pierre et Marie Curie, unpublished (May, 1980), private communications.

1480. Radford, H. E., unpublished (August 1976), private communication.

1481. Petersen, F. R., Evenson, K. M., Jennings, D. A., Scalabrin, A., *IEEE J. Quantum Electron.*, QE-16, No. 3, 319-323 (1980), New frequency measurements and laser lines of optically pumped $^{12}CH_3OH$.

1482. Dinev, S. G., Hadjichristov, G. B., Stefanov, I. L., *Opt. Comm.* 75, 273 (1990), Stimulated emission by hybrid transitions via a heteronuclear molecule.

1483. Radford, H. E. Petersen, K. M., Jennings, D. A., Mucha, J. A., *IEEE J. Quantum Electron.*, QE-13, 92-94 (1977), Heterodyne measurements of submillimeter laser spectrometer frequencies.

1484. Redon, M., Gastaud, C., Fourrier, M., *Infrared Phys.*, 20, 93-98 (1980), Far-infrared emissions in ammonia by infrared pumping using a N_2O laser.

1485. Tobin, M. S., Sattler, J. P., Wood, G. L., *Opt. Lett.*, 4, No. 11, 384-386 (1979), Optically pumped CD_3F submillimeter wave laser.

1486. Powell, H. T., Murray, J. R., Rhodes, C. K., unpublished (Oct. 1974), Laser oscillation on the green bands of xenon oxide.
1487. Humphreys, C. J., unpublished, Line list for ionized xenon.
1488. Eckardt, R., Telle, J., Haynes, L., *Conference of Laser and Electro-Optical Systems*, Tech. Digest 5 (1980), Isotopically pumped CF_4, IEEE, Piscataway, NJ.
1489. Cooper, H. G., Cheo, P. K., *Physics of Quantum Electronics Conference*, San Juan, P. R., 690-697(1966), McGraw-Hill, New York, Ion laser oscillations in sulfer.
1490. Johnson, D. E., Eden, J. G., *IEEE J. Quantum Electron.*, QE-19, 1462 (1983), Lasing on the lithium resonance line at 670.7 nm.
1491. Hemmati, H., Collins, G. J., *IEEE J. Quantum Electron.*, QE-16, 1014-1016 (1980), Atomic Sn, Sb and Ge photodissociation lasers.
1492. Xu-hui, Zh., Jian-bang, L., *Appl. Phys.* B., 29, 291-292 (1982), A hollow cathode bismuth ion laser.
1493. White, J. C., Boker, J., Henderson, D., *IEEE J. Quantum Electron.*, QE-18, 320-322 (1982), Optically pumped atomic thulium lasers.
1494. Gevasimor, V. A., Prokopev, V. E., Sokovikov, V. G., Soldatov, A. N., *Sov. J. Quantum Electron.*, 14, 426-427 (1984), New laser lines in the visible and infrared parts of the spectrum of a thulium vapor laser.
1495. Basov, N. G., Baranov, V. V., Danilychev, V. A. et al., *Sov. J. Quantum Electron.*, 15, 1004-1006 (1985), High-pressure power laser utilizing 3p-3s transitions in nei generating radiation of wavelengths 703 and 725 nm.
1496. Basov, N. G., Alexsandrov, A. Yu., Danilychev, V. A. et al., *JETP Lett.*, 41, 191-194 (1985), Intense quasi-cw lasing in the visible region in a high pressure mixture of inert gases.
1497. Bunkin, F. V., Derzhiev, V. I., Mesyats, G. A. et al., *Sov. J. Quantum Electron.*, 15, 159-160 (1985), Plasma laser emitting at the wavelength of 585 nm with penning clearing of the lower level in a dense mixtures with neon excited by an electron.
1498. Brown, E. L., Gundersen, M. A., Williams, P. F., *IEEE J. Quantum Electron.*, QE-16, 683-685 (1980), New infrared laser lines in argon, krypton and xenon.
1499. Macklin, J. J., Wood, O. R., Silfvast, W. T., *IEEE J. Quantum Electron.*, QE-18, 1832-1835 (1982), New recombination lasers in Li, Al, Ca and Cu in a segmented plasma device employing foil electrodes.
1500. Muller, W., McClelland, J. J., Hertel, I. V., *Appl. Phys.* B., 31, 131-134 (1983), Infrared laser emission study in a resonantly excited sodium vapor.
1501. Sharma, A., Bhaskar, N. D., Lu, Y. Q., Happer, W., *Appl. Phys. Lett.*, 39, 209-211 (1981), Continuous-wave mirrorless lasing in optically pumped atomic Cs and Rb vapors.
1502. Madigan, M., Hocker, L. O., Flint, J. H., Dewey, C. F., *IEEE J. Quantum Electron.*, QE-16, 1294-1296 (1980), Pressure dependence of the infrared laser lines in barium vapor.
1503. Gerck, E., Fill, E., *IEEE J. Quantum Electron.*, QE-17, 2140-2146 (1981), Blue-green atomic photodissociation lasers in Group IIb: Zn, Cd and Hg.
1504. Shigenari, T., Vesugi, F., Takuna, H., *Opt. Lett.*, 7, 362-364 (1982), Superfluorescent laser action at 404.7 nm from the transition of mercury atom by photodissociation of $HgBr_2$ by an ArF laser.
1505. Xu-hui, Zh., Jian-bang, L., Fu-Cheng, L., *Appl. Phys.* B., 29, 111-115 (1982), Laser action of C, N, O, F, Cl and Br in hollow cathode discharges.
1506. Gnadig, K., Fu-Cheng, L., *Appl. Phys.*, 25, 273-274 (1981), A hollow cathode carbon atom laser.
1507. Ozcomert, J. S., Jones, P. L., *Chem. Phys. Lett.* 169, 1 (1990), An optically pumped K_2 supersonic beam laser.
1508. Petersen, A. B., New Developments and Applications in Gas Lasers, 106(1987), Enhanced CW ion laser operation in the range 270 nm to 380 nm., SPIE.

1509. Petersen, A. B., *Conference Digest, Annual Meeting (IEEE/LEOS)*(1989), CW ion laser operation in krypton, argon and xenon at wavelengths down to 232 nm., IEEE/LEOS, Piscataway, NJ.
1510. Latush, E. L., Sém, M. F., Chebotarev, G. D., *Sov. J. Quantum. Electron.* 19, 1537 (1989), Recombination gas-discharge lasers utilizing transitions in multiply charged O III and Xe IV ions.
1511. Hirata, K., Yoshida, H., Ninomiya, H., *Appl. Phys. Lett.* 57, 1709 (1990), Amplification by stimulated emission from optically pumped nickel vapor in an ultraviolet region.
1512. Hooker, S. M., Webb, C. E., *IEEE J. Quantum Electron.* 26, 1529 (1990), F_2 pumped NO: Laser oscillation at 218 nm and prospects for new laser transitions in the 160-250 nm region.
1513. Hooker, S. M., Webb, C. E., *Opt. Lett.* 15, 437 (1990), Observation of laser oscillation in nitric oxide at 218 nm.
1514. Dinev, S. G., Hadjichristov, G. B., Marazov, O., *Appl. Phys. B* 52, 290 (1991), Amplified stimulated emission in the NaK (D$\rightarrow$X) band by high power copper vapor laser pumping.
1515. Hamada, N., Sauerbrey, R., Wilson, W. L., Tittel, F. K., Nighan, W. L., *IEEE J. Quantum Electron.*, QE-24, 1571-1578 (1988), Performance characteristics of an injection-controlled electron-beam pumped XeF (C $\rightarrow$ A) laser system.
1516. Mandl, A., Litzenberger, L. N., *Appl. Phys. Lett.*, 53, 1690-1692 (1988), Efficient, long pulse XeF (C $\rightarrow$ A) laser at moderate electron beam pump rate.
1517. Zuev, V. S., Kashnikov, G. N., Kozlov, N. P., Mamaev, S. B., Orlov, V. K., *Sov. J. Quantum Electron.*, 16, 1665-1667 (1986), Characteristics of an XeF (C-A) laser emitting visible light as a result of optical pumping by surface-discharge radiation.
1518. Chaltakov, I. V., Minkovsky, N. I., Tomov, I. V., *Opt. Commun.*, 65, 33-36 (1988), Transverse discharge pumped triatomic Xe_2F excimer laser.
1519. Loree, T. R., Showalter, R. R., Johnson, T. M., Birmingham, B. S., Hughe, *Opt. Lett.*, 11, 510-512 (1986), Lasing XeO in liquid argon.
1520. McCown, A. W., Eden, J. G., *Appl. Phys. Lett.*, 39, 371-373 (1981), ZnI (B $\rightarrow$ X) laser: 600-604 nm.
1521. McCown, A. W., Ediger, M. N., Eden, J. G., *Opt. Commun.*, 40, 190-194 (1982), Quenching kinetics and small signal gain spectrum of the ZnI photodissociation laser.
1522. Greene, D. P., Eden, J. G., *Appl. Phys. Lett.*, 42, 20-22 (1983), Discharge pumped ZnI (599-606 nm) and CdI (653-662 nm) amplifiers.
1523. Uehara, Y., Sasaki, W., Saito, S., Fujiwara, E., Kato, Y., Yamanaka, M., *Opt. Lett.*, 9, 539-541 (1984), High power argon excimer laser at 126 nm pumped by an electron beam.
1524. Wellegehausen, B., *IEEE J. Quantum Electron.*, QE-15, 1108-1130 (1979), Optically pumped CW dimer lasers.
1525. Drosch, S., Gerber, G., *J. Chem. Phys.*, 77, 123-130 (1982), Optically pumped cw molecular bismuth laser.
1526. Wodarczyk, F. J., Schlossberg, H. R., *J. Chem. Phys.*, 67, 4476-4482 (1977), An optically pumped molecular bromine laser.
1527. Diegelmann, M., Grieneisen, H. P., Hohla, K., Hu, X.-J., Krasinski, J., *Appl. Phys.*, 23, 283-287 (1980), New TEA lasers based on D' $\rightarrow$ A' transitions in halogen monofluoride compounds: ClF (284.4 nm) BrF (354.5 nm), IF (490.8 nm).
1528. Bondbey, V. E., *J. Chem. Phys.*, 65, 2296-2304 (1976), Sequential two photon excitation of the C_2 Swan transitions and C_2 relaxation dynamics in rare gas solids.

1529. Ediger, M. N., McCown, A. W., Eden, J. G., *Appl. Phys. Lett.*, 40, 99-101 (1982), CdI and CdBr photodissociation lasers at 655 and 811 nm: CdI spectrum identification and enhanced laser output with $^{114}CdI_2$.

1530. Dinev, S. G., Daniel, H.-U., Walther, H., *Opt. Commun.*, 41, 117-120 (1982), New atomic and molecular laser transitions based on photodissociation of CdI_2.

1531. Greene, D. P., Eden, J. G., *Appl. Phys. Lett.*, 43, 418-420 (1983), Lasing on the $B \rightarrow X$ band of cadmium monoiodide (CdI) and ^{114}CdI in a UV-preionized transverse discharge.

1532. Zhang, J., Cheng, B., Zhang, D., Zhao, L., Zhao, Y. and Wang, T., *Opt. Comm.* 68, 220 (1988), Atomic lead photodissociation laser at 722.9 nm.

1533. Greene, D. P., Eden, J. G., *Opt. Lett.*, 10, 59-61 (1985), Injection locking and saturation intensity of a cadmium iodide laser.

1534. Diegelmann, M., Hohla, K., Kompa, K. L., *Opt. Commun.*, 29, 334-338 (1979), Interhalogen UV laser on the 285 nm band of ClF*.

1535. Pummer, H., Egger, H., Luk, T. S., Srinivasan, T., Rhodes, C. K., *Phys. Rev.* A, 28, 795-801 (1983), Vacuum-ultraviolet stimulated emission from two photon excited molecular hydrogen.

1536. Cefalas, A. C., Skordoulis, C., Nicolaides, C. A., *Opt. Commun.*, 60, 49-54 (1986), Superfluorescent laser action around 495 nm in the blue-green band of the mercury trimer Hg_3.

1537. Bazhulin, S. P., Basov, N. G., Zuev, V. S., Leonov, Yu. S., Stoilov, Yu., *Sov. J. Quantum Electron.*, 8, 402-403 (1978), Stimulated emission at λ=502 nm as a result of prolonged optical pumping of $HgBr_2$ vapour.

1538. Bazhulin, S. P., Basov, N. G., Bugrimov, S. N., Zuev, V. S., Kamrukov, A., *Zh. TF Pis'ma Red.*, 12, 1423-1429 (1986), Photodissociation molecular laser emitting in blue-green region with energy of 3J per pulse.

1539. Greene, D. P., Killeen, K. P., Eden, J. G., *Appl. Phys. Lett.*, 48, 1175-1177 (1986), $X^2\Sigma \rightarrow B^2\Sigma$ absorption band of HgBr: optically pumped 502 nm laser.

1540. Greene, D. P., Killeen, K. P., Eden, J. G., *J. Opt. Soc. Am.* B, 3, 1282-1287 (1986), Excitation of the HgBr $B^2\Sigma +1/2 \rightarrow X^2\Sigma +1/2$ band in the ultraviolet.

1541. Ediger, M. N., McCown, A. W., Eden, J. G., *IEEE J. Quantum Electron.*, QE-19, 263-266 (1983), Spectroscopy and efficiency of the $^{200}Hg^{81}Br$ and ^{64}ZnI photodissociation lasers.

1542. Hanson, F. E., Rieger, H., Cavanaugh, D. B., *Appl. Phys. Lett.*, 43, 622-623 (1983), Relative efficiency of $^{200}Hg^{79}Br$, $Hg^{79}Br$ and HgBr electric discharge lasers.

1543. Sugii, M., Sasaki, K., *Appl. Phys. Lett.*, 48, 1633-1635 (1986), Improved performance of the discharge pumped HgBr and Hg cell lasers by adding SF_6.

1544. Bazhulin, S. P., Basov, N. G., Bugrimov, S. N., Zuev, V. S., Kamrukov, A., *Sov. J. Quantum Electron.*, 16, 990-993 (1986), Mercury-halide vapor molecular laser pumped by a wide-band optical radiation and emitting three-color visible radiation.

1545. Berry, A. J., Whitehurst, C., King, T. A., *J. Phys. D.*, 21, 855-858 (1988), The tunable operation of mixed mercury halide lasers.

1546. Bazhulin, S. P., Basov, N. G., Bugrimov, S. N., Zuev, V. S., Kamrukov, A., *Sov. J. Quantum Electron.*, 16, 836-838 (1986), Green-emitting mercury chloride laser pumped by wide band optical radiation.

1547. Bazhulin, S. P., Basov, N. G., Bugrimov, S. N., Zuev, V. S., Kamrukov, A., *Sov. J. Quantum Electron.*, 16, 663-665 (1986), Blue-violet HgI/HgI_2 laser with wide- band optical pumping by a linearly stabilized surface discharge.

1548. Ewing, J. J., Brau, C. A., *Appl. Phys. Lett.*, 27, 557-559 (1975), Laser action on the 342 nm molecular iodine band.

1549. Shaw, M. J., Edwards, C. B., O'Neill, F., Fotakis, C., Donovan, R. J., *Appl. Phys. Lett.*, 37, 346-348 (1980), Efficient laser action on the 342 nm band of molecular iodine using ArF laser pumping.

1550. Zuev, V. S., Mikheev, L. D., Shirokikh, A. P., *Sov. J. Quantum Electron.*, 12, 342-348 (1982), Investigation of an I_2(D'- A') laser pumped by wide-band radiation.

1551. Zuev, V. S., Mikheev, L. D., Shirokikh, A. P., *Sov. J. Quantum Electron.*, 13, 567-568 (1983), Permissible heating of a medium and specific ultraviolet output energy of an optically pumped I_2 laser.

1552. Sauerbrey, R., Langhoff, H., *IEEE J. Quantum Electron.*, QE-21, 179-181 (1985), Excimer ions as possible candidates for VUV and XUV lasers.

1553. Zuev, V. S., Mikheev, L. D., Startsev, A. V., Shirokikh, A. P., *Sov. J. Quantum Electron.*, 9, 1195-1196 (1979), Pulse periodic operation of an iodine ultraviolet laser pumped by radiation from quartz flashlamps.

1554. Killeen, K. P., Eden, J. G., *Appl. Phys. Lett.*, 43, 539-541 (1983), Gain on the green (504 nm) excimer band of I_2.

1555. Eden, J. G., Killeen, K. P., *Proc. SPIE,* 476, 34,35 (1984), I_2 amplifier in the green.

1556. Kaslin, V. M., Petrash, G. G., Yakushev, O. F., *Sov. J. Quantum Electron.*, 9, 639 (1979), Pulsed stimulated emission due to electronic transitions in the I_2 molecule pumped optically by a copper vapor laser.

1557. Dlabal, M. L., Eden, J. G., *Appl. Phys. Lett.*, 38, 487-491 (1981), Gain and transient absorption profiles for the iodine monofluoride 490 nm and iodine.

1558. Dlabal, M. L., Eden, J. G., *J. Appl. Phys.*, 53, 4503-4505 (1982), On the upper state formation kinetics and line tunability of the iodine monofluoride discharge laser.

1559. Dlabal, M. L., Hutchison, S. B., Eden, J. G., Verdeyen, J. T., *Appl. Phys. Lett.*, 37, 873-875 (1980), Multiline (480-496 nm) discharge-pumped iodine monofluoride laser.

1560. DeYoung, R. J., *Appl. Phys. Lett.*, 37, 690-692 (1980), Lasing characteristics of iodine-monofluoride.

1561. Dlabal, M. L., Hutchison, S. B., Eden, J. G., Verdeyen, J. T., *Opt. Lett.*, 6, 70-72 (1981), Iodine monofluoride 140 kW laser: small signal gain and operating parameters.

1562. Steigerwald, F., Emmert, F., Langhoff, H., Hammer, W., Griegel, T., *Opt. Commun.*, 56, 240-242 (1985), Observation of an ionic excimer state in CsF^+.

1563. Kubodera, S., Frey, L., Wisoff, P. J., Sauerbrey, R., *Opt. Lett.*, 13, 446-448 (1988), Emission from ionic cesium fluoride excimers excited by a laser-produced plasma.

1564. Basov, N. G., Voitik, M. G., Zuev, V. S., Kutakhov, V. P., *Sov. J. Quantum Electron.*, 15, 1455-1460 (1985), Feasibility of stimulated emission of radiation from ionic heteronuclear molecules. I. Spectroscopy.

1565. Basov, N. G., Zuev, V. S., Mikheev, L. D., Yalovoi, V. I., *Sov. J. Quantum Electron.*, 12, 674 (1982), Blue-green laser emission from IF subjected to wide-band optical pumping.

1566. Mikheev, L. D., *Izv. Akad. Nauk SSSR*, Ser. F, 1377-1386 (1987), Visible and UV photochemical lasers.

1567. Zolotarev, V. A., Kryukov, P. G., Podmar'kov, Yu. P., Frolov, M. P. et al., *Sov. J. Quantum Electron.*, 18, 643-646 (1988), Optically pumped pulsed IF (B $\rightarrow$ X) laser utilizing a CF_3I-N, F_2-He mixture.

1568. Davis, S. J., Hanko, L., Wolf, P. J., *J. Chem. Phys.*, 82, 4831-4837 (1985), Continuous wave optically pumped iodine monofluoride $B^3\Pi\ (O^+) \rightarrow X^1\Sigma^+$ laser.

1569. Davis, S. J., Hanko, L., Shea, R. F., *J. Chem. Phys.*, 78, 172-182 (1983), Iodine monofluoride $B^3\Pi\ (O^+) \rightarrow X^1\Sigma^+$ lasing from collisionally pumped states.

1570. Davis, S. J., Hanko, L., *Appl. Phys. Lett.*, 37, 692-694 (1980), Optically pumped iodine monofluoride $B^3\Pi\ (O^+) \rightarrow X^1\Sigma^+$ laser.
1571. Kurosawa, K., Sasaki, W., Fujiwara, E., Kato, Y., *IEEE J. Quantum Electron.*, QE-24, 1908-1914 (1988), High-power narrow-band operation and Raman frequency conversion of an electron beam pumped krypton excimer laser.
1572. Basov, N. G., Zuev, V. S., Kanaev, A. V., Mikheev, L. D., *Sov. J. Quantum Electron.*, 15, 1449, 1450 (1985), Lasing in optically excited Kr cell.
1573. Tittel, F. K., Smayling, M., Wilson, W. L., Marowsky, G., *Appl. Phys. Lett.*, 37, 862-864 (1980), Blue laser action by the rare gas halide trimer Kr_2F.
1574. Basov, N. G., Zuev, V. S., Kanaev, A. V., Mikheev, L. D., Stavrovskii, D., *Sov. J. Quantum Electron.*, 10, 1561-1562 (1980), Stimulated emission from the triatomic excimer Kr_2F subjected to optical pumping.
1575. Basov, N. G., Voitik, M. G., Zuev, V. S., Kutakhov, V. P., *Sov. J. Quantum Electron.*, 15, 1461-1469 (1985), Feasibility of stimulated emission of radiation from ionic heteronuclear molecules., II. Kinetics.
1576. Rajaei-Rizi, A., Bahns, J. T., Verma, K. K., Stwalley, W. C., *Appl. Phys. Lett.*, 40, 869-871 (1982), Optically pumped ring laser oscillation in the $^6Li_2$ molecule.
1577. Kaslin, V. M., Yakushev, O. F., *Sov. J. Quantum Electron.*, 12, 201-203 (1982), Optically pumped pulsed Li_2 laser.
1578. Eden, J. G., *Appl. Phys. Lett.*, 36, 393-395 (1980), 406 nm laser on the $C \rightarrow B$ band of N_2.
1579. Sauerbrey, R., Langhoff, H., *Appl. Phys.*, 22, 399-402 (1980), Lasing in an e-beam pumped Ar-N_2 mixture at 406 nm.
1580. Chou, M. S., Zawadzkas, G. A., *J. Quantum Electron.*, QE- 17, 77-81 (1981), Long-pulse N_2 lasers at 357.7, 380.5 and 405.9 nm in N_2/Ar/Ne/He mixtures.
1581. Basov, N. G., Aleksandrov, A. Y., Danilychev, V. A., Dolgikh, V. A. et al., *JETP Lett.*, 42, 47-50 (1985), Efficient high-pressure quasi CW laser using the first negative system of nitrogen.
1582. Emmert, F., Dux, R., Langhoff, H., *Appl. Phys.* B., 47, 141-148 (1988), Improved lasing properties of the He-$N+_2$ system at 391 nm and 428 nm by H_2 admixture.
1583. Burrows, M. D., Baughcum, S. L., Oldenborg, R. C., *Appl. Phys. Lett.*, 46, 22-24 (1985), Optically pumped NO ($A^2\Sigma^+ \rightarrow X^2\Pi$) ultraviolet laser.
1584. Wang, Z. G., Ma, L. A., Xia, H. R., Zhang, K. C., Cheng, I. S., *Opt. Commun.*, 58, 315-318 (1986), The generation of UV and violet diffuse band stimulated radiation in a sodium dimer.
1585. Basov, N. G., Voitik, M. G., Zuev, V. S., Klementov, A. D., Kutakhov, V., *Sov. J. Quantum Electron.*, 17, 106-107 (1987), Efficiency of rare gas-alkali ionic molecules in stimulated emission of ultraviolet and far ultraviolet radiation.
1586. Wu, C. Y., Chen, J. K., Judge, D. L., Kim, C. C., *Opt. Commun.*, 48, 28-32 (1983), Spontaneous amplified emission of a Na_2 diffuse violet band produced through two photon excitation of sodium vapor.
1587. Bahns, J. T., Stwalley, W. C., *Appl. Phys. Lett.*, 44, 826-828 (1984), Observation of gain in the violet bands of sodium vapor.
1588. Jones, P. L., Gaubatz, U., Hefter, U., Bergmann, K., Wellegehausen, B., *Appl. Phys. Lett.*, 42, 222-224 (1983), Optically pumped sodium dimer supersonic beam laser.
1589. Bahns, J. T., Verma, K. K., Rajaei-Rizi, A. R., Stwalley, W. C., *Appl. Phys. Lett.*, 42, 336-338 (1983), Optically pumped ring laser oscillation to vibrational levels near dissociation and to the continuum in Na_2.
1590. Kanorskii, S. I., Kaslin, V. M., Yakushev, O. F., *Sov. J. Quantum Electron.*, 10, 1275-1276 (1980), Optically pumped Na_2 laser.
1591. Dinev, S. G., Koprinkov, I. G., Stefanov, I. L., *Opt. Commun.*, 52, 199-203 (1984), Na_2 $b^3\Sigma_g$- $^3\Sigma_u$ excimer laser emission in the ir.

1592. Wang, Q., Wu, T., Lu, Z., Xing, D., Liu, W., Ma, Z., *Bull. Am. Phys. Soc.*, 33, 1673 (1988), Na_2 $1^3\Sigma_g$-$1^3\Sigma_u$ lasing with peak around 892.0 nm.

1593. Wellegehausen, B., Luhs, W., Topouzkhanian, A., d'Incan, J., *Appl. Phys. Lett.*, 43, 912-914 (1983), Cascade laser emission of optically pumped Na_2 molecules.

1594. Bernage, P., Niay, P., Bocquet, H., *J. Mol. Spectrosc.*, 98, 304-314 (1983), Laser transitions among the $C^1\Pi_u$, $^1\Sigma +_g$, $A^1\Sigma_u$ and $X^1\Sigma_g$ states of Na_2.

1595. Herman, P. R., Madej, A. A., Stoicheff, B. P., *Chem. Phys. Lett.*, 134, 209-213 (1987), Rovibronic spectra of Ar_2 and coupling of rotation and electronic motion.

1596. Shahdin, S., Wellegehausen, B., Ma, Z. G., *Appl. Phys.* B., 29, 195-200 (1982), Ultraviolet excited laser emission in Na_2.

1597. Kaslin, V. M., Yakushev, O. F., *Sov. J. Quantum Electron.*, 13, 1575-1579 (1983), Optically pumped gas laser using electronic transitions in the NaRb molecule.

1598. Leone, S. R., Kosnik, K. G., *Appl. Phys. Lett.*, 30, 346-348 (1977), A tunable visible and ultraviolet laser on S_2.

1599. Girardeau-Montaut, J. P., Moreau, G., *Appl. Phys. Lett.*, 36, 509-511 (1980), Optically pumped superfluorescence S_2 molecular laser.

1600. Epler, J. E., Verdeyen, J. T., *IEEE J. Quantum Electron.*, QE-19, 1686-1691 (1983), Broad-band gain in optically pumped S_2.

1601. Wellegehausen, B., Topouzkhanian, A., Effantin, C., d'Incan, J., *Opt. Commun.*, 41, 437-442 (1982), Optically pumped continuous multiline Se_2 laser.

1602. Topouzkhanian, A., Wellegehausen, B., Effantin, C., d'Incan, J. et al., *Laser Chem.*, 1, 195-209 (1983), New continuous laser emissions in Te_2.

1603. Topouzkhanian, A., Babaky, O., Verges, J., Willers, R., Wellegehausen, B, *J. Mol. Spectrosc.*, 113, 39-46 (1985), Fourier spectroscopic investigations of $^{130}Te_2$ infrared fluorescence and new optically pumped continuous laser lines.

1604. Basov, N. G., Zuev, V. S., Kanaev, A. V., Mikheev, L. D., *Sov. J. Quantum Electron.*, 15, 1289-1290 (1985), Stimulated emission from an optically pumped Xe_2 cell laser.

1605. Sauerbrey, R., Tittel, F. K., Wilson, W. L., Nighan, W. L., *IEEE J. Quantum Electron.*, QE-18, 1336-1340 (1982), Effect of nitrogen on XeF ($C \rightarrow A$) and Xe_2 cell laser performance.

1606. Nahme, H., Kessler, T., Markus, R., Chergui, M., Schwentner, N., *J. Lumin.*, 40-41, 821-822 (1988), High density excitation of rare gas crystals for stimulated emission.

1607. Gross, R. W. F., Schneider, L. E., Amimoto, S. T., *Appl. Phys. Lett.*, 53, 2365-2367 (1988), XeF laser pumped by high power sliding discharges.

1608. Zuev, V. S., Isaev, I. F., Kanaev, A. V., Mikheev, L. D., Stavrovskii, D., *Sov. J. Quantum Electron.*, 11, 221-222 (1981), Lasing as a result of a B-X transition in the excimer XeF formed as a reult of photodissociation of KrF_2 in mixtures with Xe.

1609. Zuev, V. S., Mikheev, L. D., Stavrovskii, D. B., *Sov. J. Quantum Electron.*, 14, 1174-1178 (1984), Efficiency of an optically pumped XeF laser.

1610. Shahidi, M., Jara, H., Pummer, H., Egger, H., Rhodes, C. K., *Opt. Lett.*, 10, 448-450 (1985), Optically excited XeF* excimer laser in liquid argon.

1611. Fisher, C. H., Center, R. E., Mullaney, G. J., McDaniel, J. P., *Appl. Phys. Lett.*, 35, 26-28 (1979), A 490 nm XeF electric discharge laser.

1612. Burnham, R., *Appl. Phys. Lett.*, 35, 48, 49 (1979), A discharge pumped laser on the $C \rightarrow A$ transition of XeF.

1613. Campbell, J. D., Fisher, C. H., Center, R. E., *Appl. Phys. Lett.*, 37, 348-350 (1980), Observations of gain and laser oscillation in the blue-green during direct pumping of XeF by microsecond electron beam pulses.

1614. Nighan, W. L., Nachson, Y., Tittel, F. K., Wilson, W. L., Jr.., *Appl. Phys. Lett.*, 42, 1006-1008 (1983), Optimization of electrically excited XeF (C → A) laser performance.

1615. Nighan, W. L., Tittel, F. K., Wilson, W. L., Nishida, N., Zhu, Y., Sauer, *Appl. Phys. Lett.*, 45, 947-949 (1984), Synthesis of rare gas-halide mixtures resulting in efficient XeF (C → A) laser oscillation.

1616. Tittel, F. K., Marowsky, G., Nighan, W. L., Zhu, Y. et al., *IEEE J. Quantum Electron.*, QE-22, 2168-2173 (1986), Injection-controlled tuning of an electron beam excited XeF (C → A) laser.

1617. Huestis, D. L., Hill, R. M., Eckstrom, D. J. et al., SRI International Report No. MP78-07 (May, 1978), New electronic transition laser systems.

1618. Deutsch, T. F., *Appl. Phys. Lett.* 11, 18 (1967), Laser emission from HF rotational transitions.

1619. Welling, H., Wellegehausen, B., *Laser Spectroscopy III* (1977), 365-369, Springer-Verlag, New York, Optically pumped continuous alkali dimer lasers.

1620. Niay, P., Bernage, P., *C. R. Acad. Sci.*, Ser. II, 294, 627-632 (1982), Spectroscopie de polarisation du sodium moleculaire sur des transitions $^1\Sigma_g \rightarrow A^1\Sigma_u$ situees dans l'infraroug.

1621. Miller, H. C., Yamasaki, K., Smedley J. E., Leone, S. R., *Chem. Phys. Lett.* 181, 250 (1991), An optically pumped laser on SO (B $^3\Sigma^-$ - X $^3\Sigma^-$).

1622. Sasaki, W., Kurosawa, K., Herman, P. R., Yoshida, K., Kato, Y., *Proc. Conf. on Short Wavelength Coherent Radiation*: Generation and applications of intense coherent radiation in the VUV and XUV region with electron beam pumped rare gas excimer lasers.

1623. Eden, J. G., *Proc. Int. Conf. on Lasers '82*, 346-356(1982), Metal halide dissociation lasers.

1624. Jeffers, W. Q., *AIAA J.*, (1988), Short wavelength chemical lasers.

1625. Petersen, F. R., Wells, J. S., Maki, A. G., Siemsen, K. J., *Appl. Opt.*, 20, 3635 (1981), Heterodyne frequency measurements of $^{14}CO_2$ laser hot band transitions.

1626. Drozdowicz, Z., Rudko, R. I., Kinhares, S. J., Lax, B., *IEEE J. Quantum Electron.*, QE-17, 1574 (1981), High gain 4.3-4.5 μm optically pumped CO_2 laser.

1627. Rutt, H. N., *Infrared Phys.*, 24, 535 (1984), Optically-pumped laser action in monodeuteroacetylene.

1628. Fischer, T. A., Wittig, C., *Appl. Phys. Lett.*, 41, 107 (1982), Rotationally relaxed, grating tuned laser oscillations in optically pumped C_2D_2.

1629. Kroeker, D. F., Reid, J., *Appl. Opt.*, 25, 86 (1986), Line-tunable cw ortho- and para-NH_3 lasers operating at wavelengths of 11 to 14 μm.

1630. Rolland, C., Reid, J., Gardise, B. K., *Appl. Phys. Lett.*, 44, 380 (1984), Line-tunable oscillation of a cw NH_3 laser from 10.7 to 13.3 μm.

1631. Pinson, P., Delage, A., Girard, G., Michon, M., *J. Appl. Phys.*, 52, 2634 (1981), Characteristics of two-step and two-photon-excited emissions in $^{14}NH_3$.

1632. Morrison, H. D., Reid, J., Garside, B. K., *Appl. Phys. Lett.*, 45, 321 (1984), 16-21 μm line-tunable NH_3 laser produced by two-step optical pumping.

1633. Wessel, R., Theiler, T., Keilmann, F., *IEEE J. Quantum Electron.*, QE-23, 385 (1987), Pulsed high- power mid-infrared gas lasers.

1634. Bobrovskii, A. N., Kiselev, V. P., Kozhevnikov, A. V., Kokhanskii, V. V., *Sov. J. Quantum Electron.*, 13, 1521 (1983), Two-photon optical pumping of NH_3 in a multipass cell.

1635. Akhrarov, M., Vasilev, B. I., Grasyuk, A. Z., Soskov, V. I., *Sov. J. Quantum Electron.*, 14, 572 (1984), Middle-infrared laser utilizing isotopically substituted $^{15}NH_3$ ammonia molecules.

1636. Akhrarov, M., Vasilev, B. I., Grasyuk, A. Z., Soskov, V. I. et al., *Sov. J. Quantum Electron.*, 16, 1016 (1986), $^{15}NH_3$ laser with two-photon optical pumping.
1637. Rutt, H. N., Travis, D. N., Hawkins, K. C., *Int. J. Infrared Millimeter Waves*, 5, 1201 (1984), Mid infrared and far infrared studies of the propyne optically pumped laser.
1638. Hon, J. F., Novak, J. R., *IEEE J. Quantum Electron.*, QE-11, 698 (1975), Chemically pumped hydrogen fluoride overtone laser.
1639. Bashkin, A. S., Igoshin, V. I., Leonov, Yu. S., Oraevskii, A. N. et al., *Sov. J. Quantum Electron.*, 7, 626 (1977), An investigation of a chemical laser emitting due to an overtone of the HF molecule.
1640. Bradley, L. C., Soohoo, K. L., Freed, C., *IEEE J. Quantum Electron.*, QE-22, 234 (1986), Absolute frequencies of lasing transitions in nine CO_2 isotopic species.
1641. Vedeneev, A. A., Volkov, A. Yu., Demin, A. I., Kudryavtsev, E. M., *Sov. Tech. Phys. Lett.*, 8, 110 (1982), 18.4 μm CO_2-Ne gas dynamic laser.
1642. Akimov, V. A., Volkov, A. Yu., Demin, A. I., Kudryavtsev, E. M. et al., *Sov. J. Quantum Electron.*, 13, 556 (1983), Continuous-wave CO_2-Ar gasdynamic laser emitting at 18.4 μ.
1643. Johnson, R. P., Cornelison, D., Duzcek, C. J., *Appl. Phys. Lett.*, 44, 162 (1984), 4.4 μm cascade ^{13}C^{16}O_2 laser.
1644. Johnson, R. P., *Appl. Phys. Lett.*, 44, 1119 (1984), Further characterization of the 4.4 μm $^{13}C^{16}O_2$ cascade laser.
1645. Jeffers, W. Q., unpublished (1988), Private communication.
1646. Knight, D. J. E., *Handbook of Laser Science and Technology*, Volume II, Gas Lasers, Vol. II, Tables of far-infrared cw gas laser lines, CRC Press, Boca Raton, FL.
1647. Whitford, B. G., *IEEE Trans. Instr. Meas.*, IM-29, 168 (1980), Measurement of the absolute frequencies of CO_2 laser transitions by multiplication of CO_2 laser difference frequencies.
1648. Saxon, G., *Review of Laser Engineering*, 13, 398 (1985), Free electron lasers - a survey.
1649. Davies, P. B., Jones, H., *Appl. Phys.*, 22, 53-55 (1980), New cw far-infrared molecular lasers from ClO_2, HCCF, FCN, CH_3NC, CH_3F and propynal.
1650. Taubmann, G., Jones, H., Davies, P. B., *Appl. Phys.* B., 41, 179 (1986), New cw optically pumped FIR laser lines.
1651. Dyubko, S. F., Svich, V. A., Fesenko, L. D., *Sov. Tech. Phys. Lett.*, 1, 192-193 (1975), Magnetically tuned submillimeter laser based on paramagnetic ClO_2.
1652. Bugaev, V. A., Shliteris, E. P., *Sov. J. Quantum Electron.*, 14, 1331 (1984), submillimeter lasing transitions in isotopic modifications of SO_2.
1653. Calloway, A. R., Danielewicz, E. J., *Int. J. Infrared Millimeter Waves*, 2, 933 (1981), Predicted new optically pumped FIR molecular lasers.
1654. Yoshida, H., Ninomiya, H., Takashima, N., *Appl. Phys. Lett.* 59, 1290 (1991), Laser action in a KrF laser pumped Ta vapor.
1655. Shafik, S., Crocker, D., Landsberg, B. M., Butcher, R. J., *IEEE J. Quantum Electron.*, QE-17, 115 (1981), Phosphine far-infrared CW laser transitions: optical pumping at more than 100 MHz from resonance.
1656. Jones, H., Davies, P. B., *IEEE J. Quantum Electron.*, QE-17, 13 (1981), A powerful new optically pumped FIR laser - formylfluoride.
1657. Jones, H., Davies, P. B., Lewis-Bevan, W., *Appl. Phys.* B., 30, 1 (1983), New FIR laser lines from optically pumped DCOF.
1658. Temps, F., Wagner, H. G., *Appl. Phys.* B., 29, 13 (1982), Strong far-infrared laser action in carbonyl fluoride and vinyl fluoride.
1659. Tobin, M. S., *Opt. Lett.*, 7, 322 (1982), Carbonyl fluoride (COF_2): strong new continuous-submillimeter-wave laser source, Erratum: *Opt. Lett.*, 8, 509 (1983).

1660. Tobin, M. S., Daley, T. W., *Int. J. Infrared Millimeter Waves*, 7, 1649 (1986), Heterodyne frequency measurements of COF_2, CD_3F, $^{13}CD_3F$ and C_2H_3I lasers.

1661. Davis, I. H., Pharaoh, K. I., Knight, D. J. E., *Int. J. Infrared Millimeter Waves*, 8, 765 (1987), Frequency measurements on far-infrared laser emissions of carbonyl fluoride (COF_2).

1662. Tobin, M. S., *IEEE J. Quantum Electron.*, QE-20, 5 and 985 (1984), SMMW laser emission and frequency measurements in doubly deuterated methyl fluoride (CHD_2F).

1663. Tobin, M. S., Felock, R. D., *IEEE J. Quantum Electron.*, QE-17, 825 (1981), Submillimeter wave laser emission in optically pumped methyl fluoride - $^{13}CD_3F$.

1664. Davies, P. B., Stern, D. P., *Int. J. Infrared Millimeter Waves*, 3, 909 (1982), New cw FIR laser lines from optically pumped silyl fluoride (SiH_3F).

1665. Inguscio, M., Moruzzi, G., Evenson, K. M., Jennings, D. A., *J. Appl. Phys.*, 60, R161 (1986), A review of frequency measurements of optically pumped lasers from 0.1 to 8 THz.

1666. Scalabrin, A., Tomaselli, J., Pereira, D., Vasconcellos, E. C. C., Evenson, K. M., *Int. J. Infrared Millimeter Waves*, 6, 973 (1985), Optically pumped $^{13}CH_2F_2$ laser: wavelength and frequency measurements.

1667. Vasconcellos, E. C. C., Petersen, F. R., Evenson, K. M., *Int. J. Infrared Millimeter Waves*, 2, 705 (1981), Frequencies and wavelengths from a new, efficient FIR lasing gas: CD_2F_2.

1668. Tobin, M. S., Sattler, J. P., Daley, T. W., *IEEE J. Quantum Electron.*, QE-18, 79 (1982), New SMMW laser transitions optically pumped by a tunable CO_2 waveguide laser.

1669. Anacona, J. R., Davies, P. B., Ferguson, A. H., *IEEE J. Quantum Electron.*, QE-20, 829 (1984), Optically pumped FIR laser action in chlorofluoromethane (CH_2ClF).

1670. Tobin, M. S., Felock, R. D., *Opt. Lett.*, 5, 430 (1980), Fluoroform - d optically pumped submillimeter-wave laser.

1671. Tobin, M. S., Leavitt, R. P., Daley, T. W., *J. Mol. Spectrosc.*, 101, 212 (1983), Spectroscopy of the v5 band of CDF_3 by heterodyne measurement of SMMW laser emission frequencies.

1672. Tobin, M. S., Daley, T. W., *IEEE J. Quantum Electron.*, QE-16, 592 (1980), Optically pumped $CHClF_2$ and C_2H_3I submillimeter wave lasers.

1673. Telle, J., *IEEE J. Quantum Electron.*, QE-19, 1469 (1983), Continuous wave 16 μm CF_4 laser.

1674. Vasconcellos, E. C. C., Wyss, J. C., Petersen, F. R., Evenson, K. M., *Int. J. Infrared Millimeter Waves*, 4, 40 (1983), Frequency measurements of far infrared cw lasing lines in optically pumped $CHCl_2F$.

1675. Gastaud, G., Redon, M., Fourrier, M., *Int. J. Infrared Millimeter Waves*, 8, 1069 (1987), Further new cw laser investigations in CH_3I, $^{13}CH_3I$ and CD_3I: new lines and assignments.

1676. Inguscio, M., Evenson, K. M., Peterson, F. R., Strumia, F., Vasconcellos, E. C. C., *Int. J. Infrared Millimeter Waves*, 5, 1289 (1984), A new efficient far infrared lasing molecule: $^{13}CD_3OH$.

1677. Ioli, N., Moretti, A., Strumia, F., D'Amato, F., *Int. J. Infrared Millimeter Waves*, 7, 459 (1986), $^{13}CH_3OH$ and $^{13}CD_3OH$ optically pumped FIR laser: new large offset emission and optoacoustic spectroscopy.

1678. Vasconcellos, E. C. C., Evenson, K. M., *Int. J. Infrared Millimeter Waves*, 6, 1157 (1985), New far infrared laser lines obtained by optically pumping $^{13}CD_3OD$.

1679. Landsberg, B. M., Shafik, M. S., Butcher, R. J., *IEEE J. Quantum Electron.*, QE-17, 828 (1981), CW optically pumped far-infrared emissions from acetaldehyde, vinyl chloride, and methyl isocyanide.

1680. Gilbert, B., Butcher, R. J., *IEEE J. Quantum Electron.*, QE-17, 827 (1981), New optically pumped millimeter wave cw laser lines from methyl isocyanide.

1681. Duxbury, G., Petersen, J. C., *Appl. Phys.*, B35, 127 (1984), Optically pumped submillimetre laser action in formaldoxime and ammonia.

1682. Dyubko, S. F., Fesenko, L. D., Shevyrev, A. S., Yartsev, V. I., *Sov. J. Quantum Electron.*, 11, 1248 (1981), New emission lines of methylamine and methyl alcohol molecules in optically pumped laser.

1683. Dyubko, S. F., Fesenko, L. D., Shevyrev, A. S., Yartsev, V. I., *Sov. J. Quantum Electron.*, 11, 1247 (1981), Optically pumped CH_3NO_2 and CH_3 COOD submillimeter lasers.

1684. Fourrier, M., Redon, M., *Opt. Commun.*, 64, 534 (1987), A new cw FIR lasing medium: methyl fluoroform.

1685. Bugaev, V. A., Shliteris, E. P., *Sov. J. Quantum Electron.*, 13, 150 (1983), Optically pumped molecular laser utilising C_2H_5Br and C_2H_3I halogen derivatives of ethane.

1686. Bugaev, V. A., Shliteris, E. P., Klement'ev, Yu. F., Kudryashova, V. A., *Sov. J. Quantum Electron.*, 12, 304 (1982), Laser spectroscopy, submillimeter lasing, and passive Q switching when dimethyl ether is pumped with CO_2 laser radiation.

1687. Clairon, A., Dahmani, B., Filimon, A., Rutman, J., *IEEE Trans. Instr. Meas.*, IM-34, 265 (1985), Precise frequency measurements of CO_2/OsO_4 and He-Ne/CH_4 stabilized lasers.

1688. Kantowicz, P., Palluel, P., Pontvianne, J., *Microwave J.*, 22, 57 (1979), New developments in submillimeter-wave BWOs.

1689. Evenson, K. M., Jennings, D. A., Petersen, F. R., *Appl. Phys. Lett.*, 44, 576 (1984), Tunable far-infrared spectroscopy.

1690. unknown, *Reviews of Infrared and Millimeter Waves*, Vol. 2(1984), Optically pumped far-infrared lasers, Plenum Press, New York.

1691. Ninomiya, H. and Hirata, K., *J. Appl. Phys.* 66, 2219 (1989), Laser action of optically pumped atomic titanium vapor.

1692. Knight, D. J. E., NPL Report QU 45(Feb. 1981), Ordered list of far-infrared laser lines (continuous, λ >12 μm), National Physical Laboratory, Teddington, Middlesex, England.

1693. Ninomiya, H., Hirata, K., Yoshino, S., *J. Appl. Phys.* 66, 3961 (1989), Optically pumped titanium vapor laser at 625.8 nm.

1694. Dyubko, S. F., Fesenko, L. D., *3rd Int. Conf. on Submillimetre Waves and Their Applications*, 70, Frequencies of optically pumped submillimeter lasers.

1695. Davies, P. B., Ferguson, A. H., Stern, D. P., *Proc. 3rd Int. Conf. on Infrared Physics* (July 1984), New optically pumped lasers containing silicon.

1696. Dyubko, S. F., Fesenko, L. D., Polevoy, B. I., *VIII Conf. on Quantum Electronics and Nonlinear Optics*, Poznan, Poland, Spectrum investigation of the submillimeter CF_2HCl laser radiation.

1697. Fourrier, M., Belland, P., Gastaud, C., Redon, M., *Proc. 3rd Int. Conf. on Infrared Physics*, 803(1984), New cw FIR lasing lines in optically pumped vinyl halides: CH_2CHCl, CH_2CHF, CH_2CHBr.

1698. Flynn, G. W., Feld, M. S., Feldman, B. J., *Bull. Am. Phys. Soc.*, 12, 15 (1967), New infrared-laser transition and g-values in atomic oxygen.

1699. Tobin, M. S., Submillimeter wave laser emission of optically pumped methyl fluoride - $^{13}CD_3F$, *IEEE Quantum Electron.* QE-17, 825 (1981).

1700. Tobin, M. S., Daley, T. W, *Int. J. Infrared Millimeter Waves* 7, 1649 (1986), Heterodyne frequency measurements of COF_2, CD_3F, $^{13}CD_3F$ and C_2H_2I lasers.

1701. McCombie, J., Petersen, J. C., Duxbury G., *Electronic & Electr. Opt.*, Knight, P. L., Ed. (1981).

1702. Vasconcellos, E. C. C., Jennings, D. A., *Int. J. Infrared Millimeter Waves* 6, 157 (1985), New far infrared laser lines obtained by optically pumping $^{13}CD_3OD$.

1703. Inguscio, M., Evenson, K. M., Petersen, F. R. Strumia, F., Vasconcellos, E., A new efficient far infrared lasing molecule: $^{13}CD_3OH$, *Int. J. Infrared Millimeter Waves,* 5, 1289 (1984).

1704. Scalabrin, A., Tomaselli, J., Pereira, D. et al., *Int. J. Infrared Millimeter Waves*, 6, 973 (1985), Optically pumped $^{13}CH_2F_2$ laser: wavelength and frequency measurements.

1705. Gastaud, C., Redon, M., Fourrier, M., *Int. J. Infrared Millimeter Waves* 8 1069 (1987), Further new cw laser investigations in CH_3I, $^{13}CH_3I$ and CD_3I: new lines and assignments.

1706. Pereira, D., Scalabrin, A., *Appl. Phys. B,* 44, 67 (1987), Measurement and assignment of new FIR laser lines in $^{12}CH_3OH$ and $^{13}CH_3OH$.

1707. Ioli, N., Moretti, A., Strumia, F., D'Amato, F., *Int. J. Infrared Millimeter Waves* 7, 459 (1985), $^{13}CH_3OH$ and $^{13}CD_3OH$ optically pumped FIR laser: new large offset emission and optoacoustic spectroscopy.

1708. Belland, P., Fourrier, M., Submillimeter emission lines from CD_2Cl_2 optically pumped lasers, *Int. J. Infrared Millimeter Waves,* 7, 1251 (1986).

1709. Shevirev, A. S., Dyubko, S. F., Fesenko, L. D.and Yartsev, V. I., New emission lines of a submillimeter molecular CD_2Cl_2 laser, *Sov. J. Quantum Electron.* 16, 568 (1987).

1710. Vasconcellos, E. C.C., Petersen, F. R., Evenson, K. M., *Int. J. Infrared Millimeter Waves,* 2, 705 (1981), Frequencies and wavelengths from a new, efficient FIR lasing gas: CD_2F_2.

1711. Dyubko, S. F., Fesenko, L. D., Ferguson, A. H., *3rd. Int. Conf. Submm. Waves and Appl.*, Univ. Surrey (1978).

1712. Landsberg, B. M., *Appl. Phys. B* 23, 345 (1980), New cw optically pumped FIR emission in HCOOH, D_2CO and CD_3Br.

1713. Mathias, L. E. S., Crocker, A., *Phys. Lett.* 13, 35 (1964), Stimulated emission in the far-infrared from water vapour and deuterium oxide discharges.

1714. Vasconcellos, E. C.C., Scalabrin, A., Petersen, F. R., Evenson, K. M., *Int. J. Infrared Millimeter Waves* 2, 533 (1981), New FIR laser lines and frequency meaurements in CD_3OD.

1715. Pereira, D., Vasconcellos, E. C. C., Scalabrin, A., Evenson. K. M., Petersen, F. R., Jennings, D. A., Measurement of new FIR laser lines in CD_3OD, *Int. J. Infrared Millimeter Waves* 6, 877 (1985).

1716. Fourrier, M., Kreisler, A., *Appl. Phys. B* 41, 57 (1986), Further investigation on ir pumping of CH_3OD and CD_3OD by a cw CO_2 laser.

1717. Petersen, J. C., Duxbury, G., *Appl. Phys. B* 37, 209 (1985), New submillimetre laser lines from CH_3OD and CD_3OD.

1718. Garelli, G., Ioli, N., Moretti, A., Pereira, D., Strumia, F., *Appl. Phys. B, 44*, 111 (1987), New large effect far-infrared laser lines from CD_3OH.

1719. Pereira, D., Ferrari, C. A., Scalabrin, A., *Int. J. Infrared Millimeter Waves,* 7, 1241 (1986), New optically pumped FIR laser lines in CD_3OD.

1720. Sigg, H., Bluyssen, H. J. A., Wyder, P., New laser lines with wavelengths from $\lambda = 61.7\ \mu m$ down to $\lambda = 27.2\ \mu m$ in optically pumped CH_3OH and CD_3OH, *IEEE Quantum Electron.* QE-20, 616 (1984).

1721. Tobin, M. S., Leavitt, R. P., Daley, T. W., *J. Mol. Spec.* 101, 212 (1983), Spectroscopy of the $\nu5$ band of CDF_3 by heterodyne measurement of SMMW laser emission frequencies.

1722. Tobin, M. S., Sattler, J. P., Daley, T. W., *IEEE Quantum Electron.* QE-18, 79 (1982), New SMMW laser transitions optically pumped by a tunable CO_2 waveguide laser.

1723. Lourtioz., J.-M., Ponnau, J., Morillon-Chapey, M., Deroche, J.-C., Submillimeter laser action of CW optically pumped CF_2Cl_2 (fluorocarbon 12), *Int. J. Infrared Millimeter Waves*, 2, 49 (1981).

1724. Lourtioz., J.-M., Ponnau, J., Meyer, C., Optically pumped CF_3Br FIR laser. New emission lines and tentative assignments, *Int. J. Infrared Millimeter Waves* 2, 525 (1981).

1725. Herman. H., Wiggleworth, M. J., New FIR lasers from CO_2 laser-excited 1,1-difluoethene, *Int. J. Infrared Millimeter Waves* 5, 29 (1984).

1726. Fourrier, M., Gastaud, C., Redon, M., Deroche, J. C., *Opt. Commun.* 48, 347 (1984), New CW far infrared lasing lines in optically pumped 1,1-difluoroethylene.

1727. Belland, P., Gastaud, C., Redon, M., *Appl. Phys. B,* 34, 175 (1984), Fourrier, New cw fir laser emission from CO_2 laser-pumped vinyl bromide, M..

1728. Fourrier, M., Belland, P., Mangili, D., *IEEE Quantum Electron.* QE-20, 85 (1984), New CW FIR laser action in optically pumped vinyl chloride.

1729. Taubmann. G., Jones, H., Davies, P. B., *Appl. Phys. B,* 41 179 (1986), New cw optically pumped FIR laser lines.

1730. Gastaud, C., Redon, M., Belland, P., Fourrier, M., *Int. J. Infrared Millimeter Waves* 5, 875 (1984), Far-infrared laser action in vinyl chloride, vinyl bromide, and vinyl fluoride optically pumped by a cw N_2O laser.

1731. Gastaud, C., Redon, M., Belland, P., Kreisler. A., Fourrier, M., *Int. J. Infrared Millimeter Waves* 6, 63 (1985), Optical pumping by CO_2 and N_2O lasers: new CW FIR emission in vinyl cyanide.

1732. Redon, M., Gastaud, C., Fourrier, M., *Opt. Lett.*, 9, 71 (1984), Far infrared laser lines in vinyl fluoride optically pumped by a CO_2 laser near 10 μm.

1733. Temps, F., Wagner, H. Gg., *Appl. Phys. B, 29, 13* (1982), Strong far-infrared laser action in carbonyl fluoride.

1734. Horiuchi, Y., Murai, A., *IEEE Quantum Electron.,* QE-12, 547 (1976), Far-infrared laser oscillations from H_2CO.

1735. Akitt, D. P., Yardley, J. T., *IEEE Quantum Electron.,* QE-6, 113 (1970), Far-infrared laser emission in gas discharges containing boron trihalides.

1736. Herman. H., Wiggleworth, M. J., *Int. J. Infrared Millimeter Waves,* 3, 395 (1982), Far infrared stimulated emissions from CO_2 laser excited dichloromethane.

1737. Anacona. J. R., Davies, P. B., Ferguson, A. H., *IEEE Quantum Electron.* QE-20, 829 (1984), Optically pumped FIR laser action in chlorofluoromethane (CH_2ClF).

1738. Petersen, F. R., Scalabrin, A., Evenson, K. M., *Int. J. Infrared Millimeter Waves* 1, 111 (1980), Frequencies of cw FIR laser lines from optically pumped CH_2F_2.

1739. Golby, J. A., Cross, N. R., Knight, D. J. E., *Int. J. Infrared Millimeter Waves* 7, 1309 (1986), Frequency ;;measurements on far-infrared emission from $^{12}C^{16}O$-pumped methyl chloride and from $^{12}C^{16}O$-pumped difluoromethane.

1740. Duxbury, G., Petersen. J. C., *Appl. Phys. B* 35, 127 (1984), Optically pumped submillimetre laser action in formaldoxime and ammonia.

1741. Ioli, N., Moretti, A., Pereira, D., Strumia, F., Garelli, G., *Appl. Phys. B,* 48, 299 (1987), A new efficient far-infrared optically pumped laser gas: $CH_3{}^{18}OH$.

1742. Bugaev, V. A., Shliteris, E. P., *Sov. J. Quantum Electron.* 15, 547 (1985), Use of ethyl alcohol and its deuteroderivatives CH_3CHDOH and in CH_3CD_2OH generatioon of submillimeter radiation.

1743. Fourrier, M., Redon, M., *Opt. Commun.* 64, 534 (1987), A new cw FIR lasing medium: methyl fluoroform.

1744. Bugaev, V. A., Shliteris, E. P., *Sov. J. Quantum Electron.* 13, 150 (1983), Optically pumped molecular laser utilising C_2H_2Br and C_2H_2Cl halogen derivatives of ethane.

1745. Douglas, N. G., Krug, P. A., *IEEE Quantum Electron.* QE-18, 1409 (1982), CW laser action in ethyl chloride.

1746. Müller, W. M., Flesher, G. T., Continuous wave submillimeter oscillation in H_2O_2, D_2O_2 and CH_3CN, *Appl. Phys. Lett.* 8, 217 (1966).

1747. Fourrier, M., Belland, P., Redon, M., Gastaud, C., *Proceedings Third Conference on Infrared Physics (CIRP3),* New cw FIR lasing lines in optically pumped vinyl halides: CH_2CHCl, CH_2CHF,CH_2CHBr. See, also, references 1727, 1728 and 1730.

1748. Landsberg, B. M., Shafki, M. S., Butcher, R. J., *Int. J. Infrared Millimeter Waves* 2, 49 (1981), Submillimeter laser action of cw optically pumped CH_2Cl_2 (fluorocarbon 12).

1749. Dyubkko, S. F., Fesenko, L. D., Shevyrev, A. S., Yartsev, V. I., *Sov. J. Quantum Electron.* 11, 1247 (1981), Optically pumped CH_3NO_2 and CH_3COOD submilliimeter lasers.

1750. Gilbert, B., Butcher, R. J., *IEEE Quantum Electron.* QE-17, 827 (1982), New optically pumped millimeter wave cw laser lines from methyl isocyanide.

1751. Landsberg, B. M., *Appl. Phys. B* 23, 127 (1980), New cw FIR lines from optically pumped ammonia analogues.

1752. Dyubko, S. F., Fesenko, L. D., Shevyrev, A. S., Yartsev, V. I., *Sov. J. Quantum Electron.* 11, 1248 (1981), New emission lines of methylamine and methyl alcohol molecules in optically pumped lasers.

1753. Hassler, J. C., Coleman., P. D., *Appl. Phys. Lett.* 14, 135 (1969), Far infrared lasing in H_2S, OCS, and SO_2.

1754. Bugaev, V. A., Shliteris, E. P., Klement'ev, Yu. F., Kudryashova, V. A., *Sov. J. Quantum Electron.* 12, 304 (1982), Laser spectroscopy, submillimeter lasing, and passive Q switching when dimethyl ether is pumped with CO_2 laser radiation.

1755. Deutsch, T. F., *IEEE Quantum Electron.* QE-3, 419 (1967), New infrared laser transitions HCl, HBr, DCl, and DBr.

Section 5: Other Lasers

Section 5.1

EXTREME ULTRAVIOLET AND SOFT X-RAY LASERS

Introduction to the Table

In Table 5.1.1 soft x-ray and extreme ultraviolet lasers are listing in order of increasing wavelength. The highly ionized lasing medium is specified by its oxidation state. The electron configurations of these highly ionized states are similar to those of neutral atoms with the same number of electrons. Thus Al^{10+}, for example, which has only three electrons is described as "Li-like" and Se^{24+} (with ten electrons) is described as "Ne-like". The lasing transition and references to the experiments are given in the two final columns of Table 5.1.1.

Further Reading

Attwood, D. T. and Boker, J., Eds., *Short Wavelength Coherent Radiation: Generation and Applications*, AIP Conf. Proc. No. 147, American Institute of Physics, New York (1986).

Elton, R. C., *X-Ray Lasers*, Academic Press, San Diego (1990).

Falcone, R. W. and Kirz, J., Eds., *Short Wavelength Coherent Radiation: Generation and Applications,* Optical Society of America, Washington, DC (1989).

Kapteyn, H. C., Da Silva, L. B. and Falcone, R. W., Short-wavelength lasers, *Proc. IEEE* 80, 342 (1992).

Matthews, D. and Freeman, R., Eds., *The Generation of Coherent XUV and Soft X-Ray Radiation,* Special Dedicated Volume, *J. Opt. Soc. Am.* B 4 (1987).

Matthews, D. L. and Rosen, M. D., Soft-X-Ray Lasers, *Scientific American* 25, 86 (1988).

Matthews, D. L., X-Ray Lasers, in *Handbook of Laser Science and Technology, Suppl. 1: Lasers*, CRC Press, Boca Raton, FL (1991), p. 559.

Suckewer, S. and Skinner, C. H., Soft X-Ray Lasers and Their Applications, *Science* 247, 1553 (1990).

Waynant, R. W. and Ediger, M. N., Eds., *Selected Papers on UV, VUV, and X-Ray Lasers*, SPIE Milestone Series, Vol. MS71, SPIE Optical Engineering Press, Bellingham, WA (1993).

See, also, proceedings of the x-ray laser conferences held every two years:

X-ray Lasers 1990, Proceedings of the 2nd International Colloquium on X-ray Lasers, Tallents, G.J., Ed., IOP Conf. Series 116, IOP Publishing, Bristol, U.K. (1990).

X-ray Lasers 1992, Proceedings of the 3rd International Colloquium on X-ray Lasers, Fill, E. E., Ed., IOP Conf. Series 125, IOP Publishing, Bristol, U.K. (1992).

X-ray Lasers 1994, Proceedings of the 4th International Colloquium on X-ray Lasers, Eder, D. C. and Matthews, D. L., Eds., AIP Conf. Proc. 332, American Institute of Physics, New York (1994).

X-ray Lasers 1996, Proceedings of the 5th International Colloquium on X-ray Lasers, Svanberg, S. and Wahlström, C-G., Eds., IOP Conf. Series 151, IOP Publishing, Bristol, U.K. (1996).

Table 5.1.1
Extreme Ultraviolet and Soft X-Ray Lasers Arranged in Order of Increasing Wavelength

Wavelength (nm)	Ion	Transition	Reference
3.560	Ni-like Au^{51+}	4d→4p	1,2
4.318	Ni-like W^{46+}	4d→4p	1,3
4.483	Ni-like Ta^{45+}	4d→4p	1,3
4.553	H-like Mg^{11+}	3→2	4
4.607	Co-like Ta^{46+}	4d→4p	5
5.023	Ni-like Yb^{42+}	4d→4p	1,6
5.097	Ni-like Ta^{45+}	4d→4p	1,6
5.176	Co-like Yb^{43+}	4d→4p	5
5.419	H-like Na^{10+}	3→2	4
5.611	Ni-like Yb^{42+}	4d→4p	1,6
5.77	Li-like Ca^{17+}	4f→3d	7
5.88	Ni-like Dy^{38+}	4d→4p	8
6.11	Ni-like Tb^{37+}	4d→4p	8
6.37	Ni-like Dy^{38+}	4d→4p	8
6.39	Ni-like Gd^{36+}	4d→4p	8
6.583	Ni-like Eu^{35+}	4d→4p	9
6.67	Ni-like Tb^{37+}	4d→4p	8
6.832	Ni-like Sm^{34+}	4d→4p	10
6.92	Ni-like Gd^{36+}	4d→4p	8
7.100	Ni-like Eu^{35+}	4d→4p	9
7.24	Ni-like W^{46+}	4d→4p	1,3
7.331	Ni-like Sm^{34+}	4d→4p	8,10,53
7.442	Ni-like Ta^{45+}	4d→4p	1,3
7.535	Ni-like W^{46+}	4d→4p	1,3
7.747	Ni-like Ta^{45+}	4d→4p	1,3
7.906	Ni-like Nd^{32+}	4d→4p	8,11,51
8.091	H-like F^{8+}	3→2	12
8.107	Ni-like Yb^{42+}	4d→4p	1
8.156	Ne-like Ag^{37+}	3p→3s	13
8.2	Ni-like Pr^{31+}	4d→4p	51

Table 5.1.1—*continued*
Extreme Ultraviolet and Soft X-Ray Lasers Arranged in Order of Increasing Wavelength

Wavelength (nm)	Ion	Transition	Reference
8.440	Ni-like Eu^{35+}	4d→4p	9
8.441	Ni-like Yb^{42+}	4d→4p	1,6
8.6	Ni-like Ce^{30+}	4d→4p	51
8.73	Li-like Si^{11+}	5d→3p	14
8.89	Li-like Si^{11+}	5f→3d	14
8.9	Ni-like La^{29+}	4d→4p	14
9.936	Ne-like Ag^{37+}	3p→3s	13
10.0	Ni-like Xe^{26+}	4d→4p	15
10.038	Ne-like Ag^{37+}	3p→3s	13
10.039	Ni-like Eu^{35+}	4d→4p	9
10.243	H-like O^{7+}	3→2	16
10.456	Ni-like Eu^{35+}	4d→4p	9
10.508	Ne-like Ag^{37+}	3p→3s	13
10.57	Li-like Al^{10+}	5f→3d	12,17,18
10.64	Ne-like Mo^{32+}	3p→3s	19
11.1	Ni-like Te^{24+}	4d→4p	51
11.25	Ne-like Nb^{31+}	3p→3s	20
11.473	Ni-like Sn^{22+}	4d→4p	21
11.7	Ne-like Ru^{30+}	3p→3s	22
11.8	Ne-like Ru^{30+}	3p→3s	22
11.89	Ne-like Zr^{30+}	3p→3s	20
11.910	Ni-like Sn^{22+}	4d→4p	21
12.300	Ne-like Ag^{37+}	3p→3s	13
12.989	Li-like Si^{11+}	4f→3d	17,23
12.35	Be-like Al^{9+}	5d→3p	18
13.1	Ne-like Mo^{32+}	3p→3s	19
13.27	Ne-like Mo^{32+}	3p→3s	19
13.86	Ne-like Nb^{31+}	3p→3s	20
13.94	Ne-like Mo^{32+}	3p→3s	19
14.04	Ne-like Nb^{31+}	3p→3s	20

Table 5.1.1—*continued*
Extreme Ultraviolet and Soft X-Ray Lasers Arranged in Order of Increasing Wavelength

Wavelength (nm)	Ion	Transition	Reference
14.16	Ne-like Mo^{32+}	3p→3s	19
14.3	Ni-like Ag^{19+}	4d→4p	51
14.590	Ne-like Nb^{31+}	3p→3s	20
14.66	Ne-like Zr^{30+}	3p→3s	20
14.76	Ne-like Nb^{31+}	3p→3s	20
14.86	Ne-like Zr^{30+}	3p→3s	20
15.040	Ne-like Zr^{30+}	3p→3s	20
15.466	Li-like Al^{10+}	4f→3d	12,17,18,23
15.4985	Ne-like Y^{29+}	3p→3s	24–27
15.63	Ne-like Zr^{30+}	3p→3s	21
15.71	Ne-like Y^{29+}	3p→3s	24,25
15.98	Ne-like Sr^{28+}	3p→3s	28
16.41	Ne-like Sr^{28+}	3p→3s	28
16.490	Ne-like Y^{29+}	3p→3s	25
16.65	Ne-like Sr^{28+}	3p→3s	28
16.867	Ne-like Se^{24+}	3p→3s	29
17.35	Ne-like Rb^{27+}	3p→3s	30
17.455	Ne-like Sr^{28+}	3p→3s	28
17.61	Ne-like Rb^{27+}	3p→3s	30
17.63	Ne-like Br^{25+}	3p→3s	20
17.78	Be-like Al^{9+}	4f→3d	18
18.210	H-like C^{5+}	3→2	31–33
18.243	Ne-like Se^{24+}	3p→3s	24,25,29,34
18.52	Ne-like Rb^{27+}	3p→3s	30
19.47	Ne-like Br^{25+}	3p→3s	20
19.606	Ne-like Ge^{22+}	3p→3s	35
19.78	Ne-like Br^{25+}	3p→3s	20
20.25	Ne-like Nb^{31+}	3p→3s	20
20.42	Ni-like Nb^{13+}	4d→4p	36
20.465	Ne-like Fe^{16+}	3p→3s	37

Table 5.1.1—*continued*
Extreme Ultraviolet and Soft X-Ray Lasers Arranged in Order of Increasing Wavelength

Wavelength (nm)	Ion	Transition	Reference
20.638	Ne-like Se^{24+}	3p→3s	24,25,38
20.65	Li-like S^{13+}	5g→4f	39
20.79	Ne-like Br^{25+}	3p→3s	20
20.96	Ne-like Zr^{30+}	3p→3s	20
20.978	Ne-like Se^{24+}	3p→3s	24,25
21.217	Ne-like Zn^{20+}	3p→3s	39–41
21.884	Ne-like As^{23+}	3p→3s	40
22.028	Ne-like Se^{24+}	3p→3s	24,29
22.111	Ne-like Cu^{19+}	3p→3s	35,42
22.256	Ne-like As^{23+}	3p→3s	40
22.49	Ne-like Sr^{28+}	3p→3s	28
23.11	Ne-like Ni^{18+}	3p→3s	42,43
23.224	Ne-like Ge^{22+}	3p→3s	35
23.35	Ne-like Rb^{27+}	3p→3s	30
23.626	Ne-like Ge^{22+}	3p→3s	35,44
24.02	Ne-like Cr^{14+}	3p→3s	37
24.24	Ne-like Co^{17+}	3p→3s	43
24.670	Ne-like Ga^{21+}	3p→3s	40
24.732	Ne-like Ge^{22+}	3p→3s	35
25.111	Ne-like Ga^{21+}	3p→3s	40
25.24	Ne-like Br^{25+}	3p→3s	20
25.487	Ne-like Fe^{16+}	3p→3s	37
26.1	Ne-like V^{13+}	3p→3s	45
26.232	Ne-like Zn^{20+}	3p→3s	40,41
26.294	Ne-like Se^{24+}	3p→3s	24,29
26.723	Ne-like Zn^{20+}	3p→3s	40,41
27.931	Ne-like Cu^{19+}	3p→3s	35,42,43
28.467	Ne-like Cu^{19+}	3p→3s	35,42,43
28.538	Ne-like Cr^{14+}	3p→3s	37
28.646	Ne-like Ge^{22+}	3p→3s	35

Table 5.1.1—*continued*
Extreme Ultraviolet and Soft X-Ray Lasers Arranged in Order of Increasing Wavelength

Wavelength (nm)	Ion	Transition	Reference
29.62	Ne-like Cu^{19+}	3p→3s	35,42,43
29.77	Ne-like Ni^{18+}	3p→3s	42,43
30.36	Ne-like Ni^{18+}	3p→3s	42,43
30.4	Ne-like V^{13+}	3p→3s	45
31.2	Ne-like Sc^{11+}	3p→3s	46
31.48	Ne-like Ni^{18+}	3p→3s	42,43
31.80	Ne-like Co^{17+}	3p→3s	43
32.45	Ne-like Co^{17+}	3p→3s	43
32.63	Ne-like Ti^{12+}	3p→3s	37,43,47
33.15	Ne-like Cu^{19+}	3p→3s	35,42,43
34.75	Ne-like Ni^{18+}	3p→3s	42,43
34.796	Ne-like Fe^{16+}	3p→3s	37
35.2	Ne-like Sc^{11+}	3p→3s	46
36.73	Ne-like Co^{17+}	3p→3s	43
38.3	Ne-like Ca^{10+}	3p→3s	46
38.93	Ne-like Fe^{16+}	3p→3s	37
40.235	Ne-like Cr^{14+}	3p→3s	37
41.8	Pd-like Xe^{8+}	5d→5p	52
42.1	Ne-like K^{9+}	3p→3s	46
44.077	Ne-like Cr^{14+}	3p→3s	37
45.1	Ne-like Ar^{8+}	3d→3p	48
46.9	Ne-like Ar^{8+}	3p→3s	15,48,49
47.22	Ne-like Ti^{12+}	3p→3s	43
50.76	Ne-like Ti^{12+}	3p→3s	43
52.9	Ne-like Cl^{7+}	3p→3s	46
60.1	Ne-like S^{6+}	3d→3p	50
60.8	Ne-like S^{6+}	3p→3s	50
87.4	Ne-like Si^{4+}	3p→3s	50

References

1. MacGowan, B. J., DaSilva, L. B., Fields, D. J., Fry, A.R., Keane, C. J., Koch, J. A., Matthews, D. L., Maxon, S., Mrowka, S., Osterheld, A. L., Scofield, J. H. and Shimkaveg, G., Short wavelength nickel-like x-ray laser development, *Proceedings of the 2nd International Colloquium on X-ray Lasers*, York, U.K., Sept. 1990, Inst. Phys. Conf. Series 116, 221 (1991), and Energies of nickel-like 4d to 4p lasing lines, Scofield, J.H., and MacGowan, B.J., *Physica Scripta* 46, 361 (1992).
2. MacGowan, B. J., DaSilva, L. B., Fields, D. J., Keane, C. J., Koch, J. A., London, R. A., Matthews, D. L., Maxon, S., Mrowka, S., Osterheld, A. L., Scofield, J. H., Shimkaveg, G., Trebes, J. E. and Walling, R. S., Short wavelength x-ray laser research at the Lawrence Livermore National Laboratory, *Phys. Fluids B* 4, 2326 (1992).
3. MacGowan, B. J., Maxon, S., DaSilva, L. B., Fields, D. J., Keane, C. J., Matthews, D. L., Osterheld, A. L., Scofield, J. H., Shimkaveg, G. and Stone, G. F., Demonstration of x-ray amplifiers near the carbon K edge, *Phys. Rev. Lett.* 65, 420 (1990).
4. Kato, Y., Miura, E., Tachi, T., Shiraga, H., Nishimura, H., Daido, H., Yamanaka, M., Jitsuno, T., Takagi, M., Herman, P.R., Takabe, H., Nakai, S., Yamanaka, C., Key, M.H., Tallents, G.J., Rose, S.J. and Rumsby, P.T., Observation of gain at 54.2 Å on the Balmer-alpha transition of hydrogenic sodium, *Appl. Phys. B*, 50, 247 (1990).
5. MacGowan, B. J., DaSilva, L. B., Fields, D. J., Keane, C. J., Maxon, S., Osterheld, A. L., Scofield, J. H. and Shimkaveg, G., Observation of $3d^84d$ - $3d^84p$ soft x-ray laser transitions in high-Z ions isoelectronic to Co I, *Phys. Rev. Lett.* 65, 2374 (1990).
6. MacGowan, B. J., Maxon, S., Keane, C. J., London, R. A., Matthews, D. L. and Whelan, D. A., Soft X-ray amplification at 50.3 Å in nickellike ytterbium, *J. Opt. Soc. Am. B* 5, 1858 (1988).
7. Xu, Z., Fan, P., Lin, L., Li, Y., Wang, X., Lu, P., Li, R., Han, S., Sun, L., Qian, A., Shen, B., Jiang, Z., Zhang, Z. and Zhou, J. Space- and time-resolved investigation of short wavelength x-ray laser in Li-like Ca ions, *Appl. Phys. Lett.* 63, 1023 (1993).
8. Daido, H., Kato, Y., Murai, K., Ninomiya, S., Kodama, R., Yuan, G., Oshikane, Y., Takagi, M., Takabe, H. and Koibe, F., Efficient soft x-ray lasing at 6 to 8 nm with nickel-like lanthanide ions, *Phys. Rev. Lett.* 75, 1074 (1995).
9. MacGowan, B. J., Maxon, S., Hagelstein, P. L., Keane, C. J., London, R.A., Matthews, D. L., Rosen, M. D., Scofield, J. H. and Whelan, D. A., Demonstration of soft x-ray amplification in nickel-like ions, *Phys. Rev. Lett.* 59, 2157 (1987).
10. Lewis, C. L. S., O'Neill, D. M., Neely, D., Uhomoibhi, J. O., Burge, R., Slark, G., Brown, M., Michette, A., Jaegle, P., Klisnick, A., Carillon, A., Dhez, P., Jamelot, A., Raucourt, J.P., Tallents, G.J., Krishnan, J., Dwivedi, L., Chen, H. Z., Key, M. H., Kodama, R., Norreys, P., Rose, S. J., Zhang, J., Pert, G. J. and Ramsden, S. A., Collisionally excited x-ray laser schemes: progress at Rutherford Appleton Laboratory, Proceedings of SPIE's 1991 International Symposium on Optical and Optoelectronic Applied Science and Engineering, San Diego, CA, July 1991, *SPIE Proceedings* 1551, 49 (1992).
11. Nilsen, J. and Moreno, J. C., Lasing at 7.9 nm in nickel-like neodymium, *Optics Lett.* 20, 1386 (1995).
12. Lewis, C. L. S., Corbett, R., O'Neill, D., Regan, C., Saadat, S., Chenais-Popovics, C., Tomie, T., Edwards, J., Kiehn, G. P., Smith, R., Willi, O., Carillon, A., Guennou, H., Jaeglé, P., Jamelot, G., Klisnick, A., Sureau, A., Grande, M., Hooker, C., Key, M. H., Rose, S. J., Ross, I. N., Rumsby, P. T., Pert, G. J. and Ramsden, S. A., Status of soft x-ray laser research at the Rutherford-Appleton Laboratory, *Plasma Phys. Controlled Fusion* 30, 35 (1988).

13. Desenne, D., Berthet, L., Bourgade, J-L., Bruneau J., Carillon, A., Decoster, A., Dulieu, A., Dumont, H., Jacquemot, S., Jaeglé, P., Jamelot, G., Louis-Jacquet, M., Raucourt, J-P., Reverdin, C., Thébault, J-P. and Thiell, G., X-ray amplification in Ne-like silver: Gain determination and time-resolved beam divergence measurement, in X-ray Lasers 1990, Proceedings of the 2nd International Colloquium on X-ray Lasers, York, England, edited by G.J. Tallents, *IOP Conf. Series* 116, 351 (1991; also, Fields, D.J., Walling, R.S., Shimkaveg, G., MacGowan, B. J., DaSilva, L.B., Scofield, J. H., Osterheld, A.L., Phillips, T. W., Rosen, M.D., Matthews, D. L., Goldstein W. H. and Stewart, R. E., Observation of High gain in Ne-like Ag lasers, *Phys. Rev. A* 46, 1606 (1992).

14. Xu, Z., Fan, P., Zhang, Z., Chen, S., Lin, L., Lu, P., Wang, X., Qian, A., Yu, J., Sun, L. and Wu, M., Soft x-ray lasing and its spatial characteristics in a lithium-like silicon plasma, *Appl. Phys. Lett.* 56, 2370 (1990).

15 Fiedorowicz, H., Bartnik, A., Li, Y., Lu, P. and Fill, E. E., Demonstration of soft X-ray lasing with neonlike argon and nickel-like xenon ions using a laser-irradiated gas puff target, *Phys. Rev. Lett.* 76, 415 (1996).

16. Matthews, D. L., Campbell, E. M., Estabrook, K., Hatcher, W., Kauffman, R. L., Lee, R. W. and Wang, C. L., Observation of enhanced emission of the O VIII H_α line in a recombining laser-produced plasma, *Appl. Phys. Lett.* 45, 2226 (1984).

17. Jaeglé, P., Jamelot, G., Carillon, A., Klisnick, A., Sureau, A. and Guennou, H., Soft x-ray amplification by lithium-like ions in recombining hot plasmas, *J. Opt. Soc. Am. B* 4, 563 (1987).

18. Hara, T., Kozo, A., Kusakabe, N., Yashiro, H. and Aoyagi, Y., Soft x-ray lasing in an Al plasma produced by a 6 J laser, *Jap. J. Appl. Phys.* 28, L1010 (1989).

19. MacGowan, B. J., Rosen, M. D., Eckart, M. J., Hagelstein, P. L., Matthews, D. L., Nilson, D. G., Phillips, T. W., Scofield, J. H., Shimkaveg, G., Trebes, J. E., Walling, R. S., Whitten, B. L. and Woodworth, J. G., Observation of soft X-ray amplification in neon-like molybdenum, *J. Appl. Phys.* 61, 5243 (1987).

20. Nilsen, J., Porter, J. L., MacGowan, B., Da Silva, L.B. and Moreno, J. C., Neon-like x-ray lasers of zirconium, niobium, and bromine, *J. Phys. B* 26, L243 (1993).

21. Enright, G. D., Dunn, J., Villeneuve, D. M., Maxon, S., Baldis, H. A., Osterheld, A. L., La Fontaine, B., Kieffer, J.C., Nantel, M. and Pépin, H., A search for gain in an Ni-like tin plasma, Proceedings of the 3rd International Colloquium on X-Ray Lasers, Schliersee, Germany, May 1992, X-ray lasers 1992, E.E. Fill, Ed., *Inst. Phys. Conf. Series* 125, 45 (1992).

22. Nilsen, J., Moreno, J.C., Koch, J. A., Scofield, J.H., MacGowan, B. J. and Da Silva, L. B., Hyperfine splittings, prepulse technique, and other new results for collisional excitation neon-like x-ray lasers, *AIP Conference Proceedings 332 - X-ray Lasers 1994*, Eder, D. C. and Matthews, D. L., Eds., American Institute of Physics, New York (1994), p. 271.

23. Kim, D., Skinner, C.H., Wouters, A., Valeo, E., Voorhees, D. and Suckewer, S., Soft x-ray amplification in lithium-like Al XI (154Å) and Si XII (129Å), *J. Opt. Soc. Am. B* 6, 115 (1989).

24. Matthews, D. L., Hagelstein, P. L., Rosen, M. D., Eckart, M. J., Ceglio, N. M., Hazi, A. U., Medecki, H., MacGowan, B. J., Trebes, J. E., Whitten, B. L., Campbell, E. M., Hatcher, C. W., Hawryluk, A. M., Kauffman, R. L., Pleasance, L. D., Rambach, G., Scofield, J. H., Stone, G. and Weaver, T. A., Demonstration of a soft x-ray amplifier, *Phys. Rev. Lett.* 54, 110 (1985).

25. Matthews, D., Brown, S., Eckart, M., MacGowan, B., Nilson, D., Rosen, M., Shimkaveg, G., Stewart, R., Trebes, J. and Woodworth, J., Status of the Nova x-ray laser experiments, in Proceedings of the O.S.A. Topical Meeting on Short Wavelength Coherent Radiation: Generation and Applications, Monterey, California, Attwood, D. and Bokor, J., Eds., *AIP Conference Proceedings* 147, 117 (1986).

26. Da Silva, L. B., MacGowan, B. J., Mrowka, S., Koch, J. A, London, R. A., Matthews, D. L. and Underwood, J. H., Power measurements of a saturated yttrium x-ray laser, *Optics Lett.* 18, 1174 (1993).
27. Koch, J. A., Lee, R.W., Nilsen, J., Moreno, J. C., MacGowan, B. J. and Da Silva, L. B., X-ray lasers as sources for resonance-fluorescence experiments, *Appl. Phys. B* 58, 7 (1994).
28. Keane, C. J., Matthews, D. L., Rosen, M. D., Phillips, T. W., Whitten, B. L., MacGowan, B. J., Louis-Jacquet, M., Bourgade, J. L., DeCoster, A., Jacquemot, S., Naccache, D. and Thiell, G., Study of soft x-ray amplification in laser produced strontium plasma, *Phys. Rev. A* 42, 2327 (1990).
29. Eckart, M. J., Scofield, J. H. and Hazi, A. U., XUV emission features from the Livermore soft x-ray laser experiments, Proceedings of the International Colloquium on UV and X-ray Spectroscopy, Beaulieu-sur-Mer, France, September 1987, *J. Phys. Paris* 49, C1-361 (1988).
30. Nilsen, J., Porter, J. L., Da Silva, L. B. and MacGowan, B. J. 17-nm rubidium-ion x-ray laser, *Optics Lett.*. 17, 1518 (1992).
31. Jacoby, D., Pert, G., Shorrock, L. and Tallents, G. J., Observations of gains in the extreme ultraviolet, *Phys. Rev. B* 15, 3557 (1982).
32. Suckewer, S., Skinner, C. H., Milchberg, H., Keane, C., and Voorhees, D., Amplification of stimulated soft x-ray emission in a confined plasma column, *Phys. Rev. Lett.* 55, 1004 (1986).
33. Chenais-Popovics, C., Corbett, R., Hooker, C. J., Key, M. H., Kiehn, G. P., Lewis, C. L. S., Pert, G. J., Regan, C., Rose, S. J., Sadaat, S., Smith, R., Tomie, T. and Willi, O., Laser amplification at 18.2 nm in recombining plasma from a laser-irradiated carbon fiber, *Phys. Rev. Lett.* 59, 2161 (1987).
34. Nilsen, J. and Moreno, J.C., Nearly monochromatic lasing at 182 angstroms in neon-like selenium, *Phys. Rev. Lett.* 74, 3376 (1995).
35. Lee, T. N., McLean, E. A. and Elton, R. C., Soft X-ray lasing in neon-like germanium and copper plasmas, *Phys. Rev. Lett.* 59, 1185 (1987).
36. Basu, S., Hagelstein, P. L., Goodberlet, J. G., Muendel, M. H. and Kaushik, S., Amplication in Ni-like Nb at 204.2 Å pumped by a table-top laser, *Appl. Phys. B*, 57, 303 (1993).
37. Nilsen, J., MacGowan, B. J., Da Silva, L. B. and Moreno, J. C., Prepulse technique for producing low-Z Ne-like XUV lasers, *Phys. Rev. A* 48, 4682 (1993).
38. Koch, J. A, MacGowan, B. J., Da Silva, L. B., Matthews, D. L., Underwood, J. H., Batson, P. J. and Mrowka, S., Observation of gain-narrowing and saturation behavior in Se x-ray laser line profile, *Phys. Rev. Lett.* 68, 3291 (1992).
39. Jaeglé, P., Carillon, A., Gauthe, B., Goedtkindt, P., Guennou, H., Jamelot, G., Klisnick, A., Moller, C., Rus, B., Sureau, A. and Zeitonn, P., Lasing near 200Å with neon-like zinc and lithium-like sulfur, *Appl. Phys. B* 57, 313 (1993).
40. Lee, T. N., McLean, E. A., Stamper, J. A., Griem, H. R. and Manka, C. K., Laser driven soft x-ray laser experiments at NRL, *Bull. Am. Phys. Soc.* 33, 1920 (1988).
41. Fill, E. E., Li, Y., Schloëgl, D., Steingruber, J. and Nilsen, J., Sensitivity of lasing in neon-like zinc at 21.2 nm to the use of the prepulse technique, *Optics Lett.* 20, 374, 1995).
42. Nilsen, J., Moreno, J. C., MacGowan, B. J. and Koch, J. A., First observation lasing at 231Å in neon-like nickel using the prepulse technique, *Appl. Phys. B* 57, 309 (1993).
43. Nilsen, J., MacGowan, B. J., Da Silva, L. B., Moreno, J.C, Koch, J. A. and Scofield, J. H., Reinterpretation of the neon-like titanium laser experiments, *Opt. Eng.* 33, 2687 (1994).

44. Carillon, A., Chen, H. Z., Dhez, P., Dwivedi, L., Jacoby, J., Jaeglé, P., Jamelot, G., Zhang, J., Key, M. H., Kidd, A., Klisnick, A., Kodama, R., Krishnan, J., Lewis, C.L.S., Neely, D., Norreys, P., O'Neill, D., Pert, G.J., Ramsden, S. A., Raucourt, J. P., Tallents, G. J. and Uhomoibhi, J., Saturation and near-diffraction-limited operation of an XUV laser at 23.6 nm, *Phys. Rev. Lett. 68*, 2917 (1992).
45. Li, Y., Pretzler, G. and Fill, E. E., Observation of lasing on the two J = 0-1, 3p-3s transitions at 26.1 and 30.4 nm in neonlike vanadium, *Optics Lett.* 20, 1026 (1995).
46. Li, Y., Pretzler, G. and Fill, E.E., Neon-like ion lasers in the extreme ultraviolet region, *Phys. Rev. A* 52, R3433 (1995).
47. Boehly, T., Russotto, M., Craxton, R. S., Epstein, R., Yaakobi, B., Da Silva, L. B., Nilsen, J., Chandler, E. A., Fields, D. J., MacGowan, B. J., Matthews, D. L., Scofield, J. H. and Shimkaveg, G., Demonstration of a narrow divergence x-ray laser in neon-like titanium, *Phys. Rev. A* 42, 6962 (1990).
48. Nilsen, J., Fiedorowicz, H., Bartnik, A., Li, Y., Lu, P. and Fill, E. E., Self photopumped neonlike X-ray laser, *Optics Lett.* 21, 408 (1996).
49. Rocca, J. J., Shlyaptsev, V., Tomasel, F. G., Cortazar, O.D., Hartshorn, D. and Chilla, J. L. A., Demonstration of a discharge pumped table top soft x-ray laser, *Phys. Rev. Lett.* 73, 2192 (1994); Rocca, J. J., Tomasel, F. G., Marconi, M. C., Shlyaptsev, V. N., Chilla, J. L. A., Szapiro, S. T. and Guidice, G., Discharge-pumped soft x-ray laser in neon-like argon, *Phys. Plasmas* 2, 2547 (1995).
50. Li, Y., Lu, P., Pretzler, G. and Fill, E. E., Lasing in neonlike sulfur and silicon, *Opt. Commun.* 133, 196 (1997).
51. Daido H., Ninomiya S., Imani, T., Kodama, R., Takagi, M., Kato Y., Murai K., Zhang, J., You, Y. and Gu, Y., Nickellike soft-x-ray lasing at the wavelengths between 14 and 7.9 nm, *Optics Lett.* 21, 958 (1996).
52. Lemoff, B. E., Yin, G. Y., Gordon III, C. L., Barty, C. P. J. and Harris, S. E., Demonstration of a 10-Hz femtosecond-pulse-driven XUV laser at 41.8 nm in Xe IX, *Phys. Rev. Lett.* 74, 1574 (1995).
53. Zhang, J., MacPhee, A. G., Lin, J. et al., A saturated x-ray laser beam at 7 nanometers, *Science* 276, 1097 (1997).

Section 5.2

FREE ELECTRON LASERS

Introduction to the Table

Free electron lasers are arranged in order of increasing wavelength in Table 5.2.1. When a range of wavelengths was reported, lasers are listed in order of the lowest lasing wavelength. For each laser the operating configuration (oscillator, amplifier, single pass, amplified spontaneous emission) and the accelerator (rf linac, induction linac, storage ring, electrostatic, pulse line, microtron, modulator, ignition coil) used are given, together with the primary reference or references to laser action. The operating and performance characteristics of these lasers are summarized in recent reviews by Colson and by Freund and Granatstein (see Further Reading below).

Further Reading

Brau, C. A., *Free-Electron Lasers*, Academic Press, Boston (1990).

Colson, W. B., Short wavelength free electron lasers in 1996, *Nucl. Instr. and Meth.* A 393, 6 (1997).

Colson, W. B., Pellegrini, C. and Remeri, A., Eds., Free Electron Lasers, in *Laser Handbook*, Vol. 6, North Holland, Amsterdam (1990).

Colson, W. B. and Prosnitz, D., Free Electron Lasers, in *Handbook of Laser Science and Technology, Suppl. 1: Lasers*, CRC Press, Boca Raton, FL (1991), p. 515.

Couprie, M. E., Storage rings FELs, *Nucl. Instr. and Meth.* A 393, 13 (1997).

Freund, H. P. and Autonsen, T. M., Jr., *Principles of Free Electron Lasers*, 2nd edition, Chapman and Hall, London (1996).

Freund, H. P. and Granatstein, V. L., Long wavelength free electron lasers in 1996, *Nucl. Instr. and Meth.* A 393, 9 (1997).

Freund, H. P. and Parker, R. K., Free-Electron Lasers, in *Encyclopedia of Physical Science and Technology*, Academic Press, San Diego (1991), p. 49.

Granatstein, V. L., Parker, R. K. and Spangle, P. A., Millimeter and Submillimeter Lasers, in *Handbook of Laser Science and Technology, Vol. I: Lasers and Masers*, CRC Press, Boca Raton, FL (1981), p. 441.

Luchini, P. and Motz, H., *Undulators and Free-Electron Lasers*, Oxford University Press, Oxford (1990).

Marshall, T. C., *Free Electron Lasers*, Macmillan, New York (1985).

Poole, M. W., FEL Sources: Present and future prospects, *Rev. Sci. Instrum.* 63, 1528 (1992).

Prosnitz, D., Free Electron Lasers, in *Handbook of Laser Science and Technology, Vol. I: Lasers and Masers*, CRC Press, Boca Raton, FL (1981), p. 425.

See, also, Proceedings of the International Free Electron Laser Conferences in *Nucl. Instr. and Meth.* A 272 (1988), A 285 (1989), A 296 (1990), A 304 (1991), A 318 (1992), A 331 (1993), A 341 (1994), A 358 (1995), A 375 (1996), A 393 (1997).

Table 5.2.1
Free Electron Lasers Arranged in Order of Wavelength

Wavelength (μm)	Configuration	Accelerator	Reference
0.244–0.69	oscillator	storage ring	1
0.25	oscillator	rf linac	2
0.3	oscillator	storage ring	3
0.35	oscillator	storage ring	4
0.35	oscillator	storage ring	5
0.37	oscillator	rf linac	6
0.46–0.68	oscillator	storage ring	7
0.5	oscillator	rf linac	2
0.5145	amplifier	storage ring	8
0.525	oscillator	rf linac	9
0.598	oscillator	storage ring	10
0.63	oscillator	rf linac	11
0.662	oscillator	rf linac	12
1.57	oscillator	rf linac	13
1.88	oscillator	rf linac	14
1.9–8.1	oscillator	rf linac	15
2–100	oscillator	rf linac	33
2.2–9.6	oscillator	rf linac	16
3	amplifier	rf linac	17
3	oscillator	rf linac	18
3–10	oscillator	rf linac	19
3–53	oscillator	rf linac	73
3.04	oscillator	rf linac	20

Table 5.2.1—*continued*
Free Electron Lasers Arranged in Order of Wavelength

Wavelength (μm)	Configuration	Accelerator	Reference
4–6	oscillator	rf linac	21
4–40	oscillator	rf linac	22
4.2	oscillator	rf linac	23
5	oscillator	rf linac	24
5–35	oscillator	rf linac	25
5.5	oscillator	rf linac	14
8	oscillator	rf linac	26
10	oscillator	rf linac	27
10.6	amplifier	induction linac	28
12–21	oscillator	rf linac	29
15	oscillator	rf linac	30
16–110	oscillator	rf linac	25
19–65	oscillator	rf linac	32
20	oscillator	rf linac	31
40	oscillator	rf linac	34
40	oscillator	rf linac	35
43	oscillator	rf linac	36
47	oscillator	rf linac	74
60	oscillator	electrostatic	37
65	oscillator	rf linac	75
80–200	oscillator	rf linac	38
120–800	oscillator	electrostatic	39
340	oscillator	electrostatic	37
400	single pass	electrostatic	40
400	oscillator	electrostatic	41
640	–	electrostatic	42
1000	oscillator	electrostatic	43
1500	ASE	electrostatic	44, 45
2000	amplifier	pulse line	46
2000	oscillator	microtron	47
2000	amplifier	induction linac	48
2100–2600	oscillator	mircotron	49
≤2500	oscillator	electrostatic	50
2700	oscillator	rf linac	51
3000	ASE	pulse line	52

Table 5.2.1—*continued*
Free Electron Lasers Arranged in Order of Wavelength

Wavelength (μm)	Configuration	Accelerator	Reference
3000	oscillator	pulse line	53
3200	ASE	electrostatic	54
3500	amplifier	pulse line	55
4000	oscillator	pulse line	56
6000	amplifier	induction linac	57
6700	oscillator	pulse line	58
8000	amplifier	pulse line	59
8000	ASE	pulse line	60
8000	amplifier	induction linac	61
8000	oscillator	induction linac	62
8000	amplifier	pulse line	63
8700	amplifier	induction linac	64, 65
26,000	oscillator	pulse line	66
30,000	ASE	pulse line	67
30,000	oscillator	modulator	68
32,000	ASE	induction linac	69
35,000	oscillator	modulator	70
68,000	ASE	electrostatic	71
300,000	oscillator	ignition coil	72

(S)ASE – (self) amplified spontaneous emission

References

1. Drobyazko, I. B., Kulipanov, G.N., Litvinenko, V. N., Pinoyev, I. V., Popik, V. M., Silvestrov, I. G., Skrinsky, A. N., Sokolov, A. S. and Vinokurov, N. A., Lasing in visible and ultraviolet regions in optical kylstron installed on the VEPP-3 storage ring, *Nucl. Instr. and Meth. A* 282, 424 (1989).
2. Batchelor, K., Ben-Zvi, I., Fernow, R. C. et al., Status of the visible free-electron laser at the Brookhaven Accelerator Test Facility, *Nucl. Instr. and Meth. A* 318, 159 (1992).
3. Takano, S., Hana, H. and Isoyama, G., Lasing of a free electron laser in the visible on the UVSOR storage ring, *Nucl. Instr. and Meth. A*331, 20 (1993; Hama, H., Yamozaki, J. and Isoyama, G., FEL experiment on the UVSOR storage ring, *Nucl. Instr. and Meth. A* 341, 12 (1994).
4. Hara, T., Hama, H. and Isoyama, G., Lasing of a free electron laser in the visible on the UVSOR storage ring, *Nucl. Instr. and Meth. A* 341, 21 (1994).

5. Yamazaki, T. Yamada, K., Sugiyama, S. et al., First lasing of the NIJI-IV storage-ring free-electron laser, *Nucl. Instr. and Meth. A* 331, 27 (1993), Yamazaki, T. et al., Present status of the NIJI-IV free-electron laser, *Nucl. Instr. and Meth. A* 341, ABS 3 (1994).
6. O'Shea, P. G., Bender, S. C., Byrd, D. A. et al., Demonstration of ultraviolet lasing with a low energy electron laser, *Nucl. Instr. and Meth. A* 341, 7 (1994).
7. Billardon, M., Elleaume, P., Lapierre, Y., Ortego, J.M., Bazin, C., Bergher, M., Marilleau, J. and Petroff, Y., The Orsay storage ring free-electron laser - new results, *Nucl. Instr. and Meth.* 250, 26 (1986).
8. Barbini, B. R., Vignola, G., Trillo, S., Boni, R., DeSimone, S., Faini, S., Guiducci, S., Preger, M., Serio, M., Spataro, B., Tazzari, S., Tazzioli, F., Vescovi, M., Cattoni, A., Sanelli, C., Castellano, M., Cavallo, N., Cevenini, F., Masullo, M. R., Patteri, P., Rinzivillo, R., Solimeno, S. and Cutolo, A., *J. Phys.* (Paris), 44, C1-1 (1985).
9. Edighoffer, J. A., Neil, G. R., Fornaca, S., Thompson, H. R., Smith, T. I., Schwettman, H. A., Hess, C. E., Frisch, J. and Rohtagi, R., Visible free-electron-laser oscillator (constant and tapered wiggler), *Appl. Phys. Lett.* 52, 1569 (1988).
10. Yamazaki, T., Yamada, K., Sugiyama, S., Ohgaki, N., Tomimasu, T., Noguchi, T., Mikado, T., Chiwakei, M. and Suzuki, R., Lasing in visible of a storage-ring free electron laser at ETL, *Nucl. Instr. and Meth. A* 309, 343 (1991).
11. Third harmonic; see reference 14
12. Shoffstall, D. et al., (at 1989 U. S. Particle Accelerator Conference, Chicago, Illinois, 1989).
13. Edighoffer, J. A., Neil, G., R., Hess, C. E., Smith, T. I., Fornaca, S. W. and Schwettman, H. A., Variable-wiggler free-electron laser oscillation. *Phys. Rev. Lett.* 52, 344 (1984).
14. Kobayashi, A., Saeki, K., Oshita, E. et al., *Nucl. Instr. and Meth. A* 375, 317 (1996).
15. Benson, S. et al. (at 1989 U. S. Particle Accelerator Conference, Chicago, Illinois, 1989).
16. Brau, C., The Vanderbilt University Free Electron Laser Center, *Nucl. Instr. and Meth. A* 318, 38 (1992).
17. Bhowmik, A., Curtin, M. S., McMillin, W. A., Benson, S. V., Madey, J. M. J., Richardson, B. A. and Vintro, L., First operation of the Rocketdyne/Stanford free electron laser, *Nucl. Instr. and Meth. A* 272 (1988).
18. Benson, S. V. et al., The Stanford Mark III infrared free electron laser, *Nucl. Instr. and Meth. A* 250, 39 (1986).
19. Smith, T. and Marziali, A., Feedback stabilization of the SCA/FEL wavelength, *Nucl. Instr. and Meth. A* 331, 59 (1993; Smith, T. et al., *Proc. SPIE* 1854, 23 (1993).
20. Bhowmik, A., Curtin, M. A., McMullin, M. A., Benson, S., Richman, B. A. and Vintro, L., First operation of the Rocketdyne/Stanford free-electron laser. *Nucl. Inst. and Meth. A* 272, 10 (1988).
21. Nguyen, D. C. et al., Initial performance of Los Alamos Advanced Free Electron Laser, *Nucl. Instr. and Meth. A* 341, 29 (1994).
22. Warren., R., Sollid, J. E. and Feldman, D. W., Near-ideal lasing with a uniform wiggler, *Nucl. Instr. and Meth. A* 285, 1 (1989).
23. Feinstein, J., Fisher, A. S., Reid, M. B., Ho, A., Ozcan, M., Dulman, H. D., and Pantell, R. H., Experimental results on a gas-loaded free-electron laser, *Phys. Rev. Lett.* 60, 18 (1988).
24. Auerhammer, J. et al., First observation of amplification of spontaneous emission achieved with the Darmstadt IR-FEL, *Nucl. Instr. and Meth. A* 341, 63 (1994).

65. Orzechowski, T. J., Anderson, B. R., Clark, J. C., Fawley, W. M., Paul, A. C., Prosnitz, D., Scharlemann, E. T., Yarema, S. M., Hopkins, D. B., Sessler, A. M. and Wurtele, J. S., High-efficiency extraction of microwave radiation from a tapered wiggler free-electron laser, *Phys. Rev. Lett.* 57, 2172 (1986).
66. Mizuno, T., Ohtsuki, T., Ohshima, T. and Saito, H., Experimental mode analysis of a circular free electron laser, *Nucl. Instr. and Meth. A* 358, 131 (1995).
67. Wang, M. C. (personal communication–H. P. Freund).
68. Einat, M., Jerby, E. and Shahadi, A., Dielectric-loaded free-electron maser in a stripline structrure, *Nucl. Instr. and Meth.* A 375, 21 (1996).
69. Saito, K., Takeyama, K., Ozaki, T., Kishiro, J., Ebihara, K. and Hiramatsu, S., X-band prebunched FEL amplifier, *Nucl. Instr. and Meth.* A 375, 237 (1996).
70. Al'Shamma'a, A., Stuart, R. A. and Lucas, J., A wiggler magnet for a CW-FEM, *Nucl. Instr. and Meth.* A 375, 424(1996).
71. Cohen, M., Eichenbaum, A., Kleinman, H., Arbel, M., Yakover, I.M. and Gover, A., Report of first masing and single-mode locking in a prebunched beam FEM oscillator, *Nucl. Instr. and Meth.* A 375, 17(1996).
72. Drori, R., Jerby, E., Shahadi, A., Einat, M. and Sheinin, M., Free-electron maser operation at the 1 GHz/1 keV regime, *Nucl. Instr. and Meth.* A 375, 186 (1996).
73. Ortega, J. M., Berset, J. M., Chaput, R. et al., Activities of the CLIO infrared facility, *Nucl. Instr. and Meth.* A 375, 618 (1996).
74. Asakawa, M. et al. *Proc. 15th Annual Meeting of the Laser Society of Japan*, Osaka, Japan, January 19-20, 1995.
75. Miyamoto, S., *Nucl. Instr. and Meth.* A (in press).

45. Gilgenbach, R. M., Marshall, T. C. and Schlesinger, S. P., Spectral properties of stimulated Raman radiation from an intense relativistic electron beam, *Phys. Fluids* 22, 971 (1979).
46. Dodd, J. W. and Marshall, T. C., "Spiking" radiation in the Columbia free electron laser, *IEEE Trans. Plasma Sci.* 18, 447 (1990).
47. Doria, A., Gallerano, G. P., Giovenale, E., Kimmitt, M. F. and Messina, G., The ENEA F-CUBE facility: trends in RF driven compact FELs and related diagnostics, *Nucl. Instr. and Meth.* A 375, ABS 11 (1996).
48. Throop, A. L., Orzechowski, T. J., Anderson, B. R., Chambers, F. W., Clark, J. C., Fawley, W. M., Jong, R. A., Paul, A. C., Prosnitz, D., Scharlemann, E. T., Steve, R. D., Westenskow, G. A. and Yarema, S. M., Experimental characteristics of high-gain free-electron laser amplifier operating at 8-mm and 2-mm wavelengths, presented at AIAA 19th Fluid Dynamics & Lasers Conf., June 8, 1987, American Institute of Aeronautics and Astronautics, Honolulu, Hawaii.
49. Ciocci, F., Bartolini, R., Doria, A., Gallerano, G. P., Giovenale, E., Kimmitt, M. F., Mesina, G. and Renieri, A., Operation of a compact free-electron laser in the millimeter-wave region with a bunched electron beam, *Phys. Rev. Lett.* 70, 928, 993).
50. Ramian, G., The new UCSB free-electron lasers, *Nucl. Instr. and Meth.* A 318, 225 (1992).
51. Asakawa, M., Sakamoto, N., Inoue, N., Yamamoto, T., Mima, K., Nakai, S., Chen, J., Fujita, M., Imasaki, K., Yamanaka, C., Agari, T., Asakuma, T., Ohigashi, N. and Tsunawaki, Y., Experimental study of a waveguide free-electron laser using the coherent synchrotron radiation emitted from electron bunches, *Appl. Phys. Lett.* 64, 1601 (1994).
52. Renz, G. (personal communication–H. P. Freund).
53. Pasour, J. (personal communication–H. P. Freund).
54. Zhukov, P. G., Ivanov, V. S., Rabinovich, M. S., Raizer, M. D. and Ruchadze, A. A., Stimulated Compton scattering from relativistic electron beam, *Proceedings of the Third International Topical Conference on Higher Power Electron and Ion Beam Research and Technology*, Vol. 1, Novosibirsh (1979), p. 705.
55. Cheng, S., Granatstein, V. L., Destler, W. W., Levush, B., Rodgers, J. and Antonsen, Jr., T. M., FEL with applications to magnetic fusion research, *Nucl. Instr. and Meth.* A 375, 160 (1996).
56. Ginzburg, N.S., (personal communication–H. P. Freund).
57. Shiho, M. (personal communication–H. P. Freund).
58. Peskov, N. Yu., Bratman, V. L., Ginzburg, G. G., Denisov, N. S., Kol'chugin, B. D., Samsonov, S. V. and Volkov, A. B., Experimental study of a high-current FEM with a broadband microwave system, *Nucl. Instr. and Meth.* A 375, 377 (1996).
59. Rullier, J. L., Devin, A., Gardelle, J., Labrouche, J. and Le Taillandier, P., Strong coupling operation of a FEL amplifer with an axial magnetic field, *Nucl. Instr. and Meth.*, A 358, 118 (1995).
60. van der Slot, P. J. M. and Witteman, W. J., Energy and frequency measurements on the Twente Raman free-electron laser, *Nucl. Instr. and Meth.* A331, 140 (1993).
61. Fanbao, M. (personal communication–H. P. Freund).
62. Kiminsky, A. K., Bogachenkov, V. A., Ginzburg, N. S. et al., presented at the 17th Int. Free Electron Laser Conf., New York, *Nucl. Instr. and Meth.* A 375, (1996).
63. Conde, M. E. and Bekefi, G., Experimental study of a 33.3-GHz free-electron-laser amplifier with a reversed axial guide magnetic field, *Phys. Rev. Lett.* 67, 3082 (1991).
64. Orzechowski, T. J., Anderson, B. R., Fawley, W. M., Prosnitz, D., Scharlemann, E. T., Varema, S. M., Hopkins, D. B., Paul, A. C., Sessler, A. M. and Wurtele, J. S., Microwave radiation from a high-gain free-electron laser amplifier, *Phys. Rev. Lett.* 54, 889 (1985).

65. Orzechowski, T. J., Anderson, B. R., Clark, J. C., Fawley, W. M., Paul, A. C., Prosnitz, D., Scharlemann, E. T., Yarema, S. M., Hopkins, D. B., Sessler, A. M. and Wurtele, J. S., High-efficiency extraction of microwave radiation from a tapered wiggler free-electron laser, *Phys. Rev. Lett.* 57, 2172 (1986).
66. Mizuno, T., Ohtsuki, T., Ohshima, T. and Saito, H., Experimental mode analysis of a circular free electron laser, *Nucl. Instr. and Meth. A* 358, 131 (1995).
67. Wang, M. C. (personal communication–H. P. Freund).
68. Einat, M., Jerby, E. and Shahadi, A., Dielectric-loaded free-electron maser in a stripline structrure, *Nucl. Instr. and Meth.* A 375, 21 (1996).
69. Saito, K., Takeyama, K., Ozaki, T., Kishiro, J., Ebihara, K. and Hiramatsu, S., X-band prebunched FEL amplifier, *Nucl. Instr. and Meth.* A 375, 237 (1996).
70. Al'Shamma'a, A., Stuart, R. A. and Lucas, J., A wiggler magnet for a CW-FEM, *Nucl. Instr. and Meth.* A 375, 424(1996).
71. Cohen, M., Eichenbaum, A., Kleinman, H., Arbel, M., Yakover, I.M. and Gover, A., Report of first masing and single-mode locking in a prebunched beam FEM oscillator, *Nucl. Instr. and Meth.* A 375, 17(1996).
72. Drori, R., Jerby, E., Shahadi, A., Einat, M. and Sheinin, M., Free-electron maser operation at the 1 GHz/1 keV regime, *Nucl. Instr. and Meth.* A 375, 186 (1996).
73. Ortega, J. M., Berset, J. M., Chaput, R. et al., Activities of the CLIO infrared facility, *Nucl. Instr. and Meth.* A 375, 618 (1996).
74. Asakawa, M. et al. *Proc. 15th Annual Meeting of the Laser Society of Japan*, Osaka, Japan, January 19-20, 1995.
75. Miyamoto, S., *Nucl. Instr. and Meth.* A (in press).

Section 5.3

NUCLEAR PUMPED LASERS

Introduction to the Table

Table 5.3.1 lists nuclear pumped lasers in order of increasing wavelength for both reactor-pumped lasers and nuclear-device-pumped lasers. The lasing medium (gas or gas mixtures) is given in the second column and the nuclear reaction giving rise to excitation of the lasing medium is given in the third column. Primary references are included in the final column. This table was adapted from a table provided by G. H. Miley and E. G. Batyrbekov (private communication and to be published).

Further Reading

Lipinski, R. J. et al., Survey and comparison of mission for a nuclear-reactor-pumped laser, *Proc. Nuclear Technologies for Space Exploration* (NTSE-92), 564, Jackson Hole, WY (1992).

Magda, E. P., Analysis of experimental and theoretical research of nuclear-pumped lasers at the Institute of Technical Physics, *Laser and Particle Beams* 11, 469 (1993).

McArthur, D. A., Nuclear Pumped Lasers, in *Encyclopedia of Lasers and Optical Technology*, Meyers, R. A., Ed., Academic Press, San Diego (1991).

Miley, G. H., Overview of nuclear-pumped lasers, *Laser and Particle Beams* 11, 575 (1993).

Miley, G. H., DeYoung, R., McArthur, D., and Prelas, M., Fission reactor pumped lasers: history and prospects, in *50 Years With Nuclear Fission*, American Nuclear Society, New York (1989), p. 333.

Mis'kevich, A.I., Visible and near-infrared direct nuclear-pumped lasers, *Laser Physics* 1, 445 (1991).

Petra, M. and Miley, G. H., Investigation of thermal lensing in nuclear pumped lasers, *Proc. 13th International Conference on Laser Interactions and Related Plasma Phenemona*, Monterey, CA (in press).

Schneider, R. T. and Hohl, F., Nuclear-pumped lasers, *Adv. Nucl. Sci. Technol.* 16, 123 (1984).

Shaban, Y. R. and Miley, G. H., Practical, visible wavelength nuclear-pumped laser, *Laser and Particle Beams* 11, 559 (1993).

Thom, K. and Schneider, R. T., Nuclear pumped gas lasers, *AIAA Journal* 10, 400 (1972).

Table 5.3.1
Nuclear Pumped Lasers Arranged in Order of Wavelength

Wavelength (μm)	Active medium	Pumping reactions	Reference
Reactor Pumped Lasers			
0.4416	He–Cd	U^{235}(n,f)F	1
0.4416	^{3}He–Cd	^{3}He(n,p)T	2,3
0.5337	He–Cd	U^{235}(n,f)F	1
0.5337	^{3}He–Cd	^{3}He(n,p)T	2, 3
0.5378	He–Cd	U^{235}(n,f)F	1
0.5378	^{3}He–Cd	^{3}He(n,p)T	2, 3
0.5461	He–Xe–Hg–H_2	U^{235}(n,f)F	4
0.5852	He–Ne–Ar	^{3}He(n,p)T	5
0.5852	He–Ne–H_2	$B^{10}(n,\alpha)Li^7$	8
0.5853	He–Ne–Ar	U^{235}(n,f)F	1,6,7
0.615	He–Hg	$B^{10}(n,\alpha)Li^7$	9,10
0.7032	Ne–Kr(Ar)	U^{235}(n,f)F	5,7
0.7245	(He)–Ne–Kr	U^{235}(n,f)F	1
0.7245	Ne–Kr(Ar)	U^{235}(n,f)F	5,7
0.7479	He–Zn	^{3}He(n,p)T	11
0.7479	He–Zn	U^{235}(n,f)F	1
0.8066	He–Cd	U^{235}(n,f)F	1
0.8531	He–Cd	U^{235}(n,f)F	1
0.8629	Ne–N_2	$B^{10}(n,\alpha)Li^7$	12–14
0.9393	Ne–N_2	$B^{10}(n,\alpha)Li^7$	12–14
1.149	He–Ar	U^{235}(n,f)F	15
1.190	He–Ar	U^{235}(n,f)F	15
1.27	^{3}He–Ar	^{3}He(n,p)T	16,17,37
1.4300	He–Cd	U^{235}(n,f)F	1
1.45	He–CO,He–CO_2,Ar–CO_2	$B^{10}(n,\alpha)Li^7$	18–20
1.4550	Ne–CO, Ne–CO_2	$B^{10}(n,\alpha)Li^7$	18,21
1.587	^{3}He–Cd	^{3}He(n,p)T	17
1.6500	He–Cd	U^{235}(n,f)F	1
1.69	^{3}He–Ar	^{3}He(n,p)T	37
1.69	^{3}He–Ne	^{3}He(n,p)T	37

Table 5.3.1—*continued*
Nuclear Pumped Lasers Arranged in Order of Wavelength

Wavelength (μm)	Active medium	Pumping reactions	Reference
1.73	Ar–Xe	$B^{10}(n,\alpha)Li^7$	22
1.73	Ar–Xe	$U^{235}(n,f)F$	23
1.732	Ar–Xe	$U^{235}(n,f)F$	24
1.732	He–Ar–Xe	$U^{235}(n,f)F$	1,25
1.78	He–Kr	$U^{235}(n,f)F$	15,26
1.79	^{3}He–Ar	$^3He(n,p)T$	16,17,27,37
1.79	^{3}He–Ne	$^3He(n,p)T$	37
2.026	^{3}He–Xe	$^3He(n,p)T$	17,28–30
2.026	Ar–Xe	$U^{235}(n,f)F$	15
2.03	He–Ar–Xe	$U^{235}(n,f)F$	23
2.19	^{3}He–Ar	$^3He(n,p)T$	17,30
2.19	^{3}He–Kr	$^3He(n,p)T$	17
2.397	He–Ar	$U^{235}(n,f)F$	15
2.48	He–Ar–Xe	$U^{235}(n,f)F$	36
2.482	Ar–Xe	$U^{235}(n,f)F$	25
2.5	^{3}He–Kr	$^3He(n,p)T$	30
2.52	^{3}He–Ar	$^3He(n,p)T$	17,30
2.52	He–Kr	$U^{235}(n,f)F$	15,26
2.60	He–Ar–Xe	$U^{235}(n,f)F$	1,25
2.627	Ar–Xe	$U^{235}(n,f)F$	1
2.627	He–Xe	$U^{235}(n,f)F$	15,31
3.07	He–Kr	$U^{235}(n,f)F$	15,26
3.508	^{3}He–Xe	$^3He(n,p)T$	17,28
3.5080	He–Xe	$U^{235}(n,f)F$	32
3.652	^{3}He–Xe	$^3He(n,p)T$	17,28
5.1–5.6	CO	$U^{235}(n,f)F$	33,34
Nuclear Device Pumped Lasers			
0.17	Xe_2	gamma rays	35
2.7	HF	gamma rays	35

f and F denote light and heavy fission fragments emitted in the reaction of neutrons (n) with ^{235}U.

References

1. Magda, E. P., Grebyonkin, K. F. and Kryzhanovsky, V. A., Nuclear pumped lasers at the Institute of Technical Physics, *Transactions, Lasers '90*, 827, San Diego, CA (1991).
2. Dmitirev, A. B., Il'ishenko, V. S., Miskevich, A. I. et al., *Pis'ma v ZTF* 52, 2235 (1982) (in Russian)
3. Miskevich, A. I., Dmi'triev, A. B., Il'ishenko, V. S. et al., *Pis'ma v ZTF* 6, 818 (1980) (in Russian)
4. Bochkov, A. V., Kryzhanovskii, V. A., Magda, E. P. et al., Quasi-cw lasing on the 7^3S_1-6^3P_2 atomic mercury transition, *Sov. Tech. Phys. Lett.* 18, 241 (1992).
5. Copai-Gora, A. P., Miskevich, A. I. and Salamadia, B. S., *Pis'ma v ZTF* 16, 23 (1990) (in Russian).
6. Hebner G. A. and Hays, G. N., Fission-fragment-excited lasing at 585.3 nm in He/Ne/Ar gas mixtures. *Appl. Phys.* 57, 2175 (1990).
7. Voinov, A. V., Krivonosov, V. N., Mel'nikov, S. P. et al., *Sov. Phys. Dokl.* 35, 568 (1990).
8. Shaban, Y. and Miley, G. H., A practical visible wavelength nuclear-pumped laser, *Proceeding of Specialist Conference on Physics of Nuclear Induced Plasma and Problems of Nuclear Pumped Lasers*, Obninsk, Russia, Vol. 2, 241 (1993).
9. Akerman, M. A. and Miley, G. H., A helium-mercury direct nuclear pumped laser, *Appl. Phys. Lett.* 30, 409 (1977).
10. Akerman, M.A., Demonstration of the first visible wavelength DNPL, Ph.D. Thesis, Department of Nuclear Engineering, U. of Illinois at Urbana-Champaign (1976).
11. Miskevich, A. I., Copai-Gora, A. P. and Salamadia, B. S., *Pis'ma v ZTF* 16, 62 (1990) (in Russian).
12. DeYoung, R., A direct nuclear pumped neon-nitrogen laser, Ph.D. Thesis, Department of Nuclear Engineering, U. of Illinois at Urbana-Champaign (1976).
13. DeYoung, R., Wells, W. E., Miley, G. H. and Verdeyen, J. T., Direct nuclear pumped Ne-N_2 laser, *Appl. Phys. Lett.* 28, 519 (1976).
14. Cooper, G., Verdeyen, J. T., Wells, W. and Miley, G. H., The pumping mechanism for the neon-nitrogen nuclear-excited laser, *Proceedings, 3rd Conf. Uranium Plasmas and Applications*, Princeton, NJ (June 1976).
15. Voinov, A. M., Dovbych, L. E., Krivonosov, V. N. et al., *Pis'ma v ZTF* 5, 422 (1979) (in Russian).
16. DeYoung, R. J., Jalufka, N. W., Hohl, F. and Williams, M. D., Direct nuclear pumped lasers using the volumetric ^3He reaction, *Conf. on Partially Ionized and Uranium Plasmas*, 96, Princeton, NJ (1976).
17. DeYoung, R. J., Jalufka, N. W. and Hohl, F., Direct nuclear-pumped lasers using He^3(n,p)T reaction, *AIAA Journal* 16, 991 (1978).
18. Prelas, M. A., Anderson, J. H., Boody, F. P. et al. , Nuclear pumping of a neutral carbon laser, *Progress in Astronautics and Aeronautics, Radiation Energy Conversion in Space* 61, 411 (1978).
19. Prelas, M. A., Akerman, M. A., Boody, F. P. and Miley, G. H., A direct nuclear pumped 1.45μ atomic carbon laser in mixtures of He-CO and He-CO_2, *Appl. Phys. Lett.* 31, 428 (1977).
20. Prelas, M. A., Akerman, M. A., Boody, F. P. and Miley, G. H., A direct nuclear pumped 1.45μ atomic carbon laser in mixtures of He-CO and He-CO_2, *4th Workshop on Laser Interaction and Related Plasma Phenomena,* Plenum Press, New York (1976), p. 249.
21. Prelas, M. A., Anderson, J. H., Boody, F. P. et al., A nuclear pumped laser using Ne-CO and Ne-CO_2 mixtures, *30th Annual Electronics Conference* (Oct. 1977).

22. Batyrbekov, E. G., Poletaev, E. D., Suzuki, E. and Miley, G. H., $B^{10}(n,a)Li^7$ pumped Ar-Xe laser *Transactions of 11th International Conference on Laser Interactions and Related Plasma Phenomena* (Oct. 1993), p. 152.
23. Alford, W. J. and Hays, G. H., Measured laser parameters for reactor-pumped He-Ar-Xe and Ar-Xe lasers, *J. Appl. Phys.* 65, 3760 (1990).
24. Voinov, A. M., Dovbych, L. E., Krivonosov, V. N. et al., *Pis'ma v ZTF* 52, 1346 (1982) (in Russian)
25. Voinov, A. M., Zobnin, V. G., Konak, A. I. et al., *Pis'ma v ZTF* 16, 34 (1990) (in Russian)
26. Voinov, A. M., Dovbych, L. E., Krivonosov, V. N. et al., *Pis'ma v ZTF* 52, 1346 (1982). (in Russian).
27. Jalufka, N. W., DeYoung, R. J., Hohl, F. and Williams, M. D., Nuclear pumped He^3-Ar laser excited by $He^3(n,p)T$ reaction, *Appl. Phys. Lett.* 29, 188 (1976).
28. Mansfield, C. R., Bird, P. F., Davis, J. F. et al., Direct nuclear pumping of a He^3-Xe laser, *Appl. Phys. Lett.* 30, 640 (1977).
29. Jalufka, N. W., Nuclear pumped lasing of He^3-Xe at 2.63 μ, *Appl. Phys. Lett.* 39, 535 (1981).
30. DeYoung, R. J., Jalufka, N. W. and Hohl, F., Nuclear-pumped lasing of He^3-Xe and He^3-Kr, *Appl. Phys. Lett.* 30, 19 (1977).
31. Voinov, A. M., Dovbych, L. E. and Krivonosov, V. N. et al., *Doklady AN SSSR*, 245, 80 (1979) (in Russian).
32. Helmick, H. H., Fuller, J. I. and Schneider, R. T., Direct nuclear pumping of helium-xenon laser, *Appl. Phys. Lett.* 26, 327 (1975).
33. McArthur, D. A. and Tollefsrud, P. B., Observation of laser action in CO gas excited only by fission fragments, *Appl. Phys. Lett.* 26, 187 (1975).
34. McArthur, D. A., Schmidt, T. R., Tollefsrud, P. B. and Walker, J. V., Preliminary designs for large (1-MJ) reactor-driven laser systems, *IEEE International Conf. on Plasma Science*, Ann Arbor, MI (May 1975).
35. McArthur, D. A., Nuclear pumped lasers, in *Encyclopedia of Lasers and Optical Technology*, Meyers, R. A., Ed., Academic Press, San Diego (1991), p. 385.
36. Bochkov, A. V., Kryzhanovskii, V. A., Lyubimov, O., Magda, E. P. and Mukhin, S., Nuclear reactor-pumpd laser atomic xenon operated at 2.48 μm, *Laser and Particle Beams* 11, 491 (1993).
37. Voinov, A. V., Krivonosov, V. N., Mel'nikov, S. P., Mochkaev, I. N. and Sinyanskii, A. A., Quasi-cw nuclear-pumped laser utilizing atomic transitions in argon, *Sov. J. Quantum Electron.* 21, 157 (1991).

Section 5.4

NATURAL LASERS

Introduction to the Table

Natural lasers are listed by increasing wavelength (in micrometers) in Table 5.4.1. The transition and active gas molecule, the object from which the radiation was observed, and the original reference(s) are given in the subsequent columns.

Further Reading

Clegg, A. W. and Nedoluha, G. E., Eds., *Astrophysical Masers*, Springer-Verlag, Berlin (1993).

Deming, D., Espenak, F., Jennings, D., Kostiuk, T., F. Mumma, M. J. and Zipoy, D., Modeling of the 10-µm natural laser emission from the mesospheres of Mars and Venus, *Icarus* 55, 356 (1983).

Elitzur, M., *Astronomical Masers*, Kluwer, New York (1992).

Moran, J. M., Maser action in nature, in *Handbook of Laser Science and Technology, Suppl. 1: Lasers,* CRC Press, Boca Raton, FL (1991), p. 579.

Moran, J. M., Maser action in nature, in *Handbook of Laser Science and Technology, Vol. I: Lasers and Masers*, CRC Press, Boca Raton, FL (1982), p. 483.

Mumma, M. J., Natural lasers and masers in the solar system, in *Astrophysical Masers*, Clegg, A. W. and Nedoluha, G. E., Eds., Springer-Verlag, Berlin (1993), p. 455.

Table 5.4.1
Natural Lasers Arranged in Order of Increasing Wavelength

Wavelength (μm)	Gas - Object	Transition	Reference
10.4[a]	CO_2 – Venus	ν_3–$2\nu_2$	1,2
10.4[a]	CO_2 – Mars	ν_3–$2\nu_2$	1,2,3
52.5	H – Cygnus MWC349	H10α	4
88.8	H – Cygnus MWC349	H12α	4
169.4	H – Cygnus MWC349	H15α	4
453	H – Cygnus MWC349	H21α	5
456	H_2O – stars[b]	ν_2=1, 1_{10}–1_{01}	6
636.6	H_2O – [c]	6_{42}–5_{51}	7
682.7	H_2O – [c]	6_{43}–5_{50}	7
844.9	H_2O – VY CMa	$17_{4,13}$–$16_{7,10}$	8
850	H – Cygnus MWC349	H26α	9
885.4	CH_3OH – S231	7_{-4}–6_{-4}	8
885.4 885.5	H_2O – S269	7_3–6_3 E_1 [d] 7_{-3}–6_{-3} E_2	8
885.5	H_2O – S252	7_3–6_3	8
885.6	CH_3OH – S252	7_3–6_3	8
891.6	H_2O – VY CMa	$5_{2,3}$–$6_{1,6}$	8
922.0	para-H_2O – SFR[e]	5_{15}–4_{22}	10
933.4	ortho-H_2O – SFR[e]	10_{29}–9_{36}	11
1635	H_2O[f]	3_{13} – 2_{20}	12

(a) Predicted amplications are very small (≤ 1.1)
(b) VY CMa, R Leo, R Crt, RT Vir, W Hya, RX Boo, S CrB, U Her, VX Sgr, NML Cyg
(c) Interstellar and circumstellar water
(d) Ambiquity of the line identification
(e) SFR – star forming regions
(f) Various molecular clouds, star forming regions, and evolving stars

References

1. Johnson, M. A., Betz, A. L., McLaren, R. A., Sutton, E. C. and Townes, C. H., Nonthermal 10 micron CO_2 emission lines in the atmospheres of Mars and Venus, *Astrophys. J.* 208, L145 (1976).
2. Deming, D., Espenak, F., Jennings, D., Kostiuk, T. F., Mumma, M. J. and Zipoy, D., Observations of the 10-μm natural emission from the mesospheres of Mars and Venus, *Icarus* 55, 347 (1983).
3. Mumma, M. J., Buhl, D., Chin, G., Deming, D., Espenak, F. and Kostiuk, T., Discovery of natural gain amplification in the 10-micrometer carbon dioxide laser bands on Mars: a natural laser, *Science* 212, 45 (1981).
4. Strelnitski, V., Haas, M. R., Smith, H. A., Erickson, E. F., Colgan, S. W. J. and Hollenbach, D. J., Far-infrared hydrogen lasers in the peculiar star MWC 349A, *Science* 272, 1459 (1996).
5. Thum, C., Matthews, H. E., Harris, A. I., Tacconi, L. J., Schuster, K. F. and Martin-Pintado, Detection of H21α maser emission at 662 GHz in MWC349, *Astron. Astrophys.* 288, L25 (1994).
6. Menten, K. M. and Young, K., Discovery of strong vibrationally excited water masers at 658 GHz toward evolved stars, *Astrophys. J.* 450, L70 (1995).
7. Melnick, G. J., Submillimeter water masers, in *Astrophysical Masers*, Clegg, A. W. and Nedoluha, G. E., Eds., Springer-Verlag, Berlin (1993), p. 41.
8. Feldman, P. A., Matthews, H. E., Amano, T., Scappini, F. and Lees, R. M., Observations of new submillimeter maser lines of water and methanol, in *Astrophysical Masers*, Clegg, A. W. and Nedoluha, G. E., Eds., Springer-Verlag, Berlin (1993), p. 65.
9. Thum, C., Matthews, H. E., Martin-Pintado, J., Serabyn, E., Planesas, P. and Bachiller, R. A., Submillimeter recombination line maser in MWC 349, *Astron. Astrophys.* 283, 582 (1994).
10. Menten, K. M., Melnick, G. J., Phillips, T. G. and Neufeld, D. A., A new submillimeter water maser transition at 325 GHz, *Astrophy. J.* 363, L27 (1990).
11. Menten, K. M., Melnick, G. J. and Phillips, T. G., Submillimeter water masers, *Astrophy. J.* 350, L41 (1990).
12. Cernicharo, J., Thum, C., Hein, H., John, D., Garcia, P. and Mattioco, F., Detection of 183 GHz water vapor maser emisssion from interstellar and circumstellar sources, *Astron. Astrophys.* 231, L15 (1990).

Section 5.5

INVERSIONLESS LASERS

Introduction to the Table

Table 5.5.1 presents experiments involving either lasing or amplification in inversionless systems arranged in order of increasing wavelength. The lasers have been operated in a continuous wave mode; the amplifier experiments have measured transient gain or gain on a probe laser pulse. The lasing atomic species is given in the second column. The operative transition and the lasing without inversion (LWI) or amplification without inversion (AWI) scheme are also given. Primary references to the experiments are listed in the final column.

Further Reading

Harris, S. E., Lasers without inversion: interference of lifetime-broadened resonances, *Phys. Rev. Lett.* 62, 1033 (1989).

Khurgin. J. B. and Rosencher, E., Practical aspects of lasing without inversion in various media, *IEEE J. Quantum Electron.* 32, 1882 (1996), and Practical aspects of optically coupled inversionless lasers, *J. Opt. Soc. Am. B* 14, 1249 (1997).

Kocharovskaya, O., Amplication and lasing without inversion, *Phys. Rep.* 219, 175 (1992).

Kocharovskaya, O. and Khanin, Ya. I., Coherent amplication of an ultrashort pulse in a three-level medium without a population inversion, *JETP Lett.* 48, 630 (1988).

Kocharovskaya. O. and Mandel, P., Basic models of lasing without inversion: general form of amplification condition and problem of self-consistency, *Quantum Optics* 6, 217 (1994).

Scully, M. O. and Fleischhauer, M., Lasers without inversion, *Science* 263, 337 (1994).

Scully, M. O., Resolving conundrums in lasing without inversion via exact solutions to simple models, *Quantum Optics* 6, 203 (1994).

Scully, M. O., Zhu, S.-Y. and Gavrielides, A., Degenerate quantum-beat laser: lasing without inversion and inversion without lasing, *Phys. Rev. Lett.* 62, 2813 (1989).

See, also, Papers on Atomic Coherence and Interference, Crested Butte Workshop 1993, in *Quantum Optics* 6 (1994).

Table 5.5.1
Inversionless Lasers and Amplifiers

Wavelength (nm)	Medium	Transition	Scheme	Ref.
Lasers:				
589.76	^{23}Na atomic beam	D_1:$3^2S_{1/2} - 3^2P_{1/2}$	Λ	1
794	^{87}Rb vapor, magnetic field	D_1:$5^2S_{1/2} - 5^2P_{1/2}$	V	2
Amplifiers:				
479	^{112}Cd vapor, magnetic field	$^3S_1 - {^3P_1}$	Λ	3
571	Sm vapor, magnetic field		Λ	4,9
589	^{23}Na vapor, magnetic field	D_1:$3^2S_{1/2} - 3^2P_{1/2}$ (F=2→F=2)	Λ	7
589.0, 589.6	^{23}Na vapor magnetic field	D_1:$3^2S_{1/2} - 3^2P_{1/2}$	V	5,6
770	K vapor He buffer gas	D_1:4S(F=2) – 4P(J=3/2)	*	8
821	Ba, atomic beam	$^1D_2(2) - {^1P_1}(1)$	**	10
894	Cs vapor magnetic field	D_1: F=3,4→F"=3,4	V	11

* Four-level Raman driven amplification

** Three-level cascade system.

References

1. Padmabandu, G. G., Welch, G. R., Shubin, I. N., Fry, E. S., Nikonov, D. E., Lukin, M. D. and Scully, M. O., Laser oscillation without population inversion in a sodium atomic bean, *Phys. Rev. Lett.* 76, 2053 (1996).
2. Zibrov, A. S., Lukin, M. D., Nikonov, D. E., Hollberg, L., Scully, M. O., Velichansky, V. L. and Robinson, H. G., Experimental demonstration of laser oscillation without population inversion via quantum interference in Rb, *Phys. Rev. Lett.* 75, 1499 (1995).
3. van der Veer, W. E., van Diest, R. J. J., Dönszelmann, A. and van Linden van den Heuvell, H. B., Experimental demonstration of light amplification without population inversion, *Phys. Rev. Lett.* 70, 3243 (1993).
4. Nottelmann, A., Peters, C. and Lange, W., Inversionless amplification of picosecond pulses due to Zeeman coherence, *Phys. Rev. Lett.* 70, 1783 (1993).
5. Gao, J.-Y., Zhang, H.-Z., Cui, H.-F., Guo, X.-Z., Jiang, Y., Wang, Q.-W., Jin, G.-X. and Li, J.-S., Inversionless light amplification in sodium, *Opt. Commun.* 110, 590 (1995).

6. Gao, J., Guo, C., Guo, X., Jin, G., Wang, P., Zhao, J., Zhang, H., Jiang, Y., Wang, D. and Jiang, D., Observation of light amplification without population inversion in sodium, *Opt. Commun.* 93, 323 (1995).
7. Fry, E. S., Li, X., Nikonov, D. et al., Atomic coherence effects within the sodium D_1 line: lasing without inversion via population trapping, *Phys. Rev. Lett.* 70, 3235 (1993).
8. Kleinfeld, J. A. and Streater, A. D., Observation of gain due to coherence effects in a potassium-helium mixture, *Phys. Rev. A*, 49, R4301 (1994).
9. Lange, W., Nottelman. A. and Peters, C., Observation of inversionless amplification in Sm vapour and related experiments, *Quantum Optics* 6, 273 (1994).
10. Sellin, P. B., Wilson, G. A., Meduri, K. K. and Mossberg, T. W., Observation of inversionless gain and field-assisted lasing in a nearly ideal three-level cascade-type atomic system, *Phys. Rev. A* 54, 2402 (1996).
11. Fort, C., Cataliotti, F. S., Hänsch, T. W., Inguscio, M. and Prevedelli, M., Gain without inversion on the cesium D_1 line, *Optics Commun.* 139, 31 (1997).

Section 6: Commercial Lasers

Section 6

COMMERCIAL LASERS

Introduction to the Table

Commercial lasers are arranged in order of increasing wavelength in Table 6.1. The medium (gas, liquid, or solid), laser type, and mode of operation (pulsed or cw) are given. The data were compiled from recent (1995–1997) laser buyers' guides and manufacturer's literature may not be the only lasers available commercially nor may the lasers still be manufactured. Representative output power or energies of solid state (crystalline, glass, and polymer), semiconductor, liquid organic dye, and gas lasers are given in Tables 6.1.1, 6.2.1, 6.3.1, and 6.4.1, respectively. These latter data are also taken from recent (1995–1997) laser buyers' guides and can be expected to change due to advances in technology.

Tunable lasers are listed by their reported shortest and longest wavelengths for the pumping conditions used. Wavelengths enclosed in brackets denote the extremes of a group of discrete laser lines or chemical compositions.

Abbreviations: SH - second harmonic, TH - third harmonic, FH - fourth harmonic, FFH - fifth harmonic, R - Raman shifted, OPO - optical parametric oscillator. Acronyms and abbreviations for laser types are defined in Appendix 2.

Further Reading

Hecht, J., *The Laser Guidebook* (second edition), McGraw-Hill, New York (1992).

Hecht, J., *Understanding Lasers*, (second edition), IEEE Press, New York (1994).

Laser Focus World Buyers Guide, Pennwalt Publishing Company, Tulsa, OK.

Technology and Industry Reference, Lasers and Optronics, Morris Plains, NJ.

Optical Industry and Systems Purchasing Directory, Optical Publishing, Pittsfield, MA.

Table 6.1
Commercial Lasers Arranged in Order of Increasing Wavelength

Wavelength (μm)	Medium	Laser Type	Operation
0.157	gas	F_2 excimer	pulsed
0.193	gas	ArF excimer	pulsed
0.2–0.4	liquid	organic dyes (SH)	pulsed
0.209	solid	Nd:YLF (FFH)	pulsed
0.213	solid	Nd:YAG (FFH)	pulsed
0.22–0.39	liquid	organic dyes (SH)	cw
0.222	gas	KrCl excimer	pulsed
0.2243	gas	He-Ag^+	pulsed, cw
0.248	gas	KrF excimer	pulsed
[0.248–0.270]	gas	Ne-Cu^+	pulsed, cw
0.25–0.30	solid	Ti:sapphire (TH)	pulsed
0.263	solid	Nd:YLF (FH)	pulsed
0.266	solid	Nd:YAG (FH)	pulsed, cw
[0.282–0.292]	gas	He-Au^+	pulsed, cw
0.308	gas	XeCl excimer	pulsed
0.325	gas	He-Cd	cw
0.3324	gas	Ne ion	cw
0.3371	gas	nitrogen (N_2)	pulsed
0.347	solid	ruby (SH)	pulsed
0.351	gas	Ar ion	cw
0.351	gas	XeF excimer	pulsed
0.351	solid	Nd:YLF (TH)	pulsed
0.355	solid	Nd:glass (TH)	pulsed
0.355	solid	Nd:YAG (TH)	pulsed, cw
0.36–0.40	solid	alexandrite (SH)	pulsed
0.36–0.46	solid	Ti:sapphire (SH)	pulsed, cw
0.3713	gas	Ne	cw
0.373	gas	Ne	cw
0.38–1.0	liquid	organic dyes	pulsed, cw
0.415	solid	GaN	pulsed
0.42–0.43	solid	GaAlAs (SH)	cw
0.430	solid	Ce:$LiSrAlF_6$	cw
0.4416	gas	He-Cd	cw
0.4545	gas	Ar ion	cw
0.4579	gas	Ar ion	cw

Table 6.1—*continued*
Commercial Lasers Arranged in Order of Increasing Wavelength

Wavelength (μm)	Medium	Laser Type	Operation
[0.458–0.676]	gas	Ar–Kr ion	cw
0.473	solid	Nd:YVO_4 (SH)	cw
0.4762	gas	Kr ion	pulsed, cw
0.4765	gas	Ar ion	pulsed, cw
[0.48–0.54]	gas	Xe ion	pulsed
0.4880	gas	Ar ion	pulsed, cw
0.4965	gas	Ar ion	cw
0.5017	gas	Ar ion	pulsed, cw
0.5105	gas	Cu vapor	pulsed
0.5145	gas	Ar ion	pulsed, cw
0.5208	gas	Kr ion	pulsed, cw
0.523	solid	Nd:YLF (SH)	pulsed, cw
0.527	solid	Nd:YLF (SH)	pulsed, cw
0.527	solid	Nd:glass (SH)	pulsed
0.5287	gas	Ar ion	pulsed, cw
0.531	solid	Nd:YAB (SH)	pulsed, cw
0.5319	gas	Kr ion	pulsed, cw
0.532	solid	Nd:YAG (SH)	pulsed, cw
0.532	solid	Nd:YVO_4 (SH)	cw
0.5395	gas	Xe ion	pulsed
[0.543–3.39]	gas	He-Ne	cw
0.5435	gas	He-Ne	cw
[0.55–0.70]	solid	organic dye	pulsed
0.5782	gas	Cu vapor	pulsed
0.58–0.66	solid	Cr:Mg_2SiO_4 (SH)	pulsed
0.5941	gas	He-Ne	cw
0.6119	gas	He-Ne	cw
0.628	gas	Au vapor	pulsed, cw
[0.63–0.68]	solid	InGaAlP	cw
0.6328	gas	He-Ne	cw
0.6471	gas	Kr ion	pulsed, cw
0.67	solid	GaAsP	pulsed, cw
0.67–1.13	solid	Ti:sapphire	cw
0.6764	gas	Kr ion	pulsed, cw
0.68	solid	InGaAlP	pulsed

Table 6.1—*continued*
Commercial Lasers Arranged in Order of Increasing Wavelength

Wavelength (μm)	Medium	Laser Type	Operation
0.6943	solid	ruby	pulsed, cw
0.7–0.8	solid	alexandrite	cw
0.7–1.1	solid	Ti:sapphire	pulsed
0.72–0.82	solid	alexandrite	pulsed
0.74–0.84	solid	Cr:GSGG	pulsed
[0.75–0.85]	solid	GaAlAs	pulsed, cw
0.7525	gas	Kr ion	pulsed, cw
0.78–0.85	solid	Cr:$KZnF_3$	pulsed
[0.78–0.91]	solid	GaAlAs	pulsed
0.78–1.01	solid	Cr:LiSAF	pulsed
0.7993	gas	Kr ion	pulsed, cw
0.85	solid	Er:YLF	pulsed
0.9–4.5	liquid	organic dyes (R)	pulsed
0.904	solid	GaAs	pulsed
0.91–0.98	solid	InGaAs	pulsed, cw
0.946	solid	Nd:YAG	pulsed
1.04–1.2	solid	alexandrite (R)	pulsed
1.047	solid	Nd:YLF	pulsed, cw
1.053	solid	Nd:YLF	pulsed, cw
1.053	solid	Nd:phosphate glass	pulsed
1.053	solid	Nd:YAP (YALO)	pulsed
1.06	solid	Nd:YAB	pulsed, cw
1.061	solid	Nd: silicate glass	pulsed
1.061	solid	Nd:GSGG	pulsed
1.062	solid	Nd:GGG	pulsed
1.064	solid	Nd:YAG	pulsed, cw
1.079	solid	Nd:YAP (YAlO)	pulsed
1.09–1.27	solid	LiF (F_2^-)	pulsed
[1.1–1.55]	solid	InGaAsP	pulsed, cw
1.13–1.36	solid	Cr:Mg_2SiO_4	pulsed
1.152	gas	He-Ne	cw
1.313	solid	Nd:YLF	pulsed, cw
1.315	gas	iodine	pulsed
1.319	solid	Nd:YAG	pulsed, cw
1.321	solid	Nd: YLF	cw

Table 6.1—*continued*
Commercial Lasers Arranged in Order of Increasing Wavelength

Wavelength (μm)	Medium	Laser Type	Operation
1.335	solid	Nd: YAG	pulsed, cw
1.39	solid	Nd:YAG (R)	pulsed
1.45–1.85	solid	NaCl:OH (F_2^+)	cw
1.48–1.72	solid	NaCl:OH (F_2^+)	pulsed
1.52–1.57	solid	Er:glass	pulsed, cw
1.523	gas	He-Ne	cw
1.54	solid	Er:silica	pulsed
1.54	solid	Nd:YAG (R)	pulsed
1.75–2.50	solid	Co:MgF_2	pulsed, cw
1.85–2.16	solid	Co:MgF_2	pulsed
1.91	solid	Nd:YAG (R)	pulsed
[2–4]	gas	Xe-He	pulsed, cw
[2.006–2.025]	solid	Tm:YAG	pulsed, cw
2.019, 2.033	solid	Tm:LuAG	pulsed, cw
[2.048–2.069]	solid	Ho: YLF	pulsed
2.088	solid	Ho: YSGG	pulsed
2.09	solid	Ho,Tm,Cr:YAG	pulsed, cw
[2.088–2.091]	solid	Ho:YAG	pulsed, cw
2.127	solid	Ho:YAG	cw
2.30–2.50	solid	KCl:Na (F_B)	cw
2.45–2.80	solid	KCl:Li (F_A)	cw
2.52–2.90	solid	KCl:Li (F_A)	pulsed
[2.6–3.0]	gas	HF (chemical)	pulsed, cw
2.70–3.30	solid	RbCl:Li (F_A)	cw
2.73–3.18	solid	RbCl:Li (F_A)	pulsed
2.796	solid	Er,Cr:YSGG	pulsed
2.9–3.6	solid	Pb salts (77 K)	cw
2.90	solid	Er:YAG	pulsed, cw
2.94	solid	Er:YAG	pulsed, cw
3.3–27	solid	Pb salts	pulsed, cw
3.391	gas	He-Ne	cw
[3.6–4.0]	gas	DF (chemical)	pulsed, cw
[5–7]	gas	CO	cw
[9.2–11.4]	gas	CO_2	pulsed, cw

Table 6.1—*continued*
Commercial Lasers Arranged in Order of Increasing Wavelength

Wavelength (μm)	Medium	Laser Type	Operation
10.6	gas	CO_2	pulsed, cw
10.65	gas	N_2O	pulsed, cw
[10.3–11.1]	gas	N_2O	cw
[37–1224]	gas	methanol (CH_3OH)	pulsed, cw
[40–1200]	gas	various molecules[(a)]	pulsed, cw
[496,1222]	gas	methyl fluoride (CH_3F)	pulsed, cw

(a) See footnote to Table 6.4.1 for examples of specific molecules.

Section 6.1

SOLID STATE LASERS

Introduction to the Table

Commercial solid state lasers, mode of operation (cw or pulsed), principal wavelengths, and representative outputs are given in Table 6.1.1.

Abbreviations: SH - second harmonic, TH - third harmonic, FH - fourth harmonic, FFH - fifth harmonic, R - Raman shifted, SD - self doubled, DP – diode pumped.

Table 6.1.1
Commercial Solid State Lasers

Laser Type	Operation	Principal wavelengths (μm)	Output
Lanthanide Lasers:			
Nd: YAB [$YAl_3(BO_3)_4$]	cw	0.53 (SH)	10 mW
	cw	1.06	10–200 W
	cw (DP)	1.06	0.1–1 W
Nd:YAG ($Y_3Al_5O_{12}$)	pulsed	0.213 (FFH)	4–15 mJ
	cw	0.266 (FH)	0.02–0.6 W
	pulsed	0.266 (FH)	1–300 mJ
	cw	0.355 (TH)	0.01–1.5 W
	pulsed	0.355 (TH)	1–800 mJ
	cw	0.532 (SH)	0.1–60 W
	cw (DP)	0.532 (SH)	0.1–0.5 W
	pulsed	0.532 (SH)	0.1–100 J
	pulsed (DP)	0.532 (SH)	0.001–0.1 J
	cw	0.946	10 mW
	cw (multimode)	1.064	1–3000 W
	cw (TEM_{00})	1.064	0.1–60 W
	cw (DP)	1.064	1 mW–20 W
	pulsed (multi.)	1.064	0.1–2000 J
	pulsed (TEM_{00})	1.064	1–2.5 J
	pulsed (DP)	1.064	0.1–250 mJ
	cw	1.319	0.2–100 W
	cw (DP)	1.319	0.2–2 W
	pulsed	1.319	1–5 J

Table 6.1.1—*continued*
Commercial Solid State Lasers

Laser Type	Operation	Principal wavelengths (μm)	Output
Nd:YLF ($LiYF_4$)	pulsed	0.209 (FFH)	0.2 mJ
	pulsed	0.263 (FH)	0.2–2 mJ
	pulsed	0.351 (TH)	0.3–2 mJ
	pulsed	0.523 (SH)	0.02–15 mJ
	pulsed	0.527 (SH)	1–15 mJ
	cw	1.047	0.5–6 W
	cw (DP)	1.047	0.5–2 W
	pulsed	1.047	0.5 J
	pulsed (DP)	1.047	0.01–0.15 J
	cw	1.053	2–45 W
	cw (DP)	1.053	0.5–5 W
	pulsed	1.053	0.1–10 J
	pulsed (DP)	1.053	≤ 1 mJ
	cw	1.313	1.5–3 W
	cw (DP)	1.313	0.04–0.8 W
	pulsed (DP)	1.313	≤ 10^{-5} J
	cw	1.321	40-200 mW
Nd:YVO_4	pulsed (DP)	0.355 (TH)	30 mW
	cw	0.473 (SH)	1–100 mW
	cw (DP)	0.473 (SH)	20 mW
	cw	0.532 (SH)	0.01–5 W
	cw (DP)	0.532 (SH)	10–50 mW
	pulsed	1.064	≤ 150 mJ
	cw	1.064	2.5–10 W
Nd:GGG ($Gd_3Ga_5O_{12}$)	pulsed	1.062	14 J
Nd:YAP or YALO ($YAlO_3$)	cw, pulsed	1.079	≤ 60 W
Nd,Cr:GSGG ($Gd_3Sc_2Ga_3O_{12}$)	pulsed	1.061	0.5–40 J
Nd:glass (phosphate)	pulsed	0.263 (FH)	0.04–4 J
	pulsed	0.351 (TH)	0.1–8 J
	pulsed	0.527 (SH)	0.2–22 J
	pulsed	1.054	1.0–80 J
Nd:glass (silicate)	pulsed	0.26 (FH)	0.1–0.8 J
	pulsed	0.35 (TH)	0.3–2 J
	pulsed	0.53 (SH)	0.1–5 J
	pulsed	1.06	0.2–20 J

Table 6.1.1—*continued*
Commercial Solid State Lasers

Laser Type	Operation	Principal wavelengths (μm)	Output
Ho:YLF	cw	2.048–2.069	0.05–1 W
Ho:YAG ($Y_3Al_5O_{12}$)	cw	2.088–2.091	0.05–1 W
	pulsed	2.1	1–5 J
Ho:YSGG ($Y_3Sc_2Ga_3O_{12}$)	pulsed	2.088	3 J
Ho,Tm,Cr:YAG ($Y_3Al_5O_{12}$)	cw	2.09	0.05–1 W
	pulsed	2.09	0.5–2 J
Er:glass	pulsed (DP)	1.54	1.2 J
Er:glass (fiber)	cw	1.52–1.57	30 mW
	cw (DP)	1.54	50 mW
	pulsed	1.54	0.5–10 mJ
Er:YAG ($Y_3Al_5O_{12}$)	cw	2.90, 2.94	2–10 W
	pulsed	2.90, 2.94	1–4 J
Er:YSGG ($Y_3Sc_2Ga_3O_{12}$)	pulsed	2.79	2 J
Tm:LuAG ($Lu_3Al_5O_{12}$)	pulsed (DP)	2.019, 2.033	0.01 J
	cw	2.019, 2.033	0.05–1 W
Tm:YAG ($Y_3Al_5O_{12}$)	pulsed	2.01	2 J
	cw	2.006–2.025	0.05–1 W
Transition Metal Lasers:			
Ruby ($Cr:Al_2O_3$)	pulsed (SH)	0.347	0.1–0.3 J
	cw	0.6943	7 W
	pulsed	0.6943	0.3–100 J
Alexandrite ($Cr:BeAl_2O_4$)	pulsed	0.36–0.4 (SH)	10–500 mJ
	cw	0.7–0.8	0.1–2 W
	pulsed	0.72–0.82	10 mJ–3 J

Table 6.1.1—*continued*
Commercial Solid State Lasers

Laser Type	Operation	Principal wavelengths (μm)	Output
Ti: sapphire ($Ti:Al_2O_3$)	pulsed	0.25–0.30 (TH)	50 μJ–1 mJ
	cw	0.36–0.46 (SH)	0.01-0.2 mW
	pulsed	0.36–0.45 (SH)	0.3–25 mJ
	cw	0.67–1.13	0.25–5 W
	pulsed	0.7–1.1	10 mJ–3J
Cobalt perovskite ($Co:MgF_2$)	pulsed	1.75–2.5	20–25 mJ
Chromium:LiSAF ($Cr:LiSrAlF_6$)	pulsed	0.78–1.01	2 mJ
Chromium fluoride ($Cr:KZnF_3$)	cw	0.78–0.85	1 W
	pulsed	0.78–0.85	10 mJ
Forsterite ($Cr:Mg_2SiO_4$)	pulsed	0.58–0.66 (SH)	2 mJ
	pulsed	1.13–1.36	0.1–20 J
Color Center Lasers:			
LiF (F_2^+)	pulsed	1.09–1.27	≤ 50 mJ
NaCl:OH (F_2^-)	cw	1.45–1.85	0.35 W
	pulsed	1.48–1.72	0.1 J
KCl:Na (F_B)	cw	2.30–2.55	1–100 mW
KCl:Li (F_A)	cw	2.45–2.80	100 mW
	pulsed	2.52–2.90	15 mJ
RbCl:Li(F_A)	cw	2.70–3.30	10 mW
	pulsed	2.73–3.18	15 mJ
Organic Dye Laser			
polymeric host	pulsed	0.55–0.70[(a)]	≤ 150 mJ

(a) Tunable; several different polymer rods are needed to cover the wavelength range indicated.

Section 6.2

SEMICONDUCTOR LASERS

Introduction to the Table

Commercial semiconductor lasers (single diode and arrays), mode of operation (pulsed or cw), principal wavelengths, and representative outputs are given in Table 6.2.1.

Table 6.2.1
Commercial Semiconductor Lasers

Laser Material	Operation	Principal wavelengths (μm)	Output
GaN	pulsed	0.415	20 nJ
GaAlAs	cw	0.42, 0.43 (SH)	0.4–4.0 W
InGaAlP	cw	[0.63–0.68]	1–500 mW
	pulsed	0.68	≤ 10 J
GaAsP	cw	0.67	1–10 mW
	pulsed	0.67	3–10 J
GaAlAs	cw	[0.75–0.85]	1–200 mW
	pulsed	[0.75–0.85]	1–500 mJ
GaAlAs (array)	cw	[0.75–0.85]	10 W–40 W
	pulsed	[0.78–0.91]	0.1–30 J
GaAs	pulsed	0.904	≤ 0.8 J
GaAs (array)	pulsed	0.904	≤ 5 J
InGaAs	cw	0.905–0.98	0.02–1 W
	pulsed	0.905–0.98	10^{-6}–1 J
InGaAs (array)	cw	0.91–0.98	30 W
InGaAsP	cw	1.27–1.33	0.1–3.0 W
	cw	1.52–1.58	0.5–100 mW
	pulsed	1.06–1.55	0.2 mJ
	pulsed	1.55	10^{-3}–0.6 J
InGaAsP (array)	pulsed	1.55	2.5 J
Pb salts	cw	3.3–27	0.1–25 mW
	pulsed	3.3–25	< 2 J
Pb salts (77 K)	cw	2.9–3.6	1–5 mW

Section 6.3

DYE LASERS

Introduction to the Table

Wavelength ranges for commercial laser dyes, pump sources, and representative outputs are given in Table 6.3.1.

Lasing of organic dyes is dependent on the solvent, dye concentration, pumping source and rate, and other operating conditions. Relative energy outputs and tuning curves that may be obtained from commercially available pump sources and dyes are shown in Figures 6.3.1 – 6.3.12 (figures courtesy of Richard N. Steppel). The information is provided only as a guide and may not necessarily be extrapolated to systems other than those cited.

Table 6.3.1
Commercial Dye Lasers Arranged in Order of Wavelength

Wavelength (μm)	Pump source	Output
CW Lasers:		
0.22–0.39 (SH)	Ar ion laser	0.01 W
0.38–1.0	Ar ion laser	0.1–2 W
Pulsed Lasers:		
0.2–0.4 (SH)	Nd:YAG, excimer lasers	1–60 mJ
0.25–0.4 (SH)	coaxial flashlamp	0.1–0.9 J
0.3–0.9	linear flashlamp	0.5–3 J
0.32–1.0	excimer laser	10–150 mJ
0.36–0.95	nitrogen laser	0.1–150 mJ
0.4–1.0	Nd:YAG laser	5–200 mJ
0.44–0.8	coaxial flashlamp	< 1–30 J
0.53–0.9	Cu vapor laser	0.1– 2 mJ
0.695–0.905	Ti:sapphire laser	≤ 0.15 J
0.9–4.5 (R)	Nd:YAG laser	1–10 mJ

R - Raman shifted, SH - second harmonic.

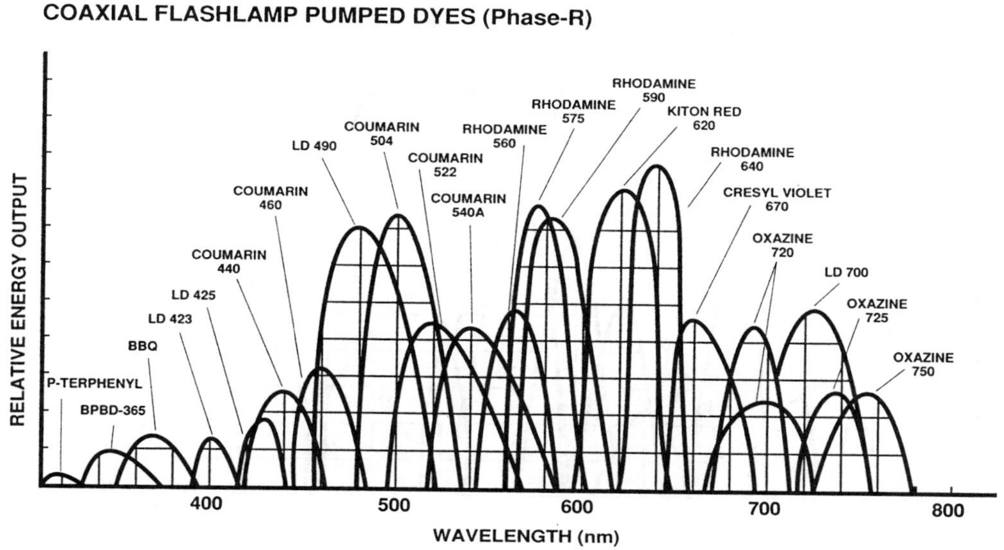

FIGURE 6.3.1 Tuning curves and relative energy outputs of various coaxial flashlamp pumped dyes. Data courtesy of Phase-R Corp., Box G-2, Old Bay Road, New Durham, NH.

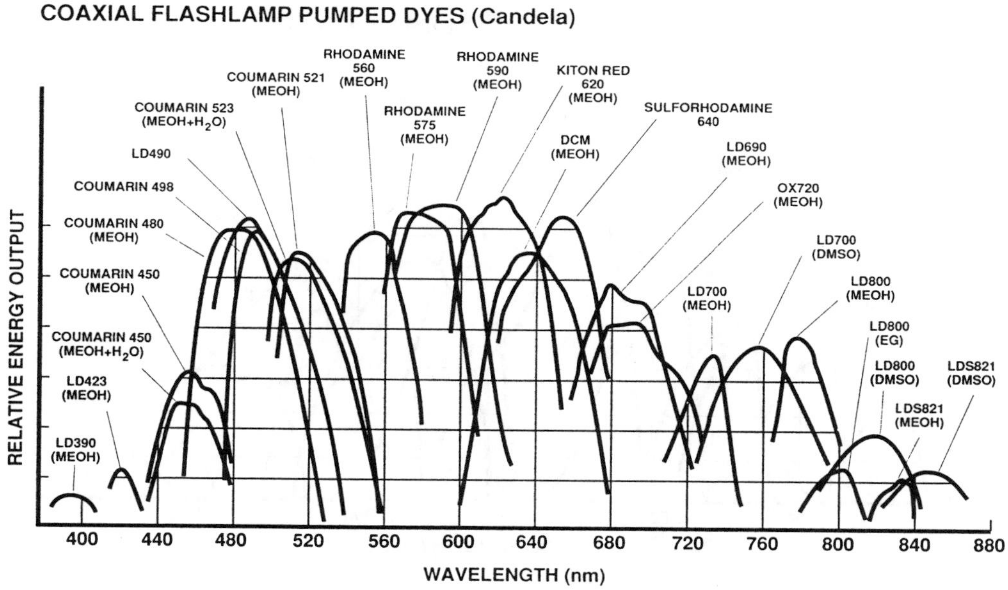

FIGURE 6.3.2 Tuning curves and relative energy outputs of various coaxial flashlamp pumped dyes. Data courtesy of Candela Laser Corp., 530 Boston Post Road, Wayland, MA.

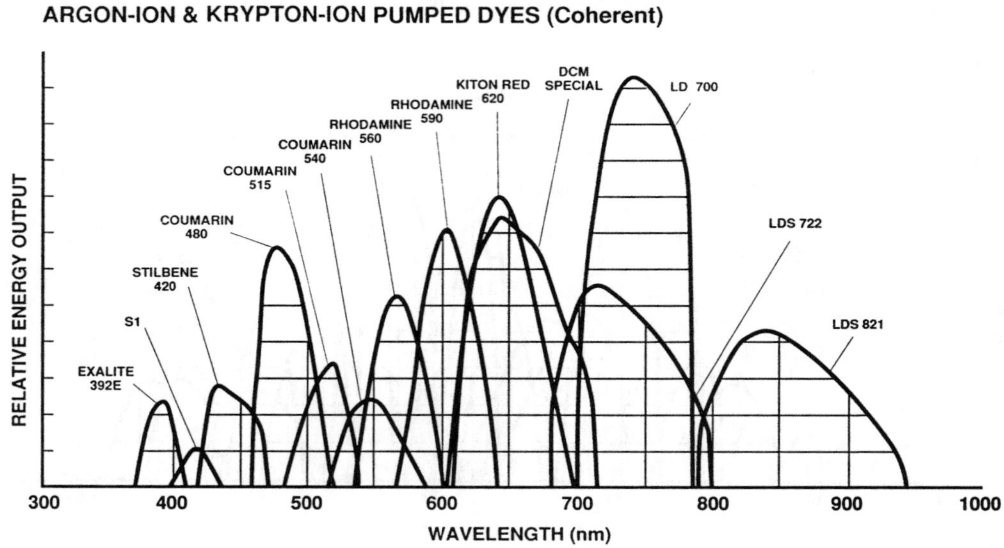

FIGURE 6.3.3 Tuning curves and relative energy outputs of various argon-ion and krypton-ion laser pumped dyes. Data courtesy of Coherent Inc., 3210 Porter Drive, Palo Alto, CA.

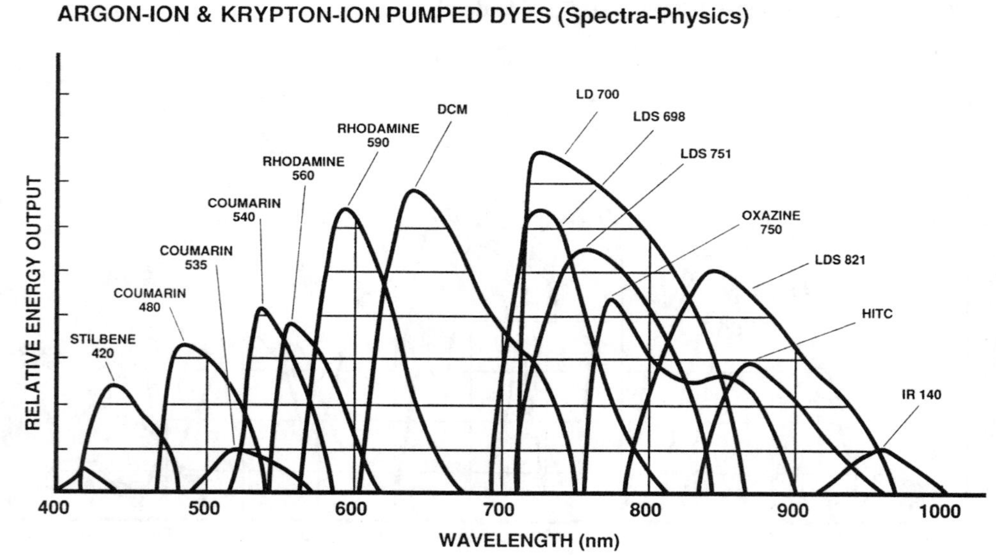

FIGURE 6.3.4 Tuning curves and relative energy outputs of various argon-ion and krypton-ion laser pumped dyes. Data Courtesy of Spectra-Physics Inc., 1250 Middlefield Road, Mountain View, CA.

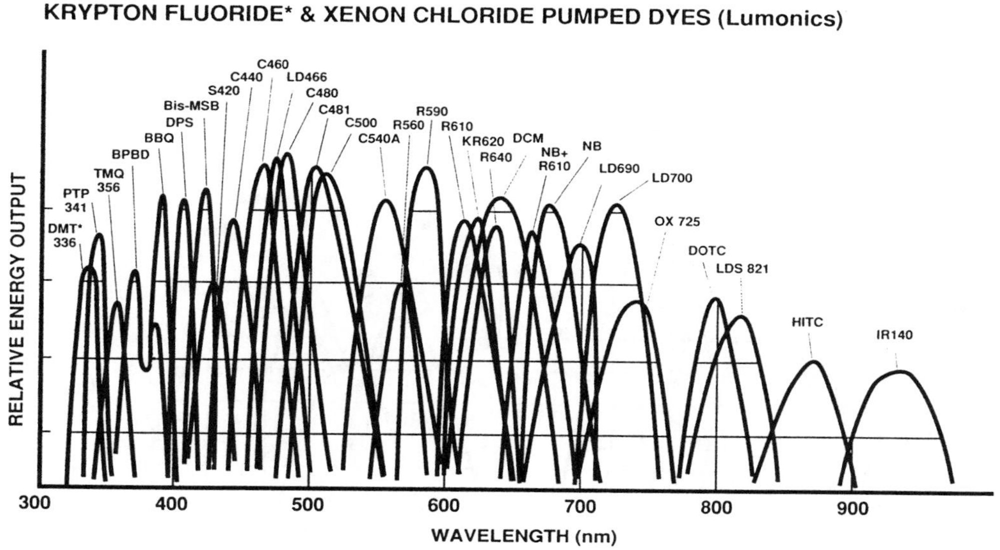

FIGURE 6.3.5 Tuning curves and relative energy outputs of various krypton fluoride and xenon chloride laser pumped dyes. Data courtesy of Lumonics, Inc., 105 Schneider Road, Kanata (Ottawa), Ontario, Canada.

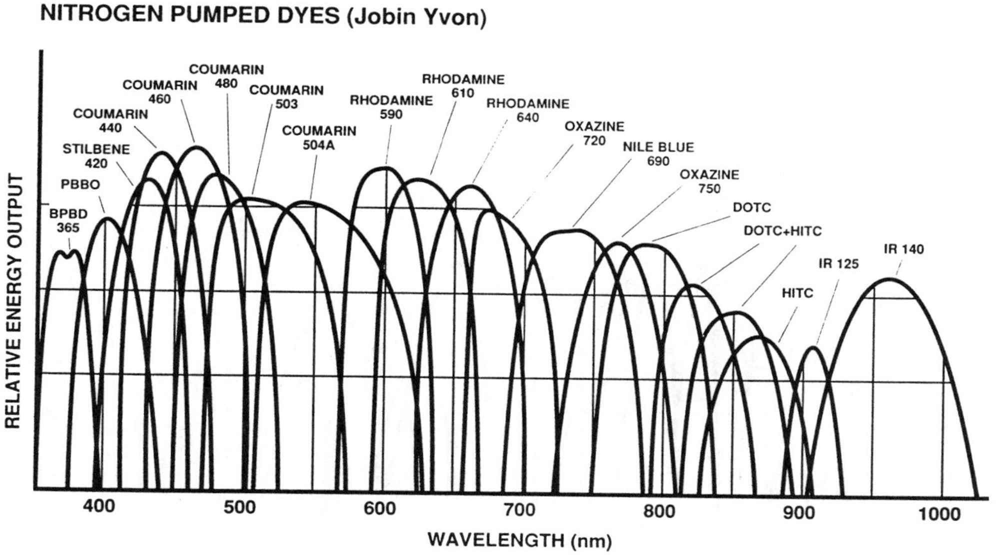

FIGURE 6.3.6 Tuning curves and relative energy outputs of various nitrogen laser pumped dyes. Data courtesy of Jobin Yvon, 16-18, rue du Canal B. P. 118, 91163 Longjumeau Cedex, France.

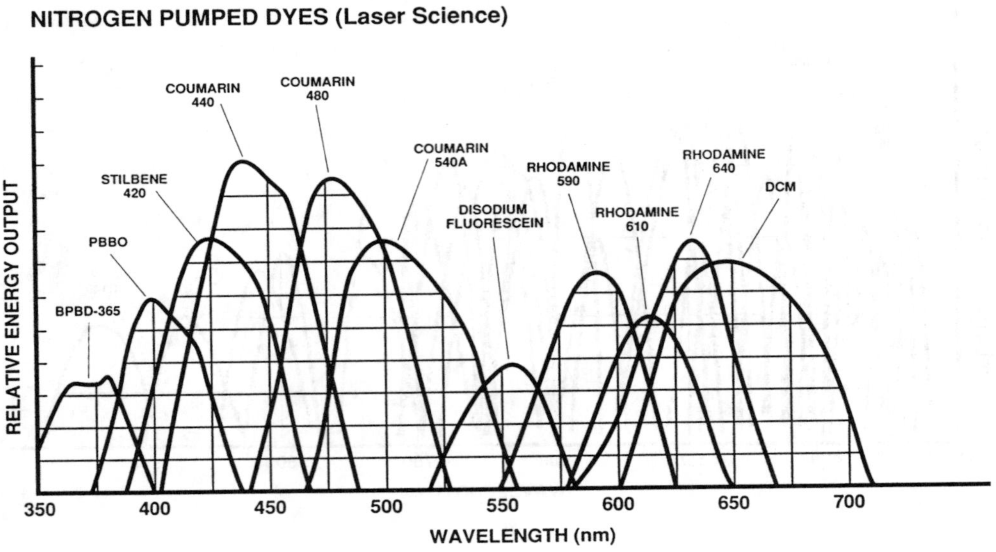

FIGURE 6.3.7 Tuning curves and relative energy outputs of various nitrogen laser pumped dyes. Data courtesy of Laser Science, Inc., 26 Landsdowne Street, Cambridge, MA.

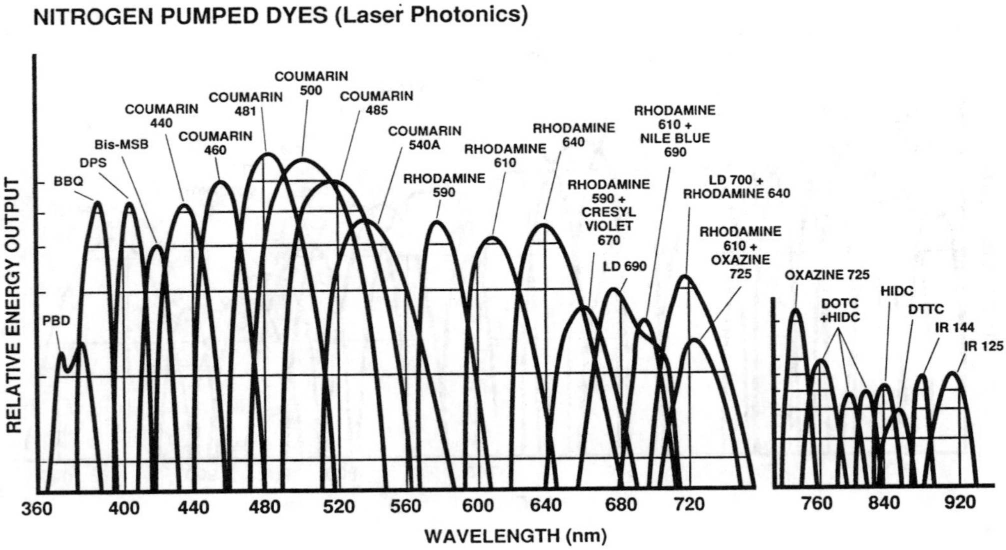

FIGURE 6.3.8 Tuning curves and relative energy outputs of various nitrogen laser pumped dyes. Data courtesy of Laser Photonics, Inc., 12351 Research Parkway, Orlando, FL.

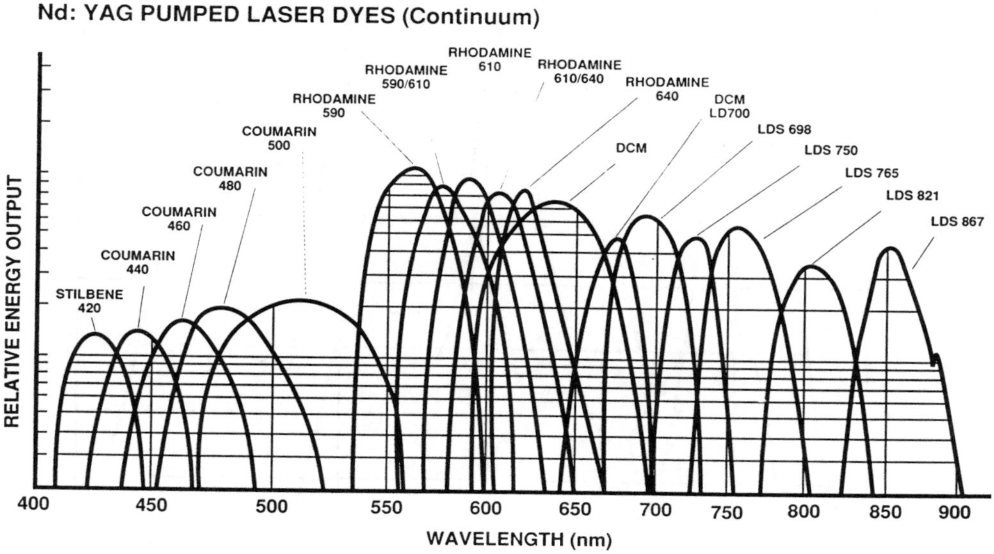

FIGURE 6.3.9 Tuning curves and relative energy outputs of various Nd:YAG laser pumped dyes. Data courtesy of Continuum, 3150 Central Expressway, Santa Clara, CA.

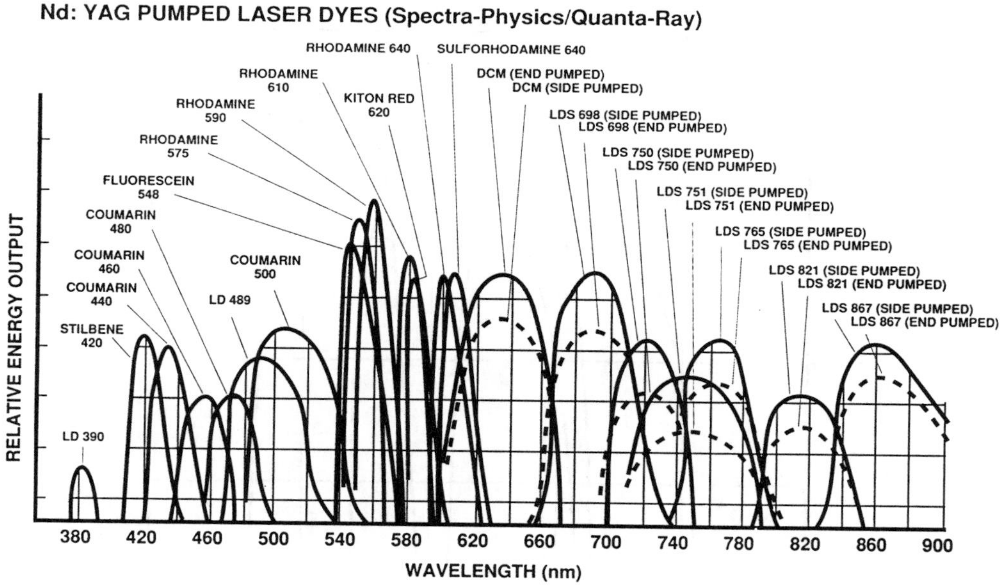

FIGURE 6.3.10 Tuning curves and relative energy outputs of various Nd:YAG laser pumped dyes. Data courtesy of Spectra-Physics/Quanta-Ray, 1250 Middlefield Road, Mountain View, CA.

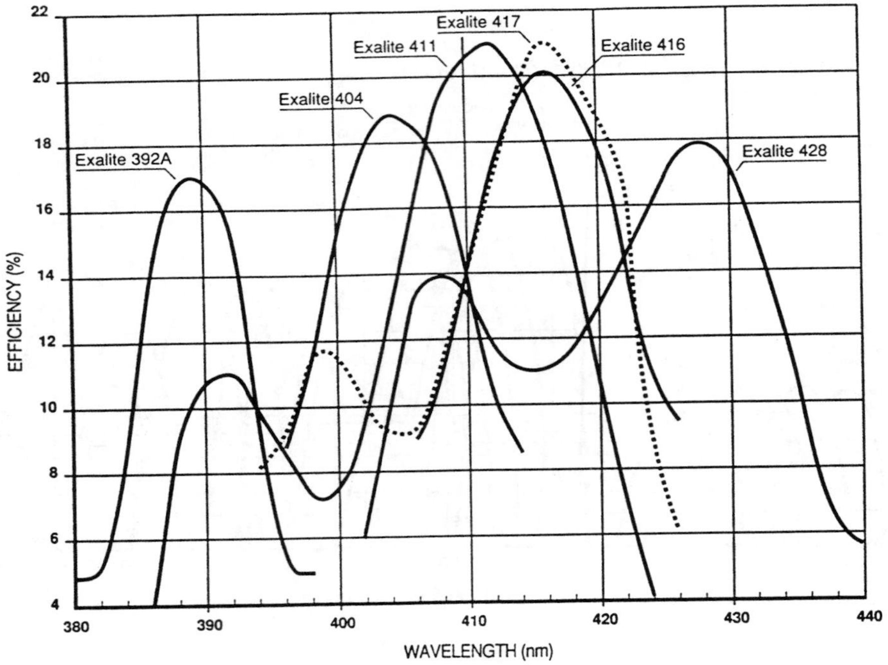

FIGURE 6.3.11 Tuning curves and efficiency of Exalite laser dyes (Exciton, Inc.) for Nd:YAG pumping at 355 nm. Data courtesy of Spectra-Physics/Quanta-Ray, 1250 Middlefield Road, Mountain View, CA.

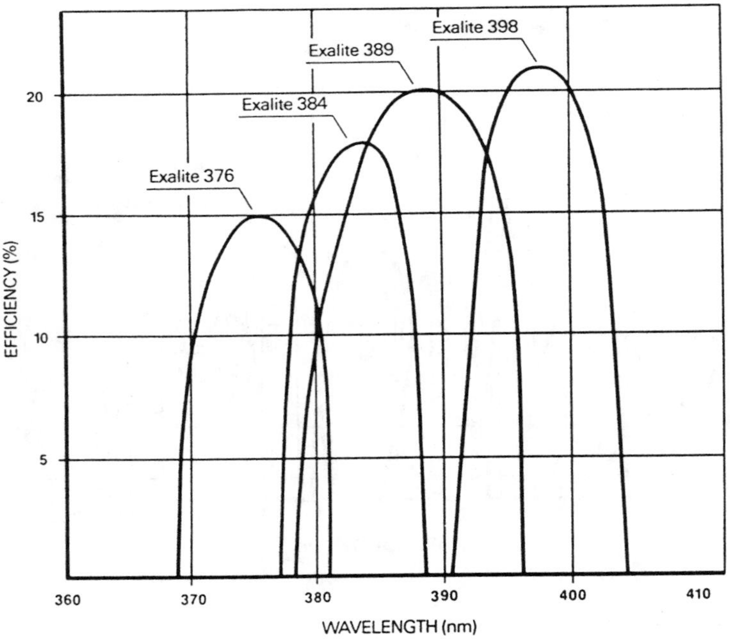

FIGURE 6.3.12 Tuning curves and efficiency of Exalite laser dyes (Exciton, Inc.) for Nd:YAG pumping at 355 nm. Data courtesy of Lumonics, Inc., 105 Schneider Road, Kanata (Ottawa), Ontario, Canada.

Section 6.4

GAS LASERS

Introduction to the Table

Commercial gas laser types, mode of operation (cw or pulsed), lasing wavelengths, and representative outputs are given in Table 6.4.1.

Table 6.4.1
Commercial Gas Lasers Arranged in Order of Wavelength

Laser Type	Operation	Wavelength(s) (μm)	Output
Helium-neon (He-Ne)	cw	0.5435	0.1–3 mW
	cw	0.5941	0.5–7 mW
	cw	0.6119	0.5–7 mW
	cw	0.6328	0.5–50 mW
	cw	1.152	1–13 mW
	cw	1.523	0.5–1 mW
	cw	3.391	1–40 mW
Helium-cadmium (He-Cd)	cw	0.325	1–100 mW
	cw	0.4416	10–200 mW
Helium-silver ($He-Ag^+$)	cw	0.2243	1 mW
	pulsed	0.2243	0.1 J
Helium-gold ($He-Au^+$)	cw	[0.282–0.292]	3 mW
	pulsed	[0.282–0.292]	0.3 J
Neon-copper ($Ne-Cu^+$)	cw	[0.248–0.270]	3 mW
	pulsed	[0.248–0.270]	0.3 J
Xenon-helium (Xe-He)	cw	2–4 μm	1–600 mW
	pulsed	2–4 μm	0.5 J
Molecular Lasers:			
Carbon dioxide (CO_2)[(a)]	cw	10.6	1 W–10 kW
	pulsed	10.6 (other lines from 9.2 to 11.4)	100 mJ–3 kJ

Table 6.4.1—*continued*
Commercial Gas Lasers Arranged in Order of Wavelength

Laser Type	Operation	Wavelength(s) (μm)	Output
Carbon monoxide (CO)	cw, pulsed	several lines between 5 and 7 μm	1–35 W
Iodine (I_2)	pulsed	1.315	≤ 1–3 J
Nitrogen (N_2)	pulsed	0.3371	0.1–10 mJ
Nitrous oxide (N_2O)	cw	10.65 (other lines 10.3 to 11.1)	15 W
	pulsed	10.65 (other lines 10.3 to 11.1)	1 mJ
Metal Vapor Lasers:			
Copper (Cu)	cw	0.5105, 0.5782	100 W
	pulsed	0.5105, 0.5782	1–20 mJ
Gold (Au)	cw	0.628	2 W
	pulsed	0.628	0.2–0.6 mJ
Ion Lasers:			
Neon (Ne^+)	cw	0.3324 (other lines–0.3392, 0.3378, 0.3345, 0.3713, 0.373)	1 W
Argon (Ar^+)	cw, pulsed	0.4880, 0.5145 (other lines–0.351, 0.4545, 0.4579, 0.4765, 0.4965, 0.5017, 0.5287)	5 mW–50 W
Krypton (Kr^+)	cw, pulsed	0.6471 (other lines–0.3375, 0.3564,0.4762, 0.5208, 0.5309, 0.5682, 0.6764, 0.7525, 0.7993)	0.1–6 W
Argon-Krypton (Ar^+-Kr^+)	cw	many lines between 0.34–0.80	1–3 W
		several lines between 0.458–0.676	0.2–10 W
Xenon (Xe^{3+})	pulsed	0.5395	0.6 J
Excimer Lasers:			
Fluorine (F_2)	pulsed	0.157	1–60 mJ

Table 6.4.1—*continued*
Commercial Gas Lasers Arranged in Order of Wavelength

Laser Type	Operation	Wavelength(s) (μm)	Output
Argon fluoride (ArF)	pulsed	0.193	3–700 mJ
Krypton chloride (KrCl)	pulsed	0.222	0.3–1.2 J
Krypton fluoride (KrF)	pulsed	0.248	5 mJ–2 J
Xenon chloride (XeCl)	pulsed	0.308	0.1–0.3 J
Xenon fluoride (XeF)	pulsed	0.351	2 mJ–0.5 J
Chemical Lasers:			
Hydrogen fluoride (HF)	cw	2.6-3.0	2–1000 W
	pulsed		50 mJ–3 J
Deuterium fluoride (DF)	cw	3.6-4.0	1–100 W
	pulsed	3.6-4.0	30 mJ–3 J
Far Infrared Lasers:			
Methanol (CH_3OH)	pulsed, cw	37.9, 70.5, 96.5, 118, 571, 699 other lines from 37 to 1224 μm[(b)]	< 1 W
Methyl fluoride (CH_3F)	pulsed, cw	496, 1222 μm	< 1 W
Other molecules[(b)]	cw	lines from ~40 to 1000 μm	0.1–1 W
	pulsed	lines from ~40 to 1200 μm	≤750 mJ

(a) Operating configurations include axial gas flow (20 W–5 kW), transverse gas flow (500 W–15 kW), sealed tube (3 W–100 W), TEA (tranverse excited, atmospheric pressure), and waveguide (0.1–50 W).

(b) Methanol (fully deuterated) (CD_3OD): 41.0, 184, 229, 255 μm.
Methylamine (CH_3NH_2): 147.8 μm, other lines from 100 to 351 μm.
Methyl iodide (fully deuterated) (CD_3I): 461, 520 μm; other lines from 272 to 1550 μm.
Formic acid (HCOOH): 432.6 μm, other lines from 134 to 1213 μm.
Difluoromethane (CH_2F_2): 375, 889, 1018 μm.

Appendices

Appendices

APPENDIX 1

Abbreviations, Acronyms, Initialisms, and Common Names for Types and Structures of Lasers and Amplifiers

APM — additive pulse mode-locked (laser)
ASE — amplified spontaneous emission
ASRL — anti-Stokes Raman laser
AWI — amplification without inversion
BC — buried crescent (laser)
BEL — bound-electron laser
BFA — Brillouin fiber amplifier
BGSL — broken-gap superlattice (laser structure)
BH — buried heterostructure (laser)
BIG — bundle-integrated-guide (laser)
BOG — buried optical guide (laser)
BRS — buried ridge structure (laser)
BVSIS — buried V-groove substrate inner stripe (laser)
C^3 — cleaved coupled cavity (laser)
CBH — circular buried heterostructure (laser)
CC-CDH — current confined constricted double heterostructure (laser)
CCGSE — concentric-circle grating surface emitting (laser)
CCL — color center laser
CDH — constricted double heterostructure (laser)
CEL — correlated emission laser
CMBH — capped-mesa buried-heterostructure (laser)
COIL — chemical oxygen-iodine laser
CPA — chirped pulse amplification
CPM — colliding-pulse mode-locked (laser)
CPM — corrugation-pitch-modulated (laser)
CRL — compact rugged laser
CSP — channeled-substrate planar (laser)
CVL — copper vapor laser
CW — continuous wave (laser)
D^3 — directly doubled diode (laser system)
DASAR — darkness amplification by stimulated absorption of radiation
DBR — distributed Bragg reflector (laser)
DC — direct current (continuous output)
DCFL — double clad fiber laser
DCPBH — double channel planar buried heterostructure (laser)
DDS — deep-diffuse stripe (laser)

Appendix 1—*continued*

Abbreviations, Acronyms, Initialisms, and Common Names for Types and Structures of Lasers and Amplifiers

DD-WGM	—	dye-doped whispering galley mode (laser)
DFB	—	distributed feedback (laser)
DFC	—	distributed forward coupled (laser)
DH	—	double-heterostructure (laser)
DPL	—	diode-pumped laser
DPSSL	—	diode-pumped solid-state laser
DQW	—	double quantum well (laser)
DR	—	distributed reflector (laser)
DS	—	diffused stripe (laser)
DSM	—	dynamic-single-mode (laser)
ECDL	—	external cavity diode laser
EDFA	—	erbium-doped fiber amplifier
EEDL	—	edge-emitting diode laser
EEL	—	edge-emitting laser
EFA	—	erbium fiber amplifier
EML	—	electroabsorption modulated laser
ESA-FEL	—	electrostatic-accelerator free-electron laser
ETDL	—	energy transfer dye laser
ETU	—	energy transfer upconversion (laser)
excimer	—	excited dimer (laser)
F-(center)	—	Farbe, German word for color (laser)
FCSEL	—	folded-cavity surface-emitting laser
FEDL	—	flashlamp-excited dye laser
FEL	—	free electron laser
FEM	—	free-electron maser
FG-ECL	—	fiber-grating external-cavity laser
FSFL	—	frequency-shifted feedback laser
GC	—	grating coupled (laser)
GDL	—	gas dynamic laser
GRINSCH	—	graded-index separate confinement heterojunction (laser)
GSE	—	grating-surface-emitting (laser)
GVL	—	gold vapor laser
HAP	—	high average power (laser)
HDL	—	homodyne laser
HEL	—	high enery laser
HENE	—	He-Ne (gas laser)
HPP	—	high peak power (laser)
HRO	—	heteroepitaxial ridge overgrown (laser)

Appendix 1—*continued*

Abbreviations, Acronyms, Initialisms, and Common Names for Types and Structures of Lasers and Amplifiers

ICF	—	inertial confinement fusion (laser)
LASER	—	light amplification by stimulated emission of radiation
LD	—	laser diode
LEC	—	long external cavity (laser)
LM-MQW	—	lattice-matched multiple quantum well (laser)
LWI	—	lasing without inversion
MASELA	—	matrix-addressable surface-emitting-laser array
MASER	—	microwave amplification by stimulated emission of radiation
MCS	—	modified channeled substrate (laser)
MDC	—	mirror dispersion controlled (oscillator)
MDR	—	morphology-dependent resonance (laser)
MIH	—	monolithically-integrated hybrid (laser)
MO	—	master oscillator
MOFPA	—	master oscillator-fiber power amplifier
MOPA	—	master oscillator power amplifier
MOPO	—	master oscillator power oscillator
MPL	—	microgun-pumped laser
MQB	—	multi-quantum barrier (laser)
MQW	—	multiple quantum well (laser)
MVL	—	metal vapor laser
NDFA	—	neodymium fiber amplifier
NDPL	—	nuclear device pumped laser
NPL	—	nuclear-pumped laser
OCL	—	optical confinement layer (laser)
OPA	—	optical parametric amplifier
OPAL	—	optical parametric amplifier laser
OPO	—	optical parametric oscillator
OPOL	—	optical parametric oscillator laser
OPPO	—	optical parametric power oscillator
OPS	—	optically pumped semiconductor (laser)
PBC	—	p-type buried crescent (laser)
PBC	—	planar buried crescent (laser)
PBH	—	planar buried heterostructure (laser)
PCSEL	—	planar cavity surface-emitting laser
PDFFA	—	praseodymium-doped fluoride fiber amplifier
PIL	—	photolylic iodine laser
PINSCH	—	periodic-index separate-confinement heterostructure (laser)

Appendix 1—*continued*

Abbreviations, Acronyms, Initialisms, and Common Names for Types and Structures of Lasers and Amplifiers

PLC	—	planar lightwave circuit (laser)
POFA	—	polymer optical fiber amplifier
POWA	—	planar optical waveguide amplifier
PRFA	—	praseodymium fiber amplifier
QB	—	quantum box (laser)
QC	—	quantum cascade (laser)
QCL	—	quantum cascade laser
QD	—	quantum dot (laser)
QF	—	quantum film (laser)
QW	—	quantum well (laser)
QW	—	quantum wire (laser)
QWH	—	quantum well heterostructure (laser)
QWR	—	quantum well ridge (laser)
QWS	—	quantum well structure (laser)
RFA	—	Raman fiber amplifier
RGH	—	rare gas halide (laser)
RPL	—	reactor pumped laser
RS	—	Raman-shifted (laser)
RW	—	ridge waveguide (laser structure)
SASE	—	self-amplified spontaneous emission
SB-BGSL	—	strain-balanced BGSL (laser structure)
SBL	—	space based laser
SBR	—	selective buried ridge (laser structure)
SBS	—	stimulated Brillouin scattering (amplifier)
SCBH	—	separate confinement buried heterostructure (laser)
SCH	—	separate carrier heterostructure (laser)
SCH	—	separate confinement heterostructure (laser)
SCL	—	semiconductor laser
SCLA	—	semiconductor laser amplifier
SDL	—	semiconductor diode laser
SE	—	surface-emitting (laser)
SEL	—	surface-emitting laser
SELD	—	surface-emitting laser diode
SELDA	—	surface-emitting laser diode array
SIPBH	—	semi-insulating planar buried heterostructure (laser)
SL	—	superlattice (laser structure)
SLA	—	semiconductor laser amplifier
SL-MQW	—	strained-layer multiple quantum well (laser)

Appendix 1—*continued*

Abbreviations, Acronyms, Initialisms, and Common Names for Types and Structures of Lasers and Amplifiers

SLS	—	strained-layer superlattice (laser structure)
SM	—	submillimeter (laser)
SOA	—	semiconductor optical amplifier
SP-APM	—	stretched-pulse additive pulse mode-locked (laser)
SPL	—	short pulse laser
SPML	—	synchronously-pumped mode-locked (laser)
SPPO	—	synchronously pumped parametric oscillator
SQW	—	single quantum well (laser)
SRS	—	stimulated Raman scattering (amplifier)
SSL	—	serpentine superlattice (laser structure)
SSQW	—	strained single quantum well laser
SXR	—	soft x-ray (laser)
T2QWL	—	type II quantum well laser
T^3	—	tabletop-terawatt (laser)
TAL	—	thin active layer (laser)
TAPS	—	tapered stripe (laser)
TCL	—	taper coupled laser
TCSM	—	twin-channel substrate mesa (laser)
T-cubed	—	tabletop terawatt (laser)
TDL	—	tunable diode laser
TEA	—	tranverse excited atmospheric pressure (laser)
TFR	—	tightly folded resonator (laser)
TIE	—	tunable interdigital electrode (DBR laser)
TJS	—	transverse junction stripe (laser structure)
TRS	—	twin-ridge structure (laser)
TTW-SLA	—	traveling wave semiconductor laser amplifier
UCL	—	upconversion laser
VCSEL	—	vertical cavity surface-emitting laser
VECOD	—	vertical-coupled quantum dot (laser)
VLD	—	visible laser diode
VSIS	—	V-channeled substrate inner stripe (laser)
VSQW	—	variable-strained quantum well (laser)
WDM	—	wavelength-division multiplexing
WGM	—	whispering galley mode (laser)
YAG	—	yttrium aluminum garnet (laser host crystal)
YDFA	—	ytterbium-doped fiber amplfier
YLF	—	yttrium lithium fluoride (laser host crystal)

Appendix 1—*continued*

Abbreviations, Acronyms, Initialisms, and Common Names for Types and Structures of Lasers and Amplifiers

YVO	—	yttrium vanadate (laser host crystal)
Z-laser	—	zone laser (self-focusing)

APPENDIX 2

Abbreviations, Acronyms, Initialisms, and Mineralogical or Common Names for Solid State Laser Materials*

alexandrite	—	chromium-doped chryoberyl ($BeAl_2O_4$)
BEL	—	lanthanum beryllate ($La_2Be_2O_5$)
BLGO	—	barium lanthanum gallate ($BaLaGa_3O_7$)
BYF	—	barium yttrium fluoride (BaY_2F_8)
CNGG	—	calcium niobium gallium garnet ($Ca_3[NbLiGa]_5O_{12}$)
colquiriite	—	lithium calcium aluminum fluoride ($LiCaAlF_6$)
CTH:YAG	—	Cr,Tm,Ho:$Y_3Al_5O_{12}$
emerald	—	chromium-doped beryl ($Be_3Al_2Si_6O_{18}$)
FAP	—	calcium fluoroapatite ($Ca_5[PO_4]_3F$)
forsterite	—	magnesium silicate (Mg_2SiO_4)
GGG	—	gadolinium gallium garnet ($Gd_3Ga_5O_{12}$)
GSGG	—	gadolinium scandium gallium garnet ($Gd_3Sc_2Ga_3O_{12}$)
HAP	—	high-average-power (laser glass)
LiCAF	—	lithium calcium aluminum fluoride ($LiCaAlF_6$)
LiSAF	—	lithium scandium aluminum fluoride ($LiScAlF_6$)
LMA	—	lanthanum magnesium hexaluminate ($LaMgAl_{11}O_{19}$)
LSB	—	lanthanum scandium borate [$LaSc_3(BO_3)_4$]
NYAB	—	neodymiun yttrium aluminum borate $Nd_xY_{1-x}Al_3(BO_3)_4$
ruby	—	Cr-doped aluminum oxide (Al_2O_3)
S–FAP	—	strontium fluoroapatite [$Sr_5(PO_4)_3F$]
silica	—	silicon dioxide (amorphous)
S–VAP	—	strontium vanadium fluoroapatite [$Sr_5(VO_4)_3F$]
Ti:sapphire	—	titanium-doped aluminum oxide (Al_2O_3)
YAB	—	yttrium aluminum borate [$YAl_3(BO_3)_4$]
YAG	—	yttrium aluminum garnet ($Y_3Al_5O_{12}$)
YALO	—	yttrium aluminate ($YAlO_3$)
YAP	—	yttrium aluminum perovskite ($YAlO_3$)
YBF	—	yttrium barium fluoride (Y_2BaF_8)
YGG	—	yttrium gallium garnet ($Y_3Ga_5O_{12}$)
YLF	—	yttrium lithium fluoride ($LiYF_4$)
YSAG	—	yttrium scandium aluminum ($Y_3Sc_2Al_3O_{12}$)
YSGG	—	yttrium scandium gallium ($Y_3Sc_2Ga_3O_{12}$)
YSO	—	yttrium silicon oxide, yttrium orthosilicate (Y_2SiO_5)
YVO	—	yttrium vanadate (YVO_4)
ZBLAN	—	Zr-Ba-La-Al-Na fluorozirconate glass

* For a more complete listing of abbreviations, acronyms, initialisms, and mineralogical or common names for laser materials, see Appendix 2 of the *Handbook of Laser Science and Technology: Supplement 2 – Optical Materials*, CRC Press, Boca Raton (1995).

APPENDIX 3

Fundamental Physical Constants

Quantity	Symbol	Value
speed of light in vacuum	c	299 792 458 m/s
permeability of vacuum	μ_0	1.256 637 061 4 N/A^2
permittivity of vacuum, $1/\mu_0c^2$	ε_0	8.854 187 817 F/m
Planck constant	h	6.626 075 5×10^{-34} J s
elementary charge	e	1.602 177 33×10^{-19} C
magnetic flux quantum, h/2e	Φ_0	2.067 834 61x10-15 Wb
electron mass	m_e	9.109 389 7×10^{-31} kg
proton mass	m_p	1.672 623 1×10^{-27} kg
fine structure constant	α	7.297 353 08×10^{-3}
inverse fine-structure constant	$1/\alpha$	137.035 989 5
Rydberg constant, $m_ec\alpha^2/2h$	R_∞	10 973 731.534 m^{-1}
Bohr radius, $\alpha/4\pi R_\infty$	a_0	0.529 177 249×10^{-10} m
Hartree energy, $e^2/(4\pi\varepsilon_0)a_0 = 2R_\infty hc$	E_h	4.359 748 2×10^{-18} J
Compton wavelength, h/m_ec	λ_C	2.426 310 58×10^{-12} m
classical electron radius, α^2a_0	r_e	2.817 940 92×10^{-15} m
Bohr magneton, $eh/4\pi m_e$	μ_B	9.274 015 4×10^{-24} J/T
nuclear magneton, $eh/4\pi m_p$	μ_N	5.050 786 6×10^{-27} J/T
electron magnetic moment	μ_e	9.284 770 1×10^{-24} J/T
magnetic moment anomaly, $\mu_e/\mu_B - 1$	a_e	1.159 653 193×10^{-3}
electron g factor, $2(1 - a_e)$	g_μ	2.002 319 304 386
proton gyromagnetic ratio	γ_p	2.675 221 28x108 $s^{-1}T^{-1}$
Avogadro constant	N_A	6.022 136 7×10^{23} mol^{-1}
Boltzmann constant, R/N_A	k	1.380 658×10^{-23} J/K
Faraday constant, N_Ae	F	96 485.309 C/mol
molar gas constant	R	8.314 510 J/mol K
Stefan-Boltzmann constant	s	5.670 51×10^{-8} $W/m^2\,K^4$

References:

Cohen, E. R. and Taylor, B. N., The 1986 adjustment of the fundamental physical constants, *Rev. Mod. Phys.* 59, 1121 (1987).

Taylor, B. N. and Cohen, E. R., Recommended values of the fundamental physical constants: a status report, *J. Res. Natl. Inst. Stand. Technol.* 95, 497 (1990).